The Star Atlas Companion

What You Need to Know about the Constellations

Philip M. Bagnall

The Star Atlas Companion

What You Need to Know about the Constellations

Philip M. Bagnall
Wallsend
UK

Front cover: Light echoes from the variable star V838 in the constellation of Monoceros as captured by the Hubble Space Telescope. Image courtesy of NASA, ESA and the Hubble Heritage Team (AURA/STScI).

Back cover: The Pleiades open cluster in Taurus. Image courtesy of Giovanni Benintende/Shutterstock.com

SPRINGER–PRAXIS BOOKS IN POPULAR ASTRONOMY
SUBJECT *ADVISORY EDITOR*: John Mason, M.B.E., B.Sc., M.Sc., Ph.D.

ISBN 978-1-4614-0829-1 ISBN 978-1-4614-0830-7 (eBook)
DOI 10.1007/978-1-4614-0830-7
Springer New York Heidelberg Dordrecht London

Library of Congress Control Number: 2012933587

Cover design: Jim Wilkie
Project copy editor: Dr John Mason
Typesetting: BookEns, Royston, Herts., UK

Printed on acid-free paper

Springer is part of Springer Science+Business Media (www.springer.com)

Contents

To my wife, Pauline

About this book

This book is about the properties of the stars that make up the 88 constellations: how far away they are, their diameters, their color, mass, luminosity, magnitude and shape, whether they are part of a binary or multiple star system, whether they have planets or are surrounded by a debris disk. The illustrations take the reader from the pinpoints of light seen through a telescope's eyepiece to an image of what the individual stars would look like as if they were only a few astronomical units away. The text supplements the images by providing additional information to put each star in context.

Acknowledgments

I am indebted to Clive Horwood at Praxis Publishing for his faith in this project and to Dr John Mason who suggested changes that improved the original proposal immensely.

In gathering the information for this book I have drawn on a number of websites and I would particularly like to acknowledge the following:

The VizieR catalogue access tool and the SIMBAD database operated by CDS, Strasbourg, France
The WEBDA database of open star clusters at the University of Vienna
The Extrasolar Planets Encyclopedia at www.exoplanet.eu
The Students for the Exploration and Development of Space (SEDS) website
Jim Kaler's *Stars* website at the University of Illinois which often helped to clarify a confusing jumble of data
A.A.Tokovinin's *Multiple Star Catalogue* at the National Optical Astronomy Observatory's Cerro Tololo Inter-American Observatory, and
The SAO/NASA Astrophysics Data System.

In addition I have often referred to Richard Dibon-Smith's *StarList 2000* (Wiley), Wil Tirion, Barry Rappaport and George Lovi's *Uranometria 2000.0* (Willmann-Bell Inc.) and, of course, Ian Ridpath's *Norton's Star Atlas* (Dutton).

These websites and books provided the raw data I needed to be able to write up the descriptions and construct the images for *The Star Atlas Companion* and it is unlikely that this book could have been written without them. I hope that I have interpreted the information correctly but any errors are all mine.

Philip M. Bagnall
October 2011

Introduction

That most famous of all star atlases, *Norton*'s, is now in its Twentieth Edition and, under the expert editorship of Ian Ridpath, is still the amateur astronomer's favorite. Over the past decade or so it has been joined by a number of new kids on the block such as *Uranometria 2000.0* and *Sky Atlas 2000.0* and several planetarium-type software programs. These road maps of the night sky are invaluable for finding your way around the constellations and locating a particular star but they all suffer from the same problem.

If you were to look at a real road map of, say, England then you should easily be able to find your way between Cambridge in the south of the country and Corbridge in the north. But there is nothing on a road map to indicate that Cambridge is one of the world's great university cities while Corbridge is a small market town. It is not until you actually get there that you realize the difference. And that is the problem with the current crop of star atlases: they do not reveal what the individual stars look like. Yet there is an almost infinite variety of colors, diameters, shapes, luminosities, distances, magnitudes and so on. Some constellation books go a little further and will describe particular stars, but unless you have a really good imagination and an unprecedented grasp of stellar dimensions then even this information is difficult to assemble into a realistic mental picture. *The Star Atlas Companion* overcomes this problem by graphically illustrating the various stars and star clusters.

This book came about when I was preparing a talk on the size of the Universe. No matter where I looked there was a dire shortage of images I could use to illustrate the talk. So I created my own. Having done that it occurred to me that I could write a book that would supplement those star atlases currently on the market, and which would help amateur astronomers gain a better idea of what they are actually looking at.

It seemed like a good idea at the time. Gather some information on the physical properties of the stars, write a brief description of each and draw a set of diagrams to illustrate their appearance, distances and so on. Then things started to get complicated.

First, there are few absolutes to work with. Stellar theory ties together a star's diameter, its mass, luminosity, magnitude and distance as well as a number of other parameters. In some cases stars fit the theoretical model perfectly. More often than not, however, some of the figures just don't add up. In addition there are usually several sets of contradictory data for most stars. So what do we do? Do we adjust the figures so they all fit nicely together? Or do we just use the data we have even though some of it does not appear to be right? We have to go with the latter purely on the basis that we should tweak our theories to match observations, not doctor the observations to match the theories.

The second problem is the sheer number of stars. Just limiting the selection to those stars that we can see without optical aid still runs to more than 3,000 stars – far too many to include in a book such as this. So I have had to be selective. I have mostly restricted the selection to stars that are brighter than magnitude +5.5. People with good eyesight can see stars that are a full magnitude fainter than this if their sky is sufficiently dark, but the fact is that most readers will live in urban areas where light pollution has drowned out the fainter stars. I have tried to include the more interesting stars – those that are unusual or which add to the cosmic story – at the expense of the more mundane stars if, indeed, any star could truly be considered mundane. But if I have missed a particular favorite of yours then do let me know and should this book run to a second edition then I will certainly consider including it.

Occasionally I have broken my own rules. Some classes of variable stars, for example, are named after prototypes that are not visible to the naked eye. I have included some of these prototypes to help the reader make sense of the differences between the various types. In addition, many of the apparently single stars turn out to be part of binary or multiple star systems when viewed with even modest optical aid. It would be plain daft not to include their companions.

Finally, we have chosen to illustrate this book with black and white images but, of course, we live in a highly colorful Universe. I hope this limitation will not detract from the purpose of this book: to visualize our small corner of the Galaxy.

Making sense of the data

This chapter covers the language and symbols we will use throughout this book and describes some of the problems and gray areas, which the reader needs to bear in mind.

Star names and designations

Astronomy has a language all of its own: a mixture of Greek and Latin, mathematical and chemical symbols and countless abbreviations. If you are a newcomer to astronomy then the descriptions of the stars may look like they came out of a book on black magic, but once you understand the shorthand it all makes sense.

The Greek alphabet is commonly used in astronomy, with many stars in each constellation labeled using lower-case Greek letters. Johann Bayer (1572-1625) introduced this Greek letter system as a way of identifying stars in his star atlas *Uranometria*, published in 1603. The Greek (Bayer) letters are used with the Latin genitive case of the constellation name. Table 1 shows the Greek alphabet in full. Note that there are two sigmas, a lowercase σ and an upper case Σ; the latter is used to identify certain double stars.

Of course, Bayer was not the only astronomer to draw up a catalog and other designations are routinely used throughout this book. These include Flamsteed numbers, such as 1 Andromedae and 20 Scorpii. John Flamsteed (1646–1720), Britain's first Astronomer Royal and Director of the Royal Greenwich Observatory, cataloged the stars using numbers instead of letters. A century later Friedrich Argelander (1799–1875) compiled his 324,188 star *Bonner Durchmusterung* catalog and identified those stars that varied in brightness by using the double letters RR-RZ, SR-SZ and so on all the way up to ZZ. Stars are sometimes still given their BD number. Another important catalog was compiled by the American astronomer Henry Draper (HD) and you will find many references to HD stars in this book. Other commonly used catalog numbers are those of the Smithsonian Astrophysical Observatory, SAO, and the Tycho catalog, TYC, which was compiled from the data sent back by the Hipparcos satellite that mapped the celestial sphere in unprecedented detail. In the wider literature you will find reference to HR, the *Harvard Revised Catalog*, ADS, *Aitken Double Star Catalog* and WDS, the *Washington Double Star Catalog* although these are not used here.

Far older than any catalog are the traditional names of the stars such as Sirius meaning 'sparkling' and the more sinister sounding Algol, 'the Demon Star.' I

Table 1: The Greek alphabet

α	alpha
β	beta
γ	gamma
δ	delta
ε	epsilon
ζ	zeta
η	eta
θ	theta
ι	iota
κ	kappa
λ	lambda
μ	mu
ν	nu
ξ	xi
ο	omicron
π	pi
ρ	rho
σ	sigma
τ	tau
υ	upsilon
φ	phi
χ	chi
ψ	psi
ω	omega
Σ	Sigma

have included some of the better known names, but I have left many out as this book is not really about how the stars got their names. I have however highlighted a couple of dubious names that have recently come into existence; the result of schoolboy pranks spreading across the Internet.

Natural yardsticks

On a galactic scale the usual forms of measurement, such as the kilometer for distance and the kilogram for mass, become woefully inadequate. Instead we have to use, for example, the size of the Earth, the Sun and the Solar System as our yardsticks otherwise the numbers become just too big to grasp. Table 2 summarizes the yardsticks we will be using in this book.

Table 2 introduces a couple of new symbols: ⊕ for the Earth and ⊙ for the Sun and these are used extensively throughout the text.

I will often use words in place of numbers simply because a text full of numbers is a very dry read. I have avoided the use of 'billion' for several reasons. One is that a billion means different things to different people. An American billion is one thousand million (10^9) but a British billion is a million million (10^{12}), so there is obvious room for confusion. My second, more important reason is people's grasp of big numbers. Most people know how big a thousand is and can imagine how big a million is, but a billion is totally incomprehensible. The third reason is because of some of the comparisons we will be making. Stars like the Sun will typically live for around 10,000 million years but some very massive stars will last only a few tens of millions of years. That's fine because we are comparing tens of millions with thousands of millions. But once we start throwing billions into the mix the comparison is not so obvious. This book is all about perspective, comparing the stars to our own Sun and its planetary system so readers can get a real feel for how big the Galaxy and its components are. Standardizing all the measurements makes such comparisons easier.

Finally, the metric system is used throughout. Except for the word 'yardstick', which doesn't sound right when converted to a 'meterstick'!

Distance measurement

Professional astronomers use the parsec, a shortened term for parallax second, to measure distance. A parsec is the distance at which the mean radius of the Earth's orbit (i.e. 1 AU) subtends an angle of one arc second. It is abbreviated as pc. Amateur astronomers almost invariably use the light year, which is the distance traveled by a beam of light in one year, equivalent to 9,460,500,000,000 km or, in words, about 10 million million kilometers. This is easier to understand and is the standard form of distance measurement we will be using. A parsec is equivalent to 3.2615 ly or 30,855,420,750,000 km (about 31 million million km). We will often use '10 pc' as this is the distance at which stars are 'placed' in order to calculate their absolute magnitudes (see section on Magnitude below).

Whenever we measure something there is an element of error, and gauging the distances to the stars is no different. Up to about 60 ly the error is relatively small. By the time we reach 100 ly the error is typically in the order of 1 ly. In other words a star that is said to be 100 ly away is, in reality, probably between 99 and 101 ly. At 1,000 ly the error is commonly about 165 ly while at 2,000 ly it can be around 500 ly.

Stellar diameter, $D_\odot$

When we wish to describe how big a star is it is convenient to compare it to the size of the Sun. So a star might be described as being 'four times bigger than the Sun.' Professional astronomers tend to use solar radii or $R_\odot$ when making comparisons, and for good mathematical reasons. But the rest of us see the Sun and stars as whole spheres or disks and naturally think in terms of solar diameters

Table 2: Useful yardsticks

The yardstick	The measurement	
Earth, ⊕	Diameter: $D_{\oplus}$ 12,756 km Mass: $M_{\oplus}$ 5.97 × 10^{24} kg Volume: 1.083 × 10^{12} km^3	
Jupiter, J	Diameter: D_J 142,800 km or 11.2 Earths Mass: M_J 1.9 × 10^{27} kg or 317.9 Earths	
Sun, ⊙	Diameter: $D_{\odot}$ 1,392,000 km or 10^9 Earths Mass: $M_{\odot}$ 1.989 × 10^{30} kg or 333,000 Earths Volume: $V_{\odot}$ 1.412 × 10^{18} km^3 or 1,303,782 Earths Apparent magnitude, m_V: -26.8 Absolute magnitude, M_V: +4.8 Luminosity: 3.83 × 10^{26} watts	
Astronomical Unit, AU	The average distance from the Sun to Earth: 149,597,870 km	
Orbits of the planets	Mean radius	Mean diameter
Mercury	0.4 AU	0.8 AU
Venus	0.7 AU	1.4 AU
Earth	1.0 AU	2.0 AU
Mars	1.5 AU	3.0 AU
Jupiter	5.2 AU	10.4 AU
Saturn	9.5 AU	19.0 AU
Uranus	19.2 AU	38.4 AU
Neptune	30.1 AU	30.2 AU
Pluto*	39.5 AU	79.0 AU
Light Year, ly	9,460,500,000,000 km or 63,239 AU	
Parsec, pc	30,855,420,750,000 km or 3.2615 ly or 206,254 AU	

*Now relegated to the role of a 'dwarf planet' but still a useful yardstick.

or $D_{\odot}$. In reality, of course, there is no difference: a star that is four times bigger than the Sun has both a radius of 4 $R_{\odot}$ and a diameter of 4 $D_{\odot}$. For convenience we will use solar diameters as our yardstick.

Measuring the diameter of a star is also not without its problems. Direct measurements can be tricky as a bright star can appear larger than it actually is while a dim star can appear smaller; the result of limb darkening. In most cases the diameter is implied from theory. Not all stars are perfect spheres. A rapidly rotating star will distort into an oblate spheroid; gravity pulling its poles in towards the center while its equatorial region bulges outwards. The polar regions become hotter and brighter, because the gases are compressed, while the equator cools and darkens – a phenomenon known as the Von Zeipel effect. As yet we do not really know what such stars actually look like. The equatorial band may appear quite dark because of its contrast against the much lighter background, but if it was possible to view the band in isolation then it is likely to appear bright.

Closely orbiting binary stars can gravitationally distort one another into teardrop shapes and, depending on how we on Earth view their orbit, can present us with an ever changing cross-section as we sometimes see a perfect sphere and sometimes a stretched out limb.

All these variations conspire to make the process of measuring a star's diameter just a little bit tricky and it is not unusual to find that different methods produce different results.

Stellar mass, $M_{\odot}$

Again it is convenient to compare the mass of a star with the mass of the Sun, $M_{\odot}$. Mass is often implied from stellar theory, but binary stars offer a more accurate means of estimating mass because of the relationship between the stars' masses and their orbital characteristics.

Stellar luminosity, $L_{\odot}$

Also compared to the Sun's luminosity, $L_{\odot}$. Luminosity is the total outflow of power but it can be quite deceptive. The brightness of a star is of course related to its luminosity, but very hot stars radiate much of their energy in the ultraviolet part of the spectrum while red giants, for example, radiate mainly in the infrared, both of which are invisible to the human eye. Luminosity is frustratingly difficult to illustrate, even if we just consider luminosity in the visible part of the spectrum. The Sun is blindingly luminous. A star that is 10 $L_{\odot}$ is still blindingly luminous.

There are nine luminosity classes (Table 3), which together with their spectral class, help us to define a star. The Sun, for example, is a G2 V indicating it is a yellow Main Sequence dwarf (see also section on Spectral class below).

Magnitude, M_v and m_v

Magnitude is a measure of the relative brightness of a star. It comes in two flavors: apparent and absolute.

Apparent magnitude, as the name suggests, is how bright a star (or other celestial object) appears. There are various types of apparent magnitude. How bright a star appears to the human eye is sometimes termed the visual apparent

Table 3: Luminosity classes of stars

Code	Luminosity Class
O	Very luminous supergiants
Ia	Bright supergiants
Ib	Supergiants
II	Bright giants
III	Normal giants
IV	Subgiants
V	Main Sequence dwarfs
VI	Subdwarfs
VII	White dwarfs

magnitude, m_v, while its image on a photograph is the photographic magnitude, m_{pg}. This book uses m_v. Somewhat paradoxically the brighter a star is, the smaller the numerical value of the magnitude. The faintest stars visible in clear dark skies to those with good eyesight are usually around the sixth magnitude (m_v +6). Most urban dwellers, however, can only see down to the fifth magnitude, and those who live in brightly lit cities are lucky if they can see stars of the first magnitude. Spica (α Virginis) and Pollux (β Geminorum) are both first magnitude stars. Capella (α Aurigae) is a zero magnitude star. Then we start to go into 'negative' magnitude values: Sirius (α Canis Majoris) is the brightest star in the night sky at a sparkling m_v -1.6. Of course the Sun, the Moon and some of the planets are brighter still: Venus at her most brilliant is m_v -4.4, the full Moon is m_v -12.7 and the Sun a stunning m_v -26.8. Some astronomy writers drop the + sign for positive values, but I have always used it and will do so throughout this book.

Absolute magnitude, or M_v, is how bright a star would look at a standard distance of 32.615 ly or 10 pc. The star 34 Cygni has an absolute magnitude of M_v -8.9 although its apparent magnitude is a more modest m_v +4.81. By comparison our Sun placed at a distance of 10 pc would shine at just M_v +4.8, barely visible to the naked eye.

It is a great pity that we cannot see the stars in terms of absolute magnitude. The sky would be a very different and very impressive canopy with some stars being clearly visible in broad daylight. Constellations like Canis Major would be particularly impressive. Hopefully this book will go some way to revealing what that sky would look like.

Variable stars

All stars are likely to be variable, but some are more variable than others. There are stars like the Sun that are variable over many centuries or millennia. At the other extreme there are stars that vary in brightness over a fraction of a day.

Some stars are microvariable, fluctuating by thousandths of a magnitude while others can change by several magnitudes. One night they are shining brightly in the sky while some time later there is no sign of them, having dropped to below naked eye visibility. The well-known variable χ Cygni can be as bright as m_v +3.3 but for most of the time it is invisible, fading to m_v +14.2 at minimum. On average, it takes 408 days to go from maximum magnitude to minimum and back again. This is termed its 'period.'

There is a bewildering array of variable stars. However, they can mostly be placed into five categories:

Eruptive: Caused by flaring and other violent events in the star's chromosphere and corona. There is an ejection of mass and strong stellar winds.
Pulsating: The star expands and contracts like a heart. In 'radially' pulsating cases the star remains roughly spherical while in the 'non-radially' pulsating type the star can distort, deviating from a spherical shape.
Rotating: Variations in brightness are due to the star having dark or perhaps light spots on its surface, similar to sunspots, or activity in the atmosphere caused by the rotation of the magnetic pole being out of sync with the polar axis.
Cataclysmic: Caused by explosions in the star's surface layers (nova) or deeper (supernova). They are usually associated with close binary systems, the proximity of the two stars disrupting each other. Material is often transferred from the larger, cooler star to the smaller, hotter dwarf and forms an accretion disk.
Close binary eclipsing systems: Due to two stars periodically passing in front of one another. The exact light curve depends on several factors such as the line of sight and the closeness of the two stars which may cause them to distort.

The variable 'types' tend to be named after the first star to display a particular characteristic. The table below lists some of the main types of variable and is based on the work of N.N. Samus of the Moscow Institute of Astronomy and O.V. Durlevich of the Sternberg Astronomical Institute, also in Moscow.

Category	Type	Notes
Eruptive	FU Orionis	Gradual increase in brightness by up to 6 magnitudes over several months. Stay at maximum brightness for long periods or decline by 1-2 magnitudes.
	γ Cassiopeiae	Irregular. Rapidly rotating B-class giants or sub-giants.
	R Coronae Borealis	High luminosity stars that are both eruptive and pulsating. Slowly fade by up to 9 magnitudes over 30 to several hundred days.

Category	Type	Notes
Eruptive (cont'd)	RS Canum Venaticorum	Close binary systems that interact with one another causing eruptions in the stars' chromospheres. Some RS CVn stars are also eclipsing variables.
	S Doradûs	High luminosity B and F-class stars. Irregular changes of up to 7 magnitudes.
	UV Ceti	K and M-class stars showing sudden flaring of up to 6 magnitudes in just a few seconds before returning to normal over several minutes.
	WR	Wolf-Rayet variables. Changes of up to 0.1 magnitude probably caused by the star suddenly ejecting mass.
Pulsating	α Cygni	Occur in B and A-class stars. Multiple overlapping variable cycles lasting from days to weeks.
	β Cep	β Cephei or β Canis Majoris type. Occur in O8 to B6 class stars. Light varies by up to 0.3 magnitude between 0.1 to 0.6 days.
	Cepheids	High luminosity stars that change from F-class at maximum brightness to G or K-class at minimum. Periods are 1-135 days.
	W Virginis	Vary in brightness between 0.3 to 1.2 magnitudes over 0.8 to 35 days.
	δ Cephei	Occur in F and G-class stars. Brightness changes between 0.2 to 2 magnitudes over 1 to 50 days.
	δ Scuti	Occur in A and F-class stars. Brightness changes by 0.2 magnitude over 0.05 to 0.2 days.
	Lb	Slow irregular variables. Tend to be giants of K, M, C and S-class.
	Lc	Irregular supergiant variables. Brightness changes by about 1 magnitude.
	M	Mira type variables named after o Ceti. Occur in M, C and S-class stars. Brightness changes are in the range of 2.5 to 10 magnitudes over 100 to 600 days.
	RR Lyrae	Sometimes called Short Period Cepheids. Occur in A and F-class stars. Vary in brightness between 0.5 to 1.5 magnitudes over 0.05 to 1 day.

Category	Type	Notes
Pulsating (cont'd)	RV Tauri	Occur in F and G-class stars at maximum magnitude and K and M-class at minimum. Changes of up to 4 magnitudes over 30-150 days.
	SR	Semi-regular variables. Giants or supergiants of intermediate to late spectral types. Can be microvariable to up to 2.5 magnitudes in the range 20 to 200 days.
	SRa	Semi-regular variables of M, C or S spectral class giants. Changes in brightness are usually less than 2.5 magnitudes over 35 to 1,200 days.
	SRb	Semi-regular variables of M, C or S spectral class giants. Poorly defined periods but in the range of 20 to 2,300 days.
	SRc	Semi-regular variables of F, G or K spectral class giants and supergiants. Brightness varies by 0.1 to 4 magnitudes over a period of 30 to 1,100 days.
	SRd	Semi-regular variables of M, C or S spectral class supergiants.
	ZZ Ceti	White dwarfs. Magnitude range is 0.001 to 0.2 over 0.5 to 25 minutes.
Rotating	α^2 Canum Venaticorum	Occur in B8 to A7 class stars. Magnitude range is 0.01 to 0.1 over 0.5 to more than 160 days.
	BY Draconis	Dwarfs of K and M-class. Magnitude changes of up to 0.5 mag with periods of up to 120 days. Variability probably due to spots, chromospheric activity and flaring
	FK Comae Berenices	Giants of G-K class. Rapid rotators. Brightness fluctuates by up to several tenths of a magnitude.
	PSR	Pulsars. Over a period of 0.004 to 4 seconds their magnitude can vary by up to 0.8 magnitude.
	SX Arietis	Also called helium variables. Affects B-class stars.

Category	Type	Notes
Rotating (cont'd)		Variable period of about a day during which magnitude varies by about 0.1 mag.
	Ell	Ellipsoidal variables. Occur in close binary systems. Variable period matches orbital period. Amplitude of 0.1 magnitude.
Cataclysmic	SS Cygni	Increase in brightness by 1-2 magnitudes over 1-2 days before returning to normal. Period can be anywhere from 10 to several thousand days.
	SU Ursae Majoris	Two types: normal and supermaxima. Normal – similar to SS Cygni. Supermaxima – Brighter by 2 magnitudes, and five times longer.
	Z Camel-opardalis	Brighten by 2 to 5 magnitudes over 10 to 40 days without returning to normal but remaining at a magnitude between minimum and maximum.
Close binary eclipsing systems	E	The orbital plane of the two stars is close to the observer's line of sight and the stars periodically eclipse each other.
	EA	Algol or β Persei type. Occur in B, A and F classes. Magnitude range is 0.1 to 3 mags. Period from 0.1 to 10,000+ days.
	EB	β Lyrae type. Occur in O, B and A classes. Magnitude range is 0.1 to 1.5 mags. Period from 0.5 to 200 days.
	EW	W Ursae Majoris type. Occur in F, G and K classes. Magnitude range is 0.1 to 1 mag. Period from 0.1 to 1 day.

Star color

What color is the Sun? That's an easy question. It is obviously yellow. Astronomers class the Sun as a yellow dwarf and even very young children will often paint a bright yellow Sun in a cloudless blue sky. Yet if you look at the Sun through thin cloud when it is high in the sky then it looks very definitely white. In fact, it only looks yellow when it is close to the horizon - but then again, it can also look orange or red. This coloration is partly due to the dust in the atmosphere filtering out all but the yellow, orange and red light. So why do we think the Sun is yellow?

The problem lies not with the Sun but in the way we perceive color and brightness. Our eyes have evolved to be efficient at detecting light and less efficient at detecting color. The Sun is so bright that its brilliance bleaches out its true color. If you could travel in space away from the Sun then you would reach a point where the Sun looks a very definite yellow. Travel much farther out, however, and the color becomes feeble and the Sun again looks white.

The Sun is in fact yellow in the conventional, every day meaning of the word: it radiates much of its energy in the yellow part of the visible spectrum. In this book we have used shades of gray to indicate the different colors.

Finally, the obligatory warning: never look directly at the Sun and never look through a telescope or a binocular at the Sun. It can permanently damage your eyesight.

Spectral class

The intricacies of the spectral classification of stars are beyond the scope of this book, but there is certainly no shortage of websites that provide the detail that some readers may need. The spectral classification of a star is related to several factors such as its composition, age and luminosity. The interplay between these and other factors are illustrated by the elegant Hertzsprung-Russell Diagram (see Figure 1) of which there are several variations. The band of stars running diagonally across the diagram is the Main Sequence and in here, at a point that corresponds to spectral class G2, a temperature of 5,770 K and a luminosity of 1 $L_{\odot}$ is, of course, the Sun, a useful reference point. The Main Sequence is not an evolutionary pathway; stars do not start at the bottom and climb to the top like some sort of game of cosmic snakes and ladders. Their evolution is far more complex, perhaps leaving and rejoining the Main Sequence several times during their lives.

There are ten spectral classes:

Class	Color	Temperature	Naked eye Abundance
WR	Blue	~50,000 K	<1%
O	Blue	28,000 K to 50,000 K	1%
B	Blue to bluish-white	10,000 K to 28,000 K	22%
A	Bluish-white to pale yellow	7,500 K to 10,000 K	20%
F	Pale yellow to yellow	6,000 K to 7,500 K	12%
G	Yellow to yellowish-orange	5,000 K to 6,000 K	13%
K	Yellowish-orange to reddish-orange	3,500 K to 5,000 K	23%

M	Reddish-orange to deep red	2,000 K to 3,400 K	7%
S	Deep red	2,000 K to 3,500 K	<1%
C	Deep red	<2,000 K	<1%

As mentioned earlier there are few absolutes in this book. Each class has subdivisions running from 0 to 9. The Sun is a G2 with a 'surface' temperature of about 5,770 K. The starting point for the G-class in the table above, G0, indicates a temperature of 6,000 K, but this is for Main Sequence stars similar to the Sun. Giant stars of G0 class are a couple of hundred degrees cooler at about 5,800 K while supergiants are cooler again at 5,600 K.

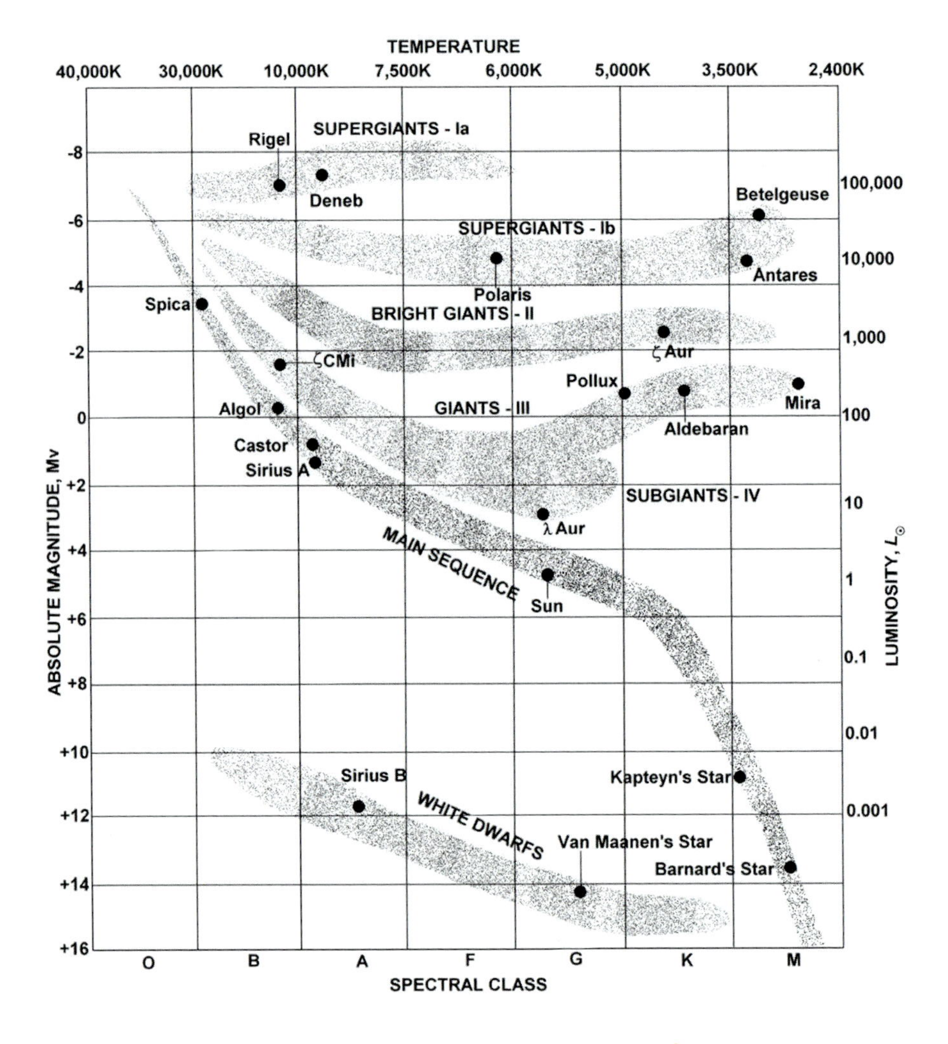

Figure 1: The Hertzsprung-Russell Diagram

Some of the spectral codes are supplemented with additional information such as A5p, meaning the star has a peculiar spectrum, or A5m indicating it is metal rich.

The abundance column suggests that M-class stars are rare at just 7%. In fact M-class stars make up 80% of the Main Sequence – we just don't see them with the naked eye because they tend to be too faint. G-class stars like the Sun account for just 3.5% of the Main Sequence. So when we look at the stars that make up the constellations we are looking at a biased sample, distorted by the limitations of our own eyes.

The WR class is Wolf-Rayet stars; highly evolved objects that are shedding huge amounts of material into space. S-class are zirconium rich while the closely related C-class are carbon-rich stars.

Temperature

In the previous section we introduced temperature. We use the Kelvin measurement scale, which is abbreviated to K. The Kelvin scale starts at Absolute Zero, equivalent to -273.15°C or -459.67°F. Hence the Sun's temperature of 5,770 K works out at 5,497°C.

Temperature is given as 'surface temperature.' This is an odd concept, since stars do not have a solid surface in the same way as the Earth. To all intents and purposes the 'surface' is the visible sphere.

Just above a star's surface the temperature can rise to a few million Kelvin, while at its core the temperature is typically from about five million to several tens of millions of K.

Radial velocity, and the motion of stars through the Galaxy

Many stars, like the Sun, have almost circular orbits. They stay at about the same distance from the galactic center for the whole of their lives although they do bob up and down passing through the galactic arms. Some stars have elongated orbits. A notable example is ε Andromedae whose orbit takes it from within just 14,000 ly of the galactic center, past the Sun's orbit at 30,000 ly and out to 35,000 ly before turning around and heading back towards the center. There is some variation in the direction in which stars are moving and how fast they move through the Galaxy, with some stars heading towards us, some away from us and some staying at roughly the same distance. Add to this the Earth gyrating on its axis and we end up with a dynamic sky full of stars heading in all different directions and at differing speeds. Fortunately with the Galaxy being so vast the movement of an individual star is almost imperceptible: the constellations look much the same now as they did to ancient civilizations 2,000, 3,000 and even 5,000 years ago. However over a period of 25,000 years or more the constellations will appear to change shape. Some of the diagrams in this book that show these changes are based on the assumption that the stars' direction of travel and velocity will not change over the next 250 centuries. In some cases that will be true but not in all, so the diagrams are an idealized situation.

Rotational velocity

Nothing is particularly easy to measure on a galactic scale, but rotational velocity brings its own challenges. A star may appear to be rotating at a certain velocity but this can be affected by the direction of the rotational axis. If the star's rotation axis is perpendicular to our line of sight (i.e. the star is face on to us) then the measured velocity can be fairly accurate. But if the rotational axis is angled, especially towards us so that we are looking at the star's pole, then it can cause errors with the star appearing to spin more slowly than it actually is. We do not usually know in which direction the axis is pointing. As a general rule of thumb, stars slow down as they age, so a fast spinning star often indicates a stellar youngster.

Rotational period

Stars display 'differential rotation.' In other words, the equator and poles spin at different speeds with a gradation between the two. The Sun rotates at 2 km/s at the equator and so the equatorial region takes about 25 days to spin once while nearer the poles it takes about 34 days. Some stars rotate more slowly than the Sun but most are faster. Rotational periods of naked eye stars usually range from half a day to several decades for hypergiants.

Surface features

Observation of the Sun will often reveal dark areas known as sunspots caused by local disturbances in the Sun's magnetic field. It is likely that similar spots are to be found on some stars, but precisely which stars is open to debate. The consensus is that spots will appear on stars of up to a few solar masses in the spectral groups F, G, K and M. In addition to spots, some stars may have a blotchy appearance with large light and dark patches. The illustrations in this book are my idea of what such stars will look like. I may be completely wrong – it is always a precarious business basing predictions on a sample size of one! Certainly, one of the things that images from the Hubble Space Telescope have taught us is that the Universe is full of surprises.

Open and globular clusters

Open clusters consist of tens to hundreds of stars that are gravitationally bound together and which are moving through the galactic disk in the same direction. When someone is swamped with information we often say that 'they cannot see the wood for the trees.' It is likewise true that you sometimes 'cannot see the cluster for the stars.' In order to photograph clusters, astronomers take long duration exposures which unfortunately tend to also pick up all the background stars, so the actual cluster can often merge with the background making it difficult to see. To overcome this problem, I have used the WEBDA database of star clusters at the University of Vienna and redrawn the clusters without the background star 'noise.' This method is not without its problems since not all the stars in the database are genuine members of the clusters. I have also limited the number of stars to those brighter than twelfth magnitude, which is what can often be seen through a small telescope or a good pair of binoculars.

Not all clusters are real. Some are just chance alignments and are really 'asterisms.' A number of asterisms are identified in this book.

Globular clusters contain from several thousand up to a million stars. They tend to be more spherical and larger than open clusters and much farther away, residing in the galactic halo. They are not as well mapped as some of the open clusters but are included in this book for completeness.

Exoplanets

We are at an exciting time in our discovery of planets orbiting stars other than our own Sun. Technological advancements are making it possible to detect smaller planets and, perhaps by the time you read these words, Earth-sized planets will have been discovered.

A year ago (at the time of writing) there were 400-odd exoplanets. Now there are more than 1,200, such is the rate of discovery. I have included a few of the most interesting planets; to include them all would require another book which would be out of date by the time it reaches Amazon. The tables containing planetary system information use a number of symbols and abbreviations:

q = Periastron distance in AU (i.e. the closest the planet gets to the parent star in its orbit)
Q = Apastron distance in AU (i.e. the farthest the point in the orbit)
P = Orbital period in days (d) or years (y)
$\mathbf{M_J}$ = Jupiter masses
$\mathbf{M_\oplus}$ = Earth masses

Constellation details

Each constellation is presented as a separate section and headed with a table that gives some basic information about the constellation. The example below is for Aquila.

Constellation:	Aquila	**Hemisphere**	Equatorial
Translation:	The Eagle	**Area**:	652 deg^2
Genitive:	Aquilae	**% of sky**:	1.580%
Abbreviation:	Aql	**Size ranking**:	22nd

The eight headings are:

- **Constellation:** The Latin name of the constellation
- **Translation:** The English translation of the constellation's name
- **Genitive:** Used with individual stars. So the alpha star is α Aquilae not α Aquila.
- **Abbreviation:** An internationally agreed 3-letter code for the constellation. Hence the alpha star can be written as α Aql.
- **Hemisphere:** Indicates whether the constellation resides entirely within the Northern or Southern Celestial Hemisphere. If a constellation straddles the border between the two it is shown as Equatorial.
- **Area:** The area of the constellation in square degrees.
- **% of sky:** The percentage of the celestial sphere covered by the constellation.
- **Size ranking:** Where the constellation fits in a league table of size.

Illustrations

Each constellation includes a number of standard diagrams:

- **Apparent magnitude** which also acts as an identifier for the individual stars.
- **Absolute magnitude** which shows what the constellation would look like if the stars were all at a distance of 10 pc.
- **Distance** of the individual stars from the Sun.
- **An information rich diagram** that shows the relative sizes of each of the stars and which specifies their characteristics. Some surface features are also shown.

Some constellations include additional illustrations to show, for example, how the constellation outline will distort over time, binary orbits, etc.

The Constellations

Andromeda

Constellation:	Andromeda	**Hemisphere:**	Northern
Translation:	Princess Andromeda	**Area:**	722 deg^2
Genitive:	Andromedae	**% of sky:**	1.750%
Abbreviation:	And	**Size ranking:**	19th

In mythology Andromeda was the daughter of Cepheus, King of Ethiopia, and Queen Cassiopeia.

α **Andromedae** or Alpheratz is unusual in that it also forms one of the corner stars of another constellation, Pegasus, and is sometimes referred to as δ Pegasi. It has an apparent magnitude of m_v +2.02 but is slightly variable, dimming to m_v +2.06 with a period of 23^h 11^m 22^s. It is classed as an α CV rotating variable, the fluctuating magnitude being due to the presence of mercury clouds in the star's equatorial region. This, and an abundance of manganese, makes it a chemically peculiar star. α And is accompanied by an unseen spectroscopic companion in a 96.7 day orbit.

β **Andromedae** or Mirach may possibly be a semi-regular variable fluctuating between m_v +2.01 and +2.10 magnitudes. A red giant lying at a distance of almost 200 light years, it would fill the orbit of Mercury. Despite its 85 $D_\odot$ its mass is probably around 3.7 $M_\odot$. Like α And, β has a companion in a 1,700 AU orbit. It is quite dim shining at just 14th magnitude.

To the unaided eye γ **Andromedae** appears to be a single star of m_v +2.26 but it actually has four components. Lying at a distance of 355 ly the primary is an 83 $D_\odot$ K3 with a mass of 3.3 $M_\odot$. Just 9.6″ away lies the blue +4.84 magnitude γ^{Ba} And; a B8 dwarf which is locked in a 2.67 day orbit with a B9 dwarf, γ^{Bb} And. Their masses come in at 4.5 and 3.8 $M_\odot$ respectively. The Ba-Bb pair are also in orbit with a 3.2 $M_\odot$ A0, γ^C And, of which little is known. The orbit takes 61.1 years to run full circle, although circle is perhaps not the right word, the orbit being quite eccentric with the three stars coming as close as 13 AU and separating by as much as 52 AU. The Ba-Bb-C trio take 6,606 years to orbit the primary star.

δ **Andromedae** is a triple star system about 101 ly from Earth. The primary component, δ^{Aa} **Andromedae,** is an orange K3 giant, 14 times larger than the Sun and 4.6 times as massive. The second component is a spectroscopic binary,

δ^{Ab} **Andromedae,** a 0.5 $M_{\odot}$ white dwarf whose elliptical orbit brings it to within 9 AU of the primary and then out to about 28 AU. The orbital period is likely thought to be 55.2 years. This binary pair is also believed to be enveloped in a dust shell. In orbit some 900 AU away, 28.7″ on the celestial sphere, is δ^{B} **Andromedae**, an M2 red dwarf of 0.37 $M_{\odot}$ which takes at least 11,377 years to complete an orbit.

ε **Andromedae** is going places. Whilst our own Sun stays at about a steady 28,000 ly from the center of the Galaxy, ε And's orbit is highly eccentric, taking it from just 14,000 ly from the galactic center out to 35,000 ly before turning around and heading back in. Currently, it is hurtling towards us at 83.6 km/s putting it in the swiftest 1% of stars visible to the naked eye. At just 650 million years it is a young star compared to our own 4,500 million year old Sun and although it is now a yellow G8 giant, 8.7 times larger than the Sun, it actually started life as a B9 dwarf. It rotates at about 4 km/s (Sun = 2 km/s) taking 110 days to rotate once.

Spinning ten times faster at 40 km/s is the orange K1 giant ζ **Andromedae** which is 15 $D_{\odot}$. This eclipsing binary has a period of 17.77 days during which its magnitude varies between m_v +3.92 and +4.14. It is classed as an EB or β Lyrae type variable. The binary component, ζ^{B} **Andromedae**, lies 32.6″ directly north of the primary but at magnitude +15.3 is far below naked eye visibility.

Another spectroscopic binary is η **Andromedae,** a G8 star lying at a distance of 243 ly but travelling towards the Solar System at 10.3 km/s. Believed to be just 800 million years old η And is nearly 8 times larger than the Sun and is 35 times more luminous. Its companion is almost an identical twin. The primary has a mass of between 2.25 and 2.95 solar masses whilst the secondary is between 1.99 and 2.61: they both have a temperature of around 4,900 K.

θ **Andromedae** is an A2 Main Sequence white dwarf somewhat less than 2 $D_{\odot}$ across. Another binary the separation between the two is just 0.1″. It lies 253 ly away and is thought to be variable between m_v +4.58 and +4.62.

At 6.5 $D_{\odot}$ λ **Andromedae** is certainly not the largest star in the sky, but it is set to grow. Believed to be about 1,000 million years old its helium core is long dead and the star is gradually swelling. It is the brightest semi-regular variable known and is a spectroscopic binary with an orbital period of 20.5212 days. The presence of the companion causes the two stars to interact tidally, leading to increased rotational velocities and significant magnetic activity. The net result is the appearance of massive starspots or very bright regions that cause the star's magnitude to appear to fluctuate by about a tenth as it rotates once every 47 days. It is therefore classed as an RS CVn variable. The corona is also affected by these tidal forces, raising the temperature to between 10 and 40 million K. By comparison, our Sun's corona is a feeble 2 million K.

λ **Andromedae** is also accompanied by no less than three M-class dwarfs. At a little over 1,300 AU a 13th magnitude companion orbits the primary every 25,000 years. A further 4,300 AU on is another binary system which orbit each other with a period of 96,000 years but, together, take about 200,000 years to circle λ And proper.

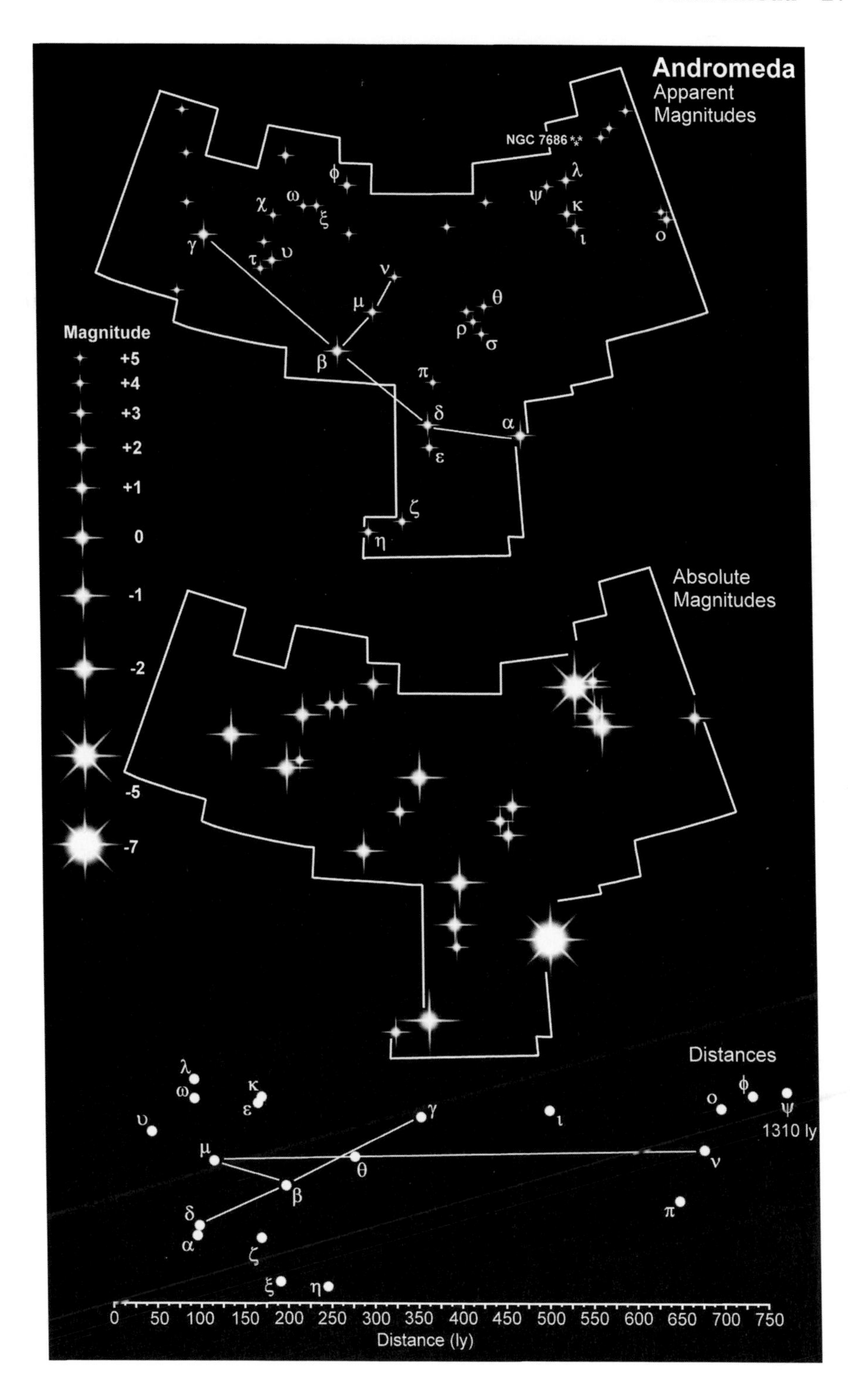

Andromeda
Apparent Magnitudes
NGC 7686
Magnitude
+5
+4
+3
+2
+1
0
-1
-2
-5
-7
Absolute Magnitudes
Distances
1310 ly
0 50 100 150 200 250 300 350 400 450 500 550 600 650 700 750
Distance (ly)

μ **Andromedae** is a useful pointer for finding **M31**, the Great Andromeda Nebula in older texts (now the Andremada Galaxy), which is the farthest object visible to the unaided eye at 2,500,000 ly. It is estimated to be 141,000 ly in diameter. Just follow a line from β And through μ until you come to a faint, tiny smudge on the celestial sphere. Other than that, this A5 white star is unremarkable, lying at a distance of about 136 ly.

At about 679 ly from our own Sun lies ν **Andromedae,** a bluish white B5 star just 3.4 times larger than the Sun but 1,700 times more luminous. A spectroscopic binary its companion is an F8 dwarf that lies just 13.92 million km away: ν^1 And itself is 3.48 million km in diameter. The result is that these two stars have an orbital period of just 4.2828 days. The primary rotates on its axis at about 80 km/s at the equator while the companion spins at a more leisurely 11 km/s. Both stars take about 2 days to complete one spin so they are locked in a gravitational embrace that sees the same hemispheres continually facing one another, in much the same way that the Moon always presents the same face towards Earth. At about 80 million years old ν^1 And is beginning to swell and will eventually envelope its less massive (1.1 $M_\odot$) companion.

The 4th magnitude **o Andromedae** is a member of a four star system. A cataclysmic or eruptive variable with a magnitude range of m_v +3.58 to +3.78, **o^{Aa} Andromedae** is a 6.6 $D_\odot$, 5.9 $M_\odot$ B6 with a 3.1 $M_\odot$ spectroscopic companion, **o^{Ab} Andromedae,** in an 8.3 year orbit. A second binary pair consists of a m_v +6.13 A2 with a mass of 3.7 $M_\odot$, **o^{Ba} Andromedae,** and the 2.9 $M_\odot$ star **o^{Bb} Andromedae,** the pair being in a 33.01 day orbit. The B-pair are in a 65.61 year long orbit with the A-pair.

π^A Andromedae is a hot bluish-white B5 star nearly five times as large as the Sun and more massive by a factor of 5.8. A small telescope or binocular will reveal a 7th magnitude companion, **π^B Andromedae,** an A6 dwarf which is in a long period orbit of at least 153,730 years. Unseen is a third member, **π^{Ab} Andromedae,** which is almost identical to the B5 primary. The two are in a 143.61 day long orbit during which their separation varies between 0.6 AU and 2.1 AU.

Wobbling noticeably through the Galaxy 43 ly away is the yellowish-white **υ Andromedae.** Just over one and a half times the size of the Sun and 3.4 times more luminous it is somewhat younger than our own star by about 1,200 million years and rotates at about 9 km/s, taking 9 days to spin on its axis. The reason it wobbles, however, is because it has a planetary system.

In 1996 two astronomers at San Francisco State University, R. Paul Butler and Geoffrey Marcy, discovered that the star's radial velocity varied rhythmically with time, giving the appearance of the star wobbling. They concluded that this must be due to the presence of a large planet gravitationally tugging the star off course. Calling the planet υ Andromedae b they estimated that the planet was about 69% the mass of Jupiter, lay at a distance of just 8.9 million kilometers, and orbited the star in just 4.6 days. However, further work showed that this single planet model could not fully account for the star's wandering path. Instead, a three planet system was proposed. According to this revised model, υ

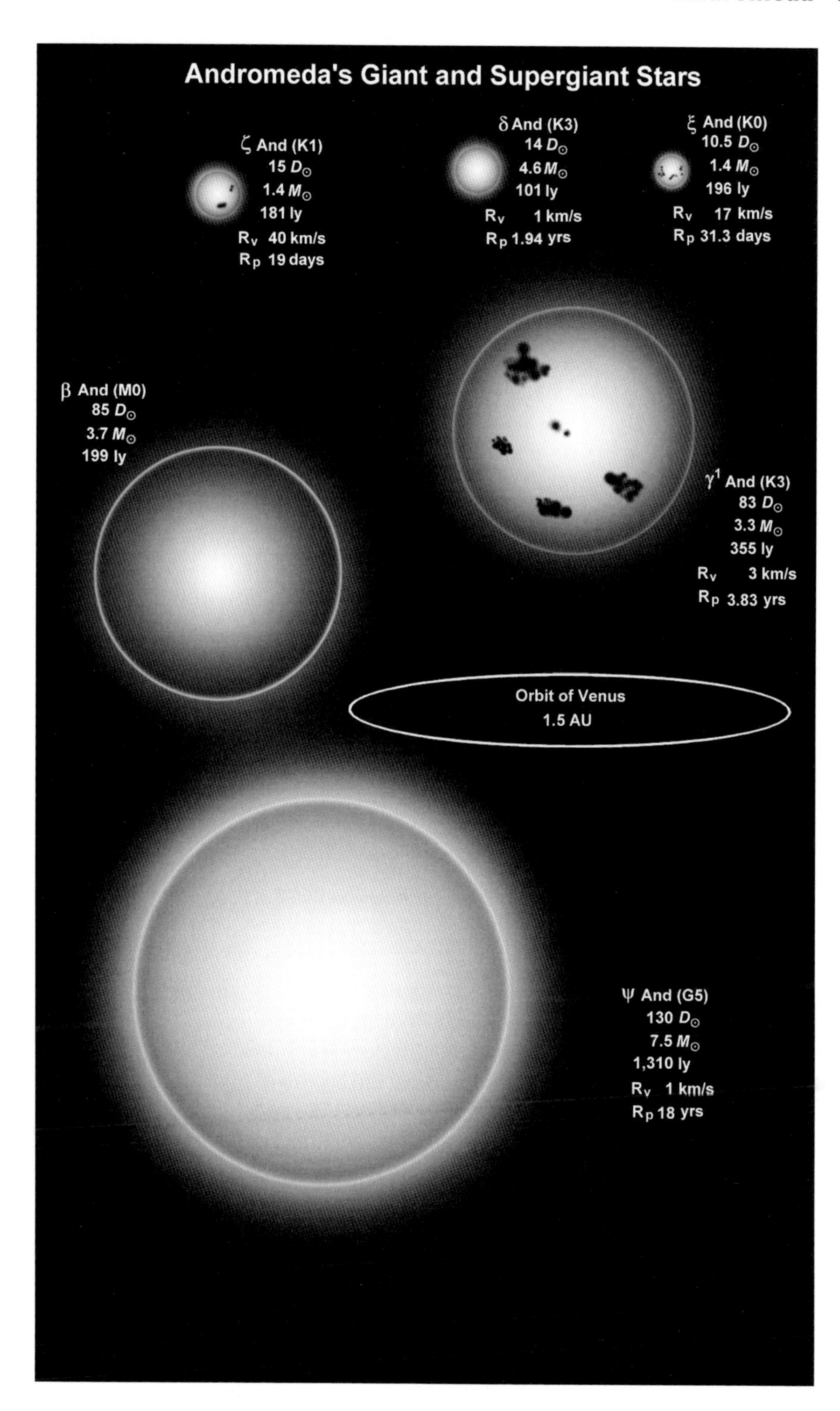
Andromeda's Giant and Supergiant Stars
ζ And (K1)
15 D⊙
1.4 M⊙
181 ly
Rv 40 km/s
Rp 19 days
δ And (K3)
14 D⊙
4.6 M⊙
101 ly
Rv 1 km/s
Rp 1.94 yrs
ξ And (K0)
10.5 D⊙
1.4 M⊙
196 ly
Rv 17 km/s
Rp 31.3 days
β And (M0)
85 D⊙
3.7 M⊙
199 ly
γ1 And (K3)
83 D⊙
3.3 M⊙
355 ly
Rv 3 km/s
Rp 3.83 yrs
Orbit of Venus
1.5 AU
ψ And (G5)
130 D⊙
7.5 M⊙
1,310 ly
Rv 1 km/s
Rp 18 yrs

Andromedae c has a mass of almost two Jupiters and orbits 0.8 AU from the star, taking 241 days to complete an orbit. The third planet, υ Andromedae d, is more massive again at about 4 Jovian masses, and takes 1,290 days to complete its year, lying at about 2.5 AU from the star. This outermost planet resides in what is termed the 'habitable zone'. This is a fairly narrow torus of space around a star in which a planet could theoretically support life, if we use Earth as a model of a typical living planet. However, what Nature tells us time and time again is that life has the amazing ability to survive and even flourish is the most hostile of environments, so the idea that life is restricted to such zones is, perhaps, somewhat naive. υ Andromedae c's orbit also changes from being highly elliptical to almost circular and back again every 13,400 years.

υ Andromedae b poses a bit of a mystery for atmospheric scientists. If an imaginary line is drawn from the center of a star to the center of a planet then the hottest part of the planet will be where the line intersects the planet's surface – the sub-stellar point. If the planet has an atmosphere this hot spot can drift by up to 20°. In the case of υ And b though the hot spot is 80° from the sub-stellar point, almost at right angles to the star. No one is really sure why this is happening but it seems likely that we will find other examples of drifting hot spots among extrasolar planets. One of the other consequences of having such a massive planet as υ And b so close to a star is that it disrupts the gaseous outer layers of the star leading to enhanced activity in the chromosphere. The table below summarizes what we think we know about this planetary system.

Planetary system in Andromeda

Star	$D_{\odot}$	Spectral class	ly	m_v	Planet	Minimum mass	q	Q	P
υ And	1.8	F8	43	+4.10	υ And b	0.69 M_J	0.058	0.060	4.62 d
					υ And c	1.98 M_J	0.646	1.018	241 d
					υ And d	3.95 M_J	1.849	3.180	3.53 y

Almost 2,000 times more luminous than the Sun, ϕ^A **Andromedae** is another interesting binary. Bluish-white it is a B-emission star indicating that it has a circumstellar disk of solid material. B-emission stars tend to rotate at very high speeds, up to about 300 km/s, but ϕ^A And seems to be spinning at a more modest 80 km/s. The most likely explanation is that its pole is tilted towards us so we do not get a true indication of its rotational speed. Its companion is also a bluish-white star in an eccentric orbit that takes it between 80 and 140 AU. It takes 371.6 years to complete an orbit.

Marking the furthest boundary of Andromeda at 1,310 light years is ψ **Andromedae.** To the unaided eye it is an unremarkable white 5th magnitude object. But bring it to 10 parsecs (32.6 ly) of Earth and it would appear to be a brilliant, magnitude M_v -4.5 yellow star. This G5 supergiant is 1,300 $L_{\odot}$ and is heading towards us at 25 km/s.

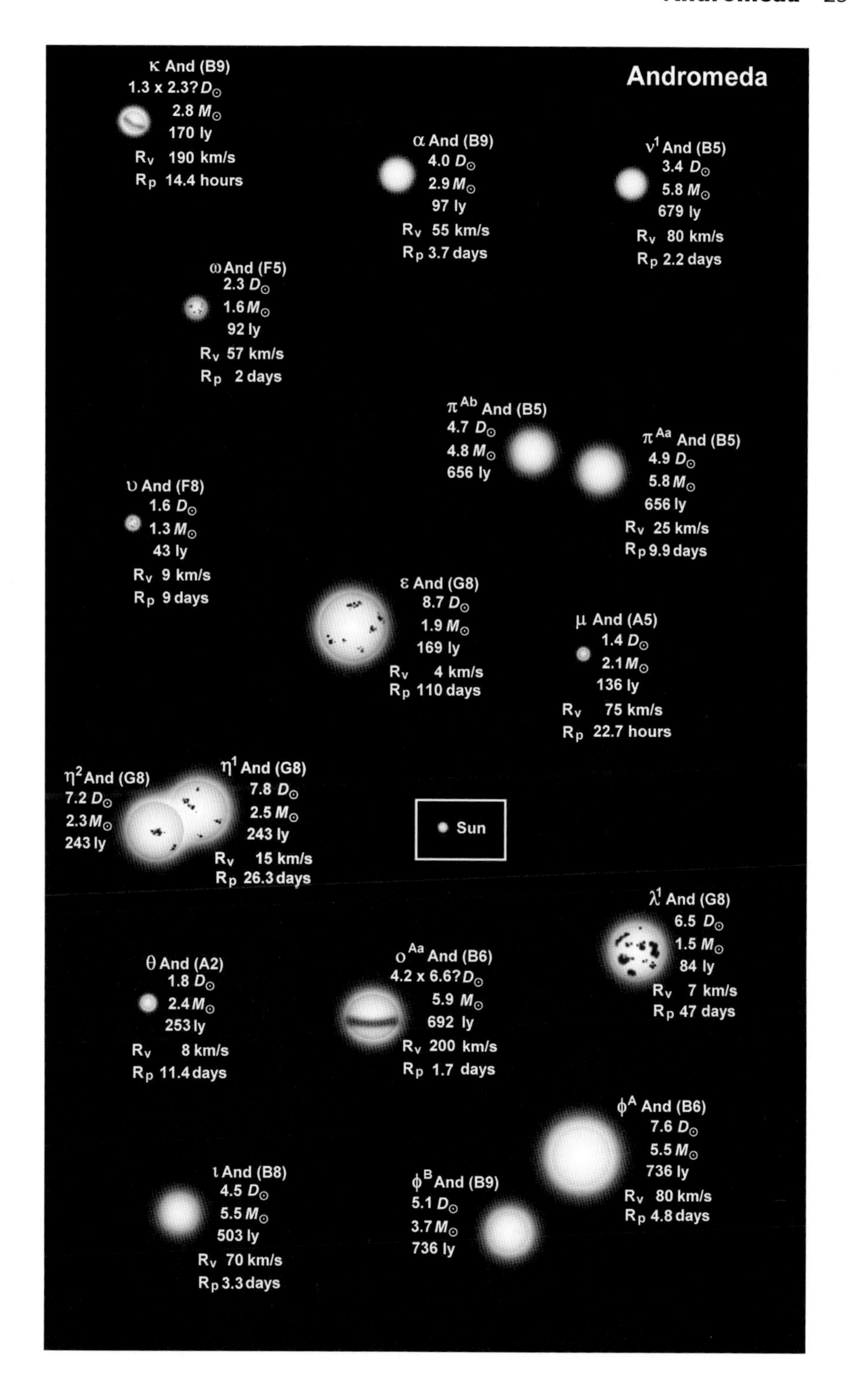

Andromeda
κ And (B9)
1.3 x 2.3? $D_\odot$
2.8 $M_\odot$
170 ly
R_V 190 km/s
R_P 14.4 hours
α And (B9)
4.0 $D_\odot$
2.9 $M_\odot$
97 ly
R_V 55 km/s
R_P 3.7 days
$ν^1$ And (B5)
3.4 $D_\odot$
5.8 $M_\odot$
679 ly
R_V 80 km/s
R_P 2.2 days
ω And (F5)
2.3 $D_\odot$
1.6 $M_\odot$
92 ly
R_V 57 km/s
R_P 2 days
$π^{Ab}$ And (B5)
4.7 $D_\odot$
4.8 $M_\odot$
656 ly
$π^{Aa}$ And (B5)
4.9 $D_\odot$
5.8 $M_\odot$
656 ly
R_V 25 km/s
R_P 9.9 days
υ And (F8)
1.6 $D_\odot$
1.3 $M_\odot$
43 ly
R_V 9 km/s
R_P 9 days
ε And (G8)
8.7 $D_\odot$
1.9 $M_\odot$
169 ly
R_V 4 km/s
R_P 110 days
μ And (A5)
1.4 $D_\odot$
2.1 $M_\odot$
136 ly
R_V 75 km/s
R_P 22.7 hours
$η^2$ And (G8)
7.2 $D_\odot$
2.3 $M_\odot$
243 ly
$η^1$ And (G8)
7.8 $D_\odot$
2.5 $M_\odot$
243 ly
R_V 15 km/s
R_P 26.3 days
Sun
$λ^1$ And (G8)
6.5 $D_\odot$
1.5 $M_\odot$
84 ly
R_V 7 km/s
R_P 47 days
θ And (A2)
1.8 $D_\odot$
2.4 $M_\odot$
253 ly
R_V 8 km/s
R_P 11.4 days
o^{Aa} And (B6)
4.2 x 6.6? $D_\odot$
5.9 $M_\odot$
692 ly
R_V 200 km/s
R_P 1.7 days
$φ^A$ And (B6)
7.6 $D_\odot$
5.5 $M_\odot$
736 ly
R_V 80 km/s
R_P 4.8 days
ι And (B8)
4.5 $D_\odot$
5.5 $M_\odot$
503 ly
R_V 70 km/s
R_P 3.3 days
$φ^B$ And (B9)
5.1 $D_\odot$
3.7 $M_\odot$
736 ly

The closest star to us in Andromeda is **HD 10307** at just 41 ly, marginally beating υ And by just 2 light years. HD 10307 is a close cousin of the Sun. As far as we can tell it is about the same size, but about one-third more luminous and 1,400 million years older. It is also yellow, belonging to spectral class G1.5 (Sun = G2). However where HD 10307 differs significantly from our own star is that it is a binary.[1] Its companion is a red dwarf with an orbital period of just 20 years and an elliptical path that takes it to within 4.2 AU (about 1 AU closer than Jupiter is to the Sun) and out to 10.5 AU (or about 1 AU farther than Saturn is from the Sun).

According to Margaret Turnbull, an astrobiologist at the University of Arizona in Tucson, a planetary system around HD 10307 would be one of the most promising local candidates for life. Turnbull, along with colleague Jill Tarter, have whittled down the 118,218 stars listed in the Hipparcos Catalog to just 17,129 that could have living planets. Their *Catalog of Nearby Habitable Systems* – or *HabCat* – was used to identify stars to which messages could be sent. On 6 July 2003 a message was transmitted to HD 10307 by the 70 m Eupatoria Planetary Radar. Known as Cosmic Call 2 it will arrive in September 2044. Assuming someone is at home we may get a reply around Christmas 2085.

About 5° south west of γ And is the large scattered open cluster **NGC 752**. About a degree across the cluster lies at a distance of 1,300 ly and is about 51 ly in diameter. To the naked eye it appears as a m_v +5.7 fuzzy 'star', but a decent binocular will reveal the individual members, perhaps 100 stars in total. Of these about 60% are brighter than m_v +9.0. The central 0.5° has 90 stars brighter than 12th magnitude but a third are not actually members of the cluster including the three brightest stars: HD 11885, a m_v +7.12 G7 at 584 ly; the A3 HD 12027 a little closer at 553 ly and m_v +8.24, and HD 11720, a m_v +8.07 of unknown distance. About 80% of the cluster members are F-class sub-giants on their way to gianthood. NGC 752 is deficient in smaller, low-mass stars which may have escaped during the 1,100 million year history of the cluster.

Open clusters in Andromeda

Name	Size arc min	Size ly	Distance ly	Age million yrs	Brightest star in region*	No. stars m_v >+12*	Apparent magnitude m_v
NGC 752	117′	51	1,500	1,122	56 And m_v +5.67	94	+5.7
NGC 7686	15	?	?	?	HD 221246 m_v +6.17	21	+5.6

*May not be a cluster member.

[1] University of California Professor Richard Muller, in his book *Nemesis: The Death Star*, proposed that the Sun was also a binary. Its companion disrupted the Oort Cloud every 26 million years, hurling comets into the inner Solar System, some of which struck Earth and were responsible for mass extinctions. The problem with the hypothesis is that the orbit of Nemesis would take it half way to α Centauri and would be inherently unstable.

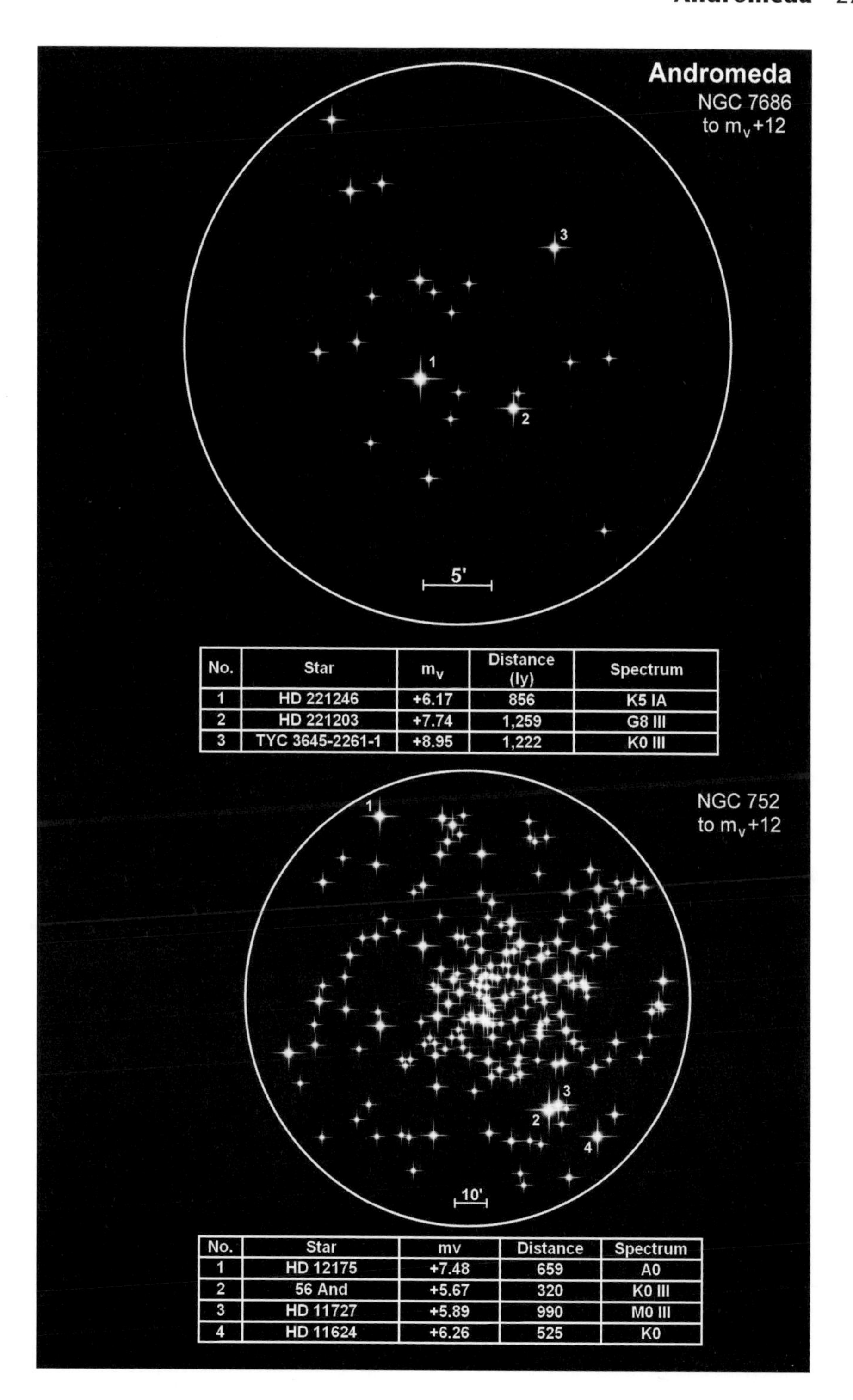

No.	Star	m_V	Distance (ly)	Spectrum
1	HD 221246	+6.17	856	K5 IA
2	HD 221203	+7.74	1,259	G8 III
3	TYC 3645-2261-1	+8.95	1,222	K0 III

No.	Star	mv	Distance	Spectrum
1	HD 12175	+7.48	659	A0
2	56 And	+5.67	320	K0 III
3	HD 11727	+5.89	990	M0 III
4	HD 11624	+6.26	525	K0

Antlia

Constellation:	Antlia	**Hemisphere:**	Southern
Translation:	The Air Pump	**Area:**	239 deg^2
Genitive:	Antliae	**% of sky:**	0.579%
Abbreviation:	Ant	**Size ranking:**	62nd

This faint constellation was introduced by Abbé de La Caille in his *Coelum Australe Stelliferum,* published in 1763, to commemorate Robert Boyle's mechanical pump.

α **Antliae** is an orange K4 giant about 367 ly from Earth but heading away from us at 12 km/s. Some 28 times larger than the Sun, with a luminosity 480 times as great, this 4th magnitude star has an absolute magnitude of M_V -0.4.

ε **Antliae** could be α Ant's twin. It too is an orange giant of spectral class K3, 28 times larger than the Sun but far more luminous at 592 $L_\odot$.It is also a lot farther from us than α Antliae at 700 ly and is travelling at twice the speed but in much the same direction, which raises the question of whether the two stars have a common birthplace.

At 1.6 $D_\odot$ and a mere 106 ly away, η **Antliae** belongs to spectral class F1 and is about 7 times brighter than our own star. Its mass is estimated to be 1.5 $M_\odot$ and it is heading away from us at 30 km/s.

Just 10% larger than the Sun, the A8 θ^A **Antliae** is 135 times more luminous and lies at 384 ly. It is a binary, its whitish-yellow F8 dwarf companion, θ^B **Antliae,** taking 18.32 years to complete an orbit.

ι **Antliae** is the only star in this constellation that is heading towards us – but only just – at 0.2 km/s. A modest K1 giant of 11 $D_\odot$ its mass weighs in at 1.4 $M_\odot$. It is 199 light years distant.

Discovered as recently as 2000 as part of the Deep Near-Infrared Survey (DENIS), **DENIS 1048-39** could be one of the closest 'stars' to us at just 13.2 ly. It may also be a brown dwarf about 10% the size of the Sun. It belongs to spectral type M9, has a very low temperature of about 2,500K and a luminosity of just 0.00015 $L_\odot$.

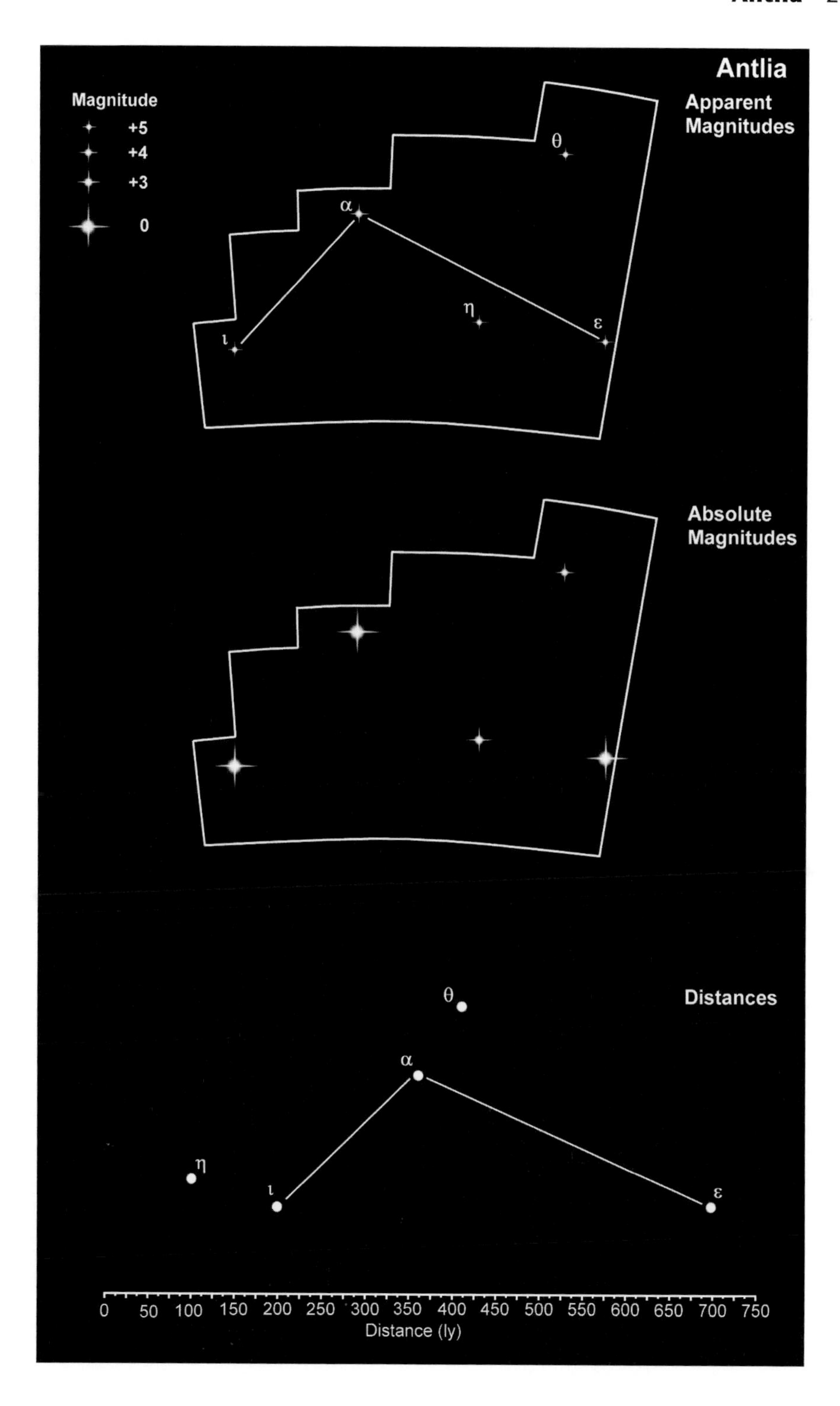

Antlia
Apparent Magnitudes
Magnitude
+5
+4
+3
0
θ
α
η
ε
ι
Absolute Magnitudes
Distances
θ
α
η
ι
ε
0 50 100 150 200 250 300 350 400 450 500 550 600 650 700 750
Distance (ly)

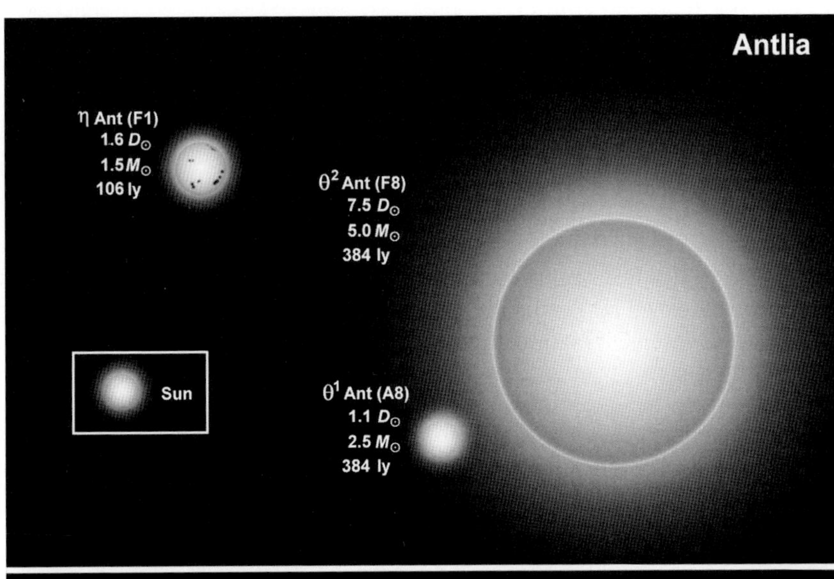
Antlia
η Ant (F1)
1.6 $D_\odot$
1.5 $M_\odot$
106 ly
θ² Ant (F8)
7.5 $D_\odot$
5.0 $M_\odot$
384 ly
Sun
θ¹ Ant (A8)
1.1 $D_\odot$
2.5 $M_\odot$
384 ly

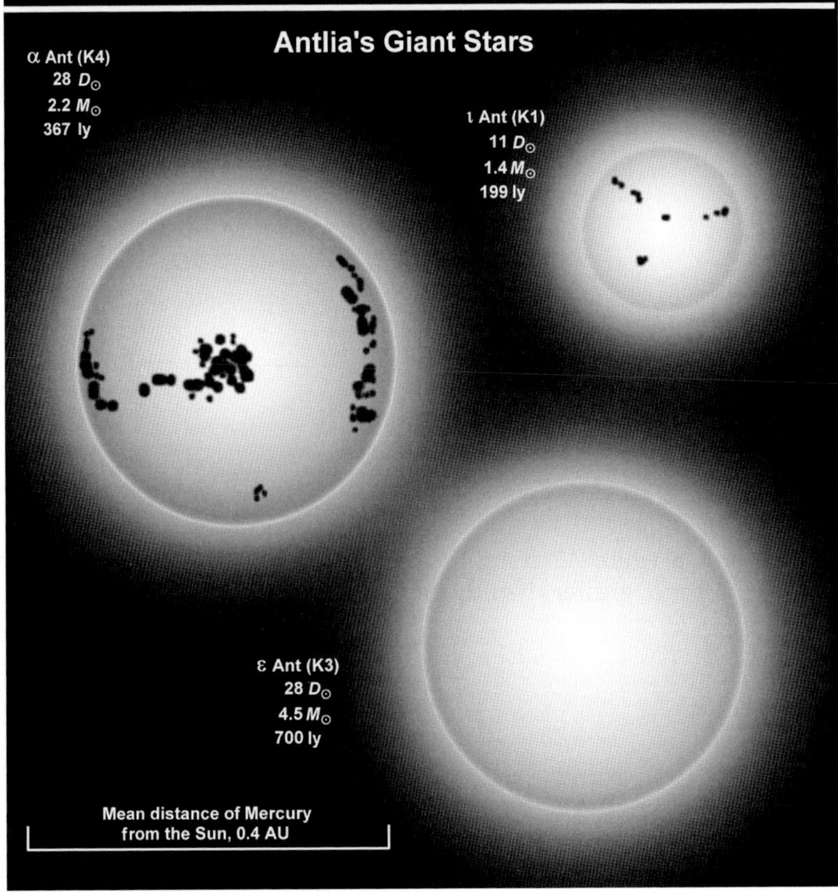
Antlia's Giant Stars
α Ant (K4)
28 $D_\odot$
2.2 $M_\odot$
367 ly
ι Ant (K1)
11 $D_\odot$
1.4 $M_\odot$
199 ly
ε Ant (K3)
28 $D_\odot$
4.5 $M_\odot$
700 ly
Mean distance of Mercury
from the Sun, 0.4 AU

Apus

Constellation:	Apus	**Hemisphere:**	Southern
Translation:	The Bird of Paradise	**Area:**	206 deg^2
Genitive:	Apodis	**% of sky:**	0.499%
Abbreviation:	Aps	**Size ranking:**	67th

Close to the South Celestial Pole, Apus first appeared about 1597 on a Dutch celestial globe created by Petrus Plancius from observations made by Frederick de Houtman and Pieter Dirkszoon Keyser. Like Antlia it is composed of only a few stars, the brightest of which is magnitude +3.81.

α **Apodis,** the brightest star in the constellation, lies between 385 and 437 ly and shines with the luminosity of 750 Suns. It is a K2.5 giant, 49 $D_\odot$ across.

β **Apodis** at magnitude m_v +4.23 is not the second but the third brightest star in the constellation. It also belongs to the K-spectral class, K0, and is 11 times larger than the Sun. It is a fast moving star and in 25,000 years will end up in Volans.

γ **Apodis** comes in as the second brightest. Sometimes recorded as a G8, sometimes as a K0, it is 11 $D_\odot$ and lies at a distance of 160 ly. Unlike most of the stars in Apus, γ Aps is moving away from us at 5.4 km/s.

δ^1 and δ^2 **Apodis** form a nice double just a few arc seconds apart but, in reality, are separated in space by about 100 ly. δ^1 Aps is the farthest at 766 ly. A red M5 it is 37 times larger than the Sun, 573 times more luminous and is an Lb-class pulsating variable. Its partner, δ^2 Aps, is an orange K3 and 100 ly closer to us.

The bluish ε **Apodis** is a dim m_v +5.04 but would glow at M_v -2 if placed at a distance of 10 parsecs. Currently 551 ly away and receding at 4.5 km/s this 3.9 $D_\odot$ star is as luminous as 1,614 Suns. Rotating at 240 km/s the star is distorted into an oblate spheroid.

Right on the northernmost boundary of the constellation, almost drifting into Ara, is the +4.77 magnitude ζ **Apodis.** A yellowish-orange K2 it dwarfs the Sun being 21 times larger and nearly 100 times more luminous. It has a mass of 1.4 $M_\odot$.

η **Apodis** is a white A2 star 140 ly from Earth. It is 2.2 $D_\odot$ across, has a luminosity 16 times greater than the Sun and appears as a magnitude m_v +4.90 star but, at 10 parsecs, would be magnitude M_v +1.7. It has a mass of 1.8 $M_\odot$.

θ **Apodis** is one of only a handful of naked eye late M-class red giants. An estimated 308 to 348 ly away it is 33 $D_\odot$ and over a period of 119 days swings from a faint m_v +5.50 to an invisible m_v +8.60 and back.

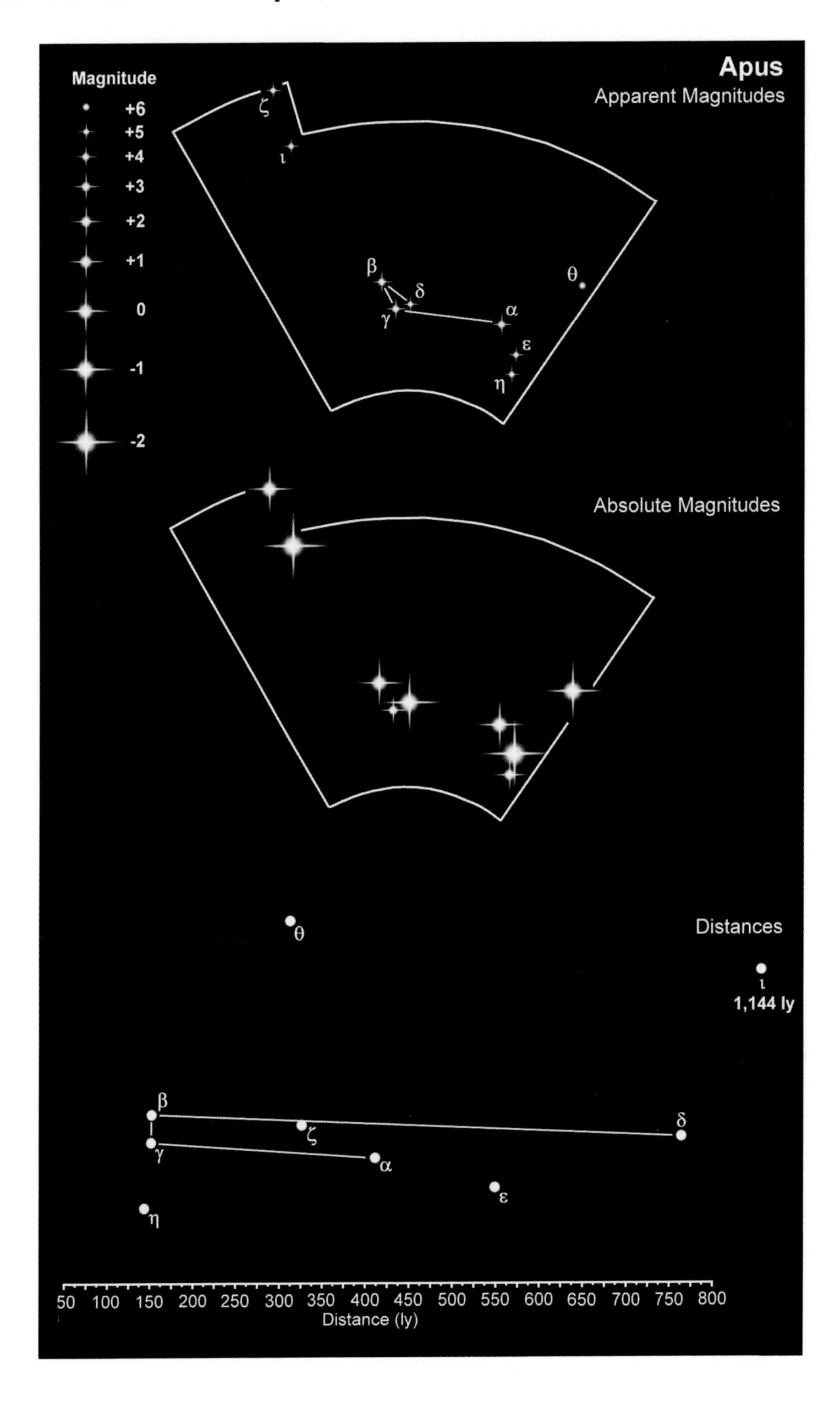
Apus
Apparent Magnitudes
Magnitude
+6
+5
+4
+3
+2
+1
0
-1
-2
ζ
ι
β
δ
γ
α
θ
ε
η
Absolute Magnitudes
Distances
θ
ι
1,144 ly
β
δ
ζ
γ
α
ε
η
50 100 150 200 250 300 350 400 450 500 550 600 650 700 750 800
Distance (ly)

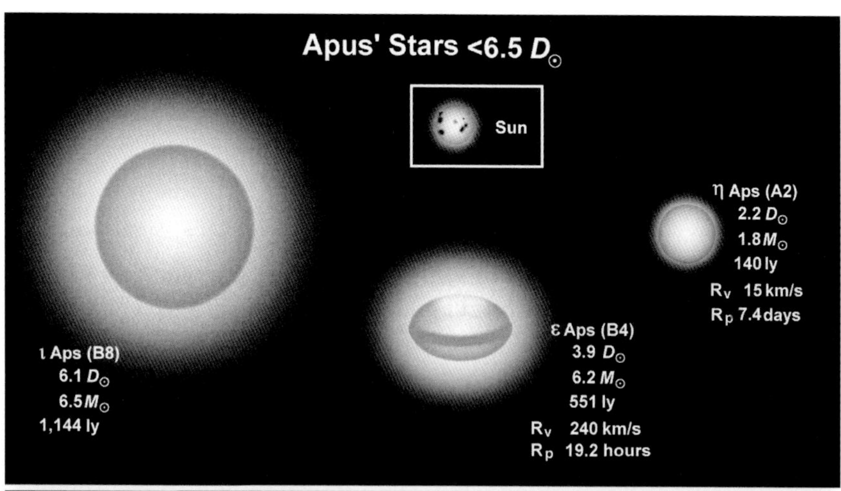
Apus' Stars <6.5 $D_\odot$
Sun
η Aps (A2)
2.2 $D_\odot$
1.8 $M_\odot$
140 ly
R_V 15 km/s
R_P 7.4 days
ε Aps (B4)
3.9 $D_\odot$
6.2 $M_\odot$
551 ly
R_V 240 km/s
R_P 19.2 hours
ι Aps (B8)
6.1 $D_\odot$
6.5 $M_\odot$
1,144 ly

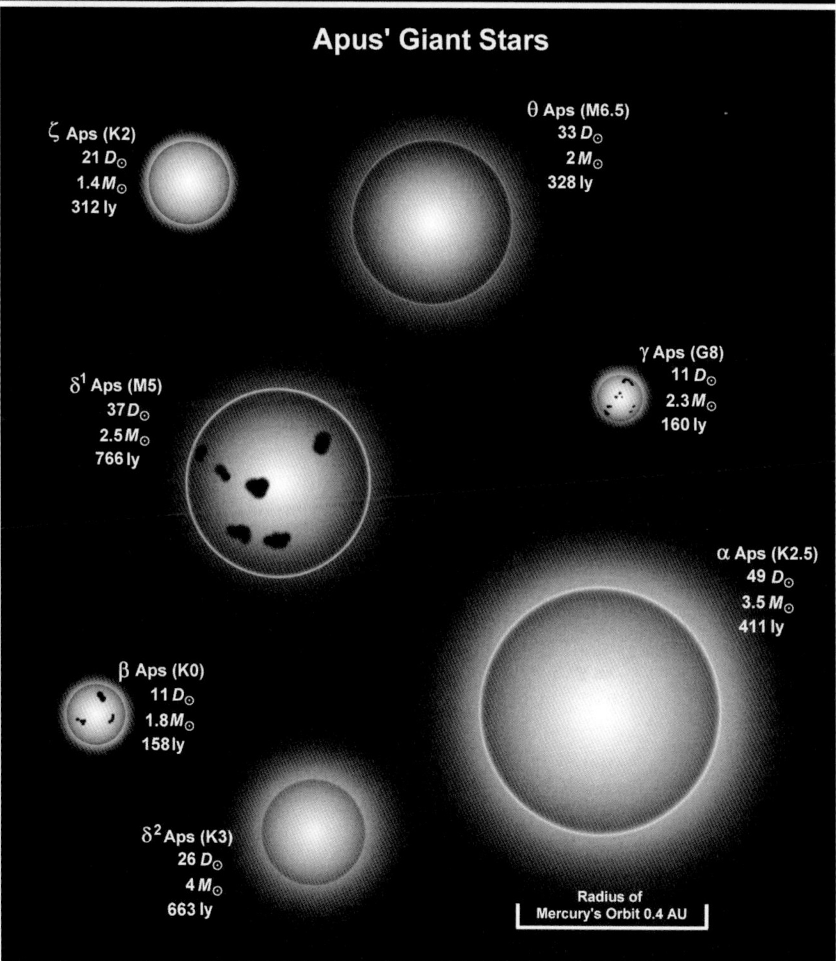
Apus' Giant Stars
θ Aps (M6.5)
33 $D_\odot$
2 $M_\odot$
328 ly
ζ Aps (K2)
21 $D_\odot$
1.4 $M_\odot$
312 ly
γ Aps (G8)
11 $D_\odot$
2.3 $M_\odot$
160 ly
δ^1 Aps (M5)
37 $D_\odot$
2.5 $M_\odot$
766 ly
α Aps (K2.5)
49 $D_\odot$
3.5 $M_\odot$
411 ly
β Aps (K0)
11 $D_\odot$
1.8 $M_\odot$
158 ly
δ^2 Aps (K3)
26 $D_\odot$
4 $M_\odot$
663 ly
Radius of
Mercury's Orbit 0.4 AU

Aquarius

Constellation:	Aquarius	**Hemisphere:**	Equatorial
Translation:	The Water Carrier	**Area:**	980 deg^2
Genitive:	Aquarii	**% of sky:**	2.376%
Abbreviation:	Aqr	**Size ranking:**	10th

One of the 12 Zodiacal constellations it depicts a man pouring water from a vase. The Sun enters Aquarius on 16 February and leaves on 11 March.

With a modest magnitude of m_v +2.93 α **Aquarii** or Sadalmelik is quite deceptive. At 110 $D_\odot$ it is by far the largest star in the constellation with a luminosity 3,000 times greater than the Sun. Placed in the center of our own Solar System α Aqr would more than fill the orbit of Mercury and would swallow up 1,331,000 Suns. This G2 yellow supergiant has an absolute magnitude of M_v -4.5, making it about as bright as Venus at her most brilliant. It lies at a distance of 759 ly.

About 20 solar diameters smaller than α Aqr is **β Aquarii** or Sadalsuud; the 'Luckiest of the Lucky'. It is actually slightly brighter than α Aqr at magnitude m_v +2.87, but 1,000 times less luminous. Even so, the two stars would have an absolute magnitude of M_v -4.5 at 10 parsecs from Earth. Like α, β Aqr is a yellow supergiant, this time a G0, and is travelling away from us at 6.7 km/s. α Aqr, by comparison, is moving at a slightly faster 7.2 km/s and a common origin for both stars seems likely. A third star, ε Pegasi, may be part of the same group.

One of the smallest star in Aquarius γ **Aquarii,** a bluish-white A0 star, 2.4 times larger than the Sun but more than 55 $L_\odot$. Its Arabic name, Sadachbia, means 'Lucky Star of Hidden Things' which is somewhat intriguing as it is actually a spectroscopic binary. Virtually nothing is known of its companion other than it lies about 0.4 AU from γ Aqr and orbits it once every 58 days.

The white 3rd magnitude δ **Aquarii** is also known as Skat but, like so many original names, its meaning is lost in time. It lies at 160 ly and is 4.3 $D_\odot$ across.

There is some evidence that ε **Aquarii** has an accretion disk that has not, as yet, accreted into any planets sufficiently large to be detected from Earth. An A1.5 star almost five times the size of the Sun its Arabic name, Al Bali, means 'The Good Fortune of the Swallower'. Its mass is 2.8 $M_\odot$.

Most books on astronomy will tell you that ζ **Aquarii** is a double: a fine pair of brilliant white F2 stars, ζ^1 and ζ^2 Aqr, that have an orbital period of about 760 years. In fact, there are three stars in the system but the determination of their orbits has been fraught with problems including misidentification of the various components, erroneous assumptions and a third star which, because it is so much fainter than the ζ^1 and ζ^2, is effectively invisible and we have had to rely on, for example, infrared detection methods. The two main stars appear to have an orbit which brings them as close together as, perhaps, 90-95 AU but which also separates them by as much as 200-220 AU at their furthest point. The orbital period could well be 760 years but could also be as long as 856 years or as short as

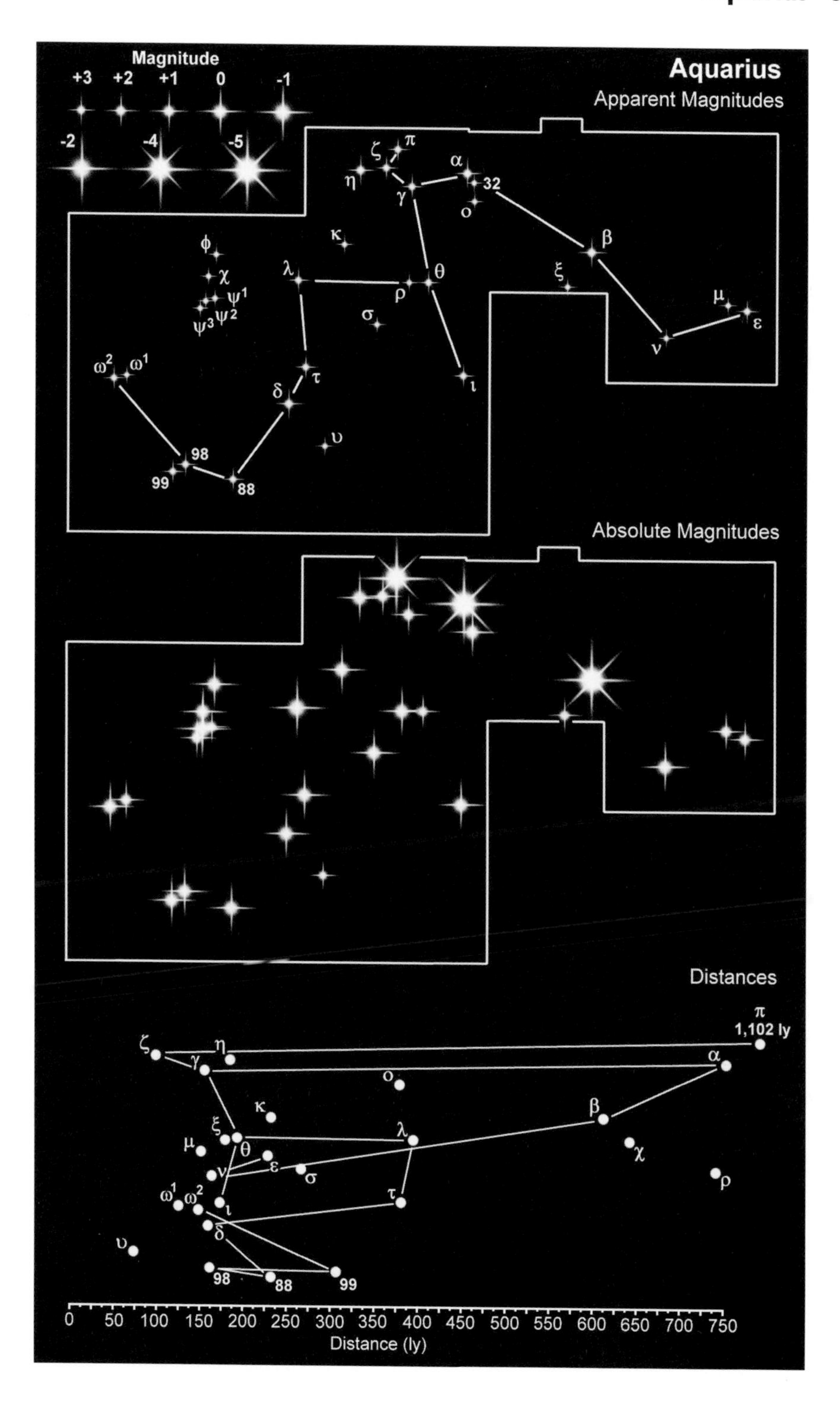

Magnitude
+3 +2 +1 0 -1
-2 -4 -5
Aquarius
Apparent Magnitudes
ζ π η α 32 γ ο κ β φ χ ψ1 ψ3 ψ2 λ θ ρ ξ μ ε σ ν ω2 ω1 τ δ ι υ 98 99 88
Absolute Magnitudes
Distances
π
1,102 ly
0 50 100 150 200 250 300 350 400 450 500 550 600 650 700 750
Distance (ly)

486.7 years. The third star, which was discovered as long ago as 1942 by K.Strand, seems to have a 25.7 y orbital period and has often been associated with ζ^2 Aqr but may, in fact, be in orbit around ζ^1. These uncertainties aside, what we *think* we know about ζ^1 is that it is 4.2 $D_\odot$ in size and is 16 times more luminous than the Sun. ζ^2 is slightly smaller, at 3.4 $D_\odot$, but is considerably more luminous at 31 $L_\odot$. The third star is very much a mystery. They are all heading away from us at just over 24 km/s. One thing is for certain, however, they are migrant stars. Until November 2003 they resided in the Southern Hemisphere but precession – the gradual gyration of the Celestial Poles – has caused the stars to drift into the Northern Hemisphere.

η **Aquarii** is 2.5 times larger than the Sun but 104 times more luminous. It appears to be a fast spinner, rotating on its axis once every 10.4 hours which equates to a rotational velocity of 291 km/s. Its high spin speed pushes the poles in towards the center of the star causing its equator to bulge producing an oblate spheroid. Our Sun rotates at just 2 km/s and is almost a perfect sphere. Its magnitude varies between m_v +3.48 and +3.77.

More than a dozen times larger than the Sun and 60 times more luminous θ **Aquarii** is a yellow-going-on-orange G8 star at 191 ly. It weighs in at 2.2 $M_\odot$ and has a lifespan of less than 2,000 million years, just 20% that of our own Sun. Spinning at 4 km/s, twice the speed of the Sun, it takes 157 days to rotate once on its axis.

Thought to be an Lb pulsating variable λ **Aquarii** is 104 $D_\odot$ but has a mass of just 3 $M_\odot$. It is usually a magnitude m_v +3.76 star but brightens by about a tenth at irregular intervals. Not particularly interesting to watch, λ Aqr has spent most of its fuel and is in the final stages of its life.

At a distance of more than 1,100 ly π **Aquarii** appears as a modest 4th magnitude star, but at 10 parsecs it would rival Venus in brightness. Although only 6 times larger than the Sun it is 16,000 times more luminous. It actually varies in its apparent magnitude between m_v +4.42 and +4.70 but over a period of some 40 years. Classed as a γ Cas eruptive variable it rotates at a speed of at least 278 km/s causing the equator to bulge and the poles to be drawn in. Also known as a Be- or B-emission 'shell star' it is surrounded by a shell or an equatorial disk of gas, angled at about 60° to our line of sight. It is accompanied by a spectroscopic companion, a 2.7 $M_\odot$ star in an 84.1 day long orbit.

τ **Aquarii** is a red giant, 53 times larger than the Sun. It is currently 380 light years away and ambling away from us at a steady 1 km/s.

At magnitude m_v +5.21 υ **Aquarii** is a difficult star to spot unless the skies are sufficiently dark, but it is a useful finder star for the Helix Nebula which lies just 5′ away (see below).

A star teetering on the edge of stability is χ **Aquarii,** an aged M3, 27 times larger than the Sun. Classed as an Lb pulsating variable its magnitude changes between m_v +4.9 and +5.3 though with no apparent rhythm. χ Aqr is preparing to shed its outer gaseous layers before settling down to become a white dwarf for a few million years.

You have to have exceptionally good eyesight and dark, clear skies to be able

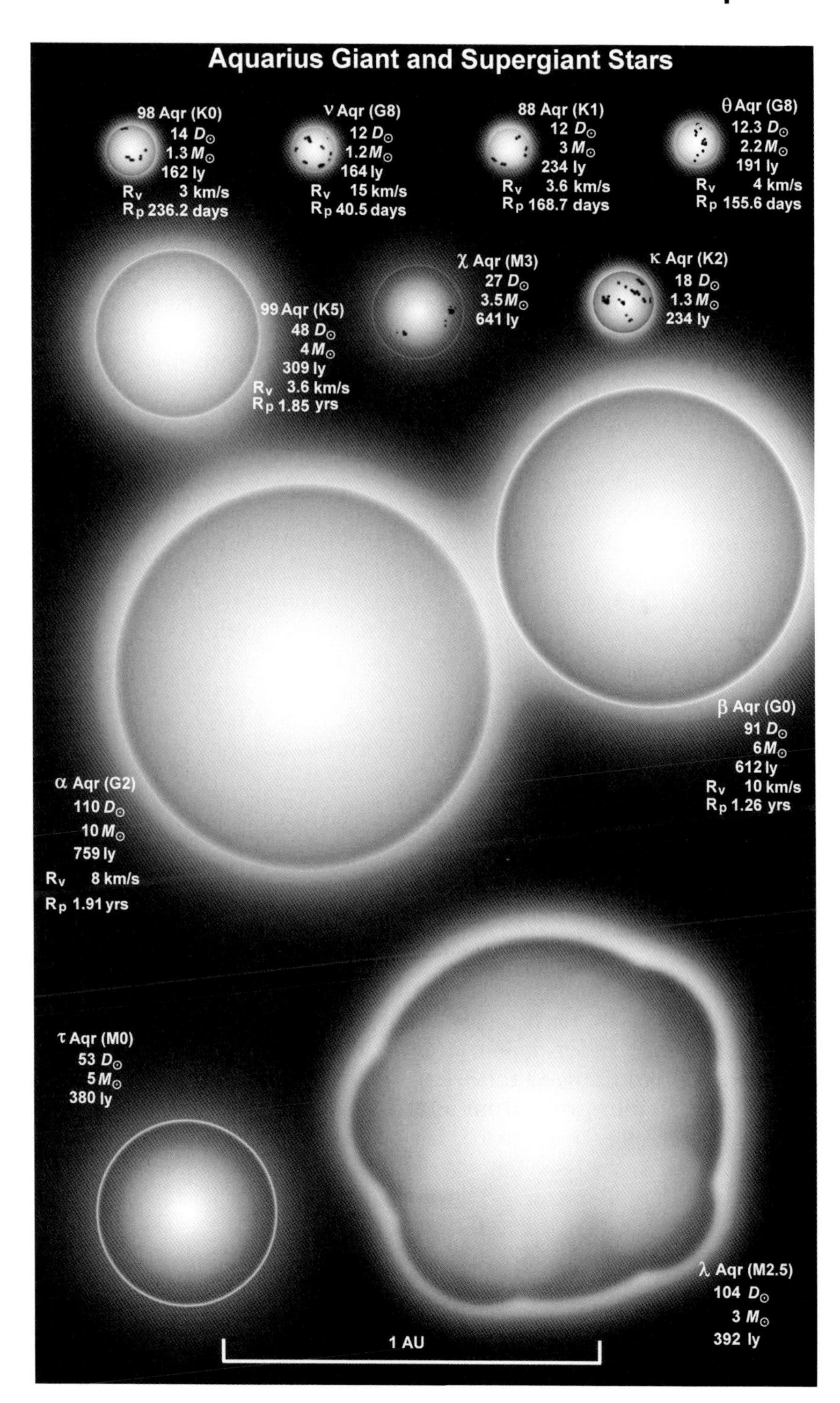

Aquarius Giant and Supergiant Stars
98 Aqr (K0)
14 $D_\odot$
1.3 $M_\odot$
162 ly
R_v 3 km/s
R_p 236.2 days
ν Aqr (G8)
12 $D_\odot$
1.2 $M_\odot$
164 ly
R_v 15 km/s
R_p 40.5 days
88 Aqr (K1)
12 $D_\odot$
3 $M_\odot$
234 ly
R_v 3.6 km/s
R_p 168.7 days
θ Aqr (G8)
12.3 $D_\odot$
2.2 $M_\odot$
191 ly
R_v 4 km/s
R_p 155.6 days
99 Aqr (K5)
48 $D_\odot$
4 $M_\odot$
309 ly
R_v 3.6 km/s
R_p 1.85 yrs
χ Aqr (M3)
27 $D_\odot$
3.5 $M_\odot$
641 ly
κ Aqr (K2)
18 $D_\odot$
1.3 $M_\odot$
234 ly
β Aqr (G0)
91 $D_\odot$
6 $M_\odot$
612 ly
R_v 10 km/s
R_p 1.26 yrs
α Aqr (G2)
110 $D_\odot$
10 $M_\odot$
759 ly
R_v 8 km/s
R_p 1.91 yrs
τ Aqr (M0)
53 $D_\odot$
5 $M_\odot$
380 ly
λ Aqr (M2.5)
104 $D_\odot$
3 $M_\odot$
392 ly
1 AU

to locate the binary **53 Aquarii.** At the limit of naked eye visibility they would not normally get a mention in this book were it not for the fact that they are almost identical to the Sun. The brighter of the two, **53^{A} Aqr**, has a visual magnitude of m_V +6.23 and is a G2 yellow dwarf only slightly smaller than our own Sun (0.94 $D_{\odot}$). Its companion, **53^{B} Aqr**, is a little smaller again at 0.91 $D_{\odot}$ and a G3. Lying at somewhere between 62.3 and 68.7 ly the two were originally thought to be unconnected. We now know they are a true pair with an orbital period of 3,500 years, their orbits bringing them to within 30 AU of one another before flinging them apart by as much as 575 AU.

Lying at a distance of 69.4 ly and with a magnitude of just m_V +6.63, **HD 210277** is invisible to the naked eye but is still of interest. A spectral class G8 yellow dwarf it is almost identical in size and luminosity to the Sun. It also has at least one planet, thought to be 1.28 M_J, in a highly elliptical orbit that takes it to within 0.6 AU of the star and then out to 1.6 AU. Its year is 1.2 Earth years. HD 210277 is surrounded by a flattened, dense dust disk similar to our Kuiper Belt stretching from 30 to 62 AU.

Gliese-Jahreiss 876 is an M3.5 red dwarf just over a third of the size of the Sun and fairly close at 15.3 ly. It seems to have three planets that range in mass from 8.4 $M_{\oplus}$ to 2.64 M_J. The closest planet to the star, GJ 876 d, is also the smallest. It orbits at an average of just 0.021 AU (3.1 million km) from the star and takes 1.94 days to complete an orbit. It is technically a 'super-Earth' with a mass of 6.36 $M_{\oplus}$ (0.02 M_J). At 0.14 AU planet GJ 876 c takes 30.3 days to orbit the star, its mass being 0.62 M_J. The largest planet, GJ 876 b, is 2.64 M_J and has an orbital period of 60.8 days. A number of other stars in Aquarius also have planetary systems (see table).

At a distance of about 695 ly the **Helix Nebula (NGC 7293)** is probably the closest planetary nebula to Earth. It is more than one-third the diameter of the full Moon and appears gray when seen through a telescope. A long exposure photograph, however, shows it to be mainly reds, yellows and blues. Although it looks like a circle of gas it is, in fact, a bubble or shell expanding outwards in all directions from the white hot dwarf in the center. The Helix marks the death of a star. When it uses up all of its fuel, a relatively small star will swell to become a red giant, typically 10 to 100 $D_{\odot}$. Eventually its outer gaseous layers will continue to expand and separate from the core which shrinks to become a white dwarf. The Helix Nebula's parent star ended its red giant stage sometime between 9,500 and 13,000 years ago. The shell is now expanding at between 30 and 40 km/s and is about 5.8 ly in diameter. The remnant white dwarf star is thought to be perhaps as small as 40,000 km across but highly luminous. Eventually it will shrink to 10,000 km diameter – slightly smaller than the Earth – and fade over thousands of millions of years to become a black dwarf. The nebula itself contains around 20,000 'knots' which look like comets. Their average length is 50 AU or about equal to the size of Pluto's orbit at its farthest point from the Sun.

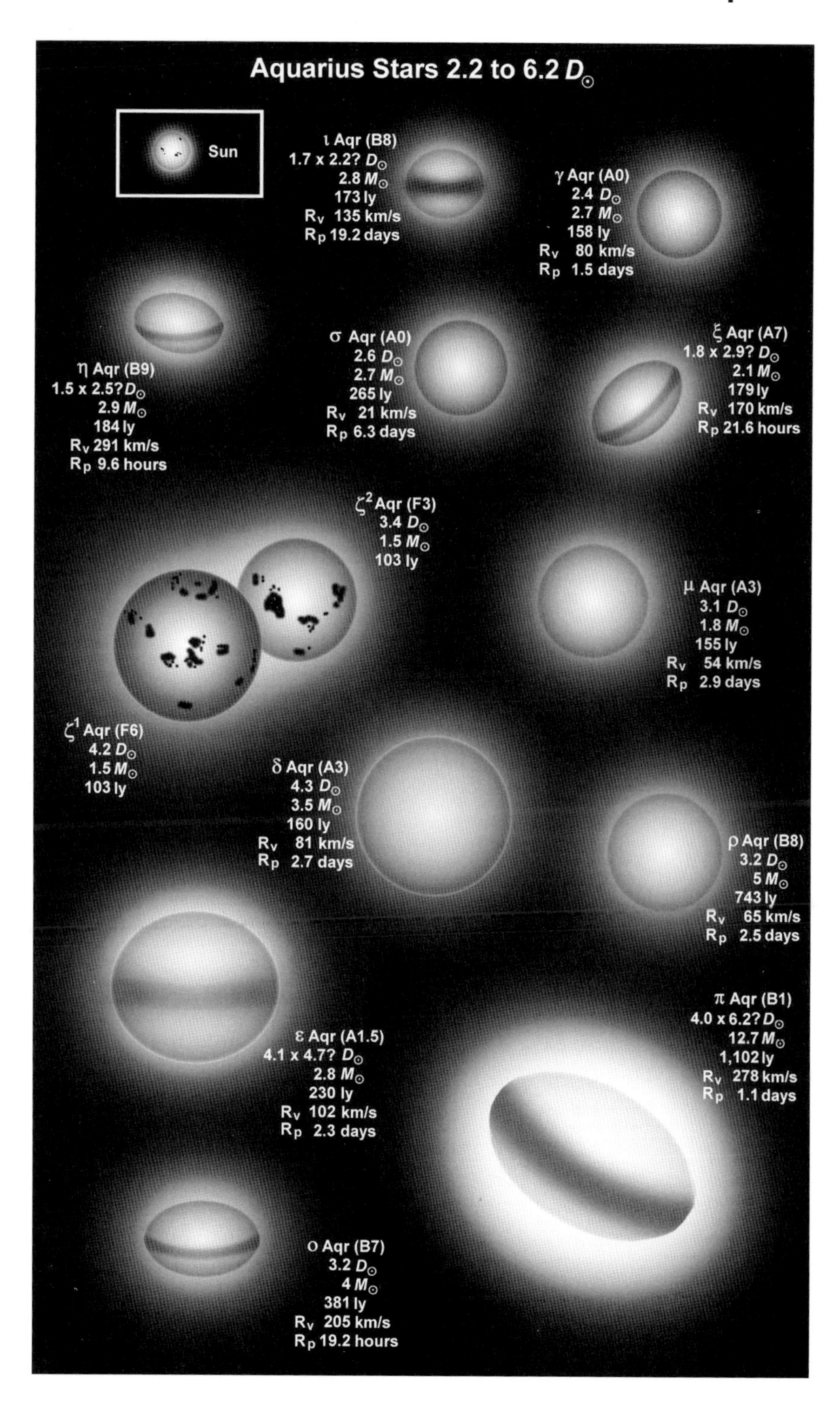
Aquarius Stars 2.2 to 6.2 $D_\odot$
Sun
ι Aqr (B8)
1.7 x 2.2? $D_\odot$
2.8 $M_\odot$
173 ly
R_V 135 km/s
R_P 19.2 days
γ Aqr (A0)
2.4 $D_\odot$
2.7 $M_\odot$
158 ly
R_V 80 km/s
R_P 1.5 days
η Aqr (B9)
1.5 x 2.5? $D_\odot$
2.9 $M_\odot$
184 ly
R_V 291 km/s
R_P 9.6 hours
σ Aqr (A0)
2.6 $D_\odot$
2.7 $M_\odot$
265 ly
R_V 21 km/s
R_P 6.3 days
ξ Aqr (A7)
1.8 x 2.9? $D_\odot$
2.1 $M_\odot$
179 ly
R_V 170 km/s
R_P 21.6 hours
ζ^2 Aqr (F3)
3.4 $D_\odot$
1.5 $M_\odot$
103 ly
μ Aqr (A3)
3.1 $D_\odot$
1.8 $M_\odot$
155 ly
R_V 54 km/s
R_P 2.9 days
ζ^1 Aqr (F6)
4.2 $D_\odot$
1.5 $M_\odot$
103 ly
δ Aqr (A3)
4.3 $D_\odot$
3.5 $M_\odot$
160 ly
R_V 81 km/s
R_P 2.7 days
ρ Aqr (B8)
3.2 $D_\odot$
5 $M_\odot$
743 ly
R_V 65 km/s
R_P 2.5 days
π Aqr (B1)
4.0 x 6.2? $D_\odot$
12.7 $M_\odot$
1,102 ly
R_V 278 km/s
R_P 1.1 days
ε Aqr (A1.5)
4.1 x 4.7? $D_\odot$
2.8 $M_\odot$
230 ly
R_V 102 km/s
R_P 2.3 days
ο Aqr (B7)
3.2 $D_\odot$
4 $M_\odot$
381 ly
R_V 205 km/s
R_P 19.2 hours

Planetary systems in Aquarius

Star	$D_{\odot}$	Spectral class	ly	m_v	Planet	Minimum mass	q	Q	P
GJ 849	0.5	M3.5	28.6	+10.4	GJ 849 b	0.82 M_J	2.209	2.491	5.18 y
GJ 876	0.4	M3.5	15.3	+10.2	GJ 876 b	2.64 M_J	0.205	0.217	60.8 d
					GJ 876 c	0.62 M_J	0.097	0.167	30.3 d
					GJ 876 d	6.36 $M_{\oplus}$	0.018	0.024	1.94 d
HD 210277	1.1	G8	69.4	+6.63	HD 210277 b	1.28 M_J	0.581	1.619	1.21 y
HD 219449	7.1	K0	148	+4.25	HD 219449 b	1.90 M_J	0.300	0.300	182 d
HD 222582	1.2	G5	136	+7.70	HD 222582 b	7.75 M_J	0.371	2.329	1.57 y

Aquila

Constellation:	Aquila	**Hemisphere:**	Equatorial
Translation:	The Eagle	**Area:**	652 deg^2
Genitive:	Aquilae	**% of sky:**	1.580%
Abbreviation:	Aql	**Size ranking:**	22nd

In mythology Aquila carried the thunderbolts of Zeus.

At magnitude m_v +0.77 **α Aquilae** is the 11th brightest star in the night sky. Or the 12th. It all depends on α Orionis – Betelgeuse – which varies between magnitudes +0.58 and +1.3. α Aql is one of the closest stars to us at 16.7 ly and also fluctuates in brightness, but quite erratically and by only a few thousandths of a magnitude making it a δ Scuti-class variable. Popularly known by its Arabic name Altair (meaning 'flying eagle') it is the 39th swiftest moving star, flying off towards Delphinus *en route* to Vulpecula, the fox. It covers a full degree in just 6,626 years. Physically it is also a fast spinner with a rotational velocity of 240 km/s, rotating once on its axis in just 10.3 hours compared to the Sun which takes more than 25 days (rotational velocity of 2 km/s). The effect of such rapid rotation is that the equator bulges outwards so the globe has turned into an oblate spheroid, 2.03 $D_\odot$ (2.83 million km) across its equator but just 1.63 $D_\odot$ (2.37 million km) pole to pole. This causes Altair to display the von Zeipel effect in which the equator darkens and the mid-latitude and polar regions develop bright hot spots. This is because the surface gravity and effective temperature are less at the equator. Altair is a white A7 Main Sequence star, 10.6 times more luminous than the Sun and has a mass of 1.79 $M_\odot$. At 1,000 million years it is less than a quarter of the age of the Sun.

β Aquilae or Alshain is a yellow G8 star three times the size of the Sun. It is also a physical binary, its companion being an m_v +11.6 red dwarf 175 AU from the primary. The main star is a lightweight 1.3 $M_\odot$ that has a temperature of 5,100 K. It is believed to spin on its axis at about 1.8 km/s completing a full rotation once every 84.4 days.

γ Aquilae or Tarazed may be a modest m_v +2.69 magnitudes but its yellowish-orange tincture betrays the fact that this is a giant K3 of 110 $D_\odot$ – just a bit larger than the orbit of Mercury. It is 461 ly distant but, at 10 parsecs, would brighten to M_v -2.3. Believed to be just 100 million years old – 1/45th the age of the Sun – γ Aql is about 2,960 times as luminous as the Sun and has a temperature of just 4,100 K.

The F0 yellowish-white **δ Aquilae** or Song lies just 50 light years from Earth and is typical of this type of star. F-class stars make up only 2% of all Main Sequence members and range in temperature from 7,200 K (F0) to 6,100 K (F8). δ Aql has a diameter of 1.7 $D_\odot$ (the average F0 is 1.5 $D_\odot$), is 8.7 times more luminous than the Sun (average is 6.5 $L_\odot$) and has a mass of 1.65 $M_\odot$ (average 1.60 $M_\odot$). It is believed to rotate at 30 km/s, taking just 2.9 days to spin once on its axis. δ Aql is a triple star system. Its closest companion, δ^C Aql, orbits the primary in just 3.77 hours. δ^B Aql is farther out and has an orbital period of 3.42 years.

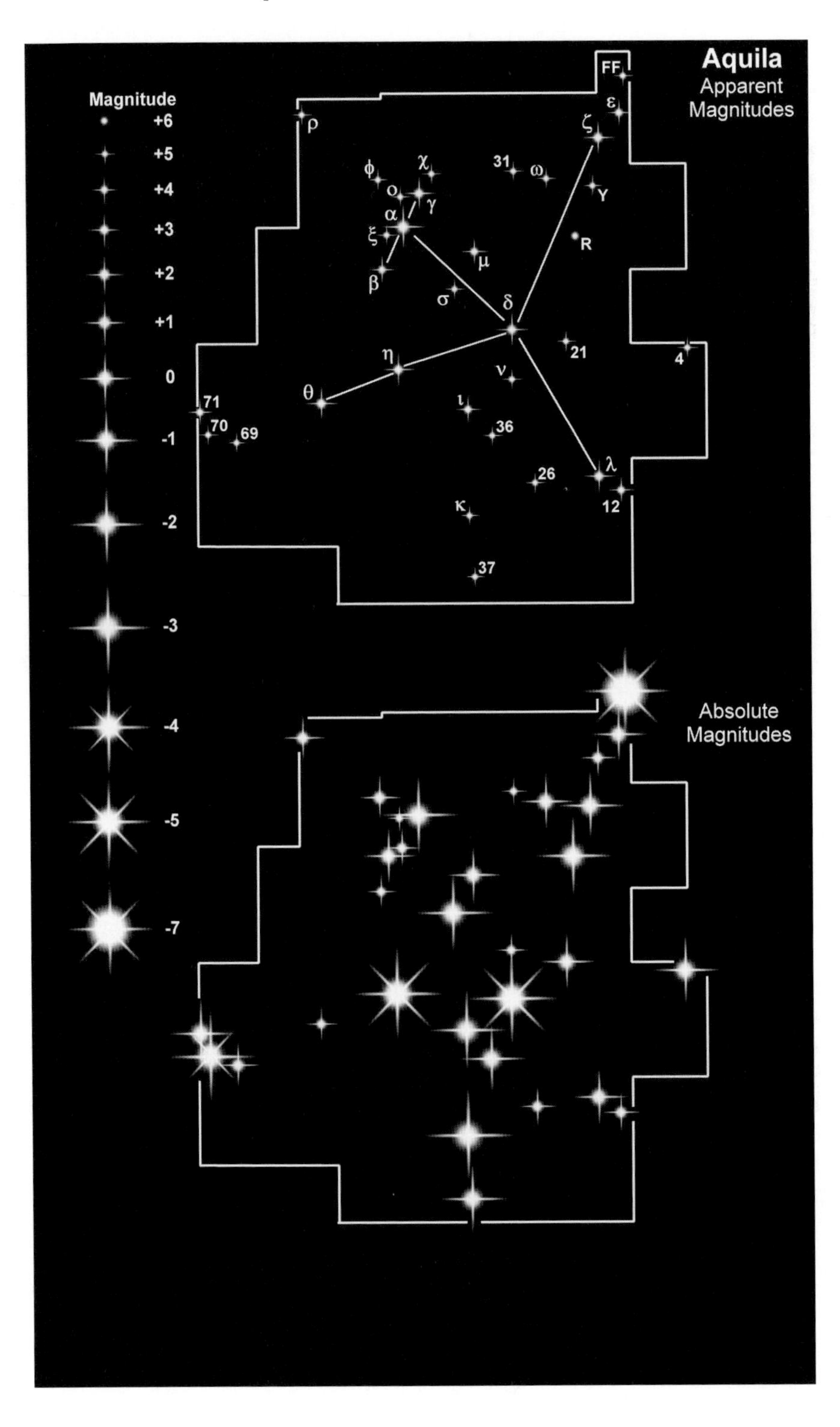
Aquila
Apparent
Magnitudes
Magnitude
+6
+5
+4
+3
+2
+1
0
-1
-2
-3
-4
-5
-7
FF
ε
ζ
ρ
φ
χ
31
ω
ο
γ
Y
α
ξ
R
μ
β
σ
δ
21
4
η
ν
θ
71
70
69
ι
36
26
λ
12
κ
37
Absolute
Magnitudes

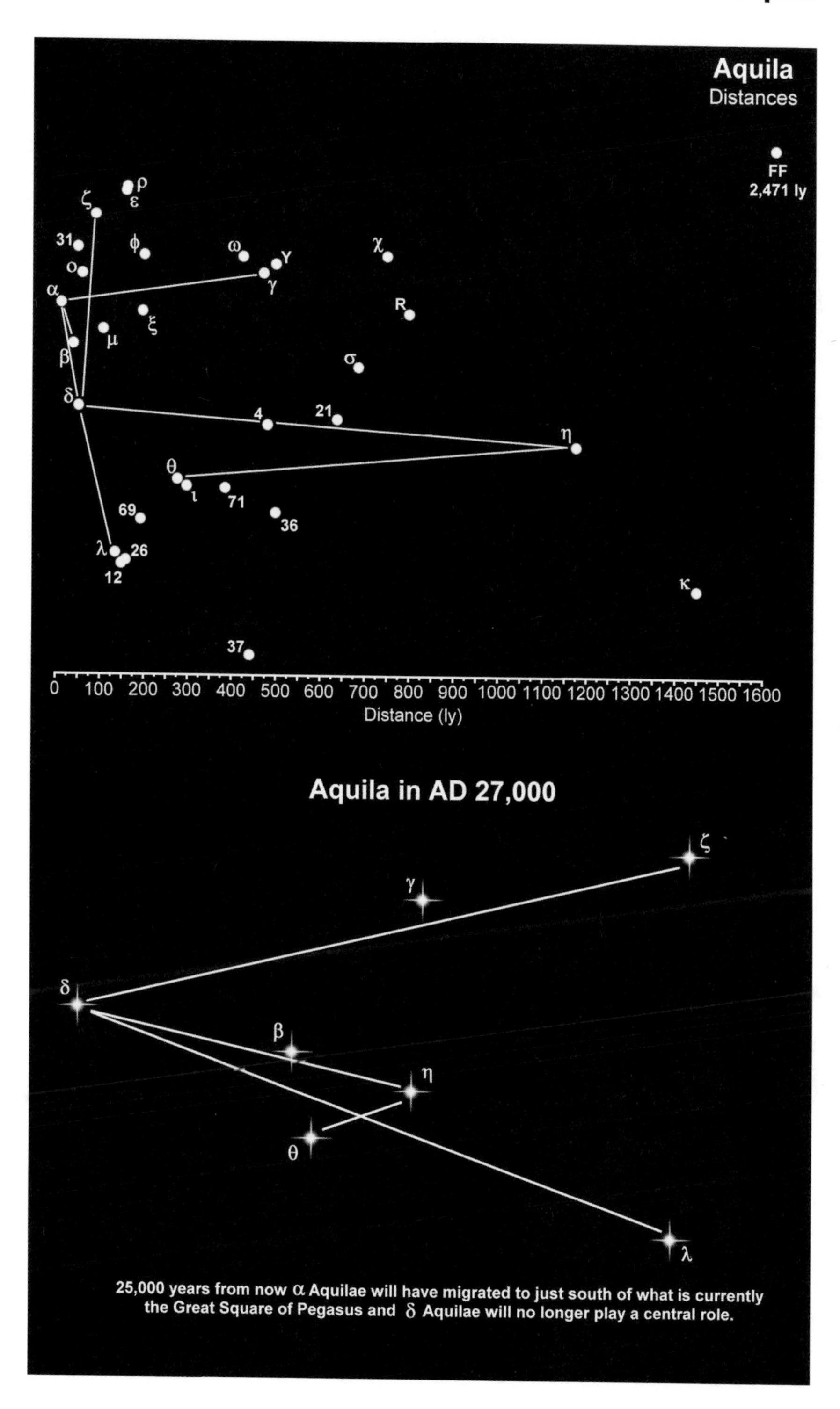
Aquila
Distances
FF
2,471 ly
ρ
ε
ζ
31
φ
ω
Y
χ
ο
γ
α
R
ξ
μ
β
σ
δ
4
21
η
θ
ι
71
69
36
λ
26
12
κ
37
0 100 200 300 400 500 600 700 800 900 1000 1100 1200 1300 1400 1500 1600
Distance (ly)
Aquila in AD 27,000
ζ
γ
δ
β
η
θ
λ
25,000 years from now α Aquilae will have migrated to just south of what is currently the Great Square of Pegasus and δ Aquilae will no longer play a central role.

ε **Aquilae** is a barium-rich K1 star, 17 $D_{\odot}$ across and with a luminosity of 44.4 $L_{\odot}$. It may have a companion in a 3.5 AU orbit with a 1,270 day period and a second in a long period orbit of 110,000 years, taking it out as far as 3,700 AU from the primary star. ε Aql is heading towards us at a rate of knots – 48 km/s – putting it among the top 2% of the fastest moving stars.

Like α Aql, ζ **Aquilae** is a fast spinning young star – but this one is heading for the record books with a rotational velocity of at least 300 km/s. It is also an A-class star, this time an A0 compared to the A7 of α Aql, with a diameter of 2.8 $D_{\odot}$ and a lot more luminous: 35 $L_{\odot}$. It also has two faint companions, although it is not clear whether these are really associated or just line-of-site illusions. They both appear to be M-class dwarfs, one lying at a distance of about 125 AU and with an orbital period of 800 years, the other more than 6,000 AU distant and taking at least 250,000 years to complete a full orbit.

η **Aquilae** or Bezek is one of the brighter pulsating Cepheid variables – also called a Cδ – and varies in magnitude from m_v +3.48 to +4.39 over a period of $7^d\ 4^h\ 14^m\ 22^s$. The outer layers of the star expand and contract as regular as clockwork over a week, with the temperature changing from a high of 6,200K to a low of 5,300K. Along with changes in magnitude and temperature, η Aquilae's spectrum also changes from F6 to G0, yellowing in color as it does so. There is some disagreement as to the actual size of η Aql with some authorities estimating an average 60 $D_{\odot}$ and others as large as 100 $D_{\odot}$. In reality, of course, the diameter will vary by perhaps as much as 30% as the star pulsates. Another consequence of the pulsations is that its radial velocity – the speed at which a star is either moving towards or away from us – swings between positive and negative, appearing to move towards us as the star expands and then traveling away from us as the star contracts. Overall, though, it is heading in our direction at 14.8 km/s. Strangely enough, η Aql was the very first 'Cepheid' to be discovered.

η Aquilae is not the only Cepheid variable in this constellation. **FF Aquilae** varies by just 3/10th of a magnitude over $4^d\ 11^h\ 18^m\ 7^s$ but is a far more impressive object than η with a diameter of 219 $D_{\odot}$ – the size of the Earth's orbit – and a luminosity of around 3,400 Suns. At 10 parsecs it would reach a brilliant M_v of -6.5.

θ Aquilae is a double and possibly a triple star. The visual component, θ^A Aql, is a white hot B9.5, rather more than two solar diameters across and 327 times more luminous. Lying at 287 ly it appears as a +3.2 magnitude star. Hidden almost behind it is θ^B Aql, a 5th magnitude companion in an orbit that varies between 0.1 and 0.4 AU, the stars circling their common center of gravity in just 17.1 days. The system is probably very young, on a cosmic timescale, being around 200 million years old. The third, rather tenuous component, θ^C Aql, is a 13th magnitude object in a 300,000 year orbit that is separated from the other two by 10,000 AU. It is possible that θ^C Aql is just a line-of-sight illusion and is not physically related.

The faint +4.67 magnitude **ν Aquilae** marks the furthest boundary of the constellation at 11,649 ly, though it could be as far as 20,461 ly or as near as 2,837 ly, such is the margin of error in our estimation of its distance. One thing is for certain, though, with a diameter of 67 $D_{\odot}$ ν Aql is a supergiant F2 with a

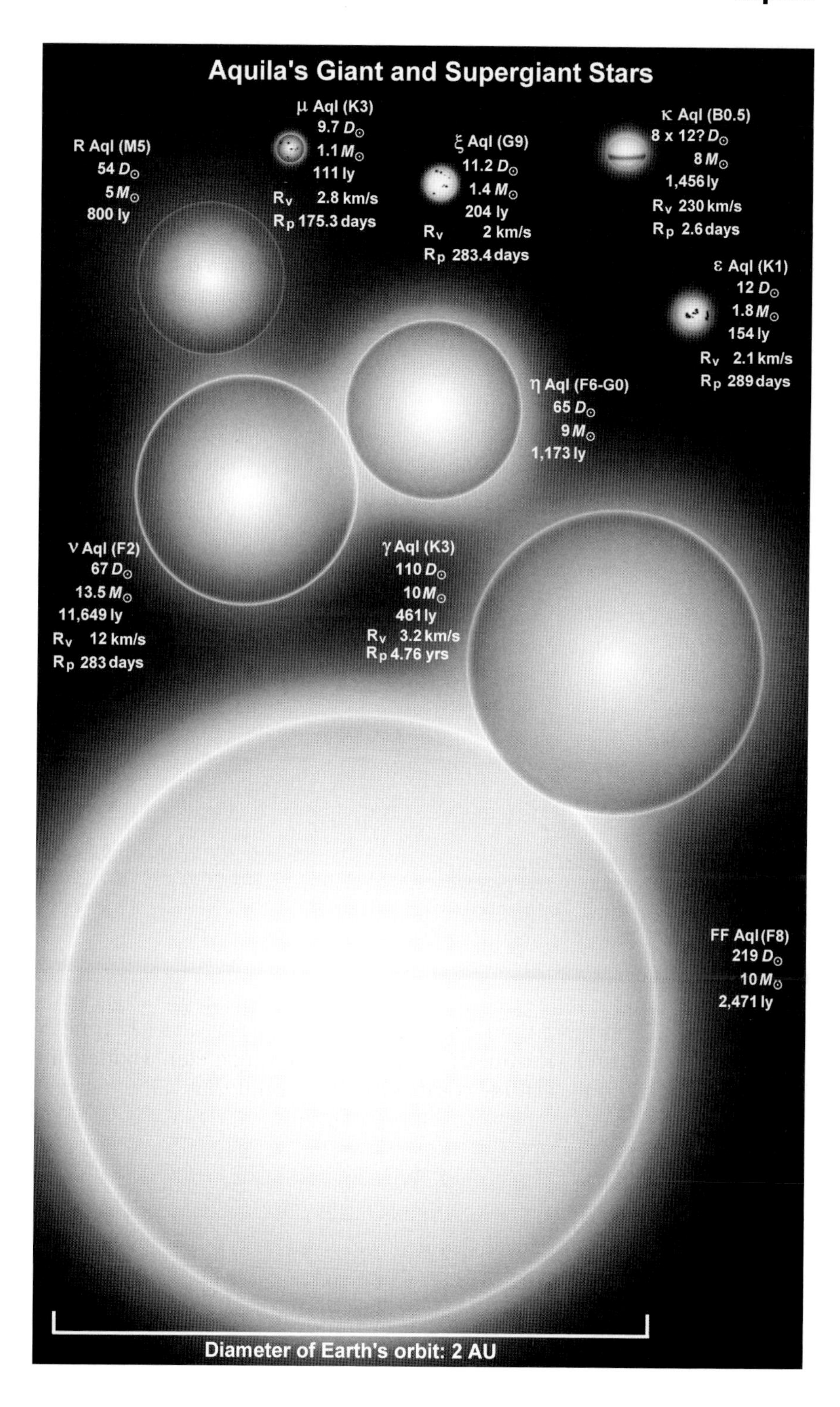
Aquila's Giant and Supergiant Stars
R Aql (M5)
54 $D_{\odot}$
5 $M_{\odot}$
800 ly
μ Aql (K3)
9.7 $D_{\odot}$
1.1 $M_{\odot}$
111 ly
R_V 2.8 km/s
R_P 175.3 days
ξ Aql (G9)
11.2 $D_{\odot}$
1.4 $M_{\odot}$
204 ly
R_V 2 km/s
R_P 283.4 days
κ Aql (B0.5)
8 x 12? $D_{\odot}$
8 $M_{\odot}$
1,456 ly
R_V 230 km/s
R_P 2.6 days
ε Aql (K1)
12 $D_{\odot}$
1.8 $M_{\odot}$
154 ly
R_V 2.1 km/s
R_P 289 days
η Aql (F6-G0)
65 $D_{\odot}$
9 $M_{\odot}$
1,173 ly
ν Aql (F2)
67 $D_{\odot}$
13.5 $M_{\odot}$
11,649 ly
R_V 12 km/s
R_P 283 days
γ Aql (K3)
110 $D_{\odot}$
10 $M_{\odot}$
461 ly
R_V 3.2 km/s
R_P 4.76 yrs
FF Aql (F8)
219 $D_{\odot}$
10 $M_{\odot}$
2,471 ly
Diameter of Earth's orbit: 2 AU

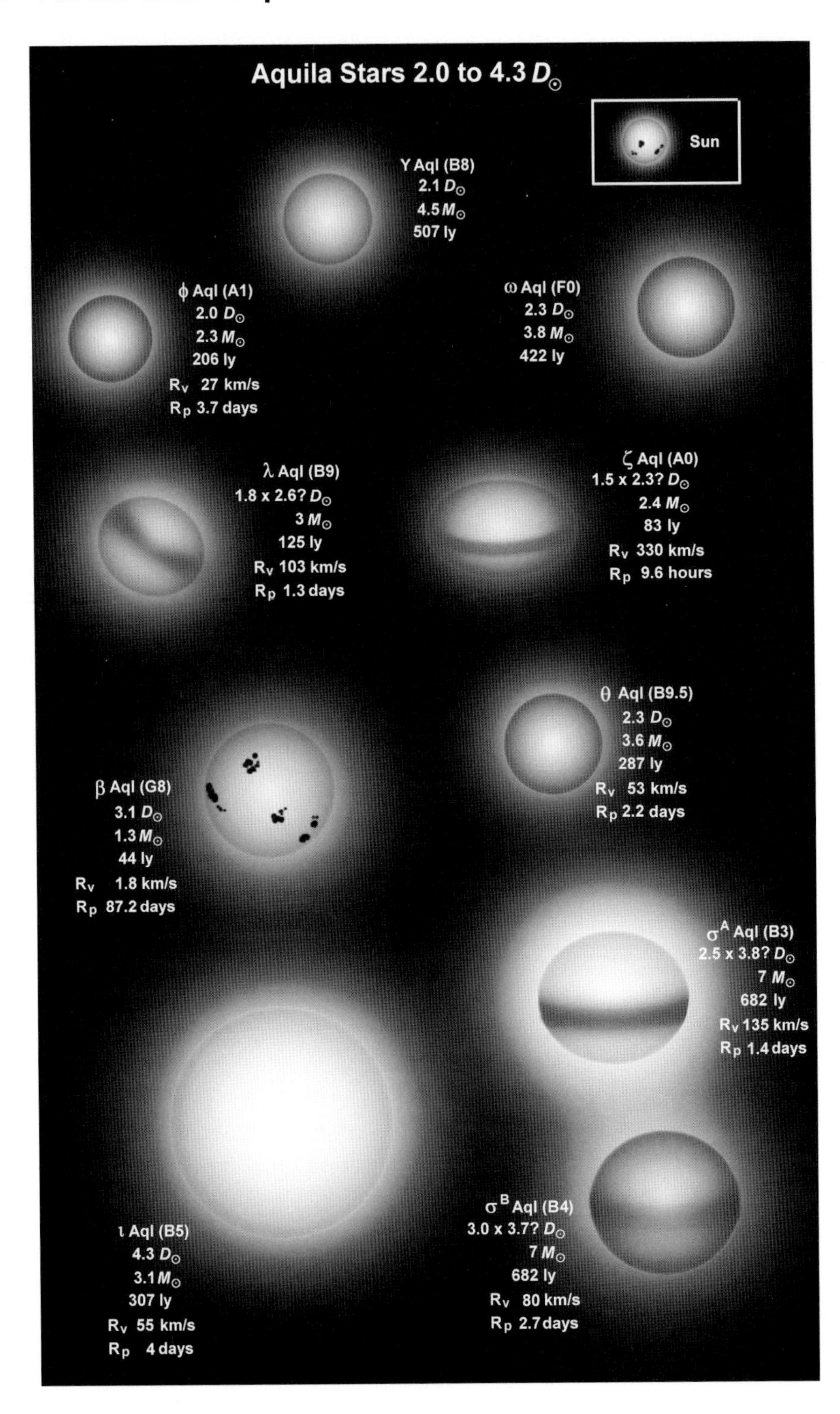
Aquila Stars 2.0 to 4.3 $D_\odot$
Sun
Y Aql (B8)
2.1 $D_\odot$
4.5 $M_\odot$
507 ly
ϕ Aql (A1)
2.0 $D_\odot$
2.3 $M_\odot$
206 ly
R_V 27 km/s
R_P 3.7 days
ω Aql (F0)
2.3 $D_\odot$
3.8 $M_\odot$
422 ly
λ Aql (B9)
1.8 x 2.6? $D_\odot$
3 $M_\odot$
125 ly
R_V 103 km/s
R_P 1.3 days
ζ Aql (A0)
1.5 x 2.3? $D_\odot$
2.4 $M_\odot$
83 ly
R_V 330 km/s
R_P 9.6 hours
θ Aql (B9.5)
2.3 $D_\odot$
3.6 $M_\odot$
287 ly
R_V 53 km/s
R_P 2.2 days
β Aql (G8)
3.1 $D_\odot$
1.3 $M_\odot$
44 ly
R_V 1.8 km/s
R_P 87.2 days
σ^A Aql (B3)
2.5 x 3.8? $D_\odot$
7 $M_\odot$
682 ly
R_V 135 km/s
R_P 1.4 days
ι Aql (B5)
4.3 $D_\odot$
3.1 $M_\odot$
307 ly
R_V 55 km/s
R_P 4 days
σ^B Aql (B4)
3.0 x 3.7? $D_\odot$
7 $M_\odot$
682 ly
R_V 80 km/s
R_P 2.7 days

luminosity of between 110,900 and 169,330 $L_{\odot}$. If the mid range is correct then ν Aql would shine as a M_V -4.6 star at 10 pc.

At first glance **o Aquilae** is not much different from our Sun. It's an F8 (the Sun is a G2), is slightly larger at 1.3 $D_{\odot}$ and nearly three times more luminous – usually. But in 1998 o Aql suddenly brightened. This unusual event was due to a superflare: a coronal mass ejection of material up to 10 million times more energetic than the flares observed on the Sun. No one is absolutely certain why superflares happen – only nine stars are known to exhibit this phenomenon – but it may be caused by the interaction of the star's magnetic field with that of a giant planet. Both F and G-type stars can release superflares, which is potentially catastrophic for life on Earth with our G2 Sun. So far, however, there is no real evidence that the Sun undergoes flaring on this scale. Researchers believe that superflares occur once every 100 years or so.

ρ Aquilae is a wandering star. Having started in Aquila it has now migrated to neighboring Delphinus, crossing the border in 1999. It is very young, somewhere between 50 and 150 million years and, as might be expected, is buried in a cloud of dust which reduces it magnitude by about 15%. It is probably too early for planets to have been formed within the dust cloud although there is some spectroscopic evidence to suggest a planet, though this has never been independently confirmed. An A2, it is twice the size of the Sun and 19 times more luminous. It takes about 14 hours to spin once on its axis at 165 km/s.

σ Aquilae is an EB eclipsing binary. Consisting of two blue stars (a B3 and a B4) separated by just 0.07 AU – 10.5 million kilometers – they eclipse one another every $1^d\ 22^h\ 48^m\ 23^s$. The two stars seem to be similar in size at 3.7 and 3.8 $D_{\odot}$ but they are also fast spinners at about 80 and 135 km/s. As a result they are oblate, bulging at their equators. Couple this with the fact that the orbit is somewhat inclined to our line of sight and it all means that the two stars never completely pass in front of one another but just partially eclipse their partners. The rotational speeds also mean that the two stars are face locked, the same hemispheres always pointing towards one another.

31 Aquilae is one of the most rapidly moving naked eye stars. It is hurtling towards us at 100 km/s making it the 8th fastest star in the sky. Perhaps gravitationally kicked out of its original galactic orbit it brings with it a legacy of its neighborhood in the form of an unusually rich mix of elements including iron, sulfur, silicon, magnesium, oxygen and carbon. A yellow G8, it is 1.7 $D_{\odot}$ and 1.59 $L_{\odot}$ and currently lies at just 49 light years away. It will be in our neighborhood in just 53.7 million years.

A Mira-type pulsating variable lurks half way along the right hand wing of the Eagle. Known simply as **R Aquilae** it is a red giant, 54 $D_{\odot}$ across and 800 light years away in what is called the Great Rift in the Milky Way. For most of its 284.2 day cycle it is invisible to the unaided eye, dipping to 12th magnitude and cooling from a 3,100K M5 to a 2,600K M8 before brightening again to magnitude +5.5. There is no guarantee that, during any particular cycle, R Aql will brighten enough to be seen without optical aid. Mira-type variables are somewhat erratic in their behavior and R Aquilae's cycle also seems to be shortening: 100 years ago it was about 350 days. Well worth recording when it is visible.

Two other naked eye variables are on display in Aquila, but their fluctuations are barely noticeable. **Y Aquilae** (or **18 Aql** if you prefer the Flamsteed number) changes from magnitude +5.02 to +5.06 with a period of $1^d\ 7^h\ 15^m\ 16^s$. A bluish white dwarf, twice the size of the Sun, it is an elliptical eclipsing E-binary. **V1288 Aquilae** (**21 Aql**) is an α CV rotating variable. Another B8 it is 1.6 $D_\odot$ and dips from magnitude +5.16 to +5.06 with a period of $1^d\ 17^h\ 31^m\ 12^s$.

Lurking in the middle of the constellation is Van Biesbroeck's star, a red dwarf of such low mass – just 0.08 $M_\odot$ – it is barely a star, tottering on the brink of being a brown dwarf. It was discovered in 1944 by George van Biesbroeck and is quite close to us at 18.72 ly. Often just referred to as **VB 10** (it is also better known as **V1298 Aquilae**) it is thought to have a planet, one of 7 stars in the constellation to harbor planets but all too dim to be seen with the naked eye (see table). VB10 is just 13 times the mass of its planet.

Planetary systems in Aquila

Star	$D_\odot$	Spectral class	ly	m_v	Planet	Minimum mass	q	Q	P
CoRoT-3	1.6	F3	2,200	+13.3	CoRoT-3 b	4.25 M_J	0.057	0.057	21.7 d
CoRoT-6	1.0	F5	?	+13.9	CoRoT-6 b	3.30 M_J	0.077	0.094	8.9 d
HD 179079	1.5	G5	210	+7.96	HD 179079 b	0.08 M_J	0.097	0.123	14.5 d
HD 183263	1.2	G2	180	+7.86	HD 183263 b	3.69 M_J	0.942	2.098	1.74 y
					HD 183263 c	3.82 M_J	4.143	4.358	8.08 y
HD 192263	0.8	K2	65	+7.79	HD 192263 b	0.72 M_J	0.150	0.150	24.4 d
HD 192699	4.3	G8	220	+6.44	HD 192699 b	2.50 M_J	0.987	1.333	351.5 d
V1298 Aql	0.1	M8	18.72	+17.3	VB 10 b	6.40 M_J	0.007	0.713	271.5 d

Open clusters in Aquila

Name	Size arc min	Size ly	Distance ly	Age million yrs	Brightest star in region*	No. stars m_v >+12*	Apparent magnitude m_v
NGC 6709	40′	41	3,500	151	No identification m_v +7.60	84	+6.7
NGC 6755	29′	39	4,630	52.4	DM +03° 3903 m_v +10.23	25	+7.5

*May not be a cluster member.

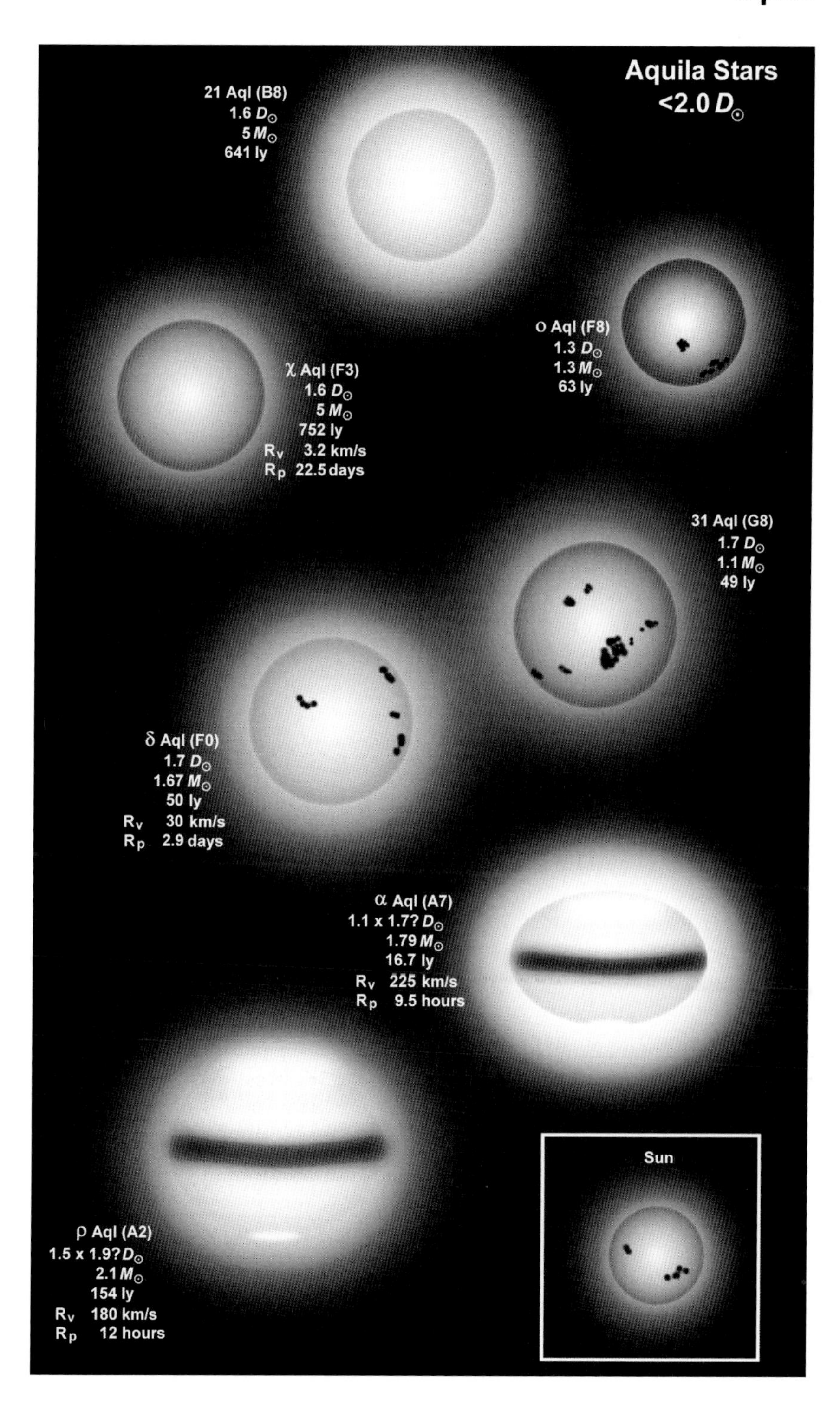
Aquila Stars
<2.0 $D_\odot$
21 Aql (B8)
1.6 $D_\odot$
5 $M_\odot$
641 ly
ο Aql (F8)
1.3 $D_\odot$
1.3 $M_\odot$
63 ly
χ Aql (F3)
1.6 $D_\odot$
5 $M_\odot$
752 ly
R_V 3.2 km/s
R_P 22.5 days
31 Aql (G8)
1.7 $D_\odot$
1.1 $M_\odot$
49 ly
δ Aql (F0)
1.7 $D_\odot$
1.67 $M_\odot$
50 ly
R_V 30 km/s
R_P 2.9 days
α Aql (A7)
1.1 x 1.7? $D_\odot$
1.79 $M_\odot$
16.7 ly
R_V 225 km/s
R_P 9.5 hours
Sun
ρ Aql (A2)
1.5 x 1.9? $D_\odot$
2.1 $M_\odot$
154 ly
R_V 180 km/s
R_P 12 hours

Ara

Constellation:	Ara	**Hemisphere:**	Southern
Translation:	The Altar	**Area:**	237 deg^2
Genitive:	Arae	**% of sky:**	0.575%
Abbreviation:	Ara	**Size ranking:**	63rd

Known to the ancient Greek and Roman scholars the constellation was named by Cicero and was included in Ptolemy's original list of 48 constellations. The Altar is associated with several Greek myths including the Centaur Chiron, Dionysus the God of Wine, and Lycaon who was turned into a wolf after sacrificing a child on the Altar.

The bluish B2 α **Arae** lies at about 242 ly from Earth and seems to want to stay there – it does not appear to be heading either towards or away from us. To all intents and purposes it appears as a normal star: 5.2 solar diameters across, 340 times more luminous than the Sun and with an absolute magnitude of M_v -1.7. But what makes α Arae interesting is its spin rate. No one is exactly sure how fast it is rotating on its axis but the figure is somewhere between 375 and 450 km/s, completing a full rotation in anywhere between 16.8 and 14 hours. The Sun spins at 2 km/s and takes 25 days to rotate just once. At such a high rotational speed the star is physically unstable, throwing off vast amounts of material from its equatorial region which now surrounds the star.

Just 100th of a magnitude fainter is **β Arae**, an orange K3 supergiant of 310 $D_\odot$ and 2,100 $L_\odot$. It lies at a distance of 603 ly but at 10 pc it would brighten to magnitude M_v -4.4 which is as bright as Venus. The star is big enough to just about fill the orbit of Mars.

At about 1,136 ly **γ Arae** is the farthest star in this constellation, and also the most luminous at 4,754 $L_\odot$. Some 26 solar diameters across this B1 giant has an absolute magnitude of M_v -3.3.

μ Arae is interesting for several reasons. First, it is a solar analog: it is about one-third larger than the Sun, a close spectral match at G5 (Sun = G2), but the same mass, 1.0 $M_\odot$, and is rather more luminous at 1.68 $L_\odot$. It is also a cosmic stone's throw away at a mere 49 ly. In fact, an observer on a planet in orbit around μ Ara would no doubt be looking at the Sun and pondering on the similarities. Which is another interesting possibility as μ Arae appears to have a planetary system of at least four bodies. The closest planet to the star is μ Arae c, a possible 'super-Earth', ten times more massive than our own planet and approaching the limits of how large a rocky planet can become. At an average of just 0.09 AU from the star – 13.5 million km – it is probably too hot to sustain life as we know it. Mercury, by comparison, is 46.4 million km during its closest approach to the Sun when the surface temperature reaches 177 K (450°C). At 0.9 AU there may be a possible gaseous giant, μ Arae d, about half the mass of Jupiter and with an orbital period of 311 days. At roughly the position that Mars is from the Sun, 1.5 AU, another gas ball, μ Arae b, probably exists, this time it is 1.7

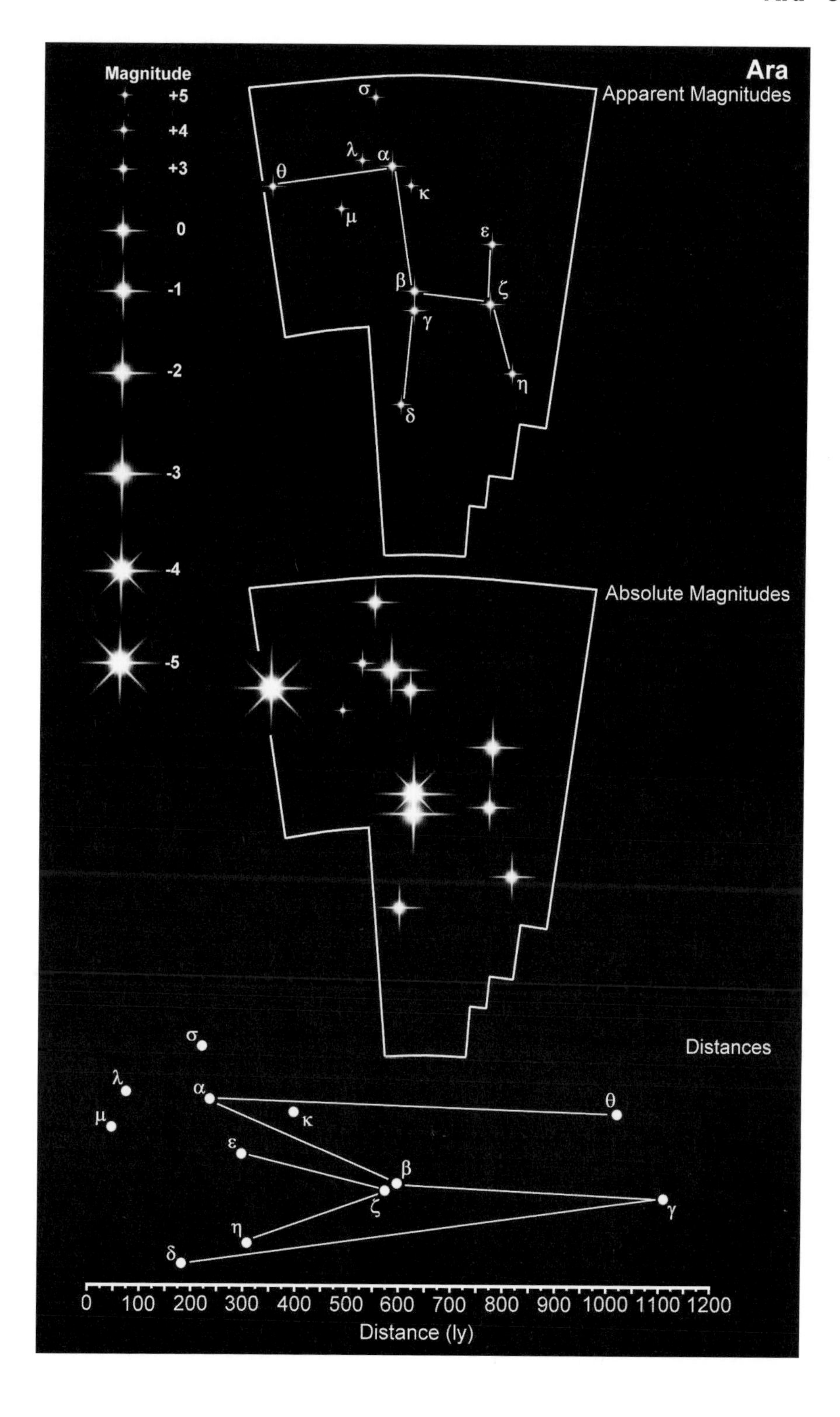
Magnitude
+5
+4
+3
0
-1
-2
-3
-4
-5
Ara
Apparent Magnitudes
σ
λ
α
θ
κ
μ
ε
β
ζ
γ
η
δ
Absolute Magnitudes
Distances
σ
λ
α
μ
κ
θ
ε
β
ζ
γ
η
δ
0 100 200 300 400 500 600 700 800 900 1000 1100 1200
Distance (ly)

Jovian masses. Still farther out at 5.2 AU – the same distance Jupiter is from the Sun – is a 1.8 M_J planet, μ Arae e, which takes 11.5 years to orbit the star (Jupiter 11.9 years). At an estimated 6,340 million years, μ Arae is considered to be somewhat older than our own Solar System which is 4,560 million years old.

In addition to μ Arae, the constellation boasts several other stars that appear to have grown planets. **HD 154857** is an invisible 7th magnitude star some 223 light years away. A yellow G5 dwarf, not unlike the Sun, it is 2.4 times larger, 1.2 times more massive but much older at 8,980 million years. In 2004 a 1.8 M_J planet, HD154857 b, was discovered to be lying at a distance of 1.2 AU. A second as yet unconfirmed planet may also exist. Close by **HD 154672** is another yellow dwarf, an 8th magnitude G3, 1.06 solar masses with a diameter of 1.3 $D_\odot$. Its planet orbits the star in just 164 days at an average distance of 0.6 AU. It is at least 5 Jovian masses. The system lies at a distance of 215 ly and is thought to be about 9,300 million years old. **GJ 674** is an M2.5 red dwarf, just 15 light years distant, and a faint 0.0032 $L_\odot$. A little less than half the diameter of the Sun, 0.42 $D_\odot$, and with only one-third of its mass, the star is a 9th magnitude object. In 2007 a 0.04 M_J planet was discovered at just 0.04 AU – only 6 million km – from the star. It takes 4.7 days to complete one orbit.

Ara has several open clusters ranging from just 6 million to 1,172 million years old and one globular cluster of about 13,400 million years old.

Planetary systems in Ara

Star	$D_\odot$	Spectral class	ly	m_v	Planet	Minimum mass	q	Q	P
μ Ara	1.25	G3	50	+5.15	HD 160691 b	1.68 M_J	1.31	1.69	1.76 y
					HD 160691 c	0.033 M_J	0.075	0.107	9.64 d
					HD 160691 d	0.522 M_J	0.860	0.982	310.6 d
					HD 160691 e	1.814 M_J	4.72	5.75	11.5 y
GJ 674	0.42	M2.5	14.8	+9.38	GJ 674 b	0.037 M_J	0.031	0.047	4.7 d
GJ 676	?	M0	54.5	+16.1	GJ 676 b	4 M_J	?	?	2.7? y
HD 152079	1.0?	G6	278	+9.18	HD 152079 b	3 M_J	1.28	5.12	5.74 y
HD 154672	1.27	G3	215	+8.22	HD 154672 b	5.02 M_J	0.234	0.966	163.9 d
HD 154857	2.42	G5	223	+7.25	HD 154857 b	1.8 M_J	0.636	1.764	1.12 y
HD 156411	1.2?	F8	180	+6.67	HD 156411 b	0.75 M_J	?	?	2.31 y

Ara Stars 1.3 to 5.2 $D_\odot$
σ Ara (A0)
1.9 $D_\odot$
3 $M_\odot$
230 ly
λ Ara (F4)
1.6 $D_\odot$
1.4 $M_\odot$
71 ly
R_V 15 km/s
R_P 5.4 days
Sun
μ Ara (G5)
1.3 $D_\odot$
1 $M_\odot$
48 ly
R_V 3.8 km/s
R_P 17.3 days
α Ara (B2)
2.8 x 5.2? $D_\odot$
4.4 $M_\odot$
242 ly
R_V 410 km/s
R_P 14.4 hours
δ Ara (B8)
2.5 x 3? $D_\odot$
3.3 $M_\odot$
187 ly
R_V 255 km/s
R_P 14.4 hours
Ara's Giant and Supergiant Stars
θ Ara (B2)
12 x 15? $D_\odot$
10 $M_\odot$
1,013 ly
R_V 117 km/s
R_P 6.5 days
γ Ara (B1)
18 x 26? $D_\odot$
11 $M_\odot$
1,136 ly
R_V 245 km/s
R_P 5.4 days
κ Ara (G8)
14 $D_\odot$
3.5 $M_\odot$
398 ly
ε Ara (K3)
28 $D_\odot$
4 $M_\odot$
304 ly
η Ara (K5)
30 $D_\odot$
5 $M_\odot$
313 ly
ζ Ara (K3)
48 $D_\odot$
8 $M_\odot$
574 ly
β Ara (K3)
310 $D_\odot$
9 $M_\odot$
603 ly
R_V 5.4 km/s
R_P 7.96 yrs
Orbit of Mars, 3.1 AU

Open and globular clusters in Ara

Name	Size arc min	Size ly	Distance ly	Age million yrs	Brightest star in region*	No. stars m_v >+12*	Apparent magnitude m_v
IC 4651	24′	20	2,900	1,140	DM -49° 11417 m_v +8.97	45	+6.9
NGC 6193	33′	36	3,770	6	HD 150135 m_v +7.60	67	+5.2
NGC 6200	22′	43	6,700	8.5	HD 150627 m_v +9.17	25	+7.4
NGC 6208	12.4′	11	3,060	1,172	DM -53° 8183 m_v +996	14	+7.2
NGC 6250	15.8	13	2,820	26	HD 152917 m_v +996	14	+5.9
NGC 6397	26′	55	7,200	13,400	Globular Cluster		+5.9

*May not be a cluster member.

Aries

Constellation:	Aries	**Hemisphere:**	Northern
Translation:	The Ram	**Area:**	441 deg^2
Genitive:	Arietis	**% of sky:**	1.069 %
Abbreviation:	Ari	**Size ranking:**	39th

Aries is associated with the Argonauts' quest for the Golden Fleece. It is one of the Zodiacal constellations, the Sun entering it on 18 April and leaving on 14 May.

α **Arietis** is magnitude m_v +2.0 star some 66 ly from Earth. A K2 it is 14.7 times larger than the Sun and is gradually dying. Its rotational velocity is slightly less than the Sun's at 1.8 km/s so it takes more than a year – 413 days – to turn once on its polar axis. It is sometimes called by its Arabic name, Hamal, which means 'the lamb' although this name was originally applied to the entire constellation.

At a distance of 60 ly β **Arietis** or Sheratan is the closest of the stars in Aries. A white A5 dwarf about twice the size and mass of the Sun it is rather more luminous at 24 $L_\odot$. It is a binary system, its companion being a Sun-like star with an estimated mass of 1.02 $M_\odot$ in a highly elongated orbit that brings the two stars to within 0.08 AU (12 million km) of one another before separating them by 1.2 AU. The orbital period is 107 days. The companion is optically invisible and can only be detected by other methods.

γ **Arietis** or Mesartim is also a binary star lying at a distance of 204 ly. γ^1 **Arietis** is the smaller of the two, a 1.2 $D_\odot$ white B9, 56 times more luminous than the Sun which gives a visual magnitude of m_v +4.59. γ^2 **Arietis** is somewhat larger at 1.9 $D_\odot$ and just makes it into the next spectral class of A1. It is actually an α CV rotating variable, fluctuating between m_v +4.62 and +4.66 over a period of $2^d14^h\ 38^m$. One of the first binary systems to be discovered, accidentally by Robert Hooke in 1664 who was following a comet, the pair are 7.5″ apart and are easily separated with a binocular although, in reality, they are at least 500 AU apart and take a minimum of 5,000 years to complete one orbit.

One of the more distant stars in Aries at 659 ly is the yellowish-orange dwarf ι **Arietis**. With a luminosity of 300 $L_\odot$ this 1.5 solar diameter star is a faint m_v +5.11. It is another spectroscopic binary system. Virtually nothing is known about its companion except that it is in a 1,568 day long orbit (4.3 years).

Apparently close to α Ari but actually about three times farther away κ **Arietis** is a Main Sequence A2 star, 26 times more luminous than the Sun. About 2.3 $D_\odot$ its apparent magnitude is m_v +5.02. An unseen companion lurks close by in a 15.3 day orbit.

There is a confusion of taus in Aries. There is a τ^{1A} Arietis, a τ^{1B} Arietis, a τ^{2A} Arietis and a τ^{2B} Arietis. The Tau 1s have nothing to do with the Tau 2s and the Tau 2s have nothing to do with each other. The Tau 1s are an eclipsing binary system at 620 ly. At a barely visible magnitude of m_v +5.50 τ^{1A} **Arietis** is a B5, 2.6 $D_\odot$ star with an 8th magnitude component, τ^{1B} **Arietis.** Close enough on

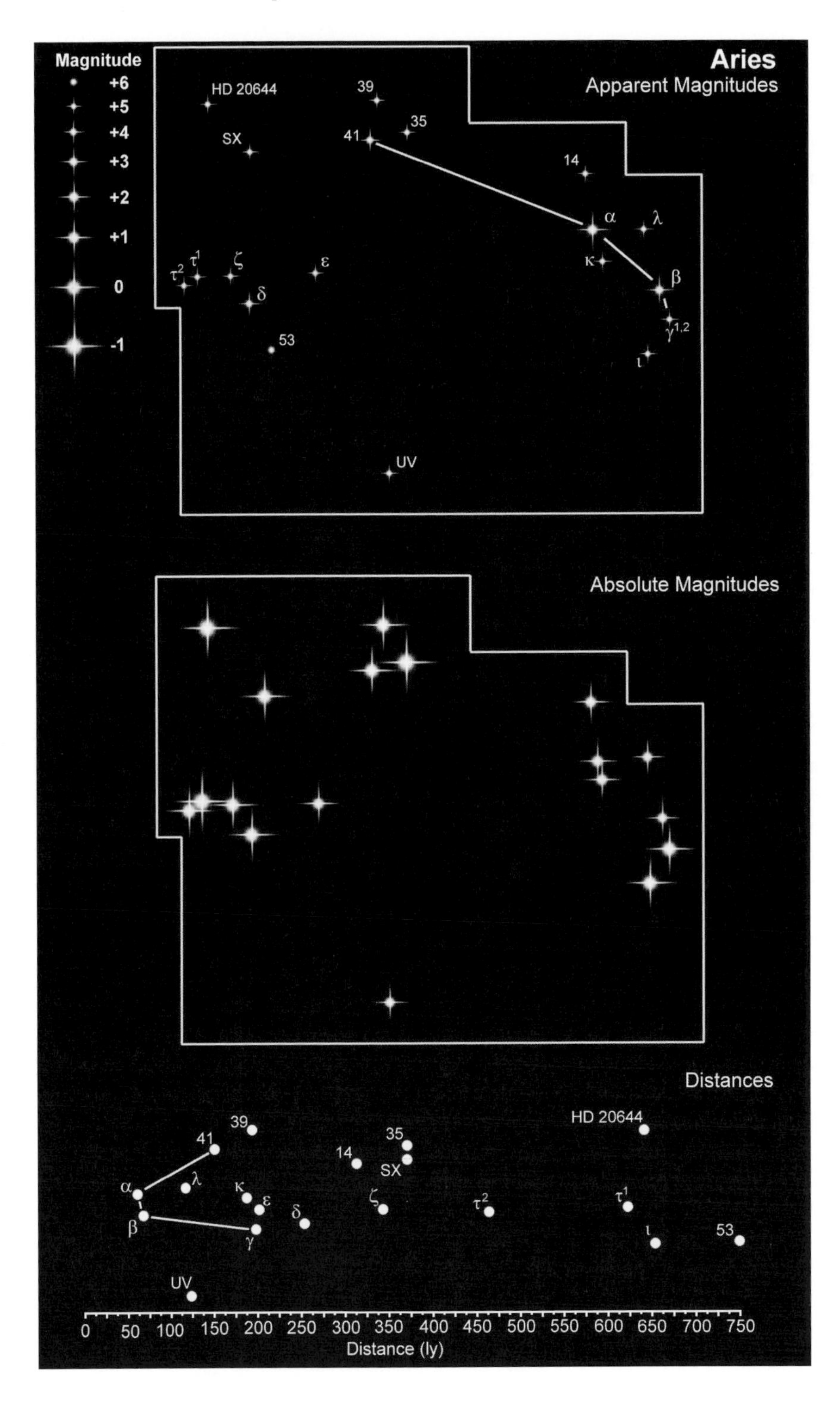
Magnitude
+6
+5
+4
+3
+2
+1
0
-1
Aries
Apparent Magnitudes
HD 20644
39
35
41
SX
14
α
λ
κ
β
γ1,2
ι
τ2
τ1
ζ
ε
δ
53
UV
Absolute Magnitudes
Distances
HD 20644
39
41
35
14
SX
α
λ
κ
ε
δ
ζ
τ2
τ1
β
γ
ι
53
UV
0 50 100 150 200 250 300 350 400 450 500 550 600 650 700 750
Distance (ly)

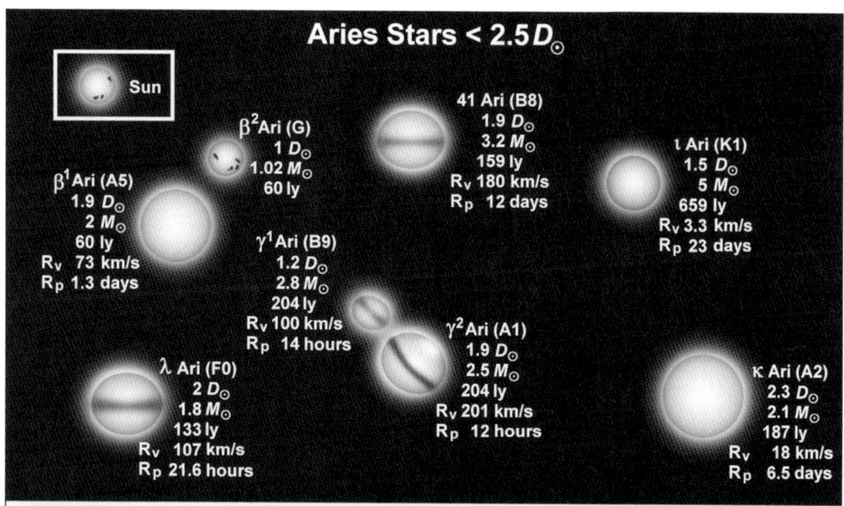
Aries Stars < 2.5$D_\odot$
Sun
41 Ari (B8)
1.9 $D_\odot$
3.2 $M_\odot$
159 ly
R_v 180 km/s
R_p 12 days
β^2Ari (G)
1 $D_\odot$
1.02 $M_\odot$
60 ly
ι Ari (K1)
1.5 $D_\odot$
5 $M_\odot$
659 ly
R_v 3.3 km/s
R_p 23 days
β^1Ari (A5)
1.9 $D_\odot$
2 $M_\odot$
60 ly
R_v 73 km/s
R_p 1.3 days
γ^1Ari (B9)
1.2 $D_\odot$
2.8 $M_\odot$
204 ly
R_v 100 km/s
R_p 14 hours
γ^2Ari (A1)
1.9 $D_\odot$
2.5 $M_\odot$
204 ly
R_v 201 km/s
R_p 12 hours
λ Ari (F0)
2 $D_\odot$
1.8 $M_\odot$
133 ly
R_v 107 km/s
R_p 21.6 hours
κ Ari (A2)
2.3 $D_\odot$
2.1 $M_\odot$
187 ly
R_v 18 km/s
R_p 6.5 days

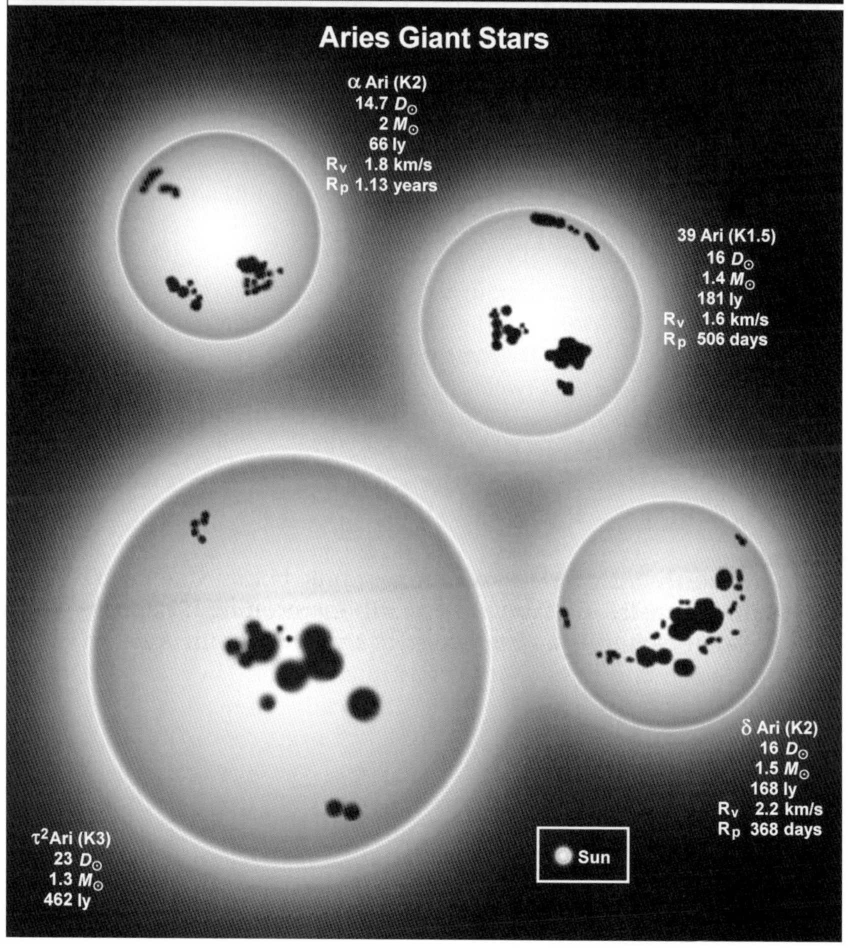
Aries Giant Stars
α Ari (K2)
14.7 $D_\odot$
2 $M_\odot$
66 ly
R_v 1.8 km/s
R_p 1.13 years
39 Ari (K1.5)
16 $D_\odot$
1.4 $M_\odot$
181 ly
R_v 1.6 km/s
R_p 506 days
δ Ari (K2)
16 $D_\odot$
1.5 $M_\odot$
168 ly
R_v 2.2 km/s
R_p 368 days
τ^2Ari (K3)
23 $D_\odot$
1.3 $M_\odot$
462 ly
Sun

the sky and faint enough to have once been mistaken as its natural companion, τ^{2A} **Arietis** is actually a lot closer at 462 ly. Sometimes listed as τ^{2B} **Arietis** an 8th magnitude 'companion' is simply a line-of-sight coincidence.

SX Arietis is the prototype of a class of variable star that bears its name and which are also known as 'helium variables'. These are Main Sequence B-class stars whose strong magnetic fields, and He I and Si III (helium and silicon) spectral lines vary in intensity. The magnetic and magnitude fluctuations coincide with their rotational periods. SX Ari itself varies between m_v +5.50 and +5.81 with a period of 17^h 28^m. It has a very uncertain diameter of 2.7 $D_\odot$ and lies 372 ly away.

Technically the brightest star in Aries is **35 Arietis.** Although it has a modest apparent magnitude of m_v +4.64 its appearance belies the fact that it is a 2.3 $D_\odot$ bluish B3, 370 ly from Earth but with a luminosity of 836 Suns. At the standard 10 pc it would actually brighten to M_v -1.7. It is another binary, its spectroscopic partner orbiting the primary every 490 days, and may even be a triple star system with the third component orbiting one of the other two stars in just 2.6 days, although this has never been confirmed.

Despite not having a Bayer designation **41 Arietis** is actually the third brightest star in Aries with a magnitude of m_v +3.59 – but it was not always that way. The star was once part of a now long defunct constellation called Musca Borealis, the 'Fly of the North', and when the constellation disappeared, along with a number of others, the star found itself in Aries like some displaced cosmic refugee. A bluish-white B8, it is not a particularly big object, just 1.4 $D_\odot$, but with a luminosity of 71 Suns it would be M_v -0.2 at 10 pc. It is also a fast spinner, 180 km/s, completing one turn in about 9.5 hours.

53 Arietis is an interloper. Originally part of the Trapezium Cluster of the Orion Nebula, about 2.7 million years ago the shockwave from a supernova, or a gravitational entanglement between two binary systems, kicked the star out of its own neighborhood towards Aries. It was not alone. AE Aurigae and μ Columbae appear to have been ejected at the same time. At m_v +6.12 it is invisible to all but those with the keenest of eyes and who are blessed with dark skies.

The honor of the largest star goes to **HD 20644**, a 43 $D_\odot$ K4 giant at a distance of 641 ly.

Just to the north of β Arietis is a 9th magnitude binary system, **HIP 8920** (or **SAO 75016).** Normally too faint to be worth mentioning it happens to be very interesting. Some 300 ly away the two stars are virtually identical to the Sun in terms of size, mass and luminosity, and probably age. They are also very close, orbiting one another in just 3.42 days. The really interesting thing about this system however, is that it seems to be bathed in a very dense, hot dust cloud. The Solar System has its own cloud of course, caused by countless collisions between asteroids and the disintegration of comets as they approach the Sun. We can see evidence of the cloud in the form of the Zodiacal Light, Zodiacal Band and Gegenschien, and meteors. But HIP 8920's cloud is more dense by a factor of about one million. This poses a mystery. The age and complexity of the binary system should have swept the area clean of dust thousands of millions of years ago by either absorbing and vaporizing the dust or ejecting it from the region sending it deep into interstellar space. The fact that there is so much dust still in the system tends to suggest that it is a recent addition, perhaps caused by the collision of two Earth-like rocky planets. Planet formation in a binary system can be tricky because of the gravitational tidal forces. What HIP 8920 perhaps demonstrates is that, even when planets do form under such conditions, they exist in a chaotic, hostile environment that can ultimately destroy them. As the celebrated science writer Nigel Calder pointed out several decades ago, we live in a violent Universe.

Auriga

Constellation:	Auriga	**Hemisphere:**	Northern
Translation:	The Charioteer	**Area:**	657 deg^2
Genitive:	Aurigae	**% of sky:**	1.593%
Abbreviation:	Aur	**Size ranking:**	21st

Auriga is one of Ptolemy's 48 constellations and contains a number of interesting and unusual stars.

The third brightest star in the Northern Celestial Hemisphere and the sixth over all, α **Aurigae** is one of the few stars better known by its ancient name Capella. There are a number of slightly different interpretations of the name Capella – 'she-goat', 'goat star', 'little she goat' – but what they all allude to is a goat being carried over the shoulder of Auriga the Charioteer. To the observer Capella appears as a yellow m_v +0.08 star, but there is more to Capella than meets the eye. A little more than 100 years ago astronomers at the Lick Observatory discovered that α Aur is a binary system but the spectral signature of the two stars proved difficult to interpret. Today's view is that the main component is a G8 giant, 13.6 $D_{\odot}$ across, 3.0 $M_{\odot}$ and 93 times more luminous than the Sun. It rotates at a modest 3 km/s taking about 230 days to complete one turn. Its companion is also a yellow star, somewhat paler due to it being a G0 class and, at 6,000K, some 10% hotter but similar in mass at 2.5 $M_{\odot}$ although smaller at 8.3 $D_{\odot}$ across. It rotates 12 times faster at 36 km/s taking just 11.7 days to spin around once. The two stars are in a near-circular orbit that keeps them separated by 0.72 AU – 100 million km – with an orbital period of about 104 days. The alignment of the orbit is such that the two stars do not eclipse one another as seen from Earth.

Further studies of Capella have revealed it to be not a binary but a *double* binary arrangement. At about 10,000 AU farther away is a pair of red dwarfs gravitationally bound to themselves and to Capella. They seem to be of similar size, about half as big as the Sun, although one is 0.5 $M_{\odot}$ while the other is a featherweight 0.2 $M_{\odot}$. They orbit one another every 388 years and Capella every 373,000 years.

The whole system is just 42 light years away as the cosmic crow flies and belongs to what is known as the Hyades Moving Group, a cluster of stars all heading in the same direction and associated with Hyades in Taurus.

β **Aurigae** or Menkarlina is also a binary but a noticeable one in that it is an Algol-type or EA eclipsing binary. The two stars are almost identical A2 class about 4.3 $D_{\odot}$ across and separated by 0.08 AU – just 11.6 million km – in a near circular orbit that takes $3^d\ 23^h$ to complete. As a result, the pair partially eclipse one another, one occluding the other by about 25% every 47 hours. The components are so close together that they are gravitationally distorted into tear-drop shapes. There is also a third component, a 14th magnitude red dwarf at 350 AU. The system is believed to be part of the Sirius Supercluster, 82 light years from Earth.

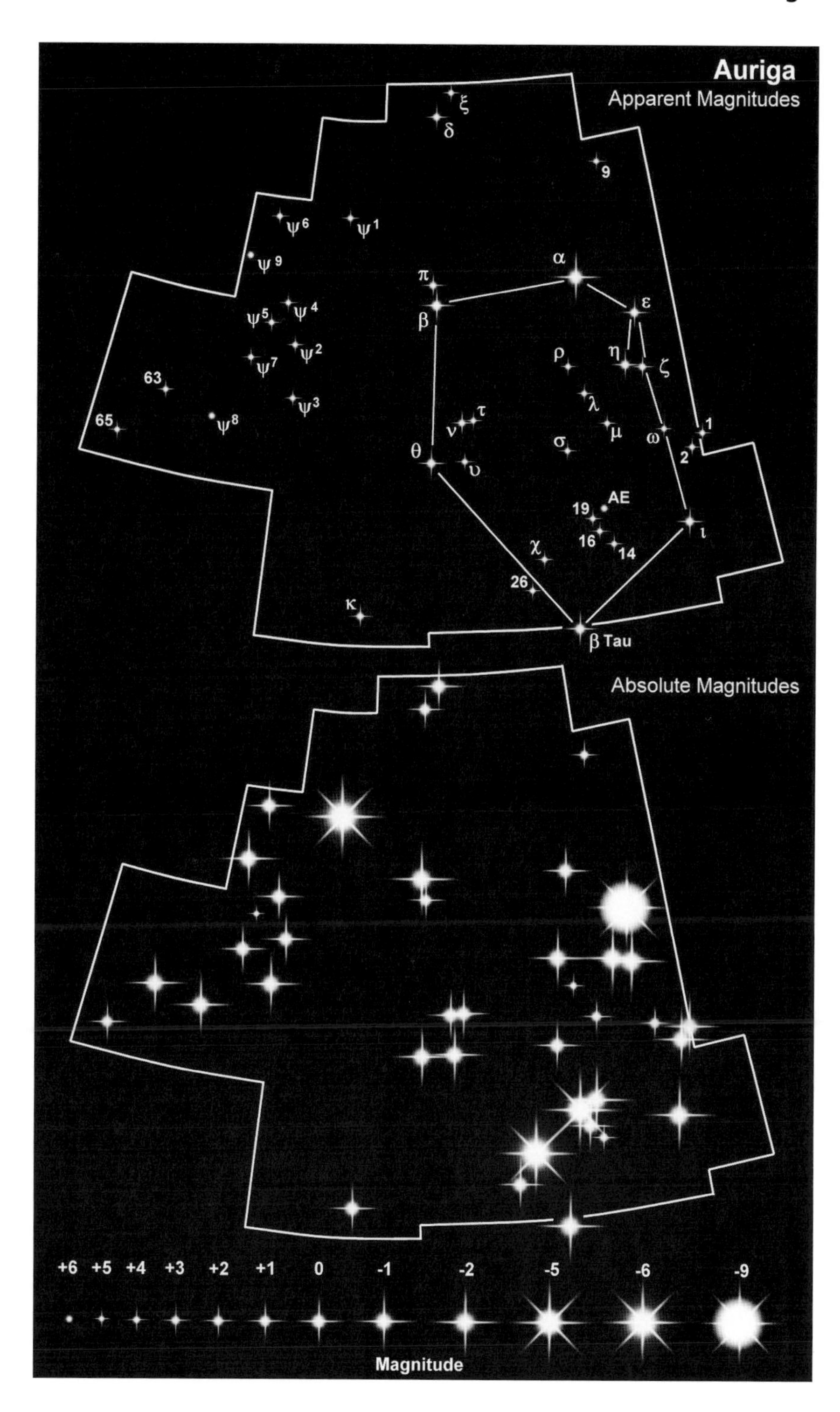
Auriga
Apparent Magnitudes
ξ
δ
9
ψ6
ψ1
ψ9
α
π
β
ε
ψ5
ψ4
ψ2
ψ7
ρ
η
ζ
63
ψ3
λ
65
ψ8
τ
ν
μ
ω
1
θ
σ
2
υ
AE
19
ι
16
14
χ
26
κ
β Tau
Absolute Magnitudes
+6
+5
+4
+3
+2
+1
0
-1
-2
-5
-6
-9
Magnitude

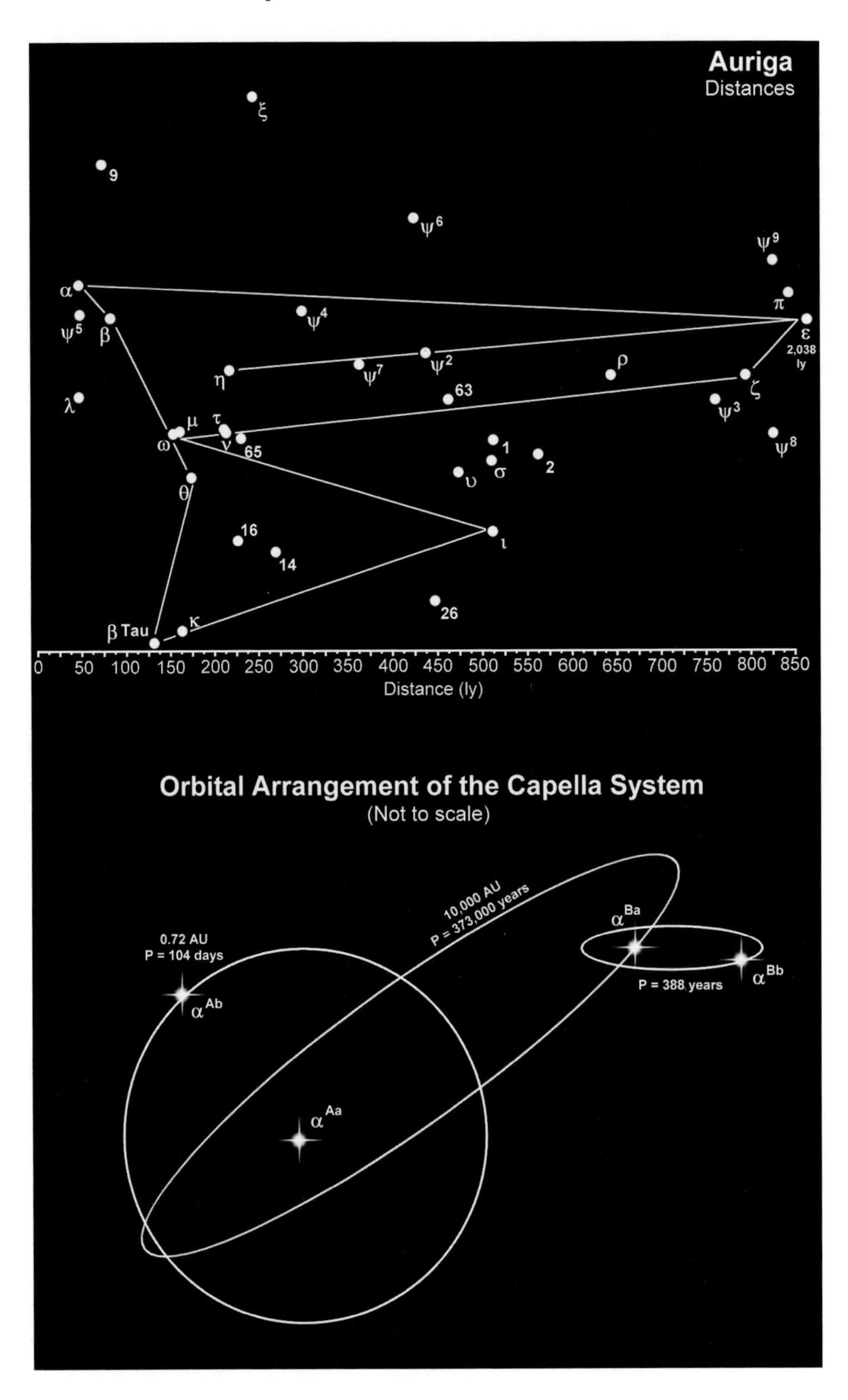

Auriga
Distances
Distance (ly)
0 50 100 150 200 250 300 350 400 450 500 550 600 650 700 750 800 850
ε 2,038 ly
β Tau
Orbital Arrangement of the Capella System
(Not to scale)
0.72 AU
P = 104 days
10,000 AU
P = 373,000 years
P = 388 years

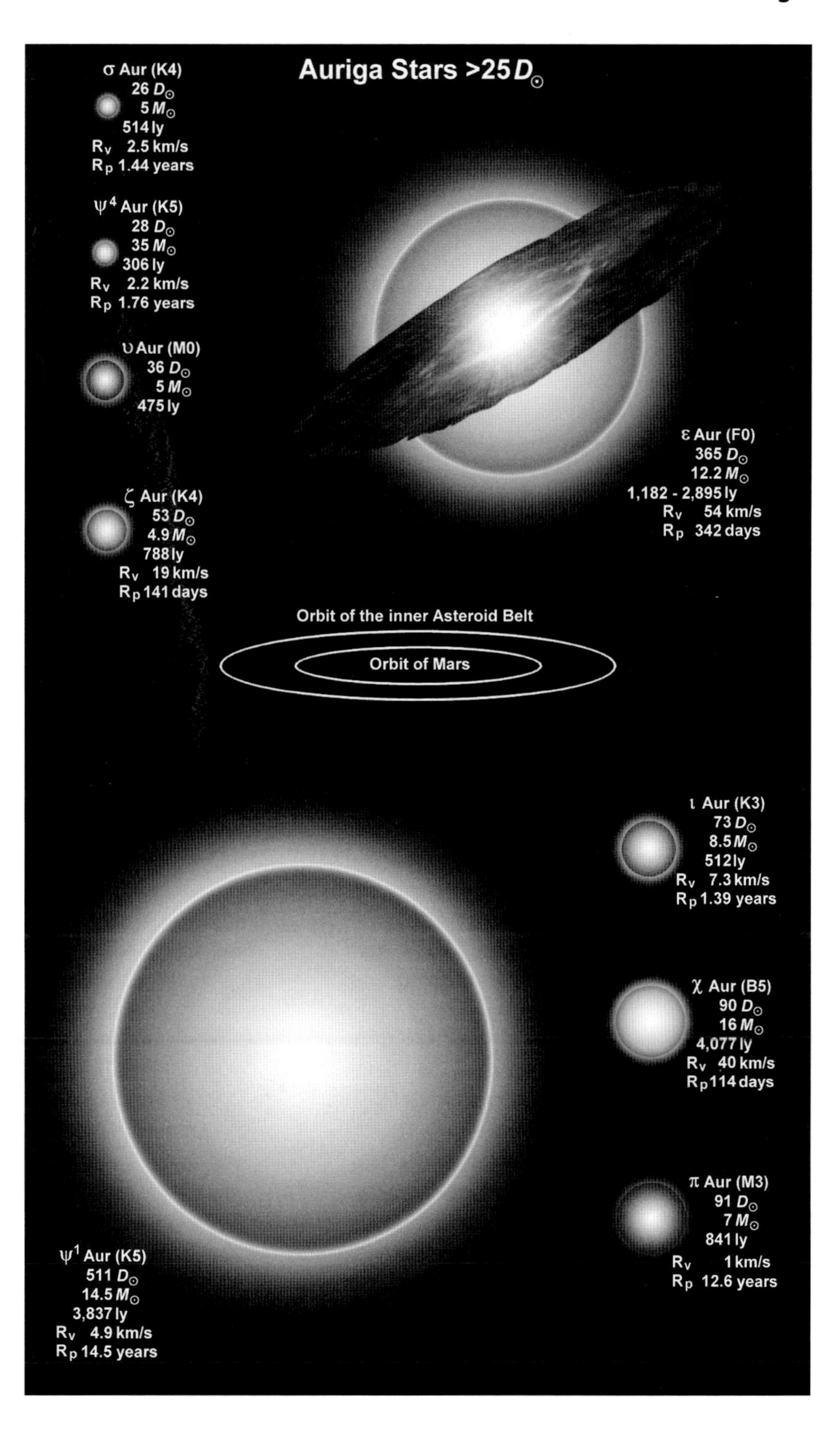
Auriga Stars >25 D☉
σ Aur (K4)
26 D☉
5 M☉
514 ly
Rv 2.5 km/s
Rp 1.44 years
ψ4 Aur (K5)
28 D☉
35 M☉
306 ly
Rv 2.2 km/s
Rp 1.76 years
υ Aur (M0)
36 D☉
5 M☉
475 ly
ζ Aur (K4)
53 D☉
4.9 M☉
788 ly
Rv 19 km/s
Rp 141 days
ε Aur (F0)
365 D☉
12.2 M☉
1,182 - 2,895 ly
Rv 54 km/s
Rp 342 days
Orbit of the inner Asteroid Belt
Orbit of Mars
ι Aur (K3)
73 D☉
8.5 M☉
512 ly
Rv 7.3 km/s
Rp 1.39 years
χ Aur (B5)
90 D☉
16 M☉
4,077 ly
Rv 40 km/s
Rp 114 days
π Aur (M3)
91 D☉
7 M☉
841 ly
Rv 1 km/s
Rp 12.6 years
ψ1 Aur (K5)
511 D☉
14.5 M☉
3,837 ly
Rv 4.9 km/s
Rp 14.5 years

When Johann Bayer compiled his *Uranometria* star atlas he arranged the stars in roughly magnitude order, so the brightest was α, the second brightest β and so on. Something seems to have gone terribly drastically when it came to Auriga. γ Aur does not exist, so naturally the third brightest should then be assigned to the next Greek letter, δ, except that δ Aurigae is actually the 8th brightest star. The third brightest is θ Aurigae which should have been assigned to the 8th brightest. Similarly ι Aur should be the 9th brightest but is, in fact, the 4th. Confusing, isn't it? So what did happen to γ Aurigae? Well, Auriga is one of the few constellations that shares a star with one of its neighbors. Nowadays γ Aurigae is called β Tauri in much the same way that α Andromedae used to be called δ Pegasi where it marks one of the corners of the Great Square of Pegasus.

The enigmatic ε **Aurigae** has puzzled astronomers for almost 200 years. It is the joint 19th largest star visible to the naked eye with a diameter of a staggering 365 $D_{\odot}$ (3.4 AU). It exists somewhere between 1,181 and 2,895 ly from Earth and has a luminosity of at least 19,600 Suns and perhaps as much as 21,600 $L_{\odot}$ and would be one of the brightest stars in the night sky at 10 pc with an M_v of -8.5. An F0 class star with a temperature of about 7,200K it appears as a washed-out yellow. And it is variable. For most of the time it is at magnitude m_v +2.92 but every 9,892 days – that's 27.1 years – its brightness dips by almost a full magnitude to m_v +3.83 and stays there for anywhere between 640 and 730 days (1.75 to 2 years). Well, almost. About midway through its dim period it brightens slightly. The suspicion is that ε Aur is an exceptionally long period Algol eclipsing variable. The problem is that the companion star that supposedly passes in front of ε Aur, thus causing it to dim, appears to be invisible! It has been suggested that the companion star is shrouded in a cloud of dust. The dust forms a torus or doughnut with a hole in the middle that allows ε Aur to brighten half way through the eclipse. Another theory was that the companion was semi-transparent. It has even been suggested that a black hole is in orbit around the primary, although this has largely been dismissed. Observations by the Spitzer infra-red space telescope suggest that the companion star is surrounding by a cloud of gravel size particles rather than the expected dust.

Another Algol or EA eclipsing binary is ζ **Aurigae**. Robert Burnham, of *Burnham's Celestial Handbook* fame, thought that this system was 1,200 light years from Earth and that the main star was at least 160 $D_{\odot}$ and possibly as large as 300 $D_{\odot}$. Latest measurements have reduced the distance by about a third to 788 ly and the diameter to a 'mere' 53 $D_{\odot}$. With a maximum magnitude of m_v +3.70 ζ Aur is a K4 giant with a temperature of about 4,000 K. Its companion is believed to be a hot 15,000 K blue dwarf, 3.9 solar diameters across. Their orbital separation swings between 2.5 and 6 AU with the result that one star eclipses the other every 972.16 days (2 years 241 days) when the magnitude drops to m_v +3.97.

η **Aurigae** is a moderately fast spinning B3 blue star. With a diameter of 3.7 Suns and a rotational velocity of 95 km/s it takes just under two days to complete one revolution.

At a distance of 173 ly **θ Aurigae** is both a rotating variable and a binary *and* a

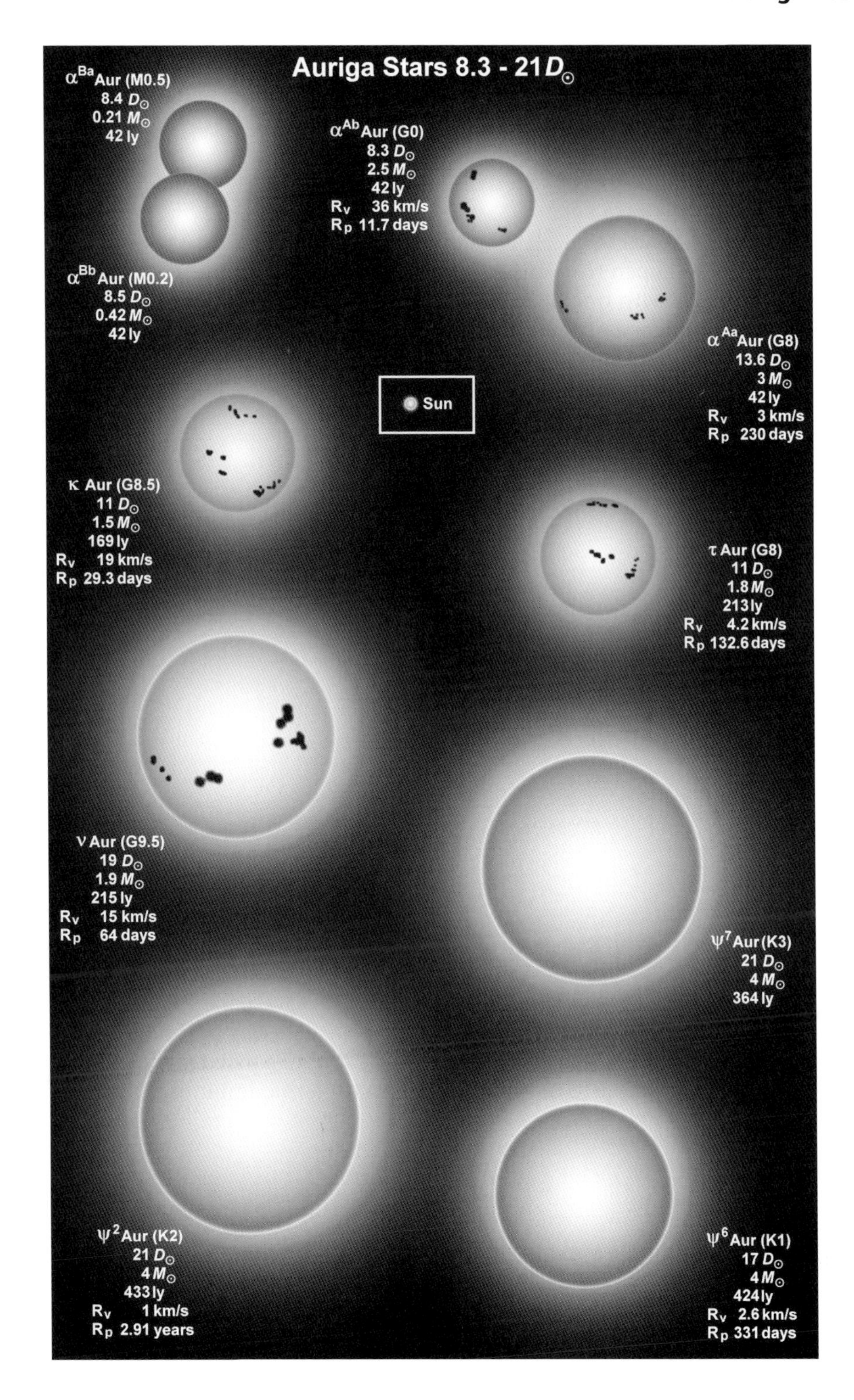

Auriga Stars 8.3 - 21$D_\odot$
α^{Ba}Aur (M0.5)
8.4 $D_\odot$
0.21 $M_\odot$
42 ly
α^{Ab}Aur (G0)
8.3 $D_\odot$
2.5 $M_\odot$
42 ly
R_v 36 km/s
R_p 11.7 days
α^{Bb}Aur (M0.2)
8.5 $D_\odot$
0.42 $M_\odot$
42 ly
α^{Aa}Aur (G8)
13.6 $D_\odot$
3 $M_\odot$
42 ly
R_v 3 km/s
R_p 230 days
Sun
κ Aur (G8.5)
11 $D_\odot$
1.5 $M_\odot$
169 ly
R_v 19 km/s
R_p 29.3 days
τ Aur (G8)
11 $D_\odot$
1.8 $M_\odot$
213 ly
R_v 4.2 km/s
R_p 132.6 days
ν Aur (G9.5)
19 $D_\odot$
1.9 $M_\odot$
215 ly
R_v 15 km/s
R_p 64 days
ψ^7Aur (K3)
21 $D_\odot$
4 $M_\odot$
364 ly
ψ^2Aur (K2)
21 $D_\odot$
4 $M_\odot$
433 ly
R_v 1 km/s
R_p 2.91 years
ψ^6Aur (K1)
17 $D_\odot$
4 $M_\odot$
424 ly
R_v 2.6 km/s
R_p 331 days

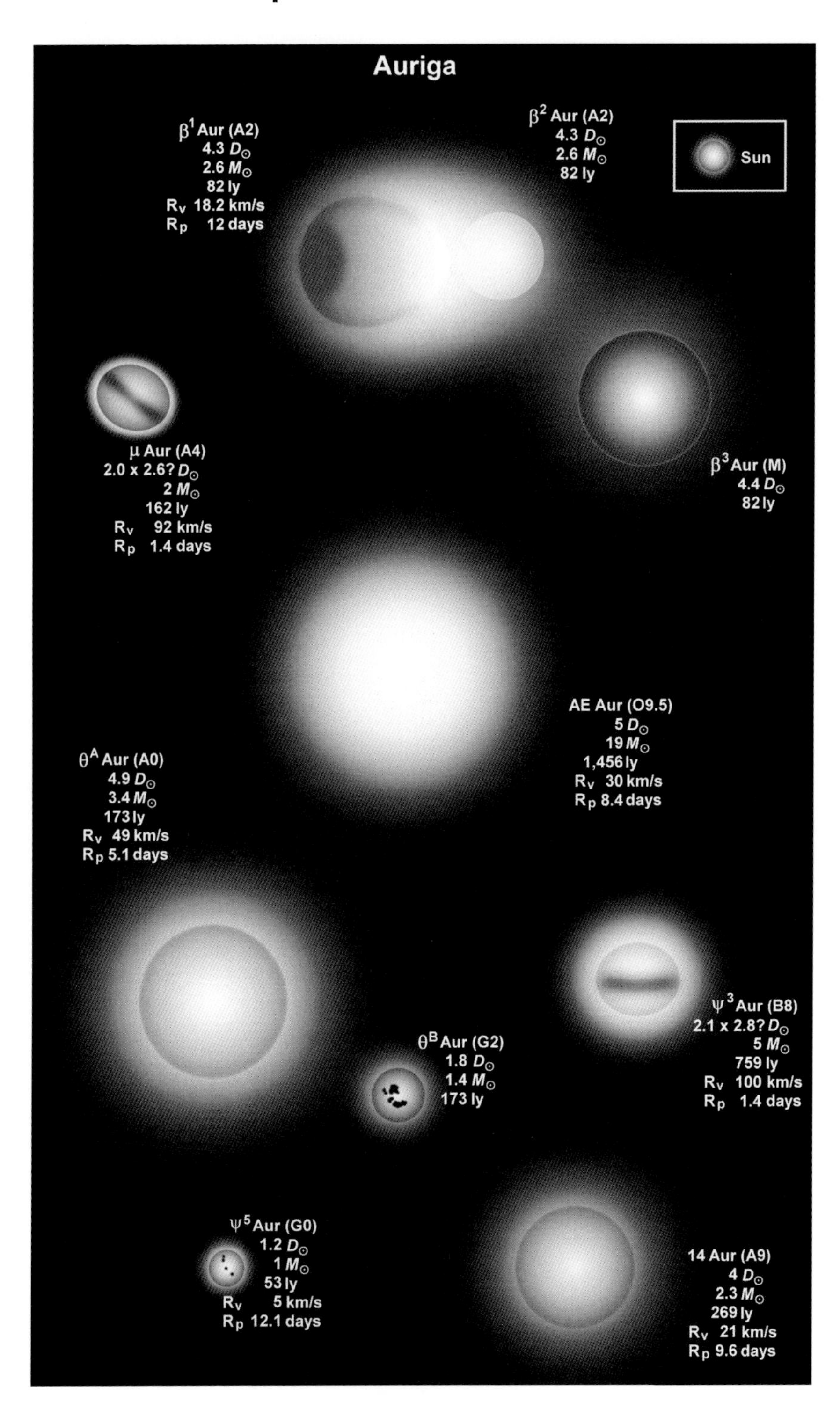

Auriga
β1 Aur (A2)
4.3 D⊙
2.6 M⊙
82 ly
Rv 18.2 km/s
Rp 12 days
β2 Aur (A2)
4.3 D⊙
2.6 M⊙
82 ly
Sun
μ Aur (A4)
2.0 x 2.6? D⊙
2 M⊙
162 ly
Rv 92 km/s
Rp 1.4 days
β3 Aur (M)
4.4 D⊙
82 ly
AE Aur (O9.5)
5 D⊙
19 M⊙
1,456 ly
Rv 30 km/s
Rp 8.4 days
θA Aur (A0)
4.9 D⊙
3.4 M⊙
173 ly
Rv 49 km/s
Rp 5.1 days
ψ3 Aur (B8)
2.1 x 2.8? D⊙
5 M⊙
759 ly
Rv 100 km/s
Rp 1.4 days
θB Aur (G2)
1.8 D⊙
1.4 M⊙
173 ly
ψ5 Aur (G0)
1.2 D⊙
1 M⊙
53 ly
Rv 5 km/s
Rp 12.1 days
14 Aur (A9)
4 D⊙
2.3 M⊙
269 ly
Rv 21 km/s
Rp 9.6 days

chemically peculiar star. Nearly five times as large as the Sun but 207 times as luminous θ Aur is a bluish-white A0 dwarf with a temperature of around 9,500 K. Sometimes referred to as a 'silicon star' its spectrum reveals high levels of silicon, chromium and iron which seem to be concentrated in pools that affect the star's magnetic field, causing it to fluctuate in strength. As with the Earth, the magnetic poles are not aligned with the rotational poles. Its magnitude varies slightly between m_v +2.62 and +2.70 over a period of $1^d\ 8^h\ 58^m$ and so it is classed as an α CV variable. Its companion is a m_v +7.1 G2 solar analog known as θ^B Aurigae.

Traditionally, θ Aur and ι **Aurigae** have been drawn linked to β Tauri, making Auriga one of only three constellations to share a star. ι Aur itself is a 73 $D_\odot$ K3 supergiant, 512 ly from Earth, and shining at m_v +2.66.

λ **Aurigae** is currently the closest of all the stars in Auriga but it is also moving away from us at an impressive 65.3 km/s making it one of the fastest naked eye stars in the sky receding from the Sun. Its space velocity is even higher at 89 km/s. A G1.5 dwarf of 1.2 $D_\odot$ and 1.2 $M_\odot$ it has a visual magnitude of m_v +4.7 which would brighten by almost a full magnitude to M_v +3.8 at 10 pc.

π **Aurigae** belongs to a relatively rare class of stars known as Lc pulsating variables. Only a dozen can be seen without optical aid. They are all K or M class and pulsate without any detectable regularity. An aging supergiant, 91 times larger than the Sun, π Aur varies between m_v +4.24 and +4.34. An M3 it lies at a distance of 841 ly.

Bringing up the rear is χ **Aurigae**, a 90 $D_\odot$ bluish B5 supergiant, twice as hot as the Sun at 12,000 K. It has a magnitude of m_v +4.74 but that is mainly because of its distance of 4,077 ly. With a luminosity of 16,100 $L_\odot$ it would have an absolute magnitude of M_v -6.3 at the standard distance of 10 pc.

There is a plethora of ψ stars in Auriga – no less than 9 – and none of them related! ψ^1 **Aurigae**, like π Aur, is an Lc irregular pulsating K5 variable at a distance of 3,837 ly. It is also huge. At a staggering 511 $D_\odot$ this red supergiant is 4.8 AU across, larger than the orbit of Mars by almost a full AU. From Earth it appears as a faint m_v +4.75 but at 10 pc it would have an absolute magnitude of M_v -5.7. ψ^2 **Aurigae** is a more modest 21 $D_\odot$ orange K2 at 433 ly. ψ^3 **Aurigae** is smaller again, just 2.8 $D_\odot$ and a fast spinning bluish-white Main Sequence star. ψ^4 **Aurigae** is similar to ψ^2 Aurigae: a bit larger at 28 $D_\odot$ it is a K5 and somewhat closer at 306 ly but unlikely to be related with ψ^2 Aur heading away from us at 16.9 km/s while ψ^4 Aur is hurtling towards us at 73.1 km/s making it the 14th fastest naked eye star heading our way. The Sun-like ψ^5 **Aurigae** is also journeying towards us, at 23.7 km/s. It is 20% larger than the Sun, a G0 type with a luminosity of 1.74 $L_\odot$ and lies at 53 ly. The two stars, ψ^6 and ψ^7 **Aurigae** are both orange giants, K1 and K3 respectively, with diameters of 17 and 21 $D_\odot$ but again, heading in opposite directions. Finally, ψ^8 and ψ^9 **Aurigae** are both similar distances from the Earth at 827 ly and both B-class – a B9.5 and a B8 – but they are probably not related, the former heading towards us at 26.6 km/s while the latter is traveling deeper into space at 41.1 km/s. ψ^9 is perhaps the most interesting of the two being a 3.7 $D_\odot$ Be emission star with a circumstellar disk. They are both 6th magnitude stars.

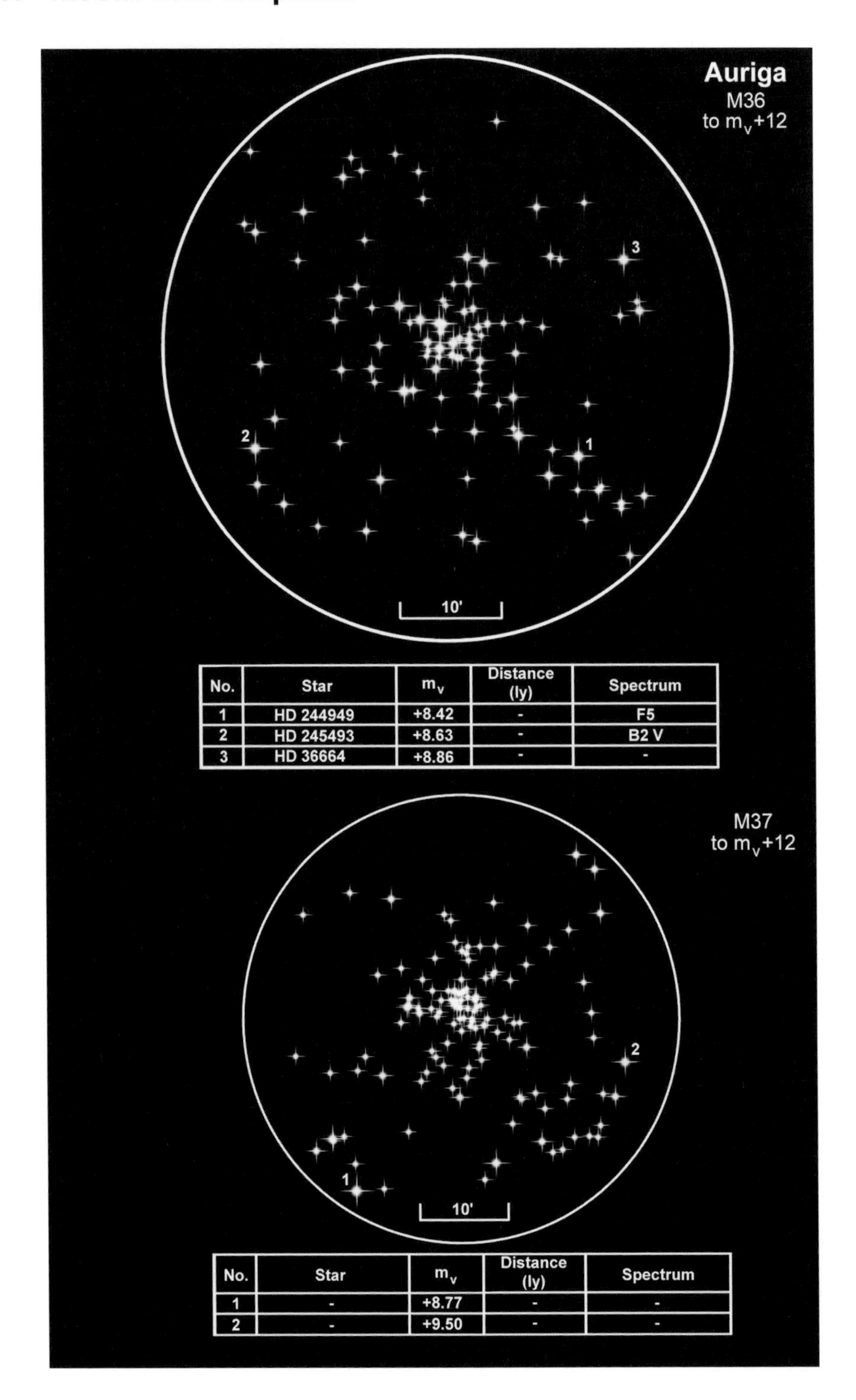

No.	Star	m_v	Distance (ly)	Spectrum
1	HD 244949	+8.42	-	F5
2	HD 245493	+8.63	-	B2 V
3	HD 36664	+8.86	-	-

No.	Star	m_v	Distance (ly)	Spectrum
1	-	+8.77	-	-
2	-	+9.50	-	-

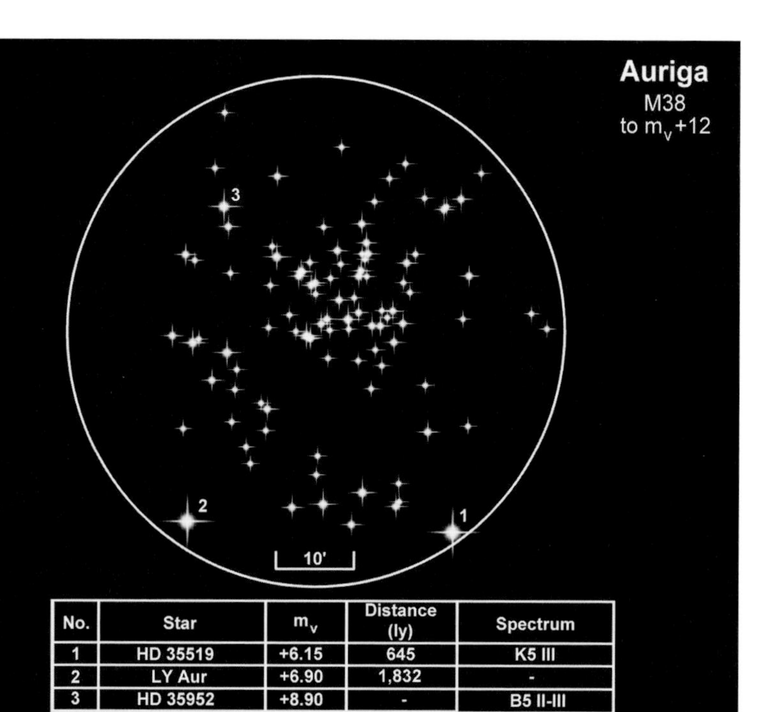

No.	Star	m_v	Distance (ly)	Spectrum
1	HD 35519	+6.15	645	K5 III
2	LY Aur	+6.90	1,832	-
3	HD 35952	+8.90	-	B5 II-III

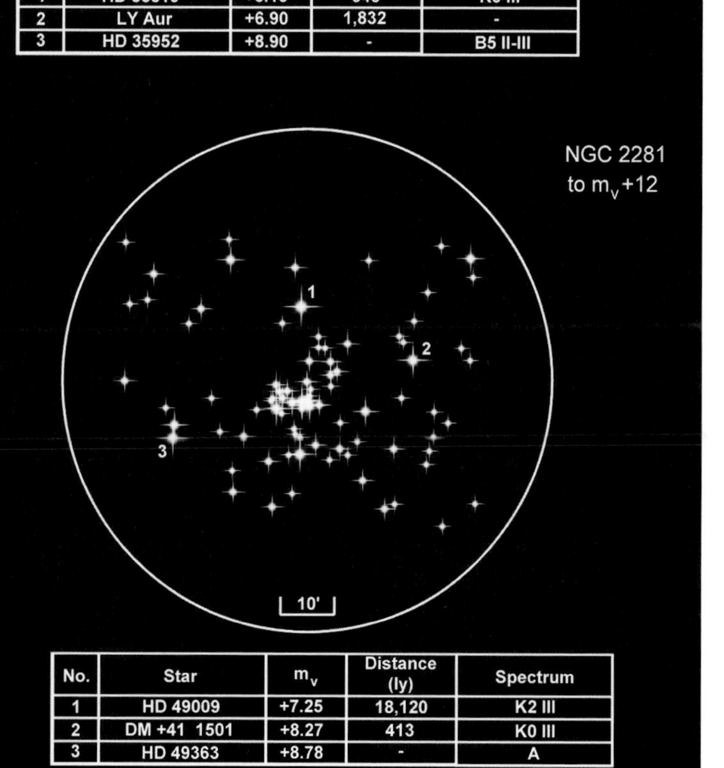

No.	Star	m_v	Distance (ly)	Spectrum
1	HD 49009	+7.25	18,120	K2 III
2	DM +41 1501	+8.27	413	K0 III
3	HD 49363	+8.78	-	A

1 Aurigae is now usually known as **HD 30504**, the star having migrated across the border into Perseus.

14 (or **KW) Aurigae** belongs to a class of variables now known as δ Scuti types but previously referred to as 'dwarf Cepheids'. An A9 yellowish-white sub-giant, 4 $D_{\odot}$, it fluctuates between magnitudes m_v +5.01 and 5.08 over $2^h\ 7^m$.

There is nothing particularly unusual about **26 Aurigae**. It has an apparent magnitude of m_v +5.42 and is a B9.5 dwarf of 3.4 $D_{\odot}$. Its importance is that it is only 2° from the galactic anticenter: the point in the sky that is directly opposite the center of our Galaxy. Draw an imaginary line between it and 136 Tauri, m_v +4.55, and the anticenter lies midway between the two.

Several stars in Auriga have planetary systems. One of the most interesting planets is **WASP-12 b**. At just 3.3 million km from the parent star, a 1.57 $D_{\odot}$ G0, it is a scorching 1,700 K. This coupled with immense tidal forces means the atmosphere from this 1.41 M_J hot Jupiter is slowly being stripped away and siphoned off into the star. About 871 ly away the planet is in a slightly eccentric 1.09 day long orbit during which it varies from 0.022 AU (3.3 million km) to 0.024 AU (3.6 million km) and is slowly spiraling inwards. It will fall into the star in about 10 million years.

At about 166 ly distance **GD 66** or **V361 Aurigae** has long been known to be a 16th magnitude variable. It is also a pulsar of 0.64 $M_{\odot}$ and appears to be moving through space in a series of small circles. This motion, detected by regular variations in the star's pulse rate, suggests the presence of a 2.11 M_J planet in an orbit of 2.356 AU. Searches have failed to directly detect the planet but it is thought to have an orbital period of about 4.52 years.

With a magnitude of just m_v +7.06 **AB Aurigae** is too faint to be seen without a binocular but this A0 class star, about twice the size of the Sun, has a couple of secrets. First, it is a strong source of X-rays which appear to emanate from just above the star. These seem to be due to material being ejected by both hemispheres and magnetically attracted to one of the poles where high speed collisions generate enough energy to release X-rays. Second, it appears to have a structured dust ring. Stretching out to about 80 AU the ring is fairly compact and contains dust particles that are larger than those found in interstellar space, suggesting they have begun to accrete. There is then a gap with a second ring extending to about 170 AU. This gap is likely to be due to the presence of a protoplanet, similar in size to the Solar System's largest asteroids (about 1,000 km across), which sweeps clean a channel through the dust cloud. With AB Aur we could be witnessing the birth of a planetary system.

About a third of the distance from ι to θ Aurigae is a star in the wrong place. **AE Aurigae** is a runaway: a star that has been ejected from its original orbit either by the pressure wave from a nearby supernova or from the collision between two binary systems. The event took place about 2.7 million years ago in the Trapezium Cluster of the Orion Nebula and launched AE Aur, 53 Arietis and μ Columbae into new trajectories. AE Aur is a blue O9.5 – almost a B0 – of 5 $D_{\odot}$ and 19 $M_{\odot}$ with a temperature of about 32,000 K. It is an Orion-type variable fluctuating by 3/10th of a magnitude from m_v +5.78 to +6.08. It retains much of the energy from the supernova explosion, slicing through space at 59.1 km/s away from us.

Auriga houses three Messier objects: the open clusters M36, M37 and M38. **M36** is the youngest of the trio at about 29 million years and lies at a distance of 4,300 ly. Not much farther at 4,500 ly **M37** contains the greatest number of stars brighter than 12th magnitude, some 130 in total, although probably not all are actual members of the cluster. **M38** is the closest at 3,500 ly. **NGC 1857** contains just a handful of bright stars while **NGC 2281**, the brightest (m_v +5.4), largest (1°) and closest (1,800 ly) is also the oldest at 358 million years.

Open clusters in Auriga

Name	Size arc min	Size ly	Distance ly	Age million yrs	Brightest star in region*	No. stars m_v >+12*	Apparent magnitude m_v
M 36 (NGC 1960)	12′	15	4,300	29	HD 244949 m_v +8.42	103	+6.3
M 37 (NGC 2029)	24′	31.4	4,500	346	DM +32° 1113 m_v +9.16	130	+6.2
M 38 (NGC 1912)	20′	20.3	3,500	290	HD 35519 m_v +6.15	109	+7.4
NGC 1664	38′	43	3,900	286	HD 30650 m_v +7.53	65	+7.6
NGC 1778	14′	19.5	4,800	143	HD 32800 m_v +9.03	19	+7.7
NGC 1857	12′	65.6	18,800	133	HD 34545 m_v +7.38	5	+7.0
NGC 1893	35′	109	10,700	11	HD 242908 m_v +9.05	45	+7.5
NGC 2281	61′	32	1,800	358	HD 49009 m_v +7.25	91	+5.4

*May not be a cluster member.

Boötes

Constellation:	Boötes	**Hemisphere:**	Northern
Translation:	The Herdsman	**Area:**	907 deg^2
Genitive:	Boötis	**% of sky:**	2.199%
Abbreviation:	Boo	**Size ranking:**	13th

Home of Arcturus, the brightest star in the Northern Celestial Hemisphere, this constellation was imagined to be a man herding the two bears, Ursa Major and Ursa Minor.

When the World Fair opened in Chicago in 1933 to celebrate a *Century of Progress* the organisers chose **α Boötis** to launch the event. At that time it was thought that the star – better known as Arcturus – was 40 light years from Earth. A similar fair had been held 40 years earlier in 1893 so it seemed appropriate that light traveling for four decades from Arcturus should be involved in the celebrations. The starlight was focused by a telescope onto a photoelectric cell which activated a switch that turned on the floodlights. Today we know that Arcturus is just 36.7 ly from Earth but this slight discrepancy should not detract from what was an imaginative and innovative idea. α Boo – note how the **ö** is replaced by a common or garden **o** – is a K1.5 giant, 23 solar diameters across, with a visual magnitude of m_v -0.04. Because it is already so close to the Sun its absolute magnitude would not change much at M_v -0.2. It is the brightest star in the Northern Celestial Hemisphere and the fourth brightest star in the entire sky after Sirius, Canopus and α Centauri. In 1635 it became the first extra-solar star to be observed in daylight. In reality it is 89 times more luminous than the Sun. Arcturus is in a highly inclined orbit that takes it above and below the main plane of the Galaxy. It is almost at its closest point to us and getting closer by 5.4 km/s. In a few thousand years it will cross the plane of the Galaxy and head off in the direction of Virgo, fading as it does so until 500,000 years from now it will be invisible to the naked eye. Seismic studies suggest it wobbles like a jelly. It may, or may not, have a companion which may be a star or an 11.7 M_J gaseous planet.

More of a warm yellow than a bright yellow **β Boötis** or Nekkar belongs to a similar spectral class as the Sun – a G8 as opposed to a G2 – but is over a dozen times larger and 148 times as luminous. It is also regarded as a flare star having been observed to emit a bright X-ray flare in 1993. These flares are thought to be caused by the collapse of magnetic field loops. Flare stars are not particularly uncommon: the closest to us include Proxima Centauri (4 ly), Barnard's Star (6 ly in Ophiuchus), Wolf 359 (7.7 ly in Leo) and the unforgettable 11th magnitude **TVLM513-46546** which is also in Boötes at a distance of just 35 ly. β Boo is not so close at 219 ly. It is also unusual in that it contains higher levels of barium than normal. Such concentrations are thought to have been caused by the star cannibalizing a neighbor, leaving behind a dead white dwarf but, so far, no such companion has been found.

A δ Scuti type variable that changes in magnitude from m_v +3.02 to +3.07 over

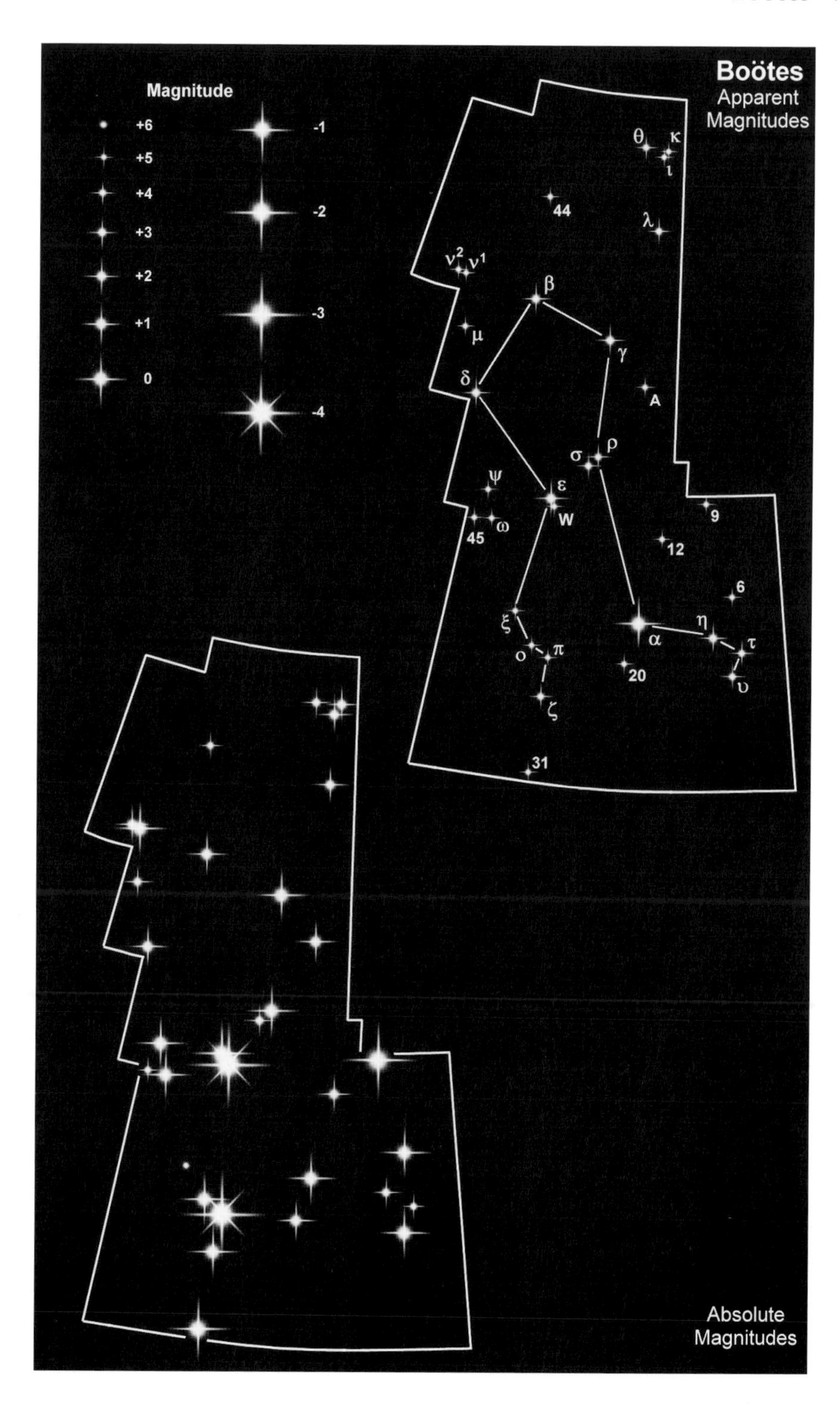
Boötes
Apparent
Magnitudes
Magnitude
+6
+5
+4
+3
+2
+1
0
-1
-2
-3
-4
θ
κ
ι
44
λ
ν2
ν1
β
μ
γ
δ
A
σ
ρ
ψ
ε
W
9
45
ω
12
6
ξ
η
α
τ
ο
π
20
υ
ζ
31
Absolute
Magnitudes

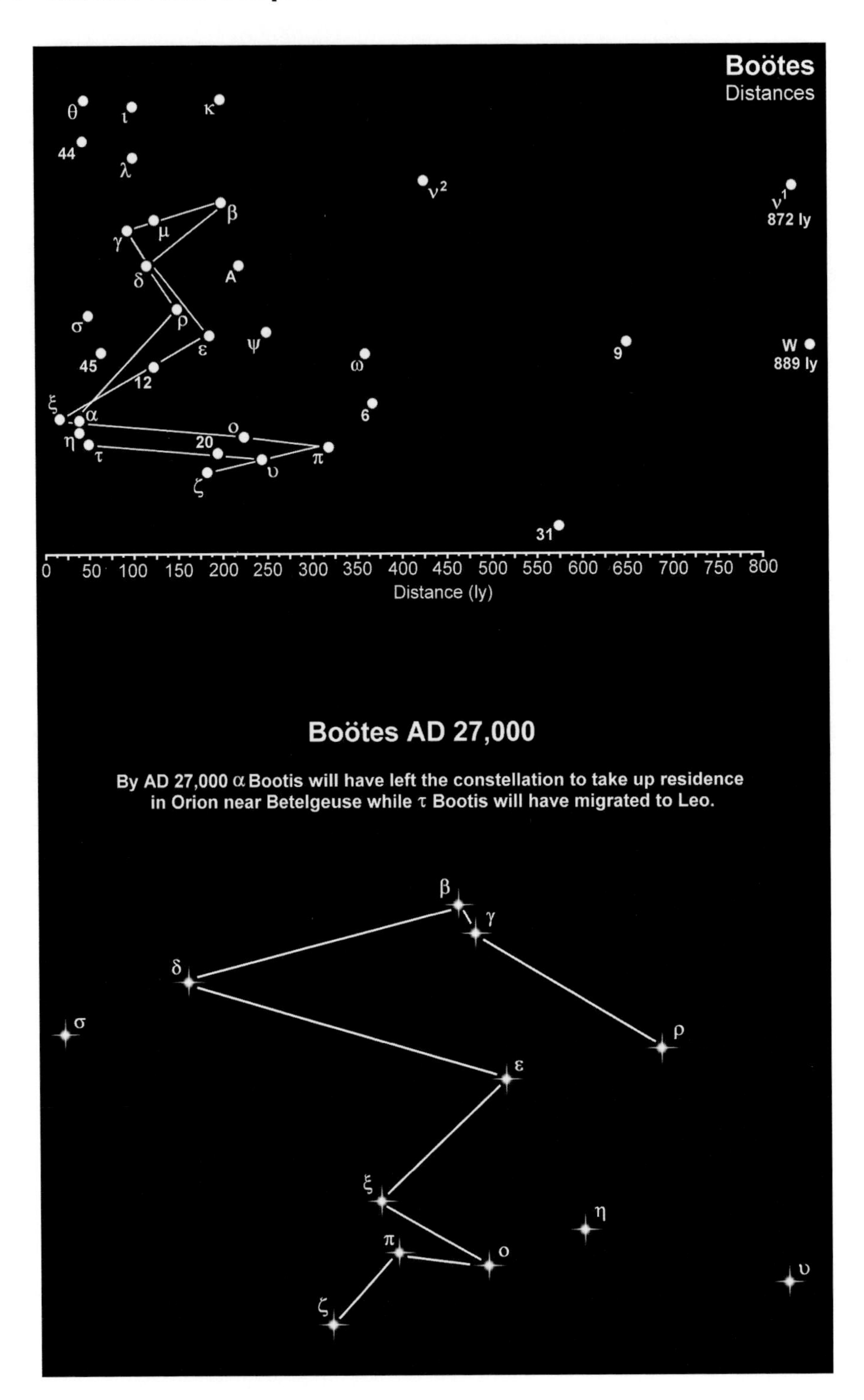

Boötes
Distances
θ
ι
κ
44
λ
ν2
ν1
872 ly
β
γ
μ
δ
A
ρ
σ
ε
ψ
45
ω
12
9
W
889 ly
ξ
α
6
η
ο
20
τ
π
υ
ζ
31
0 50 100 150 200 250 300 350 400 450 500 550 600 650 700 750 800
Distance (ly)
Boötes AD 27,000
By AD 27,000 α Bootis will have left the constellation to take up residence in Orion near Betelgeuse while τ Bootis will have migrated to Leo.
β
γ
δ
σ
ρ
ε
ξ
η
π
ο
υ
ζ

a period of 6^h 58^m γ **Boötis** or Seginus is an A7 white star which, at 3.2 $D_\odot$ is a bit on the large side: 70% of naked eye A7 type stars are less than 2 $D_\odot$. It is a moderately fast spinner at 128 km/s taking just over a day – 30.4 hours – to complete one rotation. γ Boo marks Boötes' left shoulder.

Boötes' right shoulder is indicated by δ **Boötis,** a yellowish-orange G8 giant, 11 $D_\odot$ and 1.2 $M_\odot$. It is a binary, its companion an invisible m_v +8.7 solar analog about 10% larger, more massive and more luminous than the Sun. The two stars are in a long period orbit of about 120,000 years being separated by at least 3,800 AU.

ε **Boötis** is regarded as one of the most striking binaries but you need a fairly decent telescope to separate them. The primary, ε^A Boötis is a +2.46 magnitude orange giant. Just 2.8″ away is ε^B Boötis, a blue A2, of magnitude 4.80. If you know how to work out the formula that relates distance to magnitude then ε^A Boo should be at 210 ly and ε^B Boo at 520 ly. In fact, they are both around 210 ly and separated by about 185 AU, giving an orbital period in excess of 1,000 years.

Another binary ζ **Boötis, also known as Σ1865**, consists of an almost identical pair of A3 Main Sequence stars, about 3.6 $D_\odot$ and 3.3 $D_\odot$ The most interesting feature about this system, however, is the highly elongated orbit that brings the stars to within 1.4 AU of one another before separating them by 64 AU – twice the distance of Neptune from the Sun – with a period of 123.5 years.

Rather more than 2 solar diameters across η **Boötis** appears as a m_v +2.7 yellowish-white star that also has an absolute magnitude of M_V +2.7 due to the fact that it lies at 10 parsecs: the standard distance used to calculated absolute magnitudes.

In the top right hand, north western corner of the constellation's boundary is a small triangle of stars which were traditionally regarded as the outstretched fingers of the Herdsman. For some strange reason Johann Bayer referred to them as three donkeys: Asellus Primus, Secundus and Tertius. They are not actually related. θ **Boötis is the** "First Little Donkey", an F7 yellowish-white star of the 4th magnitude and lying at a distance of some 47 ly – the closest of the donkeys. θ Boo seems to have an 11th magnitude companion: an M2.5 red dwarf, only about 30% as massive as the Sun and a mere 3% as luminous. The two are in an unstable orbit which separates by at least 1,000 AU producing an orbital period of more than 25,000 years. When at its farthest point from the primary star the red dwarf is at risk of being gravitationally tugged from its orbit by other passing stars.

ι **Boötis** is Asellus Secundus, the "Second Little Donkey", twice as far away and an A9 Main Sequence white, a little larger than the Sun at 1.2 $D_\odot$. It is a δ Scuti variable flickering between m_v +4.73 and +4.78 without any discernable regularity.

The "Third Little Donkey", or Asellus Tertius, is twice as far again and is also a δ Scuti variable. Listed as κ^2 **Boötis** its magnitude varies between m_v +4.50 and +4.58 but is as regular as clockwork: 1^h 48^m 47^s. Its fainter companion, the m_v +6.7 κ^1 **Boötis,** is an F1 dwarf in a 6,000 year long orbit.

At first glance there is nothing unusual about λ **Boötis.** It seems to be a typical Main Sequence A0 blue-white star, 1.7 $D_\odot$ and 15 $L_\odot$ lying at a distance of 97 ly.

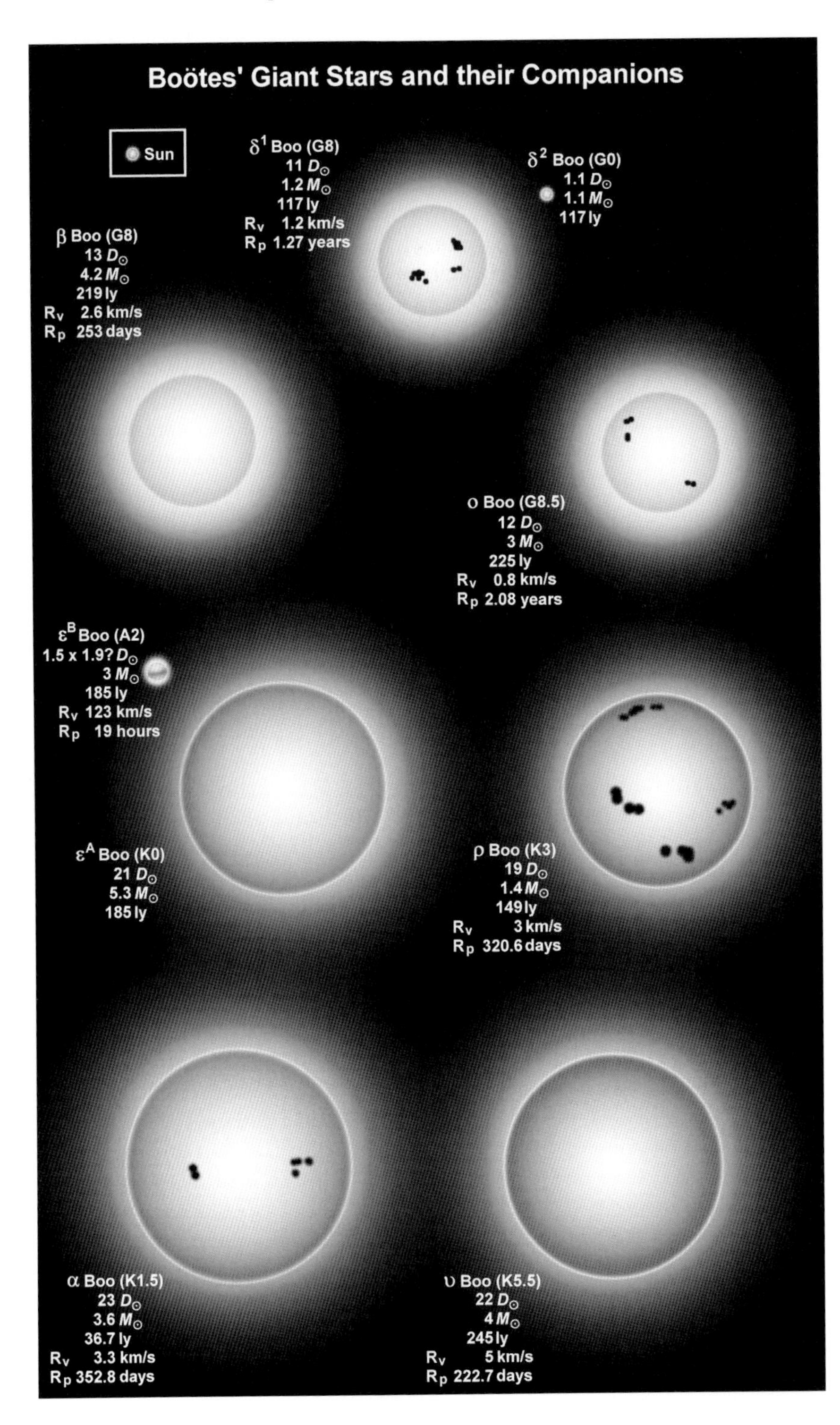
Boötes' Giant Stars and their Companions
Sun
δ¹ Boo (G8)
11 D☉
1.2 M☉
117 ly
Rv 1.2 km/s
Rp 1.27 years
δ² Boo (G0)
1.1 D☉
1.1 M☉
117 ly
β Boo (G8)
13 D☉
4.2 M☉
219 ly
Rv 2.6 km/s
Rp 253 days
ο Boo (G8.5)
12 D☉
3 M☉
225 ly
Rv 0.8 km/s
Rp 2.08 years
ε^B Boo (A2)
1.5 x 1.9? D☉
3 M☉
185 ly
Rv 123 km/s
Rp 19 hours
ε^A Boo (K0)
21 D☉
5.3 M☉
185 ly
ρ Boo (K3)
19 D☉
1.4 M☉
149 ly
Rv 3 km/s
Rp 320.6 days
α Boo (K1.5)
23 D☉
3.6 M☉
36.7 ly
Rv 3.3 km/s
Rp 352.8 days
υ Boo (K5.5)
22 D☉
4 M☉
245 ly
Rv 5 km/s
Rp 222.7 days

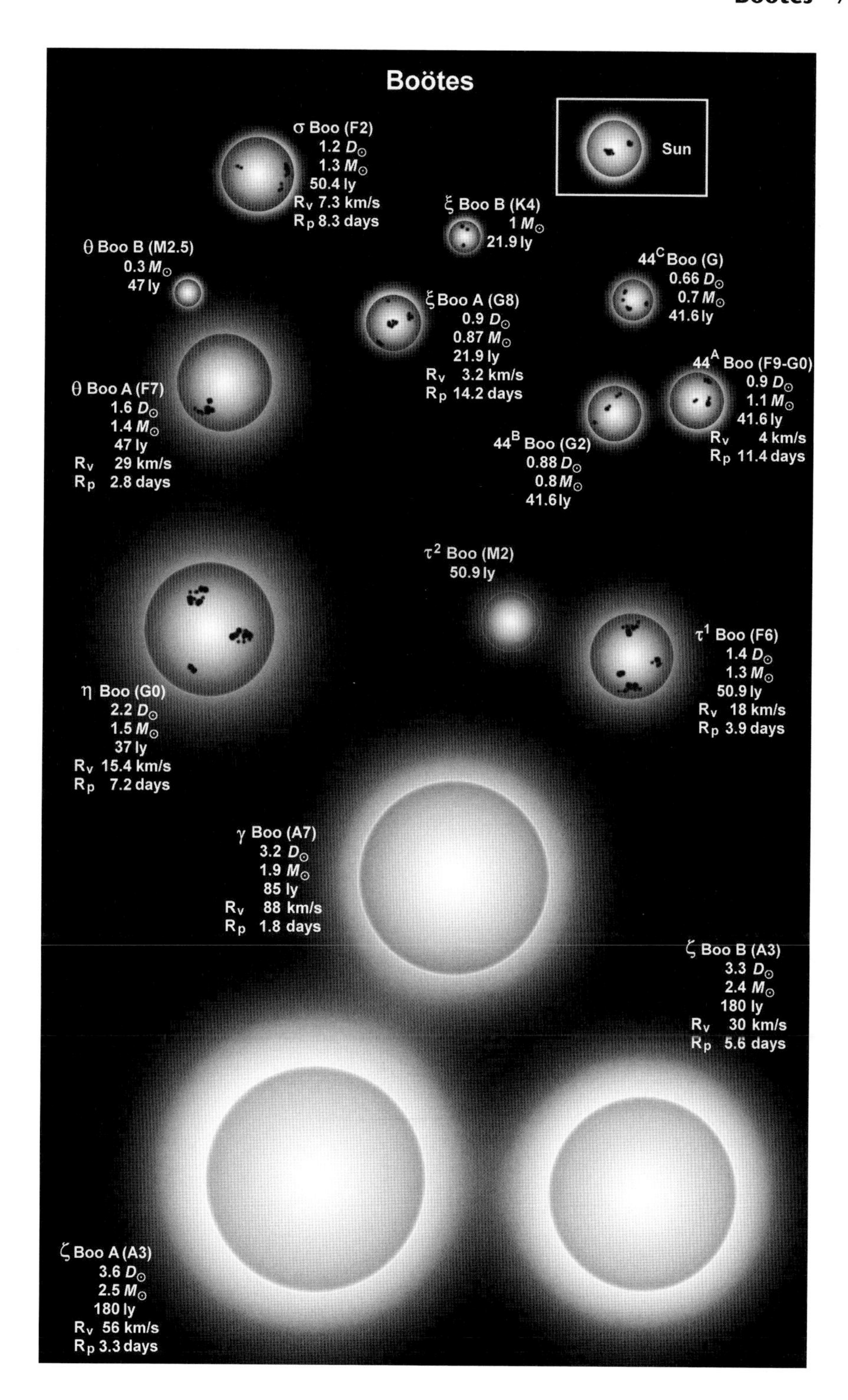

Boötes
σ Boo (F2)
1.2 $D_{\odot}$
1.3 $M_{\odot}$
50.4 ly
R_v 7.3 km/s
R_p 8.3 days
Sun
ξ Boo B (K4)
1 $M_{\odot}$
21.9 ly
θ Boo B (M2.5)
0.3 $M_{\odot}$
47 ly
44^C Boo (G)
0.66 $D_{\odot}$
0.7 $M_{\odot}$
41.6 ly
ξ Boo A (G8)
0.9 $D_{\odot}$
0.87 $M_{\odot}$
21.9 ly
R_v 3.2 km/s
R_p 14.2 days
44^A Boo (F9-G0)
0.9 $D_{\odot}$
1.1 $M_{\odot}$
41.6 ly
R_v 4 km/s
R_p 11.4 days
θ Boo A (F7)
1.6 $D_{\odot}$
1.4 $M_{\odot}$
47 ly
R_v 29 km/s
R_p 2.8 days
44^B Boo (G2)
0.88 $D_{\odot}$
0.8 $M_{\odot}$
41.6 ly
τ^2 Boo (M2)
50.9 ly
τ^1 Boo (F6)
1.4 $D_{\odot}$
1.3 $M_{\odot}$
50.9 ly
R_v 18 km/s
R_p 3.9 days
η Boo (G0)
2.2 $D_{\odot}$
1.5 $M_{\odot}$
37 ly
R_v 15.4 km/s
R_p 7.2 days
γ Boo (A7)
3.2 $D_{\odot}$
1.9 $M_{\odot}$
85 ly
R_v 88 km/s
R_p 1.8 days
ζ Boo B (A3)
3.3 $D_{\odot}$
2.4 $M_{\odot}$
180 ly
R_v 30 km/s
R_p 5.6 days
ζ Boo A (A3)
3.6 $D_{\odot}$
2.5 $M_{\odot}$
180 ly
R_v 56 km/s
R_p 3.3 days

But its spectrum reveals it to be depleted in metals. The most likely explanation is that the star accreted from an interstellar cloud that was metal poor. This can happen when the cloud is composed of gas and dust particles that readily absorb metallic atoms, effectively removing them from the cloud. The metal-free gas coalesces to form the star while the resulting radiation pressure blows away the metal-rich dust particles. Stars with this composition – known as λ Boötis type stars – are quite rare with only about 50 having so far been identified.

μ Boötis is a quadruple system some 121 ly from Earth. The primary, **μ^{Aa} Boo**, is somewhat larger than the Sun at 1.4 $D_\odot$ and 2 $M_\odot$ but is considerably more luminous at 21.3 $L_\odot$. It belongs to spectral class F1. Its closest companion is an A3 of 1.5 $M_\odot$ in a 3.75 year long orbit. Some 4,000 AU away in a 125,000 year orbit is a pair of G0 solar clones, **μ^B** and **μ^C Boo**, that also orbit each other at a distance of about 54 AU and with a period of 257 years.

ξ Boötis is one of only two naked eye BY Draconis rotating variables, the other being ε Hydri. It is also the 19th closest naked eye star at a distance of just 21.9 ly. Its magnitude varies between m_v +4.52 and m_v +4.67 over a period of 10^d 3^h 17^m. ξ Boo is a dwarf G8 star a bit smaller than the Sun at 0.9 $D_\odot$ and not nearly as luminous at 0.49 $L_\odot$. It also has a companion, a m_v +6.78 K4 dwarf in a highly elliptical orbit that reaches out as far as 50.5 AU (1 AU farther than Pluto at its maximum distance from the Sun) but which also brings the two stars to within 16.4 AU – rather closer than Uranus is to the Sun. The orbital period is 151.6 years. Such orbital arrangements tend to disrupt any debris cloud that may have remained from when the stars were formed, preventing the birth of planets. The system makes a fine pair in a small telescope or binocular, their contrasting colors having been variously described as 'yellow and orange', 'yellow and reddish-violet' and 'orange and purple'.

Another binary system is **π Boötis** though rather farther away at 317 ly. **π^1 Boo** is a 4.7 $D_\odot$, 11,000 K white B9 which appears as a +4.89 magnitude star but which, at 10 pc, would brighten to M_v -3.6. Its partner, **π^2 Boo**, is a 3,000 K cooler A6, but somewhat larger at 13 $D_\odot$. With a visual magnitude of +5.82 it too would brighten at 10 pc to M_v -2.7.

ρ and **σ Boötis** which appear close together on Boötes' belt could easily be mistaken for another binary system – but nothing could be further from the truth. **ρ Boo** lies at a distance of 149 ly and is a 19 $D_\odot$ K3 heading towards us at 13.7 km/s. **σ Boo** is much closer at 50.4 ly and a dwarf F2, just 1.2 $D_\odot$ and traveling away from us at a barely detectable 0.2 km/s.

τ Boötis is a similar distance, 50.9 ly, and is a binary system that also contains a planet. The +4.49 magnitude primary is an F6 star 1.4 solar diameters across and just over three times as luminous as the Sun. Apart from the Sun it is the only other star for which we have evidence of a magnetic field reversal. Its companion is an 11th magnitude M2 red dwarf which orbits at a distance of between 100 and 240 AU and with an orbital period of 750 to 2,600 years. In 1997 a single planet believed to be at least 3.87 Jovian masses was discovered orbiting around the visible F7 star at a distance of just 0.05 AU (7.5 million km) taking 3.31 days to complete a single orbit.

W Boötis or, if you prefer its Flamsteed number, **34 Boo**, is a semi-regular Type-b (SRb) pulsating red giant, a huge 130 Suns across – nearly as big as the orbit of Venus. Belonging to spectral class M3 it fluctuates between m_v +4.73 and +5.4 over periods of 25, 33 and 450 days. Superimposed on these periods are further minor fluctuations of a few hundredths of a magnitude with 2, 4.5 and 35.2 day periodicity. An altogether complex variable.

44 Boötis, which is also known as **Σ 1909** and **HD 133640**, is one of only a few naked eye stars to belong to the EW or W UMa eclipsing variables. 44 Boo is a complex triple star system with the primary bordering on the F9-G0 spectral class. The EW element consists of 44^B Boo and 44^C Boo which orbit the primary star every 206 to 220 years during which they approach to within 22 AU (slightly more than the average distance of Uranus from the Sun) and sweep out to 75 AU. 44^B Boo is a dim, 0.54 $L_\odot$, G2 about 0.88 $D_\odot$. Its kissing companion, 44^C Boo, is also thought to be a G-class but somewhat smaller at about 0.66 $D_\odot$ and even less luminous. The two are just 0.008 AU (1.2 million km) apart and are severely distorted into merging teardrops. The whole system appears as a m_v +4.76 star. The other naked eye EW variables include ε Coronae Australis and γ Doradus.

Caelum

Constellation:	Caelum	**Hemisphere:**	Southern
Translation:	The Chisel	**Area:**	125 deg^2
Genitive:	Caeli	**% of sky:**	0.303%
Abbreviation:	Cae	**Size ranking:**	81st

Introduced in the 1763 by the French astronomer Nicolas Louis de Lacaille in his *Coelum Australe Stelliferum* the constellation consists of just four faint naked eye stars.

α **Caeli** is the brightest and closest of this quartet of stars. At a distance of 65.7 ly the star is a +4.45 magnitude object about 50% larger than the Sun and 5.5 times more luminous. It is a yellowish-white F2 showing a hint of fluctuation in brightness and may, in fact, be a δ Scuti-type variable. It is so poorly studied however, that its variability remains to be confirmed. What we do know though is that it has a m_v +12.5 companion: a red dwarf of spectral class M0.5, less than half the diameter of the Sun and with a similarly low mass but prone to sudden brightening by a magnitude or more making it a UV Ceti variable. Its increase in magnitude is caused by flaring, the brightening taking only a few seconds to reach maximum before fading back to normal within an hour. It is thought that α Cae may be part of the Ursa Major Moving Group.

At first glance **β Caeli** could easily be α Caeli's twin. It is almost the same size – α Cae is 1.5 $D_\odot$, β Cae is 1.6 $D_\odot$ – almost the same luminosity (5.45 and 5.98 $L_\odot$), almost the same spectral class (F2 and F3) but 24 ly apart (66 and 90 ly) and heading in opposite directions with α Cae slowly edging its way towards us at just 0.6 km/s while β Cae is receding at 26.8 km/s. The big difference is the rotational velocity. α Cae spins at 40 km/s while β puts in a much higher 185 km/s causing it to bulge at the equator.

γ^1 **Caeli** is the largest star in the constellation at 13 $D_\odot$ and about 185 ly away. γ^2 **Caeli**, which is sometimes called **X Caeli**, is believed to be 334 ly away, is an F1 dwarf, 1.8 $D_\odot$ and with a magnitude of m_v +6.28 is too faint for most people to see except under very clear, very dark skies. The two stars are not related.

Marking the furthest limit of Caelum at 711 ly, give or take 10%, is **δ Caeli**, a hot 15,000 K bluish-white sub-giant, 3.9 $D_\odot$ across 2,600 times more luminous than the Sun. Its visual magnitude is a faint m_v +5.05 which would increase to M_v -1.7 at 10 pc, rivaling Sirius. In 25,000 years from now δ Cae will be the only star to remain in the constellation, barely moving from its current location while the others will have migrated into neighboring constellations.

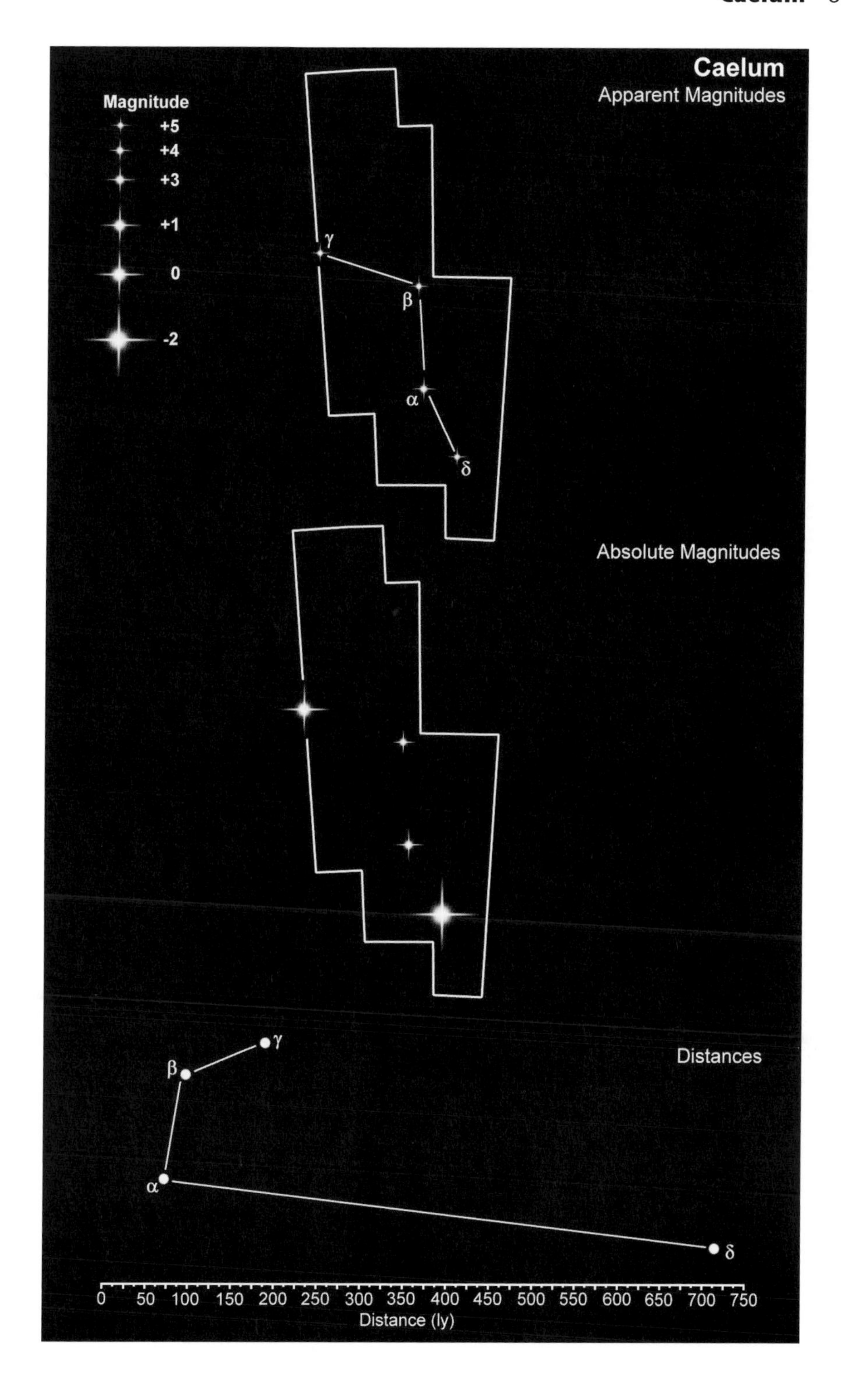
Caelum
Apparent Magnitudes
Magnitude
+5
+4
+3
+1
0
-2
γ
β
α
δ
Absolute Magnitudes
Distances
β
γ
α
δ
0 50 100 150 200 250 300 350 400 450 500 550 600 650 700 750
Distance (ly)

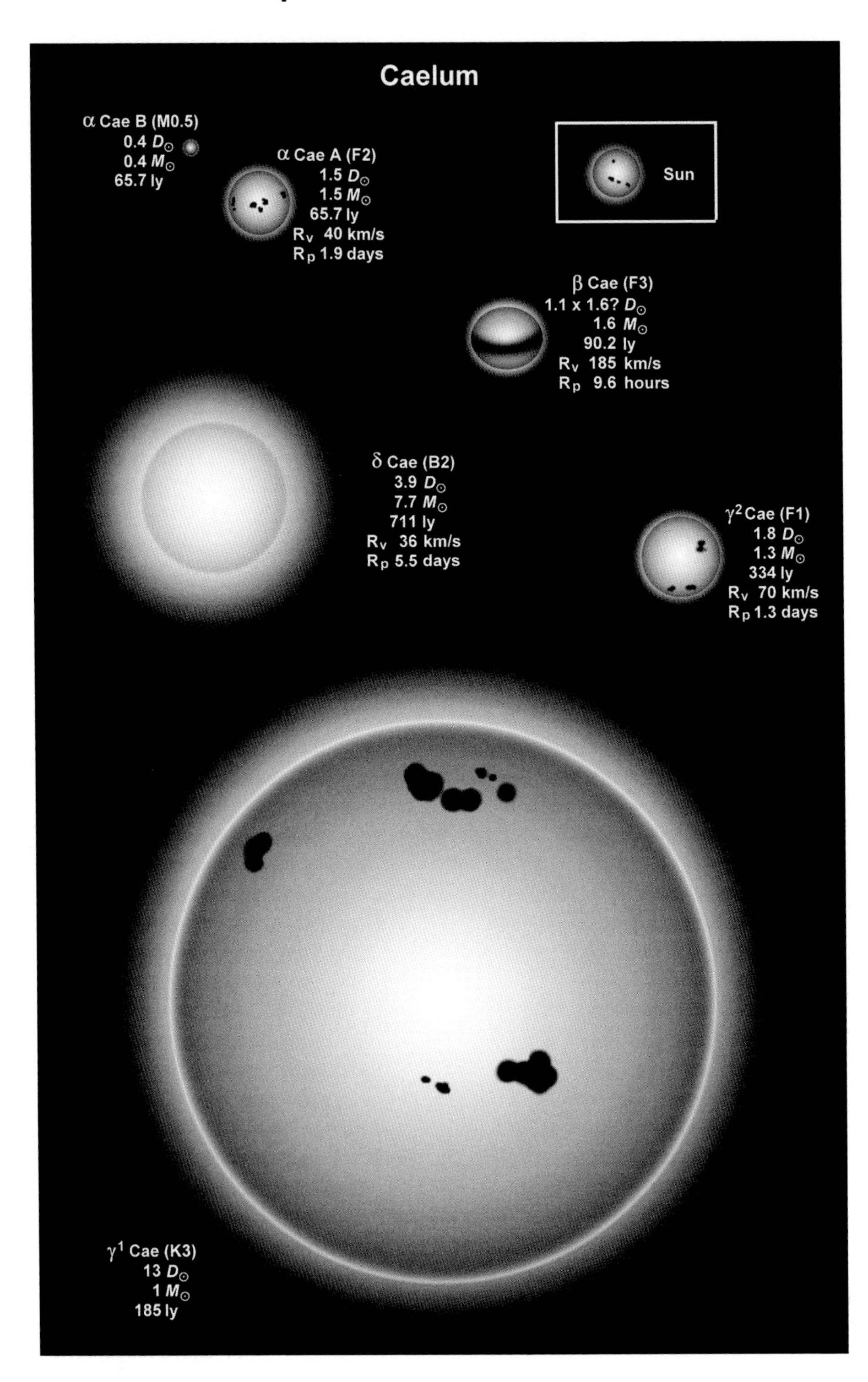
Caelum
α Cae B (M0.5)
0.4 $D_\odot$
0.4 $M_\odot$
65.7 ly
α Cae A (F2)
1.5 $D_\odot$
1.5 $M_\odot$
65.7 ly
R_V 40 km/s
R_P 1.9 days
Sun
β Cae (F3)
1.1 x 1.6? $D_\odot$
1.6 $M_\odot$
90.2 ly
R_V 185 km/s
R_P 9.6 hours
δ Cae (B2)
3.9 $D_\odot$
7.7 $M_\odot$
711 ly
R_V 36 km/s
R_P 5.5 days
γ^2 Cae (F1)
1.8 $D_\odot$
1.3 $M_\odot$
334 ly
R_V 70 km/s
R_P 1.3 days
γ^1 Cae (K3)
13 $D_\odot$
1 $M_\odot$
185 ly

Camelopardalis

Constellation:	Camelopardalis	**Hemisphere:**	Northern
Translation:	The Giraffe	**Area:**	757 deg^2
Genitive:	Camelopardalis	**% of sky:**	1.835%
Abbreviation:	Cam	**Size ranking:**	18th

Although the Northern Celestial Hemisphere contains some magnificent stellar sights the region between Auriga and Ursa Minor is not one of them, being almost devoid of any bright stars. In his work *Usus Astronomicus Planisphœrium Argentinœ*, published in 1624, Jacob Bartsch created the image of a giraffe in this region, a rather unusual choice for such a northern constellation. There is some evidence that Petrus Plancius may have earlier suggested the figure. Most modern day writers make only a passing comment on this large but faint group of stars yet, as we shall see, it is full of surprises.

α Camelopardalis is not actually the brightest star in the constellation. At m_v +4.29 it is beaten into second place by β Cam which comes in at m_v +4.03. α Cam, however, is hiding its light under a bushel or to be more accurate, behind a dust cloud without which it would be a full magnitude brighter. And it does not stop there. At an estimated 6,940 ly away α Cam marks the outermost limit of the constellation. Bring it to within 10 pc and it would glow as a magnificent M_v -6.2. It is quite a rare star: an O9 – actually an O9.5 so almost a B0 – of which less than a dozen are visible to the naked eye. It is big, some 52 $D_\odot$ across and immensely luminous at a staggering 1.8 million $L_\odot$ making it one of the most luminous naked eye stars in the entire sky. With an estimated mass of 61 $M_\odot$ its lifespan is less than half a million years. Its temperature is 29,300 K.

Not to be outshone **β Camelopardalis** also has a story to tell. Just 40 million years old it is a mere infant in the celestial nursery it calls home, almost 1,000 ly from Earth. It is a yellowish-white G1 supergiant, twice as big as α Cam, and has an absolute magnitude similar to Venus, M_v -4.50 although we see it as a dim m_v +4.03. In 1967 it was observed to suddenly flare, which was a bonus for the astronomers who were actually observing a meteor shower. At a distance of 25,000 AU a pair of stars pirouette their way around β Cam taking more than a million years to complete an orbit. Very little is known about the pair other than one is an A-class and the other a suspected F-class.

γ Camelopardalis is an unremarkable Main Sequence white dwarf – if a star can ever be regarded as being 'unremarkable'. It is more than twice as big as the Sun, 123 times more luminous and is 335 ly away.

After γ Cam there is – nothing! Well at least as far as Johann Bayer was concerned because he found the constellation so boring he only assigned Greek letters to just three stars. Which is a pity because **BE Camelopardalis**, with a diameter of 314 $D_\odot$ is one of the largest red giants in the sky. Almost as far away as β Cam this enormous star is as wide as the orbit of Mars and could swallow up 31 million Suns. It is also an Lc class pulsating variable, one of just a dozen visible

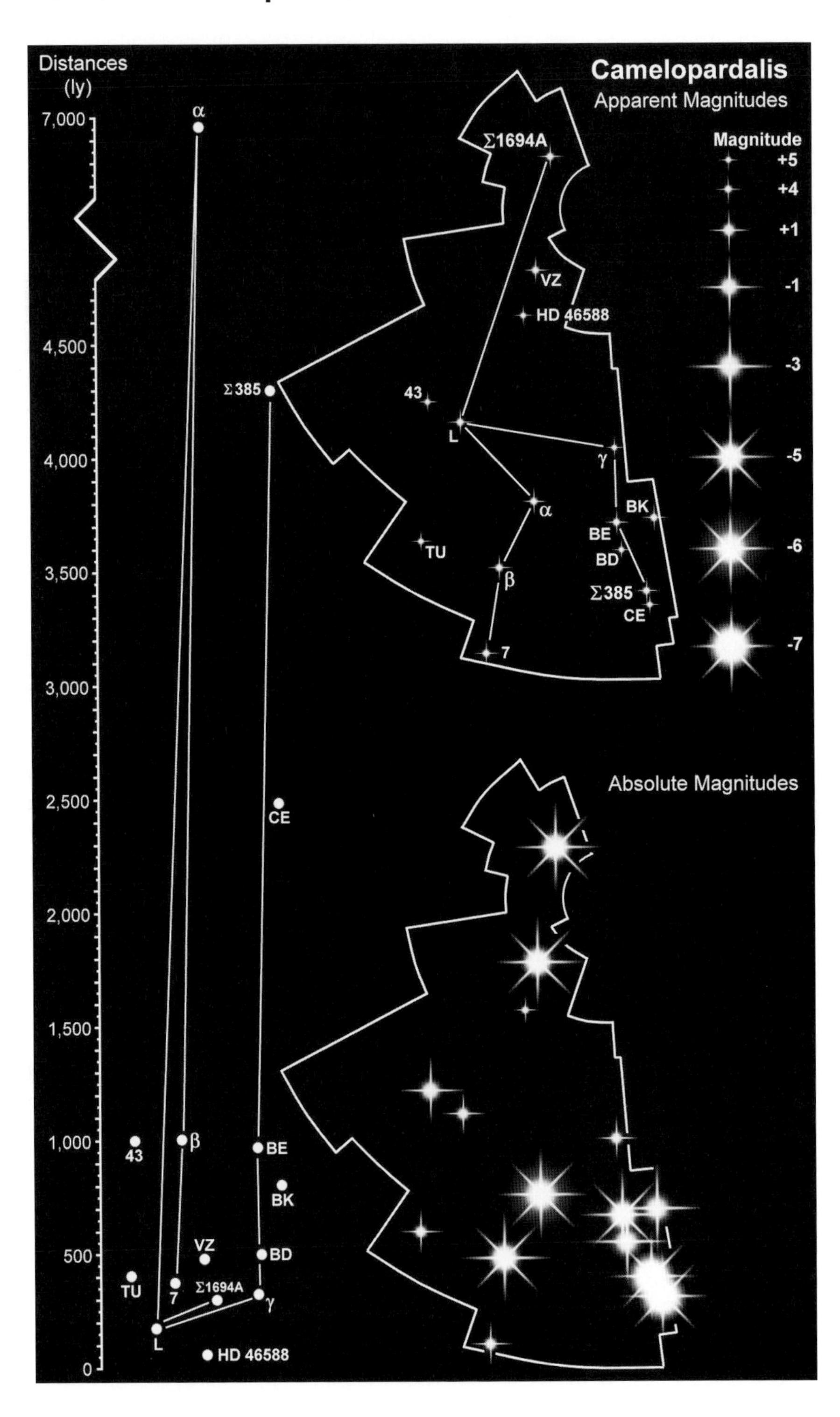

Distances (ly)
7,000
4,500
4,000
3,500
3,000
2,500
2,000
1,500
1,000
500
0
α
Σ385
CE
β
BE
BK
43
VZ
BD
TU
7
Σ1694A
γ
L
HD 46588
Camelopardalis
Apparent Magnitudes
Magnitude
+5
+4
+1
-1
-3
-5
-6
-7
Absolute Magnitudes

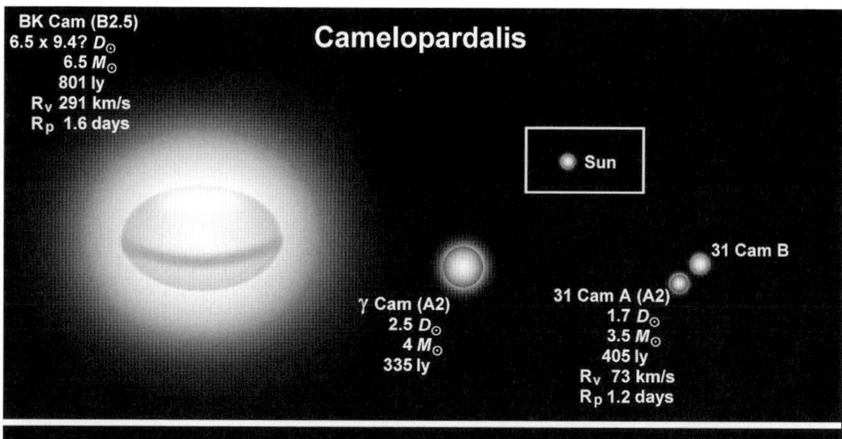
Camelopardalis
BK Cam (B2.5)
6.5 x 9.4? $D_\odot$
6.5 $M_\odot$
801 ly
R_V 291 km/s
R_P 1.6 days
Sun
γ Cam (A2)
2.5 $D_\odot$
4 $M_\odot$
335 ly
31 Cam B
31 Cam A (A2)
1.7 $D_\odot$
3.5 $M_\odot$
405 ly
R_V 73 km/s
R_P 1.2 days

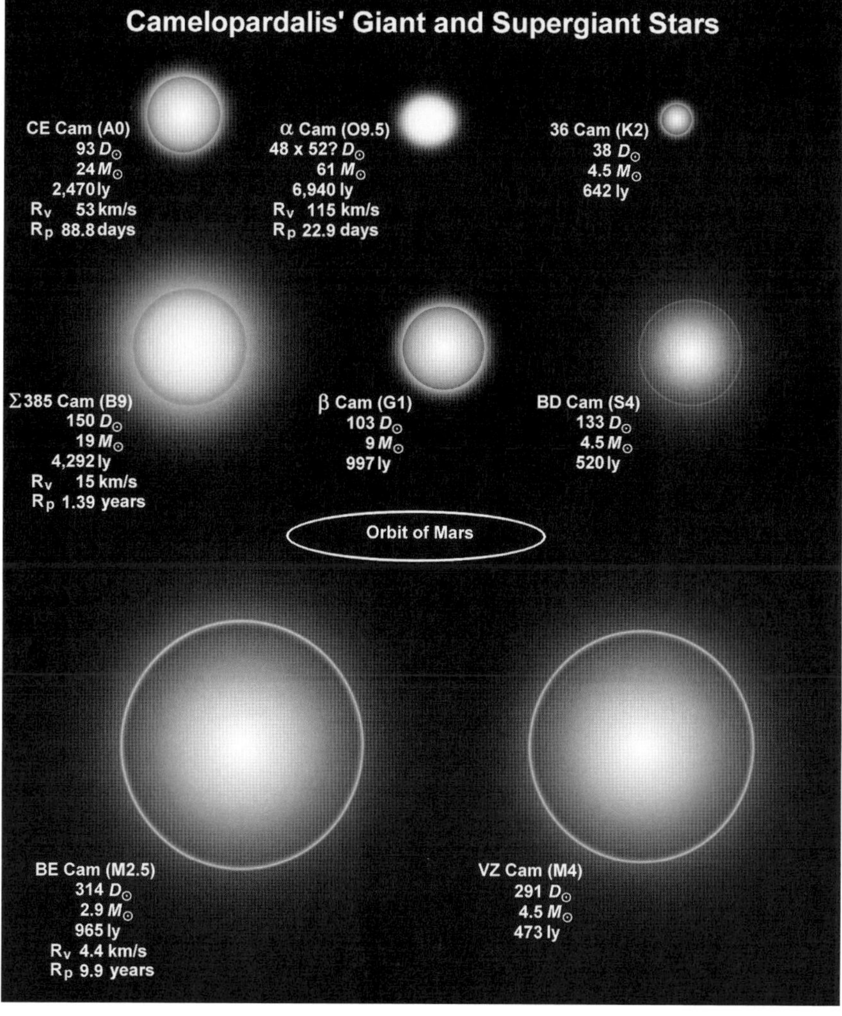
Camelopardalis' Giant and Supergiant Stars
CE Cam (A0)
93 $D_\odot$
24 $M_\odot$
2,470 ly
R_V 53 km/s
R_P 88.8 days
α Cam (O9.5)
48 x 52? $D_\odot$
61 $M_\odot$
6,940 ly
R_V 115 km/s
R_P 22.9 days
36 Cam (K2)
38 $D_\odot$
4.5 $M_\odot$
642 ly
Σ385 Cam (B9)
150 $D_\odot$
19 $M_\odot$
4,292 ly
R_V 15 km/s
R_P 1.39 years
β Cam (G1)
103 $D_\odot$
9 $M_\odot$
997 ly
BD Cam (S4)
133 $D_\odot$
4.5 $M_\odot$
520 ly
Orbit of Mars
BE Cam (M2.5)
314 $D_\odot$
2.9 $M_\odot$
965 ly
R_V 4.4 km/s
R_P 9.9 years
VZ Cam (M4)
291 $D_\odot$
4.5 $M_\odot$
473 ly

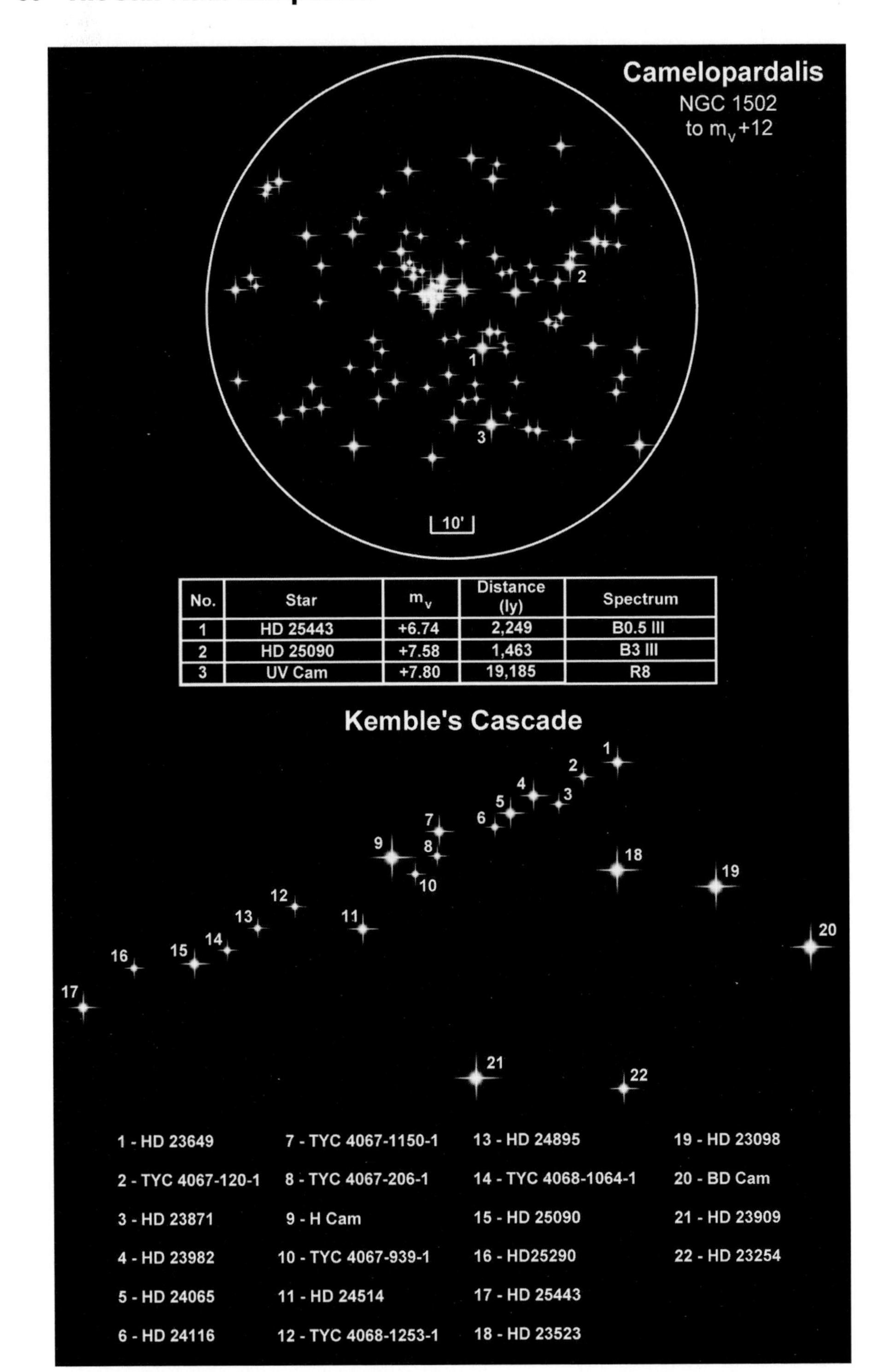

No.	Star	m_v	Distance (ly)	Spectrum
1	HD 25443	+6.74	2,249	B0.5 III
2	HD 25090	+7.58	1,463	B3 III
3	UV Cam	+7.80	19,185	R8

without optical aid. Its magnitude varies between m_v +4.35 and m_v +4.48 with no discernable period.

Camelopardalis contains a real gem. At 520 ly the 133 solar diameter **BD Camelopardalis** belongs to the rare S-type spectral group. Often called zirconium stars because of the presence of zirconium oxide in their spectral signature, S-class stars are cool, very red giants. BD Cam is an S4 with a temperature of around 2,500 K and is one of only two naked eye S-type stars, the other being o^1 Orionis. It has a magnitude of m_v +5.04. Like BE Cam it is a long period variable.

BD Cam is a good finder star for what has come to be known as Kemble's Cascade. Moving eastwards you will find **HD 23089**, m_v +4.8, and **HD 23523**, m_v +5.85, and then "...a beautiful cascade of faint stars tumbling from the north-east down to the open cluster NGC 1502", as Fr. Lucian Kemble, a Franciscan friar and amateur astronomer described it in a letter to Walter Scott Houston, the great popularizer of astronomy who wrote the Deep-Sky Wonders column for *Sky & Telescope* for nearly 50 years. The table below includes details of Kemble's Cascade.

The 9.4 $D_\odot$ **BK Camelopardalis** is a bluish B2.5 giant with a surface temperature of around 22,000 K. Some 633 times more luminous than the Sun it lies at a distance of between 700 and 900 ly and is a γ Cas eruptive variable changing in brightness by a 10th of a magnitude between m_v +4.78 and m_v +4.89.

VZ Camelopardalis is another red supergiant of 291 $D_\odot$ across and a pulsating variable. Over a period of $23^d\ 16^h\ 48^m$ its magnitude changes between m_v +4.80 and m_v +4.96.

Σ 1694A is an A1, 8,500 K bluish-white giant that spins on its axis at 275 km/s. As it is 20 $D_\odot$ across it takes just 3.7 days to spin once, the high rotational rate causing the poles to be drawn in and the equator to bulge.

The title of brightest star in the constellation is shared between **Σ385** and **CE Cam.** Although they both appear faint at m_v +4.20 and m_v +4.60 respectively, at 10 pc they would brighten to M_v -7.1 both being highly luminous bluish-white supergiants.

The smallest star is the 1.1 $D_\odot$ **HD 46588**, a yellowish-white F7 dwarf which, at 58.2 ly, is also the closest star to us in the constellation.

31 Camelopardalis (also known as **TU Cam**) is another variable, this time of the EB eclipsing binary type. A fast spinning, 76 km/s, A2 at more than 400 ly its range is between m_v +5.12 and +5.29 with a period of $2^d\ 22^h\ 23^m\ 52^s$. It is a spectroscopic binary system.

Despite being a bright open cluster **Stock 23** almost disappeared from the history books. The cluster was discovered by the German astronomer Jürgen Stock in 1956/57 but because it was so bright he assumed it had already been discovered and did not investigate it further. Then in 1977 an amateur astronomer, John Pazmino, observing from New Jersey rediscovered the cluster and it has since become known as Pazmino's Cluster. In fact it would appear not to be a true gravitationally bound cluster but an asterism.

Kemble's Cascade listed from NW to SE

Star	m_v	ly	RA	Dec
HD 23649	+7.92	?	$03^h 50^m 39^s$	+63° 42′ 26″
TYC 4067-120-1	+9.37	?	$03^h 51^m 42^s$	+63° 36′ 30″
HD 23871	+8.92	?	$05^h 52^m 15^s$	+63° 29′ 06″
HD 23982	+8.07	1,708	$03^h 53^m 13^s$	+63° 29′ 00″
HD 24065	+7.90	?	$03^h 53^m 46^s$	+63° 23′ 24″
HD 24116	+8.52	?	$03^h 54^m 16^s$	+63° 19′ 04″
TYC 4067-1150-1	+8.28	1,117	$03^h 56^m 07^s$	+63° 13′ 46″
TYC 4067-206-1	+8.69	324	$03^h 55^m 58^s$	+63° 08′ 23″
H Cam	+5.03	338	$03^h 55^m 26^s$	+63° 04′ 20″
TYC 4067-939-1	+8.74	?	$03^h 56^m 30^s$	+63° 02′ 41″
HD 24514	+7.97	317	$03^h 57^m 40^s$	+62° 46′ 39″
TYC 4068-1253-1	+8.97	?	$04^h 00^m 10^s$	+62° 46′ 00″
HD 24895	+8.47	?	$04^h 01^m 14^s$	+62° 38′ 18″
TYC 4068-1064-1	+8.45	4,236	$04^h 01^m 58^s$	+62° 30′ 51″
HD 25090	+7.35	1,463	$04^h 22^m 55^s$	+62° 25′ 17″
HD 25290	+7.69	?	$04^h 04^m 53^s$	+62° 19′ 20″
HD 25443	+6.77	2,249	$04^h 06^m 08^s$	+62° 06′ 07″

Open clusters in Camelopardalis

Name	Size arc min	Size ly	Distance ly	Age million yrs	Brightest star in region*	No. stars m_v >+12*	Apparent magnitude m_v
NGC 1502	100′	78.6	2,700	11	HD 25443 m_v +6.74	108	+5.7
Stock 23	20′	7.2	1,240	32	HD 20134 m_v +7.47	26	+6.5

*May not be a cluster member.

Cancer

Constellation:	Cancer	**Hemisphere:**	Northern
Translation:	The Crab	**Area:**	506 deg^2
Genitive:	Cancri	**% of sky:**	1.227%
Abbreviation:	Cnc	**Size ranking:**	31st

Sandwiched between the brilliant twins of Gemini and the unmistakable Leo this Zodiacal constellation of faint stars is easy to miss. It represents a crab sent by the goddess Hera to distract Hercules in his battle with Hydra. The crab nipped Hercules' toe but was immediately crushed underfoot and subsequently placed in the heavens. The Sun passes through Cancer between the 20th July and the 10th August.

Sometimes known by its proper name of Acubens, 'the claw', α **Cancri** is a triple star system 174 ly away. The primary component is a 1.8 $D_\odot$ bluish-white A5 of 2.1 $M_\odot$ and m_v +4.25. It is known to have a companion which is believed to be 1.2 $M_\odot$ and of a slightly later spectral class, perhaps an A7 or A8. The two are in a 6.75 year orbit. The third star is probably an M-class red dwarf of 0.38 $M_\odot$ and averaging 600 AU leading to an orbital period of around 7,700 years. α Cnc is considered to be a chemically peculiar star. The presence of a twin places a drag on the star causing it to rotate more slowly than normally for A-class stars. As a result there is less turbulence within the star and so elements like zirconium, strontium and barium separate out and pool near the surface.

At magnitude m_v +3.52 β **Cancri** or Altarf is actually the brightest star in the constellation (α Cnc is the 4th brightest). A giant K4, 48 times as big as the Sun it is more than 660 times as luminous and is 290 ly away. Rotating at 17 km/s it would take 143 days to complete just one revolution. It appears to have a red dwarf 14th magnitude companion at a distance of no less than 2,600 AU resulting in an orbital period of at least 75,000 years.

We previously came across three donkeys in Boötes – Asellus Primus, Asellus Secundus and Ascellus Tertius. Cancer has a further two of which γ **Cancri** is Asellus Borealis – the 'Northern Donkey'. An estimated 159 ly from Earth it is an A1 moderately fast spinner, believed to be in the order of 80 km/s. As the star is just 1.8 $D_\odot$ across it therefore spins once on its axis in 27.3 hours.

Asellus Australis, the 'Southern Donkey', is the star δ **Cancri** which is no relation to γ Cnc. It belongs to spectral class K0, is 11 $D_\odot$ across and lies at a distance of 136 ly. Eratosthenes (*c.* 276-196 BC), the Greek geographer and astronomer who was the first person to measure the circumference of the Earth, recorded that the two donkeys were rode into battle by the gods Dionysus and Silenus against the Titans. The animals made so much noise braying that they frightened the Titans and the gods won. As a reward the donkeys were placed in the heavens.

ζ **Cancri** is a quadruple or possibly a five star system. ζ^A **Cancri** which, at m_v +5.25, is the only one of the four that can just be seen without optical aid is a yellowish-white F8, 2.3 $D_\odot$ and 1.18 $M_\odot$. At a distance that varies between 15 and 29 AU orbits ζ^z **Cancri**, a G0 and about one solar diameter smaller. The orbital period is 59.6 years. ζ^C **Cancri** is similarly a G0 with a mass of slightly less than one Sun. It is in a 17.3 year orbit with ζ^D **Cancri** which could be a 0.92 $M_\odot$ white dwarf or possible a pair of identical M2 red dwarfs of 0.46 $M_\odot$ each. The AB pair are in a 1,115 year long orbit with the CD pair (or should that be the CD^aD^b triple?) during which the separation varies between 150 and 244 AU.

ι **Cancri** is a wide double, the two stars appearing 30.6″ apart in the sky. In reality they are separated by 2,800 AU resulting in an orbital period of 65,000 years. The brighter star, ι^A **Cancri**, is a +4.02 magnitude yellow G8 but much larger than the Sun at 20 $D_\odot$. Its partner is a 7th magnitude A3 Main Sequence white dwarf of 1.7 $D_\odot$. The system lies at a distance of 298 ly and is believed to be only 260 million years old.

Normally too faint to be included in this book **55 Cancri** – otherwise known as ρ^1 **Cancri** – is a G8 yellow dwarf, about 1.15 $D_\odot$ across but only 0.57 $L_\odot$ and shines at m_v +5.95. It has a companion, a 13th magnitude M6 red dwarf about 0.27 $D_\odot$. The red dwarf orbits at least 1,000 AU from the primary and takes a minimum of 30,000 years to complete one single orbit. This is important because 55 Cnc also has its own planetary system. In a more compact binary system the gravitational forces between the two stars could inhibit the formation of planets. Consisting of at least five planets the closest to 55 Cnc has a mass of just 3.4% that of Jupiter, or 11 Earth masses, and lies 0.038 AU (5.7 million km) from the star. With an orbital velocity of 165 km/s – nearly 600,000 km/h – planet **55 Cancri e** takes just 2.82 days to complete one orbit. It is likely to be a terrestrial type rocky planet. The largest of the five planets is **55 Cancri d** with a minimum mass of 3.8 M_J and orbiting at a distance of 5.9 AU, a slightly larger orbit than Jupiter. Its year is 14.2 Earth years long. The table below lists details for all the planets.

Despite counting among their ranks some of the most intelligent people in the world, when relied upon to develop a cataloging system astronomers fail miserably. 55 Cancri is a case in point. Binary and multiple star systems tend to be cataloged in one of three different ways. Suppose the α star in the hypothetical constellation Con is a binary, then it will usually be listed as:

α^1 Con and α^2 Con, or
α^A Con and α^B Con, or
α Con A and α Con B

55 Cancri is sometimes referred to as 55 Cnc A and 55 Cnc B, but then along came its planetary system and we suddenly have a planet listed as 55 Cnc b – you can no doubt see the scope for confusion.

In the early years of extrasolar planetary discovery the naming system was a right mess. Some planets were named after the proper name of the star, such as Fomalhaut b, others were named after the Bayer or Flamsteed systems (e.g. υ And b and 47 UMa b) and some were given their Harvard Revised (HR) or Henry Draper (HD) catalog numbers. So what should perhaps have been called ι Hor b

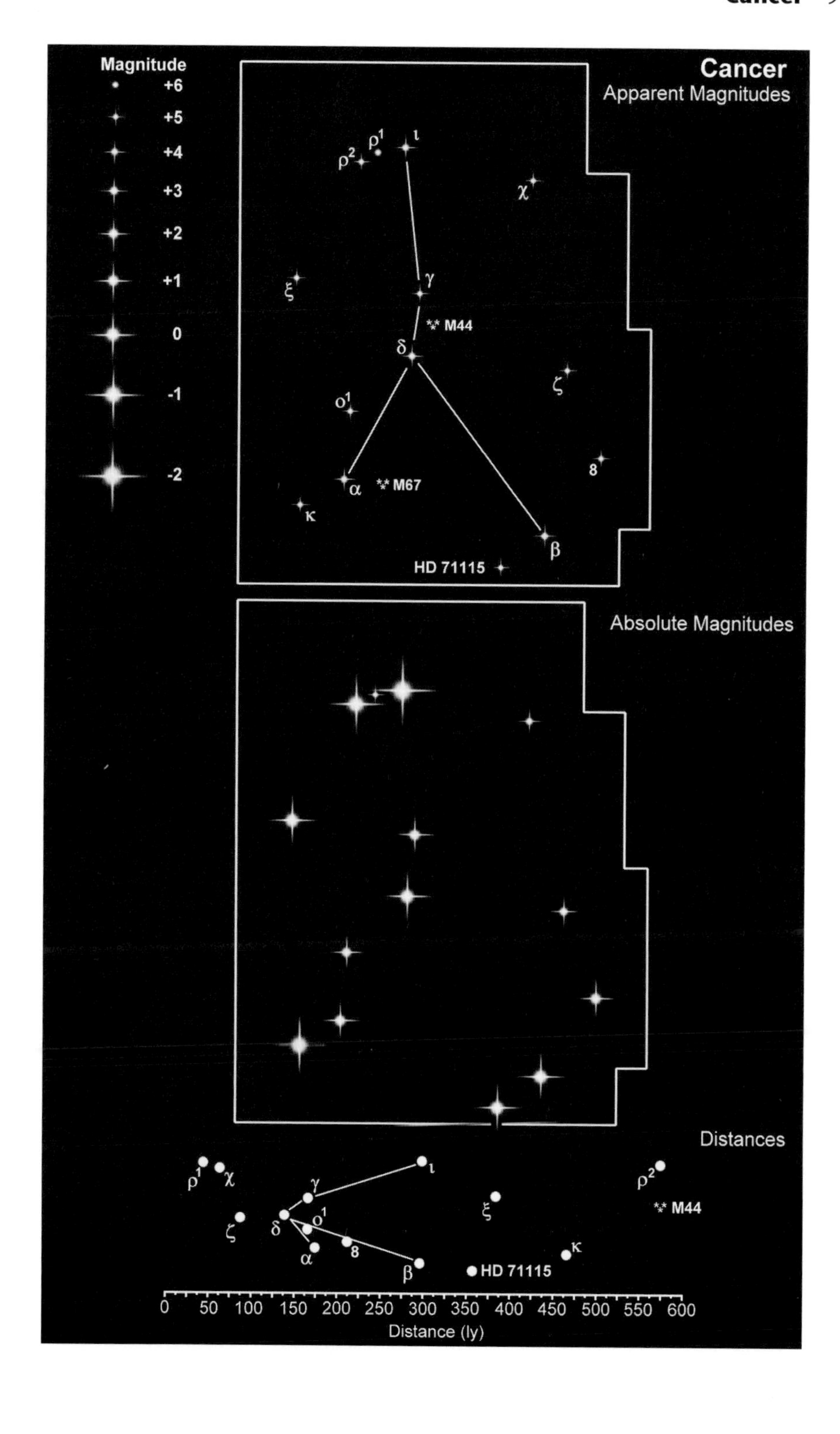
Magnitude
+6
+5
+4
+3
+2
+1
0
-1
-2
Cancer
Apparent Magnitudes
ρ2 ρ1 ι
χ
ξ
γ
M44
δ
ζ
o1
8
α
M67
κ
β
HD 71115
Absolute Magnitudes
Distances
ρ1 χ
ι
γ
ρ2
ξ
M44
ζ
δ
o1
α
8
κ
β
HD 71115
0 50 100 150 200 250 300 350 400 450 500 550 600
Distance (ly)

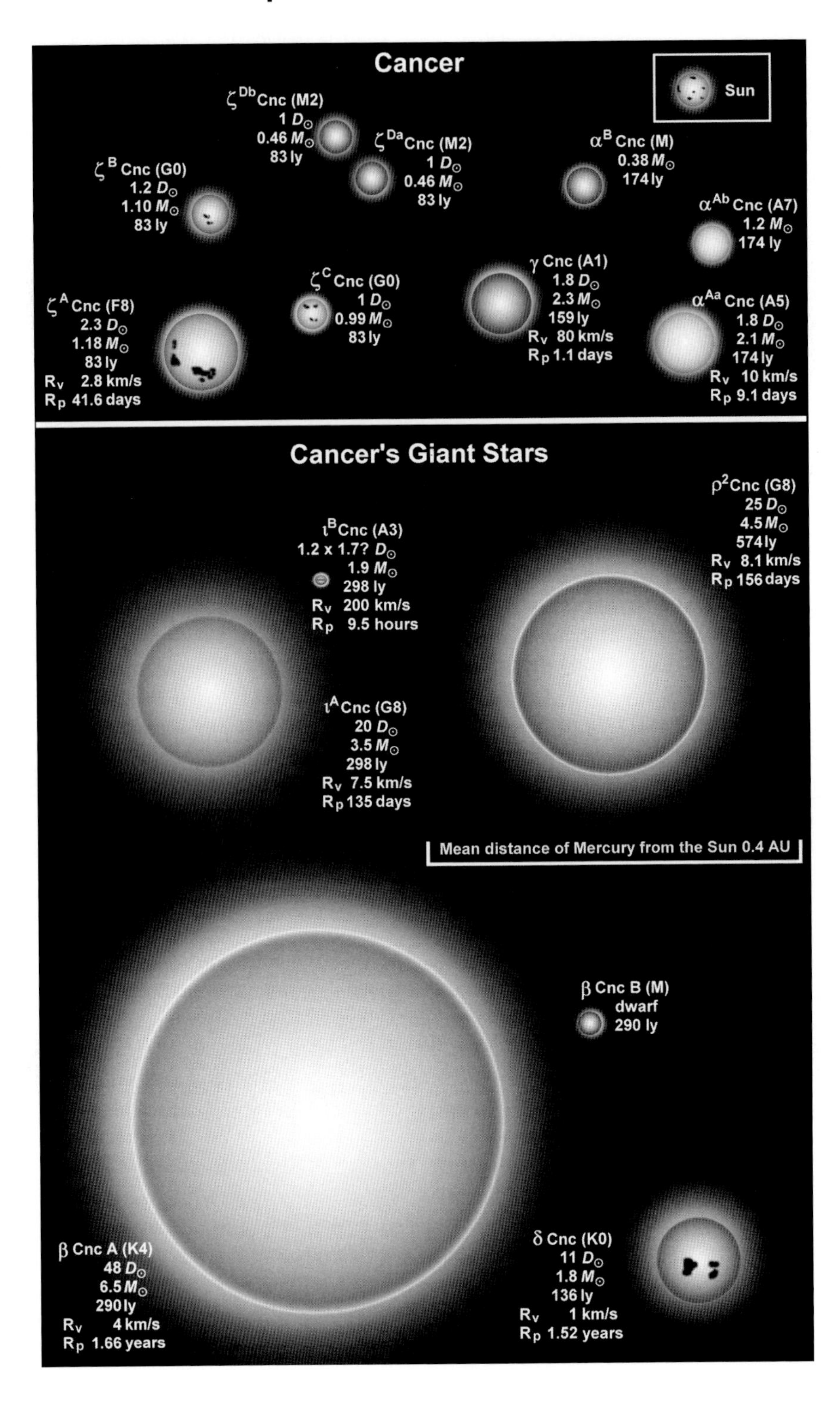
Cancer
Sun
ζDb Cnc (M2)
1 D⊙
0.46 M⊙
83 ly
ζDa Cnc (M2)
1 D⊙
0.46 M⊙
83 ly
αB Cnc (M)
0.38 M⊙
174 ly
ζB Cnc (G0)
1.2 D⊙
1.10 M⊙
83 ly
αAb Cnc (A7)
1.2 M⊙
174 ly
γ Cnc (A1)
1.8 D⊙
2.3 M⊙
159 ly
Rv 80 km/s
Rp 1.1 days
ζC Cnc (G0)
1 D⊙
0.99 M⊙
83 ly
ζA Cnc (F8)
2.3 D⊙
1.18 M⊙
83 ly
Rv 2.8 km/s
Rp 41.6 days
αAa Cnc (A5)
1.8 D⊙
2.1 M⊙
174 ly
Rv 10 km/s
Rp 9.1 days
Cancer's Giant Stars
ρ2Cnc (G8)
25 D⊙
4.5 M⊙
574 ly
Rv 8.1 km/s
Rp 156 days
ιB Cnc (A3)
1.2 x 1.7? D⊙
1.9 M⊙
298 ly
Rv 200 km/s
Rp 9.5 hours
ιA Cnc (G8)
20 D⊙
3.5 M⊙
298 ly
Rv 7.5 km/s
Rp 135 days
Mean distance of Mercury from the Sun 0.4 AU
β Cnc B (M)
dwarf
290 ly
β Cnc A (K4)
48 D⊙
6.5 M⊙
290 ly
Rv 4 km/s
Rp 1.66 years
δ Cnc (K0)
11 D⊙
1.8 M⊙
136 ly
Rv 1 km/s
Rp 1.52 years

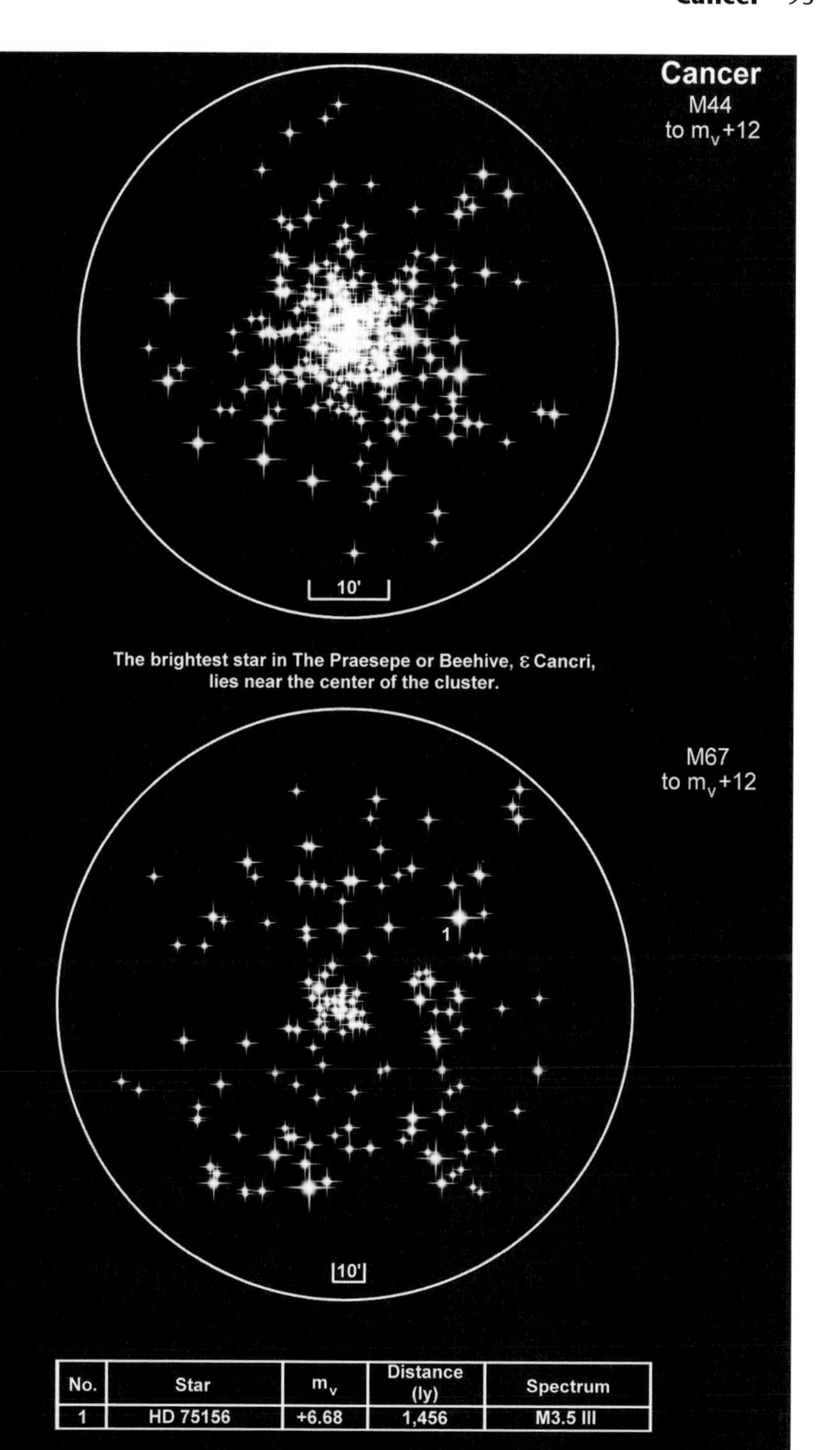

The brightest star in The Praesepe or Beehive, ε Cancri, lies near the center of the cluster.

No.	Star	m_v	Distance (ly)	Spectrum
1	HD 75156	+6.68	1,456	M3.5 III

became HR 810 b. Other catalog numbers were used, such as Bonner Durchmusterung (BD), and today we have stars and planets named after planet-hunting projects, so we find CoRoTs, HATs, Keplers, MOAs, OGLEs, TrESs, WASPs and XOs. There is certainly merit in naming stars and planets after such projects as success in this area can lever in much needed funds, but the cataloging system for the rest has been haphazard. And it is not just planets that have been poorly cataloged: comets, asteroids and meteorites have all been subject to a confusing mix of naming regimes.

To the naked eye it looks little more than a gray smudge on the celestial canvas but when Galileo turned his primitive telescope towards this tiny region in the middle of Cancer he could clearly make out many individual stars. Today we know that there are at least 350 stars and perhaps as many as 1,000 in this open cluster that is popularly known as the **Beehive** and **Praesepe** and which Charles Messier cataloged as **M44** (or if you prefer something a little more modern, **NGC 2632**). It is estimated to be about 730 million years old, which means it is of a similar age to the Hyades open cluster in Taurus and a common origin is at least feasible. Information on individual stars within M44 is patchy but the cluster appears to be 16 ly across.

Planetary systems in Cancer

Star	$D_\odot$	Spectral class	ly	m_v	Planet	Minimum mass	q	Q	P
ρ^1 Cnc	01.15	G8	42.5	+5.95	55 Cancri b	0.82 M_J	0.113	0.117	14.7 d
					55 Cancri c	0.17 M_J	0.22	0.26	44.3 d
					55 Cancri d	3.80 M_J	5.63	5.91	14.2 d
					55 Cancri e	0.03 M_J	0.035	0.041	2.82 d
					55 Cancri f	0.14 M_J	0.63	0.84	260 d

Open clusters in Cancer

Name	Size arc min	Size ly	Distance ly	Age million yrs	Brightest star in region*	No. stars m_v >+12*	Apparent magnitude m_v
M44 (NGC 2632)	95′	16	577	730	ε Cnc m_v +6.30	330	+3.7
M67 (NGC 2682)	30′	25.7	2,950	3,200	HD 75156 m_v +6.68	150	+6.1

*May not be a cluster member.

Canes Venatici

Constellation:	Canes Venatici	**Hemisphere:**	Northern
Translation:	The Hunting Dogs	**Area:**	465 deg^2
Genitive:	Canum Venaticorum	**% of sky:**	1.127%
Abbreviation:	CVn	**Size ranking:**	38th

Originally part of Ursa Major, Johannes Hevelius created the Hunting Dogs as a separate constellation in 1687 in his *Prodromus Atronomiæ* as Chara and Asterion, the mythical hounds of the herdsman Boötes.

The α^1 and α^2 **Canum Venaticorum** binary system lies at a distance of 110 ly. α^1 CVn is a yellowish-white F0 dwarf of 1.5 $D_\odot$ and 3.2 $L_\odot$ and has a visual magnitude of m_v +5.49. Its companion, α^2 CVn, is a much brighter star at m_v +2.84 and altogether more interesting. Around 2.6 $D_\odot$ and 83 times more luminous than the Sun this white A0 dwarf is the prototype for α^2 CV rotating variables. Sometimes known by it proper name of Cor Caroli (meaning the 'King's Heart' in honour of the executed King Charles I of England), α^2 CVn fluctuates in brightness between m_v +2.84 and +2.98 with a period of $5^d\ 11^h\ 16^m$. It is believed that its variability is caused by the star's immensely strong magnetic field, 1,500 times more intense than the Sun's, which may cause vast pools of elements like mercury and silicon to form in the upper layers of the star producing a surface with a blotchy appearance. As the star spins on its axis the dark blotches come into view and the star appears to darken. The two stars are separated by about 650 AU and take more than 8,000 years to complete one orbit. They represent the hound Asterion.

The other dog, Chara, is marked by β **Canum Venaticorum**, a solar analog G0 star, the same size and mass as the Sun but a little more luminous at 1.13 $L_\odot$ and the closest star in the constellation at just 27 ly. There is some dispute over its age. It spins at 3 km/s; 50% faster than the Sun and so turns once on its axis in about 16 days, which would suggest it is somewhat younger than the Sun. But Sun-like stars tend to become brighter with age so the higher luminosity would indicate a slightly older star.

For most of the time **Y Canum Venaticorum** is an impossible to see faint star, dipping down to 10th magnitude, but every 157 days it can brighten to m_v +4.99 though often only attains m_v +5.30. Yet despite its faintness it is worth looking out for because it is one of the reddest stars in the sky. At 711 ly it is the farthest star we are going to consider in this constellation. It is also the largest at 277 $D_\odot$ which, put into context, means its diameter is midway between the orbit of Earth and the orbit of Mars. Given the name of La Superba by Fr. Angelo Secchi in the 19th Century it belongs to that rare spectral class C7 indicating it is richer in carbon than in oxygen, which is odd for a red supergiant. La Superba is dying. It is losing mass at a rate of a million times faster than the Sun and is now surrounded by a vast dust shell 2.5 ly in diameter. Its upper layers are enriched in carbon by-products that filter out any light from the blue end of the spectrum

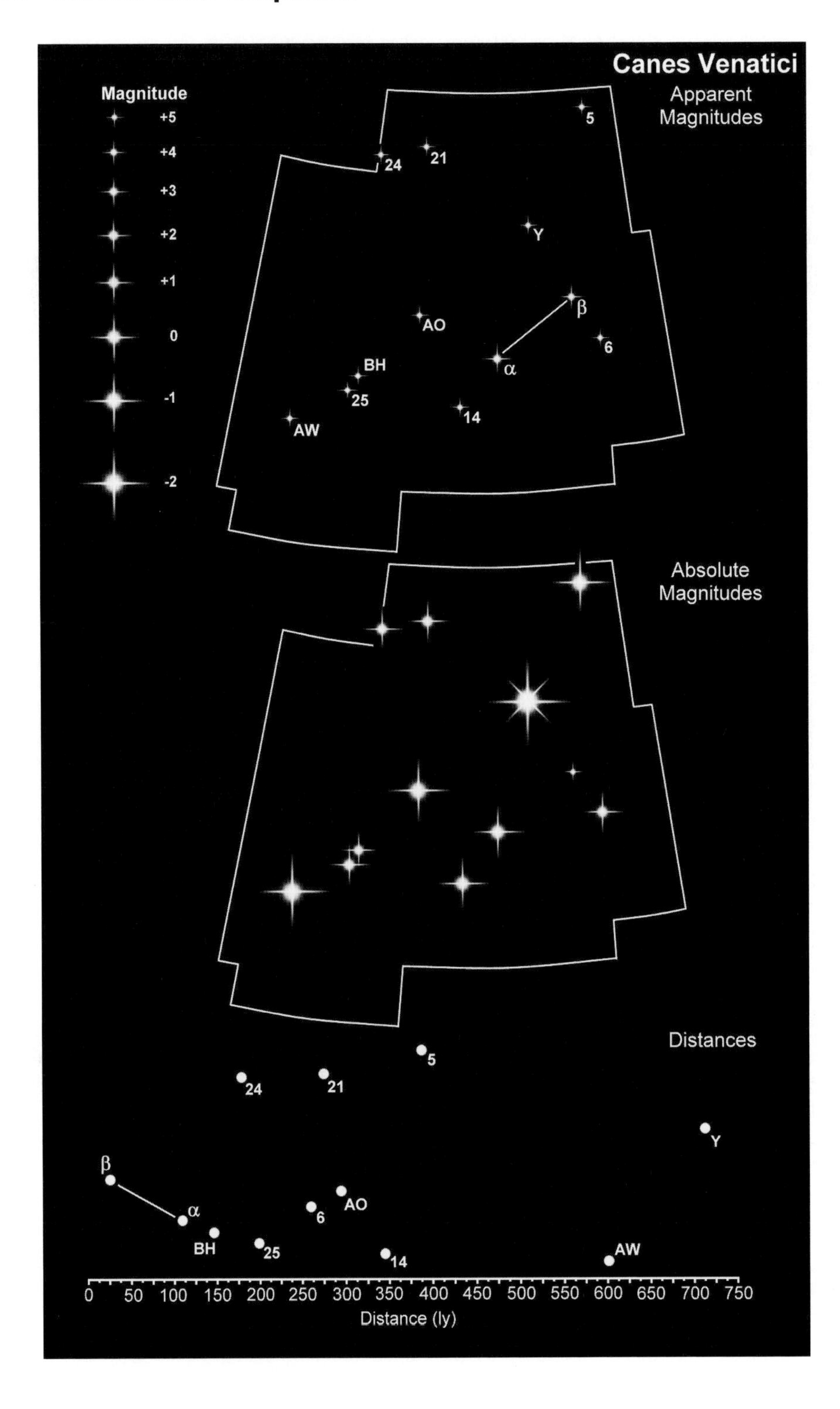
Canes Venatici
Apparent Magnitudes
Magnitude
+5
+4
+3
+2
+1
0
-1
-2
5
24
21
Y
β
AO
6
BH
α
25
14
AW
Absolute Magnitudes
Distances
5
24
21
Y
β
α
AO
6
BH
25
14
AW
0 50 100 150 200 250 300 350 400 450 500 550 600 650 700 750
Distance (ly)

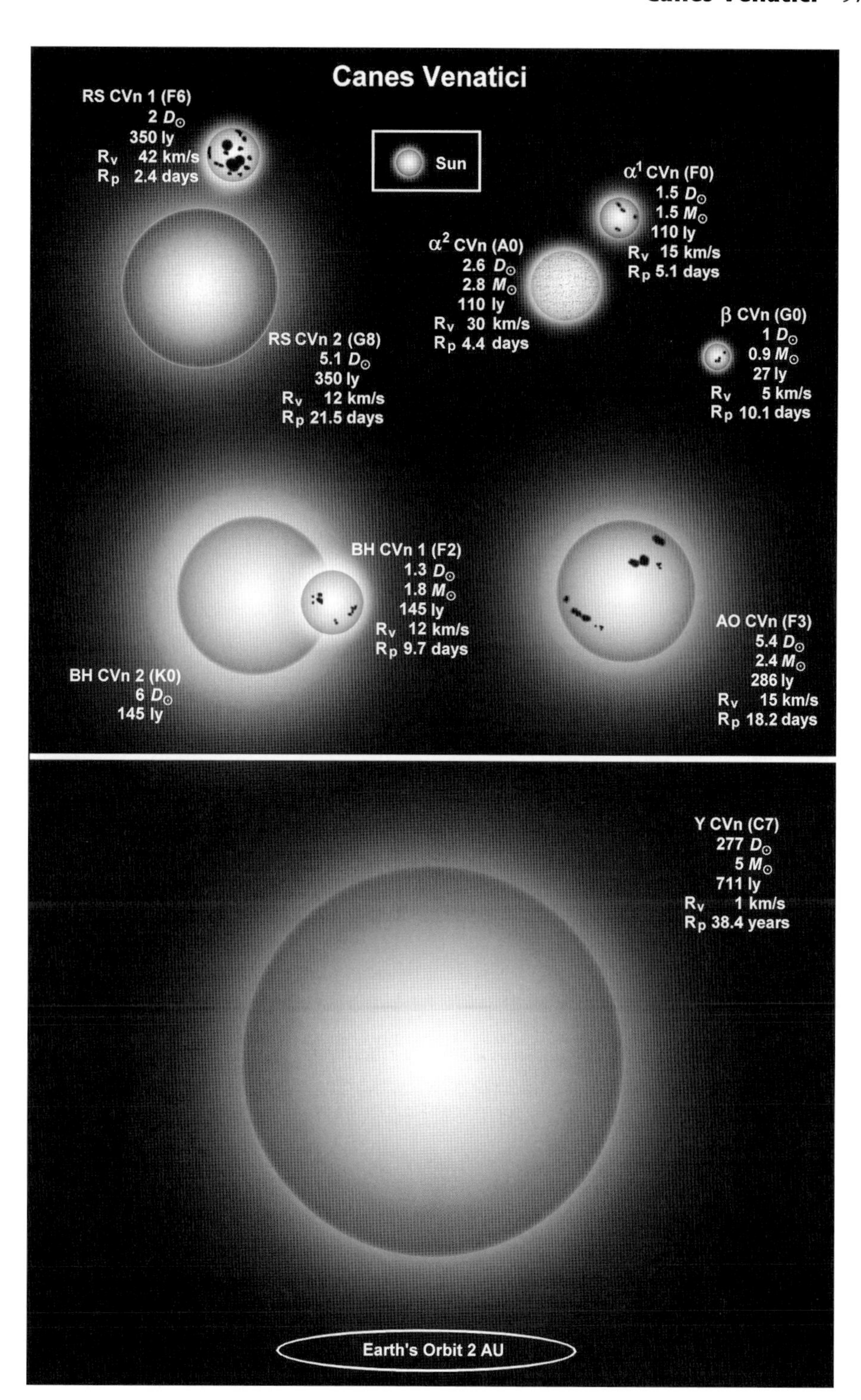
Canes Venatici
RS CVn 1 (F6)
2 $D_{\odot}$
350 ly
R_v 42 km/s
R_p 2.4 days
Sun
α^1 CVn (F0)
1.5 $D_{\odot}$
1.5 $M_{\odot}$
110 ly
R_v 15 km/s
R_p 5.1 days
α^2 CVn (A0)
2.6 $D_{\odot}$
2.8 $M_{\odot}$
110 ly
R_v 30 km/s
R_p 4.4 days
β CVn (G0)
1 $D_{\odot}$
0.9 $M_{\odot}$
27 ly
R_v 5 km/s
R_p 10.1 days
RS CVn 2 (G8)
5.1 $D_{\odot}$
350 ly
R_v 12 km/s
R_p 21.5 days
BH CVn 1 (F2)
1.3 $D_{\odot}$
1.8 $M_{\odot}$
145 ly
R_v 12 km/s
R_p 9.7 days
AO CVn (F3)
5.4 $D_{\odot}$
2.4 $M_{\odot}$
286 ly
R_v 15 km/s
R_p 18.2 days
BH CVn 2 (K0)
6 $D_{\odot}$
145 ly
Y CVn (C7)
277 $D_{\odot}$
5 $M_{\odot}$
711 ly
R_v 1 km/s
R_p 38.4 years
Earth's Orbit 2 AU

that may be produced deeper within the star. In the not too distant future, on a cosmic timescale, it will shed its outer atmosphere to form a new planetary nebula and a vanishing white dwarf.

RS Canum Venaticorum is also well beyond naked eye visibility at m_v +8.2 to +9.3 but we will consider it here because it is the prototype of the RS CVn variable stars of which there are about ten that can be seen without optical aid. RS CVn is actually a binary system consisting of an F6 and a G8 subgiant. Although the G8 star is more than twice as big it tends to radiate much of its energy in the infrared so it is the F6 star that appears the brightest of the two. As with all variables of this type it is a close binary, the two stars being tidally locked together so that their axial spin periods are equal to their orbital period. This makes the F6 component a fast spinner and prone to massive starspot activity. The theory is that its spin produces a dynamo effect that converts rotational energy into magnetic energy resulting in starspot activity that can cover as much as 20% of the star's surface. Compare this to the Sun's coverage that amounts to, at most, only a few percent. RS CVn stars are also known to flare. This combination of size, different magnitudes, giant starspots and huge flares produces a complex variability. As the two stars eclipse one another there is first a brief decrease in magnitude of about m_v 0.5 as the brightest star is eclipsed and then, 2½ days later, the magnitude again drops by about m_v 0.1 as the dimmer star is eclipsed. The presence of giant starspots on the F6 star will cause it to dim further while the occasional flare will make the whole system appear brighter. RS CVn lies at a distance of some 350 ly.

By a strange coincidence, Canum has *two* RS CVn variables. The other is **BH Canum Venaticorum**, a similar combination: a 2.3 $D_\odot$ F2 with a 6 $D_\odot$ K0 but much closer at 145 ly. The magnitude varies between m_v +4.94 and +5.01 with a period of $2^d\ 14^h\ 43^m$.

AO Canum Venaticorum is another variable, this one of the δ Sct variety. Yellowish-white it oscillates between m_v +4.70 and +4.75 with a period of $2^d\ 55^m$.

Globular cluster in Canes Venatici

Name	Size arc min	Size ly	Distance ly	Age million yrs	Apparent magnitude m_v
M3 (NGC 5272)	9′	90	33,900	8,000	+6.2

Canis Major

Constellation:	Canis Major	**Hemisphere:**	Southern
Translation:	The Greater Dog	**Area:**	380 deg^2
Genitive:	Canis Majoris	**% of sky:**	0.921%
Abbreviation:	CMa	**Size ranking:**	43rd

Canis Major depicts one of the two dogs belonging to Orion. It was one of Ptolemy's 48 constellations and contains the brightest star in the night sky, Sirius.

The brightest and arguably the most famous of all stars, there is nothing particularly unusual about α **Canis Majoris** or, to give it its more popular name, Sirius. It is a typical Main Sequence A1 white dwarf of 1.8 $D_{\odot}$ and 26 $L_{\odot}$. It appears very white when high in the sky, but when observed near the horizon it can flash different colors due to air turbulence. The reason it is so bright, at m_v -1.46, is mainly because it is only 8.6 ly away, making it the second closest star after the α Centauri system. If it were placed at 10 pc it would barely make M_v +1.41. Often called the Dog Star (although Sirius actually means 'sparkling') it is accompanied by an 8th magnitude white dwarf, **Sirius B** (nicknamed 'The Pup'), of spectral type DA2. Sirius B has a mass almost equal to that of the Sun, 0.978 $M_{\odot}$, but is a tiny 0.0084 $D_{\odot}$ making it a good 1,000 km smaller than Earth. Nevertheless, its temperature is a whopping 25,600 K – more than four times hotter than the visible surface of the Sun. The two stars are in an orbit that brings then to within 8.1 AU of one another before they separate to 31.5 AU, the orbital period being 50.1 years. Sirius B was probably, at one stage in its history, an early B-class star of up to 5 $D_{\odot}$, 100 $M_{\odot}$ and 1,000 $L_{\odot}$ making it considerably more luminous than Sirius itself. It spent the first 100 million years as a Main Sequence star before swelling to a red giant for about 25 million years and then, some 124 million years ago, evolved into a white dwarf. The system is so close that its proper motion across the sky is significant. In 25,000 years the stars will be in Fornax.

α CMa may be part of the Sirius Supercluster, about 100 stars that all have similar characteristics (e.g. spectrum, size, direction, etc). The stars in the supercluster are all aged about 225 to 250 million years, are generally less than 2 $D_{\odot}$ and lie within 500 ly of Earth with most being 180 to 280 ly away.

Side by side β **Canis Majoris** or Mirzam is a far more impressive star than Sirius. It is larger (9 $D_{\odot}$), it is hotter (25,800 K compared to 8,500 K) and it is considerably more luminous at 3,155 $L_{\odot}$. It is also much farther away at about 500 ly which is why it shines in our sky at a modest 2nd magnitude. But at 10 pc it would rival Venus at M_v -4.8. In fact the constellation is full of very luminous stars some of which will be part of the Collinder 121 open cluster (see below). β CMa is also the brightest β Cepheid variable which, just to confuse matters, are also known as β CMa variables. The variability is caused by the star pulsating with maximum brightness occurring during contraction of the star. In reality the pattern of pulsations is complex with the greatest magnitude swinging from m_v

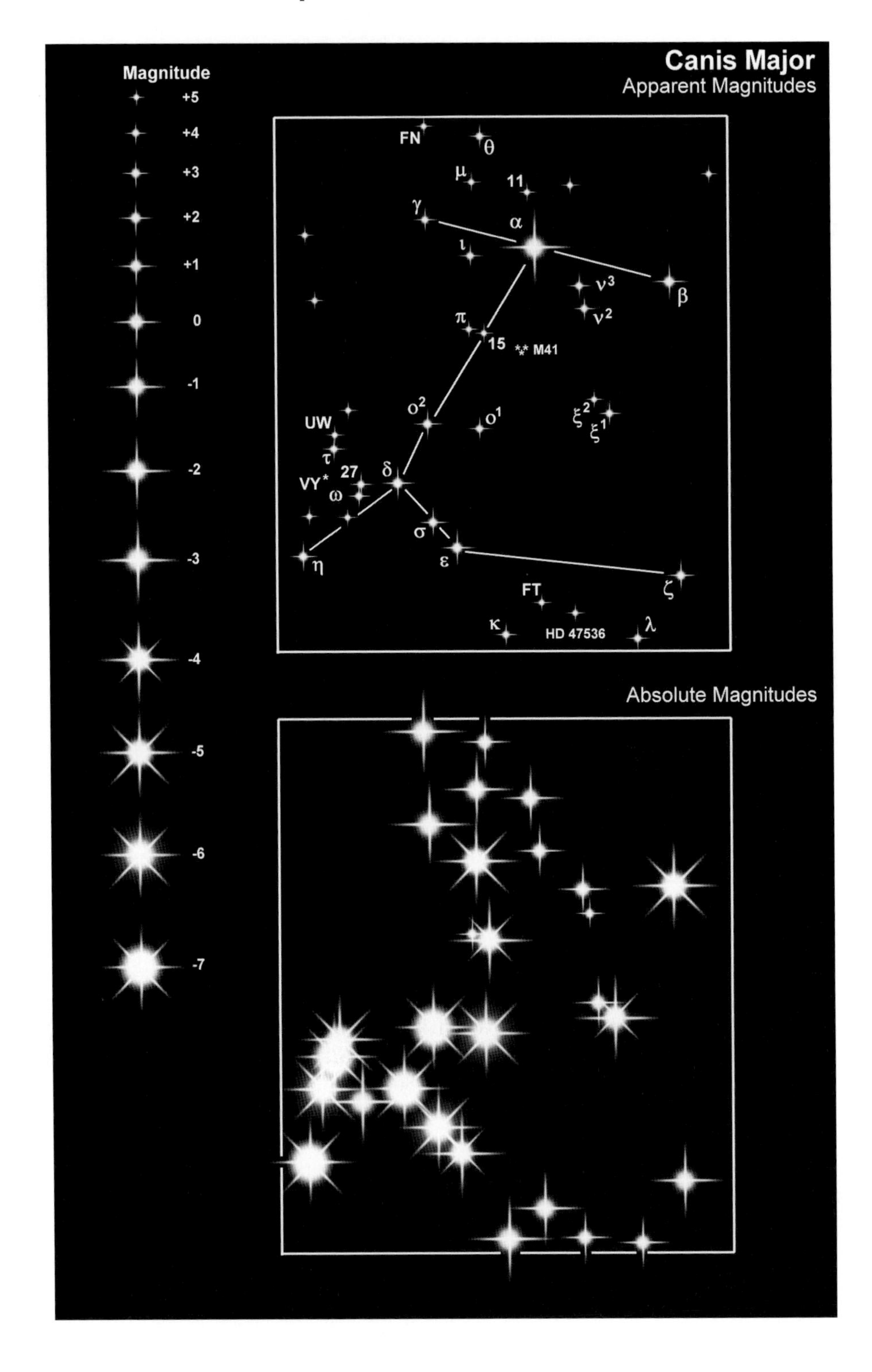
Canis Major
Apparent Magnitudes
Magnitude
+5
+4
+3
+2
+1
0
-1
-2
-3
-4
-5
-6
-7
FN
θ
μ
11
γ
α
ι
ν3
β
ν2
π
15
M41
o2
o1
ξ2
ξ1
UW
τ
27
δ
VY*
ω
σ
ε
η
ζ
FT
κ
HD 47536
λ
Absolute Magnitudes

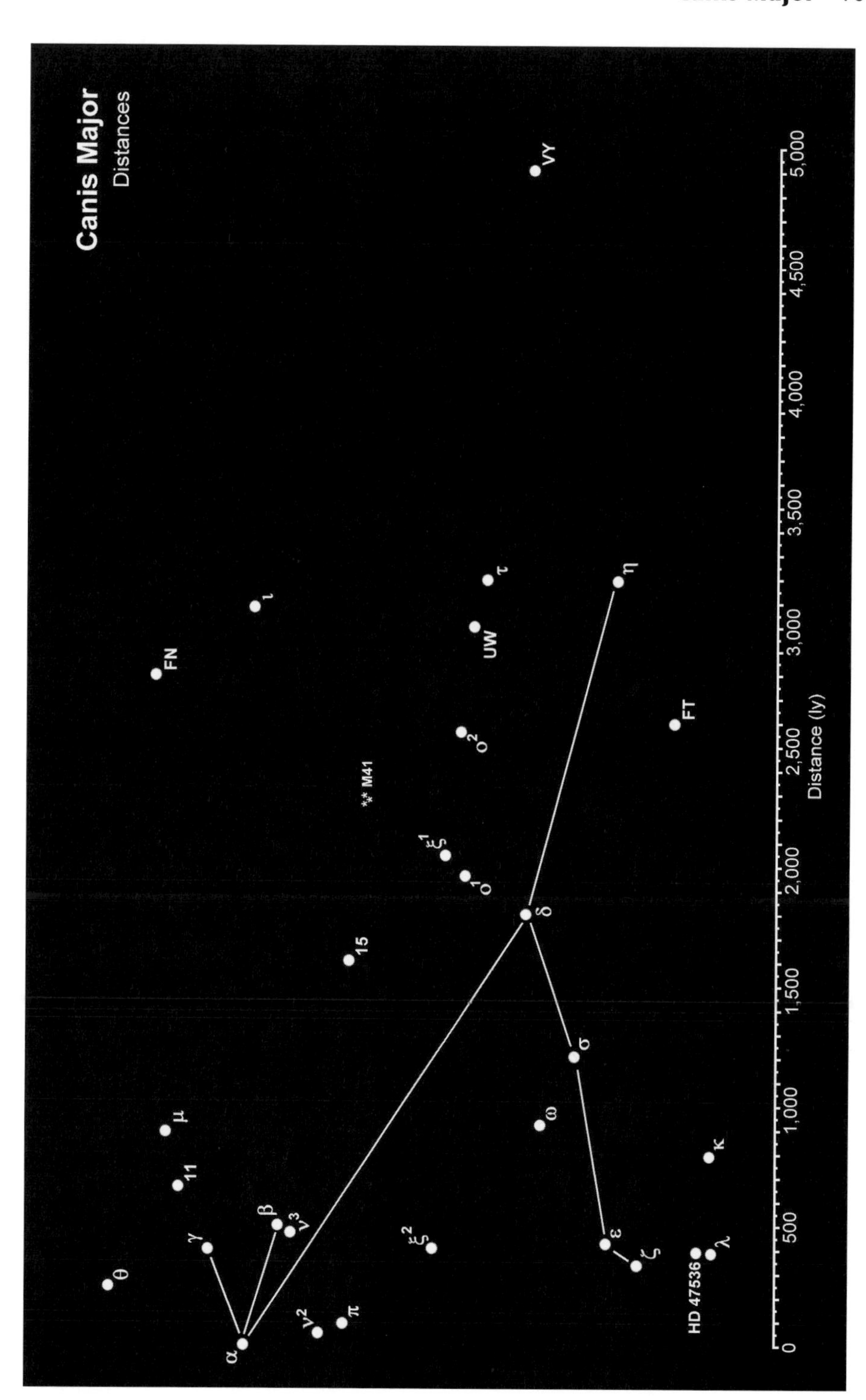

Canis Major
Distances
Distance (ly)
0
500
1,000
1,500
2,000
2,500
3,000
3,500
4,000
4,500
5,000
α
β
γ
δ
ε
ζ
η
θ
ι
κ
λ
μ
ν2
ν3
ξ1
ξ2
ο1
ο2
π
σ
τ
ω
11
15
FN
FT
UW
VY
HD 47536
M41

+1.93 to +2.00 having a period of 6^h but set against a background of almost imperceptible changes in brightness.

γ **Canis Majoris** or Muliphen presents us with a bit of a mystery. While compiling his *Uranometria* star atlas Johann Bayer called the brightest star α, the second brightest β and so on. Occasionally he got it wrong, particularly where two stars had very similar magnitudes, but with γ CMa he got it badly wrong. Following Bayer's system it should be the third brightest star in the constellation, but it isn't. Nor is it the fourth. Or the fifth. In fact it is the 14th brightest! It could be that he just overlooked it. Or it could be something else. There is a report that the star actually disappeared between 1670 and 1693. Ancient chronicles are awash with stellar enigmas, most of them little more than fantasy, but this one may have some substance to it. γ CMa is known to have high levels of manganese and mercury in its near-surface regions, rather like α Andromedae which is classed as an α CV variable because of the effects which the mercury clouds have on its brightness. Could it be that γ CMa also darkened so much in the past that it dipped below naked eye magnitude? Was it actually the third brightest star in the constellation in Bayer's time, then faded significantly, and never fully recovered to its former glory? We simply do not know.

δ **Canis Majoris** or Wezea is a supernova just waiting to happen. At 365 $D_{\odot}$ – 3.4 AU across – it has a luminosity of more than 45,000 Suns. It is one of a trio of stars within the constellation that would reach an impressive absolute magnitude of M_V -7, the others being η and τ. A yellowish F8 it is less than 10 million years old but is already depleted in hydrogen. Within the next 100 millennia it will swell to a red supergiant while its core collapse will instigate helium fusion. Eventually the core will explode seeding interstellar space with iron and other heavy elements that stars like the Sun are just not powerful enough to synthesize.

About 4.5 million years ago ε **Canis Major** was the brightest star in the night sky with a magnitude of m_V -4.4, comparable to Venus. The reason it is no longer so is that it is traveling directly away from us and at that time it would have been at most 21 ly distant. At 22,000 K this brilliant white B2 star is almost four times the temperature of the Sun but more than 3,500 times as luminous. It is calculated to be 11 $D_{\odot}$ across and is currently 431 ly from Earth. With a visual magnitude of m_V +1.51 it is the second brightest star in Canis Major, not the 5th as the designation suggests. It has an 8th magnitude companion at an orbital distance of at least 900 AU resulting in a 7,500 year long period.

η **Canis Majoris** along with τ Canis Majoris mark the outposts of the constellation, the two being an equal 3,198 ly, or at least as equal as we can tell. Forty-three times as large as the Sun η CMa appears to be surrounded by a dust shell, an indication that we are looking at a star that has shed a significant amount of its mass and is in its dying throes. Like δ CMa it is likely to end its life as a supernova.

Believed to be the 9th fastest naked eye star in the sky of those heading away from us is θ **Canis Majoris**. Apart from its incredible speed of 97.3 km/s it is an otherwise normal K4 orange giant of 37 $D_{\odot}$. It rotates at about the same speed as the Sun leading to rotational period of 892 days (2.4 years).

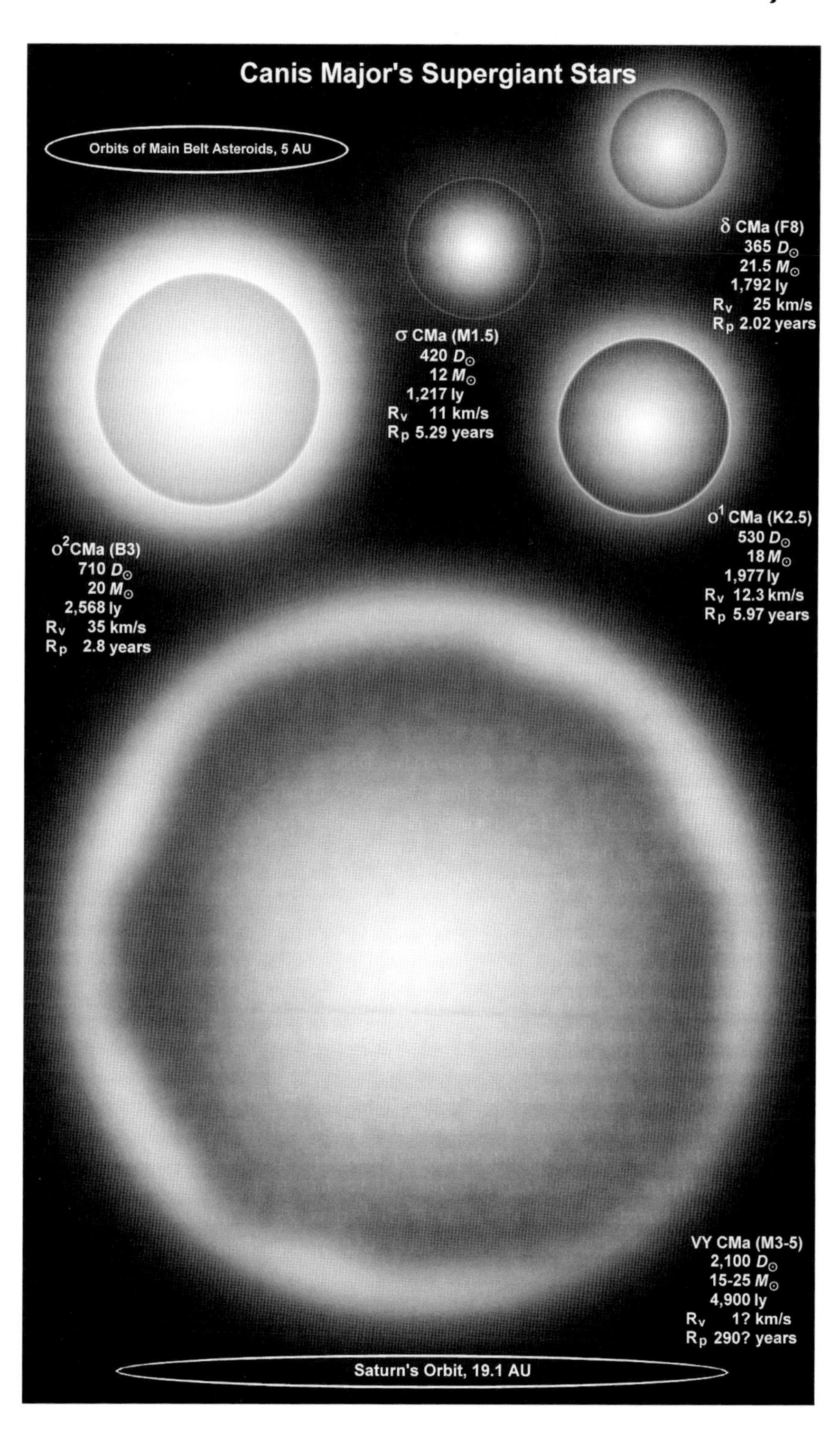
Canis Major's Supergiant Stars
Orbits of Main Belt Asteroids, 5 AU
δ CMa (F8)
365 $D_{\odot}$
21.5 $M_{\odot}$
1,792 ly
R_V 25 km/s
R_P 2.02 years
σ CMa (M1.5)
420 $D_{\odot}$
12 $M_{\odot}$
1,217 ly
R_V 11 km/s
R_P 5.29 years
o^1 CMa (K2.5)
530 $D_{\odot}$
18 $M_{\odot}$
1,977 ly
R_V 12.3 km/s
R_P 5.97 years
o^2CMa (B3)
710 $D_{\odot}$
20 $M_{\odot}$
2,568 ly
R_V 35 km/s
R_P 2.8 years
VY CMa (M3-5)
2,100 $D_{\odot}$
15-25 $M_{\odot}$
4,900 ly
R_V 1? km/s
R_P 290? years
Saturn's Orbit, 19.1 AU

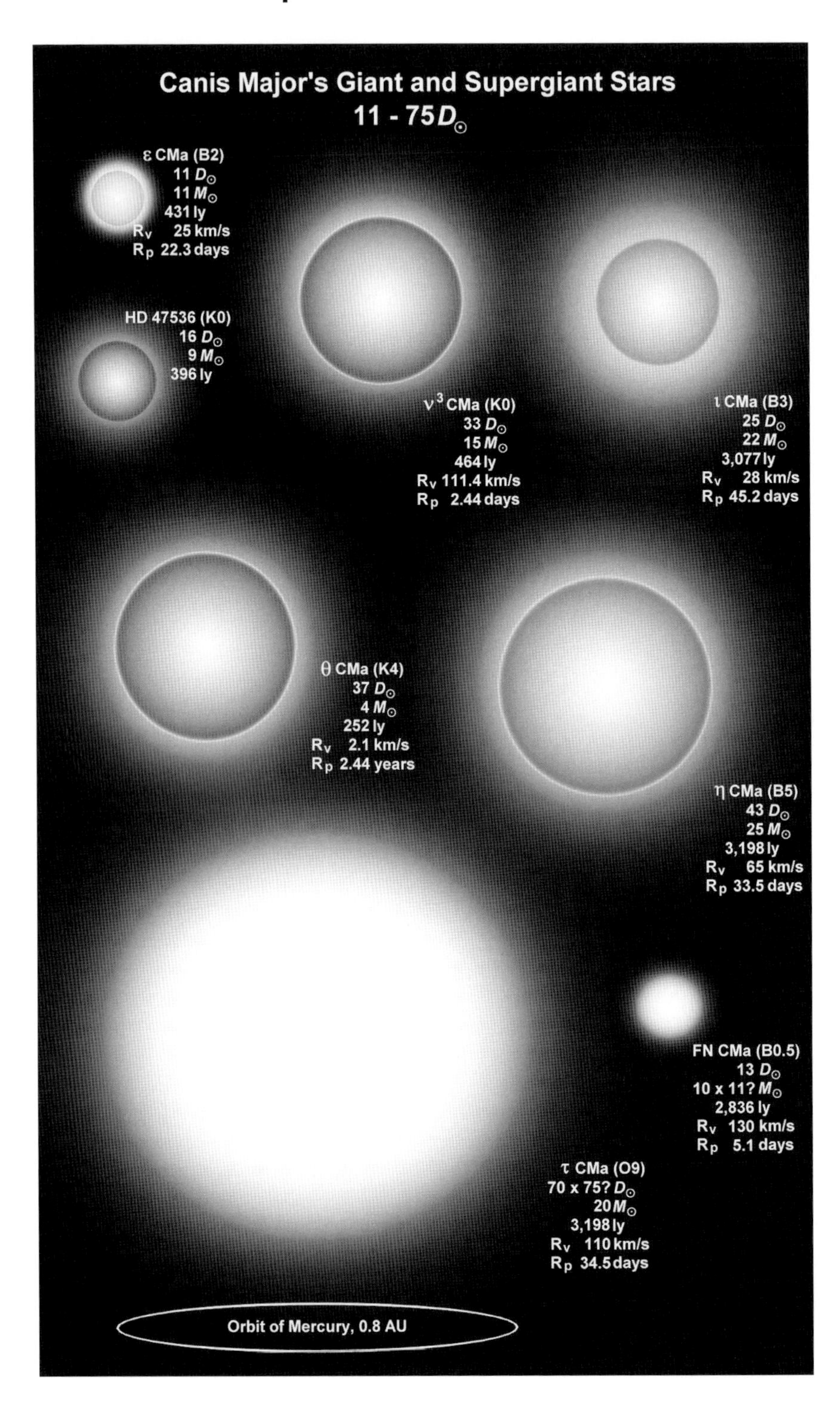
Canis Major's Giant and Supergiant Stars
11 - 75$D_\odot$
ε CMa (B2)
11 $D_\odot$
11 $M_\odot$
431 ly
R_v 25 km/s
R_p 22.3 days
HD 47536 (K0)
16 $D_\odot$
9 $M_\odot$
396 ly
ν^3 CMa (K0)
33 $D_\odot$
15 $M_\odot$
464 ly
R_v 111.4 km/s
R_p 2.44 days
ι CMa (B3)
25 $D_\odot$
22 $M_\odot$
3,077 ly
R_v 28 km/s
R_p 45.2 days
θ CMa (K4)
37 $D_\odot$
4 $M_\odot$
252 ly
R_v 2.1 km/s
R_p 2.44 years
η CMa (B5)
43 $D_\odot$
25 $M_\odot$
3,198 ly
R_v 65 km/s
R_p 33.5 days
FN CMa (B0.5)
13 $D_\odot$
10 x 11? $M_\odot$
2,836 ly
R_v 130 km/s
R_p 5.1 days
τ CMa (O9)
70 x 75? $D_\odot$
20 $M_\odot$
3,198 ly
R_v 110 km/s
R_p 34.5 days
Orbit of Mercury, 0.8 AU

Like β CMa, ι **Canis Majoris** is also a β Cepheid. A B3 giant almost at the edge of the constellation at 3,077 ly its magnitude changes between m_v +4.36 and +4.40 with a period of $1^h\ 55^m$ making it one of the faster pulsators in its class. It is also one of the largest at 25 $D_\odot$. **FN Canis Majoris** is another large β Cepheid at 13 $D_\odot$. Some 85% of the naked eye β Cepheids are less than 5 $D_\odot$.

You may think that κ and λ **Canis Majoris** are related. They are in more or less the same patch of sky and are both blue with not a huge difference in magnitude, m_v +3.78 and +4.45 respectively. However, κ CMa is about twice the distance of λ CMa and is travelling more slowly at 14 km/s compared to 41 km/s so a common origin is unlikely. They are also quite different in size with κ measuring perhaps 6.5 × 8.3 $D_\odot$ because of its high rotational velocity of 210 km/s, while λ is a more modest 1.1 × 1.5 $D_\odot$. κ CMa is a γ Cassiopeiae eruptive variable, one of many, which jumps between m_v +3.78 and +3.97 over no set period.

Canis Major harbors a number of accidental associations. You could be forgiven for thinking that ν^2 and ν^3 **Canis Majoris** are part of a wide binary system. They are both early K-class stars of about the same magnitude but actually separated in space by 400 ly, as the cosmic crow flies. Similarly, although ξ^1 and ξ^2 **Canis Majoris** are close neighbors on the celestial sphere there is more than 1,600 ly between them. And there is about 600 ly of space between o^1 and o^2 **Canis Majoris**. o^1 CMa is a K2.5 with an impressive diameter of 530 Suns. It is also an Lc (large supergiant) pulsating variable, its magnitude ranging from m_v +3.78 to +3.99. o^2 on the other hand is an enormous 710 $D_\odot$ B3 making it one of the largest stars in the constellation. Translated into Solar System language, if the Sun was replaced by o^1 CMa it would swallow up Mars and everything up to the Asteroid Belt. o^2 CMa would reach out to 3.3 AU, towards the farthest edge of the Belt.

Another Lc variable is σ **Canis Majoris**. Although not quite as big as o^1 CMa its 420 $D_\odot$ still covers a full 4 AU. It belongs to the M spectral group (M1.5) and lies at a distance of 1,217 ly.

At about 3,200 ly τ **Canis Majoris** appears as a m_v +4.40 bluish-white star but it is deceptive. An O9 supergiant, 75 $D_\odot$ across, it has a luminosity in excess of 13,500 Suns and at 10 pc would easily attain M_v -7. But it is remarkable for a couple of other reasons. First it is a complex but unstable five star system. Second, all the stars seem to be part of a larger open cluster that contains about 40 members and which is known as **NGC 2362**. All the members of this cluster are either O or B class and very young, probably less than 1 million years old. Despite this the cluster appears to be devoid of any nebulosity.

FT Canis Majoris – also called **10 CMa** and **HD 48917** – is a luminous γ Cas eruptive variable at 4,197 $L_\odot$. It is, however, relatively small at 2.3 $D_\odot$ and lies at a distance of more than 2,600 ly so we see it as a faint star fluctuating between m_v +5.13 and +5.44.

VY Canis Majoris is possibly the largest star in our visible Galaxy. An M3 to M5 red hypergiant this 10th magnitude star is estimated to be between 1,800 and 2,100 $D_\odot$. That works out at 16.7 to 19.5 AU across (Saturn's orbit is 19 AU in diameter). It is at least 200,000 times more luminous than the Sun and could be

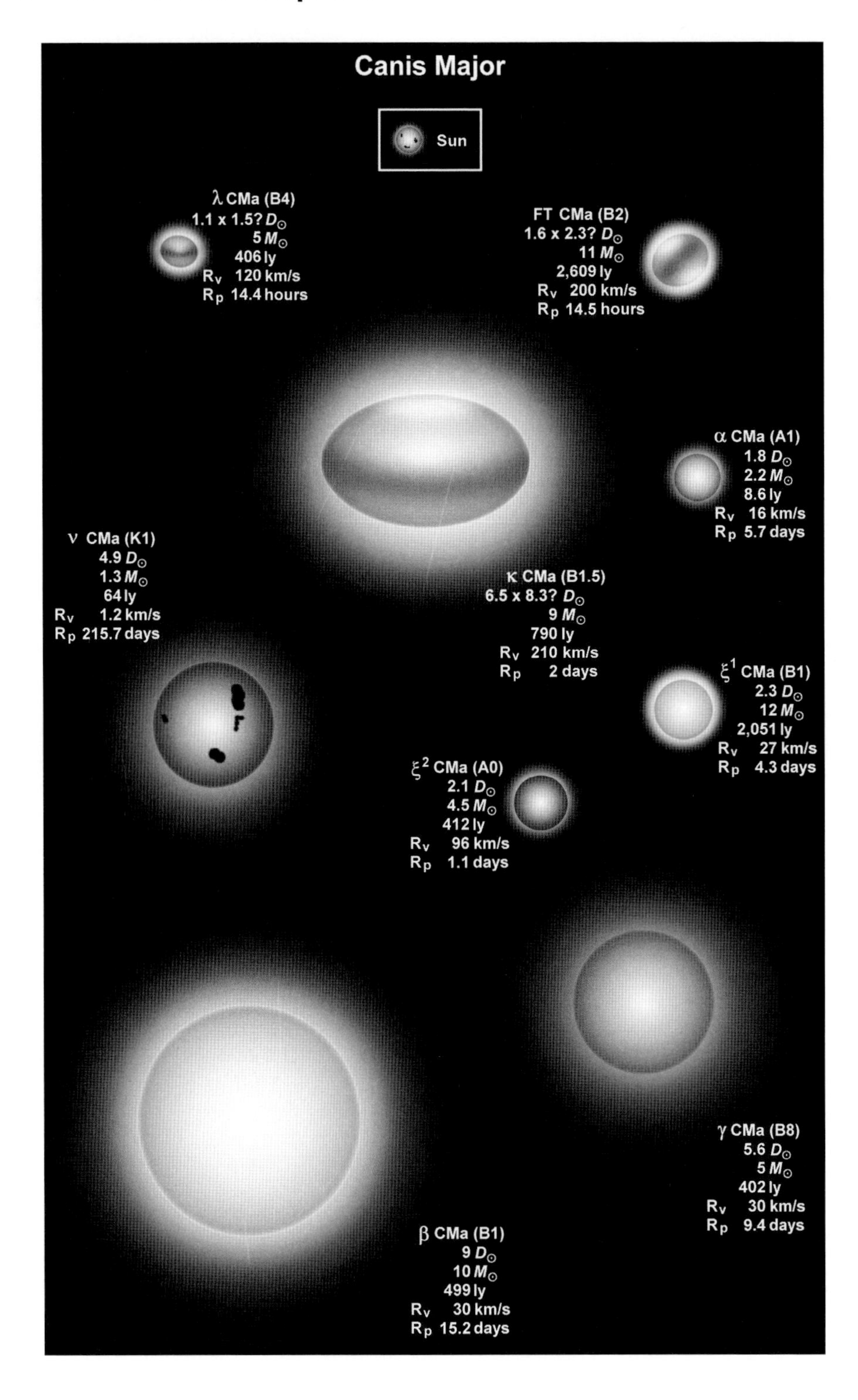

Canis Major
Sun
λ CMa (B4)
1.1 x 1.5? D☉
5 M☉
406 ly
Rv 120 km/s
Rp 14.4 hours
FT CMa (B2)
1.6 x 2.3? D☉
11 M☉
2,609 ly
Rv 200 km/s
Rp 14.5 hours
α CMa (A1)
1.8 D☉
2.2 M☉
8.6 ly
Rv 16 km/s
Rp 5.7 days
ν CMa (K1)
4.9 D☉
1.3 M☉
64 ly
Rv 1.2 km/s
Rp 215.7 days
κ CMa (B1.5)
6.5 x 8.3? D☉
9 M☉
790 ly
Rv 210 km/s
Rp 2 days
ξ1 CMa (B1)
2.3 D☉
12 M☉
2,051 ly
Rv 27 km/s
Rp 4.3 days
ξ2 CMa (A0)
2.1 D☉
4.5 M☉
412 ly
Rv 96 km/s
Rp 1.1 days
γ CMa (B8)
5.6 D☉
5 M☉
402 ly
Rv 30 km/s
Rp 9.4 days
β CMa (B1)
9 D☉
10 M☉
499 ly
Rv 30 km/s
Rp 15.2 days

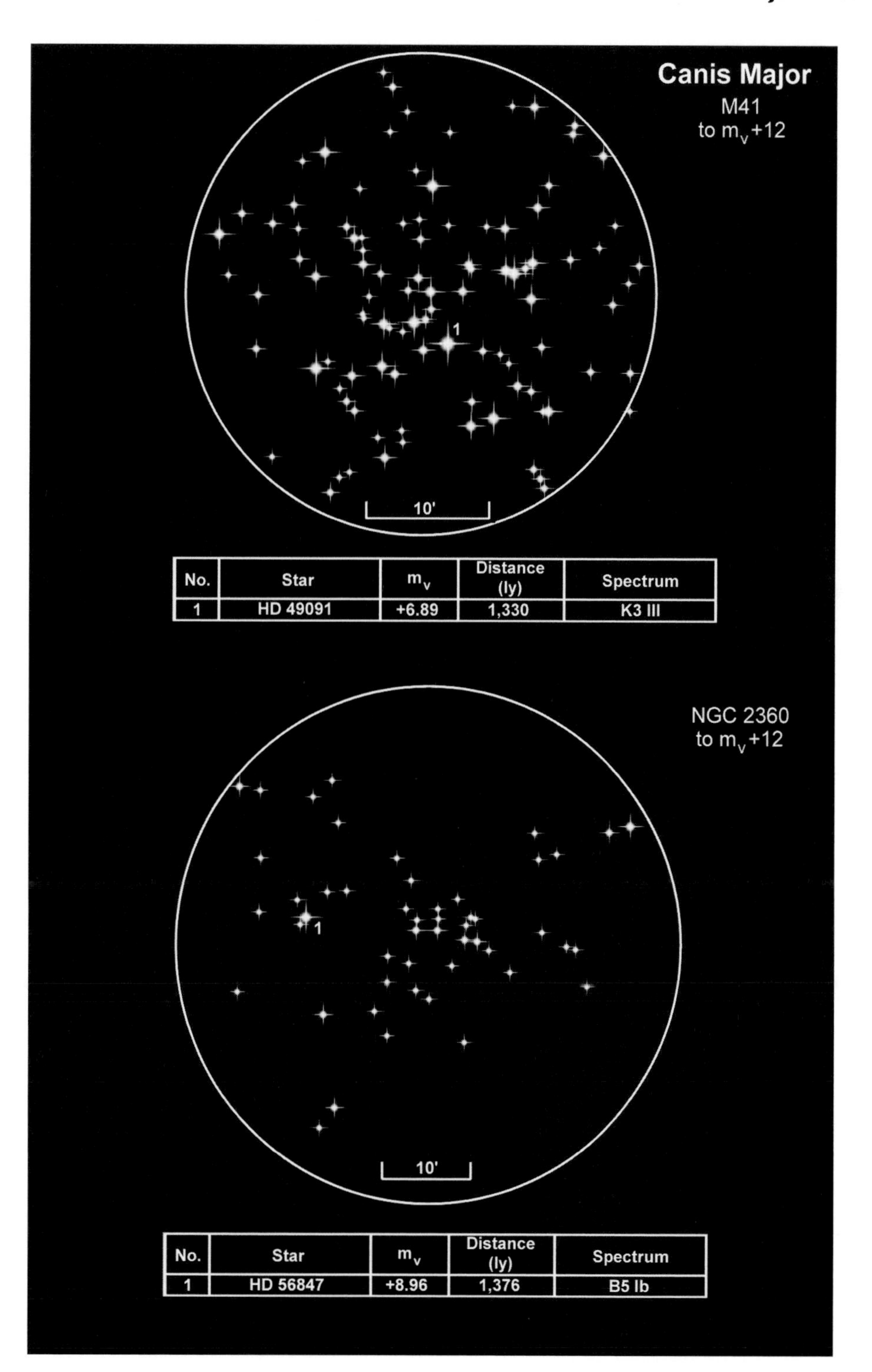

No.	Star	m_v	Distance (ly)	Spectrum
1	HD 49091	+6.89	1,330	K3 III

No.	Star	m_v	Distance (ly)	Spectrum
1	HD 56847	+8.96	1,376	B5 Ib

as much as 560,000 $L_\odot$ but its mass is put at just 15 to 25 $M_\odot$. It is estimated to be 4,900 ly away.

Collinder 121 is an open star cluster that may, or may not include γ, δ, ζ^1, η, ι o^1 and o^2 CMa. Most of the stars in the cluster are B class, occasionally A class, with a few K and M.

Open clusters in Canis Major

Name	Size arc min	Size ly	Distance ly	Age million yrs	Brightest star in region*	No. stars m_v >+12*	Apparent magnitude m_v
Collinder 121	80′	36	1,550	280	δ CMa m_v +1.84	47	+4.5
Collinder 140	42′	16	1,300	35	HD 58535 m_v +5.34	75	+3.5
M 41 (NGC 2287)	38′	25	2,250	240	HD 49091 m_v +6.89	129	+4.5
NGC 2345	12′	26	7,350	70	HD 54411 m_v +9.16	10	+7.7
NGC 2354	18′	17	13,300	135	DM -25° 4236 m_v +9.76	36	+6.5
NGC 2360	20′	36	6,200	560	HD 56847 m_v +8.96	51	+7.2
NGC 2362	39′	51	4,500	8	τ CMa m_v +4.40	57	+4.1
NGC 2367	13′	25	6,500	5.5	HD 57370 m_v +8.74	12	+7.9
NGC 2374	10′	14	4,800	290	TYC 5407-41-1 m_v +10.65	15	+8.0
NGC 2384	15′	30	6,900	8	HD 58509 m_v +8.57	28	+7.4

*May not be a cluster member.

Canis Minor

Constellation:	Canis Minor	**Hemisphere:**	Equatorial
Translation:	The Lesser Dog	**Area:**	183 deg^2
Genitive:	Canis Minoris	**% of sky:**	0.444%
Abbreviation:	CMi	**Size ranking:**	71st

Canis Minor depicts the smaller of the two hunting dogs belonging to Orion.

The seventh brightest star in the night sky α **Canis Minoris**, aka Procyon, is only 11.4 ly away. A yellowish-white F5 it is 2.0 $D_\odot$ and 6.5 $L_\odot$. Although it has a visual magnitude of m_v +0.38 this drops to just M_v +2.64 at 10 pc. As the star rises before Sirius the Greeks called it Pro Tu Kynos meaning 'Before the Great Dog' and the current spelling is simply a corruption. Perhaps the most interesting thing about the star is that it is a binary. Its companion, Procyon B, at m_v +10.82 is almost impossible to see in the glare of the primary. A white dwarf of spectral class DA it orbits Procyon A with a period of 40.82 years, coming as close as 8.9 AU – a little closer than Saturn is to the Sun – before heading out to 21 AU – a little farther than Uranus. It is a typical white dwarf: 1.35 $D_\odot$ but a paltry 0.00049 $L_\odot$. Even so it is still massive with a cubic centimeter weighing in at 300 kg! Eventually Procyon A will follow its tiny neighbor's evolutionary path and also finish its cosmic life as a white dwarf.

Rather larger than Procyon at 3.9 $D_\odot$ β **Canis Minoris**, or Gomeisa, is a Main Sequence B8 dwarf with a surface temperature of around 12,000 K. It is a γ Cas rotating variable with a magnitude swing of m_v +2.84 to +2.92. Like most B-class stars it is a fast spinner, rotating at a speed of at least 230 km/s which means it will revolve once in 20.6 hours. Gomeisa is enveloped in a thin dusty nebula.

δ^1 and η **Canis Minoris** could easily be mistaken for twins at a first glance. They are both Main Sequence pale yellow F0 dwarfs of almost identical magnitude. δ^1 at 11 $D_\odot$ is significantly larger than η though which comes in at 1.9 $D_\odot$ and more than twice the distance at 788 ly. However, δ^1 CMi is considerably more luminous at 379 $L_\odot$ compared to 76.2 $L_\odot$. It is also the faster spinner at 61 km/s taking 9.1 days to complete a revolution while η CMi spins once in 1.5 days with a rotational velocity of 54 km/s. While δ^1 is a lone star – δ^2 and δ^3 are unrelated and faint – η CMi has a companion in an orbit of at least 440 AU and with a period of more than 5,000 years.

γ **Canis Minoris** looks like an ordinary run-of-the-mill K3 giant – one of countless others – but its 25 $D_\odot$ hides a companion (probably also a K-class) which is just 2 $D_\odot$ smaller.

ζ **Canis Minoris** is almost the smallest star in the constellation. Perhaps not surprisingly it is another B8 but only 90% the size of the Sun and with a relatively slow rotational velocity of 45 km/s (the average for all B8 types is 151 km/s).

The 5th magnitude ε **Canis Minoris** lies at the farthest boundary of the constellation, at least 988 ly away and possibly as far as 1,196 ly. Although it belongs to the same spectral class as the Sun, G6.5, it is a giant at 13 solar

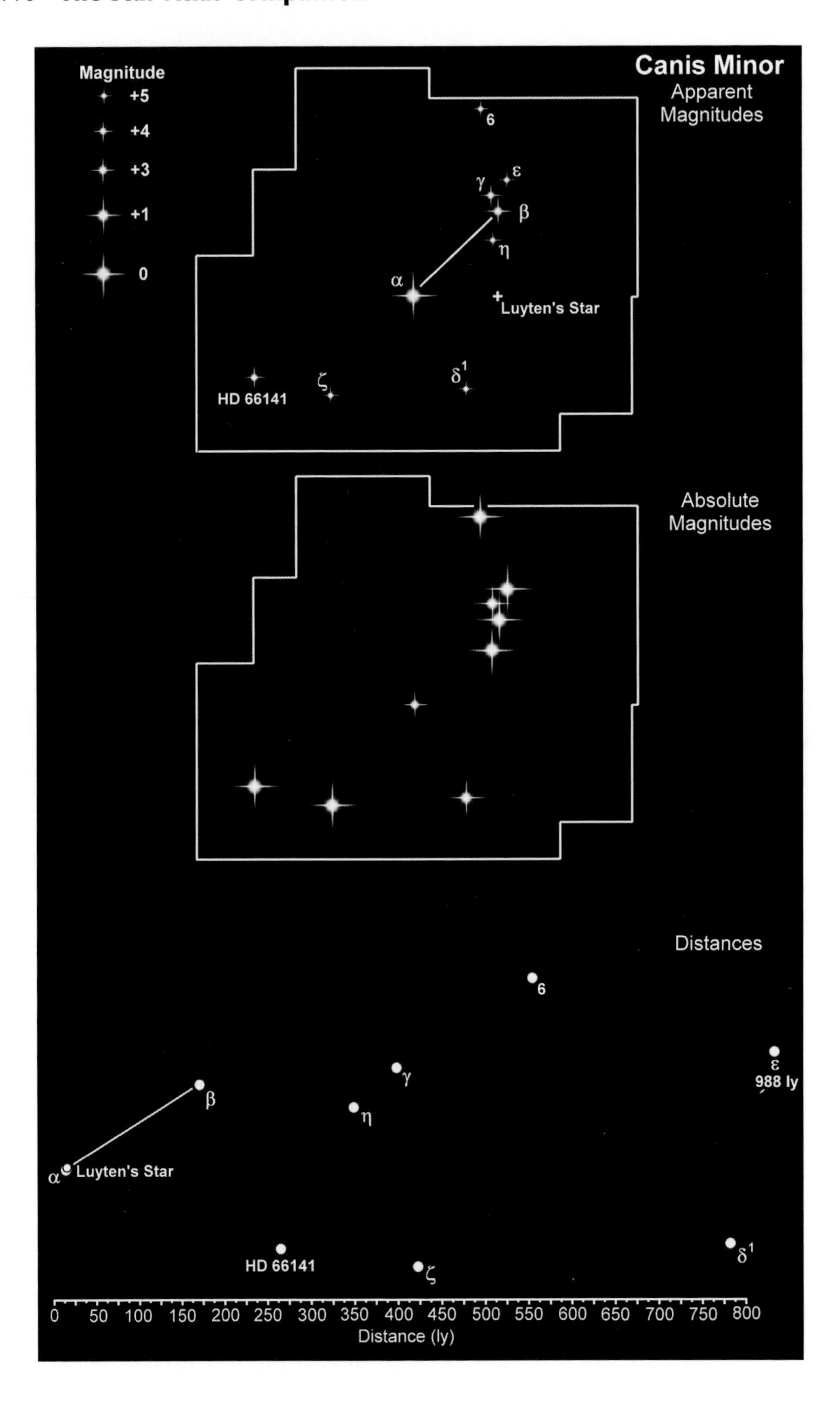

Canis Minor
Apparent Magnitudes
Magnitude
+5
+4
+3
+1
0
6
γ
ε
β
η
α
Luyten's Star
HD 66141
ζ
δ1
Absolute Magnitudes
Distances
6
ε
988 ly
γ
β
η
α
Luyten's Star
HD 66141
ζ
δ1
0 50 100 150 200 250 300 350 400 450 500 550 600 650 700 750 800
Distance (ly)

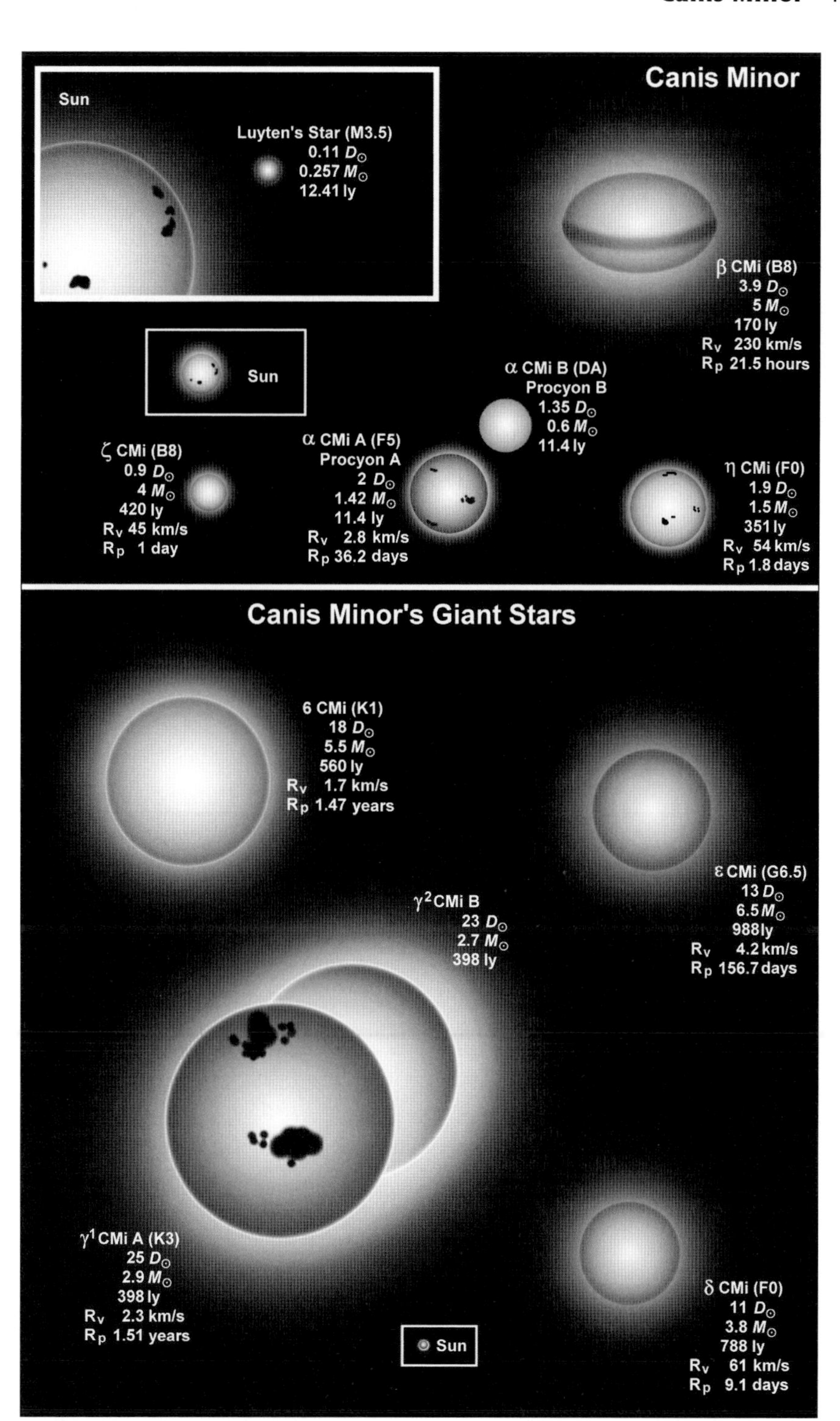
Canis Minor
Sun
Luyten's Star (M3.5)
0.11 $D_\odot$
0.257 $M_\odot$
12.41 ly
β CMi (B8)
3.9 $D_\odot$
5 $M_\odot$
170 ly
R_v 230 km/s
R_p 21.5 hours
Sun
α CMi B (DA)
Procyon B
1.35 $D_\odot$
0.6 $M_\odot$
11.4 ly
ζ CMi (B8)
0.9 $D_\odot$
4 $M_\odot$
420 ly
R_v 45 km/s
R_p 1 day
α CMi A (F5)
Procyon A
2 $D_\odot$
1.42 $M_\odot$
11.4 ly
R_v 2.8 km/s
R_p 36.2 days
η CMi (F0)
1.9 $D_\odot$
1.5 $M_\odot$
351 ly
R_v 54 km/s
R_p 1.8 days
Canis Minor's Giant Stars
6 CMi (K1)
18 $D_\odot$
5.5 $M_\odot$
560 ly
R_v 1.7 km/s
R_p 1.47 years
ε CMi (G6.5)
13 $D_\odot$
6.5 $M_\odot$
988 ly
R_v 4.2 km/s
R_p 156.7 days
γ^2 CMi B
23 $D_\odot$
2.7 $M_\odot$
398 ly
γ^1 CMi A (K3)
25 $D_\odot$
2.9 $M_\odot$
398 ly
R_v 2.3 km/s
R_p 1.51 years
Sun
δ CMi (F0)
11 $D_\odot$
3.8 $M_\odot$
788 ly
R_v 61 km/s
R_p 9.1 days

diameters and considerably more luminous by a factor of 750. Whereas the Sun has a life expectancy of 10,000 million years ε CMi will die after just 1,000 million years. In the cosmic longevity stakes, small is good.

One of the largest stars in the constellation is the K1 giant simply known as **6 Canis Minoris**. It wades in at 18 $D_{\odot}$ or 25 million km. It rotates slightly slower than the Sun at 1.7 km/s (Sun = 2 km/s) and so takes 1.5 years to complete a single rotation.

Luyten's Star (aka HIP 36208) is named after Willem Jacob Luyten (1899-1994) who discovered the star has a high proper motion rushing across the sky at 3.738″ per year. It is an M3.5 red dwarf, just 0.11 $D_{\odot}$, 0.257 $M_{\odot}$ and 0.0004 $L_{\odot}$. It is currently 12.41 ly from Earth but reaches only a dim m_v +9.85. It wobbles, suggesting it has a companion, but thus far nothing has been found. For some strange reason - probably because of a web encyclopedia - many writers put this star in Monoceros.

Capricornus

Constellation:	Capricornus	**Hemisphere:**	Southern
Translation:	The Sea Goat	**Area:**	414 deg^2
Genitive:	Capricorni	**% of sky:**	1.004%
Abbreviation:	Cap	**Size ranking:**	40th

The smallest of the Zodiacal constellations, the Sun passes through Capriconus between 19 January and 16 February.

With about 3,000 stars clearly visible to the naked eye it should not be surprising that coincidences abound. The two α stars of Capricornus are a typical example. **α¹ Capricorni** is a yellow G3 with a visual magnitude of m_v +4.23. Just 6.3′ away **α² Capricorni** is also a G-class but more of a yellowish-orange G8.5 and about half a mag brighter at m_v +3.57. It would be natural to assume they are related but it is simply not the case. α^1 lies at a distance of 687 ly whereas α^2 is much closer at 109 ly. They are also vastly different in size. α^2 is a 12 $D_\odot$ giant which is 34 times more luminous than the Sun and drifting away from us at a leisurely 0.7 km/s. α^1 is a 130 $D_\odot$ 730 $L_\odot$ supergiant galloping towards us, if indeed sea goats do gallop, at a steady 26 km/s.

β Capricorni is also known as Dabih which translates into 'The Lucky One of the Slaughterers' and conjures up a vision of a goat escaping a gruesome end although, in truth, the real meaning is lost in time. β Cap itself appears to be a complex 5-star system with the possibility of there being more components. The star visible to the naked eye is a K0 giant, 35 $D_\odot$ across and some 344 ly distant. Its spectrum suggests the presence of a B8 companion in orbit 4 AU away with a period of 1,374 days. It also has a companion that orbits at just 0.1 AU taking 8.678 days to complete a single orbit. These two orbiters are too small, too faint and too close to the primary to be seen visually and we therefore have to rely on other methods, such as spectroscopy, to detect them. Get the binocular out though and you will see a second – or is that a fourth? – star at 3.4′ separation with a visual magnitude of m_v +6.1. It is a B9.5 sub-giant which orbits β Cap at a distance of 21,000 AU and with a period in excess of 1 million years. It was originally thought that the two were separated by 9,400 AU but improved measurements have lead to revising the figure upwards. It is another star displaying evidence of mercury and manganese 'clouds' in its upper layers that may lead to brightness variations. Just to complicate matters further, orbiting this star is another at 30 AU believed to be an F-class dwarf.

There is some disagreement whether **γ Capricorni** is a 3 $D_\odot$ F0 or a 4.3 $D_\odot$ A7. Traditionally it has always been regarded as an F0 but the latest studies suggest it is an A7 and an α CVn rotating variable. The situation will perhaps become clearer when more detailed research has been undertaken.

δ Capricorni is just as confusing as γ Cap. From 39 ly its magnitude fluctuates between m_v +2.81 and +3.05 over a period of $24^h\ 33^m$ and has long been regarded as an Algol-type (EA) eclipsing binary, but it has been suggested

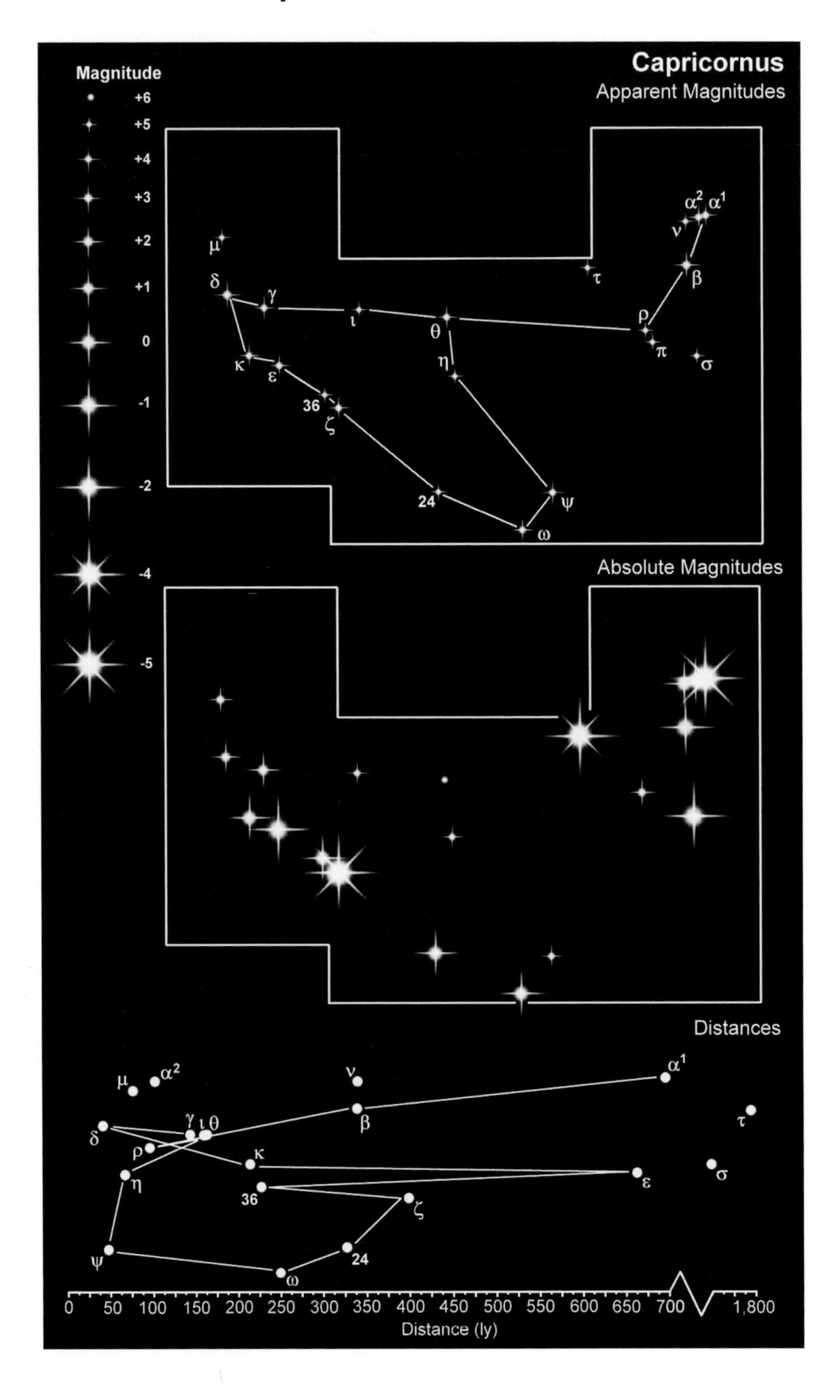
Capricornus
Apparent Magnitudes
Magnitude
+6
+5
+4
+3
+2
+1
0
-1
-2
-4
-5
Absolute Magnitudes
Distances
Distance (ly)
0 50 100 150 200 250 300 350 400 450 500 550 600 650 700 1,800

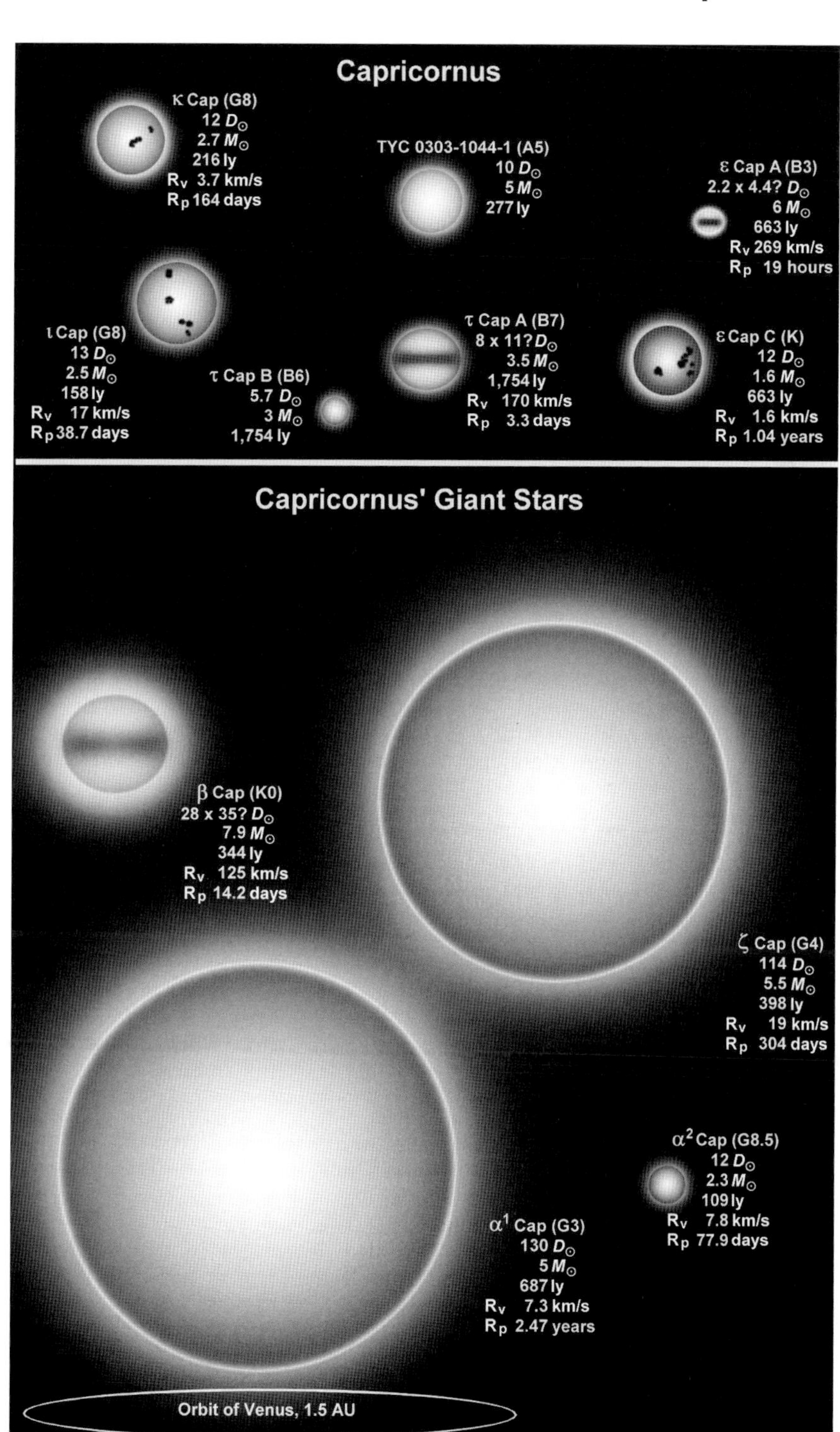
Capricornus
κ Cap (G8)
12 $D_\odot$
2.7 $M_\odot$
216 ly
R_v 3.7 km/s
R_p 164 days
TYC 0303-1044-1 (A5)
10 $D_\odot$
5 $M_\odot$
277 ly
ε Cap A (B3)
2.2 x 4.4? $D_\odot$
6 $M_\odot$
663 ly
R_v 269 km/s
R_p 19 hours
ι Cap (G8)
13 $D_\odot$
2.5 $M_\odot$
158 ly
R_v 17 km/s
R_p 38.7 days
τ Cap B (B6)
5.7 $D_\odot$
3 $M_\odot$
1,754 ly
τ Cap A (B7)
8 x 11? $D_\odot$
3.5 $M_\odot$
1,754 ly
R_v 170 km/s
R_p 3.3 days
ε Cap C (K)
12 $D_\odot$
1.6 $M_\odot$
663 ly
R_v 1.6 km/s
R_p 1.04 years
Capricornus' Giant Stars
β Cap (K0)
28 x 35? $D_\odot$
7.9 $M_\odot$
344 ly
R_v 125 km/s
R_p 14.2 days
ζ Cap (G4)
114 $D_\odot$
5.5 $M_\odot$
398 ly
R_v 19 km/s
R_p 304 days
α² Cap (G8.5)
12 $D_\odot$
2.3 $M_\odot$
109 ly
R_v 7.8 km/s
R_p 77.9 days
α¹ Cap (G3)
130 $D_\odot$
5 $M_\odot$
687 ly
R_v 7.3 km/s
R_p 2.47 years
Orbit of Venus, 1.5 AU

recently that it should be reclassified as either an RS CVn eclipsing variable or a δ Scuti pulsator. δ Cap has an odd spectral signature suggesting that, like some form of cosmic cannibal, it is feeding off a dying companion. This would lend weight to the call to reclassify the star as an RS CVn. On the other hand it hints at a slight pulsation, perhaps indicating it is about to enter its own death phase, and so could be a δ Scuti variable. As with γ Cap only time will tell. A m_V +2.85 star just 42″ to the north west is the unrelated **TYC 6363-1044-1**, an A5 lying much farther away at 277 ly.

ε **Capricorni** is a triple star system 663 ly from Earth. It is also regarded as being a γ Cas eruptive variable fluctuating between m_V +4.48 and +4.72. A fast spinning B3 at 269 km/s, it bulges to 4.4 $D_\odot$ at its equator, its polar diameter probably only around 3.3 $D_\odot$. One component, that has only been resolved spectroscopically orbits at a distance of 1 AU while the second component is at least 13,000 AU away and is a K-class giant with a m_V +9.5. The system bathes in a decretion disk of material ejected from the primary.

ζ **Capricorni** is a barium-rich star. During its early history ζ Cap had a more massive, older companion which has since exhausted most of its fuel and is now a white dwarf in a 6.5 year orbit at 6 AU. As its companion began to die it transferred some of its mass to ζ, adulterating its natural composition. Nowadays ζ Cap is a G4 supergiant, 114 $D_\odot$ across and 386 $L_\odot$. It barely makes 4th magnitude but would easily match Venus at 10 pc.

We have come across situations before where seemingly related stars were nothing more than a coincidence, but ι and κ **Capricorni** may be the genuine articles. At 13 and 12 $D_\odot$ respectively the stars are of similar size and both are G8s. They are now more than 42 light years apart but heading in opposite directions. It could be that a wayward star gave one of them a gravitational tug and separated the twins forever.

At 1,754 ly τ **Capricorni** marks for us the outer limits of the constellation. A rapidly rotating B7 (170 km/s compared to the average 160 km/s) it is 11 $D_\odot$, but it is significantly more luminous than the Sun by a factor of 2900. It is also a binary twin, its companion a warmer B6, the pair orbiting their common center of gravity once every 200 days.

Carina

Constellation:	Carina	**Hemisphere:**	Southern
Translation:	The Keel	**Area:**	494 deg^2
Genitive:	Carinae	**% of sky:**	1.197%
Abbreviation:	Car	**Size ranking:**	34th

Carina, Puppis (The Stern) and Vela (The Sail) were originally one constellation, Argo Navis, but were split by Nicolas Louis de Lacaille in the 18th Century. They represent the ship of the Argonauts. Carina lies in a very rich part of the Milky Way that includes Canopus, the second brightest star in the night sky, the Southern Pleiades, a brilliant flare star and several open clusters.

α **Carinae**, better known as Canopus, is the second brightest star after Sirius yet it has remained an enigma until fairly recently. It has not been particularly well studied, being too far south for many of the northern based telescopes, and the nature of this type of star – a rare F0 bright giant – is not well understood. The launch of the Hipparcos satellite provided astronomers with a much better understanding of α Car. It is now regarded as having a diameter of 71 $D_{\odot}$ and a luminosity of 13,600 $L_{\odot}$. If Canopus and Sirius were placed at 10 pc from Earth Canopus would appear far brighter at M_v -2.4 compared to the M_v +1.42 of Sirius. Its distance, which had previously been quoted as being anywhere between 100 to 1,200 ly is now believed to be 313 ly. It has a mass estimated at 8.6 $M_{\odot}$ and a temperature of 7,350 K. What astronomers are still uncertain about is whether it is evolving into a red giant or if it has already been through that phase and is now shrinking again. It is known to be a strong X-ray emitter suggesting it has an intense magnetic field that heats the surrounding corona to 15 million K. However its rotational period, which drives the magnetic field, has proven elusive possibly because its rotational pole is pointing directly towards us. Canopus is one of the key stars used for the navigation of spacecraft.

The otherwise unremarkable β **Carinae** together with θ, υ and ω **Carinae** make up the Diamond Cross asterism which is sometimes mistaken for Crux. A second False Cross, which looks even more like Crux, is made up of ι and ε Carinae and δ and κ Velorum.

No one is really sure whether ε **Carinae** is an eclipsing binary. The Yale *Bright Star Catalog* queries a 785 day period and variable star observers also doubt it is variable. It is certainly an orbital binary system. One star is an orange K3, possibly up to 70 $D_{\odot}$, while the other is a blue B3, probably no more than 6 $D_{\odot}$. They are at most only 4 AU apart and that is part of the problem: they are so close together we are unable to tell which of the stars we are seeing. Like many stars in the Southern Hemisphere ε Car could benefit from more detailed study. One interesting fact about the star is how it got its name, Avior. D.H. Sadler recounts in his book, *A Personal History of H.M. Nautical Almanac Office*, that the name was invented by HMNAO in the late 1930s. The Nautical Almanac Office was preparing *The Air Almanac*, a book containing details of 57 bright stars that pilots

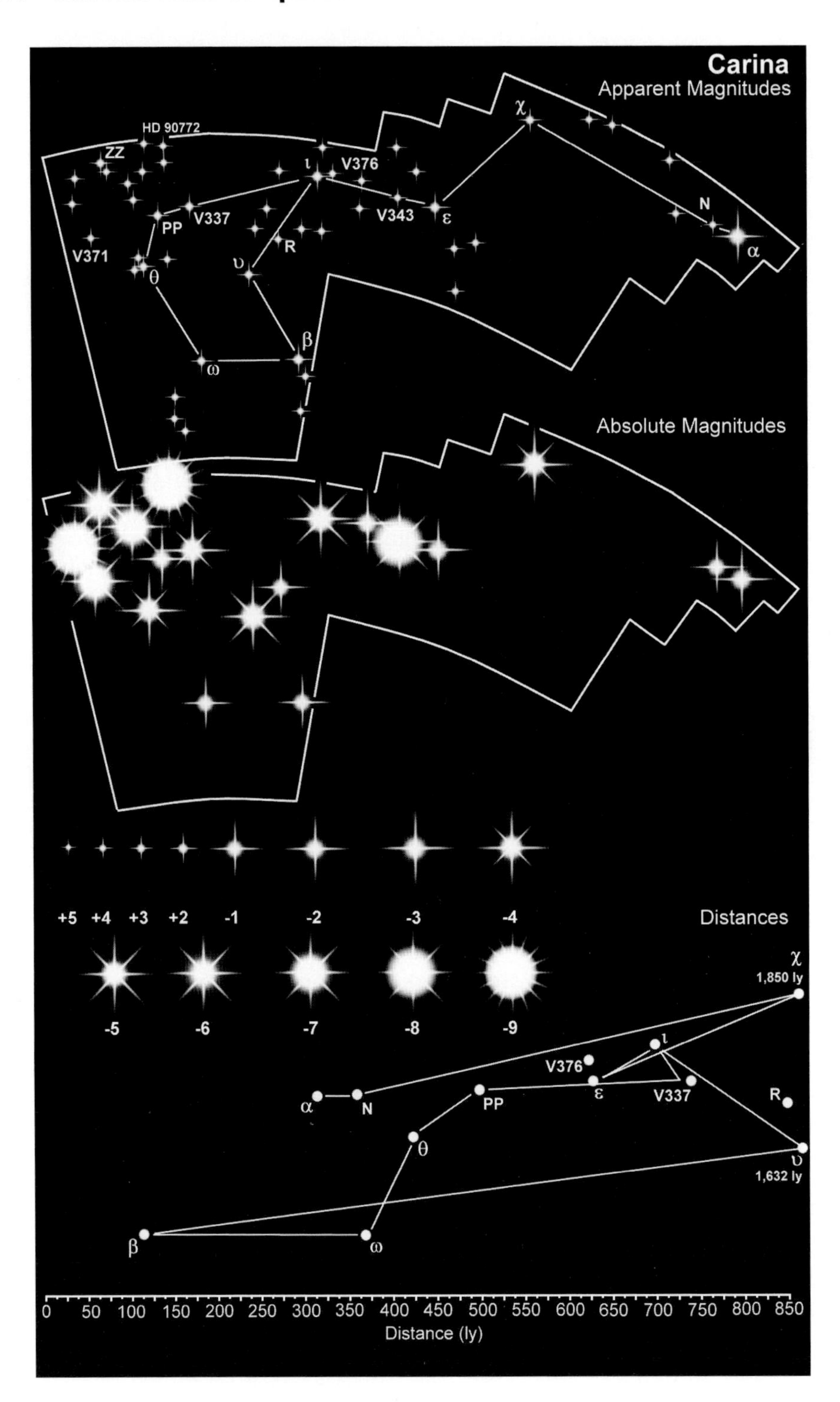

Carina
Apparent Magnitudes
HD 90772
ZZ
ι
V376
V337
PP
V343
ε
χ
N
α
V371
θ
υ
R
β
ω
Absolute Magnitudes
+5 +4 +3 +2 -1 -2 -3 -4
-5 -6 -7 -8 -9
Distances
χ
1,850 ly
ι
V376
ε
V337
R
α
N
PP
θ
υ
1,632 ly
β
ω
0 50 100 150 200 250 300 350 400 450 500 550 600 650 700 750 800 850
Distance (ly)

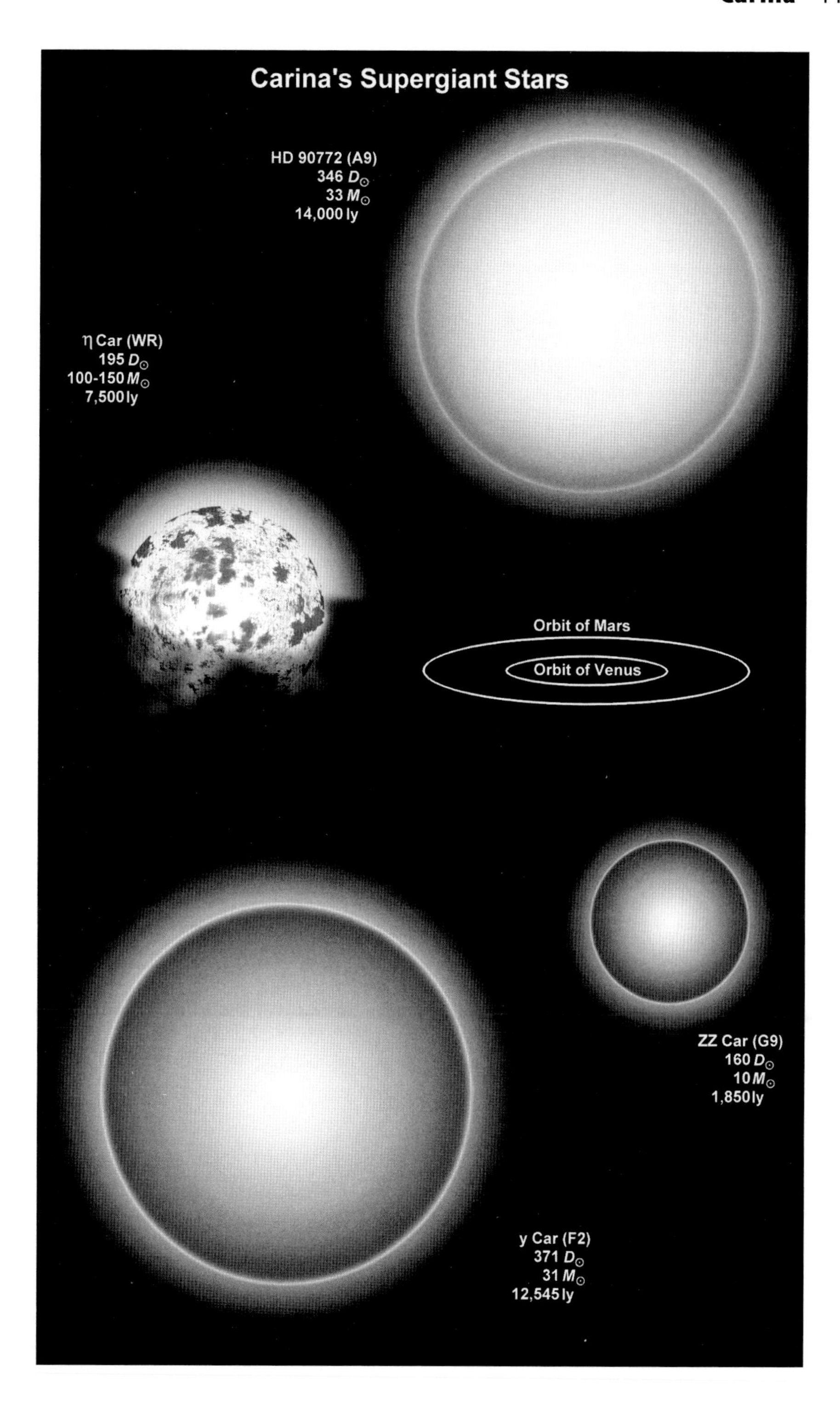
Carina's Supergiant Stars
HD 90772 (A9)
346 $D_\odot$
33 $M_\odot$
14,000 ly
η Car (WR)
195 $D_\odot$
100-150 $M_\odot$
7,500 ly
Orbit of Mars
Orbit of Venus
ZZ Car (G9)
160 $D_\odot$
10 $M_\odot$
1,850 ly
y Car (F2)
371 $D_\odot$
31 $M_\odot$
12,545 ly

could use for navigational purposes. Only two of the stars did not already have names – ε Car and α Pavonis – so they called them Avior (from the French *avis* for bird) and Peacock (the English translation of Pavo).

η **Carinae** is one of the most peculiar stars in the sky. In 1677 Edmond Halley (of Halley's Comet fame) recorded the star as being of the 4th magnitude. By 1730 it had brightened to 2nd magnitude but returned to 4th by 1782. In April 1843 it suddenly brightened to m_v -0.8 making it the second brightest star in the night sky. By 1890 it had faded to 6th magnitude and continued to do so until the end of the century when it appeared to stabilize at m_v +8. Since 1943 it has gradually brightened showing a sudden but relatively small burst in 1998-99. At the time of writing it is m_v +4.5 and brightening. No one is absolutely sure what is happening to η Car. The initial brilliant burst would suggest a large star going supernova, but it has clearly survived and may be what has been termed a 'supernova impostor'. It is a very large star at up to 195 $D_\odot$ though by no means the largest – there are another 52 naked eye stars that are larger. Its mass is somewhere between 100 and 150 $M_\odot$. However, it is its luminosity that is really staggering: some 5 million times more luminous than the Sun, and perhaps only surpassed by HD 135591 in Circinus. The variation in magnitude may be attributed to the behavior of the resulting dust cloud following its outburst in 1843, what astronomer Jim Kaler calls the 'Great Eruption'. During the Great Eruption the star ejected a Sun's worth of material. As the cloud expanded at 400 km/s it cooled and dust grains condensed from it. This made the cloud rather more opaque and blocked out much of η Carinae's light. As the cloud continued to spread out into interstellar space it thinned out, allowing more starlight to reach us.

η Car is thought to be a Wolf-Rayet star and may belong to the rare Luminous Blue Variable (LBV) or S Doradûs variable class. There is some observational evidence to indicate it is also a binary with an orbital period of 5.52 years. The pair are now imbedded in a giant gas and dust cloud 11 ly across and which has been named the Homunculus Nebula. It looks likely that η Car will explode in the not too distant future, perhaps not tomorrow but almost certainly within the next million years. And it is bubbling up to be quite an explosion. Not so much a supernova, more a super-duper nova (though professional astronomers would prefer we call it a hypernova).

θ Carinae is the brightest star in the open cluster **IC 2602** which has about 60 members, although less than 20 have been researched to any degree. θ Car is typical of the stars in the cluster. Often referred to as the Southern Pleiades IC 2602 contains half a dozen stars brighter than m_v +6 although one of these, **HD 93163**, is not actually a cluster member lying at a distance of 1,010 ly: cluster stars are found in the range of 428 to 488 ly. **HD 93607** seems a bit on the large size at 5.6 $D_\odot$ but there could be several reasons for this, not least of which could be observational error. It has been said that θ Car has the obscure proper name Vathorz Posterior which is supposed to be a combination of Old Norse and Latin meaning 'trailing (star) of the waterline (of the ship Argo)'. Quite remarkable for a star that is invisible to anyone much farther north of the Tropic of Cancer! It is, of course, a schoolboy joke that has been circulating

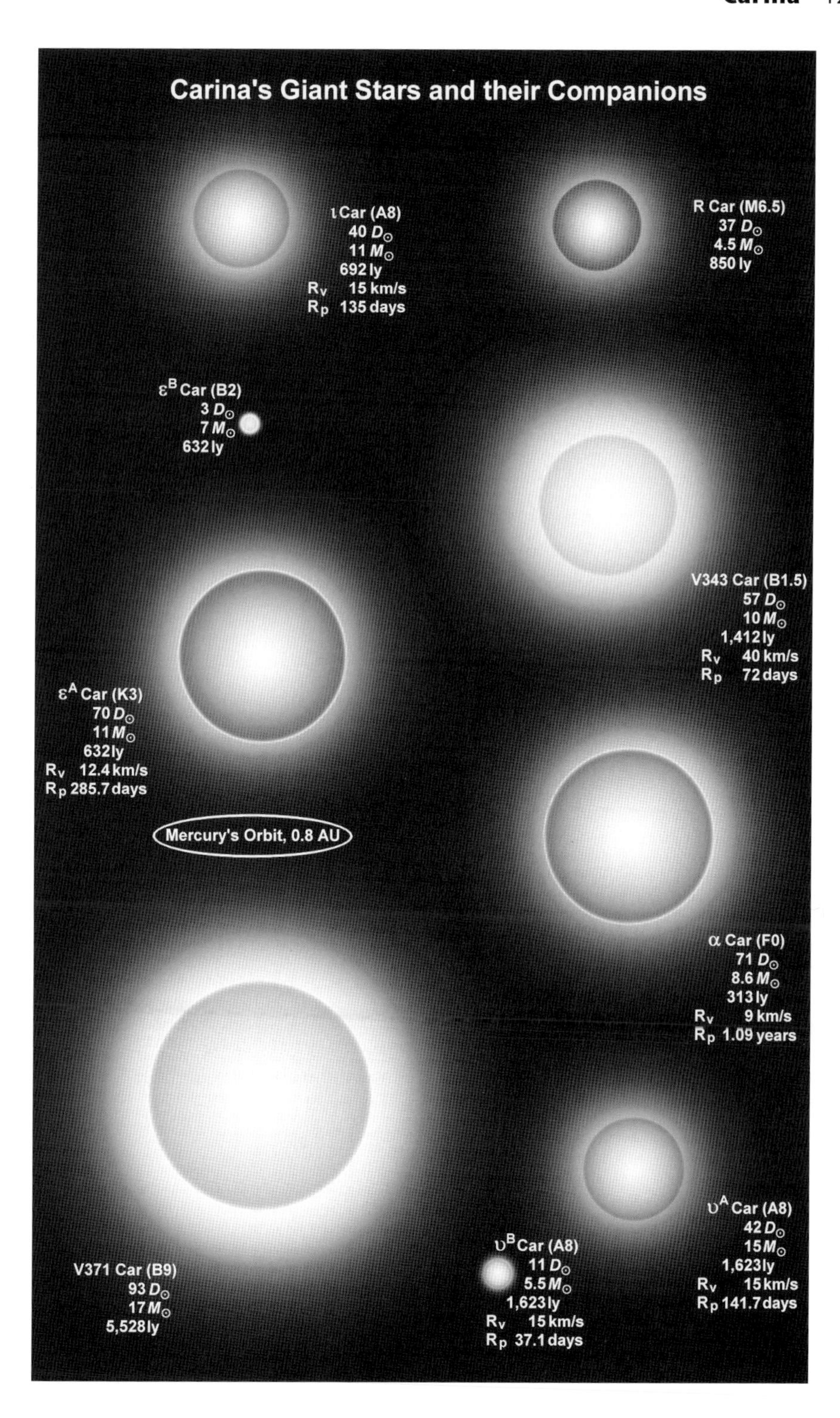
Carina's Giant Stars and their Companions
ι Car (A8)
40 D⊙
11 M⊙
692 ly
Rv 15 km/s
Rp 135 days
R Car (M6.5)
37 D⊙
4.5 M⊙
850 ly
εB Car (B2)
3 D⊙
7 M⊙
632 ly
V343 Car (B1.5)
57 D⊙
10 M⊙
1,412 ly
Rv 40 km/s
Rp 72 days
εA Car (K3)
70 D⊙
11 M⊙
632 ly
Rv 12.4 km/s
Rp 285.7 days
Mercury's Orbit, 0.8 AU
α Car (F0)
71 D⊙
8.6 M⊙
313 ly
Rv 9 km/s
Rp 1.09 years
υA Car (A8)
42 D⊙
15 M⊙
1,623 ly
Rv 15 km/s
Rp 141.7 days
υB Car (A8)
11 D⊙
5.5 M⊙
1,623 ly
Rv 15 km/s
Rp 37.1 days
V371 Car (B9)
93 D⊙
17 M⊙
5,528 ly

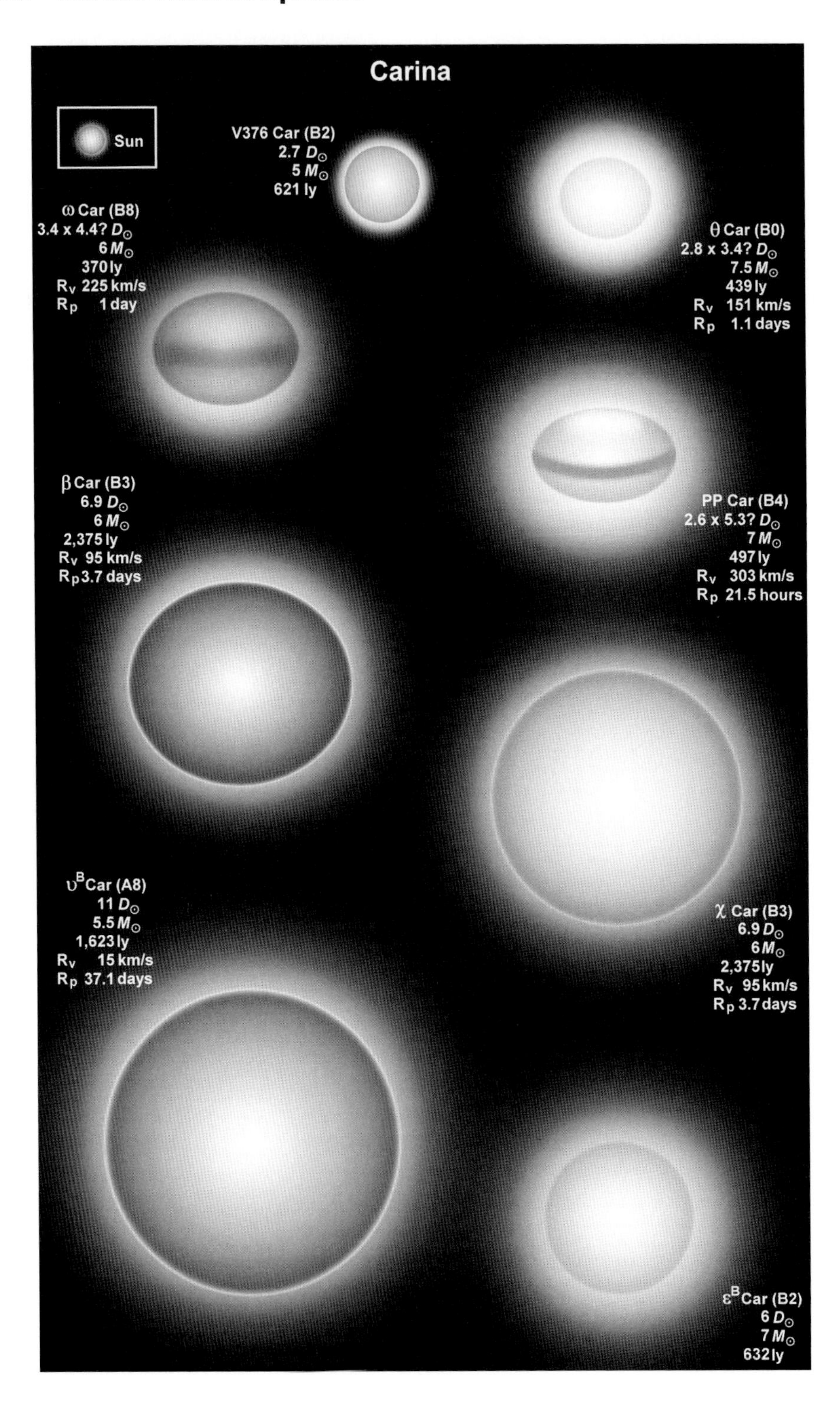

Carina
Sun
V376 Car (B2)
2.7 $D_\odot$
5 $M_\odot$
621 ly
ω Car (B8)
3.4 x 4.4? $D_\odot$
6 $M_\odot$
370 ly
R_v 225 km/s
R_p 1 day
θ Car (B0)
2.8 x 3.4? $D_\odot$
7.5 $M_\odot$
439 ly
R_v 151 km/s
R_p 1.1 days
β Car (B3)
6.9 $D_\odot$
6 $M_\odot$
2,375 ly
R_v 95 km/s
R_p 3.7 days
PP Car (B4)
2.6 x 5.3? $D_\odot$
7 $M_\odot$
497 ly
R_v 303 km/s
R_p 21.5 hours
$υ^B$ Car (A8)
11 $D_\odot$
5.5 $M_\odot$
1,623 ly
R_v 15 km/s
R_p 37.1 days
χ Car (B3)
6.9 $D_\odot$
6 $M_\odot$
2,375 ly
R_v 95 km/s
R_p 3.7 days
$ε^B$ Car (B2)
6 $D_\odot$
7 $M_\odot$
632 ly

around the Internet for far too long: Vathorz Posterior = a fat whore's backside.

Supergiant A8 and A9 class stars are a relative rarity. There are only three visible to the naked eye, and they are all in Carina. **ι Carinae** is the smallest at a 'mere' 40 $D_{\odot}$. At 692 ly it appears as m_v +2.22 star but would brighten to M_v -4.7 at 10 pc. A couple of solar diameters larger **υ^{A} Carinae** is almost 1,000 ly further away. Its companion, **υ^{B} Carinae**, is also an A8 but much smaller at just 11 $D_{\odot}$ and is all but invisible at m_v +6.03. But it is **HD 90772** that is truly the hypergiant at 346 $D_{\odot}$ – slightly larger than the orbit of Mars. This A9 class star lies at 14,000 ly and is 200,000 times more luminous than the Sun. A δ Cepheid variable, sometimes identified as **V399**, it has a period of 58.82 days during which its magnitude fluctuates between m_v +4.64 and +4.71. At 10 pc it would be the joint 2nd brightest star in the night sky at M_v -8.5, outshone only by 34 Cygni (M_v -8.9) and on par with ε Aurigae, ϕ Cassiopeiae and another Carina star, **y Carinae**, a 371 $D_{\odot}$ F2 at 12,545 ly and which has a luminosity of 168,598 $L_{\odot}$.

The 3 $D_{\odot}$ **PP Carinae** is a γ Cassiopeiae eruptive variable. A blue B4 with a temperature of around 17,000 K it is another fast spinner rotating at 303 km/s taking just 12 hours to complete one full turn.

R Carinae is a long period pulsating red giant of 37 $D_{\odot}$. Over a period of 308.71 days it goes from a respectable m_v +3.9 to +10.5 and back again.

V371 is an α Cygni type pulsating variable that varies by just 700th of a magnitude between m_v +5.12 and +5.19. At 93 $D_{\odot}$ it is a supergiant B9 of 20,470 $L_{\odot}$. It appears so faint only because it is so far away – more than 5,500 ly – but would have an absolute magnitude of M_v -7.1.

Carina is home to a several record breaking β Cepheid variables. **V343** is the largest of the naked eye β Cepheids at 57 $D_{\odot}$ (85% are less than 5 $D_{\odot}$ across). **V376** has the shortest period at 29^{m} 57.1^{s} while **χ Carinae** has the largest amplitude, 0.44, varying between m_v +3.46 and +3.90.

It might sound as though it would send you to sleep but **ZZ Carinae** is an exceptional δ Cepheid pulsating variable. Whereas most δ Cepheids have periods of less than 10 days (they average 6.7 days) ZZ Car has the longest of all the naked eye Cepheids spanning 35^{d} 12^{h} 52^{m} during which its magnitude goes from m_v +3.28 to +4.18. At 1,850 ly this yellow G5 supergiant pulses between 160 $D_{\odot}$ (222.7 million km) at about the time of magnitude minima, and 195 $D_{\odot}$ (271.4 million km) several days before magnitude maxima. Its luminosity is around 14,000 $L_{\odot}$ and, combined with a mass of 10 $M_{\odot}$, ZZ Car is a candidate exploding star.

The red giant **HD 66342** lies within the open cluster **NGC 2516**, sometimes referred to as the Diamond Cluster. This particular star is m_v +5.20 but the combined magnitudes of the 100 stars in the cluster give it the appearance of a 3rd magnitude cloud. HD 66342 lies at a distance of 1,045 ly, some 250 ly from the cluster's center. A second star just about visible to the naked eye, **HD 66194**, otherwise known as **V374**, is a γ Cas variable which changes from m_v +5.84 at maximum to +5.72 over a period of a few days.

NGC 3532 is nicknamed the Wishing Well, the stars obviously reminding someone of the coins at the bottom of a wishing well. There are no particularly bright stars but the cluster overall is about m_v +3.

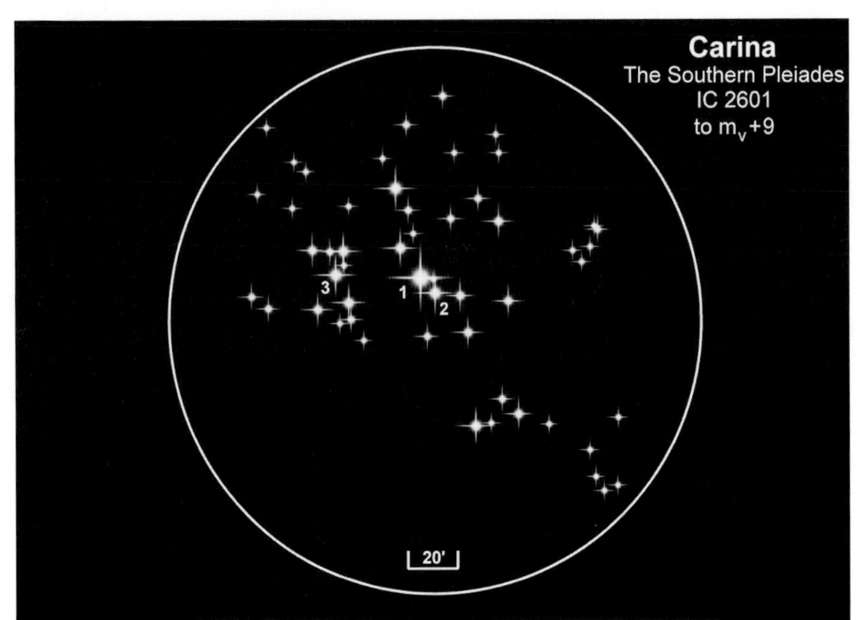

No.	Star	m_v	Distance (ly)	Spectrum
1	θ Car	+2.78	439	B0 V
2	HD 92938	+4.81	456	B4 V
3	HD 93607	+4.87	449	B3 IV

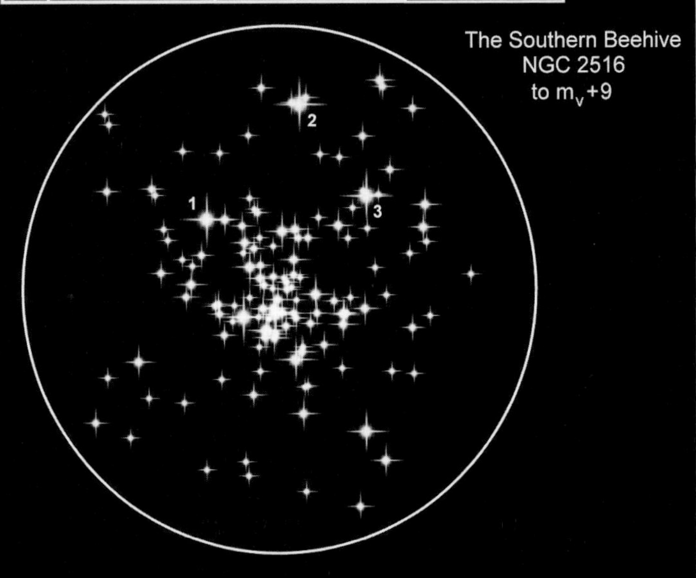

No.	Star	m_v	Distance (ly)	Spectrum
1	V460 Car	+5.18	1,045	M1.5 IIa
2	HD 65907	+5.60	52.8	F9.5 V
3	HD 65662	+5.75	1,245	K3 III

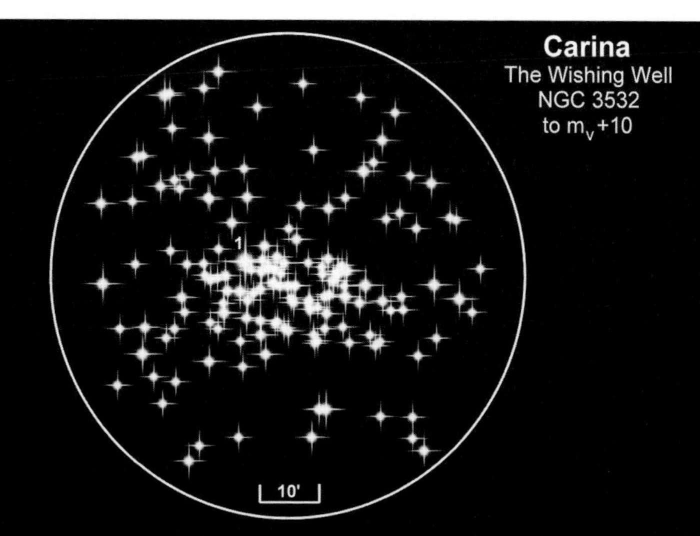

No.	Star	m_v	Distance (ly)	Spectrum
1	HD 96544	+6.02	1,226	K2 II-III

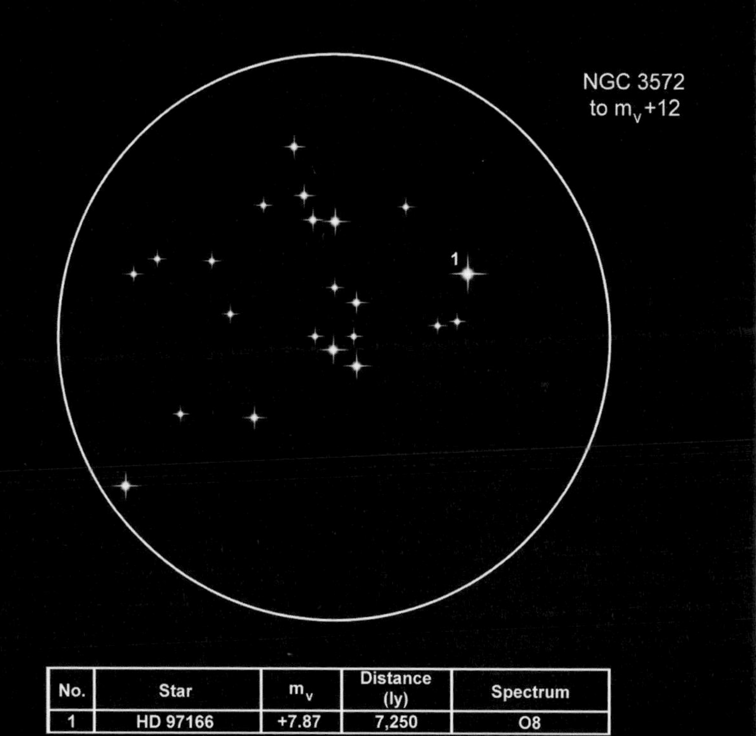

No.	Star	m_v	Distance (ly)	Spectrum
1	HD 97166	+7.87	7,250	O8

Open and globular clusters in Carina

Name	Size arc min	Size ly	Distance ly	Age million yrs	Brightest star in region*	No. stars m_v >+12*	Apparent magnitude m_v
Collinder 228	40′	84	7,200	7	QZ Car m_v +6.42	124	+4.4
Collinder 240	25′	37	5,100	15	HD 97534 m_v +4.6	12	+3.9
IC 2581	15′	35	8,000	14	V399 Car m_v +4.71	39	+4.3
IC 2602 Southern Pleiades	135′	21	525	32	θ Car m_v +2.78	146	+1.9
NGC 2516 Southern Beehive	30′	11	1,300	115	V460 Car m_v +5.18	80	+3.8
NGC 3114	50′	44	3,000	124	HD 87283 m_v +5.93	221	+4.2
NGC 3247	8′	12	5,100	121	HD 90552 m_v +9.82	12	+7.6
NGC 3293	17′	38	7,600	10	HD 91969 m_v +6.52	68	+4.7
NGC 3324	10′	23	7,600	6	HD 92207 m_v +5.45	22	+6.7
NGC 3532 Wishing Well	63′	140	1,600	310	HD 96544 m_v +6.02	514	+3.0
NGC 3572	8′	16	6,500	8	HD 97166 m_v +7.87	22	+6.6
NGC 3590	13′	21	5,400	17	HD 97581 m_v +8.85	41	+8.2
Trumpler 14	9′	24	9,000	7	HD 93129 m_v +6.97	56	+5.5
Trumpler 16	21′	54	8,800	6	η Car m_v +6.20	87	+5.0
NGC 2808	14′	128	31,300	12,500	Globular cluster		+7.8

*May not be a cluster member.

Cassiopeia

Constellation:	Cassiopeia	**Hemisphere:**	Northern
Translation:	Queen Cassiopeia	**Area:**	598 deg^2
Genitive:	Cassiopeiae	**% of sky:**	1.450%
Abbreviation:	Cas	**Size ranking:**	25th

Queen Cassiopeia was the wife of King Cepheus and the mother of Andromeda. The constellation is set against the rich backdrop of the Milky Way. The five main stars form a distinctive letter W or M.

Not the brightest but the second brightest, **α Cassiopeiae** or Shedir is a m_v +2.22 orange star, 855 times more luminous than the Sun. At 10 pc it would brighten to M_v -0.9. It has a rotational velocity of 5 km/s so its 41 $D_\odot$ takes more than a year – 415 days – to revolve just once. At one time it was believed to be variable but that suggestion has now been dismissed.

At m_v +2.25 **β Cassiopeiae** or Chaph is the third brightest δ Scuti pulsating variable. Its magnitude varies by an almost unnoticeable 0.06 magnitudes ranging from m_v +2.31 to a maximum +2.25 with a period of 2^h 30^m. Some 2.9 times the diameter of the Sun this F2 star is actually a binary although precious little is known about its +13.6 magnitude companion other than it orbits once every 27 days. β Cas spins at 70 km/s taking 2.1 days to complete a revolution.

Currently the brightest star in the constellation at m_v +1.60, **γ Cassiopeiae** or Cih may well have been a more modest 3rd magnitude object at the time Bayer compiled his *Uranometria* star atlas in 1603, hence its γ designation. γ Cas lends its name to the type of eruptive variable stars that can brighten suddenly and unpredictably. In γ Cassiopeiae's case the difference in brightness is quite large at 1.4 magnitudes – the largest of the 37 known naked eye stars – but most vary by between 0.11 and 0.19 magnitudes. With the exception of ζ Ophiuchi which varies by just 0.02 magnitudes and is an O9 spectral class star, all of the γ Cas variables are B-type stars covering the entire B-class spectrum although most lie in the B1 to B5 range. They tend to be smaller stars – two-thirds are less than 5 $D_\odot$ across – and fast spinners. γ Cas itself is one of the larger of the breed at 23 $D_\odot$ and very luminous at 40,000 $L_\odot$. It spins on its axis at 300 km/s so that, despite its substantial size, it takes just under 4 days to turn once. Its high rotational speed also means that the star is an oblate spheroid, considerably fatter at the equator than at the poles, which ejects huge amounts of material into an orbiting disk. A bright X-ray source, though no one really understands why, today's γ Cas will be tomorrow's supernova.

δ Cassiopeiae is an Algol or EA-type eclipsing binary with a period of 759 days. At almost 100 ly its companion is unseen but as it passes in front of δ Cas its magnitude dips from m_v +2.68 to +2.76.

ε Cassiopeiae is a bit of a mystery. It belongs to spectral class B3 and appears to be surrounded by a circumstellar disk of gas spreading out from its equator. So far so good: this matches the profile of a typical fast spinning B-class star which is

Magnitude
+6
+5
+4
+3
+2
+1
0
-1
-2
-3
-4
-5
-7
-8
-9
Cassiopeia
Apparent Magnitudes
HD 19275
50
48
ι
ω
ψ
52
ε
κ
4
V566
1
δ
γ
β
AR
V509
τ
χ
υ2
υ1
ρ
φ
η
σ
α
θ
λ
μ
ζ
R
ν
ξ
ο
π
Absolute Magnitudes

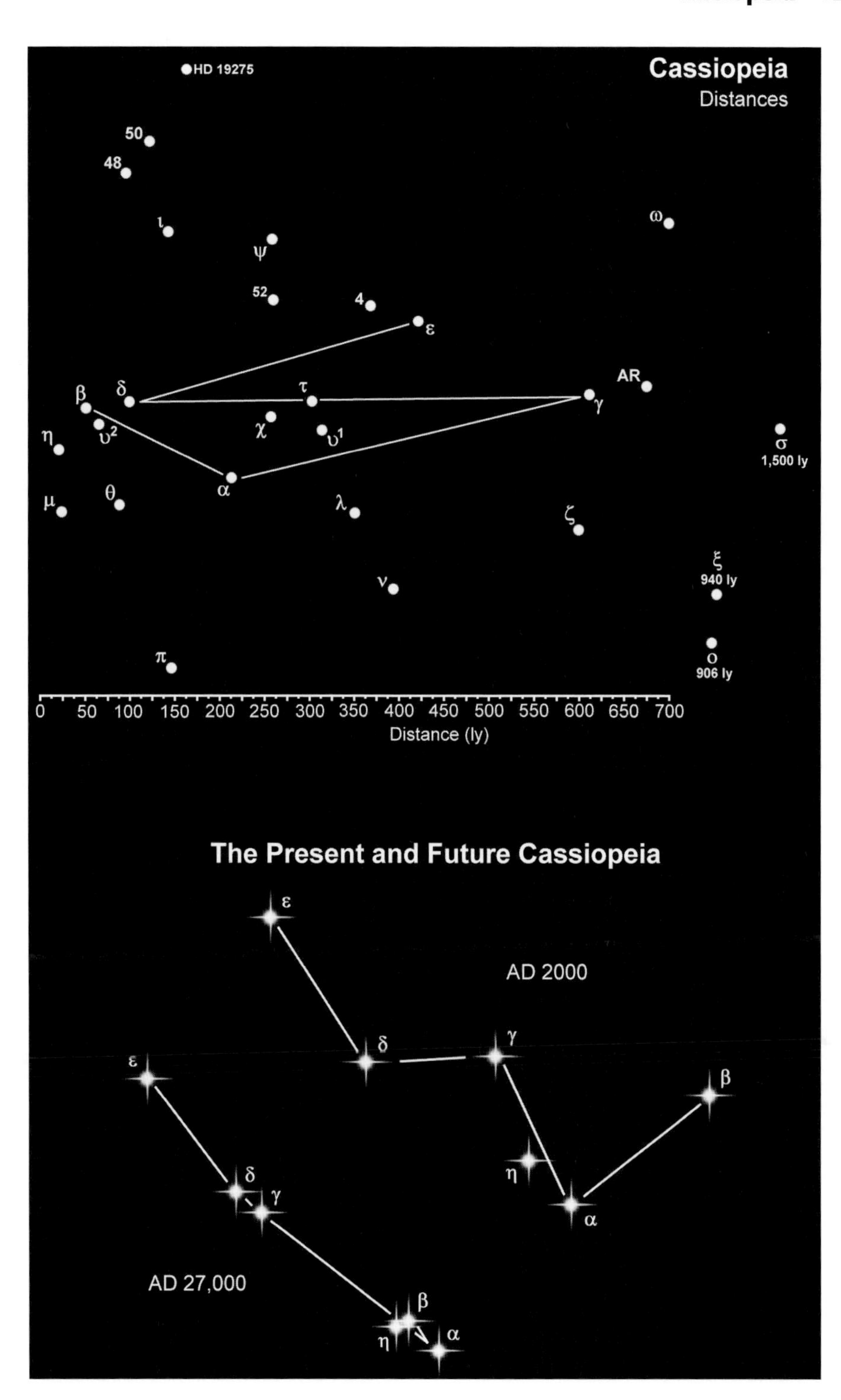
Cassiopeia
Distances
HD 19275
50
48
ι
ψ
52
4
ω
ε
AR
β
δ
τ
γ
η
υ2
χ
υ1
σ
1,500 ly
α
μ
θ
λ
ζ
ξ
940 ly
ν
π
ο
906 ly
0 50 100 150 200 250 300 350 400 450 500 550 600 650 700
Distance (ly)
The Present and Future Cassiopeia
ε
AD 2000
δ
γ
ε
β
η
δ
γ
α
AD 27,000
β
η
α

throwing off material into space. Except it is *not* a fast spinner. We would expect a rotational velocity of 200 to 250 km/s, perhaps even faster, but ε Cas is turning at a modest 50 km/s. Exactly what is going on is anyone's guess.

Although ζ **Cassiopeiae** is not usually listed as a variable it does actually belong to a rare category known as 53 Persei or Slowly Pulsating B stars (SPBs). You have to have some pretty sensitive equipment to detect the variation though as it amounts to just one thousandth of a magnitude with a period of 1.56 days. These are complex stars with some layers of the star expanding while others contract. Although difficult to accurately monitor from ground based sites (their subtle magnitude changes are often masked by our turbulent atmosphere) astronomers believe that SPBs can give us a unique insight into the inner structure and workings of variable stars.

η **Cassiopeiae** is one of those stars best seen with a small telescope or binocular. It is another solar analog, the same size as the Sun, 19% more luminous and belonging to spectral class G0 or G3. It is also fairly close to us at just 19.4 ly and appears as a m_v +3.46 star. But modest optical aid will also reveal that it is a binary, its companion being a +7.51 magnitude K7 dwarf just 11″ away (PA 307°). In reality, the two stars are in an orbit that brings them to within 36 AU of one another before separating them by 107 AU with an orbital period of around 480 years. The stars should appear as yellow and reddish-orange though depending on how good your optics are (your telescope's and your eyes) they may appear gold and purple.

ι **Cassiopeiae** is also a good telescopic object. It looks like a triple star arrangement but could, in fact, consist of up to five related stars. The brightest is ι^A Cas, a white A5 rather more than twice the diameter and mass of our own Sun and 22 times more luminous. It is an α CV rotating variable with a period of $1^d\ 17^h\ 46^m$ during which its magnitude varies between m_v +4.45 and +4.53. In orbit around ι^A Cas is ι^{Aa} Cas, a G2 dwarf about 70% the mass and luminosity of the Sun. Its distance from the primary is somewhere between 5 and 18 AU and it has an orbital period of around 47-52 years. This star is invisible in small telescopes but a third star, ι^B Cas, is an easy m_v +6.91 object lying at least 100 AU from ι^A Cas and having an orbital period of 840 years, although the jury is still out as to whether it is actually in orbit or just a passing stranger. ι^B Cas is an F5 of 1.3 $M_\odot$ and 2.7 $L_\odot$. At 300 AU from ι^A Cas is ι^C Cas which is a K1 dwarf, 0.7 $M_\odot$ and 0.7 $L_\odot$. It has a magnitude of m_v +8.40. Finally, ι^C Cas itself has a companion at 0.4″ – invisible in small telescopes and often designated ι^{Cb} Cas. It is a K5 dwarf of 0.65 $M_\odot$.

At 4,129 ly κ **Cassiopeiae** is a highly luminous (420,000 $L_\odot$) giant 40 times a big as the Sun. Belonging to spectral class B1 it carries a temperature of 24,000 K: four times hotter than the Sun. Its brightness also varies by 800th of a magnitude between m_v +4.22 and +4.30, the star belonging to the α Cygni group of pulsating variables.

λ **Cassiopeiae** consists of twin blue stars, one a B7 the other a B8, in a 640 year orbit. Their magnitudes are m_v +4.77 and +5.80 and they are separated by 0.6″. Neither star is particularly big – each a couple of solar diameters across – but they are considerable more luminous at 120 $L_\odot$. They lie at a distance of 355 ly.

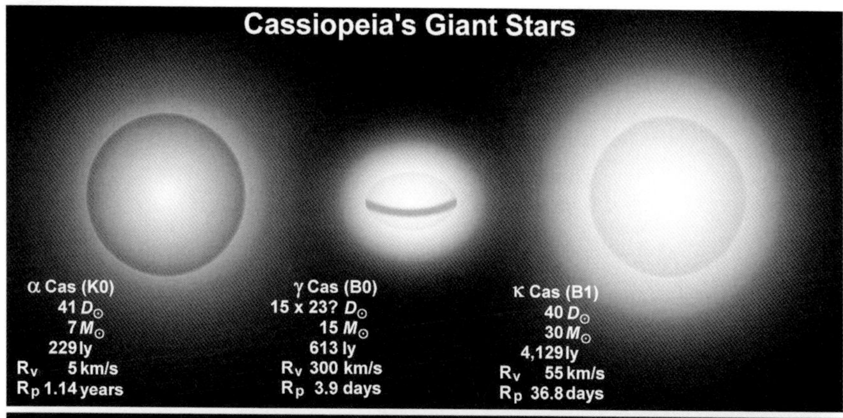
Cassiopeia's Giant Stars
α Cas (K0)
41 $D_\odot$
7 $M_\odot$
229 ly
R_V 5 km/s
R_P 1.14 years
γ Cas (B0)
15 x 23? $D_\odot$
15 $M_\odot$
613 ly
R_V 300 km/s
R_P 3.9 days
κ Cas (B1)
40 $D_\odot$
30 $M_\odot$
4,129 ly
R_V 55 km/s
R_P 36.8 days

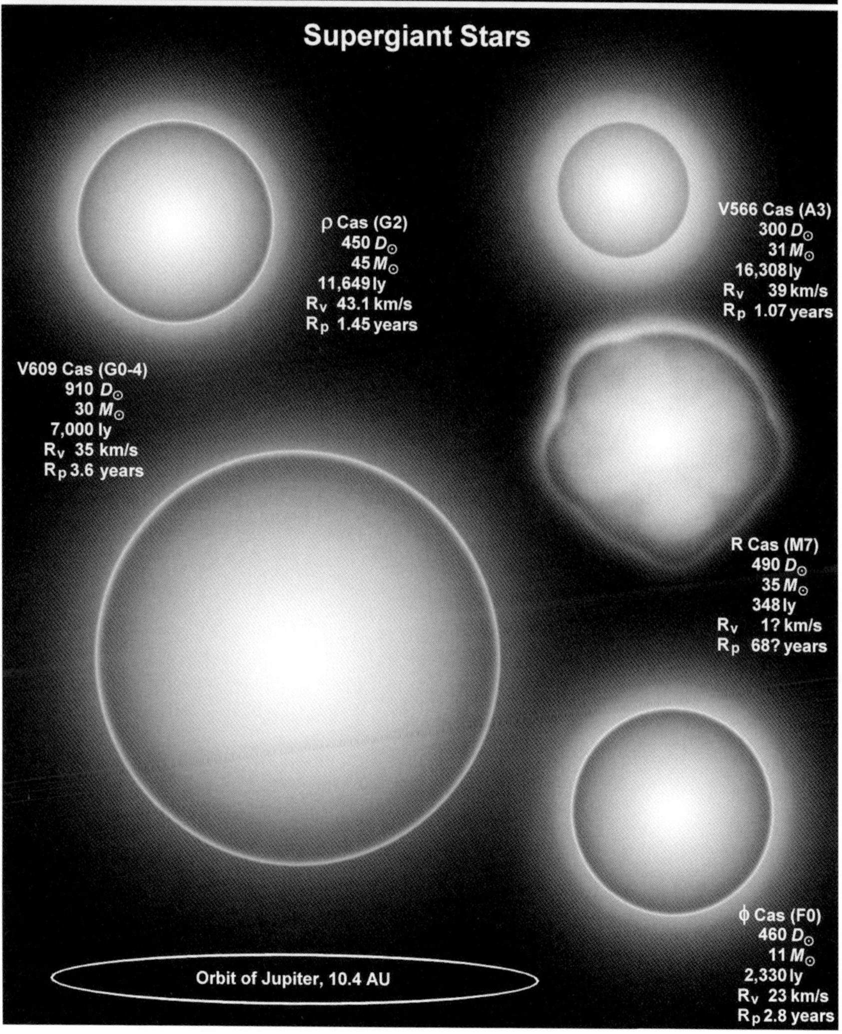
Supergiant Stars
ρ Cas (G2)
450 $D_\odot$
45 $M_\odot$
11,649 ly
R_V 43.1 km/s
R_P 1.45 years
V566 Cas (A3)
300 $D_\odot$
31 $M_\odot$
16,308 ly
R_V 39 km/s
R_P 1.07 years
V609 Cas (G0-4)
910 $D_\odot$
30 $M_\odot$
7,000 ly
R_V 35 km/s
R_P 3.6 years
R Cas (M7)
490 $D_\odot$
35 $M_\odot$
348 ly
R_V 1? km/s
R_P 68? years
ϕ Cas (F0)
460 $D_\odot$
11 $M_\odot$
2,330 ly
R_V 23 km/s
R_P 2.8 years
Orbit of Jupiter, 10.4 AU

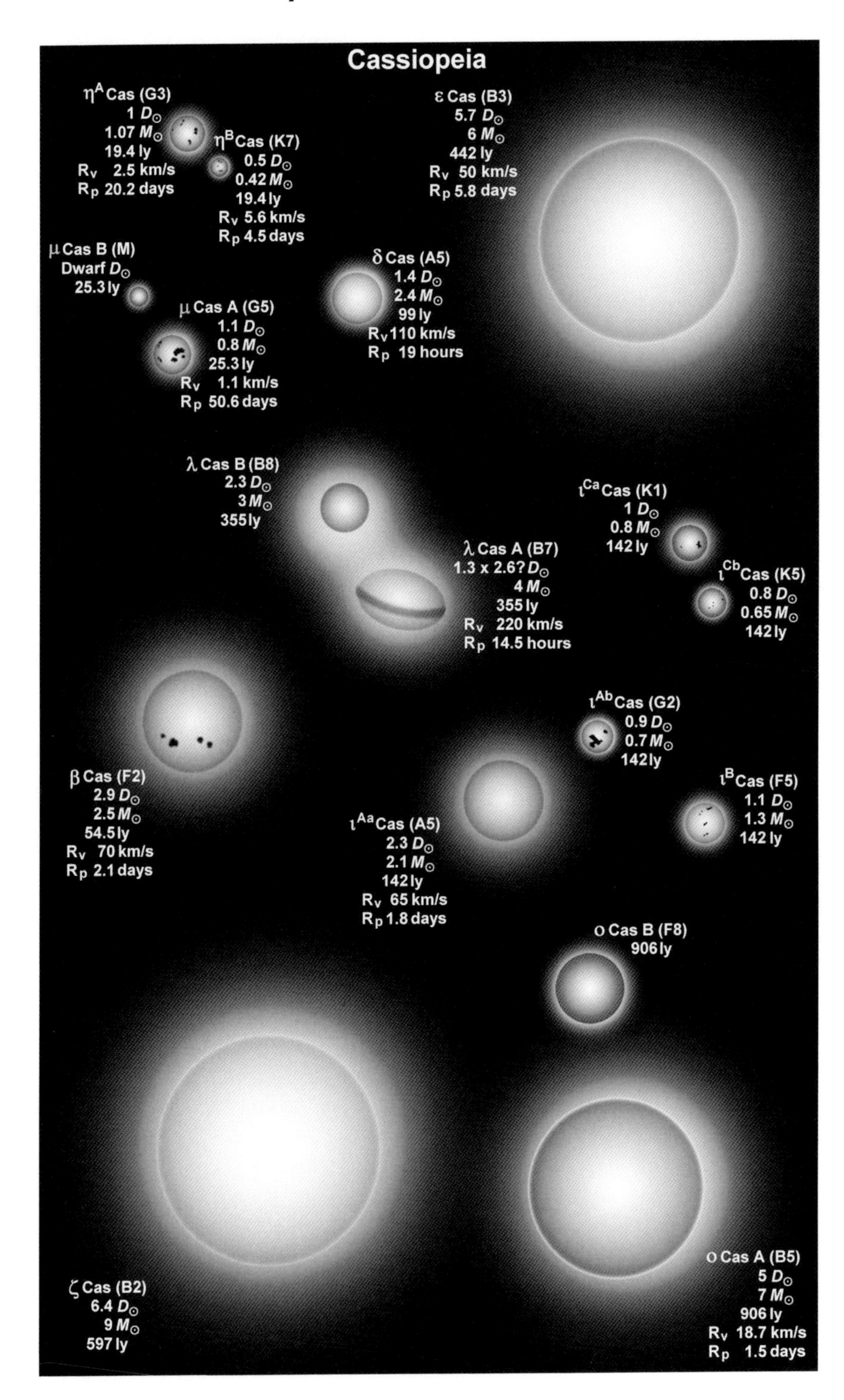
Cassiopeia
η^A Cas (G3)
1 D⊙
1.07 M⊙
19.4 ly
Rv 2.5 km/s
Rp 20.2 days
η^B Cas (K7)
0.5 D⊙
0.42 M⊙
19.4 ly
Rv 5.6 km/s
Rp 4.5 days
ε Cas (B3)
5.7 D⊙
6 M⊙
442 ly
Rv 50 km/s
Rp 5.8 days
μ Cas B (M)
Dwarf D⊙
25.3 ly
μ Cas A (G5)
1.1 D⊙
0.8 M⊙
25.3 ly
Rv 1.1 km/s
Rp 50.6 days
δ Cas (A5)
1.4 D⊙
2.4 M⊙
99 ly
Rv 110 km/s
Rp 19 hours
λ Cas B (B8)
2.3 D⊙
3 M⊙
355 ly
λ Cas A (B7)
1.3 x 2.6? D⊙
4 M⊙
355 ly
Rv 220 km/s
Rp 14.5 hours
ι^Ca Cas (K1)
1 D⊙
0.8 M⊙
142 ly
ι^Cb Cas (K5)
0.8 D⊙
0.65 M⊙
142 ly
ι^Ab Cas (G2)
0.9 D⊙
0.7 M⊙
142 ly
β Cas (F2)
2.9 D⊙
2.5 M⊙
54.5 ly
Rv 70 km/s
Rp 2.1 days
ι^Aa Cas (A5)
2.3 D⊙
2.1 M⊙
142 ly
Rv 65 km/s
Rp 1.8 days
ι^B Cas (F5)
1.1 D⊙
1.3 M⊙
142 ly
ο Cas B (F8)
906 ly
ζ Cas (B2)
6.4 D⊙
9 M⊙
597 ly
ο Cas A (B5)
5 D⊙
7 M⊙
906 ly
Rv 18.7 km/s
Rp 1.5 days

Heading towards us at 97.2 km/s μ **Cassiopeiae** is an interloper from the galactic halo. Belonging to spectral class G5 it is a 1.1 $D_{\odot}$ dwarf and is considerably less luminous than our Sun at 0.4 $L_{\odot}$ due to its very low metal content (13% to 28% that of the Sun). It also has a companion: an +11.51 magnitude M class dwarf in a 21.97 year orbit which varies from 3.3 AU at periastron to 11.9 AU at apastron. The duo travel across the sky at a rate of knots with a space velocity of 186 km/s and, in 25,000 years' time, should be in what is now Lyra. Halo stars such as μ Cas are thought to be about 10,000 million years old and may hold clues as to the composition of interstellar space at the time our Galaxy formed.

o **Cassiopeiae** is another γ Cas eruptive variable. A B5 star of 5 $D_{\odot}$ it has a rotational velocity of 155 km/s and fluctuates between a minimum m_v of +4.62 to a maximum +4.50. It also has an unseen companion, an F8 dwarf with an orbital period of 1,033 days indicating they are separated by an average of 2 AU.

Despite being 11,649 ly from Earth ρ **Cassiopeiae** is still visible as a 4th magnitude star – usually. It is one of the largest stars visible to the naked eye with a diameter of 450 $D_{\odot}$. Translated into Solar System terms this means that it has a radius of 2.1 AU; big enough to engulf Mars and reach the inner fringes of the Asteroid Belt. This hypergiant is 550,000 times more luminous than the Sun and would be a brilliant M_v -8.7 at 10 pc. It is also highly unstable, belonging to the SRd class of semi-regular pulsating variables. Its magnitude fluctuates from m_v +4.1 to +6.2 with a range of periods of 320, 510, 645 and 820 days. Astronomers are undecided as to whether it is an F8 or a G2 but as it heads towards magnitude minima it turns from yellow to red, cooling from 7,200 K to 4,000 K. ρ Cas is in its death throes and, indeed, may already have exploded as a hypernova and we are just waiting for the light to reach us. Well worth keeping an eye on this star.

R Cassiopeiae is one of only a handful of naked eye M7 red giants. Like most of the others it is a Mira-type variable, spending most of its 430.46 day period invisible without optical aid and growing as dim as m_v +13.5. When it does brighten it reaches m_v +4.70. There is considerable uncertainty over its distance, diameter and luminosity but it has been suggested that it may be 490 $D_{\odot}$.

As big as ρ and R Cas are, they are dwarfed by **V509** which is an enormous 910 $D_{\odot}$ or 8.5 AU across. Put it in the center of the Solar System and it would be just shy of reaching Jupiter by 0.7 AU. V509 resides at least 7,000 ly from Earth and could be either a G0 or G4 yellow hypergiant with a luminosity of 125,000 $L_{\odot}$. Like ρ Cas it is an SRd semi-regular pulsating variable fluctuating between m_v +4.75 and +5.50.

V566, or **6 Cas** if you prefer, is another supergiant, super-luminous star. At 300 $D_{\odot}$ it is only a third of the size of V509 but that still converts to 2.8 AU. Lying somewhere between 15,050 and 17,600 ly it is more than 166,600 times as luminous as the Sun. This A3 white hot star spins at 39 km/s taking 389 days to complete one rotation and varies in magnitude in an α CV fashion from m_v +5.45 to +5.34 though over no particular timescale.

V566 along with ρ Cas and at least 19 other stars appear to belong to a loose association called Cas OB5. This group, which is based on position, distance and

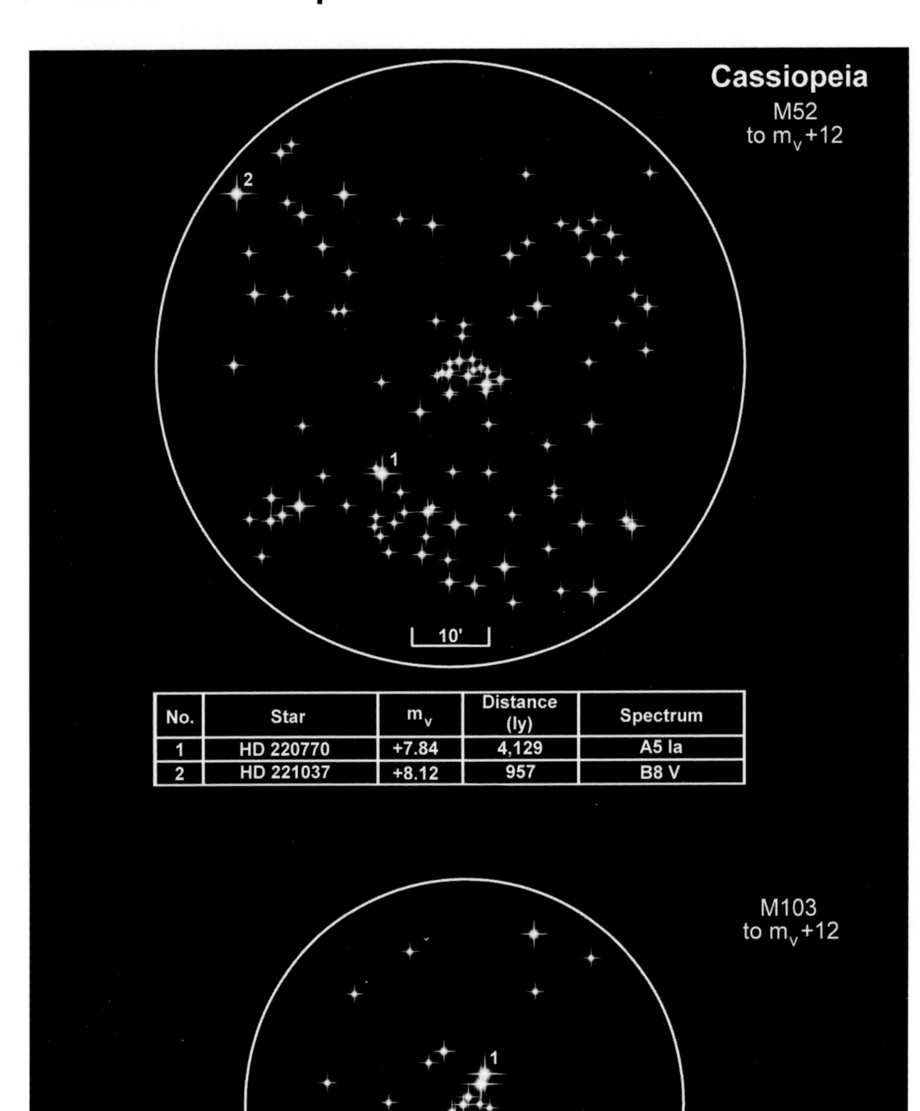

No.	Star	m_v	Distance (ly)	Spectrum
1	HD 220770	+7.84	4,129	A5 Ia
2	HD 221037	+8.12	957	B8 V

No.	Star	m_v	Distance (ly)	Spectrum
1	HD 9311	+7.28	1,337	B5 Ib

radial velocity, includes the peculiar **HD 108** which may or may not be a binary system. At m_v +7.40 it is too faint to be seen without optical aid but it has generated much debate in recent decades. Its radial velocity – how fast it travels towards the Sun – is around the 64 km/s mark but varies, suggesting the presence of a second star that gravitationally tugs at HD 108, sometimes slowing it down and sometimes speeding it up. The orbital period has been variously calculated at 1.02, 4.612, 5.7937 and 1,627.6 days. The binary component has been confirmed and rejected several times. It has also been suggested that HD 108 has undergone a supernova explosion without destroying the star and the variations in radial velocity have even been dismissed altogether with jets of material perpendicular to the line of sight being blamed for any perceived variation in radial velocity. Optical and infra-red observations suggest it is ejecting massive amounts of material while paradoxically UV studies suggest a low mass loss. Whatever the true nature of HD 108 we do know, or at least we *think* we know, that it is an O6 supergiant with a temperature of around 45,000 K.

The +5.01 magnitude **φ Cassiopeiae** is another huge star, 460 solar diameters across – just about wide enough to touch the Asteroid Belt if it were to replace the Sun – and 4,100 $L_{\odot}$. It is a rare yellowish-white F0, 2,330 ly away and through a small telescope or binocular looks as though it could be part of the open cluster NGC 457. But it isn't, the cluster being a further 6,700 ly away. Still, the cluster makes for an attractive backdrop for a star that will one day go supernova.

If you had been watching Cassiopeia in the first week of November 1572 you would have noticed a brilliant new star close to κ Cas. At magnitude m_v -4 it rivaled Venus in intensity and attracted the attention of writers and philosophers, kings and queens throughout Europe but none more so than the Danish astronomer Tycho Brahe who undertook an extensive study of the celestial visitor. Now known as Tycho's Star it is listed in modern catalogs as **SNR G120.1+01.2**, a *S*uper*N*ova *R*emnant expanding rapidly through space 7,500 ly away.

A number of stars in Cassiopeia have planetary systems although the stars are all below naked eye visibility. One planet, **HD 7924 b**, appears to be a super-Earth with a mass of at least 9.26 $M_{\oplus}$ putting it close to the upper limit for terrestrial type planets. The star **HD 17156** is a solar analog, about 45% larger than the Sun, but its sole planet orbits at a distance of just 24 million km (Mercury orbits the Sun at about 60 million km) and is likely to be a 'hot Jupiter' that formed farther away from the star but later migrated inwards.

With countless stars we should perhaps not be surprised to find a large number of coincidences but one of the most remarkable is **Kemble 2**, otherwise known as the **Mini Cassiopeia**. Although only about 10′ across this arrangement of six stars just to the east of χ Draconis is the spitting image of the real thing.

Planetary systems in Cassiopeia

Star	$D_{\odot}$	Spectral class	ly	m_v	Planet	Minimum mass	q	Q	P
HD 7924	0.83	K0	54.8	+7.21	HD 7924 b	9.26 $M_{\oplus}$	0.05	0.07	5.40 d
HD 17156	1.45	G0	255	+8.17	HD 17156 b	3.21 M_J	0.05	0.28	21.22 d
HD 240210	13.0	K3	466	+8.00	HD 240210 b	6.90 M_J	1.13	1.53	1.38 y

Open clusters in Cassiopeia

Name	Size arc min	Size ly	Distance ly	Age million yrs	Brightest star in region*	No. stars m_v >+12*	Apparent magnitude m_v
Collinder 33	36′	34	3,300	230	HD 18326 m_v +7.94	53	+5.9
Collinder 463	102′	68	2,300	235	HD 10483 m_v +8.20	76	+5.7
IC 1805	51′	92	6,200	7	HD 15557 m_v +7.39	98	+6.5
IC 1848	11′	21	6,500	7	HD 17505 m_v +7.06	7	+6.5
M 52 (NGC 7654)	50′	67	4,600	58	HD 220770 m_v +7.84	102	+6.9
M 103 (NGC 581)	25′	51	7,200	22	HD 9311 m_v +7.28	29	+7.4
NGC 225	30′	19	2,100	130	TYC 4016-97-1 m_v +9.28	39	+7.0
NGC 457	25′	58	8,000	21	ϕ Cas m_v +4.95	57	+6.4
NGC 654	63′	121	6,700	14	HD 10494 m_v +7.32	46	+6.5
NGC 659	17′	32	6,300	35	No ID m_v +10.11	13	+7.9
NGC 663	43′	80	6,400	16	TYC 4032-1211-1 m_v +8.29	55	+7.1
NGC 1027	30′	22	2,500	160	HD 16626 m_v +6.99	52	+6.7
NGC 7790	14′	38	9,600	56	TYC 4281-1230-1 m_v +10.23	7	+8.5
Stock 2	92′	26	1,000	170	HD 13122 m_v +6.67	169	+4.4
Trumpler 3	47′	17	1,300	68	HD 19661 m_v +8.77	89	+7.0

*May not be a cluster member.

Centaurus

Constellation:	Centaurus	**Hemisphere:**	Southern
Translation:	The Centaur	**Area:**	1,060 deg^2
Genitive:	Centauri	**% of sky:**	2.570%
Abbreviation:	Cen	**Size ranking:**	9th

In mythology Centaurs were creatures who were half man, half horse that resided in ancient Thessaly. They were driven out of the country when one of them attempted to abduct a bride. One Centaur is associated with Chiron, tutor to Jason of the Argonauts and Hercules. This Southern Hemisphere constellation partly overlaps the Milky Way

Despite being invisible from anywhere farther north than Miami **α Centauri** is arguably the most famous of all stars due, in no small part, to it being the closest naked eye star to the Sun at 4.39 ly. Also known as Rigil Kentaurus or simply Rigil Kent (meaning the 'Foot of the Centaur') it is not unlike our Sun: somewhat larger and more luminous by about a third but still a G2. The third brightest star in the night sky at m_v -0.01 it would fade to a barely visible M_v +4.35 at 10 pc. Estimates of its age range from 6,500 to 8,000 million years – rather older than the Sun's 4,560 million years – so considerably depleted in hydrogen and heading towards the final stages of its life as a yellow dwarf. It has one, perhaps two companions. **α^B Centauri** is a K1 dwarf about 90% the diameter and mass of the Sun. Its orbital period is around 80 years during which the two stars come as close as 11 AU (as in 1957 when they were 2″ apart) and separate by as much as 35 AU (which occurred in 1980 with a distance of 22″). At m_v +1.33 the star is relatively easy to find with a modest telescope, the duo appearing as a contrasting yellow-orange pair. **α^C Centauri** is better known as **Proxima** and may or may not be gravitationally bound to the other two. At m_v +11.05 it is by no means a bright or easy object to find lying, as it does, a good 2° from the primary star. It is slightly closer to us by about 10,000 AU but will not always be so if it is indeed in orbit around α^A Cen. Don't hold your breath waiting for it to pass behind the two brighter stars: the orbital period is in excess of 1 million years and unstable. An M5.5 red dwarf of 0.145 $D_\odot$ and 0.00005 $L_\odot$ Proxima does flare fairly regularly increasing its brightness by about a magnitude, a result of its magnetic fields collapsing. From the α Centauri system the Sun would appear as a m_v +0.5 star in Cassiopeia.

β Centauri or Agena is one of three β Cephei pulsating variables in the constellation, the other two being ε and χ. β Cen fluctuates between m_v +0.61 and +0.66 with a period of 3^h 46^m. ε **Centauri** has a somewhat longer period of 4^h 4^m during which its magnitude changes from m_v +2.29 to +2.31, while **χ Centauri** is timed at just 50.5^m and flickers between m_v +4.15 and +4.17.

The constellation includes four Be 'emission' stars – γ Cas variables – all roughly the same size and all 'shell stars' being surrounded by circumstellar disks of material. These rapidly rotating stars are thought to eject huge amounts of

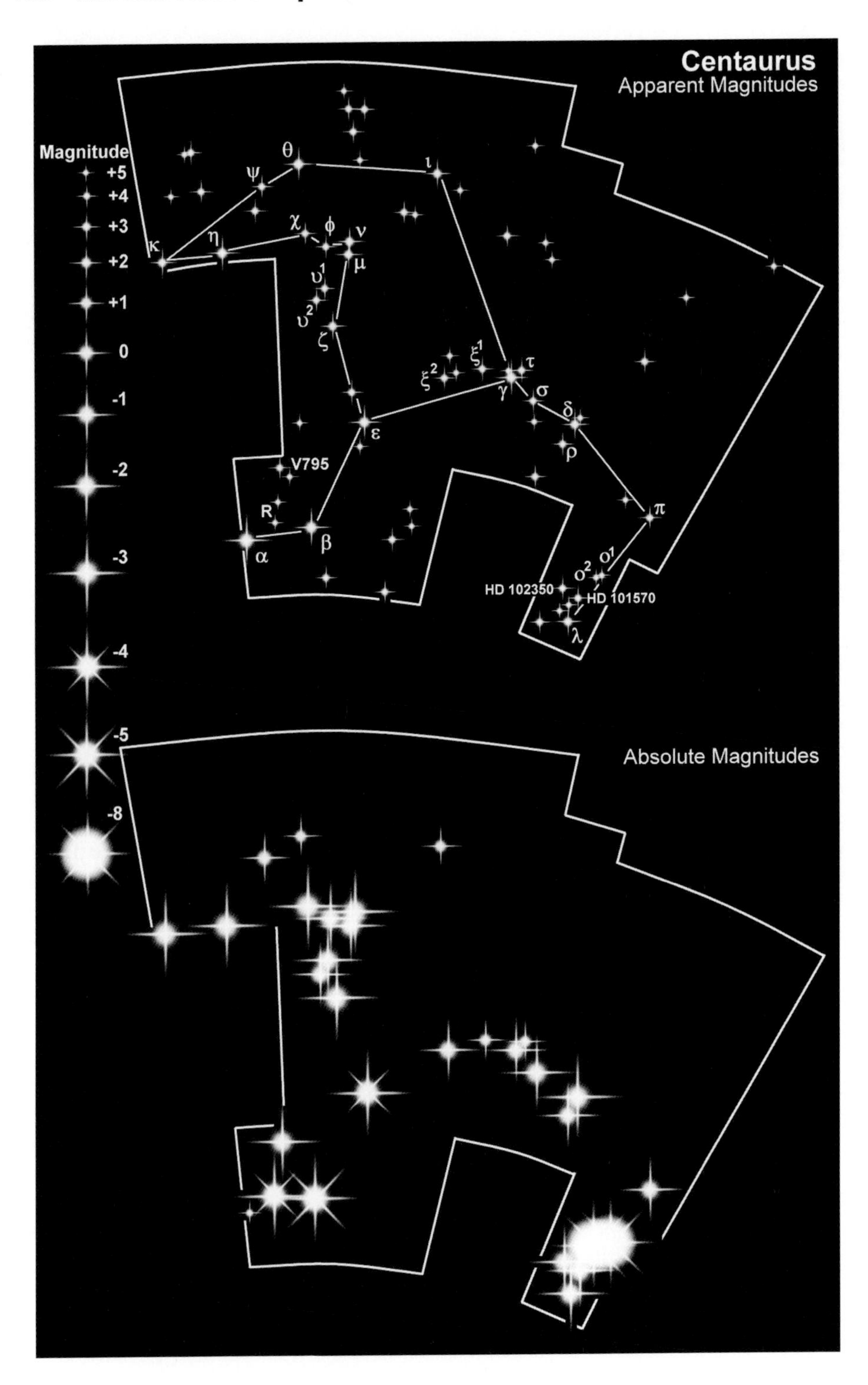
Centaurus
Apparent Magnitudes
Magnitude
+5
+4
+3
+2
+1
0
-1
-2
-3
-4
-5
-8
θ
ι
ψ
χ
ϕ
ν
κ
η
μ
υ1
υ2
ζ
ξ2
ξ1
τ
γ
σ
δ
ε
ρ
V795
R
π
α
β
o2
o1
HD 102350
HD 101570
λ
Absolute Magnitudes

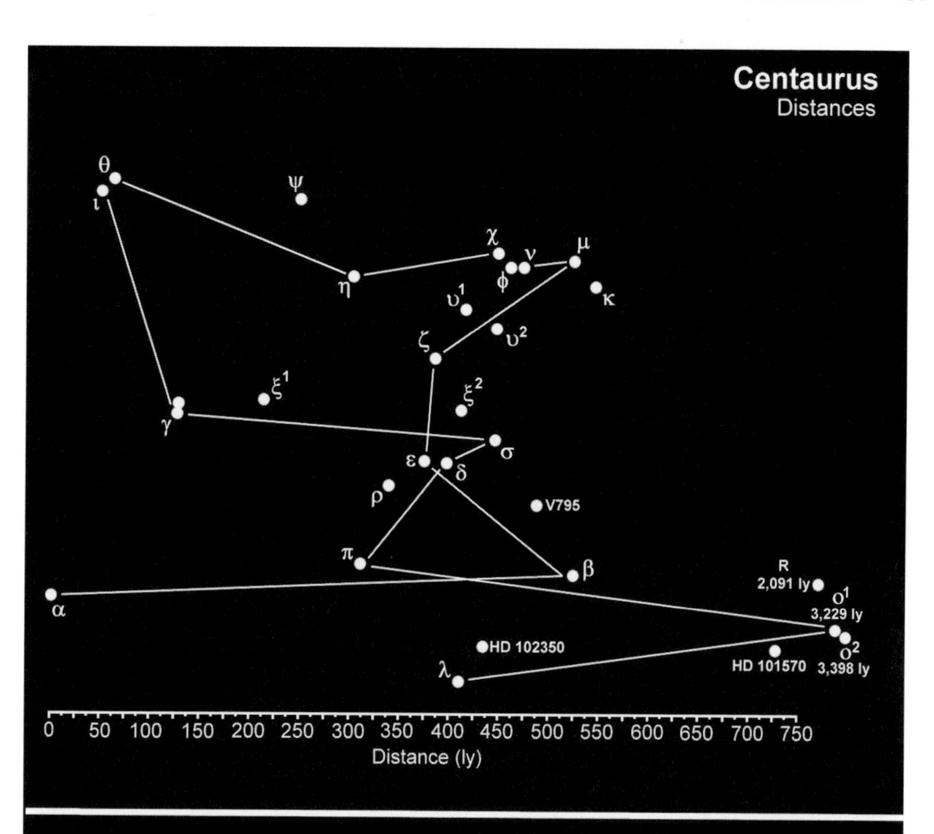

The View from α Centauri

From α Centauri the Sun would appear as a m_v+0.5 star in Cassiopeia to the south east of ε Cassiopeiae.

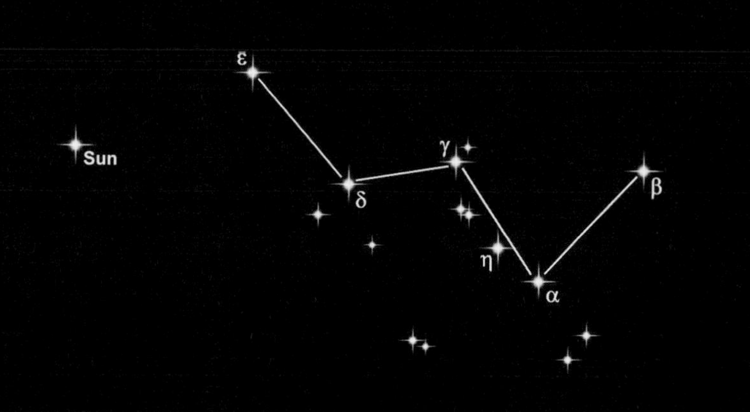

gases that eventually cool and condense to form dust particles. The table below summarizes what we know about the four stars.

γ Cas Be emission stars in Centaurus

Star	m_V min	m_V max	diff	M_V	$L_\odot$	$D_\odot$	Dist. ly	Spec	Rotational Velocity km/s	Radial Velocity km/s
δ Cen	+2.51	+2.65	0.14	-2.5	1,200	6.7	395	B2	240	11
η Cen	+2.30	+2.41	0.11	-2.9	8,200	7.2	309	B1.5	350	-0.2
μ Cen	+2.92	+3.47	0.55	-1.7	7,180	5.5	527	B2	130	9.1
V795	+4.97	+5.10	0.13	-1.7	174	3.2	486	B4	242	7

The 11 $D_\odot$ K0 giant **θ Centauri** appears to be a visitor to our corner of the Galaxy. Currently only 61 ly away it is passing us at 65 km/s. In 25,000 years' time, it will appear in the same part of the sky now occupied by ψ Velorum which, by coincidence, is also about 60 ly distant.

Also heading off towards Vela, but only making it as far as Antlia in our 25,000 year timeframe, is **ι Centauri**, an A2 Main Sequence white dwarf of 1.9 $D_\odot$. At 58.6 ly it is another fairly near star but perhaps its most interesting feature is the presence of a dust cloud, possibly the result of failed planet formation, or possibly still in the process of forming planets.

ν Centauri is a rotating ellipsoidal variable that changes in magnitude from m_V +3.38 to +3.41 without any detectable period. A hot B2 blue dwarf with a temperature of around 22,500 K it is 4.5 $D_\odot$ and rotates at 70 km/s, taking 3.3 days to complete one full turn on its axis. It is known to be a spectroscopic binary, its companion locked into an orbital period of 2.62516 days, and although nothing else is known about the secondary star its presence and close proximity may distort the shape of the primary and influence how it spins. If we could see the disk of the star from Earth it would appear to be continuously changing in shape, brightening and fading as different hemispheres are presented to us.

The two 5th magnitude stars **o^1** and **o^2 Centauri** mark the farthest boundaries of the constellation at more than 3,000 ly but are not actually related. o^1 Cen is a bright yellow G3 hypergiant, 630 $D_\odot$ and a bit cooler than our own Sun at 5,100 K but considerably more luminous by a factor of some 7,250. If dropped into the center of the Solar System o^1 Cen would swallow everything up to and including the Asteroid Belt. A semi-regular SRd pulsating variable with a period of around 200 days its magnitude varies noticeably from m_V +5.10 to an invisible m_V +6.60. If placed 10 pc from Earth it would be an impressive M_V -8.0. It is actually 3,229 ly away. Right next to it, at least on the celestial sphere, is o^2 Cen: almost identical in apparent magnitude (+5.12) but also an α Cygni variable to m_V +5.22 over a period of 46.3 days. o^2 Cen is 169 ly farther away and only one-third the size of its namesake at 210 $D_\odot$. A brilliant white A2 the pair make a for fine color contrast in a telescope with good optics.

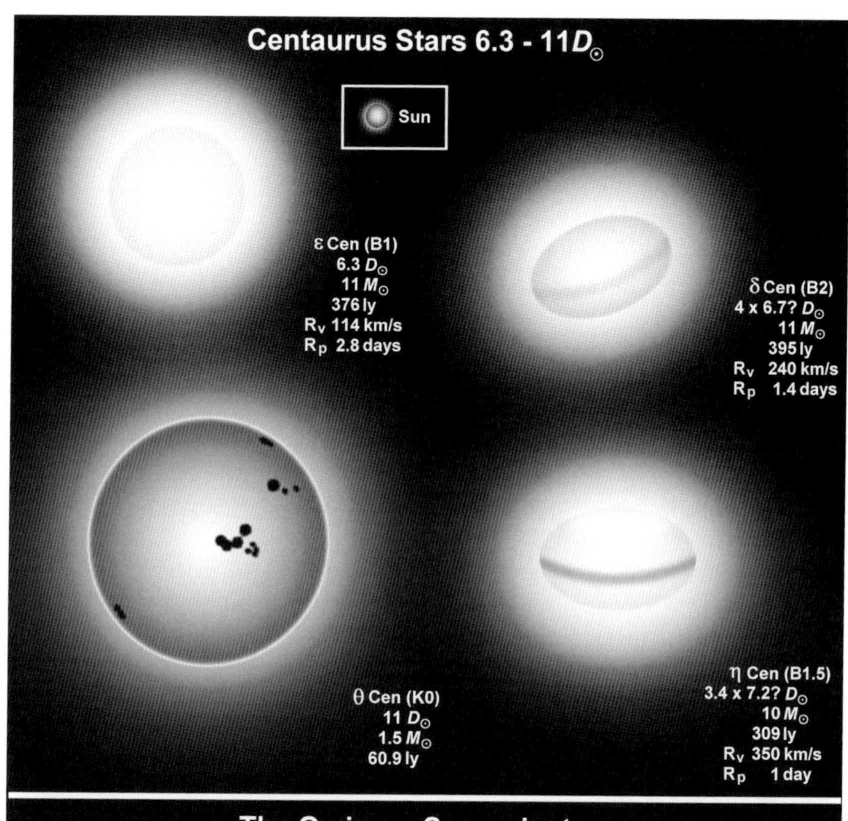
Centaurus Stars 6.3 - 11$D_{\odot}$
Sun
ε Cen (B1)
6.3 $D_{\odot}$
11 $M_{\odot}$
376 ly
R_V 114 km/s
R_P 2.8 days
δ Cen (B2)
4 x 6.7? $D_{\odot}$
11 $M_{\odot}$
395 ly
R_V 240 km/s
R_P 1.4 days
θ Cen (K0)
11 $D_{\odot}$
1.5 $M_{\odot}$
60.9 ly
η Cen (B1.5)
3.4 x 7.2? $D_{\odot}$
10 $M_{\odot}$
309 ly
R_V 350 km/s
R_P 1 day

The Omicron Supergiants
o^1 Cen (G3)
630 $D_{\odot}$
13 $M_{\odot}$
3,229 ly
o^2 Cen (A2)
210 $D_{\odot}$
13 $M_{\odot}$
3,398 ly
R_V 25 km/s
R_P 1.16 years
Mean distance of Jupiter
from the Sun, 5.2 AU

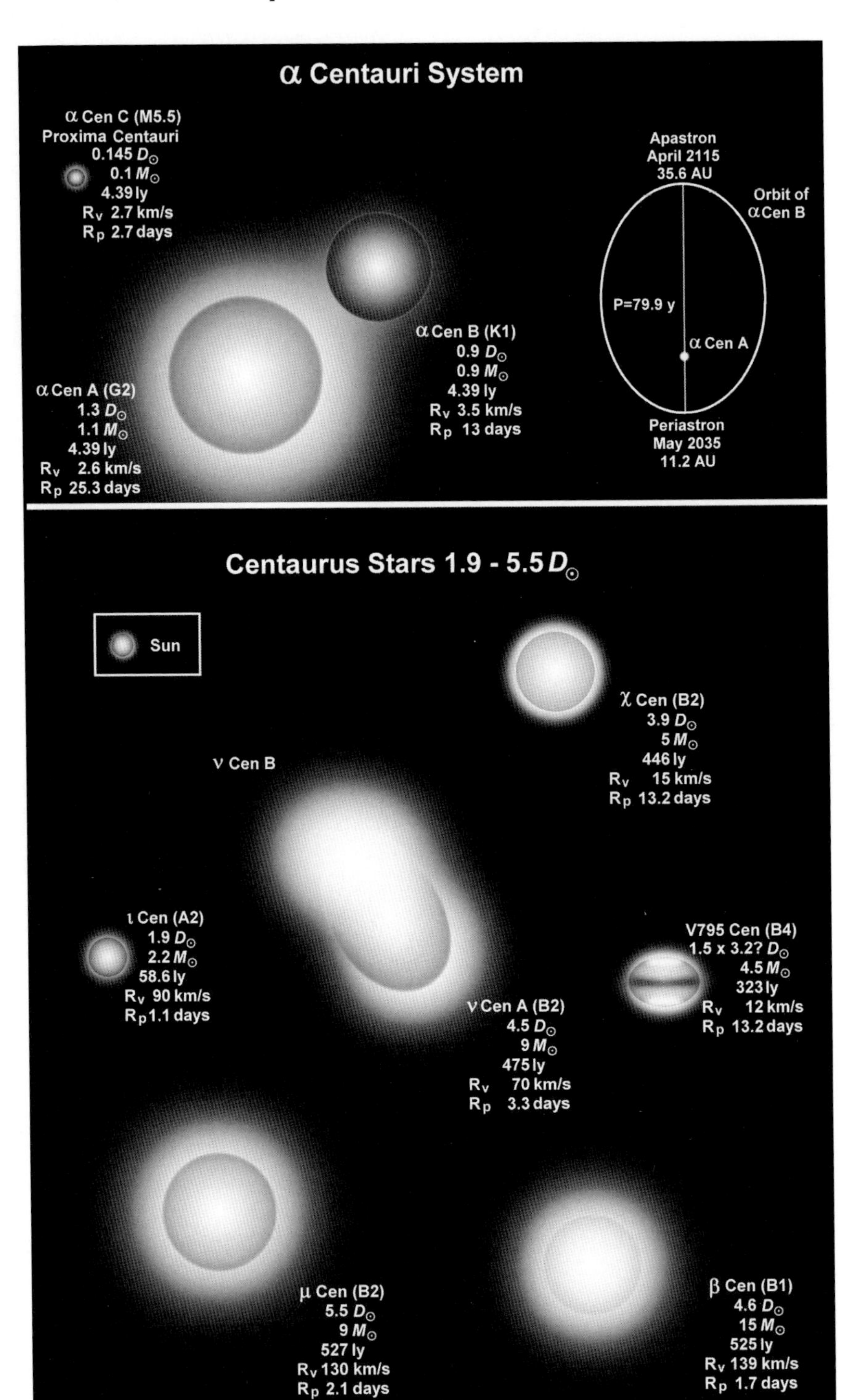
α Centauri System
α Cen C (M5.5)
Proxima Centauri
0.145 D⊙
0.1 M⊙
4.39 ly
Rv 2.7 km/s
Rp 2.7 days
Apastron
April 2115
35.6 AU
Orbit of
α Cen B
P=79.9 y
α Cen A
Periastron
May 2035
11.2 AU
α Cen B (K1)
0.9 D⊙
0.9 M⊙
4.39 ly
Rv 3.5 km/s
Rp 13 days
α Cen A (G2)
1.3 D⊙
1.1 M⊙
4.39 ly
Rv 2.6 km/s
Rp 25.3 days
Centaurus Stars 1.9 - 5.5 D⊙
Sun
χ Cen (B2)
3.9 D⊙
5 M⊙
446 ly
Rv 15 km/s
Rp 13.2 days
ν Cen B
ι Cen (A2)
1.9 D⊙
2.2 M⊙
58.6 ly
Rv 90 km/s
Rp 1.1 days
V795 Cen (B4)
1.5 x 3.2? D⊙
4.5 M⊙
323 ly
Rv 12 km/s
Rp 13.2 days
ν Cen A (B2)
4.5 D⊙
9 M⊙
475 ly
Rv 70 km/s
Rp 3.3 days
μ Cen (B2)
5.5 D⊙
9 M⊙
527 ly
Rv 130 km/s
Rp 2.1 days
β Cen (B1)
4.6 D⊙
15 M⊙
525 ly
Rv 139 km/s
Rp 1.7 days

R Centauri is a Mira-type variable with a gradually decreasing period, similar to that displayed by R Aquilae and R Hydrae. Very little is known about its physical dimensions but its magnitude can reach m_v +5.3 at maximum before falling to m_v +11.3 (often it will rise to just m_v +6.3 before dimming again). Its period is listed in the *Second Harvard Catalog of Variable Stars* as being 568.2 days. That was in 1907. Now, a century later, it is 546.2 days.

Open and globular clusters in Centaurus

Name	Size arc min	Size ly	Distance ly	Age million yrs	Brightest star in region*	No. stars m_v >+12*	Apparent magnitude m_v
IC 2944[§]	37′	62	5,800	7	HD 101545 m_v +6.30	83	+4.5
NGC 3680	30′	26	3,000	1,200	DM -42° 6983 m_v +10.06	23	+7.6
NGC 3766 Pearl Cluster	16′	27	5,700	15	HD 100943 m_v +7.11	132	+5.3
NGC 5138	12′	22	6,500	97	HD 116721 m_v +8.70	15	+7.6
NGC 5281	15′	16	3,600	14	HD 119699 m_v +6.61	42	+5.9
NGC 5316	11′	13	4,000	159	DM -61° 4106 m_v +9.38	24	+6.0
NGC 5460	51′	33	2,200	161	HD 123247 m_v +6.42	86	+5.6
NGC 5606	15′	26	5,900	12	HD 126449 m_v +8.79	25	+7.7
NGC 5617	15′	22	5,000	82	HD 126640 m_v +8.76	40	+6.3
NGC 5662	58′	37	2,200	93	HD 127753 m_v +7.07	141	+5.5
Stock 14	22′	44	7,000	11	V810 Cen m_v +5.01	38	+6.3
ω Cen (NGC 5139)	19′	87	15,800	12,000	Globular cluster		+3.9
NGC 5286	9′	94	36,000	2,000	Globular cluster		+7.3

*May not be a cluster member. [§]IC 2944 may not be a true cluster.

Cepheus

Constellation:	Cepheus	**Hemisphere:**	Northern
Translation:	King Cepheus	**Area:**	588 deg^2
Genitive:	Cephei	**% of sky:**	1.425%
Abbreviation:	Cep	**Size ranking:**	27th

The fabled King Cepheus of Ethiopia was the husband of Queen Cassiopeia and father of Andromeda. The constellation is one of the 48 listed by Ptolemy in his *Almagest* (*c.* AD 150).

Not particularly bright at m_v +2.43, especially for such a regal constellation, **α Cephei**, better known as Alderamin, is somewhat of an enigma. An A7 Main Sequence dwarf, 2.5 solar diameters across, it is supposed to have a modest rotational velocity. Instead it is spinning around at 246 km/s, behaving more like a B-class star, and emitting X-rays at levels that are on par with our own Sun. It also appears to have a magnetic field, all of which is not normal for an A7. Its very high rotational velocity – A7 stars average 113 km/s – suggests it is an oblate spheroid, probably darkened around the equator and brighter at the poles and with a noticeable temperature difference between the two. Whatever the explanation Alderamin is in transition, moving off the Main Sequence *en route* to becoming a giant. α Cep is also one of the Pole Stars. As the Earth gyrates on its axis the North Pole describes a circle on the Celestial Hemisphere, taking 25,800 years to complete a full cycle. At the moment the North Pole is pointing towards Polaris (α Ursae Minoris). In about 18,300 BC Alderamin was the Pole Star and will be again in AD 7,500. At 3° separation it never comes quite as close as Polaris at 0.4526°.

The second brightest star in Cepheus is also the standard for the β Cepheid variables. Lying at nearly 600 ly **β Cephei** or Alphirk appears as an unimpressive 3rd magnitude star but at 10 pc it would brighten to M_v -3.6. A hot, 22,700 K blue B2 with a diameter of 9 $D_\odot$ it has a luminosity of 1,430 $L_\odot$ and has two companions, both A-class stars. The closest circles the primary at a distance of about 45 AU and takes 90 years to complete a full orbit. Considerably farther out at 2,400 AU the second companion has an orbital period of at least 30,000 years. β Cep displays a complex sequence of variability during which its magnitude dips to m_v +3.27 from a maximum of +3.16. The main period of 4^h 34.3^m is caused by the star pulsating, maximum brightness occurring when the star is fully contracted. However, there are smaller magnitude changes timed at 4^h 43^m, 4h 28^m, 4^h 26^m, 4^h 53^m and 4^h 18^m in addition to 6 and 12 day variability. It appears to have a modest rotational velocity of 20 km/s (the average for this class is 142 km/s). β Cepheid type variables tend to be relatively small: 85% are less than 5 $D_\odot$ across and one-third are between 2 and 3 $D_\odot$. There are a number of exceptions, most notably τ^1 Lupi at 12 $D_\odot$, FN Canis Majoris 13 $D_\odot$, ι Canis Majoris 18 $D_\odot$, σ Scorpii 30 $D_\odot$ and V343 Carinae at 50 $D_\odot$. Three-quarters of all β Cepheids belong to the B1 or B2 spectral groups the rest belonging to B3-B6, at

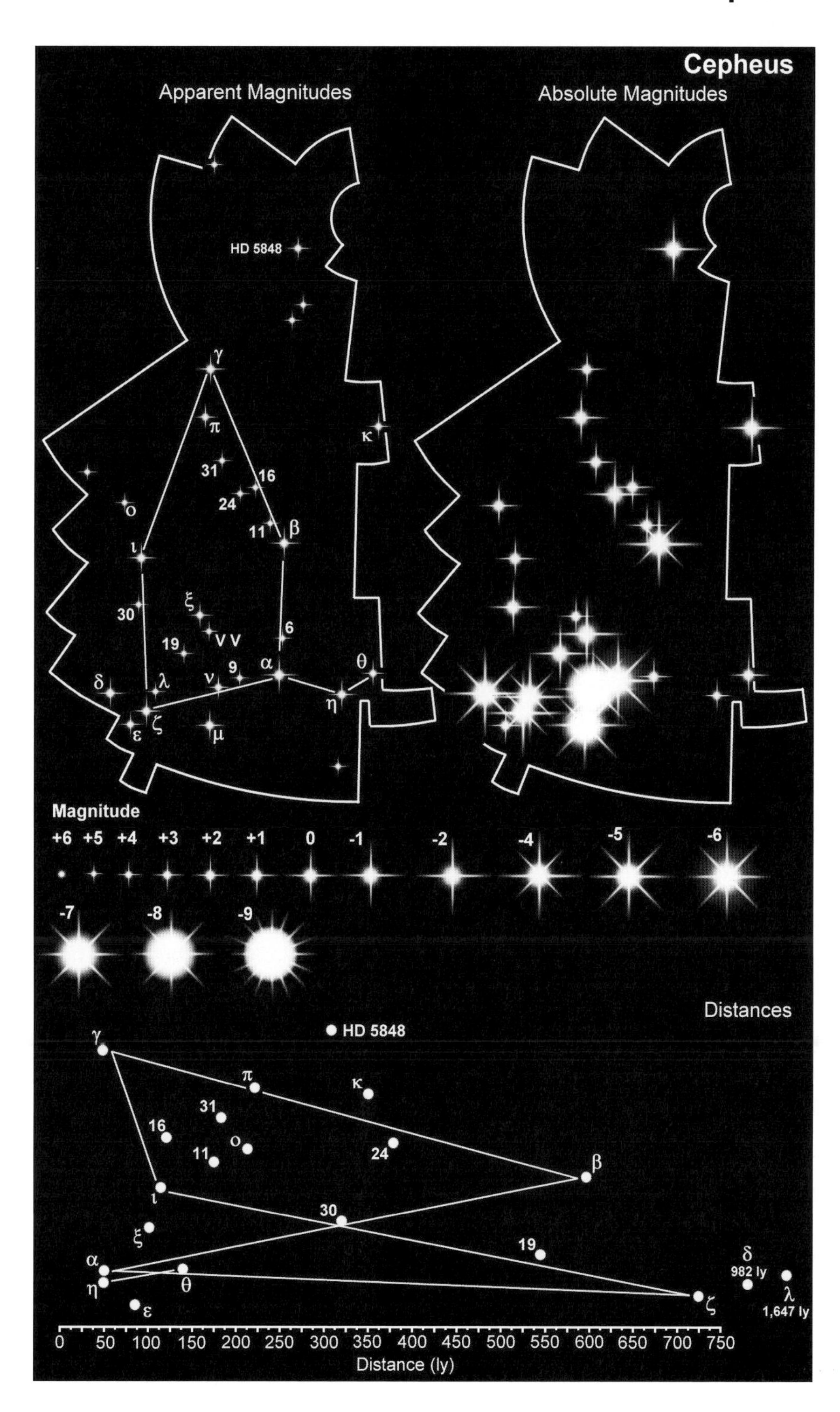
Cepheus
Apparent Magnitudes
Absolute Magnitudes
HD 5848
γ
π
κ
31
16
24
ο
11
β
ι
ξ
30
6
19
V V
9
α
θ
δ
λ
ν
η
ε
ζ
μ
Magnitude
+6 +5 +4 +3 +2 +1 0 -1 -2 -4 -5 -6
-7 -8 -9
Distances
HD 5848
γ
π
κ
31
16
ο
24
11
β
ι
30
ξ
19
δ
982 ly
α
θ
η
λ
1,647 ly
ε
ζ
0 50 100 150 200 250 300 350 400 450 500 550 600 650 700 750
Distance (ly)

least as far as naked eye stars are concerned. The variability periods are mainly between 2^h and 12^h but can be longer with o Velorum holding the record at 2^d 18^h 41^m 46^s. The shortest period among the naked eye β Cepheids is 29^m 57.1^s displayed by V376 Carinae. The amplitudes - i.e. the difference between maximum and minimum magnitudes - are usually between 0.01 and 0.08 magnitudes, the largest being χ Carinae at 0.44 (m_v +3.46 to +3.90).

γ Cephei is a 4.8 $D_\odot$ K1 evolving sub-giant 45 ly from Earth. It is important for two reasons. First, it is one of the Pole Stars and will be almost directly above the Earth's North Pole around AD 4000. Second, it is one of the relatively few binary systems that harbors a planet. γ Cephei's stellar companion is a red dwarf of 0.3 to 0.4 $M_\odot$ in an orbit that varies between 12 and 26 AU and with a period of around 62 years. The planet is at least 1.59 M_J in a less eccentric orbit that varies between 1.62 and 2.43 AU and with a period of 2.47 years. The stability of such a planetary system depends on a number of factors including the masses of the two stars, how eccentric their orbits are and how close the stars come together. In γ Cephei's case the system is relatively stable - the low mass red dwarf never gets any closer to the planet than about 9.5 AU.

δ Cephei is perhaps the most important star in the night sky. It is by no means the brightest at m_v +3.48, nor the most luminous at 1,639 $L_\odot$, and its distance is 982 ly so nothing unusual there. It is a supergiant, though a fairly modest one at 44 $D_\odot$. What is important about δ Cep is the fact that it is a variable star in which its luminosity is directly related to its pulsation period. This means that the luminosity can be found by accurately measuring the period (the periods in this type of variable are very stable and precise). Then by measuring the apparent magnitude a mathematical relationship between the magnitude and the luminosity will reveal the star's distance. It was, of course, also necessary to independently determine δ Cephei's distance which was done by the more traditional means of measuring its parallax (i.e. how the star appears to move against the much farther background stars over a period of six months). The Cepheid variables, as the class is now known, can therefore be used as yardsticks to determine distances within our own Galaxy and, because their high luminosities make them easy to detect, the distances to other galaxies. A Cepheid discovered in the famous Andromeda spiral enabled Edwin Hubble to make the first estimate of the galaxy's distance. δ Cep itself has a period of 5^d 8^h 47^m 31.9^s during which its magnitude varies from m_v +3.48 to +4.37. As the star dims its temperature plummets from 6,500 K to 5,500 K and its spectrum changes from F8 to G6, the star turning a more intense yellow. Meanwhile its diameter increases by about 15%. What appears to be happening is that the gases that make up the star's outer layers become opaque and trap heat. As the gases get hotter they expand and become more transparent, releasing the heat as a sort of natural pressure valve. Once most of the heat has escaped the outer layers contract again and the whole cycle is repeated. In orbit about δ Cep is a B7 dwarf of m_v +6.3. With a separation of at least 12,000 AU (19% of a light year) and an orbital period in excess of 2.5 million years there is every possibility that the companion will eventually be tugged out of its orbit by a passing star.

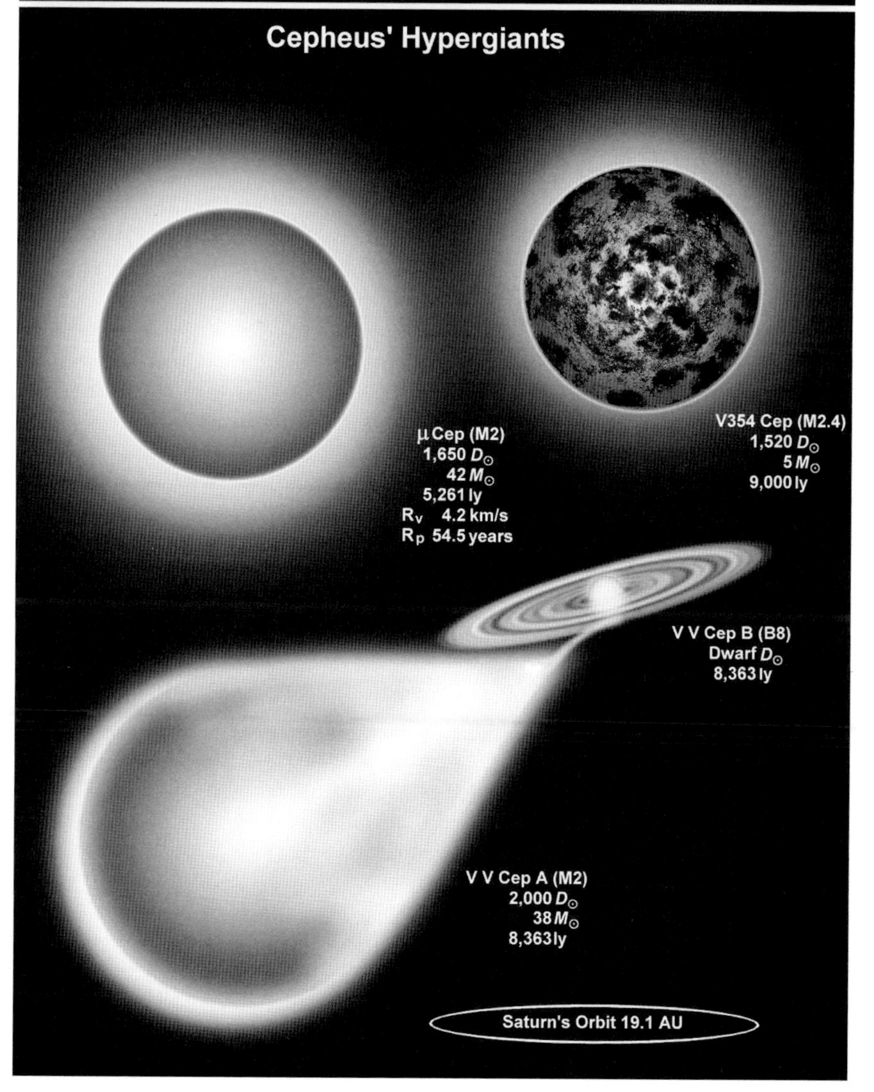

Cepheus' Supergiants
Earth's Orbit 2 AU
ζ Cep (K1.5)
230 D⊙
8.3 M⊙
726 ly
Rv 8 km/s
Rp 3.98 years
ν Cep (A2)
222 D⊙
20 M⊙
5,096 ly
Rv 49 km/s
Rp 229.3 days
Cepheus' Hypergiants
μ Cep (M2)
1,650 D⊙
42 M⊙
5,261 ly
Rv 4.2 km/s
Rp 54.5 years
V354 Cep (M2.4)
1,520 D⊙
5 M⊙
9,000 ly
V V Cep B (B8)
Dwarf D⊙
8,363 ly
V V Cep A (M2)
2,000 D⊙
38 M⊙
8,363 ly
Saturn's Orbit 19.1 AU

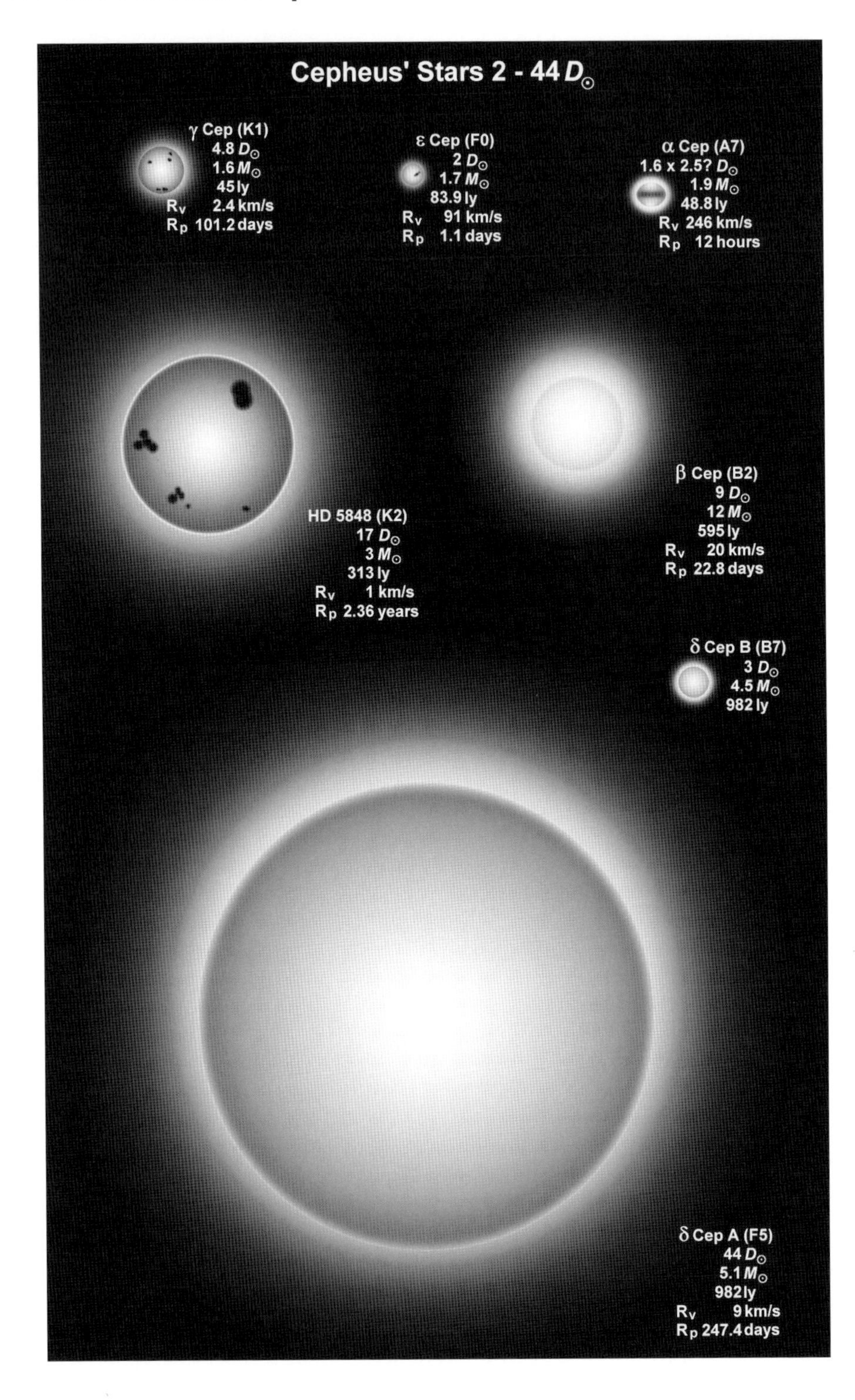
Cepheus' Stars 2 - 44 $D_{\odot}$
γ Cep (K1)
4.8 $D_{\odot}$
1.6 $M_{\odot}$
45 ly
R_V 2.4 km/s
R_P 101.2 days
ε Cep (F0)
2 $D_{\odot}$
1.7 $M_{\odot}$
83.9 ly
R_V 91 km/s
R_P 1.1 days
α Cep (A7)
1.6 x 2.5? $D_{\odot}$
1.9 $M_{\odot}$
48.8 ly
R_V 246 km/s
R_P 12 hours
β Cep (B2)
9 $D_{\odot}$
12 $M_{\odot}$
595 ly
R_V 20 km/s
R_P 22.8 days
HD 5848 (K2)
17 $D_{\odot}$
3 $M_{\odot}$
313 ly
R_V 1 km/s
R_P 2.36 years
δ Cep B (B7)
3 $D_{\odot}$
4.5 $M_{\odot}$
982 ly
δ Cep A (F5)
44 $D_{\odot}$
5.1 $M_{\odot}$
982 ly
R_V 9 km/s
R_P 247.4 days

Cepheids of this type (Classical Cepheids) – usually denoted by the symbol Cδ or δC – have periods of between 1 and 50 days and amplitudes of between 0.2 and 2.0 magnitudes. There are only a dozen naked eye Cepheids: ω Geminorum has the shortest period at $17^h\ 28^m\ 37^s$ while ZZ Carinae has the longest at $35^d\ 12^h\ 51^m\ 37^s$. Although they are all F or G class (or both) there is more variation in size with the smallest being 3 Sagittarii (28 $D_\odot$) and one of the largest being β Doradûs at 71 $D_\odot$. One of the most famous Cepheids is Polaris (α Ursae Minoris). What all Cepheids have in common is that they are dying high mass stars. Oddly enough δ Cep was not the first of the class to be recognized: that honor goes to η Aquilae, so maybe they should rightly be called Aquilids?

ε Cephei is another variable, this time belonging to the δ Scuti class. Pulsing at just under one hour – $59^m\ 53^s$ to be exact – ε Cep is the second fastest in the class, beaten only by V1644 Cygni at just under 45^m.

Cepheus is home to several supergiant stars including the 230 $D_\odot$ **ζ Cephei** of spectral class K1, and the slightly smaller **ν Cephei** at 222 $D_\odot$ which is a rare A2, the largest visible without optical aid. But compared to **μ Cephei** these two stars look like dwarfs. Sir William Herschel said of μ Cep that it is the reddest star in the sky and nicknamed it the 'Garnet Star'. What he did not know at that time was that it is also one of the largest stars visible to the naked eye at an estimated 1,650 $D_\odot$. To put that into context, it is equivalent to 15.4 AU. Replace the Sun with μ Cep and it would easily swallow up the orbit of Jupiter and stretch mid-way to Saturn. μ Cep is a semi-regular SRc variable, its magnitude fluctuating between m_v +3.43 and +5.10 over 730 days. The star is enveloped in a thick dust cloud and this, coupled with a fair amount of interstellar dust, actually reduces its brightness by 1.5 magnitudes. Larger again is **VV Cephei,** though by how much is debatable. Research puts its diameter at about 2,000 $D_\odot$. That's 18.6 AU or about as large as Saturn's orbit. Like μ Cephei it belongs to spectral group M2, its surface temperature a cool 3,500 K. It is also variable between m_v +4.80 and +5.36 with a period of 7,430 days or 20.3 years. The variability on this occasion is due to the presence of a B8 dwarf in an orbit that averages 25 AU. Although much smaller than the primary star, the dwarf has considerable density and gravitationally distorts its gigantic companion into a teardrop shape. As it does so it draws off matter into a hot circumstellar disk. A fifth gigantic star also exists in Cepheus. An estimated 1,520 $D_\odot$ **V354 Cephei** lies at a distance of about 9,000 ly and so only appears as an11th magnitude object. It's size would place it midway between the orbits of Jupiter and Saturn.

A couple of strays have found their way in and out of Cepheus. **R Cephei** is not in Cepheus at all but is well within the boundaries of Ursa Minor near λ UMi. And despite its designation, it is debatable as to whether this m_v +8.5 star is variable. What used to be called 2 Ursae Minoris was in the Little Bear but when the constellation boundaries were redrawn in the 1920s it found itself in Cepheus. It is now referred to by its other designations, most notably **HD 5848** or HR 285.

At m_v +11.5 the planetary nebula **NGC 40** is not easy to find but it represents what the Sun is likely to look like in about 6,000 million years. About 3,000 ly away and one light year across NGC 40 is the expanding shell of gas which the parent star has blown off as it began the final stages of its demise. The remnant white dwarf has a surface temperature of 50,000 K and pours heat into the nebula. A ferocious 1,000 km/s stellar wind also compresses the nebula increasing the temperature to millions of degrees.

Open clusters in Cepheus

Name	Size arc min	Size ly	Distance ly	Age million yrs	Brightest star in region*	No. stars m_v >+12*	Apparent magnitude m_v
IC 1396	90′	71	2,700	11	HD 206267 m_v +5.62	221	+3.5
NGC 188	49′	95	6,700	4,300	HD 3161 m_v +7.13	95	+8.1
NGC 7160	40′ × 90′	29 × 66	2,500	19	HD 208218 m_v +6.64	66	+6.1
NGC 7235	25′	67	9,200	12	HD 239886 m_v +6.64	25	+7.7
NGC 7380	65′	137	7,300	12	HD 215907 m_v +6.36	96	+7.2
NGC 7510	17′	33	6,800	38	HD 240221 m_v +8.75	11	+7.9

*May not be a cluster member.

Cetus

Constellation:	Cetus	**Hemisphere:**	Equatorial
Translation:	The Sea Monster	**Area:**	1,231 deg^2
Genitive:	Ceti	**% of sky:**	2.984%
Abbreviation:	Cet	**Size ranking:**	4th

Perseus was said to have rescued Andromeda from Cetus. The constellation is sometimes depicted as a sea monster and sometimes as a whale.

α Ceti or Menkar is a dying red giant. Some 77 $D_{\odot}$ across it lies at a distance of 220 ly and appears as a m_v +2.45 object making it the second brightest star in the constellation. Its magnitude fluctuates irregularly by a few percent so it is classed as an Lb variable. An M1.5 it burns at a cool 3,500 K, slightly hotter than the spots found on our own Sun, and it is heading towards us at 26 km/s.

Brighter by half a magnitude **β Ceti** or Diphda is also much closer to us at 95.8 ly. It is a yellow giant, 17 $D_{\odot}$, and is borderline G9.5 / K0. On its way to eventually becoming a red giant it is unusual in that it is a bright X-ray source, one of the brightest in our neck of the Galaxy. The theory is that rapid rotation causes strong magnetic fields which heat the corona to several million degrees generating X-rays in the process. The problem is that β Ceti rotates at just 13 km/s, a leisurely stroll in galactic terms and too slow to produce the observed effect. Its spectral signature also suggests it left the Main Sequence long ago but high levels of X-rays indicate that it is only just leaving this phase. Obviously there is more going on with β Ceti than we understand.

γ Ceti is a close double – actually a triple – and well worth taking time to observe. Sometimes described as blue and yellow, sometimes as white and yellow, the primary star, γ^A Ceti, is a +3.54 magnitude A2 dwarf of 1.7 $D_{\odot}$ and with a luminosity of 19.7 $L_{\odot}$. At 2.8″ (PA 298°) is γ^B Ceti, an F3, 1.3 $D_{\odot}$ across and with a luminosity of 1.6 $L_{\odot}$ but much fainter at m_v +6.25. In real space terms they are separated by at least 60 AU and have an orbital period of about 320 years. A decent size telescope will reveal a 10th magnitude K5 dwarf, about half the size and mass of the Sun, 14′ away. If it is part of the system then it is in an unstable orbit of at least 18,000 AU and with an orbital period of at least 1.5 million years.

The fourth magnitude **δ Ceti** now lies 19′ north of the Celestial Equator but was until 1923 in the Southern Celestial Hemisphere and has migrated due to precession. It is a typical β Cepheid pulsating variable: a B2 (the most common kind), an amplitude of 0.05 magnitude (the most common kind), an absolute magnitude of M_V -3 (the most common kind), a diameter of 6.9 $D_{\odot}$ (a bit on the large side) and a period of 3^h 52^m (the most common is around 4.5 hours). Perhaps in a parallel universe β Cepheids are known as δ Cetids. Meanwhile, in the parallel world of cyberspace δ Ceti seems to have acquired the proper name of Phycochroma meaning 'Seaweed-colored'. As δ Cet is, at best, m_v +4.05 and too faint to display any color and is, besides, a blue B2 and not seaweed green or

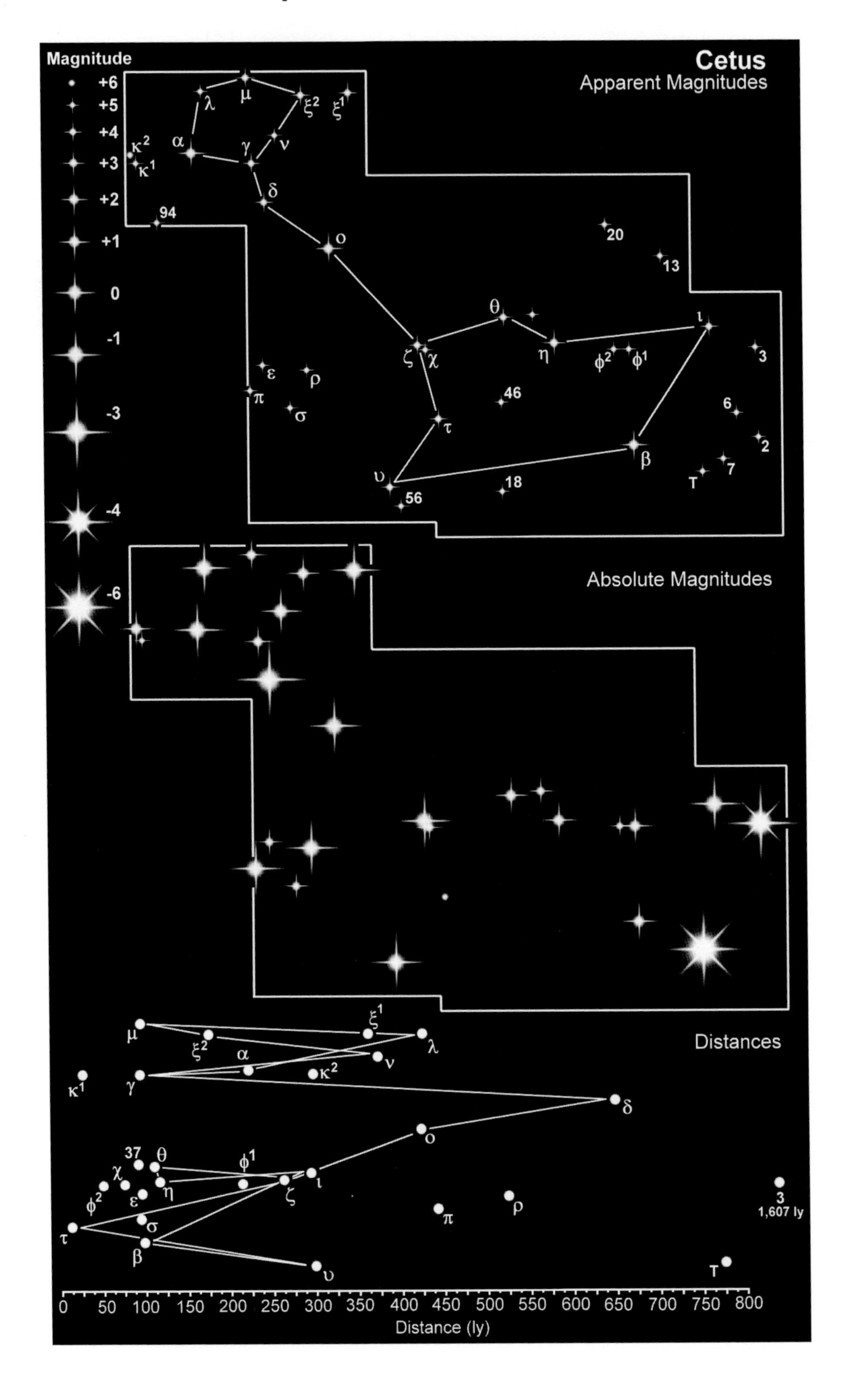
Cetus
Apparent Magnitudes
Magnitude
+6
+5
+4
+3
+2
+1
0
-1
-3
-4
-6
Absolute Magnitudes
Distances
1,607 ly
0 50 100 150 200 250 300 350 400 450 500 550 600 650 700 750 800
Distance (ly)

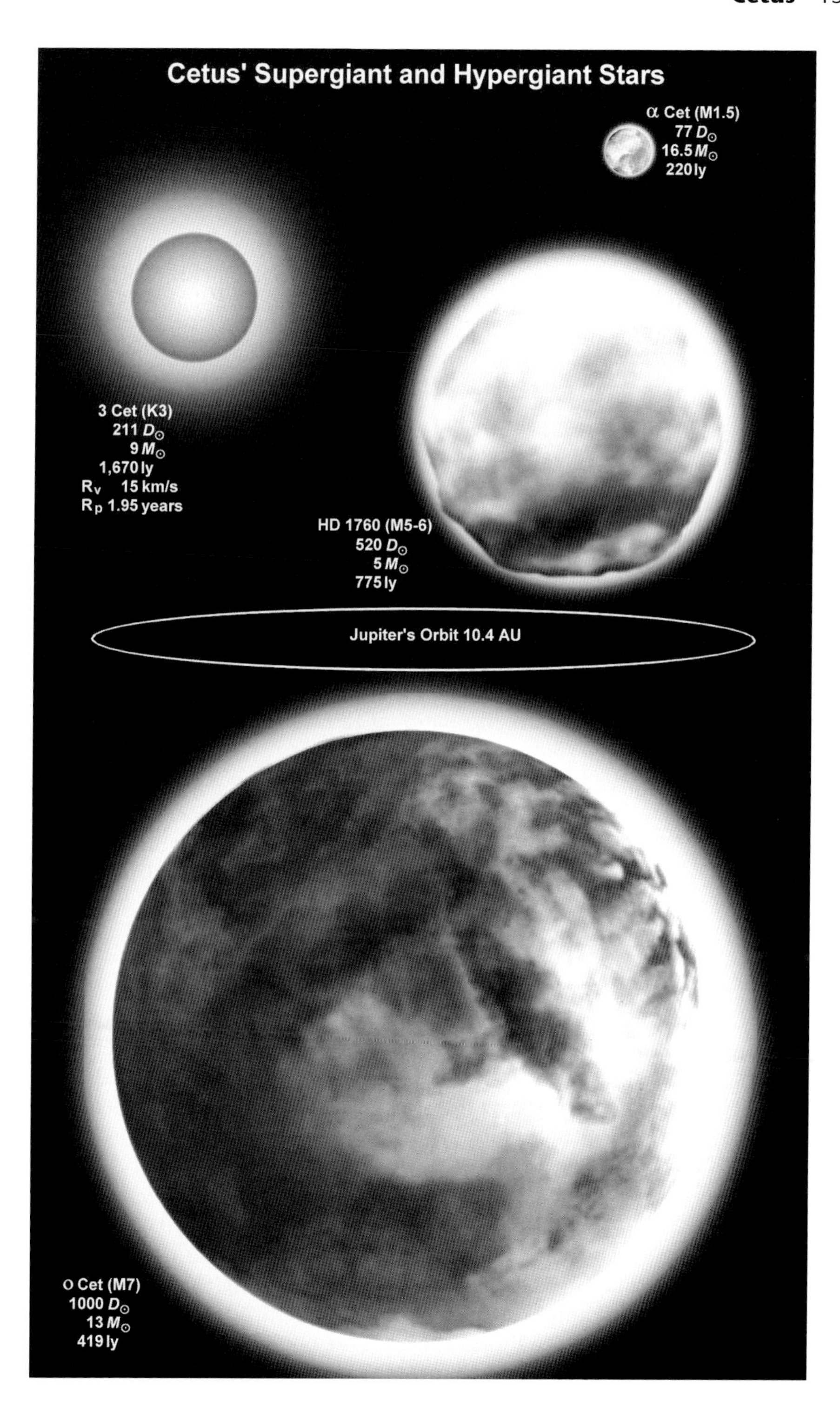
Cetus' Supergiant and Hypergiant Stars
α Cet (M1.5)
77 $D_\odot$
16.5 $M_\odot$
220 ly
3 Cet (K3)
211 $D_\odot$
9 $M_\odot$
1,670 ly
R_v 15 km/s
R_p 1.95 years
HD 1760 (M5-6)
520 $D_\odot$
5 $M_\odot$
775 ly
Jupiter's Orbit 10.4 AU
o Cet (M7)
1000 $D_\odot$
13 $M_\odot$
419 ly

brown then the likelihood of this name being real beggars belief. All mention of Phycochroma seems to track back to just one unreliable Internet source.

It is often said that looking at stars is like looking back in time. Looking at a star that is 100 ly distant means that we are seeing the star as it was a century ago, not as it is now. **κ¹ Ceti** adds an entirely new dimension. This star may well tell us what our own star looked like 3,750 million years ago. κ¹ Cet is a G5, 0.99 $D_\odot$, 0.78 $L_\odot$ star just 29.9 ly away. It has 10% more metals than the Sun but is otherwise a good solar analog. It is also young. Very young. Somewhere between 650 and 800 million years old compared to our ancient 4,560 million year old Sun. Its magnitude varies by a few percent but with two or three different periods: 9.2 and possibly 9.3 days and 8.9 days. The variability is thought to be due to large individual or large groups of dark starspots moving across the surface as the star spins. It has long been known that the Sun rotates differentially: it spins faster at the equator (one rotation taking about 25 days) than it does at the poles (about 30 days). A similar thing seems to be happening with κ¹ Cet – two or three indicators of differential rotation. Like the Sun κ¹ Cet has a starspot cycle of 5.6 years, about half of the Sun's 11 year cycle. Where κ¹ Cet differs from the Sun is that it is known to undergo massive magnetic field collapses that lead to coronal mass ejections on a gigantic scale producing 'superflares' 100 to 10 million times more energetic than those released by the Sun. Astronomers are fairly certain that the Sun is not currently a flare star but if κ¹ Cet is really a 'young Sun' then it may be that the Sun too was once a superflare star. **κ² Ceti** is also a G-class (G8.5) but totally unrelated at ten times the distance. Its apparent magnitude is m_v +5.69.

μ Ceti is the brightest of a quartet of stars 84.3 ly from Earth. An F0 of 1.7 $D_\odot$ it is a suspected δ Scuti variable. Not an awful lot is known about its three companions – they may not all be part of the same system – except that they appear to be a G2, a G3 and a G6 dwarf. This small cluster is part of the Hyades Moving Group.

o Ceti is much better known as Mira, the star that gives its name to the Mira-type variables. When it is at its brightest it can reach 2nd magnitude, although 3rd is the norm, at which point it is of spectral class M5. But as it fades to 8th, 9th or even 10th magnitude it changes to an M9, the cycle taking 331.96 days. Lying at a distance of 419 ly the diameter of Mira swings between 500 and 1,000 $D_\odot$ as it pulsates, struggling to find some sort of equilibrium but failing as it progresses to its inevitable death when it will puff off its outer layers and leave behind a white dwarf. Hurtling through the interstellar medium at 79 km/s Mira leaves behind a shockwave 15 ly long. It is not alone however. As if to serve as a constant reminder of its fate a white dwarf is on orbit around Mira at a distance of about 70 AU and with an orbital period of around 400 years. Matter ejected from Mira is forming a disk around the tiny companion star.

Literally thousands of Mira-types have been located with periods of between 80 and 1,000 days and amplitudes of 2.5 to 11 magnitudes. They are always red giants and belong to M, S (zirconium) and C (carbon) spectral classes. They sit in what is known as the Asymptotic Giant Branch (AGB) of the Herztsprung-Russell

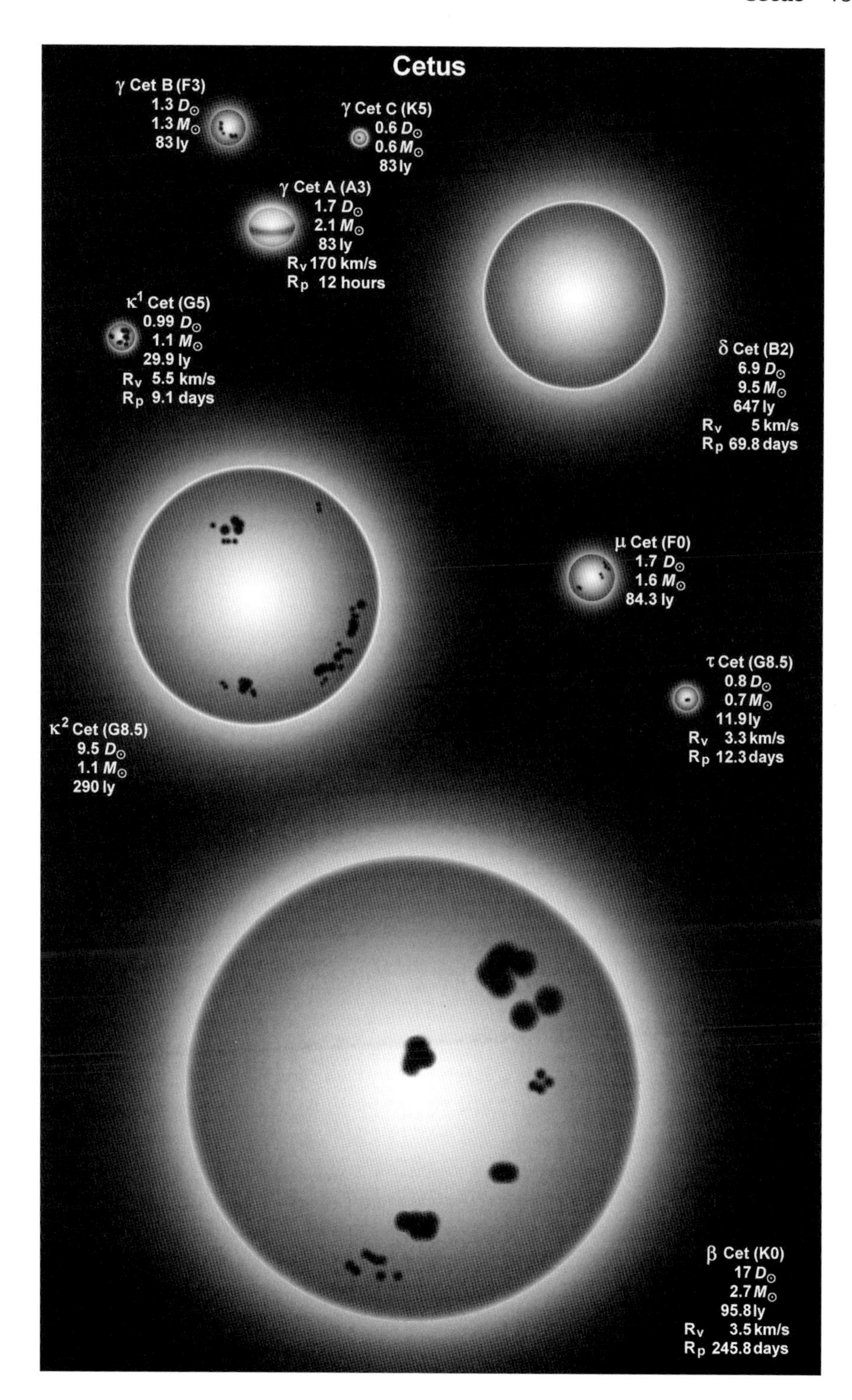
Cetus
γ Cet B (F3)
1.3 $D_\odot$
1.3 $M_\odot$
83 ly
γ Cet C (K5)
0.6 $D_\odot$
0.6 $M_\odot$
83 ly
γ Cet A (A3)
1.7 $D_\odot$
2.1 $M_\odot$
83 ly
R_V 170 km/s
R_P 12 hours
κ¹ Cet (G5)
0.99 $D_\odot$
1.1 $M_\odot$
29.9 ly
R_V 5.5 km/s
R_P 9.1 days
δ Cet (B2)
6.9 $D_\odot$
9.5 $M_\odot$
647 ly
R_V 5 km/s
R_P 69.8 days
μ Cet (F0)
1.7 $D_\odot$
1.6 $M_\odot$
84.3 ly
τ Cet (G8.5)
0.8 $D_\odot$
0.7 $M_\odot$
11.9 ly
R_V 3.3 km/s
R_P 12.3 days
κ² Cet (G8.5)
9.5 $D_\odot$
1.1 $M_\odot$
290 ly
β Cet (K0)
17 $D_\odot$
2.7 $M_\odot$
95.8 ly
R_V 3.5 km/s
R_P 245.8 days

diagram: a phase which all low to intermediate mass stars (0.6 to 10 $M_\odot$) pass through during the latter stages of their life cycle. The internal structure of such stars can be separated into four zones: an inert core of oxygen and carbon, a helium burning shell where helium is fused into carbon, a hydrogen burning shell in which hydrogen is fused into helium and an extended outer envelope that has a composition closely matching more middle aged 'normal' stars. As M-variables change size they distort into asymmetrical shapes.

τ **Ceti** holds a special place in the annals of astronomical history. In 1960 it was one of two stars chosen by Frank Drake to launch what became SETI, the Search for Extra-Terrestrial Intelligence. The other star was ε Eridani. τ Cet was chosen because of its Sun-like properties: it is a G8.5, 80% the size of the Sun but less than half as luminous. Although there are better solar analogs its real attraction is its distance – just 11.9 ly, the closest Sun-type star to us (the next closest are η Cas, δ Pav and 82 Eri all of which are 19 ly away). Despite intensive monitoring no signals were ever discovered and we are fairly certain that no planets orbit τ Cet.

Currently only 8.73 ly from Earth – the 6th closest star system – **Luyten 726-8** was even closer about 28,700 years ago at just 7.2 ly. Willem Luyten (1899-1994) discovered this binary system of red dwarfs in 1948-49. They are better known as **BL Ceti** and **UV Ceti** and are almost identical. BL Ceti is an M5.5 while UV Ceti is an M6. They both have masses of 0.1 $M_\odot$ and diameters of 0.14 $D_\odot$. BL is thought to be more luminous than UV, but only just!: 0.00006 $L_\odot$ vs. 0.00005 $L_\odot$. They orbit one another with a period of 26.5 years during which their distance varies between 2.1 and 8.8 AU. A cool 2,670 K they are normally 12th magnitude but they are also both flare stars. The most dramatic is UV Ceti which typically can brighten fivefold in less than a minute before fading back to 'normal' over 2-3 minutes. However, in 1952 it brightened by a factor of 75 in about 20 seconds. It is sometimes referred to as Luyten's Flare Star. In about 31,500 years from now it will pass within 0.93 ly of ε Eridani, a star which is suspected of having a Kuiper Belt and Oort Cloud. If that is the case then the passage of the BL-UV Ceti pair could disrupt numerous comets, sending them hurtling inwards towards ε Eridani. Any planets in orbit around the star could be in for a pounding.

Cetus contains a couple of supergiant stars. **3 Ceti** is 211 $D_\odot$ (2 AU) and marks the outermost boundary of the constellation at 1,607 ly. A K3 star it is more than 2,000 times as luminous as the Sun. At 10 pc it would rival Venus at M_v -4.4. More than twice the size of 3 Cet is **T Ceti** or **HD 1760**, an M5 red giant which, not surprisingly, is also a semi-regular SRc variable with a period of 158.9 days during which its magnitude changes from m_v +5.0 to an invisible m_v +6.9. It is the brightest star in the constellation with an absolute magnitude of M_v -6.2.

Chamaeleon

Constellation:	Chamaeleon	**Hemisphere:**	Southern
Translation:	The Chameleon	**Area:**	132 deg^2
Genitive:	Chamaeleontis	**% of sky:**	0.320%
Abbreviation:	Cha	**Size ranking:**	79th

An inconspicuous constellation near the South Celestial Pole, the Chamaeleon was the invention of the 16th Century astronomer and cartographer Petrus Plancius working on the observations of the Dutch explorers Pieter Keyser and Frederick de Houtman. Remarkably it has survived more than 400 years whilst others, such as the reindeer and the cat, have long since disappeared. The constellation is thought to be of little interest and even the great Robert Burham Jr devoted just a single page to it in his *Celestial Handbook.*

α Chamaeleontis is the brightest star in this constellation but at m_v +4.06 it is not exactly brilliant. Measurements have put the star as big as 3.0 $D_\odot$ but this has recently been revised to 1.8 $D_\odot$ with a mass of 1.6 $M_\odot$. An F5 it lies at a distance of 63.5 ly and rotates at 29 km/s taking just over 3 days to turn once in its axis.

β Chamaeleontis is a little larger than α Cha, 3 $D_\odot$, but much farther away at 271 ly. A bluish B6 it has a rotational velocity of 260 km/s, a bit on the high side for B4s which average 182 km/s but not particularly unusual. It takes 14 hours to spin once.

At 53 $D_\odot$ **γ Chamaeleontis** is the largest star in the constellation and also one of the most luminous at 296 $L_\odot$. An M0 giant it has a temperature of about 3,800 K. It may be variable and has the catalog number NSV 4913, the NSV standing for New Suspect Variable.

δ^1 and δ^2 Chamaeleontis may appear close together on the celestial sphere but are actually separated by 10 ly. δ^1 Cha is the fainter of the two at m_v +5.50 but larger at 12 $D_\odot$. It is also the closest at 354 ly. δ^2 Cha is brighter by more than a full magnitude and much smaller at 3.9 $D_\odot$. The secret, of course, is in their respective luminosities. δ^1 being a K0 comes in at 56 $L_\odot$ whereas δ^2 is a 172 $L_\odot$ B2.5. At 10 pc δ^2 Cha would far outshine δ^1 brightening to M_v -1.7 compared to just M_v +0.2. Those with good optics and color vision describe the pair as blue and yellowish-orange.

η Chamaeleontis is a bluish-white B8 dwarf that has a temperature of around 12,000 K. It is the brightest member of a 12 strong cluster lying at a distance of 316 ly making it one of the nearest open clusters to us. It was the first open cluster to be discovered through X-ray observation and is the second closest T Tauri group (TW Hya is closest at 163 ly). It is much more compact than the TW Hya association however: just 2.6 ly across compared to 60 ly. The cluster is known as Mamajek 1 after Eric Mamajek whose Australia based team discovered the association in 1999.

The constellation is home to the **Chamaeleon Dark Clouds**, one of the closest star forming regions to us at somewhere between 375 and 701 ly. The clouds contain the 'Cosmic Tornado' (**HH 49/50**) an energetic outflow of material several light years long lying 450 ly from Earth.

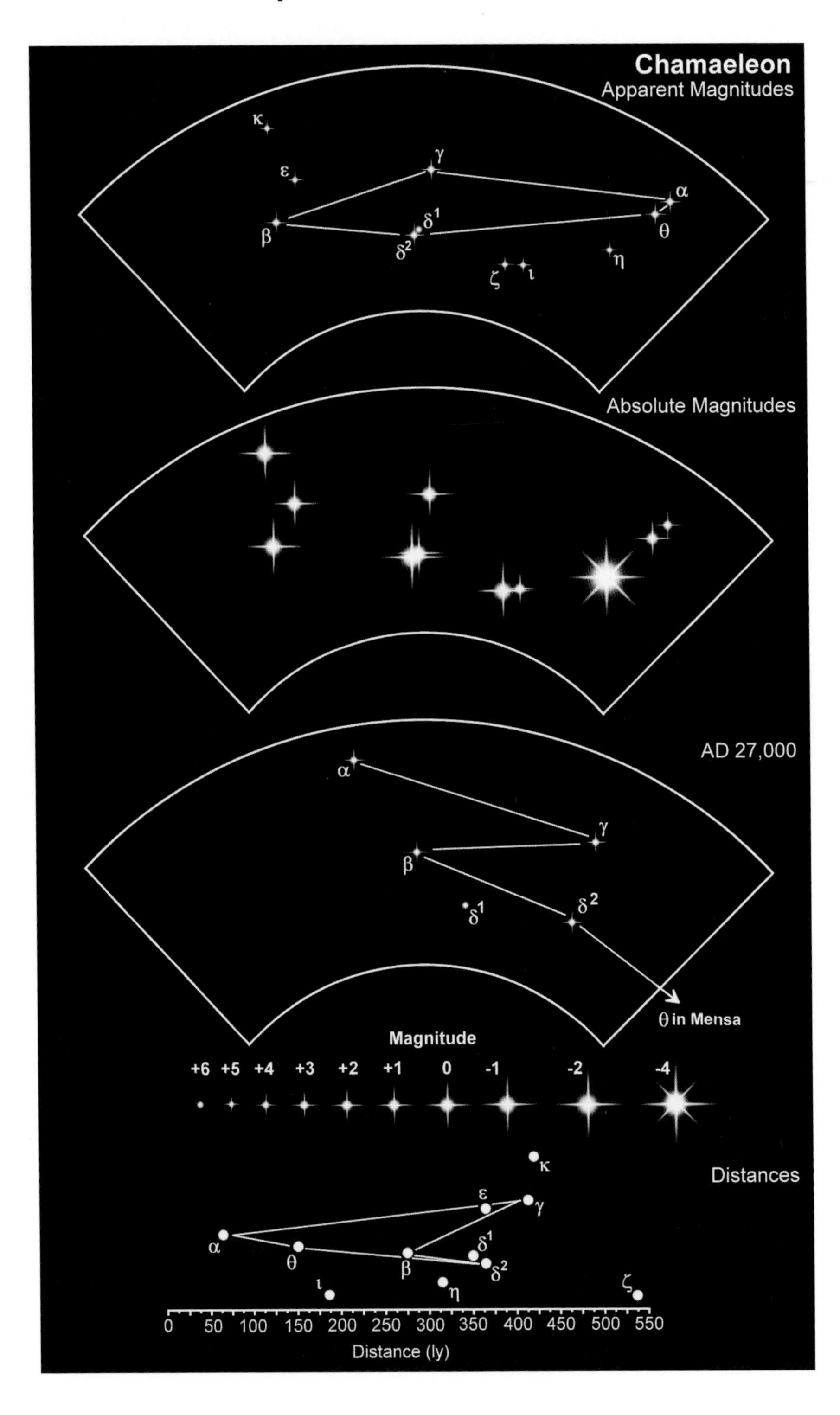
Chamaeleon
Apparent Magnitudes
κ
ε
γ
α
β
δ1
δ2
θ
η
ζ
ι
Absolute Magnitudes
AD 27,000
α
γ
β
δ1
δ2
θ in Mensa
Magnitude
+6 +5 +4 +3 +2 +1 0 -1 -2 -4
Distances
κ
ε
γ
α
θ
β
δ1
δ2
ι
η
ζ
0 50 100 150 200 250 300 350 400 450 500 550
Distance (ly)

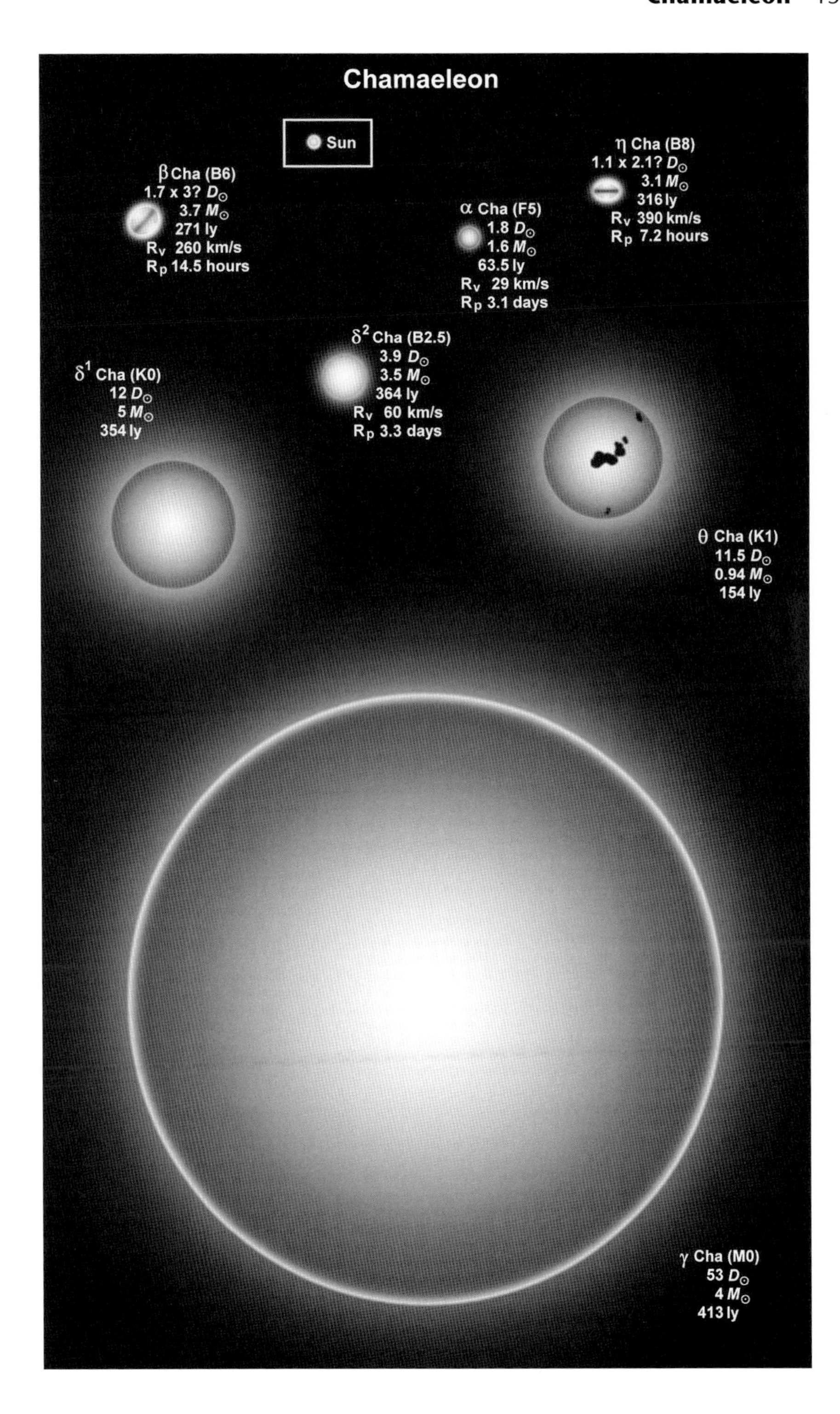
Chamaeleon
Sun
β Cha (B6)
1.7 x 3? D⊙
3.7 M⊙
271 ly
Rv 260 km/s
Rp 14.5 hours
α Cha (F5)
1.8 D⊙
1.6 M⊙
63.5 ly
Rv 29 km/s
Rp 3.1 days
η Cha (B8)
1.1 x 2.1? D⊙
3.1 M⊙
316 ly
Rv 390 km/s
Rp 7.2 hours
δ1 Cha (K0)
12 D⊙
5 M⊙
354 ly
δ2 Cha (B2.5)
3.9 D⊙
3.5 M⊙
364 ly
Rv 60 km/s
Rp 3.3 days
θ Cha (K1)
11.5 D⊙
0.94 M⊙
154 ly
γ Cha (M0)
53 D⊙
4 M⊙
413 ly

Circinus

Constellation:	Circinus	**Hemisphere:**	Southern
Translation:	The Pair of Compasses	**Area:**	93 deg^2
Genitive:	Circini	**% of sky:**	0.225%
Abbreviation:	Cir	**Size ranking:**	85th

The 18th Century astronomer Abbé La Caille seemed determined to fill the Southern skies with numerous small constellations of only a few faint stars each, including this one. The constellation lies right next to α Centauri.

α Circini belongs to the unusual spectral class A7pSrCrEu: an A-type star that is peculiar in that it has higher than normal levels of strontium, europium and chromium. The magnetic structure of the star tends to localize certain elements, including these ones. At 1.8 $D_{\odot}$ α Cir is only 53.5 ly from Earth and is an α CV rotating variable which fluctuates between m_v +3.18 and +3.21.

At 1.8 $D_{\odot}$ **β Circini** matches α Cir in size. An A3 it spins at 60 km/s – just half the average rotational velocity for this class of star – taking just 36.4 hours to complete one revolution.

The brilliant white **δ Circini** is nearly 6,400 ly from Earth and so appears as a faint 5th magnitude star in our skies. At 10 pc however it would outshine Venus at M_v -4.8. Belonging to spectral class O8.5 is has a luminosity of about 29,000 Suns. It is accompanied by a 10 solar mass partner in a 3.9 day orbit. The proximity of the two stars causes them to distort one another into ellipsoids. O8s are rare. They represent only about 6% of all the stars whose spectra have been measured and only five of these are visible without optical aid: that's just 0.17% of all naked eye stars.

θ Circini, a 5th magnitude B4, is a γ Cas variable which ranges from m_v +5.02 to +5.44 with no discernable period. Some 834 ly away it has a diameter in the region of 6.5 $D_{\odot}$ and rotates at 100 km/s taking 3.3 days to turn once on its axis.

Hipparcos measurements for **HD 135591**, a 5th magnitude O7, indicates a distance of about 163,000 ly. That would make it the most luminous star visible to the naked eye at an unbelievable 13.6 million $L_{\odot}$. At 10 pc it would have an absolute magnitude of M_v -13 – as bright as the full Moon. However, there is considerable uncertainty about the measurement and the star could be as close as 4,500 ly in which case it would only make M_v -5. HD 135591 is 102 solar diameters across and weighs in at 26 $M_{\odot}$. Like O8s, giant O7s are also very rare – only three are visible without optical aid – their life expectancy running to just a few million years.

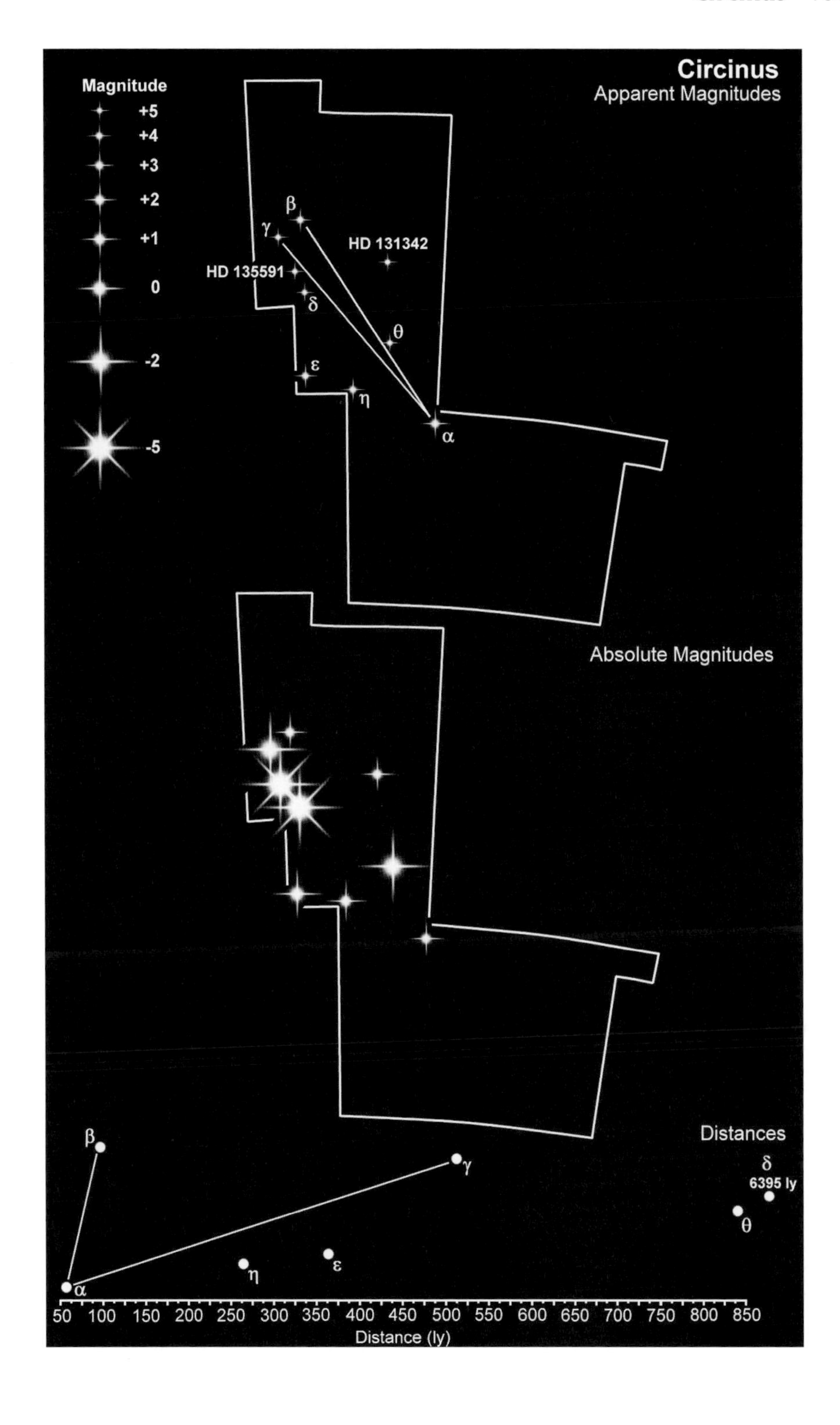
Circinus
Apparent Magnitudes
Magnitude
+5
+4
+3
+2
+1
0
-2
-5
β
γ
HD 131342
HD 135591
δ
θ
ε
η
α
Absolute Magnitudes
Distances
δ
6395 ly
θ
50 100 150 200 250 300 350 400 450 500 550 600 650 700 750 800 850
Distance (ly)

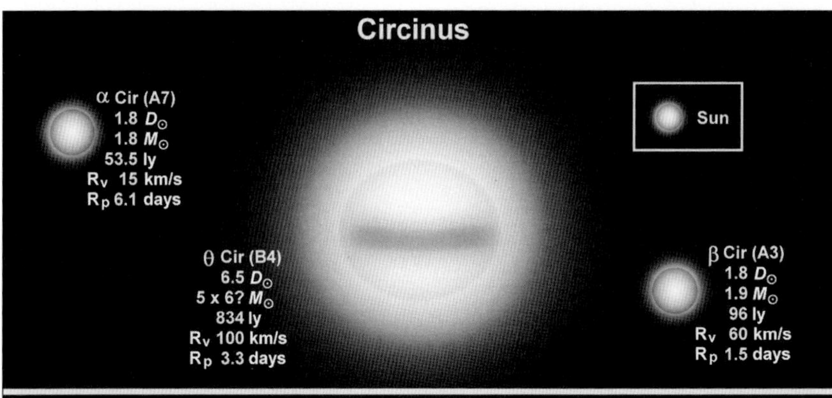
Circinus
α Cir (A7)
1.8 D⊙
1.8 M⊙
53.5 ly
Rv 15 km/s
Rp 6.1 days
Sun
θ Cir (B4)
6.5 D⊙
5 x 6? M⊙
834 ly
Rv 100 km/s
Rp 3.3 days
β Cir (A3)
1.8 D⊙
1.9 M⊙
96 ly
Rv 60 km/s
Rp 1.5 days

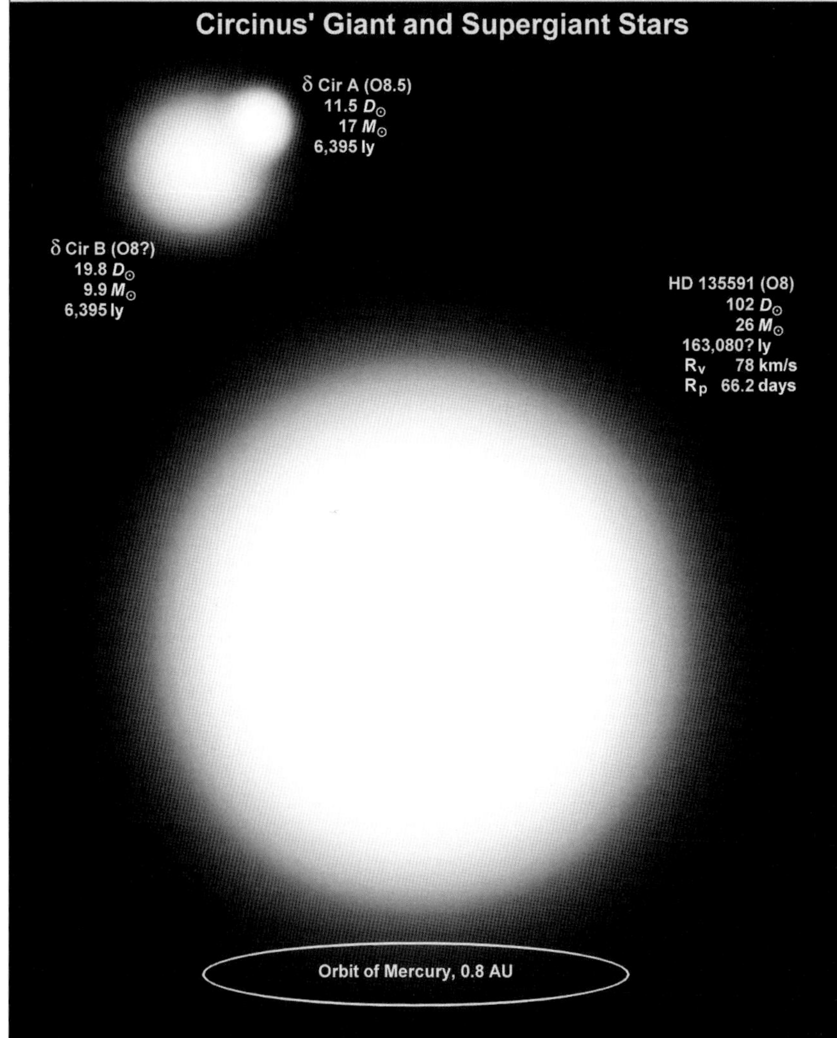
Circinus' Giant and Supergiant Stars
δ Cir A (O8.5)
11.5 D⊙
17 M⊙
6,395 ly
δ Cir B (O8?)
19.8 D⊙
9.9 M⊙
6,395 ly
HD 135591 (O8)
102 D⊙
26 M⊙
163,080? ly
Rv 78 km/s
Rp 66.2 days
Orbit of Mercury, 0.8 AU

Columba

Constellation:	Columba	**Hemisphere:**	Southern
Translation:	The Dove	**Area:**	270 deg^2
Genitive:	Columbae	**% of sky:**	0.654%
Abbreviation:	Col	**Size ranking:**	54th

Another inconspicuous constellation Columba was introduced by the Dutch explorer Petrus Plancius in the 16th Century. It is assumed to be the Biblical dove which Noah released following the Deluge and which returned with a twig in its mouth indicating the water was receding.

α **Columbae** or Phakt is a hot, fast spinning B7 star 268 ly from Earth. Its rotational velocity of 195 km/s pulls in the poles and causes the equator to bulge, ejecting a continuous stream of matter like a cosmic Catherine wheel. It is classed as a Be 'emission' star: its spectrum has prominent hydrogen emission lines. About 6.5 times larger than the Sun it takes α Col just 1.7 days to spin once on its axis.

Columba contains two stars that appear to be visitors from more distant parts of the Galaxy. The first is β **Columbae** or Wazn, an 11 $D_\odot$ K1 giant which is more of a warm yellow than orange. It is hurtling away from us at about 89 km/s – the average is 17 km/s – and has a space velocity of 112 km/s. It may well have been ejected from its original birthplace due to some cataclysmic event although it does not appear to be an interloper from the galactic halo. Halo stars tend to have a low metal content whereas β Col is significantly more enriched in metals than our own Sun indicating an origin in the Galaxy's spiral arms. It is also the closest of the Columba stars at 85.6 ly.

μ **Columbae** is also a speeding star, heading off into space at 109 km/s making it the fourth fastest star receding from us and the sixth fastest over all, at least as far as naked eye stars are concerned. Its space velocity is an estimated 150 km/s. Its origin is more certain. It appears to have been ejected from the Trapezium Cluster in the Orion Nebula, along with AE Aurigae and 53 Arietis, about 2.7 million years ago. Some 4.5 $D_\odot$ it is another B-class star (B1) with a rotational velocity of 140 km/s and is 1,294 ly away. At 10 pc it would be the M_v -4 luminary.

The 1.5 $D_\odot$ B5 λ **Columbae** is an ellipsoidal rotating variable that changes in magnitude from m_v +4.85 to +4.92 and back in $15^h\ 21.5^m$. Nothing is known of its unseen companion.

At 1,463 ly σ **Columbae** is the most distant star in the constellation. At 1.7 $D_\odot$ and 1,010 $L_\odot$ it is also by far the most luminous F2 dwarf visible without optical aid.

Globular cluster in Columba

Name	Size arc min	Size ly	Distance ly	Age million yrs	Apparent magnitude m_v
NGC 1851	11′	126	39,500	11	+7.1

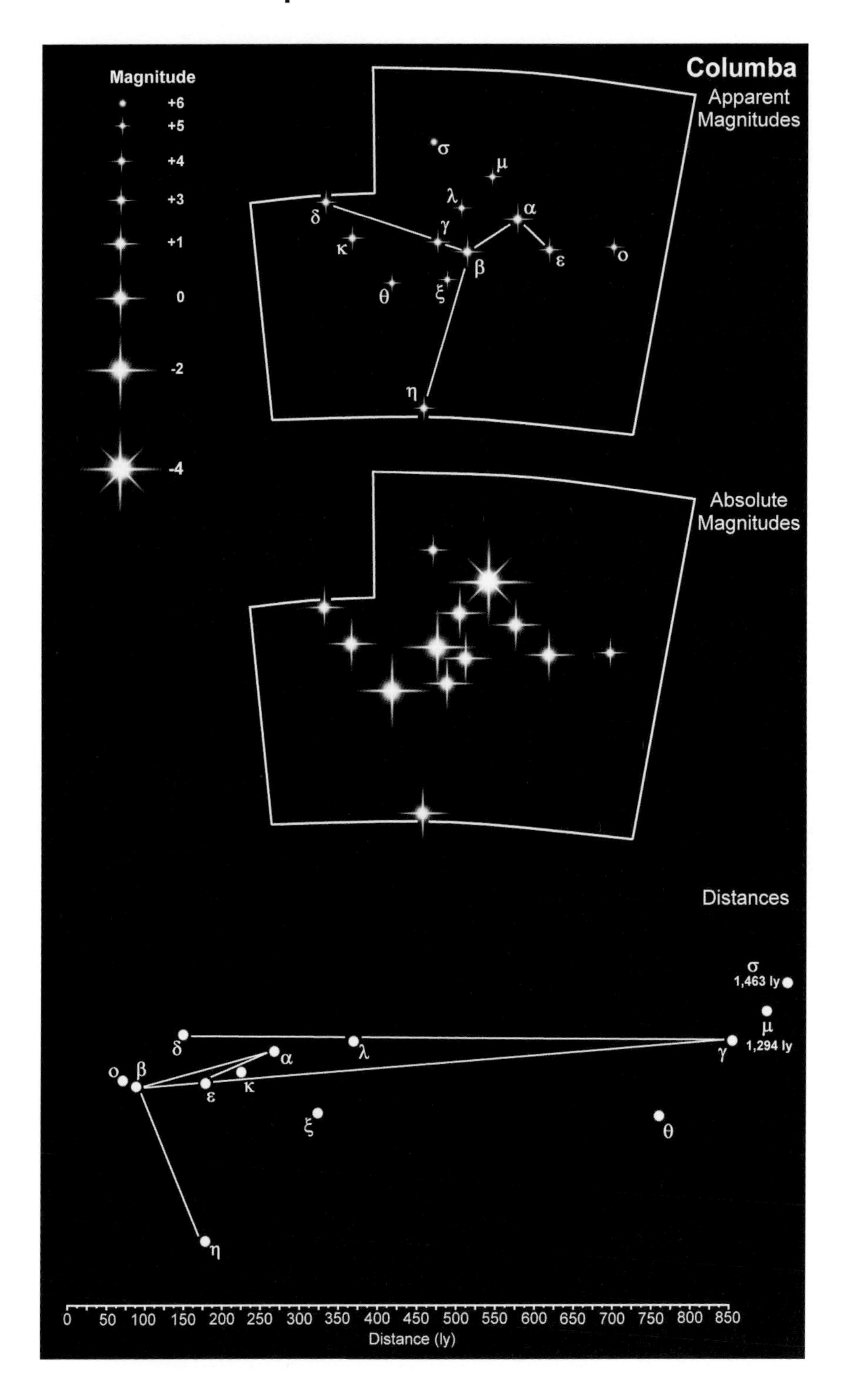
Columba
Apparent Magnitudes
Magnitude
+6
+5
+4
+3
+1
0
-2
-4
σ
μ
λ
α
δ
γ
κ
β
ε
ο
θ
ξ
η
Absolute Magnitudes
Distances
σ
1,463 ly
μ
1,294 ly
δ
λ
γ
α
ο
β
ε
κ
ξ
θ
η
0 50 100 150 200 250 300 350 400 450 500 550 600 650 700 750 800 850
Distance (ly)

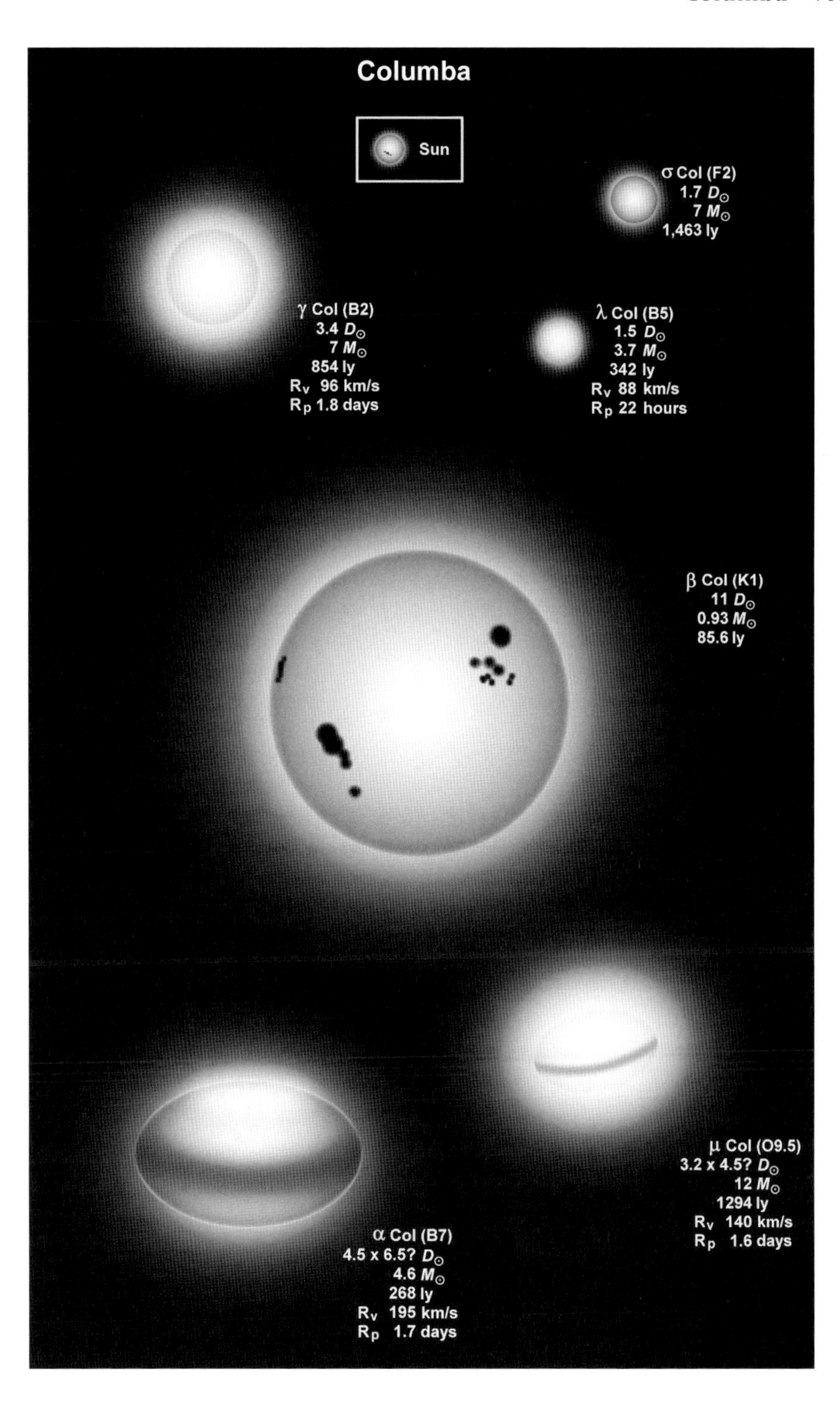
Columba
Sun
σ Col (F2)
1.7 $D_{\odot}$
7 $M_{\odot}$
1,463 ly
γ Col (B2)
3.4 $D_{\odot}$
7 $M_{\odot}$
854 ly
R_V 96 km/s
R_P 1.8 days
λ Col (B5)
1.5 $D_{\odot}$
3.7 $M_{\odot}$
342 ly
R_V 88 km/s
R_P 22 hours
β Col (K1)
11 $D_{\odot}$
0.93 $M_{\odot}$
85.6 ly
μ Col (O9.5)
3.2 x 4.5? $D_{\odot}$
12 $M_{\odot}$
1294 ly
R_V 140 km/s
R_P 1.6 days
α Col (B7)
4.5 x 6.5? $D_{\odot}$
4.6 $M_{\odot}$
268 ly
R_V 195 km/s
R_P 1.7 days

Coma Berenices

Constellation:	Coma Berenices	**Hemisphere:**	Northern
Translation:	The Hair of Berenice	**Area:**	386 deg^2
Genitive:	Comae Berenicis	**% of sky:**	0.936%
Abbreviation:	Com	**Size ranking:**	42nd

This star-poor but galaxy-rich region of the northern sky – it includes part of the Virgo cluster of galaxies – represents the hair of Queen Berenice II of Egypt which she cut off as a sacrifice to the gods for the safe return of her husband King Ptolemy III. The constellation is home to the North Galactic Pole situated at RA $12^h\ 51^m\ 26.282^s$, Dec +27° 07′ 42.01″.

α **Comae Berenicis** at m_v +4.38 is actually the second brightest star in the constellation. Possibly slightly variable but probably not, it is a binary system of what seem to be identical F5 dwarfs, each slightly larger than our own Sun. Seen edge on the stars appear to swap positions in a straight line. Although they never separate by more than 1″ in real space they can orbit as far apart as 19 AU (the same distance Uranus is from the Sun) and come as close as 6 AU (slightly farther than Jupiter), a complete orbit taking 25.87 years. The system is just 46.7 ly away. α Com's proper name is Diadem, a jewel worn in the hair.

Closer by 17 ly and brighter by 0.13 magnitudes, **β Comae Berenicis** is a solar analog of 0.95 $D_\odot$ and 1.36 $L_\odot$. Belonging to spectral group G0 the star appears as m_v +4.25 and lies at 29.9 ly. With a rotational velocity of 4.4 km/s – twice as fast as the Sun – β Com completes a full rotation in only 11 days. It shows signs of magnetic activity and appears to have a starspot cycle of 16.6 years, not unlike the solar sunspot cycle of 11 years.

At m_v +8.19 **FK Comae Berenicis** is well beyond even the most keen sighted people. It is an important star, however, in that it lends its name to a particular type of variable. FK Com is the prototype of a small group of rapidly rotating stars that fluctuate in brightness because of surface features, most likely star spots. They all belong to G or K spectral classes, they are all rapid rotators – FK Com rotates at 160 km/s and is a G5 – and they are all giants. FK Com itself varies by 0.1 magnitude with a period of 2.412 days. There is only one naked eye FK Com variable; 15 Monocerotis.

The bluish-white A3 star **13 Comae Berenicis** is an α CV rotating variable which barely fluctuates between m_v +5.15 and +5.18. About 3.5 $D_\odot$ it has a rotational velocity of 55 km/s taking just over 3 days to complete a single turn.

24 Comae Berenicis is a nice binary for a binocular. The main component is a m_v +4.99 magnitude K2, the companion a white A9 of m_v +6.56 at PA 271°. It is an optical illusion. In reality they are separated by 2,000 ly with the K2 component being closest to us at 614 ly.

Jim Kaler, Professor Emeritus of Astronomy at the University of Illinois, has unofficially named **31 Comae Berenicis** Polaris Galacticus Borealis or PolGarBol for short due to the fact that it very nearly marks the position of

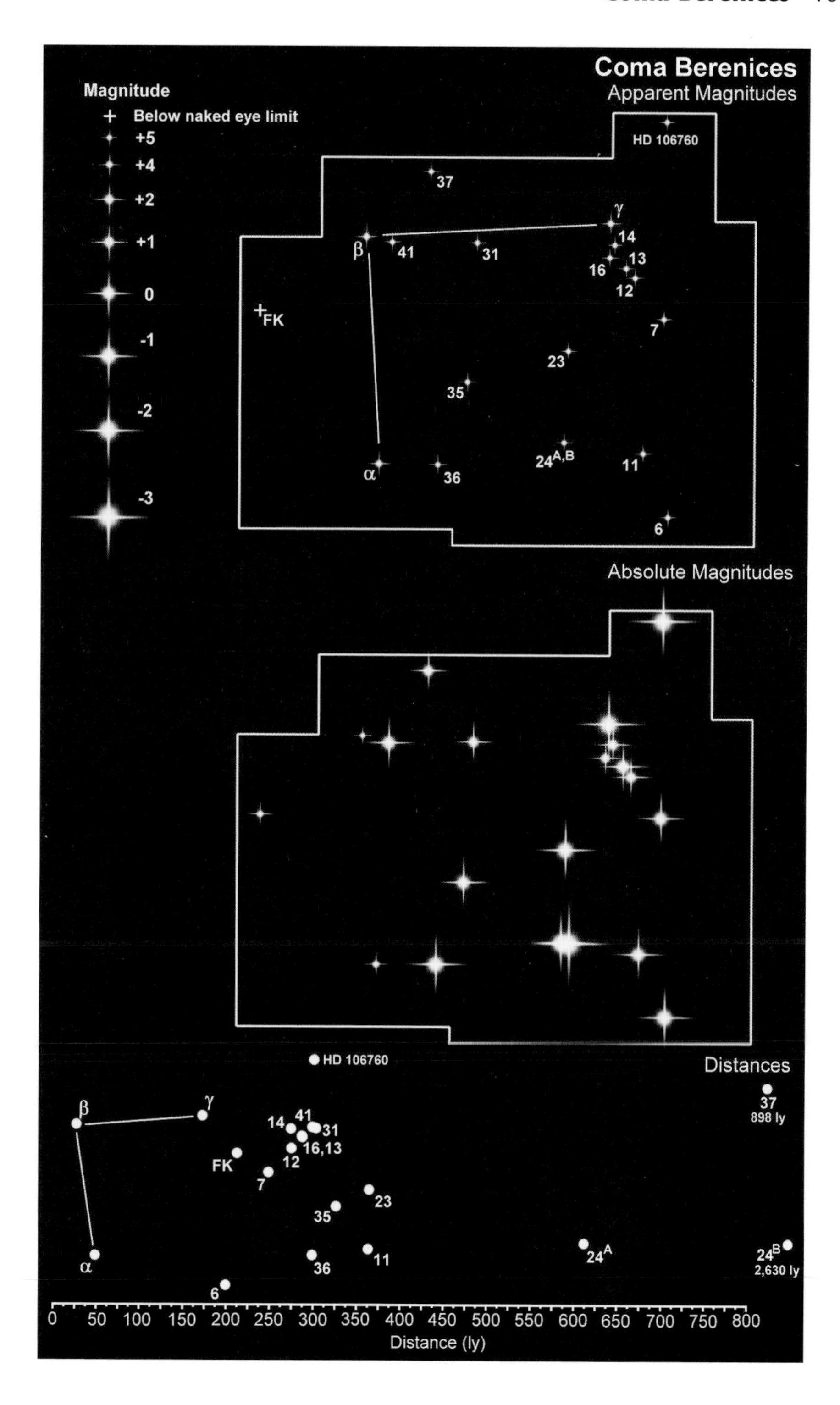
Coma Berenices
Apparent Magnitudes
Magnitude
Below naked eye limit
+5
+4
+2
+1
0
-1
-2
-3
HD 106760
37
γ
β
41
31
14
13
16
12
FK
7
23
35
α
36
24A,B
11
6
Absolute Magnitudes
Distances
HD 106760
37
898 ly
β
γ
14
41
31
16,13
FK
12
7
23
35
α
36
11
24A
24B
2,630 ly
6
0 50 100 150 200 250 300 350 400 450 500 550 600 650 700 750 800
Distance (ly)

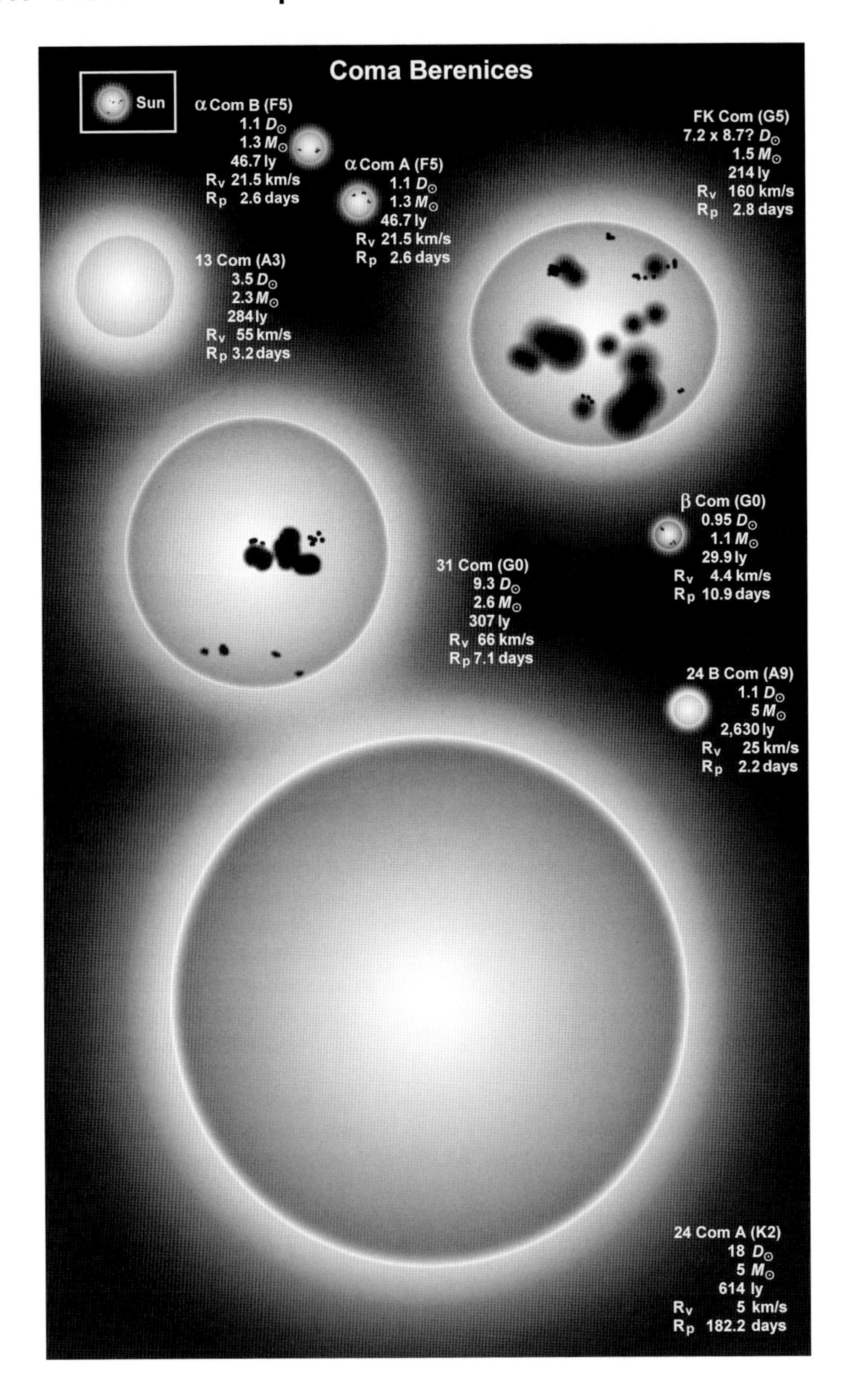
Coma Berenices
Sun
α Com B (F5)
1.1 D☉
1.3 M☉
46.7 ly
Rv 21.5 km/s
Rp 2.6 days
α Com A (F5)
1.1 D☉
1.3 M☉
46.7 ly
Rv 21.5 km/s
Rp 2.6 days
FK Com (G5)
7.2 x 8.7? D☉
1.5 M☉
214 ly
Rv 160 km/s
Rp 2.8 days
13 Com (A3)
3.5 D☉
2.3 M☉
284 ly
Rv 55 km/s
Rp 3.2 days
β Com (G0)
0.95 D☉
1.1 M☉
29.9 ly
Rv 4.4 km/s
Rp 10.9 days
31 Com (G0)
9.3 D☉
2.6 M☉
307 ly
Rv 66 km/s
Rp 7.1 days
24 B Com (A9)
1.1 D☉
5 M☉
2,630 ly
Rv 25 km/s
Rp 2.2 days
24 Com A (K2)
18 D☉
5 M☉
614 ly
Rv 5 km/s
Rp 182.2 days

the North Galactic Pole. A G0 class star, 9.3 $D_{\odot}$ and 307 ly distant it radiates with the luminosity of 77 Suns. Its rotational velocity is somewhat on the high side at 66 km/s with the result that it generates a large amount of X-rays.

The open star cluster **Melotte 111** is the third closest cluster to us at just 284 ly. Named after P.J.Melotte's catalog reference published in 1915 it was not actually recognized as a true cluster until 1938. Containing about 40 stars it has an estimated age of 400 million years. Its stars range from m_v +5 to about +10 with the A-class stars in greater abundance at greater distances and F-class stars dominating the part of the cluster closest to us, although this may be due to observational bias. The cluster is moving parallel to the Sun and so membership is determined by a combination of distance (224 to 349 ly) and radial velocities that are close to zero.

Open and globular clusters in Coma Berenices

Name	Size arc min	Size ly	Distance ly	Age million yrs	Brightest star in region*	No. stars m_v >+12*	Apparent magnitude m_v
Melotte 111	417′	38	313	450	12 Com m_v +7.80	281	+1.8
M53	13′	206	54,500	12,500	Globular cluster		+7.6
NGC 4147	4′	239	205,000	17,000	Globular cluster		+10.3
NGC 5053	10.5′	163	53,500	12,500	Globular cluster		+9.5

*May not be a cluster member.

Corona Australis

Constellation:	Corona Australis	**Hemisphere:**	Southern
Translation:	The Southern Crown	**Area:**	128 deg^2
Genitive:	Coronae Australis	**% of sky:**	0.310%
Abbreviation:	CrA	**Size ranking:**	80th

This crown of 10 stars lies at the forefeet of Sagittarius and behind the sting of Scorpius. One of Ptolemy's 48 constellations it is sometimes called Corona Austrina.

α **Coronae Australis** or Meridiana is a fast spinning A0 star 130 ly from Earth. With a rotational velocity of at least 201 km/s and a diameter of 2.2 $D_\odot$ it completes a full rotation in just 13.3 hours, compared to the Sun's 25 days. It appears to have a cool dusty disc that produces a prominent infrared signature.

The K0 giant β **Coronae Australis** is uncommonly luminous and large for its class. It measures 43 $D_\odot$ – only a couple of other K0 stars are larger – and its luminosity is a brilliant 730 $L_\odot$. Typically it should be 90 $L_\odot$. This makes it one of the most luminous naked eye K0 stars. By comparison the smaller 11 $D_\odot$ δ **Coronae Australis**, a K1, has a luminosity of 34.2 $L_\odot$.

γ **Coronae Australis** is a binary system of two identical F7 dwarfs, 1.3 $D_\odot$ and lying at a distance of 58.4 ly. Each about 4.7 times as luminous as the Sun, γ^A CrA is a m_v +4.91 while γ^B CrA comes in a tenth of a magnitude dimmer at m_v +5.01. Together they have a combined magnitude of +4.36. The two components have an orbital period of 120 years.

The F4 dwarf ε **Coronae Australis** is variable between m_v +4.74 and +5.00 with successive maxima (or minima) occurring as regular as clockwork at 14^h 11^m 39^s. Almost 100 ly from Earth ε CrA is a close binary system, so close in fact, that the two stars are almost touching. Known as an EW or W UMa eclipsing variable (after the first such system to be recorded, W Ursae Majoris) the two stars significantly distort one another as they dance their rapid orbit. Such systems are not unusual – about 1% of all stars evolve into this arrangement – but among the naked eye stars only a couple of other example exists. EWs all have periods of less than a day and belong to spectral classes F-G or later. What eventually happens to EW systems is not well understood. They may merge into one star or perhaps become unstable and tear one another apart.

Good skies and a modest telescope – at least a 75 mm (3″) but preferably a 100 mm (4″) – will help you locate the 12th magnitude star **R Coronae Australis** mid-way between and slightly to the north of a line from γ to ε CrA. A young B5 star at about 420 ly R CrA marks the position of the Corona Australis molecular cloud, the birthplace of new stars and one of the closest molecular clouds to us. At the heart of the cloud is a loose cluster of about 30 young stars showing a range of masses and at various stages of evolution but all essentially protostars.

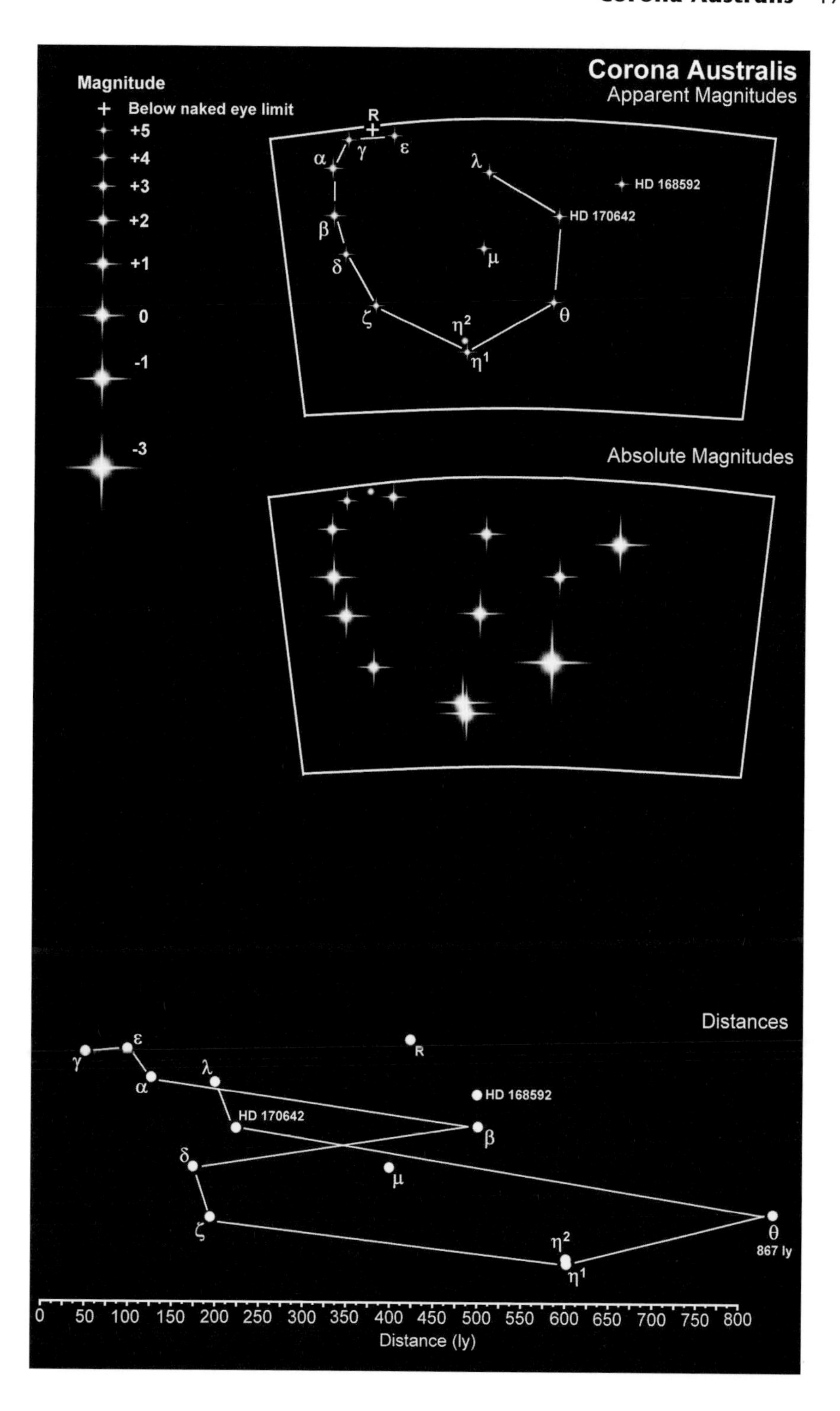

Corona Australis
Apparent Magnitudes
Magnitude
Below naked eye limit
+5
+4
+3
+2
+1
0
-1
-3
R
α
γ
ε
λ
HD 168592
HD 170642
β
μ
δ
ζ
θ
η²
η¹
Absolute Magnitudes
Distances
867 ly
0 50 100 150 200 250 300 350 400 450 500 550 600 650 700 750 800
Distance (ly)

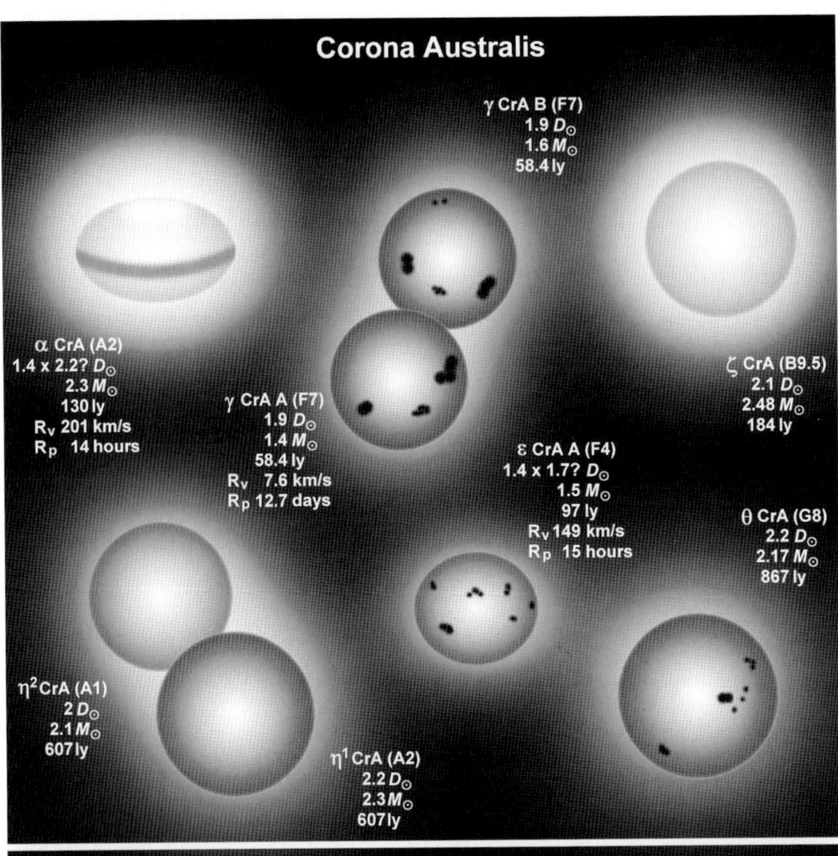
Corona Australis
γ CrA B (F7)
1.9 $D_\odot$
1.6 $M_\odot$
58.4 ly
α CrA (A2)
1.4 x 2.2? $D_\odot$
2.3 $M_\odot$
130 ly
R_V 201 km/s
R_P 14 hours
γ CrA A (F7)
1.9 $D_\odot$
1.4 $M_\odot$
58.4 ly
R_V 7.6 km/s
R_P 12.7 days
ζ CrA (B9.5)
2.1 $D_\odot$
2.48 $M_\odot$
184 ly
ε CrA A (F4)
1.4 x 1.7? $D_\odot$
1.5 $M_\odot$
97 ly
R_V 149 km/s
R_P 15 hours
θ CrA (G8)
2.2 $D_\odot$
2.17 $M_\odot$
867 ly
$η^2$ CrA (A1)
2 $D_\odot$
2.1 $M_\odot$
607 ly
$η^1$ CrA (A2)
2.2 $D_\odot$
2.3 $M_\odot$
607 ly

Corona Australis' Giant Stars
β CrA (K0)
43 $D_\odot$
4.5 $M_\odot$
508 ly
δ CrA (K1)
11 $D_\odot$
1.6 $M_\odot$
175 ly
Mean distance of Mercury
from the Sun 0.4 AU

Corona Borealis

Constellation:	Corona Borealis	**Hemisphere:**	Northern
Translation:	The Northern Crown	**Area:**	179 deg^2
Genitive:	Coronae Borealis	**% of sky:**	0.434%
Abbreviation:	CrB	**Size ranking:**	73rd

The Northern Crown is associated with several legends the most well known of which is the story of Dionysus (or Bacchus) who, wishing to prove that he was a god to Princess Ariadne of Crete, cast her crown into the heavens where the jewels turned to stars.

α Coronae Borealis is also sometimes referred to as Gemma, the 'Jewel in the Crown' but it is more usually called Alphekka. It is an Algol-type or EA variable with a period of $17^d\ 8^h\ 38^m$ during which its magnitude changes by a tenth from m_v +2.21 to +2.32 and back again. The cause of this variability is a regular eclipse by a smaller, cooler companion. The primary star is a 2.7 $D_\odot$, A0 dwarf with a temperature of 8,500 K and a luminosity of around 56 $L_\odot$. The secondary is an early G-class Sun-like star, about 0.9 $D_\odot$, in an orbit that brings them to within 0.13 AU (19.5 million km) and then out to 0.27 AU (40.4 million km). α^A CrB is surrounded by a dusty disk while α^B CrB is a bright X-ray source suggesting solar type magnetic activity. α CrB is the only star in the constellation heading away from us and is believed to be a member of the Sirius Supercluster.

The 2.5 $D_\odot$ **β Coronae Borealis** or Nusakan is a chemically peculiar star with not quite the right mix of chemical elements and about 2,000 K hotter than it should be. Such odd dwarfs are thought to have pools of elements concentrated in large and strongly magnetic starspots. This reveals itself in its variability. Classed as an α CV rotating variable its magnitude changes between m_v +3.65 and +3.72 with a period of $18^d\ 11^h\ 41^m$ between successive maxima or minima. Regarded as an F0p – 'p' for 'peculiar' – β CrB lies at a distance of 114 light years.

γ Coronae Borealis is a δ Scuti pulsating variable and a binary. Belonging to spectral class B9 this 3.4 $D_\odot$ star has a period of $43^m\ 12^s$ during which it fluctuates between m_v +3.80 and +3.86 and back again. Its 6th magnitude partner is an A3 dwarf about 1.3 $D_\odot$ with an orbital period of 92.94 years during which their separation changes from 17 AU (a little less than Uranus' distance from the Sun) to 49 AU (the same as Pluto at its furthest point).

The variability of **δ Coronae Borealis** was only recognized in 1987. Astronomers were monitoring the variable R Coronae Borealis and were using δ CrB as the control against which R CrB's magnitude could be compared. A series of strange results let to the discovery that δ CrB itself varied between m_v +4.57 and +4.69. The variability is linked to large starspot activity which appears to be cyclical: the star did not appear variable throughout the whole of 1989.

Marking one end of the Crown **θ Coronae Borealis** could be one of the fastest spinning stars visible in the night sky. At the lower end of the scale its rotational velocity has been estimated at 271 km/s but various other observations

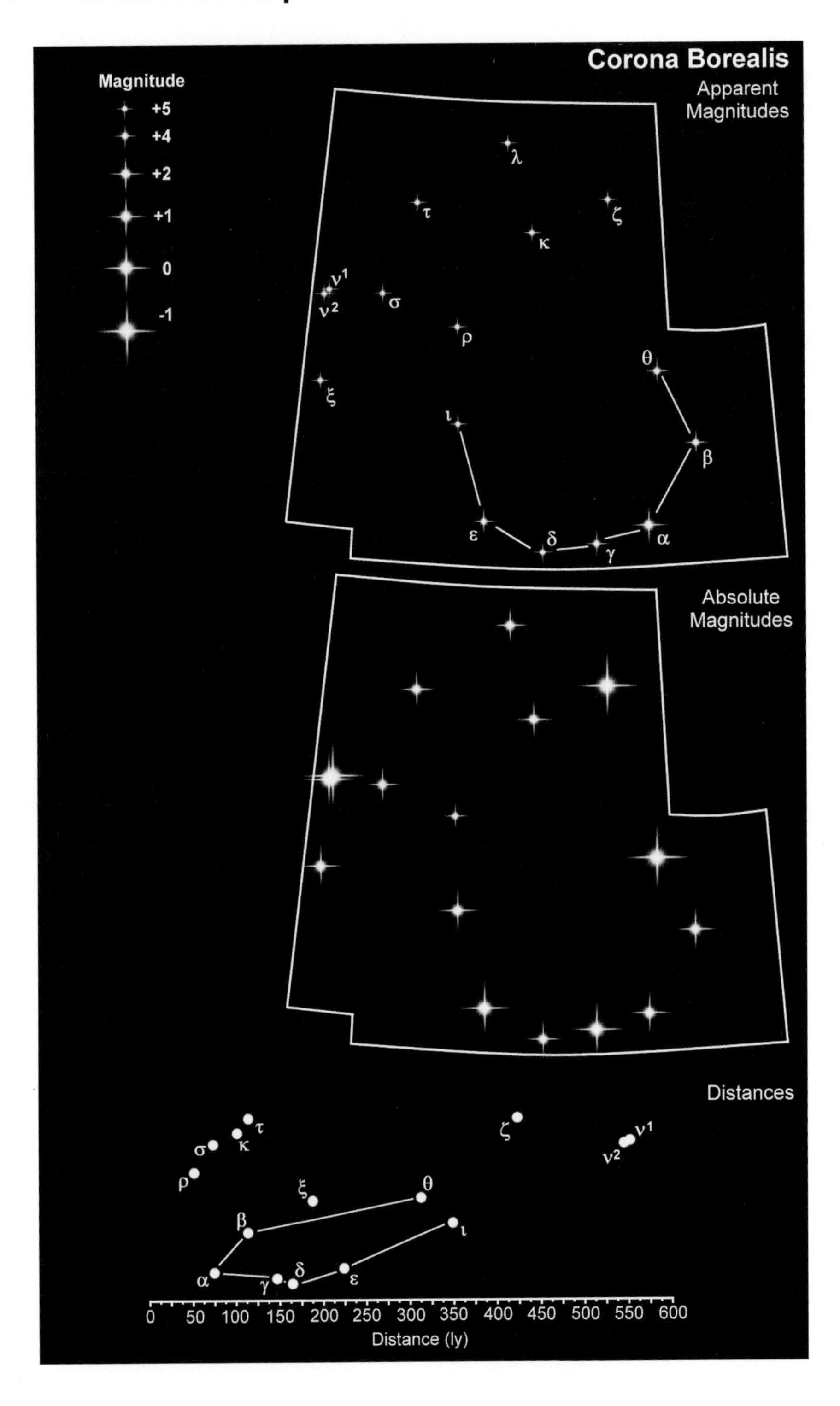
Corona Borealis
Apparent Magnitudes
Magnitude
+5
+4
+2
+1
0
-1
λ
τ
ζ
κ
ν1
ν2
σ
ρ
θ
ξ
ι
β
ε
δ
γ
α
Absolute Magnitudes
Distances
τ
ζ
σ
κ
ν1
ν2
ρ
ξ
θ
β
ι
δ
α
γ
ε
0 50 100 150 200 250 300 350 400 450 500 550 600
Distance (ly)

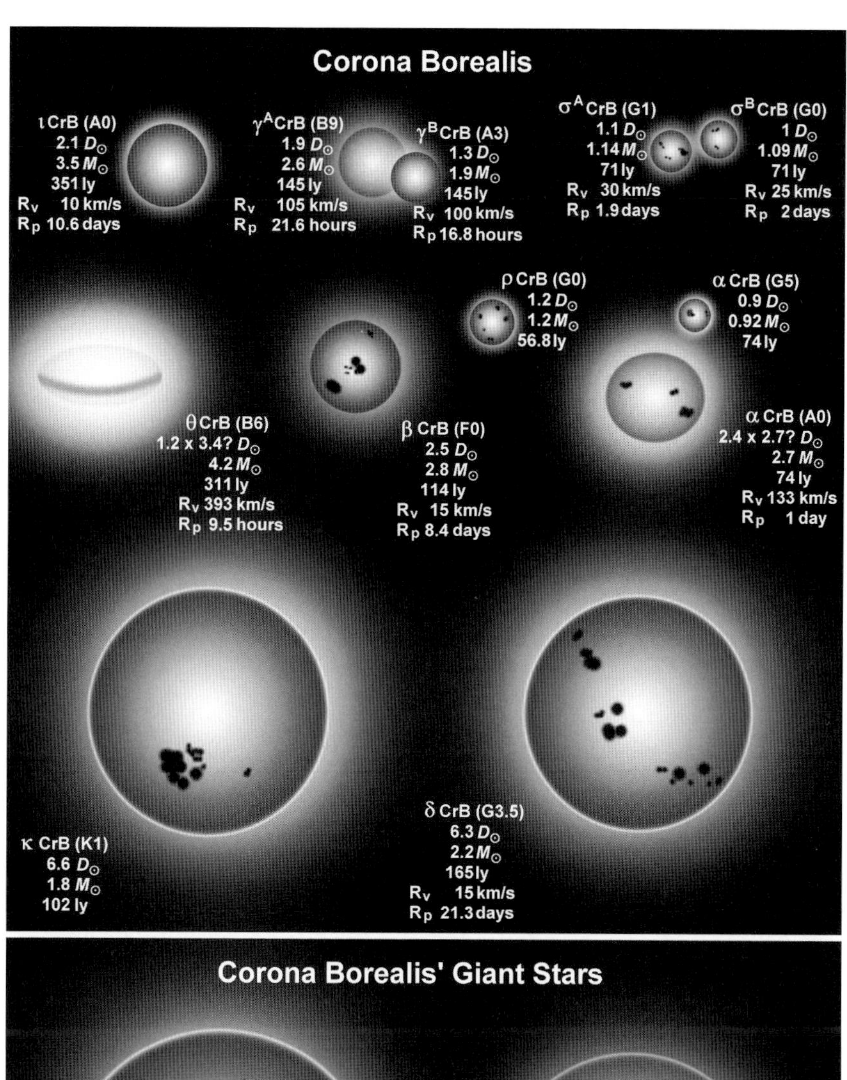
Corona Borealis
ι CrB (A0)
2.1 D☉
3.5 M☉
351 ly
Rv 10 km/s
Rp 10.6 days
γA CrB (B9)
1.9 D☉
2.6 M☉
145 ly
Rv 105 km/s
Rp 21.6 hours
γB CrB (A3)
1.3 D☉
1.9 M☉
145 ly
Rv 100 km/s
Rp 16.8 hours
σA CrB (G1)
1.1 D☉
1.14 M☉
71 ly
Rv 30 km/s
Rp 1.9 days
σB CrB (G0)
1 D☉
1.09 M☉
71 ly
Rv 25 km/s
Rp 2 days
ρ CrB (G0)
1.2 D☉
1.2 M☉
56.8 ly
α CrB (G5)
0.9 D☉
0.92 M☉
74 ly
θ CrB (B6)
1.2 x 3.4? D☉
4.2 M☉
311 ly
Rv 393 km/s
Rp 9.5 hours
β CrB (F0)
2.5 D☉
2.8 M☉
114 ly
Rv 15 km/s
Rp 8.4 days
α CrB (A0)
2.4 x 2.7? D☉
2.7 M☉
74 ly
Rv 133 km/s
Rp 1 day
κ CrB (K1)
6.6 D☉
1.8 M☉
102 ly
δ CrB (G3.5)
6.3 D☉
2.2 M☉
165 ly
Rv 15 km/s
Rp 21.3 days

Corona Borealis' Giant Stars
ν1 CrB (M2)
34 D☉
4.5 M☉
556 ly
Mean distance of Mercury from the Sun 0.4 AU
ν2 CrB (K5)
29 D☉
4 M☉
545 ly

have resulted in 320, 349, 385 and 393 km/s. To add to the confusion is the uncertainty over its size: anywhere from 1.7 $D_{\odot}$ to 3.4 $D_{\odot}$. We do know, or at least we think we know its distance – 311 ly give or take 18 ly. And we also know it is a B6. But it is a strange star. Its rapid rotation hurls vast amounts of matter from its equatorial region, creating a band of debris that radiates energy – a typical Be 'emission' star – but unlike some Be stars that flare up, θ CrB dimmed by 50% in the early 1970s and went through a series of irregular magnitude fluctuations. It also has a companion (not responsible for the dip in magnitude) which is likely to be an A2 dwarf in a 300 year orbit with a separation of at least 85 AU. The star is now reasonably settled but no one knows what it will do next, or when.

At the other end of the Crown is ι **Coronae Borealis,** a chemically peculiar A0 star about twice the size of the Sun but nearly 100 times as luminous. Not much farther than θ CrB it is 351 ly distant. It rotates at a more leisurely 10 km/s.

At 6.6 times the diameter of the Sun κ **Coronae Borealis,** a K1 about 102 ly from Earth, is on its way to becoming a true giant – which is bad news for the Jovian size planet that orbits the star. Taking 1,119 days to complete a full orbit, during which its distance from κ CrB will vary between 2.2 and 3.2 AU, the planet will be swamped with radiation as the star approaches its red giant phase and swells to 2 AU across, and then deprived of heat and light as the star shrinks to a white dwarf.

ν^1 and ν^2 **Coronae Borealis** are a faint but easy pair to separate with a binocular and even without optical aid if the sky is sufficiently dark. ν^1, the most northerly of the two, is an M2 red giant of 34 $D_{\odot}$. The other star is an orange K5 somewhat smaller at 29 $D_{\odot}$. However, they are not related: ν^1 is heading south while ν^2 is moving north. The strange thing about this pair though, is that they may be the same distance from us. ν^1 is estimated to be between 506 and 606 ly while ν^1 is between 497 and 593 ly. These are stars that will pass in the night or, if they do get too close, will gravitationally link to one another and form an orbital binary system.

Sun-like stars that have planetary systems always generate interest within the astronomical community, and ρ **Coronae Borealis** was one of the first to do so. Just 56.8 ly from Earth ρ CrB is a G0, possibly a G2 (like the Sun), a little bit larger than the Sun at 1.2 $D_{\odot}$ and rather more luminous at 1.69 $L_{\odot}$ which could be due to its greater age estimated to be 10,000 million years compared to the Sun's 4,567 million years. The star has a dusty circumstellar disk extending out to 85 AU, similar to our own Kuiper Belt, and angled at 46° to our line of sight. This may indicate the direction in which the star's poles are pointing as such disks are thought to extend from the star's equator. Imbedded in the disk is a planet with a mass of at least 1.04 Jupiters. It orbits very close to the star at an average of 0.22 AU, about half the distance of Mercury from the Sun, and completes a full orbit in about 40 says.

Whatever ρ CrB can do σ **Coronae Borealis** can do twice as well. A binary system, σ^A CrB is a G1 star about the same size as the Sun, slightly more massive at 1.14 $M_{\odot}$ but slightly less luminous at 0.85 $L_{\odot}$. With a rotational velocity of 30 km/s it spins 15 times faster than the Sun completing one rotation in just over 40 hours. Its companion, σ^B CrB, is another solar analog. A G0, again about the same size as the Sun, it has a very similar mass of 1.09 $M_\odot$, is rather more luminous at 1.17 $L_\odot$ and rotates at 25 km/s. A highly elongated orbit brings the two stars to within 31 AU (as close as Neptune gets to the Sun) and separates them by 225 AU. Situated 70.7 ly away the two stars shine at m_v +6.66 and +5.64 respectively which combined give the appearance of a m_v +5.30 star. It was once believed that two other stars were in orbit in this system but recent research has dispelled that notion. The system appears to be young: 100 million to 3,000 million years.

Corvus

Constellation:	Corvus	**Hemisphere:**	Southern
Translation:	The Crow	**Area:**	184 deg^2
Genitive:	Corvi	**% of sky:**	0.446%
Abbreviation:	Crv	**Size ranking:**	70th

Corvus is one of Ptolemy's 48 constellations and is associated with neighboring Crater and Hydra. Apollo sent Corvus to fetch a cup (Crater) containing the water of life. As the crow was returning it spotted figs on which it decided to feed, letting go of the cup and spilling the precious water. Presenting the empty cup to Apollo the crow lied that it had been attacked by the great water snake Hydra. Apollo knew this to be a lie and so cast the crow into the heavens followed by the cup, just out of the crow's reach so it would experience thirst for eternity.

When the fifth brightest star in a constellation is given the α designation then you know something is wrong. This is the case with **α Corvi**. At m_v +4.02 it is outshone by β (+2.63), γ (+2.54), δ (+2.93) and ε (+2.99). Now it is easy to understand how Johannes Bayer could not decide which one was the brightest between β and γ because the difference is just 9/100th of a magnitude, but between γ and α it is an unmistakable 1.48 magnitudes! So what's going on? α Corvi, or Alchiba to give it its proper name, is a bit of an unusual star. Previously believed to be a giant and now relegated to a dwarf F1 just 30% larger than the Sun, it burns hot but is under-luminous at 4.36 $L_\odot$ when 6 $L_\odot$ would be the norm. Perhaps α Corvi was brighter 400 years ago and has since dimmed?

At 15.5 $D_\odot$ **β Corvi** or Kraz is the second largest star in the constellation. Some 140 ly from Earth this G5 yellow bright giant has a temperature a few hundred degrees less than the Sun but is 132 times more luminous. At 10 pc it would shine at M_v -2.1. Spinning at 17 km/s β Crv takes 46 days to make a full rotation.

Lying 165 ly away is the 4.4 $D_\odot$ **γ Corvi** or Minkar. A B8 with a luminosity 350 times greater than the Sun γ Crv is a modest spinner at just 40 km/s: the average is 151 km/s although 20% of B8s have rotational velocities of 40 km/s or less. Clouds of mercury and manganese can be found in its upper layers.

δ Corvi is a fine double for a small telescope or good binocular. The stars should appear white – the A0 primary being of m_v +2.93 – and yellowish-orange, the K0 secondary having a magnitude of mv +8.51, though in his *Celestial Handbook* Robert Burnham Jr notes that they are often said to be yellowish and pale lilac. Jim Kaler on his website says they are also regarded as being pale yellow and purple. They are separated by 24.2″ at PA 214°.

By far the largest star in the constellation the K2 giant **ε Corvi** is 57 times the size of the Sun but rotates just slightly faster at 2.6 km/s. As a result it takes 1,110 days – 3 years – to turn once on its axis. Unstirred, its outermost atmosphere is no longer magnetically active.

η Corvi, a 1.5 $D_\odot$ F2 dwarf at 59.4 ly distance appears to have a belt of debris that spreads out to 200 AU from the star – four times the distance of our own

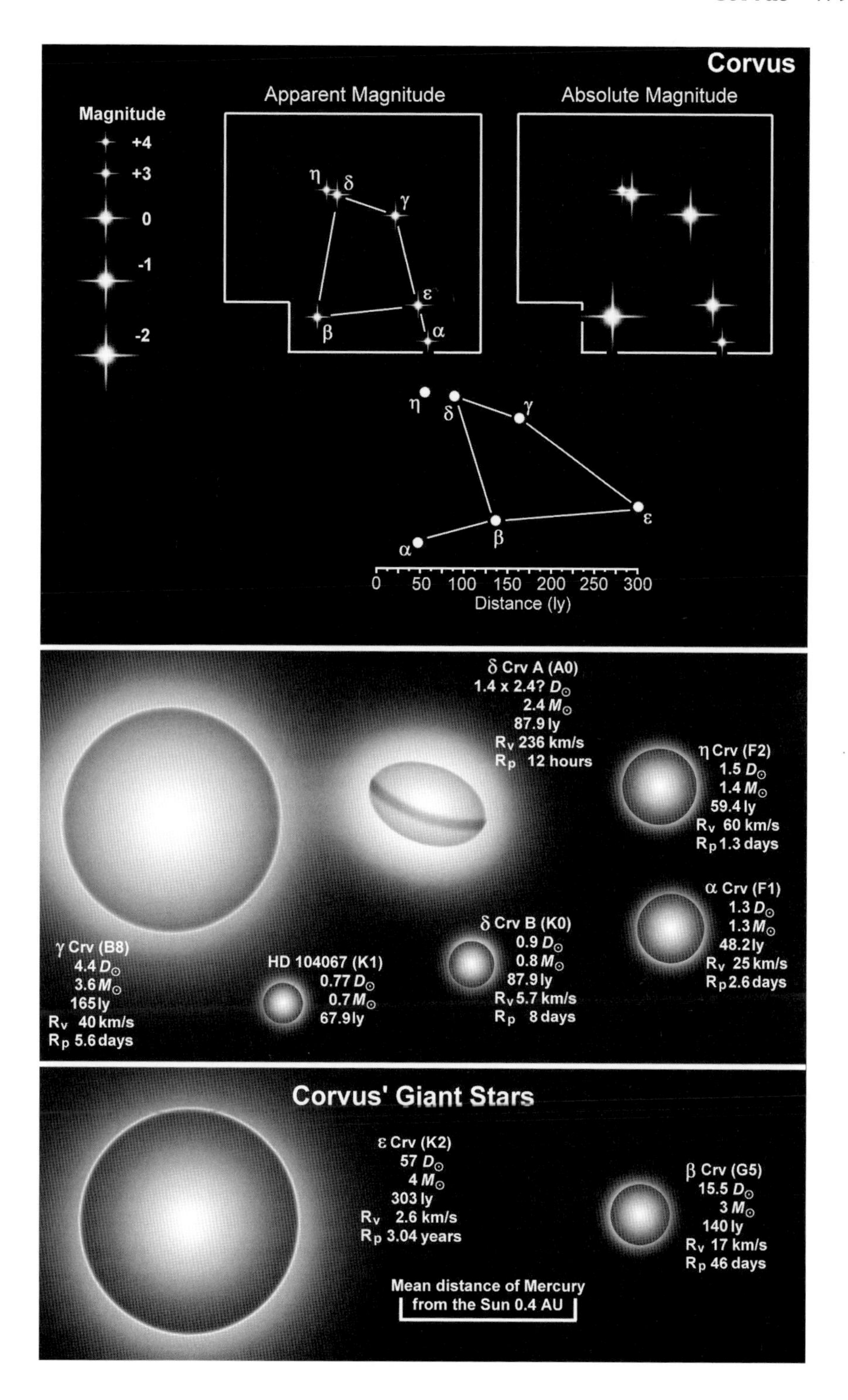
Corvus
Magnitude
+4
+3
0
-1
-2
Apparent Magnitude
Absolute Magnitude
η
δ
γ
ε
β
α
0 50 100 150 200 250 300
Distance (ly)
δ Crv A (A0)
1.4 x 2.4? $D_\odot$
2.4 $M_\odot$
87.9 ly
R_V 236 km/s
R_P 12 hours
η Crv (F2)
1.5 $D_\odot$
1.4 $M_\odot$
59.4 ly
R_V 60 km/s
R_P 1.3 days
α Crv (F1)
1.3 $D_\odot$
1.3 $M_\odot$
48.2 ly
R_V 25 km/s
R_P 2.6 days
δ Crv B (K0)
0.9 $D_\odot$
0.8 $M_\odot$
87.9 ly
R_V 5.7 km/s
R_P 8 days
γ Crv (B8)
4.4 $D_\odot$
3.6 $M_\odot$
165 ly
R_V 40 km/s
R_P 5.6 days
HD 104067 (K1)
0.77 $D_\odot$
0.7 $M_\odot$
67.9 ly
Corvus' Giant Stars
ε Crv (K2)
57 $D_\odot$
4 $M_\odot$
303 ly
R_V 2.6 km/s
R_P 3.04 years
β Crv (G5)
15.5 $D_\odot$
3 $M_\odot$
140 ly
R_V 17 km/s
R_P 46 days
Mean distance of Mercury
from the Sun 0.4 AU

Kuiper Belt. There is some evidence that the debris within the belt is not evenly distributed but tends to gather in clumps. This could be due to resonance effects with a large planet which, during its orbit, periodically sweeps the debris from some parts of the belt and ejects it in other parts. So far however, no planet has been found.

Crater

Constellation:	Crater	**Hemisphere:**	Southern
Translation:	The Cup	**Area:**	282 deg^2
Genitive:	Crateris	**% of sky:**	0.684%
Abbreviation:	Crt	**Size ranking:**	53rd

Crater is inextricably linked to Corvus and Hydra (see Corvus). This is a faint constellation that is of little interest to most observers and even Robert Burnham Jr, who was never lost for words, could find nothing to say about it in his *Celestial Handbook.*

α Crateris or Alkes is the third brightest star in the constellation, but only just. The brightest, by about half a mag, is actually γ Crt at m_V +3.56. α and β are just about neck-and-neck at +4.08 and +4.07 respectively. A K1 giant of 15 $D_\odot$ α is among the 200 fastest moving naked eye stars. Its speed relative to the Sun is in the order of 130 km/s, some three to six times faster than our local neighbors. In 25,000 years it will have left Crater altogether and taken up residence in Puppis. Spectroscopic studies have revealed that it has a very high metal content.

Another high speed star is **β Crateris** which is hurtling through space at 68 km/s but, unlike α Crt, is deficient in metals indicating it is an escapee from an older part of the Galaxy. Just 80% as big as the Sun this A1 dwarf spins at 49 km/s taking 4 days to complete one turn on its axis. It is currently 266 ly away and may be a member of the Sirius Supercluster.

The m_V +3.56 γ **Crateris** is the closest to us at 83.8 ly. Another A class, this time an A7, it is 1.8 $D_\odot$ and a binary, its companion also of the same spectral class – an A5 – with a magnitude of +9.6.

ε **Crateris** holds the record for the largest star in the constellation at 30 $D_\odot$. Perhaps not surprisingly it is a K5. It is also the farthest at 364 ly and is the second most luminous at 119 $L_\odot$.

The title for the most luminous goes to ζ **Crateris** at 122 $L_\odot$. A warm yellowish-orange this G8 giant is a dozen times larger than the Sun.

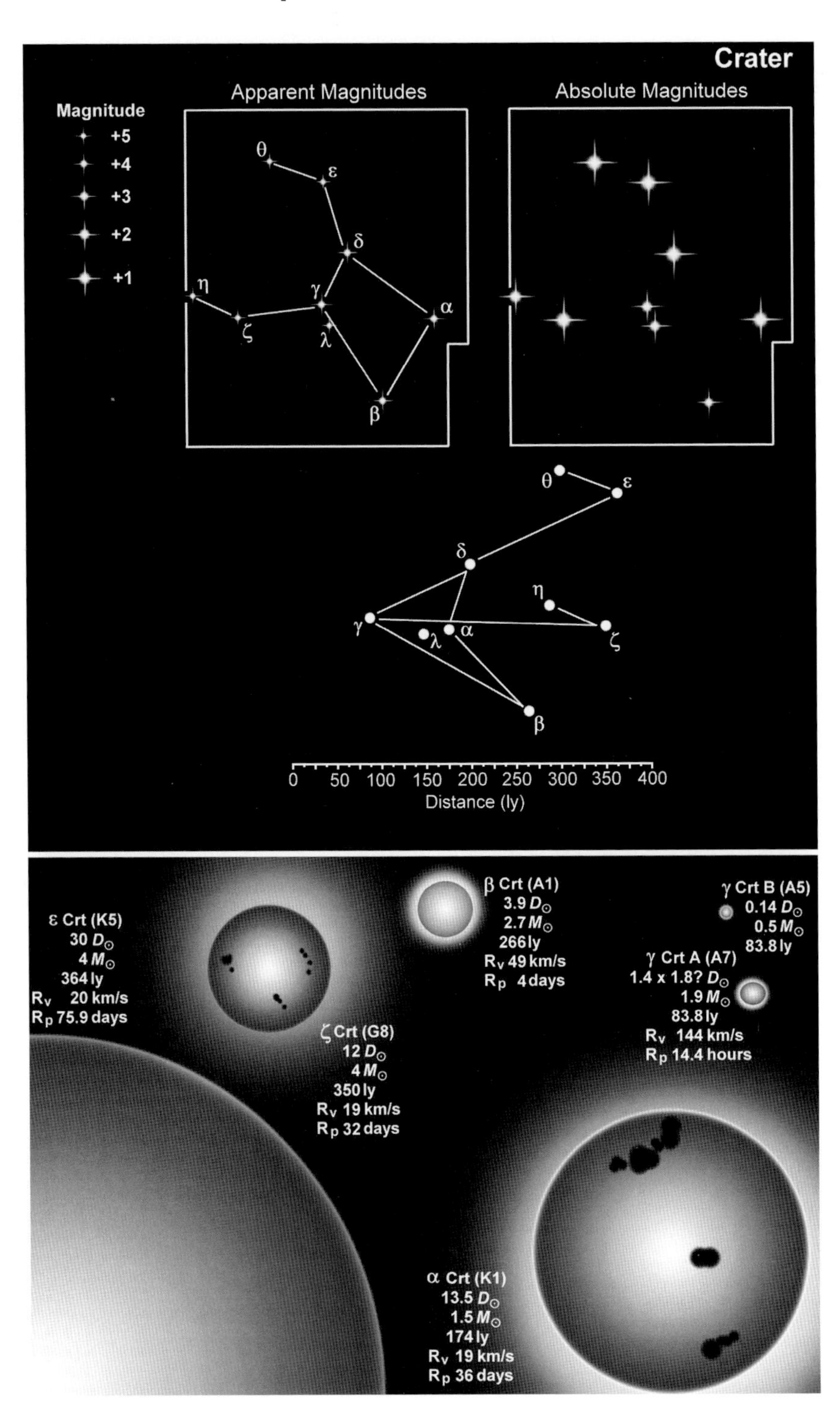
Crater
Apparent Magnitudes
Absolute Magnitudes
Magnitude
+5
+4
+3
+2
+1
0 50 100 150 200 250 300 350 400
Distance (ly)
ε Crt (K5)
30 D⊙
4 M⊙
364 ly
Rv 20 km/s
Rp 75.9 days
β Crt (A1)
3.9 D⊙
2.7 M⊙
266 ly
Rv 49 km/s
Rp 4 days
γ Crt B (A5)
0.14 D⊙
0.5 M⊙
83.8 ly
γ Crt A (A7)
1.4 x 1.8? D⊙
1.9 M⊙
83.8 ly
Rv 144 km/s
Rp 14.4 hours
ζ Crt (G8)
12 D⊙
4 M⊙
350 ly
Rv 19 km/s
Rp 32 days
α Crt (K1)
13.5 D⊙
1.5 M⊙
174 ly
Rv 19 km/s
Rp 36 days

Crux

Constellation:	Crux	**Hemisphere:**	Southern
Translation:	The Cross	**Area:**	68 deg^2
Genitive:	Crucis	**% of sky:**	0.165%
Abbreviation:	Cru	**Size ranking:**	88th

Despite being the smallest of all the constellations Crux the Cross, or more commonly the Southern Cross, is arguably the most famous group of stars helped in no small measure by its appearance on the flags of several countries including New Zealand, Australia, Samoa, Brazil and Papua New Guinea. Originally it was part of the hind legs of Centaurus but became a constellation in its own right in the late 16th Century when pioneers from Christian Europe began exploring the southern oceans. It is set against the background of the Milky Way. Fifteen out of Crux's 23 brightest stars – two-thirds – belong to spectral class B.

To the naked-eye **α Crucis** or Acrux looks like a single star of m_v +1.58 but a small telescope or binocular will reveal three stars. One of these – HD 108250, 1.5′ to the south west – is not a member of the system. The two that are members, α^1 and α^2 Crucis, are actually a threesome with α^1 Cru being a spectroscopic binary. **α^{1a} Crucis** is a 10 $D_\odot$ B0.5 of perhaps 14 $M_\odot$ and rotating at 110 km/s taking 4.6 days to spin once on its polar axis. Very little is known about the spectroscopic component **α^{1b} Crucis** except that it is in a 75.8 day long orbit which swings between 0.5 AU at periastron and 1.5 AU at apastron. Its mass is thought to be in the order of 10 $M_\odot$. **α^2 Crucis** is also a heavyweight at 13 $M_\odot$ and 9 $D_\odot$ across. Such stars last less than 20 million years. The system is 321 ly away. The interloper in this story is **HD 108250,** a B4 of unknown size but probably in the range of 1.1 to 3.2 $D_\odot$ with an estimated distance of between 326 and 526 ly. It also has a spectroscopic companion, 0.97 $M_\odot$ which is in a 1.23 day orbit.

The B0.5 **β Crucis** is a β Cepheid pulsating variable (or β CMa, if you prefer). Over a period of 5^h 30.7^m its magnitude dips from m_v +1.23 to +1.31 and back again. At somewhere between 330 and 375 ly it has a luminosity of 12,000 $L_\odot$ and a diameter of 7.1 $D_\odot$. With a temperature of 27,000 K – 4½ times greater than the Sun – it is one of the hottest B-class stars visible to the naked eye. It may or may not have one or more companions including a spectroscopic component in a 5 year orbit of 7 AU.

β Cru may be closely related to a number of other stars in Crux including ζ, λ, μ^1 and μ^2 Cru. They are all B-class and are at roughly the same distance (353 to 377 ly with a margin of error of about 20 ly), they all have very similar radial velocities (12 to 15.8 km/s) and they all have very similar proper motions of between -0.010″ to -0.015″ per year in Declination and a slightly broader -0.052″ to -0.096″ per year in Right Ascension. It would seem likely that these five stars had a common birthplace.

Apart from β Cru there are three other β Cepheids in Crux. **δ Crucis** is a 4.9 $D_\odot$

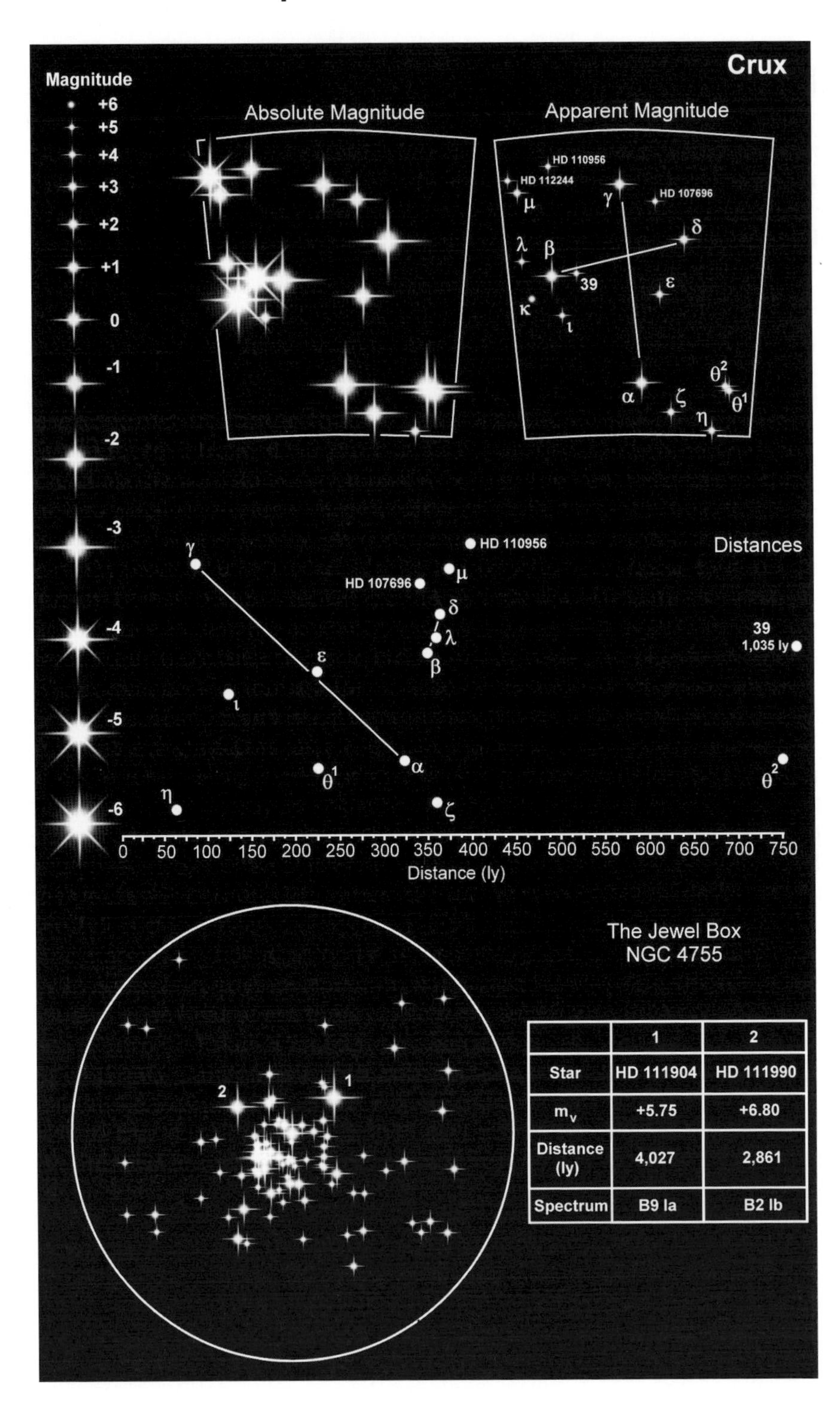

	1	2
Star	HD 111904	HD 111990
m_v	+5.75	+6.80
Distance (ly)	4,027	2,861
Spectrum	B9 Ia	B2 Ib

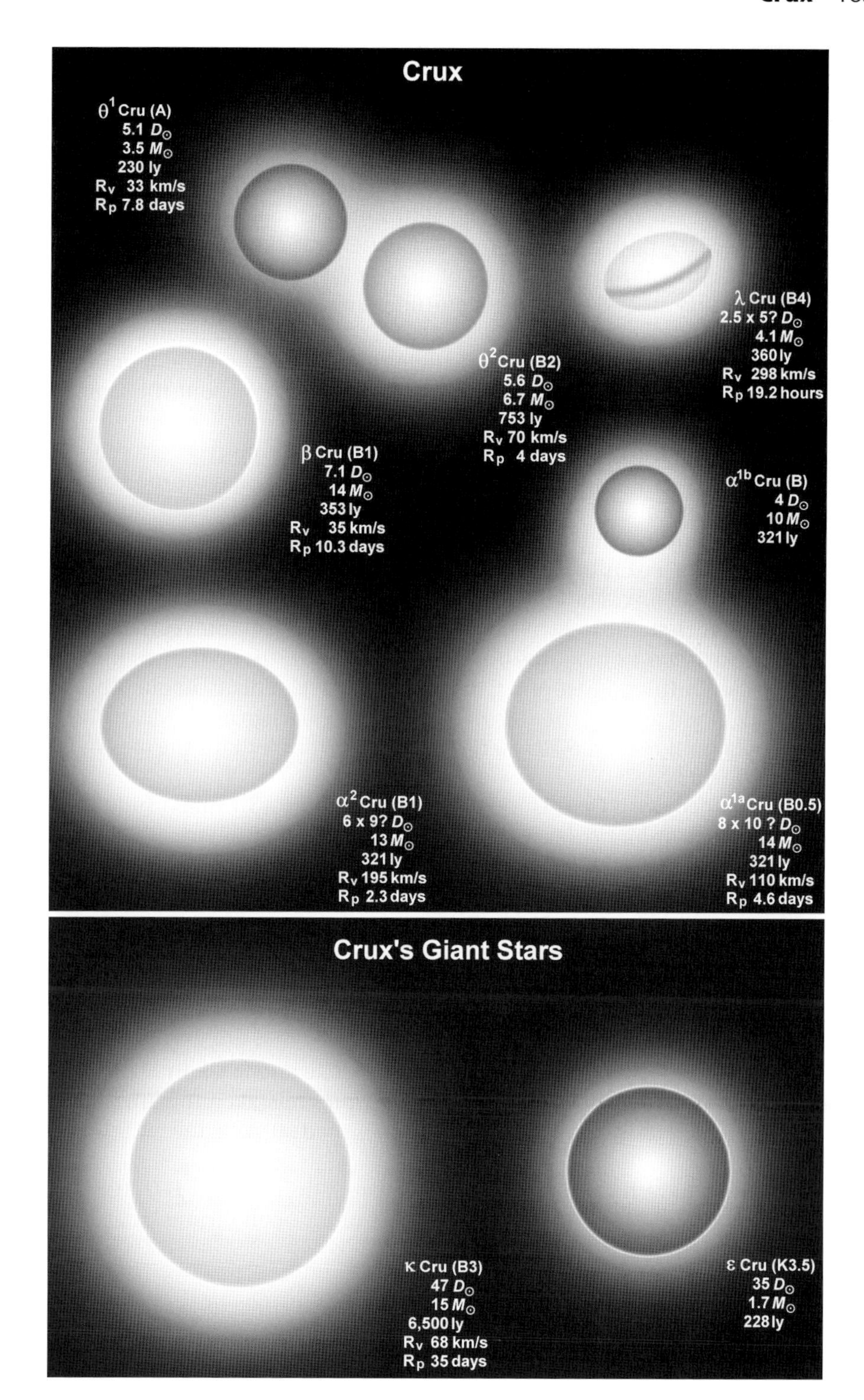
Crux
θ1 Cru (A)
5.1 D☉
3.5 M☉
230 ly
Rv 33 km/s
Rp 7.8 days
λ Cru (B4)
2.5 x 5? D☉
4.1 M☉
360 ly
Rv 298 km/s
Rp 19.2 hours
θ2 Cru (B2)
5.6 D☉
6.7 M☉
753 ly
Rv 70 km/s
Rp 4 days
β Cru (B1)
7.1 D☉
14 M☉
353 ly
Rv 35 km/s
Rp 10.3 days
α1b Cru (B)
4 D☉
10 M☉
321 ly
α2 Cru (B1)
6 x 9? D☉
13 M☉
321 ly
Rv 195 km/s
Rp 2.3 days
α1a Cru (B0.5)
8 x 10 ? D☉
14 M☉
321 ly
Rv 110 km/s
Rp 4.6 days
Crux's Giant Stars
κ Cru (B3)
47 D☉
15 M☉
6,500 ly
Rv 68 km/s
Rp 35 days
ε Cru (K3.5)
35 D☉
1.7 M☉
228 ly

B2 lying at 364 ly distance. Like most B-class stars it rotates at high speed – 135 km/s – and completes one full rotation in 1.8 days. Its variability has a period of 3^h 37.5^m during which its magnitude swings between m_v +2.78 to +2.84. λ **Crucis** is a smidgen larger at 5 $D_\odot$ and a B4 but around about the same distance – perhaps exactly the same distance – of 360 ly. Estimates of its rotational velocity are somewhat scattered between 280 and 330 km/s which means it rotates once every 18.4 to 21.7 hours. Of the four β Cepheids it has the longest period of 9^h 28^m 57^s but the shortest amplitude of m_v +4.62 to +4.64. The shortest period is that of **θ² Crucis** at 2^h 8^m 1^s. Its magnitude range is m_v +4.70 to +4.74. θ^2 Cru also has the slowest rotational velocity at 70 km/s. With a diameter of 5.6 $D_\odot$ it takes a leisurely 4 days to complete a single turn. θ^2 Cru's apparent companion, **θ¹Crucis**, is an optical illusion: it is much closer – 230 ly – about the same size at 5.1 $D_\odot$, but considerably less luminous at 95 $L_\odot$.

ε **Crucis** is the 'fifth' star of the Cross, prominently placed between α and δ Cru. Most depictions of the cross include ε except for the national flag of New Zealand. This seems strange to us simple astronomers but no doubt a vexillologist would be able to offer a perfectly sensible explanation.

The m_v +5.9 κ **Crucis** is part of **NGC 4755**, an open star cluster better known as Herschel's Jewel Box. The cluster, which is about 6,500 ly away, contains about 100 stars and is thought to be no more than about 10 million years old. Most of the stars lie within a 12 ly radius but there are a few farther out by as much as 25 ly. Just to the south and east of α Cru is the Coalsack: a dusty nebula that shows up as a dark cloud against the star rich Milky Way background. Measuring 7° × 5° the nebula is, in reality, 60 to 70 ly in diameter. It is perhaps the closest nebula to us at 500-600 ly.

To the unaided eye μ **Crucis** looks like any ordinary +3.7 magnitude star but a modest telescope or binocular will reveal its duality. The primary, μ^1, actually has a magnitude of m_v +4.03. A rather modest spinning B2, only 48 km/s, it is estimated to be 3.1 $D_\odot$ with a mass of 8 $M_\odot$ and luminosity of 2,400 $L_\odot$. Just 34.9″ away on the celestial sphere – 3,900 AU in real space – is its companion, μ^2. A touch smaller at 2.5 $D_\odot$ but significantly less luminous at 425 $L_\odot$ and, as a result, fainter at m_v +5.17 this B5 star rotates at 230 km/s. The orbital period for the pair is about 68,000 years. The system, some 377 ly distant, is thought to be very young.

Crux is hounded by a couple of imposters. β, θ, υ and ω Carinae make up the asterism of the Diamond Cross which is sometimes mistaken for Crux. A second False Cross, which looks even more like Crux, is made up of ι and ε Carinae and δ and κ Velorum.

Two planets have been located in the constellation around **HD 108147** and star number 127 in **NGC 4349** which is part of an open cluster of at least 206 stars.

Open clusters in Crux

Name	Size arc min	Size ly	Distance ly	Age million yrs	Brightest star in region*	No. stars m_v >+12*	Apparent magnitude m_v
NGC 4609	8.6′	10	4,000	78	HD 110432 m_v +5.31	21	+6.9
NGC 4755 Jewel Box	18′	34	6,500	16	HD 111904 m_v +5.75	92	+4.2

*May not be a cluster member.

Cygnus

Constellation:	Cygnus	**Hemisphere:**	Northern
Translation:	The Swan	**Area:**	804 deg^2
Genitive:	Cygni	**% of sky:**	1.949%
Abbreviation:	Cyg	**Size ranking:**	16th

The Greek god Zeus is said to have transformed himself into a swan to court Queen Leda of Sparta. The result was a pair of eggs from which sprang Castor and Pollux, the Gemini twins. Cygnus is sometimes referred to as the Northern Cross and is set against a particularly rich part of the Milky Way.

α Cygni, better known as Deneb, represents the tail of the swan (Deneb is Arabic for 'tail'). Its modest magnitude of m_v +1.21 belies the fact that it is among the most luminous stars in the sky at 54,400 $L_\odot$. There has always been great uncertainty about the distance of Deneb. The most recent calculation suggests 1,549 ly but with a considerable amount of error: anywhere between 1,338 and 1,843 ly. The extent of this error latitude means that calculating Deneb's properties is, at best, a bit of a gamble. It is a true giant of some 114 $D_\odot$ – a full Astronomical Unit across and a bit more – but with a range of between 99 and 140 $D_\odot$ depending on how accurate our estimates are. It weighs in at 20 $M_\odot$. Placed at 10 pc, the standard distance for measuring absolute magnitude, it would become a brilliant M_v -7.5. Belonging to spectral class A2 its surface temperature is around 8,500 K and it has a rotational velocity of 39 km/s, turning once on its axis once every 182 days if the star is 140 $D_\odot$ across, 148 days for 114 $D_\odot$ and 128.5 days if the 99 $D_\odot$ estimate is correct. It is very slightly variable, fluctuating between m_v +1.21 and +1.29, and is the prototype for stars that have a number of overlapping pulsation periods lasting from days to weeks. It is losing mass at a rate which is 40 million times greater than the Sun, equivalent to an Earth mass each year. In 16,000 BC Deneb was the Pole Star, 7° from the North Celestial Pole. Due to precession it will once again mark the direction of North around AD 9,800.

The fifth brightest star in Cygnus, Albireo, was strangely designated **β Cygni** by Bayer. To the unaided eye it appears as a common or garden 3rd magnitude star but with a binocular or small telescope it can be separated into a binary of contrasting blue and yellowish-orange. Lying at a distance of 386 ly the primary is a K3 giant, 45 $D_\odot$ across and with a luminosity of 650 $L_\odot$. The secondary is a B8 dwarf of 2.7 $D_\odot$ across and rotating at 220 km/s. At this speed the star rotates once in just half a day, its poles drawn in towards the center while its equatorial region bulges, spraying matter into space. No one is yet certain whether the pair are truly related: the orbit seems to be in the order of 75,000 years which is physically unstable. There is little doubt, however, that a third star in orbit around Albireo is a true companion. Too faint (m_v +5.50) and too close (an average of 60 AU) to the primary to be easily separated at the eyepiece the two stars have a 96.84 year orbital period, the fainter star being a B9.5 dwarf.

γ Cygni or Sadr is a bit of a rarity. Of the 37 naked eye F8 stars, 31 (84%) have

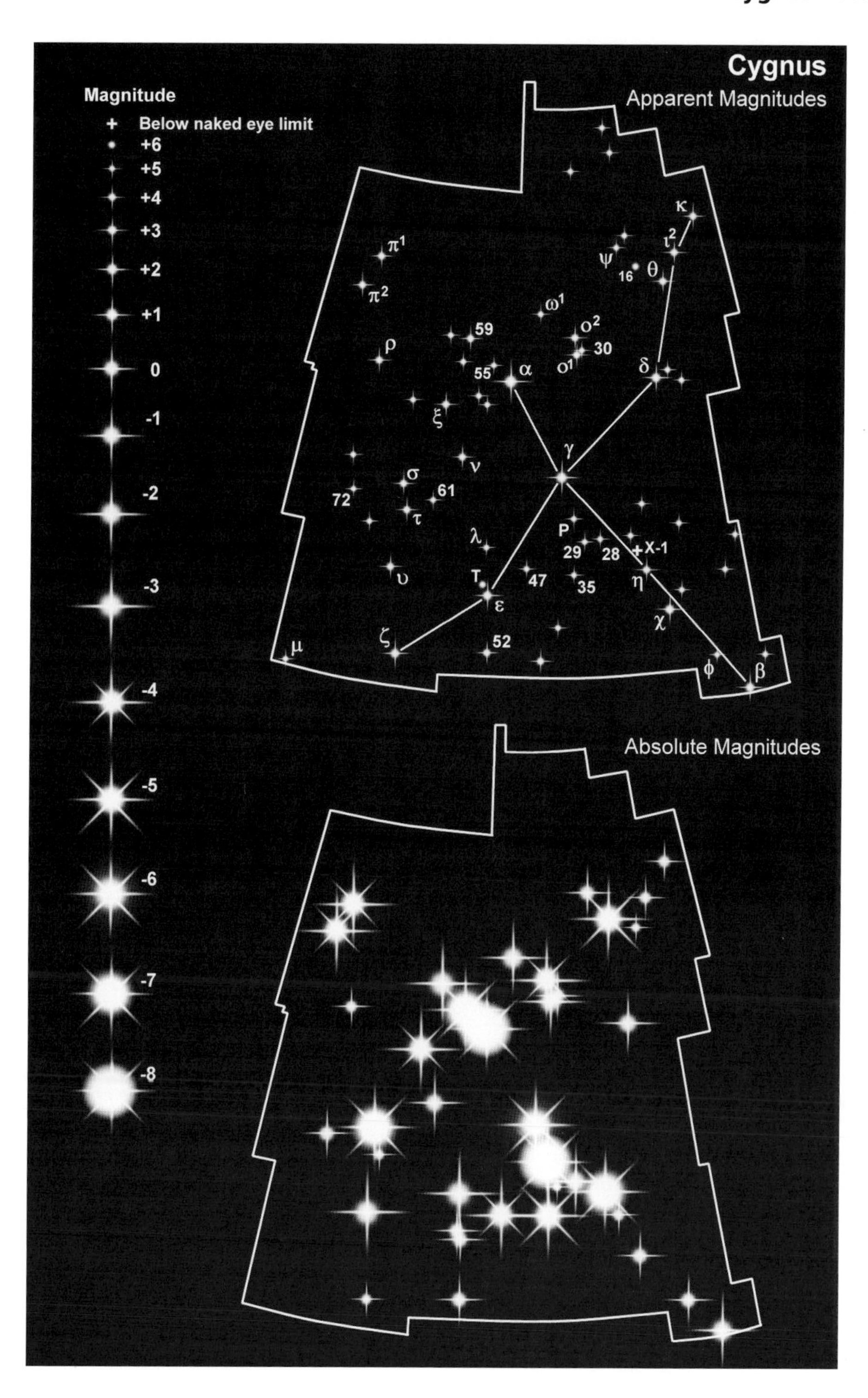
Cygnus
Apparent Magnitudes
Magnitude
Below naked eye limit
+6
+5
+4
+3
+2
+1
0
-1
-2
-3
-4
-5
-6
-7
-8
κ
ι²
ψ
16
θ
π¹
π²
ω¹
ο²
59
30
ρ
55
α
ο¹
δ
ξ
γ
ν
σ
61
72
τ
λ
P
29
28
X-1
υ
T
47
35
η
ε
χ
ζ
52
μ
φ
β
Absolute Magnitudes

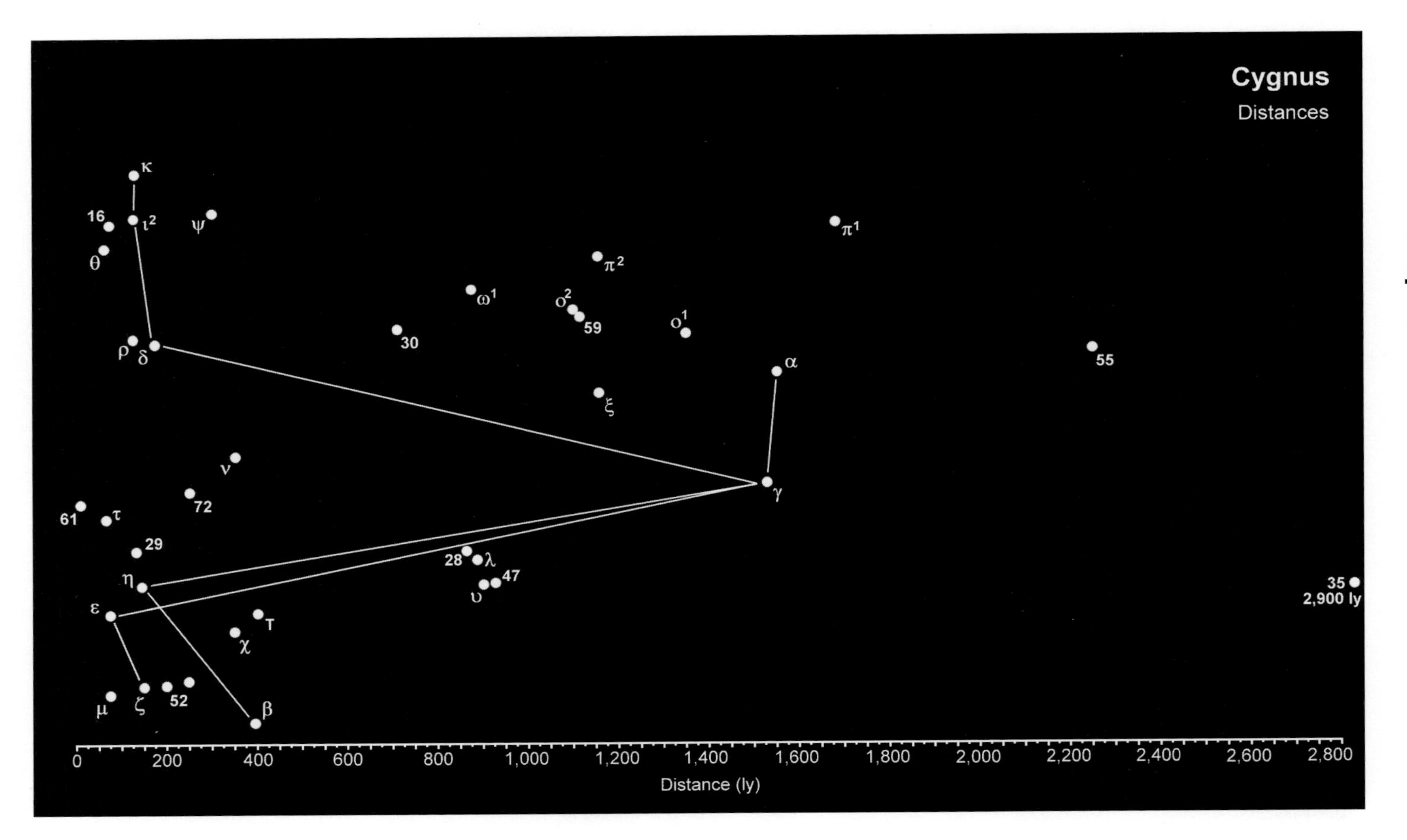
Cygnus
Distances
κ
16
ι2
ψ
θ
π1
π2
ω1
σ2
59
o1
30
ρ
δ
α
55
ξ
ν
72
γ
61
τ
29
28
λ
47
η
υ
35
2,900 ly
ε
T
χ
μ
ζ
52
β
0
200
400
600
800
1,000
1,200
1,400
1,600
1,800
2,000
2,200
2,400
2,600
2,800
Distance (ly)

diameters of less than 3.6 $D_\odot$. γ Cyg isn't one of them. This great lumbering giant is 76 $D_\odot$ across and appears to rotate at just 10 km/s. That's five times faster than the Sun but it means that it still takes more than a year – 385 days – to revolve just once. Measuring rotation is not without its problems however, the accuracy of which partly depends on whether the star is sideways on to the Earth and we are therefore taking measures from the equator, or whether the star's pole is pointing towards us which then gives the impression it is rotating more slowly than it actually is. Situated right in the middle of the cross its distance is thought to be about 1,524 ly by there is a margin of error of nearly 300 ly either way. In the longevity stakes, big is not good. Stars the size and mass of γ Cyg tend to have life cycles measured in tens of millions of years compared to the Sun which will exist for about 10,000 million years.

The third magnitude **δ Cygni** is at least a binary and possibly a triple star system. At about 171 ly the main component is a B9.5 slightly more than five solar diameters across and almost three times as massive as the Sun. With a surface temperature of 9,800 K it is much more luminous at around 154 $L_\odot$. Just 2.4″ away is an F1 dwarf, 1.5 $D_\odot$ and a faint m_v +6.55. It is much cooler at 7,300 K and has a luminosity of 5.2 $L_\odot$. The two stars' orbit is rather eccentric so that they can be separated by up to 230 AU but can also be as close as 85 AU. The orbital period is estimated to be 780 years. The third star is a difficult to spot 12th magnitude B9.5 of about half the size and mass of the Sun. It is not absolutely clear whether it is a true orbital companion. Some 2,000 years after Deneb becomes the Pole Star, δ Cyg will take its place but will be much closer at less than 2°.

At first glance the m_v +2.46 **ε Cygni** looks like it is accompanied by a fainter 5th magnitude companion in a wide orbit. Both are yellowish-orange K giants: ε Cyg is a K0 some 11 $D_\odot$ across while the other, **T Cygni,** is a K3 but more than twice as big at 23 $D_\odot$. They are, however, a line of sight coincidence. ε Cyg is 72 ly away while T Cyg is about 400 ly and totally unrelated. T Cygni is a suspected pulsating variable that fluctuates by 5/100th of a magnitude with no particular rhythm.

The diameter of **ζ Cygni** has proven difficult to pin down. From the first measurements in 1922 its size has been variously quoted as being 9.8, 14, 14.7, 26, 37 and 54 $D_\odot$ with the earliest estimate being 14 $D_\odot$ and the latest 37 $D_\odot$. Of more certainty is its distance of 151 $\pm$ 4 ly and its spectral class: a yellow G8. But even this is not pure. It contains enhanced levels of barium, a pollutant that tells of a once more massive companion that seeded ζ Cyg with the element before dying. Now a white dwarf the two stars are in a 17.8 year orbit which varies between 8 and 13 AU.

o^1 and **o^2 Cygni**, and **30 Cygni**, look like a complex five star system but the reality is much simpler. **o^1** is an EA Algol eclipsing variable. The primary star, **o^{1A} Cygni**, is a K2 bright giant of perhaps 120 $D_\odot$ and a luminosity of about 4,200 $L_\odot$. The secondary, **o^{1B} Cygni**, is a B8 dwarf in a 10.4 year long orbit. Every 3,784.3 days (10.36 years) it passes behind o^{1A} and the magnitude dips from m_v +3.77 to +3.88 for 63 days. **o^2** is also an EA Algol eclipsing variable. The primary star, **o^{2A} Cygni**, is a K3 supergiant though nearly twice the size of o^{1A} at 230 $D_\odot$ and with a much higher luminosity of about 11,100 $L_\odot$. Its secondary, **o^{2B} Cygni**, is again a B3 dwarf which eclipses every 3.14 years when the magnitude

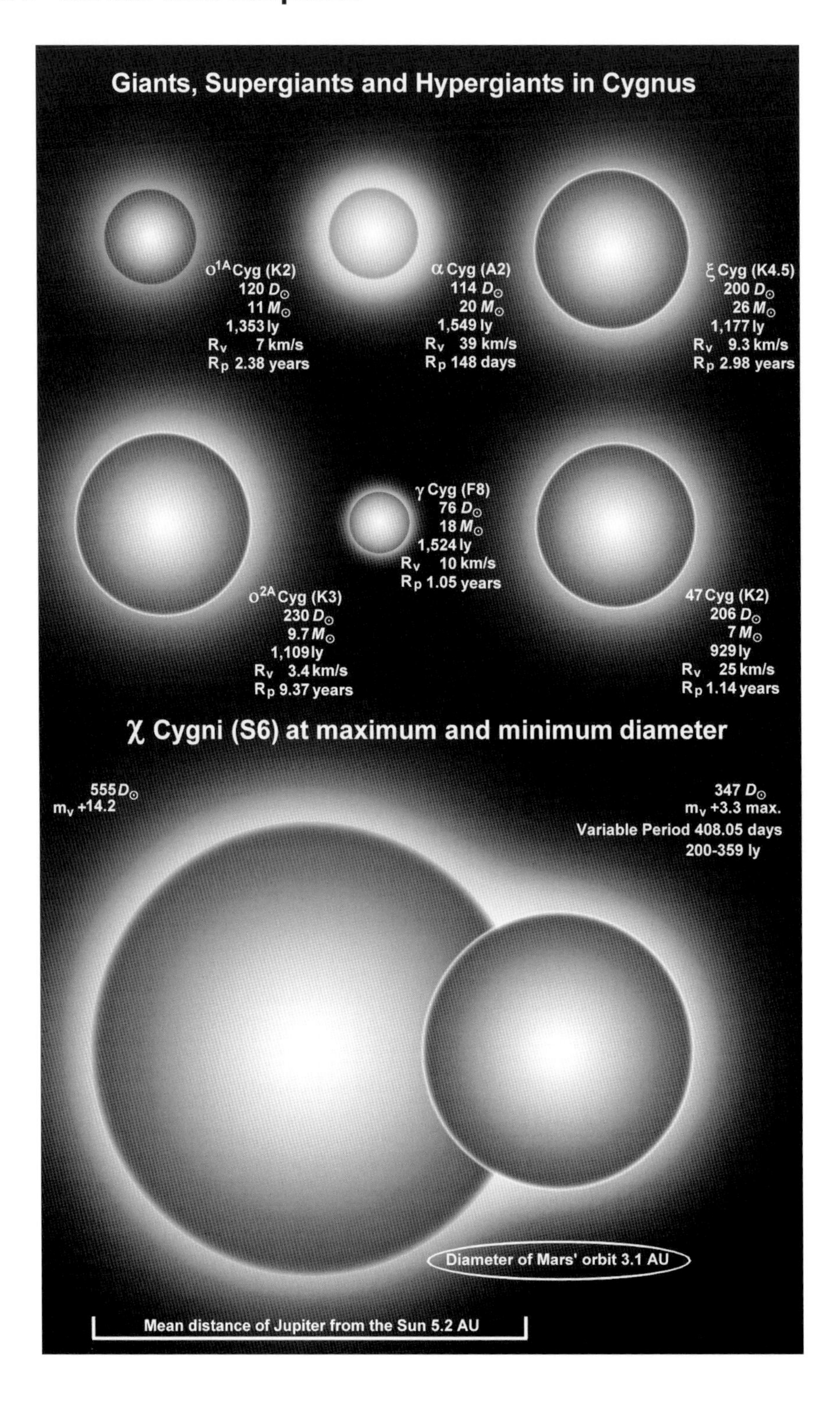
Giants, Supergiants and Hypergiants in Cygnus
o1A Cyg (K2)
120 D☉
11 M☉
1,353 ly
Rv 7 km/s
Rp 2.38 years
α Cyg (A2)
114 D☉
20 M☉
1,549 ly
Rv 39 km/s
Rp 148 days
ξ Cyg (K4.5)
200 D☉
26 M☉
1,177 ly
Rv 9.3 km/s
Rp 2.98 years
γ Cyg (F8)
76 D☉
18 M☉
1,524 ly
Rv 10 km/s
Rp 1.05 years
o2A Cyg (K3)
230 D☉
9.7 M☉
1,109 ly
Rv 3.4 km/s
Rp 9.37 years
47 Cyg (K2)
206 D☉
7 M☉
929 ly
Rv 25 km/s
Rp 1.14 years
χ Cygni (S6) at maximum and minimum diameter
555 D☉
mv +14.2
347 D☉
mv +3.3 max.
Variable Period 408.05 days
200-359 ly
Diameter of Mars' orbit 3.1 AU
Mean distance of Jupiter from the Sun 5.2 AU

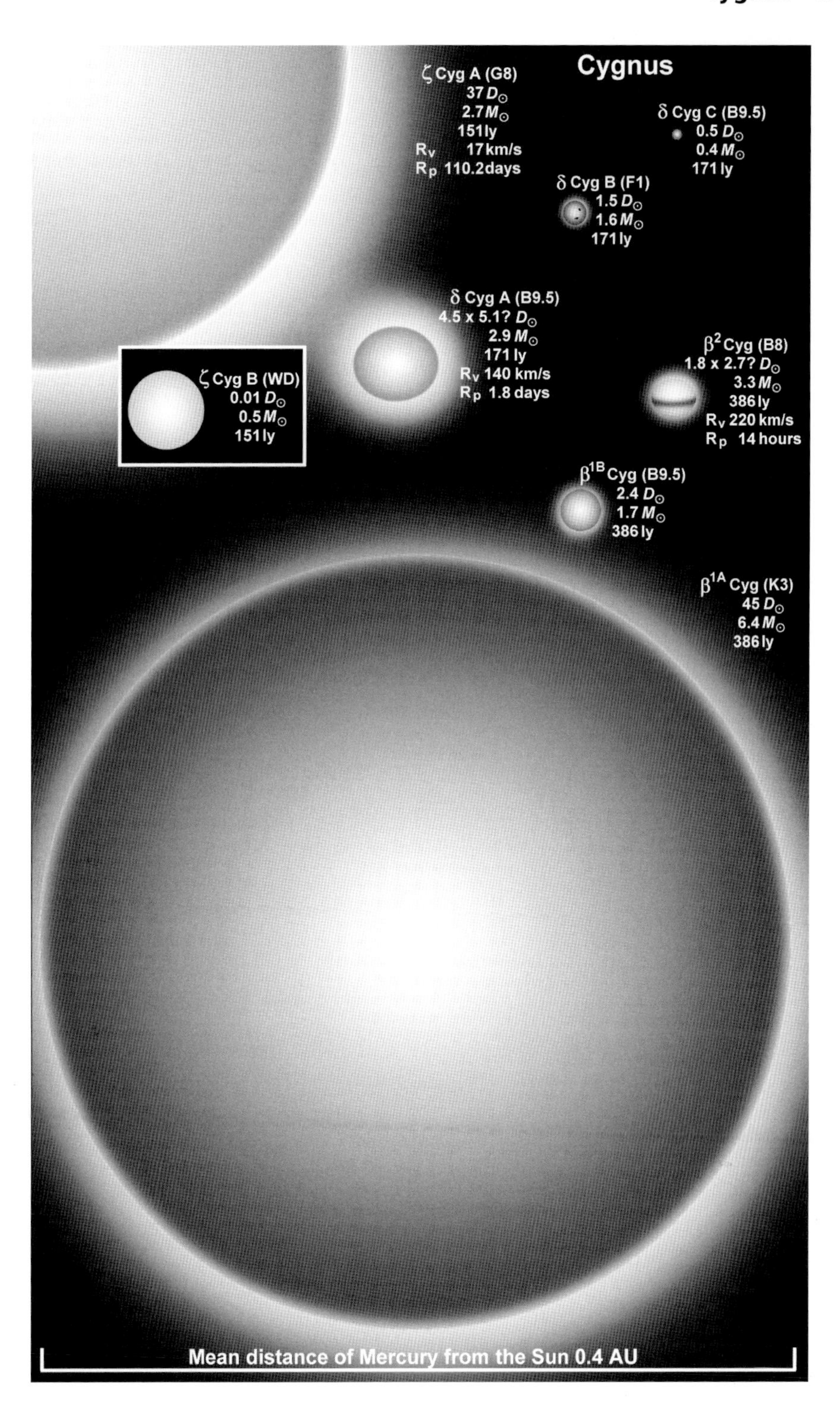
Cygnus
ζ Cyg A (G8)
37 $D_\odot$
2.7 $M_\odot$
151 ly
R_v 17 km/s
R_p 110.2 days
δ Cyg C (B9.5)
0.5 $D_\odot$
0.4 $M_\odot$
171 ly
δ Cyg B (F1)
1.5 $D_\odot$
1.6 $M_\odot$
171 ly
δ Cyg A (B9.5)
4.5 x 5.1? $D_\odot$
2.9 $M_\odot$
171 ly
R_v 140 km/s
R_p 1.8 days
ζ Cyg B (WD)
0.01 $D_\odot$
0.5 $M_\odot$
151 ly
β^2 Cyg (B8)
1.8 x 2.7? $D_\odot$
3.3 $M_\odot$
386 ly
R_v 220 km/s
R_p 14 hours
β^{1B} Cyg (B9.5)
2.4 $D_\odot$
1.7 $M_\odot$
386 ly
β^{1A} Cyg (K3)
45 $D_\odot$
6.4 $M_\odot$
386 ly
Mean distance of Mercury from the Sun 0.4 AU

drops from m_v +4.11 to +4.14. The **o²** system, however, is much closer to us at 1,109 ly while the **o¹** system is at 1,353 ly. The two are not related. Then there is 30 Cygni lurking around the two omicrons in the night sky but, in real space, is just 717 ly away and certainly not related being an A5 dwarf of just 3.3 $D_\odot$.

Cygnus contains some enormous stars. Three are in the 100+ $D_\odot$ range and three are 200+ $D_\odot$. α Cyg at 99 to 140 $D_\odot$ has already been mentioned as has o^{1A} at 120 $D_\odot$ and o^{2A} at 230 $D_\odot$ (see previous paragraph) while the barely noticeable 5th magnitude **55 Cygni** is 131 $D_\odot$ and about 2,250 ly distant. It is also an α Cyg variable cycling between m_v +4.81 and +4.87. At 929 ly can be found **47 Cygni**, 206 $D_\odot$ across and a K2 supergiant. If placed in the center of our Solar System it would just about fill the Earth's orbit. Of similar size is **ξ Cygni** at 200 $D_\odot$ across (almost 2 AU) A K4.5 ξ Cyg believed to be at a distance of 1,177 ly.

The 3rd magnitude **P Cygni** (aka **34 Cygni**) could be one of the most luminous stars in the entire sky. Belonging to the rare Luminous Blue Variable (LBV) or S Doradûs eruptive variable class of star it is thought to be around 6,272 ly distant but with a considerable amount of uncertainty that could see it as close as 3,000 ly. At 75 $D_\odot$ P Cyg is by no means the largest of stars but this 19,000 K B2 supergiant is thought to be 500,000 to 900,000 times more luminous than the Sun. It has a checkered history. It appears to have been unknown prior to 8 August 1600 when it suddenly appeared as a new 3rd magnitude star. Over the next six years it slowly faded until it again disappeared only to suddenly return in 1655 before again gradually fading to below naked eye visibility in 1662. Three years later it was back but continued to fluctuate until 1715 when it seemed to stabilize for a while at 5th magnitude. Since then it has slowly brightened so that today it is m_v +4.81. For most of the past few thousand years P Cyg has been pumping out much of its energy in the ultra-violet region of the spectrum. As the star cools more of this energy appears in the visible spectrum hence the gradual increase in magnitude. Its unpredictable eruptions eject vast amounts of material into space at a speed of 300 km/s forming a series of rings around the star. Believed to be spinning at 65 km/s it takes 58 days to complete a single rotation. P Cyg is only a fleeting resident of the Galaxy, its lifespan just a few million years after which it will explode as a supernova or hypernova and perhaps collapse to form a black hole. There is only one other confirmed naked eye LBV, ζ^1 Scorpii. η Carinae is only a suspected LBV and S Doradûs itself is at m_v +9.57, so it is well worth keeping an eye on P Cyg – you never know when she'll blow!

Another variable worth watching is **χ Cygni** – if you can find it. This Mira-type has a period of 408.05 days but for most of the time it is well below naked eye visibility, often as low as m_v +14.2. When it does put in an appearance it can reach m_v +3.3, though usually it is about 5th magnitude, but it is only then visible for a couple months. Sometimes its peak does not even attain m_v +6 before it slips back to minimum. Its diameter can shrink to 347 $D_\odot$ (3.2 AU) and swell to 555 $D_\odot$ (5.2 AU). We know that it is largest at minimum magnitude, but its distance contains a high element of uncertainty: it could be anywhere between 200 and 350 ly with the likelihood towards the farthest distance. χ Cyg belongs to the rare S6 spectral class which means that its abundance of carbon

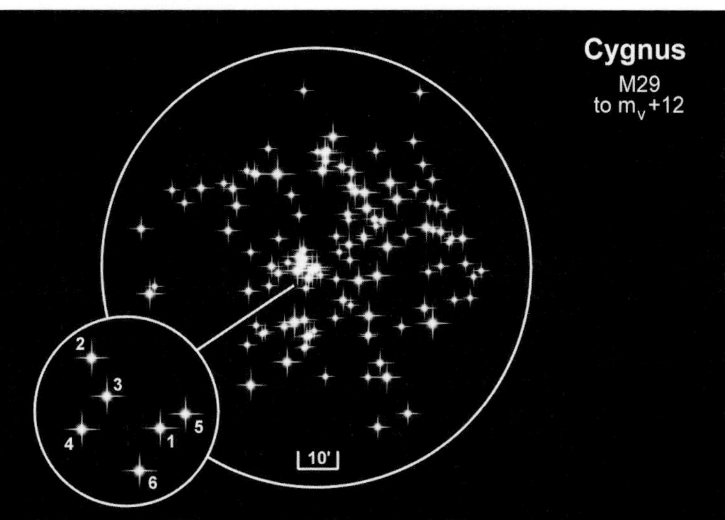

No.	Star	m_v	Distance (ly)	Spectrum
1	V2031 Cyg	+8.57	1,084	F0 III
2	HD 229238	+8.88	168	B0.5 II
3	HD 229234	+8.92	147	O9 II
4	HD 229239	+8.92	40,770	B0.5 II
5	V1322 Cyg	+9.21	-	B0 II
6	HD 229227	+9.38	4,659	B0 III

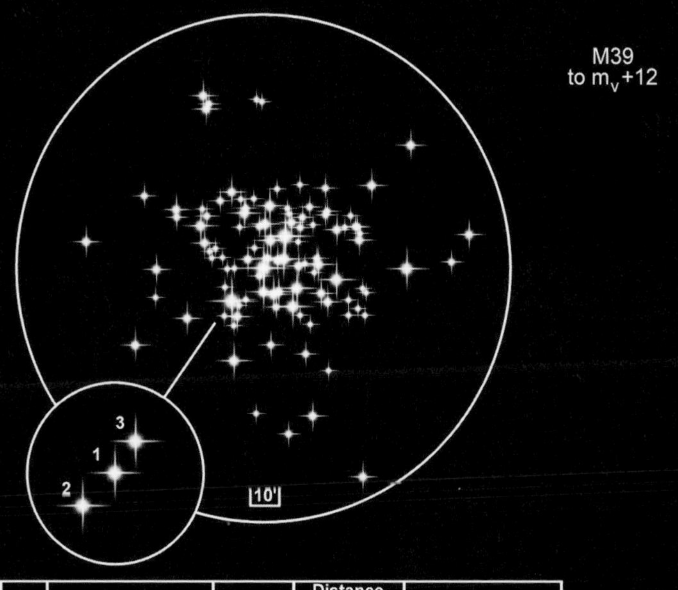

No.	Star	m_v	Distance (ly)	Spectrum
1	HD 205210	+6.57	1,177	B9.5 IV
2	HD 205331	+6.84	647	B9 IV
3	HD 205116	+6.86	943	B9.5 V

just about equals that of oxygen. Higher levels of carbon result in C-class stars. This is a planetary nebula in the making. Before too long the star will puff off its outer layers to enrich the stellar medium.

Planetary systems around binary stars are uncommon but not unknown. **16 Cygni** is one such system. The stars are nearly identical twins. **16 Cygni A** is a G1.5 dwarf of 1.4 $D_\odot$, 1.02 $M_\odot$, and 1.6 $L_\odot$. It has a rotational velocity of 26.9 km/s and spins once on its axis every 2.6 days. With a surface temperature of 5,825 K its m_v is +5.96. To the south east is **16 Cygni B**: a G2.5 dwarf, slightly smaller at 1.2 $D_\odot$, 0.97 $M_\odot$, and 1.3 $L_\odot$. It has a higher rotational velocity of 29.1 km/s and spins once on its axis every 2.1 days. With a surface temperature of 5,640 K its m_v is +6.20. The system is 70.5 ly away and is believed to be 10,200 million years old, more than twice the age of our own Solar System. The orbit is so large that the limited data available could result in several different orbital configurations. The best fit seems to be a period in excess of 18,700 years during which the two stars come within 404 AU of one another before separating by 1,350 AU. Swanning around in this system is a single planet which is in orbit around 16 Cyg B. It is a sad fact that when astronomers are asked to catalog stars, planets, comets, asteroids, meteorites and the like then any semblance of logic completely abandons them. As a result the planet is now known as 16 Cygni B b rather than, say, the Henry Draper number HD 186408 b or even 16^B Cygni b. Apart from its odd designation the planet is a typical Jovian-type, 1.68 M_J, a period of 799.5 days and an orbit that varies between 0.5 and 2.8 AU.

SX Arietis-type variables are quite rare – only four naked eye examples exist, one of which is **28 Cygni**. A 3.9 $D_\odot$ B2.5 (SX Arietis stars are all B-class) its magnitude switches between m_v +4.91 and +4.97.

29 Cygni is a double rarity. It has a very low metal content and is therefore classified as a λ Boötis type star, but it is also a δ Scuti pulsating variable with an amplitude of just 3/100th of a magnitude between m_v +4.94 and 4.97. The period is 44^m 38^s making it the fastest naked-eye δ Scuti visible.

59 Cygni is a quadruple star system. The main star is 11 $D_\odot$ and has a mass of 7.5 $M_\odot$. It is a fast spinning B1.5 with a rotational velocity that could be as high as 375 km/s, which means that it may rotate once every 1.5 days. This superhot star – 25,500 K – has a subdwarf in orbit around it. With an orbital period of 28 days it lies just a few AUs from the star. A second orbiter at 85 AU takes about 200 years to complete a circuit around the main star. A B4 dwarf it is m_v +7.5. Finally, at a distance of some 9,000 AU and with an orbital period in excess of 200,000 years is a m_v +9.4 A0 dwarf. Just to make things even more interesting the primary is a γ Cas eruptive variable with a magnitude range of m_v +4.49 to +4.88.

The German mathematician and astronomer Friedrich Wilhelm Bessel was always up for a challenge. Though he lacked a university education and any formal training in astronomy he nonetheless became the Director of the Königsberg Astronomical Observatory in Prussia at the age of 26 from where he precisely measured the positions of 50,000 stars. His greatest achievement though was in 1838 when he became the first person to measure the parallax of a star. The star in question was **61 Cygni** which he estimated was 9.8 ly away.

Today we know that it is 11.36 ly distant and the closest star to us in the Northern Celestial Hemisphere, and getting closer at 64.3 km/s. In 19 million years it is likely to be the brightest star in the sky. This faint, m_V +5.23 orange K5 is actually a binary. Its companion is also a K-class (K7) in a 722 year orbit. Hurtling through space at 105 km/s relative to the Sun these two stars are visitors from another part of the Galaxy. If they remain on the current course and speed they will end up in Cassiopeia in 25 centuries time.

Another fast moving star is **72 Cygni.** Currently 256 ly away it is heading our way at 66 km/s and moving at 91 km/s relative to the Sun. It is a 14 $D_{\odot}$ K1 giant which takes 41.7 days to turn once on its axis at 17 km/s.

The constellation is also famous for giving us the first evidence of an X-ray source being associated with a black hole. In 1964 Geiger counters aboard two Aerobee rockets detected a very strong X-ray source that was initially called Cygnus XR-1 and which is now better known as **Cygnus X-1.** Coming from 6,000 to 7,000 ly away the X-rays are caused by matter being drawn off a massive star into a black hole. As the matter is compressed its temperature rises by several million degrees generating vast amounts of X-rays and γ-rays. The star in this system originally began life just 5 million years ago as a 40 $M_{\odot}$ supergiant. Currently – well, 7,000 years ago – it had lost nearly 80% of its mass to the black hole. Its diameter was around 23 $D_{\odot}$, its temperature 30,000 K but its luminosity was a staggering 400,000 $L_{\odot}$. It shines at just m_V +8.93 but this is deceptive. Between it and us is a significant amount of interstellar dust without which the star would be visible to the naked eye. At 10 pc it would be M_V -6.7. The black hole is thought to be more than twice the size of the star at 52 $D_{\odot}$, the pair orbiting one another every 5.6 days at a distance of just 0.2 AU (30 million km).

A planet may be in orbit around **HD 188753** but it is, as yet, unconfirmed and disputed by some. HD 188753 lies at a distance of 146 ly. At m_V +7.43 it is well below naked eye visibility and belongs to the K0 spectral class, having a mass of 1.06 $M_{\odot}$. However, HD 188753 is not alone. Orbiting it once every 25.7 years at between 6.2 and 18.5 AU is a binary system consisting of a B3 and a G8 which together have a combined mass of 1.63 $M_{\odot}$. The planet is in an orbit just 0.0446 AU (10.8 million km) from the primary star and has a mass of at least 1.14 M_J but at this distance there simply is not enough available material to have created such a large planet. The theory is that this type of planet – a *hot Jupiter* – formed beyond 2.7 AU and migrated in towards the star. No one has yet come up with an undisputed mechanism to account for the migration or to explain why such planets come to a halt and do not fall into the star. Now the problem with the triple star arrangement of HD 188753 is that the gravitational forces truncate the debris disk around the star preventing it from spreading any farther than 2.7 AU. So not only is there insufficient material for the planet to have been born close to the star, but there is also a paucity of material beyond 2.7 AU. Without any plausible theory as to how the planet formed its very existence has been called into question.

Acting as a vast backdrop to **52 Cygni** is the **Veil Nebula**, otherwise known as the **Cygnus Loop**, a delicate filament of gas expanding at more than 100 km/s. It is all that remains of a supernova that exploded between 5,000 and 10,000 years ago. The star, 52 Cygni, is not related to the nebula. At 206 ly this G9.5 giant, 14 $D_{\odot}$, is much closer than the Veil which is estimated to be 1,500 ly away.

Close to α Cygni in the night sky but, in reality, about 500 ly deeper into space is **IC 5067**, the **Pelican Nebula**. Long exposure photographs show the nebula glowing red, signifying the presence of atomic hydrogen.

Messier 29 is a small but attractive cluster near to γ Cygni which can be seen with a binocular or small telescope. Containing up to 150 stars it could be as close as 3,700 ly. Its core is about 11 ly in diameter but its full diameter could be up to 52 ly. It is part of the Cygnus OB1 association. It is estimated to be about 13 million years old, its five hottest stars belonging to the B0 spectral class.

Smaller, closer and less populated is **Messier 39**. Its center is only about 1,000 ly away and contains around 100 stars but it is much older than M29 at 270 to 300 million years.

It is said by some that the **North American Nebula (NGC 7000)** can be seen by the keen sighted under very dark conditions. It is certainly popular with astrophotographers, its large size, 120′ × 100′, and proximity to Deneb (just 3° away) making it an easy target. NGC 7000 is where stars are being born. It is about 1,600 ly away.

Open clusters in Cygnus

Name	Size arc min	Size ly	Distance ly	Age million yrs	Brightest star in region*	No. stars m_v >+12*	Apparent magnitude m_v
Collinder 419	27′	19	2,400	7	HD 193322 m_v +5.96	37	+5.4
IC 4996	27′	44	5,600	9	HD 193076 m_v +7.62	51	+7.3
IC 5146	82′	66	2,800	1	TYC 3608-1659-1 m_v +9.64	12	+7.2
M29	57′	52	3,700	13	V2031 Cyg m_v +8.57	123	+6.6
M39	80′	25	1,000	278	HD 205210 m_v +6.57	106	+4.6
NGC 6819	16′	35	7,700	1,500	TYC 3140-3020-1 m_v +10.19	14	+7.3
NGC 6834	25′	49	6,700	76	HD 332845 m_v +9.74	24	+7.8
NGC 6871	30′	44	5,100	9	V1676 Cyg m_v +6.79	85	+5.2
NGC 6910	32′	34	3,700	14	HD 194279 m_v +7.01	70	+7.4
NGC 7063	30′	20	2,250	95	HD 203921 m_v +8.89	105	+7.0

*May not be a cluster member.

Delphinus

Constellation:	Delphinus	**Hemisphere:**	Northern
Translation:	The Dolphin	**Area:**	189 deg^2
Genitive:	Delphini	**% of sky:**	0.458%
Abbreviation:	Del	**Size ranking:**	69th

This small but very distinct constellation has several myths associated with it one of which is that Delphinus was the son of Triton, the sea god. A poet and singer called Arion was aboard a ship when the crew decided to rob him and throw him overboard. The gods, who used to listen to Arion's poems and songs, sent Delphinus to save him. The four brightest stars form a diamond which is sometimes referred to as 'Job's Coffin' after the Biblical character.

At m_v +3.76 α **Delphini** is the second brightest star in the constellation being beaten to first place by β Del at m_v +3.63. Lying at a distance of 241 ly this 3.6 $D_\odot$ star is a fast spinning B9 rotating at about 138 km/s. It may also a binary, its companion being an A-class apparently separated by 12 AU and in a 17.1 year long orbit, but this interpretation is disputed and the pair may just be an optical double.

Like α Del, **β Delphini** is also a double but a very definite binary with a better understood orbit due to the wider separation of the two components. Both stars are F5 subgiants that have magnitudes of m_v +4.0 and +4.9 but which together shine at m_v +3.63 technically making β Del the brightest star in the constellation. They are separated by just 0.5″ which, at a distance of 97.4 ly, translates into an average distance of 13 AU between the two components, though this will vary between 8 and 18 AU during the orbital period of 26.65 years.

Both α and β Del have peculiar proper names: Sualocin and Rotanev respectively. It seems these names first appeared in 1814 in a star catalog published by the Palermo Astronomical Observatory. Some years later the Rev'd Thomas Webb, a British astronomer, worked out that the names were the partly Latinized version of Niccolò Cacciatore, an assistant at Palermo, spelled in reverse:

> Niccolò = Nicolaus (in English) = Sualocin (in reverse).
> Cacciatore = Venator (in Latin or Hunter in English) = Rotanev (in reverse).

No one knows why the stars were so named. It could be that arrogance got the better of Cacciatore and he inserted the names into the catalog himself. Alternatively the then director of the observatory, Giuseppe Piazzi (who discovered the first asteroid, Ceres), may have honored Cacciatore by naming the stars after him. Another possibility is that someone else inserted the names as a joke. Whatever the explanation the names became established and Cacciatore succeeded Piazzi as director of the observatory.

The m_v +3.87 γ **Delphini** is also a binary although observers cannot agree

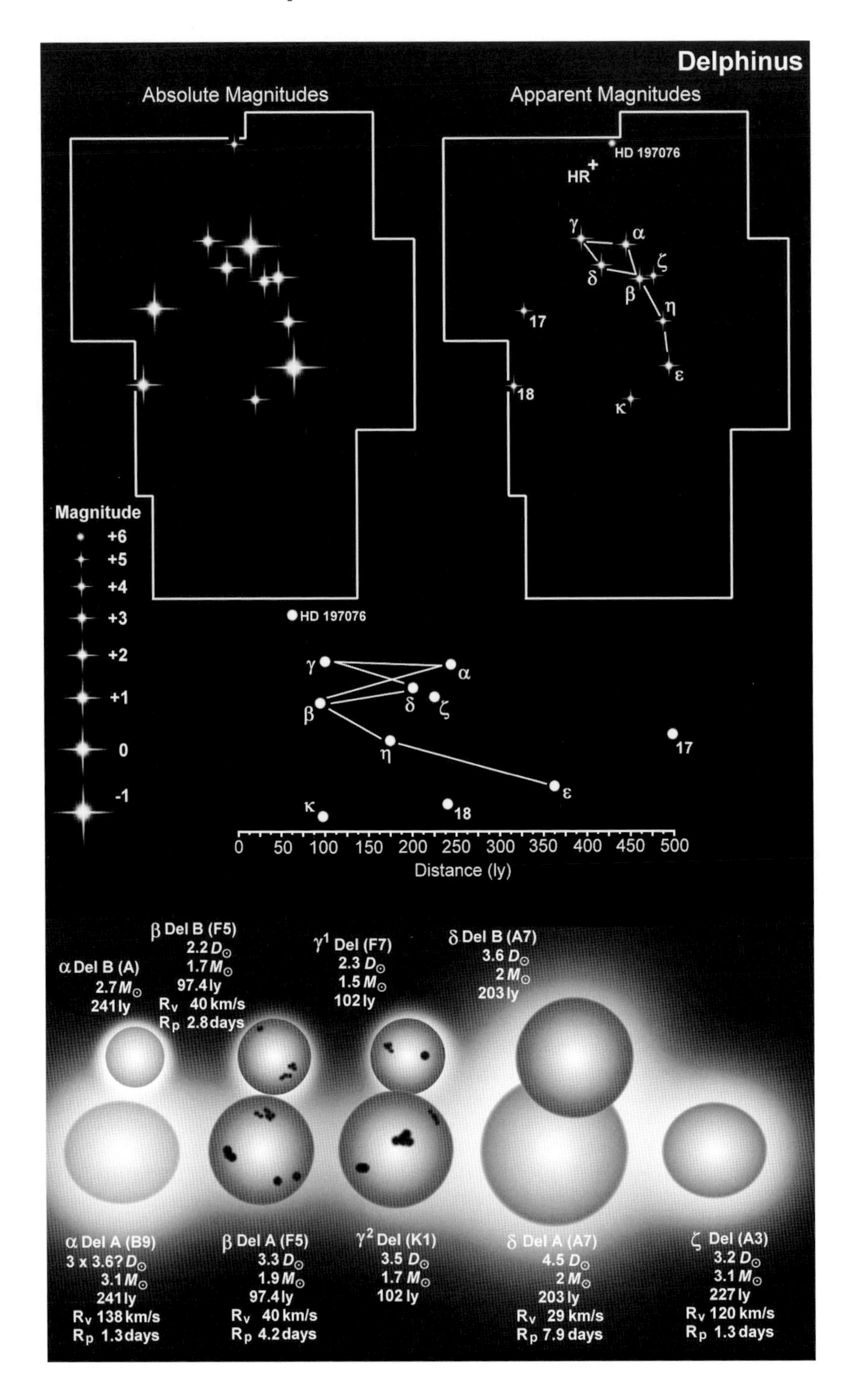

Delphinus
Absolute Magnitudes
Apparent Magnitudes
HD 197076
HR
γ
α
ζ
δ
β
η
17
18
ε
κ
Magnitude
+6
+5
+4
+3
+2
+1
0
-1
HD 197076
0 50 100 150 200 250 300 350 400 450 500
Distance (ly)
β Del B (F5)
2.2 D☉
1.7 M☉
97.4 ly
Rv 40 km/s
Rp 2.8 days
α Del B (A)
2.7 M☉
241 ly
γ1 Del (F7)
2.3 D☉
1.5 M☉
102 ly
δ Del B (A7)
3.6 D☉
2 M☉
203 ly
α Del A (B9)
3 x 3.6? D☉
3.1 M☉
241 ly
Rv 138 km/s
Rp 1.3 days
β Del A (F5)
3.3 D☉
1.9 M☉
97.4 ly
Rv 40 km/s
Rp 4.2 days
γ2 Del (K1)
3.5 D☉
1.7 M☉
102 ly
δ Del A (A7)
4.5 D☉
2 M☉
203 ly
Rv 29 km/s
Rp 7.9 days
ζ Del (A3)
3.2 D☉
3.1 M☉
227 ly
Rv 120 km/s
Rp 1.3 days

on the colors. γ^1 Del is a m_v +5.14 F7 so should be yellowish-white. γ^2 Del is a m_v +4.27 K1 and should be yellowish-orange. The brighter star is usually described as white or yellowish but the fainter has been recorded as yellow, blue, green and grayish-lilac. Lying at about 102 ly the pair are separated by 9.6″ (PA 267°) in a 3,200 year highly elliptical orbit that parts them by a maximum 600 AU but then brings them to within 40 AU of one another.

δ Delphini is a δ Scuti type variable. Its magnitude switches between m_v +4.38 and 4.49 with a period of 3^h 47.5^m. It is a 4.5 $D_\odot$ A7 – all δ Scutis are either A or F class – and is 203 ly distant. It is also a spectroscopic binary, its 3.6 $D_\odot$ companion in a 40.58 day orbit.

ζ Delphini is another A-class star, this time an A3, with a diameter of 3.2 $D_\odot$ and a mass of 3.1 $M_\odot$. It is 227 ly away.

The largest naked eye star in Delphinus is **17 Delphini** at 15 $D_\odot$. It is a K0, 144 times more luminous than the Sun and marks the farthest boundary of the constellation at somewhere between 474 and 520 ly.

The 8.5 $D_\odot$ **18 Delphini** is a bright yellow G6 giant which has a m_v of +5.48. In 2008 researchers announced the discovery of 10.3 M_J planet in an orbit that varies between 2.4 and 2.8 AU. The orbital period is about 993 days.

One star that does not appear to have any planets is **HD 197076**, a m_v +6.43 G5 of 0.96 $D_\odot$. However, it is regarded as one of the most likely candidates for life-bearing planetary systems and on 29 August 2001 a Message to Extra-terrestrial Intelligence (METI) was sent to the star. It will get there in February 2070.

On 8 July 1967 a new 5th magnitude star appeared close to the northern boundary of Delphinus. **HR Delphini**, as it is now known, is a 'slow nova'. It was discovered by the renowned British amateur astronomer and comet hunter G.E.D. Alcock. A search of photographic plates revealed that the star had previously been at magnitude +11.9 and took 30 days to brighten to naked eye visibility. Its magnitude varied over the next few months reaching a maximum of m_v +3.5 in December. It then went through a period of oscillations lasting 10 days each but by June it had become erratic and had faded to 6th magnitude. It is now at magnitude +13.4 and shows at least four shells expanding at 523, 612, 676 and 781 km/s.

Dorado

Constellation:	Dorado	**Hemisphere:**	Southern
Translation:	The Goldfish	**Area:**	179 deg^2
Genitive:	Doradûs	**% of sky:**	0.434%
Abbreviation:	Dor	**Size ranking:**	72nd

One of the 'modern' constellation constructed by the cartographer Petrus Plancius from observations made by explorers Pieter Keyser and Frederick de Houtman in the late 16th Century.

α **Doradûs** is a binary system 176 ly away. The main star is a m_v +3.26 white A0 of about 3 $M_\odot$. It is an αCV variable with a period of $2^d\ 22^h\ 48^m$ and a measured rotational velocity of 55 km/s, suggesting a diameter of 3.2 $D_\odot$. The secondary is a m_v +4.55 B9 in an elongated orbit. At close approach the two stars come within 1.9 AU while at apastron they are separated by 17.5 AU. Somewhat smaller at 1.7 $D_\odot$, it weighs in at 2.7 $M_\odot$.

β **Doradûs** could be the largest naked eye Cepheid – or perhaps not. Theoretical studies indicate a diameter of 450 $D_\odot$. That equates to 4.2 AU, so if β Dor replaced our Sun it would envelop Mercury, Venus, Earth and Mars and those minor planets on the inner edge of the Asteroid Belt. Direct measurements tell a different story however. They range from 55 to 350 $D_\odot$ (0.5 to 3.3 AU). Estimates of its distance also have a broad range from 1,040 to 7,300 ly while its luminosity has been put at between 3,000 and 125,000 $L_\odot$. On firmer ground is its variability period of $9^d\ 20^h\ 12.5^m$ and its spectrum which is F6 when the star is expanding rapidly and is at its hottest (about 6,300 K), and G5 when it is contracting rapidly and at its coolest (about 5,700 K).

A little larger than the Sun but 6.6 times more luminous γ **Doradûs** sits 66.2 ly away in the northernmost part of the constellation. Its magnitude is not stable but flickers between m_v +4.23 and +4.27 with a double period of 17.5 and 18.1 hours. Early F-class stars displaying this type of multiple periodicity are known as γ Doradûs-type variables of which about 60 are currently known. They all vary in periods between a few tenths of a day to a little over one day with amplitudes rarely exceeding 1/10th of a magnitude, the changes in brightness due to pulsations.

The B6 bluish ε **Doradûs** is just 1.8 $D_\odot$ but has a mass of 4.5 $M_\odot$. Lying at 512 ly, give or take 34 ly, it has as its backdrop the Large Magellanic Cloud (LMC), an irregular shaped galaxy some 160,000 ly away and 14,000 ly across, the third closest to us. Buried in the LMC is the Luminous Blue Variable (LBV) **S Doradûs**. At 10th magnitude it is too faint to be seen without optical aid – although it has been known to brighten to m_v +8.6 – but it is nonetheless an important star. It is believed to have a mass of at least 60 $M_\odot$ and a diameter of anywhere between 100 and 380 $D_\odot$. Its temperature could be as high as 20,000 K but it is its luminosity that is most impressive: a mind boggling 1 million Suns! Such brilliance is not without its drawbacks however. S Dor is using up its nuclear

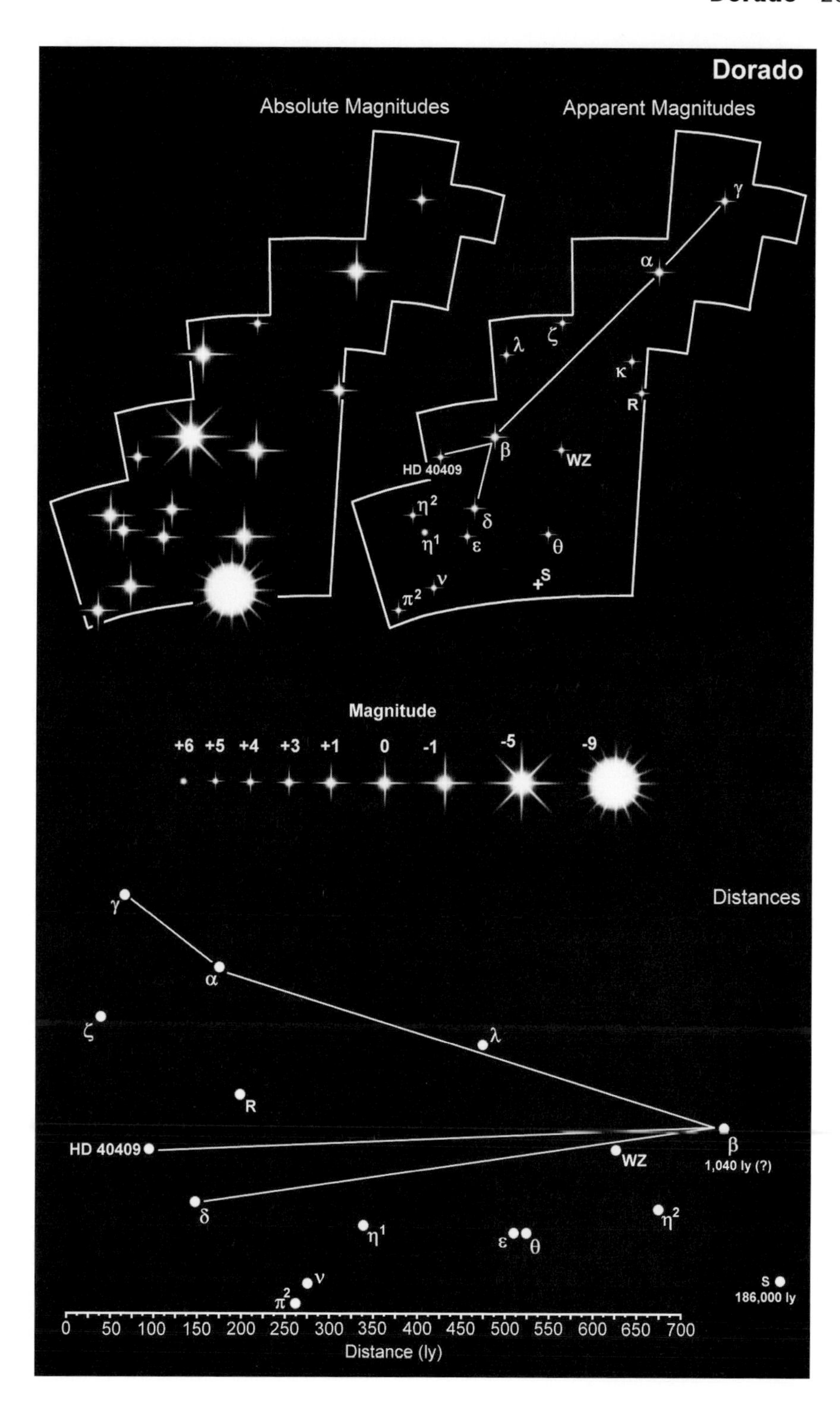
Dorado
Absolute Magnitudes
Apparent Magnitudes
L
γ
α
λ
ζ
κ
R
β
WZ
HD 40409
η2
δ
η1
ε
θ
S
π2
ν
Magnitude
+6 +5 +4 +3 +1 0 -1 -5 -9
Distances
1,040 ly (?)
186,000 ly
0 50 100 150 200 250 300 350 400 450 500 550 600 650 700
Distance (ly)

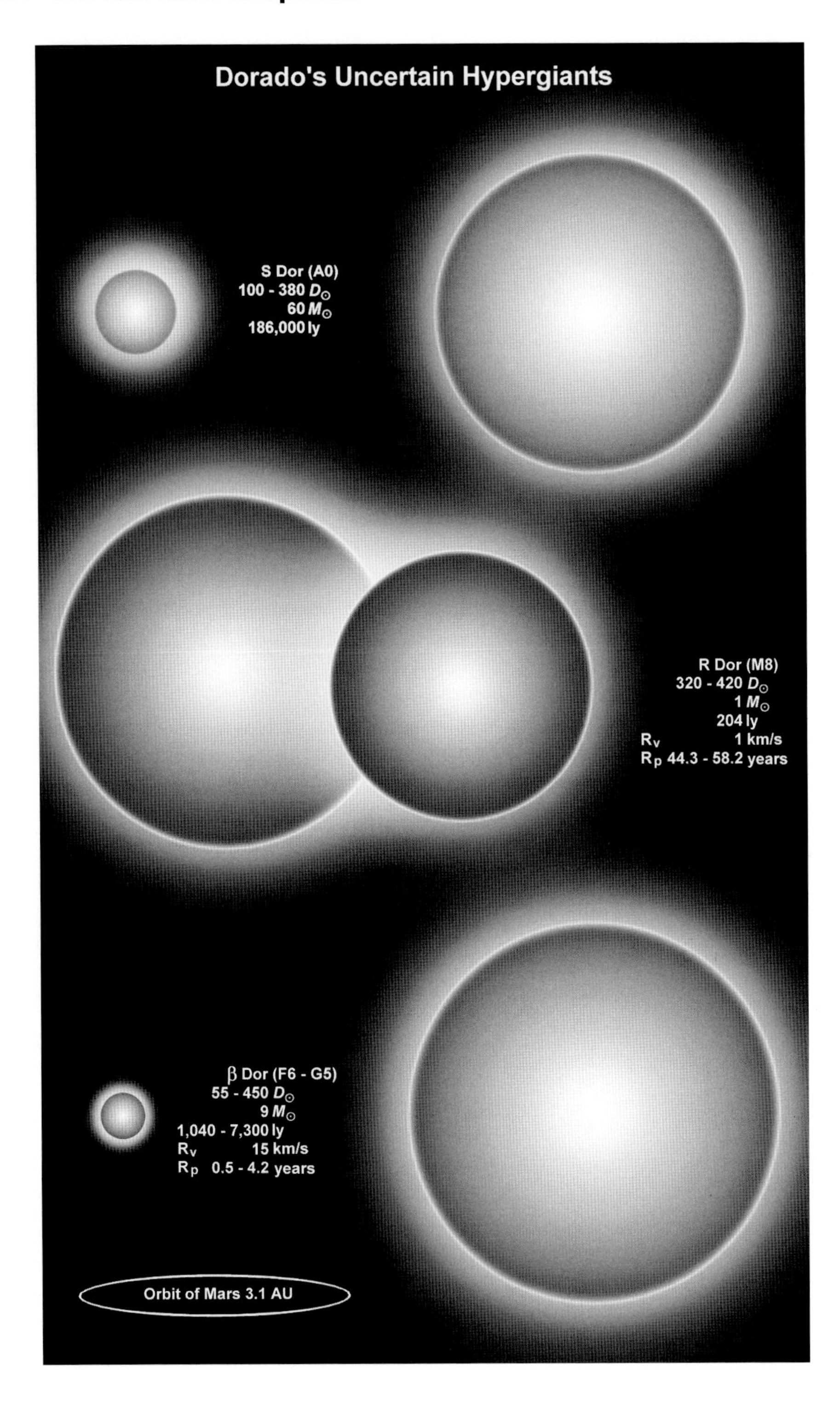
Dorado's Uncertain Hypergiants
S Dor (A0)
100 - 380 $D_{\odot}$
60 $M_{\odot}$
186,000 ly
R Dor (M8)
320 - 420 $D_{\odot}$
1 $M_{\odot}$
204 ly
R_V 1 km/s
R_P 44.3 - 58.2 years
β Dor (F6 - G5)
55 - 450 $D_{\odot}$
9 $M_{\odot}$
1,040 - 7,300 ly
R_V 15 km/s
R_P 0.5 - 4.2 years
Orbit of Mars 3.1 AU

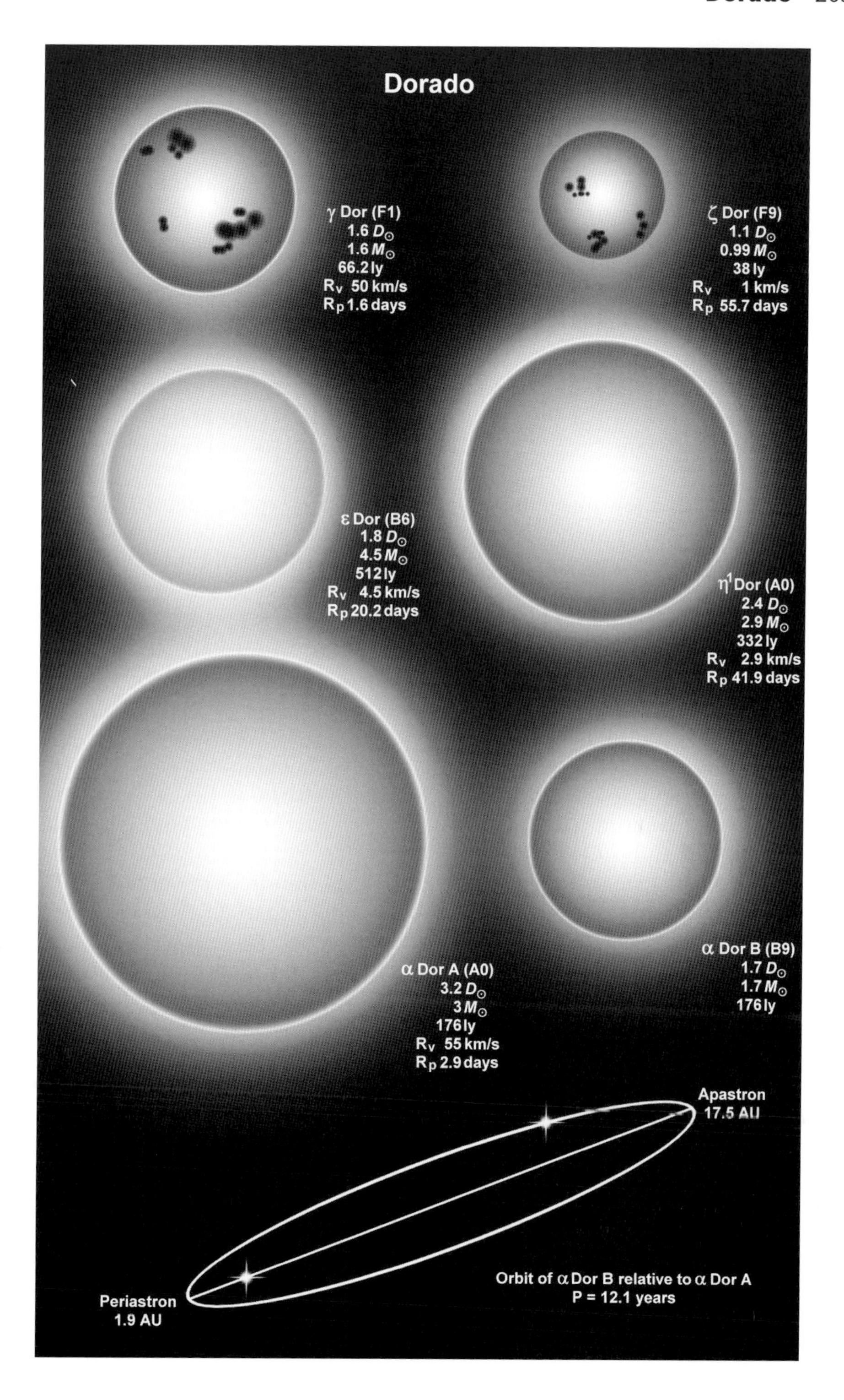
Dorado
γ Dor (F1)
1.6 $D_\odot$
1.6 $M_\odot$
66.2 ly
R_V 50 km/s
R_P 1.6 days
ζ Dor (F9)
1.1 $D_\odot$
0.99 $M_\odot$
38 ly
R_V 1 km/s
R_P 55.7 days
ε Dor (B6)
1.8 $D_\odot$
4.5 $M_\odot$
512 ly
R_V 4.5 km/s
R_P 20.2 days
η^1 Dor (A0)
2.4 $D_\odot$
2.9 $M_\odot$
332 ly
R_V 2.9 km/s
R_P 41.9 days
α Dor A (A0)
3.2 $D_\odot$
3 $M_\odot$
176 ly
R_V 55 km/s
R_P 2.9 days
α Dor B (B9)
1.7 $D_\odot$
1.7 $M_\odot$
176 ly
Apastron
17.5 AU
Periastron
1.9 AU
Orbit of α Dor B relative to α Dor A
P = 12.1 years

fuel so quickly that it will burn out in just a few million years, perhaps first evolving into a Wolf-Rayet star before ending its short life as a spectacular supernova. While it lives, the enormous radiation pressure will continuously eject vast amounts of matter into space, supplementing its self-made nebula with the occasional massive eruption. S Doradûs-type variables, as they are otherwise known, always belong to the B, A or F spectral classes and display brightness changes usually in the range of 1 to 7 magnitudes. η Carinae and P Cygni are members of this variable star type.

At 38 ly the closest of the Dorado stars is ζ **Doradûs**: a 1.1 $D_{\odot}$ F9 with an apparent magnitude of m_v +4.72 and, because of its distance of almost 10 pc, an absolute magnitude, M_v, of +4.10.

Like many stars that share a designation, η^1 and η^2 **Doradûs** are unrelated. η^1, the more southerly of the two, is a m_v +5.71 white A0 at 332 ly. η^2 is more than twice the distance at 671 ly, is brighter at m_v +5.01 and is an M2.5 red giant of 32 $D_{\odot}$.

R Doradûs is a low mass, luminous supergiant on the border with Reticulum. Approximately 200 ly away it has an estimated diameter of between 320 and 420 $D_{\odot}$. In Solar System terms if our Sun was replaced with R Dor it would engulf the orbit of Mars if the lower size estimation is correct and would stretch out to the inner edges of the Asteroid Belt if the larger diameter is correct. Despite its huge size it has a mass equivalent to just 1 $M_{\odot}$. Its luminosity is 6,500 $L_{\odot}$ but it appears dim because it radiates most of its energy in the infrared. Belonging to spectral class M8 it is a semi-regular, SRb, variable which pulsates between m_v +4.80 and +6.60 over a period of about 338 days.

The star **30 Doradûs** isn't a star at all – it is a nebula. Also cataloged as **NGC 2070** but better known as the **Tarantula Nebula** it lies at a distance of 180,000 ly and is illuminated by a 'super star cluster' of very young (1-2 million years old) giant and supergiant blue-white O-class stars, mainly O3s. This cluster, also called **RMC 136** or simply **R136**, consists of hundreds of very hot, 50,000 K individuals confined to a region of space no more than 35 ly in diameter. The total mass is in the order of 450,000 $M_{\odot}$. The nebula also contains an older cluster, Hodge 31, whose stars are 20-25 million years old. It is believed that around 40% of Hodge 31 stars have already exploded as supernovae. In 1987 a supernova was observed on the edge of the nebula which reached 3rd magnitude. Optimistically called **1987A** (a second supernova that year would have been called 1987B) its behavior was unlike other supernovae which tend to flash and then grow dimmer. Instead 1987A grew brighter as its shock wave crashed into the surrounding nebula, the compression causing a temperature rise of millions of degrees and the emission of massive amounts of X-rays and radio waves.

Draco

Constellation:	Draco	**Hemisphere:**	Northern
Translation:	The Dragon	**Area:**	1,083 deg^2
Genitive:	Draconis	**% of sky:**	2.625%
Abbreviation:	Dra	**Size ranking:**	8th

In Greek mythology Draco was the dragon Ladon who guarded the golden apples in the garden of the Hesperides, the three daughters of Atlas. As one of his 12 labors Hercules slew the dragon and carried off some of the apples to Eurystheus. A long, sprawling constellation near Ursa Major and Ursa Minor, the stars of Draco are all relatively faint.

Thuban, or α **Draconis**, is the 8th brightest star in the constellation. The difference between it and the brightest star, rather confusingly γ Dra, is 1.42 magnitudes. As a result it is not the easiest of stars to spot in the night sky although, just as α and β Ursae Majoris point to Polaris, the other two stars in the bear's quadrangle, γ and δ Ursae Majoris, point to Thuban. A 6.2 $D_{\odot}$ A0 it is 309 ly away and rotates at 15 km/s taking 21 days to complete one revolution on its axis. In 2787 BC Thuban was the Pole Star, closer to the North Celestial Pole than today's Polaris, and will be so again in about AD 20346. It is also a spectroscopic binary with a 51.42 day orbit.

G2 yellow giants are relatively rare: only a handful of naked eye examples are known (about 0.3% of visible stars). **β Draconis** or Alwaid is one of them. Some 37 $D_{\odot}$ across it lies at a distance of 362 ly and shines at m_v +2.77 but would brighten to M_v -2.1 at 10 parsecs. Like α Dra it has a rotational velocity of 13 km/s but, because of its larger size, takes much longer to rotate: 144 days. It has a 14th magnitude companion: a 0.4 $M_{\odot}$ G8 in a 4,000+ year orbit.

γ Draconis or Eltanin is the brightest star in the constellation despite being given the 3rd brightest designation. Currently at a distance of 148 ly it is a steady m_v +2.21, its K5 classification indicating its orange appearance. It is heading in our general direction at 27.6 km/s and, in 1.5 million years, will pass by at 28 ly to become one of the brightest stars in the sky. A giant at 64 $D_{\odot}$ across and with a luminosity of 600 $L_{\odot}$ Eltanin, to give the star its proper name, is a slow rotator. It spins at 3.5 km/s and so takes 2.6 years to rotate just once.

Relatively close to one another in the night sky and marking the point at which the dragon's neck curves are two G-class stars **δ** and **ε Draconis**. There is almost a magnitude between them, δ Dra being the brighter of the two at m_v +3.06 while ε Dra lags behind at +3.91, and δ is larger than ε – 13 $D_{\odot}$ and 8.8 $D_{\odot}$ respectively – the former being a G8 while the latter is a G9. δ is the closer of the two at 100 ly while ε lies at 146 ly. The big difference between them is that while δ Dra seems to be a lone star, ε Dra is a spectroscopic binary, its companion was discovered, as so many binaries were, by F.W. Struve, the founder of the Pulkovo Observatory near St. Petersburg.

η Draconis is a double star 87.7 ly from Earth. The primary is a 9.2 $D_{\odot}$ G8 with a luminosity of 47 $L_{\odot}$. Rotating at 3.7 km/s it takes nearly 126 days to complete a single revolution. Its companion is a K1 dwarf in an orbit that averages 145 AU and with a period in excess of 1,000 years.

In the tail of the dragon is a triplet of stars that are totally unassociated, other than appearing in the same small area of the night sky. The brightest of these is **κ Draconis**, a m_V +3.82 B6 'Be emission' star surrounded by a bright disk and a thick absorbing shell believed to be due to material flung off the star as it rotates at up to 200 km/s. About four times larger than the Sun it lies at a distance of 498 ly. It belongs to the γ Cas type variables switching between m_V +3.82 and +4.01 with no discernable period. Flamsteed numbered this star 5 Draconis. The one slightly to the south he named **4 Draconis** and the more northerly star was designated **6 Draconis.** 4 Dra is, officially, 83 ly deeper into space at 581 ly. However, the margin of error in the measurements means that the two stars could actually be very close neighbors. In fact, 4 Dra could actually be 4 ly closer to us than κ Dra. 4 Dra is also a variable of the pulsating variety and belonging to the Lb class. It is a red giant some 35 times larger than the Sun and sitting in spectral class M4. Its amplitude is much less than κ Dra varying between m_V +4.95 and +5.04. Because of its variability it is sometimes known as **CQ Draconis**. Then there is 6 Draconis: a 21 $D_{\odot}$ orange giant of spectral class K3. Its magnitude is rock steady at m_V +4.96. Spinning slower than the Sun at 1.7 km/s (Sun, 2 km/s) the star takes 625 days – 1.7 years – to rotate just once. It is estimated to be 546 ly away but again the margin of error means that it could well be rubbing shoulders with κ and 4 Dra (see table). On a dark, clear night with a good binocular and perfect color vision the three stars show as blue, red and orange.

The errors in measuring the distances of κ, 4 and 6 Draconis means that they may be closer neighbors than we think. They could all lie between 533 and 537 ly.

Star	Distance limits
κ Dra	< 459 to 537 ly >
4 Dra	< 533 to 629 ly >
6 Dra	< 476 to 616 ly >

The closest Draconian star to us is **σ Draconis** at 18.8 ly. A K0 dwarf just 70% the size of the Sun and only a third as luminous σ Dra is a faint m_V +4.68 which would fall by over a full magnitude at 10 pc to M_V +5.9. Its spectrum reveals it is a stranger to our corner of the Galaxy having a metal content significantly lower than the Sun.

Near the head of the dragon some 163 ly distant is the m_V +5.95 star **HD 167042**, a K1 of 4.3 $D_{\odot}$ and believed to be about 2,200 million years old (half the Sun's age). In an orbit that varies between 1.26 and 1.34 AU is a planet a little more than 1½ Jovian masses. Discovered in 2007 its year is 416 days long.

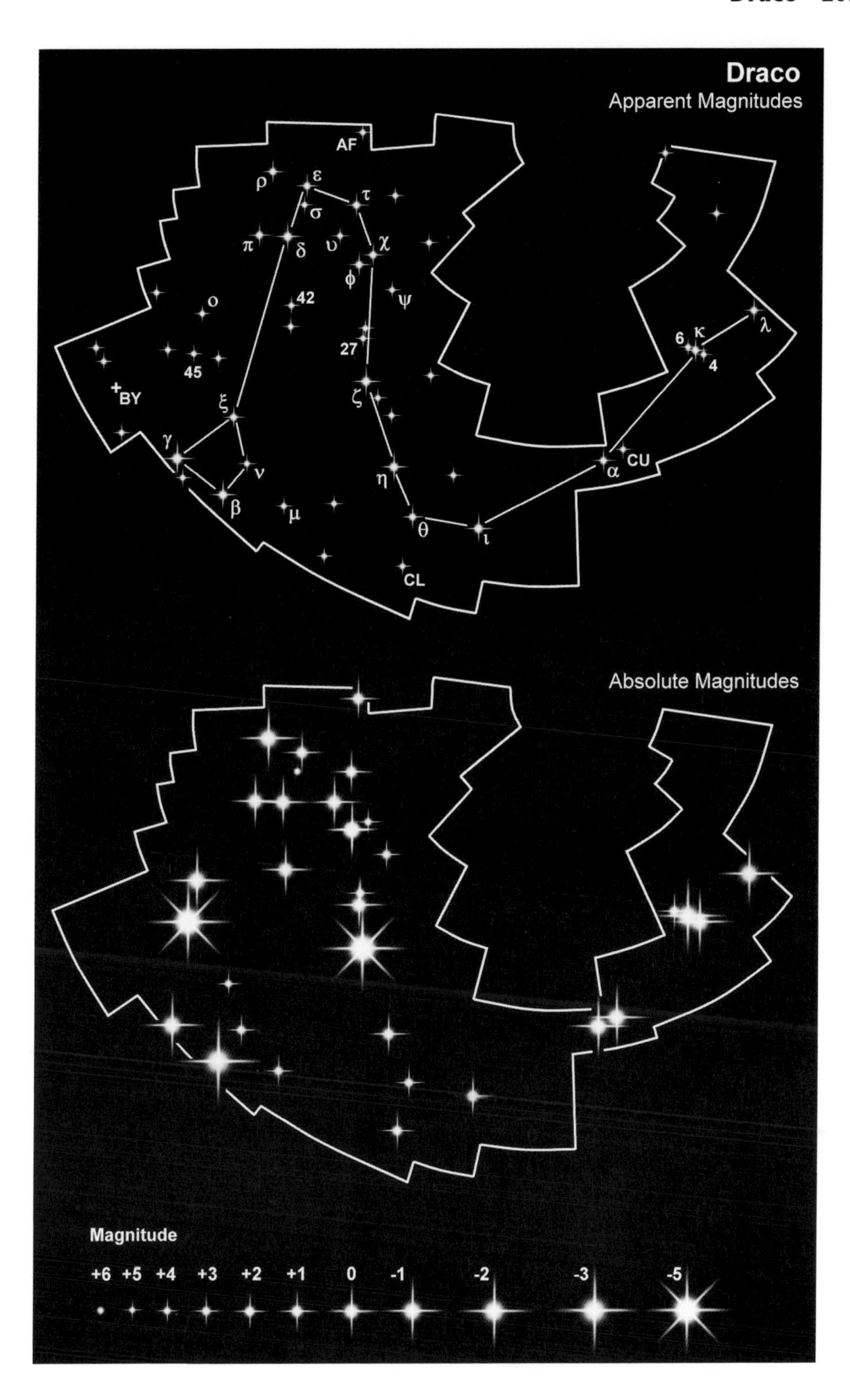
Draco
Apparent Magnitudes
AF
ρ
ε
τ
σ
π
δ
υ
χ
φ
42
ψ
ο
27
6
κ
λ
4
45
BY
ξ
ζ
γ
CU
ν
η
α
β
μ
θ
ι
CL
Absolute Magnitudes
Magnitude
+6
+5
+4
+3
+2
+1
0
-1
-2
-3
-5

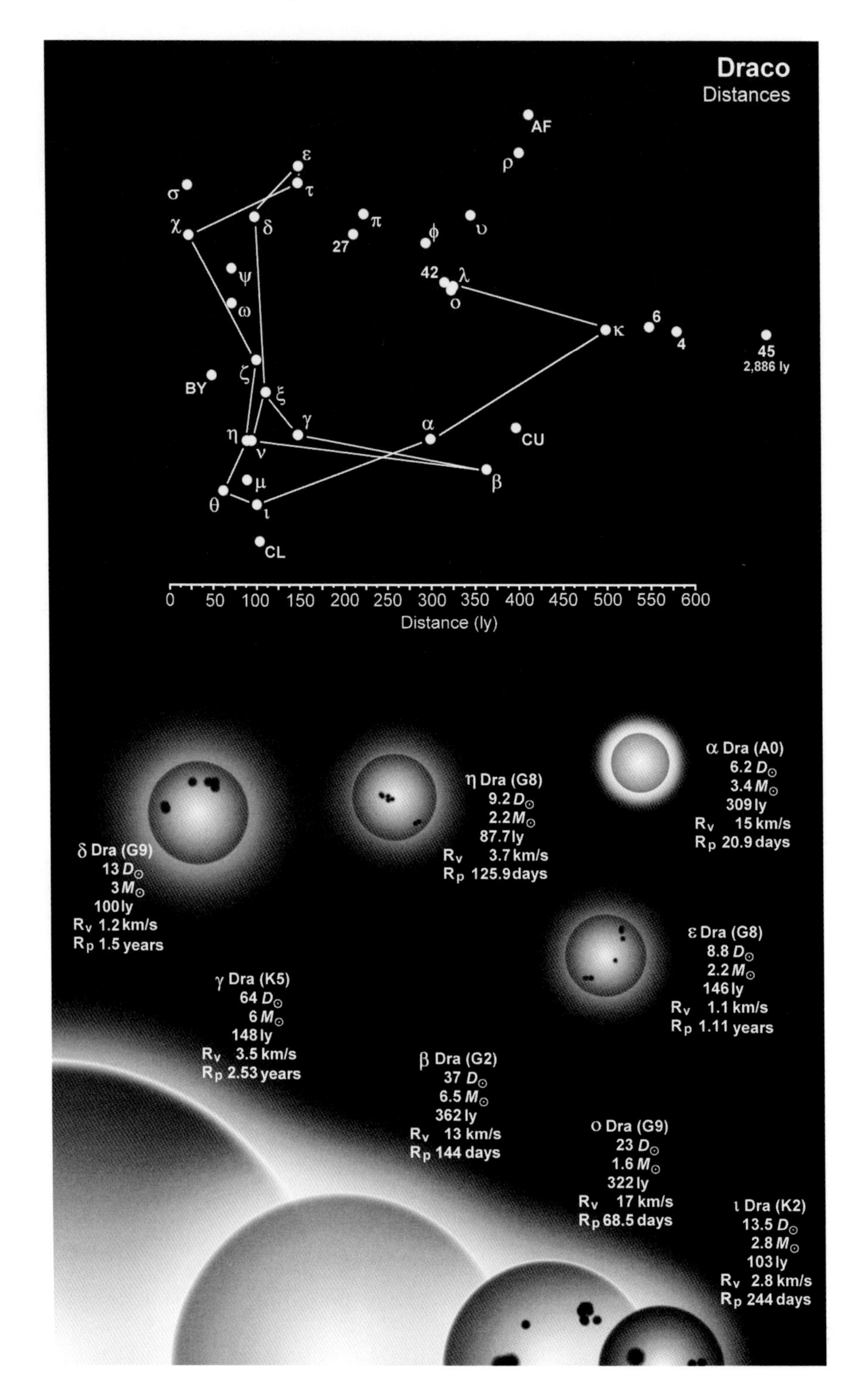

Draco
Distances
AF
ρ
ε
σ
τ
χ
δ
π
27
ϕ
υ
ψ
42
λ
ω
ο
κ
6
4
45
2,886 ly
BY
ζ
ξ
γ
α
CU
η
ν
β
μ
θ
ι
CL
0 50 100 150 200 250 300 350 400 450 500 550 600
Distance (ly)
α Dra (A0)
6.2 $D_\odot$
3.4 $M_\odot$
309 ly
R_V 15 km/s
R_P 20.9 days
η Dra (G8)
9.2 $D_\odot$
2.2 $M_\odot$
87.7 ly
R_V 3.7 km/s
R_P 125.9 days
δ Dra (G9)
13 $D_\odot$
3 $M_\odot$
100 ly
R_V 1.2 km/s
R_P 1.5 years
ε Dra (G8)
8.8 $D_\odot$
2.2 $M_\odot$
146 ly
R_V 1.1 km/s
R_P 1.11 years
γ Dra (K5)
64 $D_\odot$
6 $M_\odot$
148 ly
R_V 3.5 km/s
R_P 2.53 years
β Dra (G2)
37 $D_\odot$
6.5 $M_\odot$
362 ly
R_V 13 km/s
R_P 144 days
ο Dra (G9)
23 $D_\odot$
1.6 $M_\odot$
322 ly
R_V 17 km/s
R_P 68.5 days
ι Dra (K2)
13.5 $D_\odot$
2.8 $M_\odot$
103 ly
R_V 2.8 km/s
R_P 244 days

In 2002 **HIP 75458** was the first giant star to yield a planet. More widely known as ι **Draconis** this K2 orange star is about 13.5 $D_\odot$. The planet is not far off being a brown dwarf with a mass of nearly 9 M_J. Its orbital period is 1.4 years. A second orange giant, the 22 $D_\odot$ **42 Draconis** has a 3.9 M_J planet in a 1.3 year orbit. At an estimated age of between 7,700 and 11,250 million years this star is fast approaching the end of its life. Yet a third planetary system around an orange giant is **HD 139357**, although this system is much younger at about 3,070 million years. Its planet is estimated to be 9.76 M_J but with a high degree of uncertainty that could push its mass up to 11.91 M_J teetering on the brink of being a brown dwarf. Current views are that gaseous giants of 13 M_J are able to fuse deuterium and therefore do not behave as a planet but are undergoing processes that are more akin to stars.

HD 193664 seems to be planetless, though that may just be because our detection methods are not yet good enough to find any. The star has been identified as one of those most likely to have planetary systems that harbor life. It is a m_v +5.90 solar analog, 0.99 $D_\odot$ and 57.3 ly away.

27 Draconis is going somewhere in a hurry. Half the stars in our corner of the Galaxy have radial velocities of 10 km/s or less, either heading towards or away from us. Some 96.5% have radial velocities of up to 40 km/s. But 27 Dra is in the 1% of stars whose velocities exceed 70 km/s. 27 Dra is in fact the 33rd fastest naked eye star in the night sky, hurtling towards us at 73.5 km/s (the fastest is τ^1 Lupi closing in at a staggering 215 km/s). A 14 $D_\odot$ K0 giant it is currently 217 ly away.

The outlier is the 68 $D_\odot$ **45 Draconis** at somewhere between 2,050 and 3,720 ly. An F7 with a blinding 7,846 $L_\odot$ it has a modest visual magnitude of m_v +4.77 but an absolute magnitude of M_v -4.6, or about as bright as Venus at her most glorious.

The 8th magnitude **BY Draconis** is the prototype for a class of dwarf variable that can have periods lasting from a fraction of a day up to 120 days and whose brightness can change by up to 0.5 magnitude. The variability appears to be linked to the presence of large groups of starspots, some of which may appear in polar regions, and intense chromospheric activity. BY Dra itself is a K6 – the type only occur in K and M classes – and is believed to be 0.91 $D_\odot$ and 0.59 $M_\odot$. Lying at a distance of 53.4 ly its variability period is 5.8285 days. It is actually a triple star system. One component orbits the primary every 5.98 days and is m_v +9.26. It is slightly less massive at 0.52 $M_\odot$ and is an M1. The other member of the trio is a red sub-dwarf, an M5, of m_v +15.8. It lies at an average distance of 260 AU and takes 3,822 years to orbit the primary. It is less than half the mass at 0.21 $M_\odot$.

Just to the east of **χ Draconis** is an arrangement of six stars called Kemble 2 that look like a mini version of Cassiopeia, only much smaller and fainter (see Cassiopeia for further details).

Draco includes at least seven naked eye variable stars: κ Dra and 4 Dra are mentioned above. Details of these and the others are in the table below.

Variable stars in Draco

Star	Maximum Magnitude	Minimum Magnitude	Variable Type	Variability Period	$D_\odot$	$L_\odot$	Distance (ly)	Spectrum
κ Dra	+3.82	+4.01	γ Cas		1.4	540	498	B6
o Dra	+4.63	+4.73	RS		27	110	322	K0
ϕ Dra	+4.22	+4.26	α CV	1.71646 d	1.5	131	289	A0
AF Dra	+5.15	+5.22	α CV	20.2747 d	2.7	111	417	A0
CL Dra	+4.95	+4.97	δ Sct	0.0630 d	1.9	9.62	110	F0
CU Dra	+4.52	+4.67	Lb		34	167	392	M3.5
4 Dra	+4.95	+5.04	Lb		35	251	581	M3

Equuleus

Constellation:	Equuleus	**Hemisphere:**	Northern
Translation:	The Foal	**Area:**	72 deg^2
Genitive:	Equulei	**% of sky:**	0.175%
Abbreviation:	Equ	**Size ranking:**	87th

The second smallest and often ignored constellation. Equuleus was said to be the brother of Pegasus, which is to its north east, and is one of Ptolemy's 48 constellations. Usually drawn as a quadrangle its stars are all faint and often difficult to find.

α **Equulei**, otherwise known as Kitalpha, is the brightest star in the constellation but at m_v +3.94 is by no means bright. A G0 giant of 9.2 $D_\odot$ and 3.2 $M_\odot$ it is 68 times more luminous than the Sun. Rotating more slowly than the Sun, just 1.3 km/s compared to 2 km/s, Kitalpha takes almost a full year - 358 days - to turn on its axis. It is also a spectroscopic binary. Its companion, an A5 dwarf of 2.9 $M_\odot$, is in a near circular orbit of 0.66 AU with a period of 99 days. This is a relatively young system, 500 to 600 million years, but already the G0 giant is beginning to die. The estimated distance is 186 ly.

β **Equulei** is about 10% larger than the Sun but is an A3, its subtle bluish-white color barely detectable because of its faint magnitude, just m_v +5.15. At a distance of 360 ly β Equ has a luminosity of 83 $L_\odot$ and a fairly modest rotational velocity of 40 km/s, completing a full turn in less than 1½ days.

γ **Equulei** is an α CV rotating variable of no particular period or multiple-periods (estimates include 17.492 days, 314 days, 1,785 days and 72 years). Its magnitude fluctuates between m_v +4.58 and +4.77 as this 1.9 $D_\odot$ F1 star spins on its axis 115 ly from Earth. Seen through a small telescope or binocular γ Equ turns out to have a close optical companion. The other star, an A2 which Flamsteed numbered as **6 Equulei**, is rather deeper in space at 479 ly and is m_v +6.07.

The closest of the stars in Equuleus at 60 ly is also the most interesting. δ **Equulei** is a very close binary system with an orbital period of 5.7 years, the shortest period on record of any naked eye binary. The two components are very similar in many ways. One of the stars is an F5 with a calculated temperature of 6,600 K and a mass of 1.2 $M_\odot$. The other is a slightly cooler 6,000 K G0 with a little more mass at 1.3 $M_\odot$. Separated by about 4.5 AU the two stars are larger than the Sun in diameter at 1.7 and 1.1 $D_\odot$ respectively. The individual magnitudes come in at m_v +5.2 and +5.3 but combined they glow at m_v +4.5.

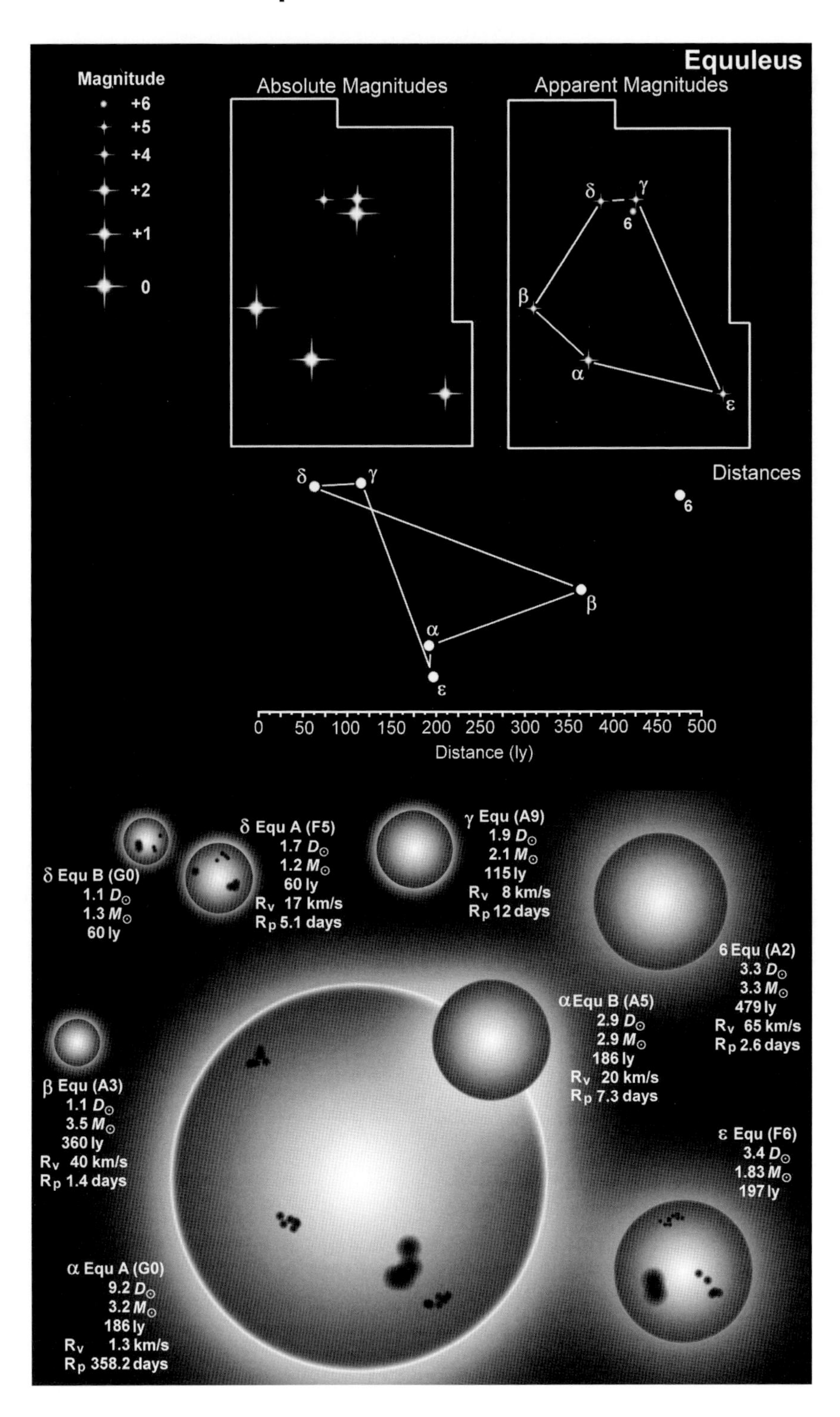

Equuleus
Magnitude
+6
+5
+4
+2
+1
0
Absolute Magnitudes
Apparent Magnitudes
δ
γ
6
β
α
ε
Distances
0 50 100 150 200 250 300 350 400 450 500
Distance (ly)
δ Equ B (G0)
1.1 D⊙
1.3 M⊙
60 ly
δ Equ A (F5)
1.7 D⊙
1.2 M⊙
60 ly
Rv 17 km/s
Rp 5.1 days
γ Equ (A9)
1.9 D⊙
2.1 M⊙
115 ly
Rv 8 km/s
Rp 12 days
6 Equ (A2)
3.3 D⊙
3.3 M⊙
479 ly
Rv 65 km/s
Rp 2.6 days
α Equ B (A5)
2.9 D⊙
2.9 M⊙
186 ly
Rv 20 km/s
Rp 7.3 days
β Equ (A3)
1.1 D⊙
3.5 M⊙
360 ly
Rv 40 km/s
Rp 1.4 days
ε Equ (F6)
3.4 D⊙
1.83 M⊙
197 ly
α Equ A (G0)
9.2 D⊙
3.2 M⊙
186 ly
Rv 1.3 km/s
Rp 358.2 days

Eridanus

Constellation:	Eridanus	**Hemisphere:**	Southern
Translation:	The Eridanus River	**Area:**	1,138 deg^2
Genitive:	Eridani	**% of sky:**	2.759%
Abbreviation:	Eri	**Size ranking:**	6th

This long, straggling constellation stretches from just south of the Celestial Equator to -60° meandering between six other constellations. It represents the river into which Phaethon, son of Helios the Sun god, fell after trying to ride his father's chariot across the sky. Tut, kids!

Meaning the 'Star at the end of the river' Achernar, or α **Eridani**, lies at the southernmost tip of the constellation and hence is hidden from most of the population of Europe and North America: essentially anyone much farther north than the Tropic of Cancer. At a very slightly variable m_v +0.46 it is the 9th brightest star in the night sky. A blue B3 it has an equatorial diameter of perhaps 12 $D_\odot$ and a polar diameter of about two-thirds that, say 8 $D_\odot$. It rotates at 250 km/s – 125 times faster than the Sun – and so spins once on its axis in just 2.3 days. Such high speeds distort the star causing it to bulge at the middle, creating a dark band around the equator and ejecting material into a disk around the star. Currently 144 ly from Earth and drifting away at 16 km/s it has a luminosity of around 5,000 Suns and would brighten to M_v -2.8 at 10 pc.

At the opposite end of the river, not far from Rigel (β Orionis) is **β Eridani** or Kursa, although some people extend the beginning of the river to include λ Eridani. One of the closer stars to us at 88.8 ly it is slightly variable between m_v +2.72 and +2.80 though does not easily fall into any recognized category. There is a report dating back to 1985 that the star suddenly brightened to zero magnitude for a couple of hours. An A3 it is about four times larger than the Sun.

Only a dozen red giants are brighter than 3rd magnitude and **γ Eridani** or Zaurak just scrapes in at m_v +2.88 to +2.96 being an Lb pulsating variable. About 69 times larger than the Sun – it could easily hold 328,509 solar globes and there would still be enough room to swing a cosmic cat – the star has a mass of only about 2 $M_\odot$. It is, nonetheless, considerably more luminous at 860 $L_\odot$. From its corner of the Galaxy 221 ly away this aging giant is no geriatric however, flying away from us at an impressive 61.7 km/s (only a few percent of stars have velocities in excess of 40 km/s). If it were heading towards us then it would reach an absolute magnitude of M_v -2 at 10 pc.

Widely listed as an RS Canum Venaticorum type variable **δ Eridani** seems to be anything but. RS CVn variability is caused by a close binary system stirring up strong magnetic fields, yet δ Eri is very much a single, isolated star. Its variable label was assigned to it in 1987 but the latest Hipparcos analysis shows it to be a rock steady m_v +3.522 and not fluctuating between +3.51 and +3.56 as so often claimed. A K1 of 11 $D_\odot$ it is only 29.5 ly away and getting closer at 6.1 km/s.

In 1960 Frank Drake used the 85-foot (26 meter) radio telescope at Green

Bank, West Virginia to search for artificial signals from two stars. One was τ Ceti, the other ε **Eridani**. Nothing was ever detected but that was only the start of the story. In 1983 IRAS (the Infra-Red Astronomy Satellite) detected a significant amount of dust circling ε Eri hinting at the possibility of a planetary system. Twelve years later ε Eri became one of the targets for Project Phoenix which sent out radio signals to those stars that were most likely to harbor planets and life. Then, in 2000, a team led by Artie Hatzes announced the discovery of a planet. What makes ε Eri so attractive is that it is only 10.4 ly from Earth, making it the 10th closest system and the 3rd closest naked eye star. The star itself is an orange K2 dwarf, about 10% smaller than the Sun at 0.895 $D_{\odot}$ and 0.83 $M_{\odot}$. It rotates on its axis once every 22.7 days and, by solar standards, is very young at just 660 million years. The debris disk is most likely the leftovers from the initial planet forming period early in the history of the system. The disk appears to have considerable structure with high concentrations of dust at 3 and 29 AU and farther out, in much the same way as the Solar System has a structured Asteroid Belt separating the rocky, terrestrial planets from the gaseous giants, and the Kuiper Belt and Oort Cloud extending beyond the farthest giant planets to perhaps half way to α Centauri. Such structure is caused by a number of processes, not least being the presence of giant planets. As giant planets successively complete one orbit after another they gravitationally sweep some areas clean while concentrating dust in others. In about 31,500 years time the BL/UV Ceti red dwarf binary will pass within 0.93 ly of ε Eri disrupting Oort Cloud, sending material into the inner ε Eri system which may impact with any planets that exist. The current status with ε Eri is that it has one confirmed and one unconfirmed planet (see table). The confirmed planet is in a highly eccentric orbit with a periastron of 1 AU and an apastron of 5.77 AU. That would be like Earth moving between our current position and that of Jupiter as we orbit the Sun. The theoretical planet, if it exists, is in a 280 year long orbit that ranges from 28 AU to 52 AU (about the distance of Neptune to almost twice its distance). It is much smaller than the confirmed planet: just 0.1 M_J or 32 $M_{\oplus}$. But what about the Phoenix signals? Well, by the time they get to the planet, are decoded by any intelligent life that may happen to live there, and a reply is sent back to us, we should expect a signal any time after 2016. Unless, of course, their government is a bureaucratic as most of ours in which case we could be in for a very, very long wait.

θ Eridani looks like an uninteresting 3rd magnitude star to the naked eye but modest optical equipment will reveal twin white stars. θ^1 Eri is a m_v +3.18 early-A class (A3) with a diameter of 5.6 $D_{\odot}$ and 106 $L_{\odot}$. Its partner, θ^2 Eri, is somewhat dimmer at m_v +4.11, belongs to the A1 spectral class, has a much smaller diameter of 1.7 $D_{\odot}$ and luminosity of 45 $L_{\odot}$. Separated by 8.2″ at PA 88° the pair lie at a distance of 161 ly. Their positions relative to one another appear to be fixed, so they are either not in orbit or their orbital period is so long as to be undetectable.

At 1,754 ly **λ Eridani** is the farthest star in the constellation. A highly luminous B2 of 38,750 $L_{\odot}$, it is 12 $D_{\odot}$ across but its rotational velocity is proving difficult to pin down with measurements ranging from 220 to 325 km/s. It is

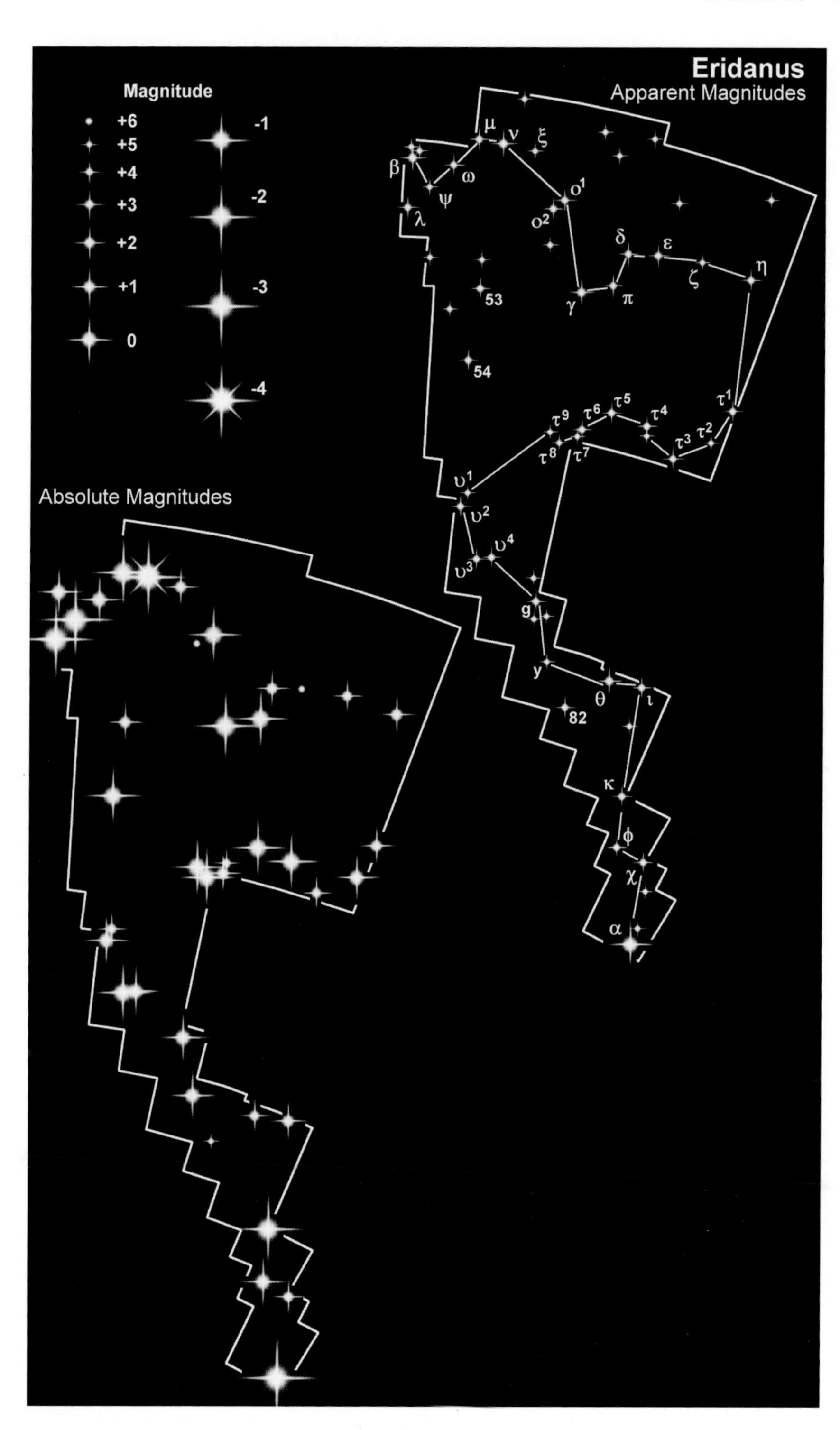
Eridanus
Apparent Magnitudes
Magnitude
+6
+5
+4
+3
+2
+1
0
-1
-2
-3
-4
Absolute Magnitudes
μ
ν
ξ
β
ω
ψ
λ
ο1
ο2
δ
ε
ζ
η
γ
π
53
54
τ1
τ2
τ3
τ4
τ5
τ6
τ7
τ8
τ9
υ1
υ2
υ3
υ4
g
y
θ
ι
82
κ
φ
χ
α

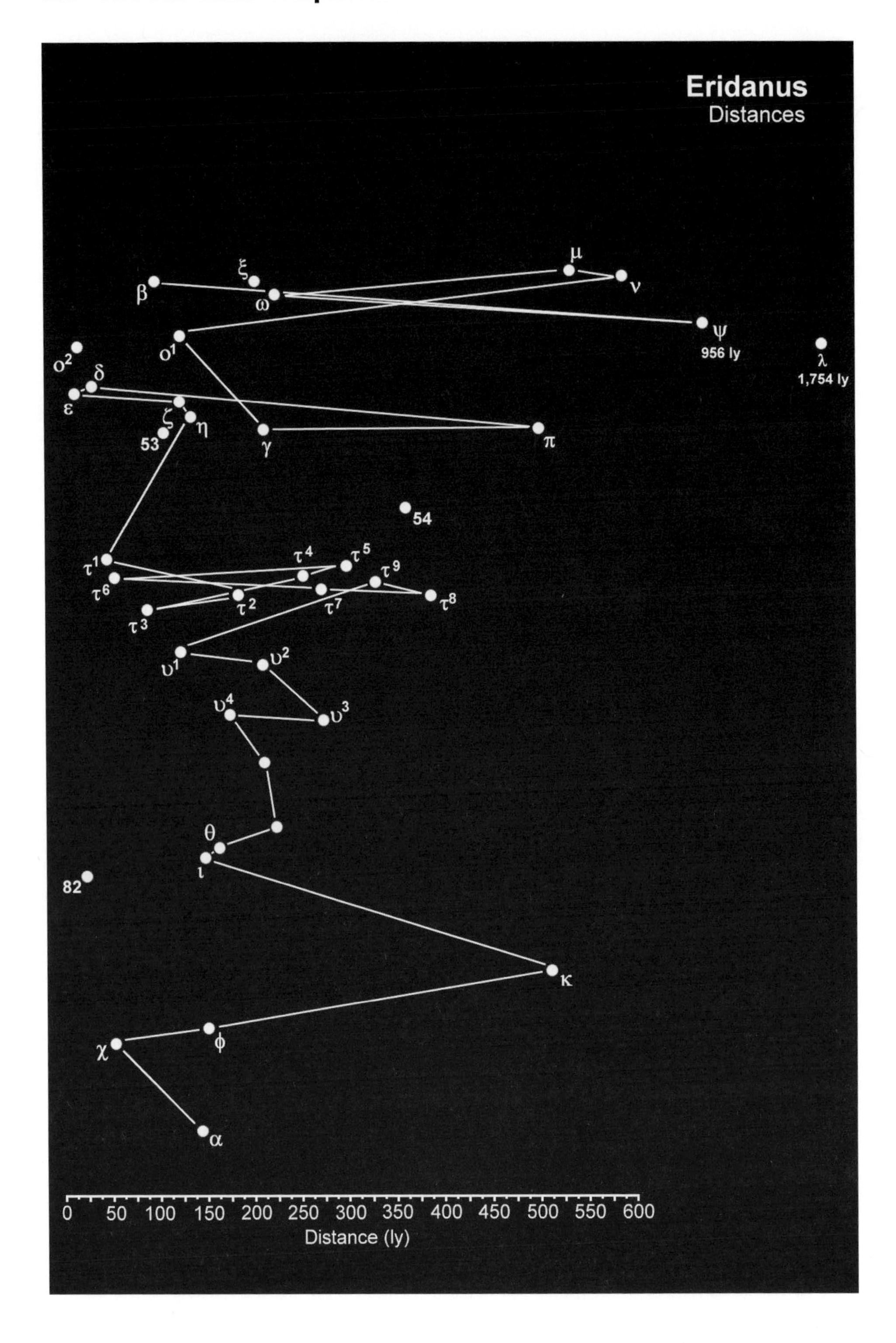

Eridanus
Distances
μ
ξ
ν
β
ω
ψ
956 ly
λ
1,754 ly
o1
o2
δ
ε
ζ
η
53
γ
π
54
τ1
τ4
τ5
τ9
τ6
τ2
τ7
τ8
τ3
υ1
υ2
υ4
υ3
θ
ι
82
κ
φ
χ
α
0 50 100 150 200 250 300 350 400 450 500 550 600
Distance (ly)

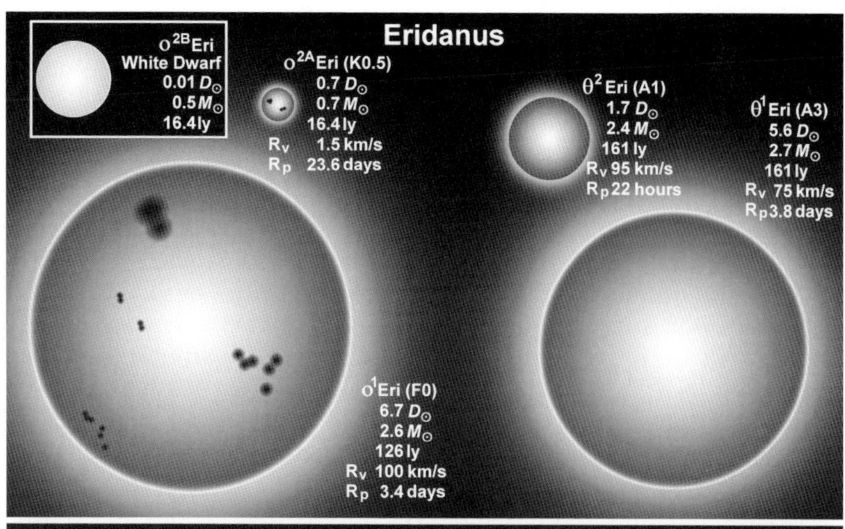
Eridanus
o^{2B}Eri
White Dwarf
0.01 $D_\odot$
0.5 $M_\odot$
16.4 ly
o^{2A}Eri (K0.5)
0.7 $D_\odot$
0.7 $M_\odot$
16.4 ly
R_v 1.5 km/s
R_p 23.6 days
θ^2 Eri (A1)
1.7 $D_\odot$
2.4 $M_\odot$
161 ly
R_v 95 km/s
R_p 22 hours
θ^1 Eri (A3)
5.6 $D_\odot$
2.7 $M_\odot$
161 ly
R_v 75 km/s
R_p 3.8 days
o^1Eri (F0)
6.7 $D_\odot$
2.6 $M_\odot$
126 ly
R_v 100 km/s
R_p 3.4 days

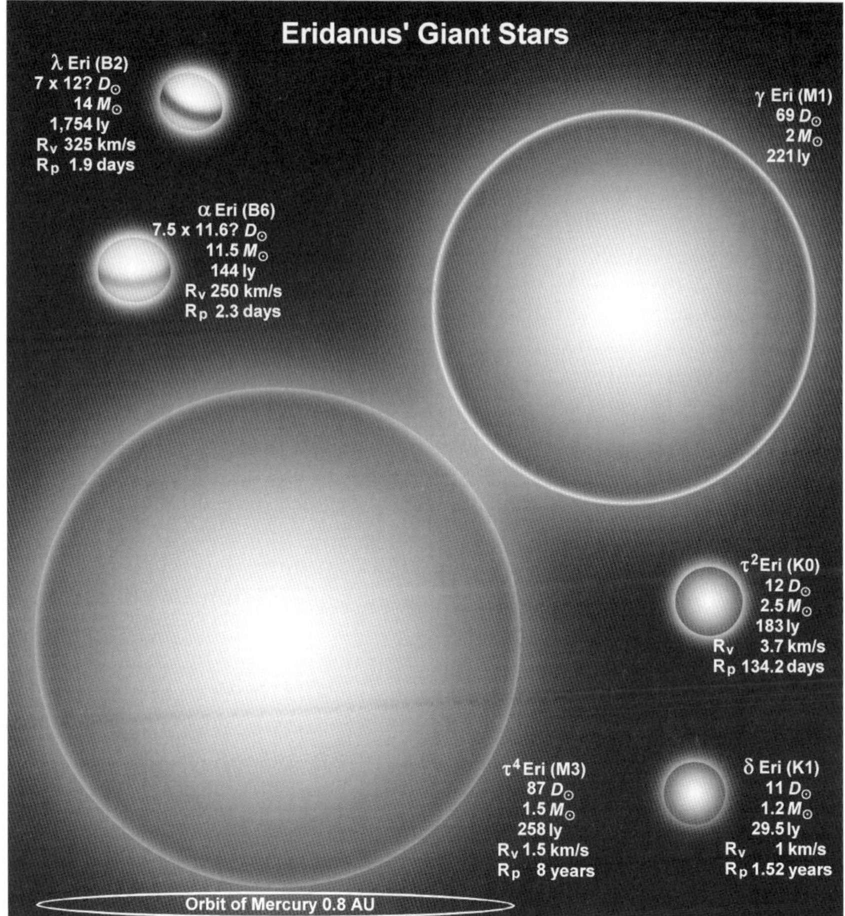
Eridanus' Giant Stars
λ Eri (B2)
7 x 12? $D_\odot$
14 $M_\odot$
1,754 ly
R_v 325 km/s
R_p 1.9 days
γ Eri (M1)
69 $D_\odot$
2 $M_\odot$
221 ly
α Eri (B6)
7.5 x 11.6? $D_\odot$
11.5 $M_\odot$
144 ly
R_v 250 km/s
R_p 2.3 days
τ^2Eri (K0)
12 $D_\odot$
2.5 $M_\odot$
183 ly
R_v 3.7 km/s
R_p 134.2 days
τ^4Eri (M3)
87 $D_\odot$
1.5 $M_\odot$
258 ly
R_v 1.5 km/s
R_p 8 years
δ Eri (K1)
11 $D_\odot$
1.2 $M_\odot$
29.5 ly
R_v 1 km/s
R_p 1.52 years
Orbit of Mercury 0.8 AU

almost certainly an oblate spheroid, its high velocity hurling material into space causing a circumstellar ring of gas and dust. It is another β Cepheid, its pulsations changing its magnitude from m_v +4.22 to +4.34 with a period of $16^h\ 50.2^m$.

Stars do not form randomly but are born in the spiral arms of the Galaxy. On this basis μ and ν **Eridani** appear to be in the wrong place, being some 300 ly below the galactic plane. They are not alone, and many bright stars including those in Orion's Belt, Rigel and a host of other stars have been displaced by, as yet, an unknown force. This grouping is known as Gould's Belt after the astronomer who researched the structure in the late 19th Century. It seems to be a spur to the Orion arm, not quite a full ring and up to 3,000 ly across. The Sun sits about 325 ly from the center of the Belt and may be part of it. As for μ and ν Eri, the former is very much run-of-the-mill B-class star lying at 532 ly from Earth. It has a spectroscopic companion, probably an A3, in a 7.36 day long orbit. ν Eri, at 587 ly, is a bit more unusual in that it is a β Cepheid pulsating variable with a period of $4^h\ 16^m$ during which its magnitude changes from m_v +3.4 to +3.6 and back. In addition it has perhaps a dozen other periods during which it fluctuates by a few thousandth of a magnitude as it vents off radiation in its hopeless battle to stabilize.

The Arabs called them Beid and Keid (meaning 'Eggs' and 'Egg Shells'), Aitken listed them as No. 3093 in his Double Star Catalog and today we call them by Bayer's designations $\mathbf{o^1}$ and $\mathbf{o^2}$ **Eridani** but the reality is that these two stars have nothing to do with one another. o^1 Eri is a 6.7 $D_\odot$ F0 some 126 ly away. Over a period of $1^h\ 51.3^m$ its magnitude dips from m_v +4.00 to +4.05 and back *à la* δ Scuti. o^2 on the other hand stares at us with an unblinking m_v +4.42 from its home just 16.4 ly away. It is a much smaller star, just 70% the size of the Sun, and a yellowish-orange K0.5. And it is not alone. In a 400 AU orbit that takes it 7,200 years to complete o^2 has a companion: a white dwarf of about half the mass of the Sun and with a feeble 0.003 $L_\odot$. Then in orbit around the white dwarf is a red dwarf with a luminosity of 0.022 $L_\odot$ and a mass of 0.16 $M_\odot$. It orbits the white dwarf once every 252 years, the orbit bringing the two dwarfs as close together as 21 AU and then separating them by 49 AU. The white dwarf component is historically important. It was discovered by Friedrich Wilhelm Herschel on 31 January 1783 and was given the spectral class A (white) by Russell, Pickering and Fleming in 1910. The spectrum was described in more detail in 1914 by Walter Adams and in 1922 Willem Luyten coined the phrase 'white dwarf' making o^2 Eri the first white dwarf to be discovered.

It is not unusual to find stars with similar designations, such as θ^1 and θ^2 Eri, but the constellation holds the record for sequential Greek numbering by having no fewer than 9 τ **Eridanis.** Apart from $\boldsymbol{\tau^4}$ **Eridanis** which is 87 $D_\odot$ and $\boldsymbol{\tau^2}$ **Eridanis** at 12 $D_\odot$ the rest are all relatively small stars – a few solar diameters or less – and they are all totally unrelated. τ^4 Eri is an Lb pulsating variable with a range of m_v +3.52 to +3.72 and no particular period. It is joined by two other variables: $\boldsymbol{\tau^8}$ **Eridanis**, an SX Arietis of m_v +4.63 to +4.66, and the αCV rotating variable $\boldsymbol{\tau^9}$ **Eridanis** which varies between m_v +4.62 and +4.67 with a period of $1^d\ 5^h\ 1.5^m$.

ψ **Eridani** is an unremarkable bluish B3 star 956 ly from Earth. What is remarkable is what it also reveals. Right next to the star is **IC 2118,** a reflection nebula better known as the Witch's Head. In long duration photographs the nebula looks very blue due in part to the blueness of the star but also due to the nebula's dust particles reflecting blue light more efficiently than any other color.

82 Eridani could be the oldest closest star in the sky. A yellowish-orange G8 it is only about 0.8 $D_{\odot}$ across and 0.59 $L_{\odot}$. It is currently just 20.9 ly away and heading away from us at 87.3 km/s, though its true velocity through the Galaxy is in the order of 129 km/s. Its orbit takes it from within 15,000 ly of the galactic center out to 35,225 ly (the Sun is about 30,000 ly from the center). In 25,000 years time it should have migrated to Puppis. About 60% as bright as the Sun it is thought to be much older at about 10,000 million years (the Sun is 4,567 million years old).

Planetary system in Eridanus

Star	$D_{\odot}$	Spectral class	ly	m_v	Planet	Minimum mass	q	Q	P
ε Eridani	0.895	K2	10.4	+3.72	ε Eridani b	1.55 M_J	1.01	5.77	6.85 y
					ε Eridani c?	0.10 M_J	28	52	280 y

Fornax

Constellation:	Fornax	**Hemisphere:**	Southern
Translation:	The Furnace	**Area:**	398 deg^2
Genitive:	Fornacis	**% of sky:**	0.965%
Abbreviation:	For	**Size ranking:**	41st

Originally called the Fornax Chemica this constellation was introduced by Abbé Nicolas Louis de Lacaille in 1756.

The brightest star in the constellation is also the closest, and a binary. α **Fornacis** is 46.4 ly from us and shines at m_v +3.30. It is about 50% larger than the Sun but rather more luminous at 4.33 $L_\odot$ and slightly more massive at 1.27 $M_\odot$. It belongs to the F6 spectral class. Its m_v +6.48 companion is a G7 of 0.75 $M_\odot$ but only 0.41 $L_\odot$. There are several published estimations of the orbital period including 155, 269, 314 and 408 years. Whatever the real value the orbit appears to be very eccentric perhaps bringing the two stars to within 10 AU at periastron and separating them by up to 100 AU at apastron.

β **Fornacis** is a B8.5 giant, 12 $D_\odot$ across and spinning at a very leisurely 1.4 km/s, taking 433.8 days (about 14 months) to complete one rotation. Like the constellation's luminary it is also a binary. Its partner seems likely to be a K2 dwarf of 0.27 $D_\odot$.

γ^2 **Fornacis** is four times larger than the Sun, give or take a cat's whisker. A bluish-white A1 lying at 555 ly it is 165 times more luminous that the Sun and spins at 135 km/s, completing a rotation in just 1.5 days. It appears as a m_v +5.38 star: its namesake, γ^1 **Fornacis,** is m_v +6.15 lying at a rather closer 363 ly.

The Sun size star κ **Fornacis** is only 71.6 ly away and is a close solar analog belonging to the G1 spectral group. It is, however, 3.21 times more luminous and 1.4 times as massive.

μ **Fornacis** is a 2.3 $D_\odot$ bluish-white B9 and rotates at a breakneck 320 km/s meaning that it completes one rotation in less than half a day. Although it has a luminosity of 60 $L_\odot$ at m_v +5.26 it may well be too faint for many people to see.

ν **Fornacis** is another B9 star but almost three times the size of the Sun. It is also an αCV variable dipping from m_v +4.68 to +4.73 and back again over a period of $1^d\ 21^h\ 21.5^m$.

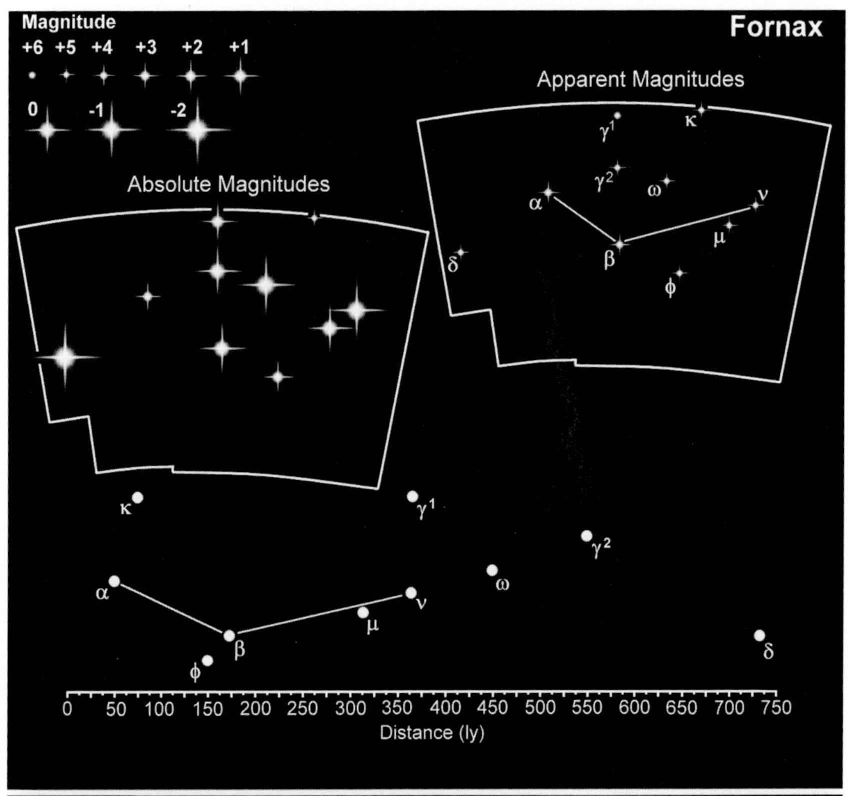
Magnitude
+6 +5 +4 +3 +2 +1
0 -1 -2
Fornax
Apparent Magnitudes
Absolute Magnitudes
γ1
κ
γ2
ω
α
ν
μ
β
δ
ϕ
0 50 100 150 200 250 300 350 400 450 500 550 600 650 700 750
Distance (ly)

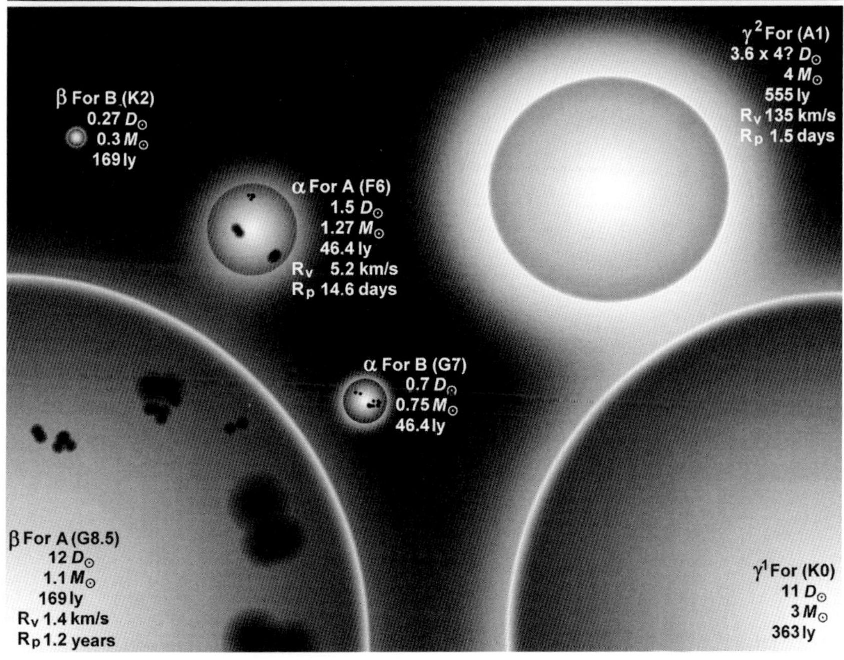
γ2 For (A1)
3.6 x 4? D⊙
4 M⊙
555 ly
Rv 135 km/s
Rp 1.5 days
β For B (K2)
0.27 D⊙
0.3 M⊙
169 ly
α For A (F6)
1.5 D⊙
1.27 M⊙
46.4 ly
Rv 5.2 km/s
Rp 14.6 days
α For B (G7)
0.7 D⊙
0.75 M⊙
46.4 ly
β For A (G8.5)
12 D⊙
1.1 M⊙
169 ly
Rv 1.4 km/s
Rp 1.2 years
γ1 For (K0)
11 D⊙
3 M⊙
363 ly

Gemini

Constellation:	Gemini	**Hemisphere:**	Northern
Translation:	The Twins	**Area:**	514 deg^2
Genitive:	Geminorum	**% of sky:**	1.246%
Abbreviation:	Gem	**Size ranking:**	30th

One of the Zodiacal constellations, the Sun enters Gemini on 21 June and leaves on 20 July. The constellation represents the twin sons of Leda and Zeus who sailed with the Argonauts, depicted by the two brightest stars Castor, the more northerly star, and Pollux. It was not only the Greeks who considered these stars to be twins: Roman, Phoenician, Arabian, Babylonian and Indian mythologies all mention the pair. Twins they may be: identical they are not.

Although it is the second brightest star in the constellation **Castor** gets the α **Geminorum** label possibly because the twins are always known as 'Castor and Pollux' and never 'Pollux and Castor'. It has been suggested that Castor was once the brightest star but there is no real evidence for this view. It is a complex sextuple system some 51.5 ly from Earth.

Castor A has a m_V of +1.98. It is 2 $D_\odot$ across and weighs in at 2.98 $M_\odot$. An A1 it is 37 $L_\odot$. Its close companion is also an A-class Main Sequence star and 2 $D_\odot$ across but is just 12 $L_\odot$. It orbits the main star every 9.2128 days in an elliptical path that varies between 0.022 and 0.065 AU (3.3 to 9.7 million km).

Castor B has a m_V of +2.88. It is 1.55 $D_\odot$ across and is an estimated 2.76 $M_\odot$. An A2 it is 13.9 $L_\odot$. Its close companion is also an A-class Main Sequence star and also 1.55 $D_\odot$ across but is just 6 $L_\odot$. It orbits Castor B every 2.9283 days at an average distance of 0.032 AU (4.8 million km). This pair orbit Castor A with a period of 445 years in an elliptical orbit that brings them to within 71 AU before parting them by 138 AU.

Castor C, which is also known as **YY Geminorum**, has a m_V of +9.1. It is a red dwarf of 0.62 $D_\odot$ across and weighs in at 0.59 $M_\odot$. It is an M0.5. Its close companion is identical, as far as we can tell. Their orbital period is $19^h\ 33^m$, the pair being separated by 0.058 AU (8.7 million km). Magnetic interaction between the two stars causes one or both to flare, hence the YY Gem designation. This pair orbit Castor A and Castor B at a distance of at least 1,000 AU and with an orbital period in excess of 14,000 years.

Pollux is not as complicated as Castor (!). While Castor is bluish-white, on a clear dark night with good seeing and good color vision, Pollux can appear yellowish-orange. That's not surprising as it is a K0 giant more than nine times the size of the Sun and the closest Gemini star at 33.7 ly. Rotating at a just 1.2 km/s it takes 383.8 days to spin once on its axis. Cataloged as β **Geminorum** Pollux has a mass of 1.86 $M_\odot$ and is one of the few giants to host a planet. Rather confusingly the planet is known by its Henry Draper catalog number of HD 62509 instead of simply β Geminorum b as ε Tauri b, γ Cephei b and others are known. It has a mass of 2.9 M_J and orbits the star every 1.6 years.

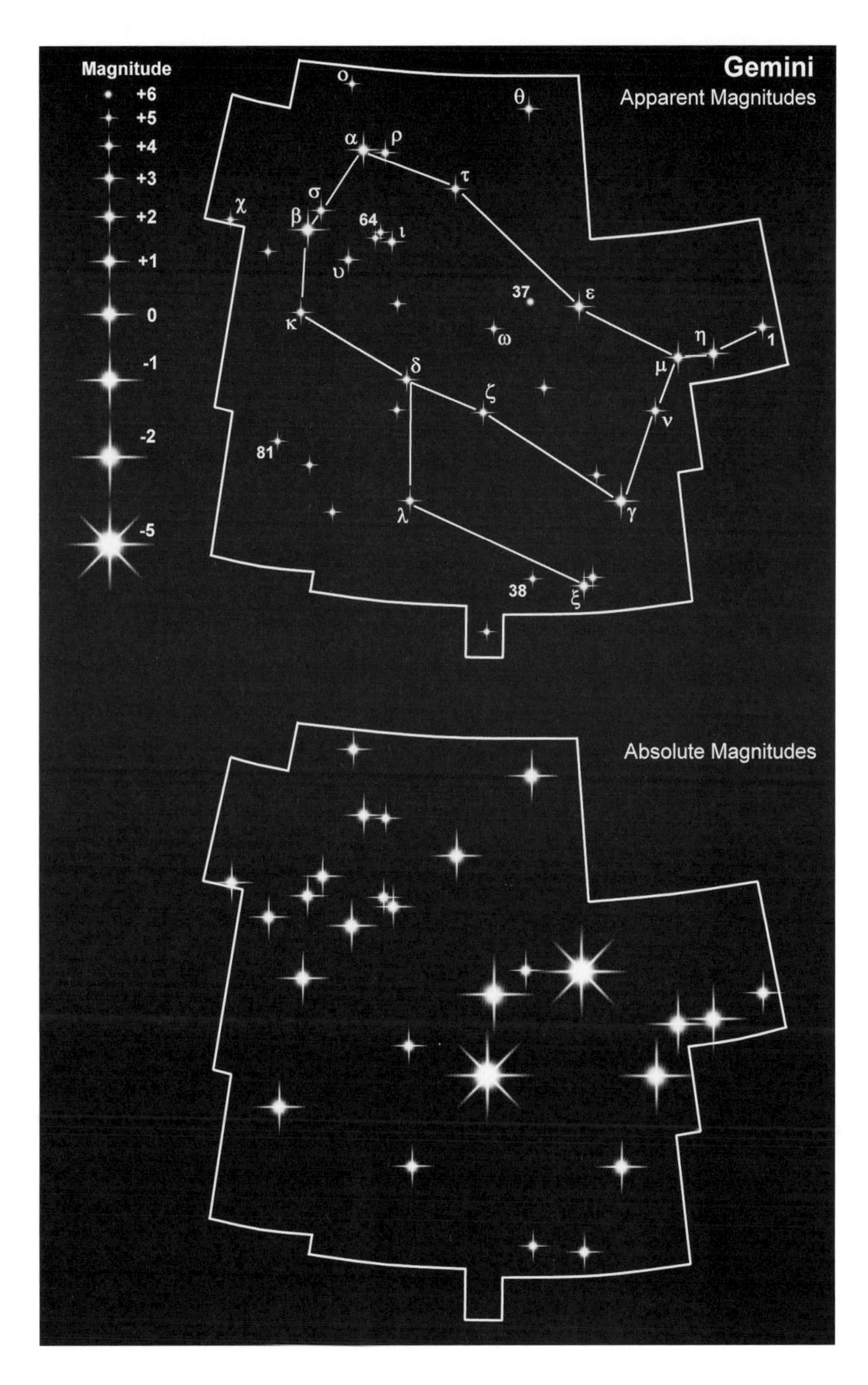
Magnitude
+6
+5
+4
+3
+2
+1
0
-1
-2
-5
Gemini
Apparent Magnitudes
ο
θ
α
ρ
τ
σ
χ
β
64
ι
υ
37
ε
κ
ω
η
1
δ
μ
ζ
ν
81
γ
λ
38
ξ
Absolute Magnitudes

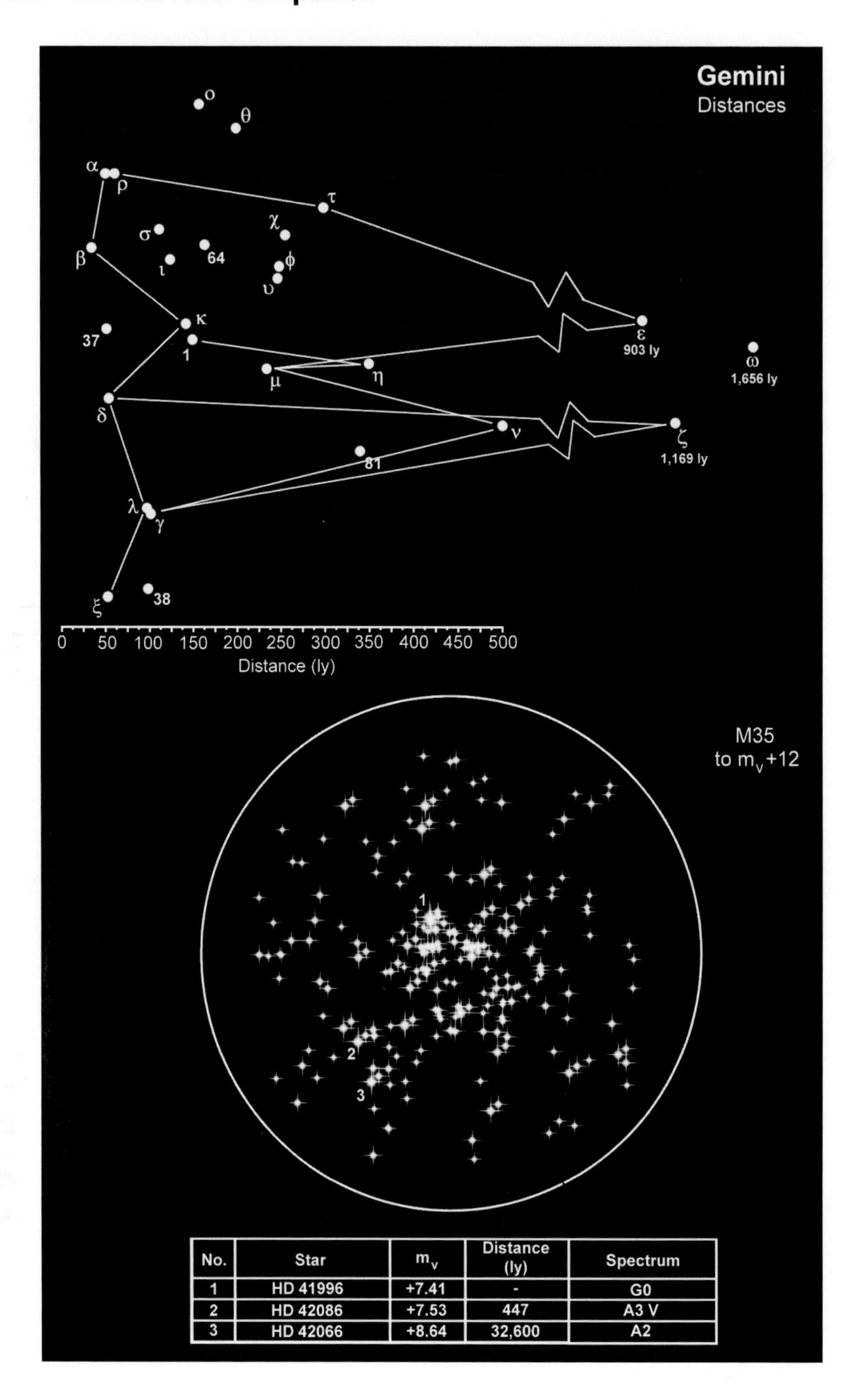

No.	Star	m_v	Distance (ly)	Spectrum
1	HD 41996	+7.41	-	G0
2	HD 42086	+7.53	447	A3 V
3	HD 42066	+8.64	32,600	A2

The m_v +2.02 γ **Geminorum** or Alhena traditionally marks the third corner of the constellation, opposite Pollux. Attempts at finding its diameter have resulted in various values of between 1.5 and 4.8 $D_\odot$ with 3.2 $D_\odot$ being the average. We have a better grasp of its distance – 105 ± 7 ly – and spectral type, A0, with its luminosity coming in at 130 ± 19 $L_\odot$. There are few absolutes in measuring the stars. We do know that it is a spectroscopic binary, its companion being a G-class whose 12.6 year eccentric orbital period brings it to within 1 AU at periastron and out to 16 AU at apastron.

Sitting just about on the Ecliptic is δ **Geminorum** which, to all intents and purposes, is a fairly ordinary star. Called Wasat by the Arabs it is an F0 class with a diameter of 1.6 $D_\odot$ and a luminosity of 10.2 $L_\odot$. Being at a distance of 58.8 ly its physical parameters result in a m_v +3.53 star that would brighten by about a magnitude to M_v +2.46 at 10 pc. It has a companion: a K3 in a 1,200 year orbit with an average separation of 100 AU. However, δ Gem's real claim to fame is its appearance on a photograph taken in 1930 by Clyde Tombaugh. One of the other stars on the plate turned out to be an unknown planet – Pluto. Alas, since its discovery Pluto has been relegated to the ranks of a dwarf planet.

At 150 $D_\odot$ ε **Geminorum** is the largest star in the constellation. If placed in the center of the Solar System the Earth would be just 45 million km from its surface. Yellowish-orange in color its 7,600 $L_\odot$ pour our way from a distance of 903 ly producing a modest +2.99 magnitudes. At 10 pc however, it would rival Venus at M_v -4.5.

Of the nine Cepheid variables visible without optical aid, ζ **Geminorum** has the longest period at $10^d\ 3^h\ 37^m\ 3.1^s$, its magnitude fluctuating between m_v +3.62 and +4.18. Lying 1,169 ly from Earth it is a 74 $D_\odot$ supergiant with a luminosity of about 2,700 $L_\odot$. At maximum magnitude it is a 6,500 K F7 changing to a 5,100 K G3 at minimum. It has an almost symmetrical light curve which is unusual for this type of variable.

Only two naked eye Semi-Regular type a (SRa) giants are known: one is GZ Pegasi and the other is η **Geminorum**, a triple star system 349 ly away. The main component is a 130 $D_\odot$ red giant M3 with a luminosity of 425 $L_\odot$ and mass of 5.5 $M_\odot$. Its magnitude varies between m_v +3.20 and +3.90 with a period of 232.9 days. At an average distance of 7 AU is a B-class companion in a 2,983 day (8.2 year) orbit. It is 2.06 $M_\odot$ and shines at m_v +5.46. Taking more than 700 years to complete its orbit is the third star in the system, an F or possibly a G-class with a mass of 1.18 $M_\odot$ and a magnitude of m_v +8.02. It averages 150 AU from the main star.

It is interesting to compare η Gem with μ **Geminorum** which is also an M3, somewhat smaller at 104 $D_\odot$ but nearly four times as luminous at 1,540 $L_\odot$. It is also variable, m_v +2.75 to +3.02 but, belonging to the Lb variable class, has no real period although a 2,000 day cycle overlapped by a 27 day cycle has been reported. It is 232 ly from Earth.

ν **Geminorum** is another complex quadruple system. The main star is a m_v +4.30 B6 with a diameter of 3.5 $D_\odot$. It has a companion in a 53.7 day orbit that is an almost circular 0.5 AU. Approaching the main star to within 1.5 AU before

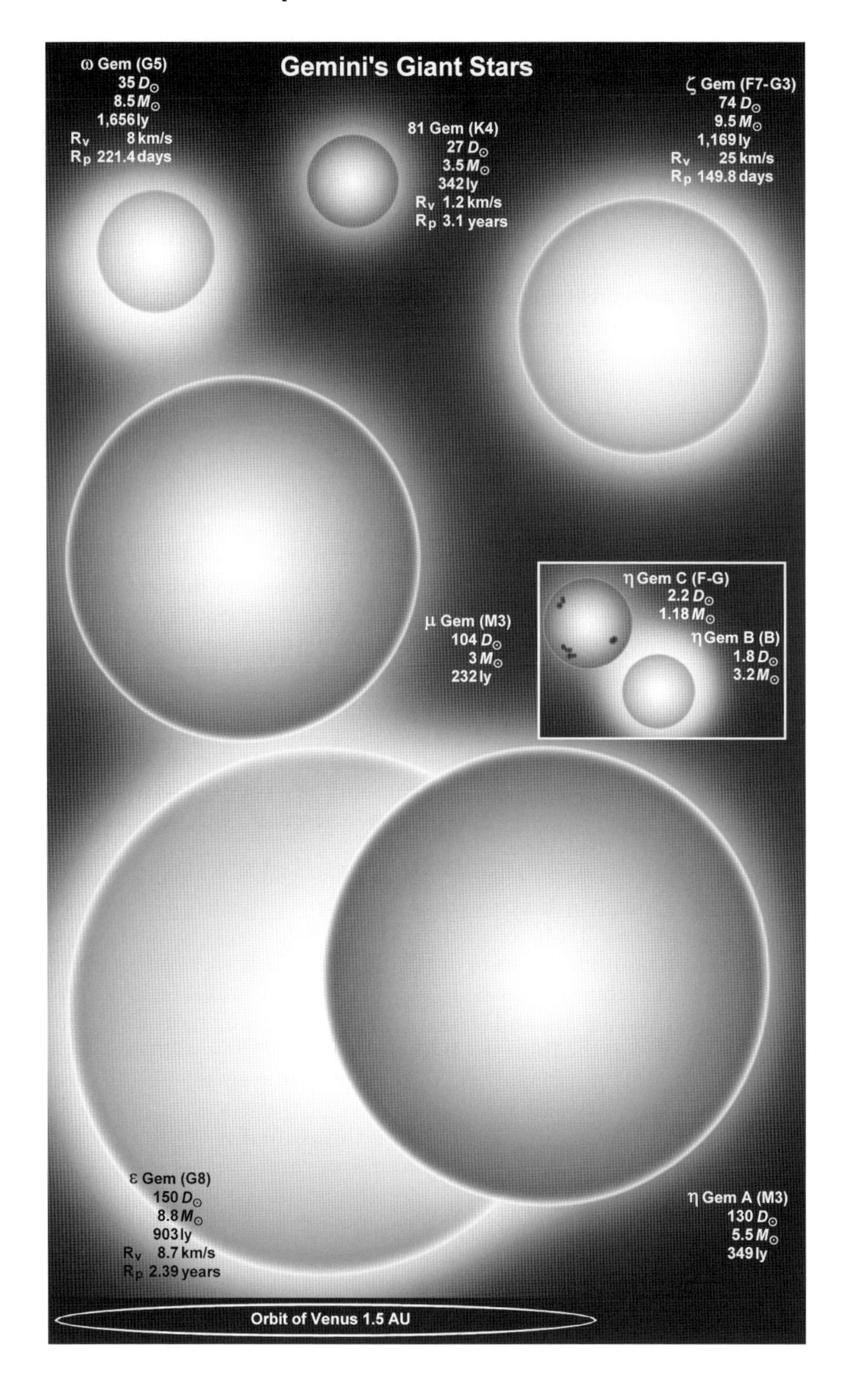
Gemini's Giant Stars
ω Gem (G5)
35 D⊙
8.5 M⊙
1,656 ly
Rv 8 km/s
Rp 221.4 days
81 Gem (K4)
27 D⊙
3.5 M⊙
342 ly
Rv 1.2 km/s
Rp 3.1 years
ζ Gem (F7-G3)
74 D⊙
9.5 M⊙
1,169 ly
Rv 25 km/s
Rp 149.8 days
η Gem C (F-G)
2.2 D⊙
1.18 M⊙
η Gem B (B)
1.8 D⊙
3.2 M⊙
μ Gem (M3)
104 D⊙
3 M⊙
232 ly
ε Gem (G8)
150 D⊙
8.8 M⊙
903 ly
Rv 8.7 km/s
Rp 2.39 years
η Gem A (M3)
130 D⊙
5.5 M⊙
349 ly
Orbit of Venus 1.5 AU

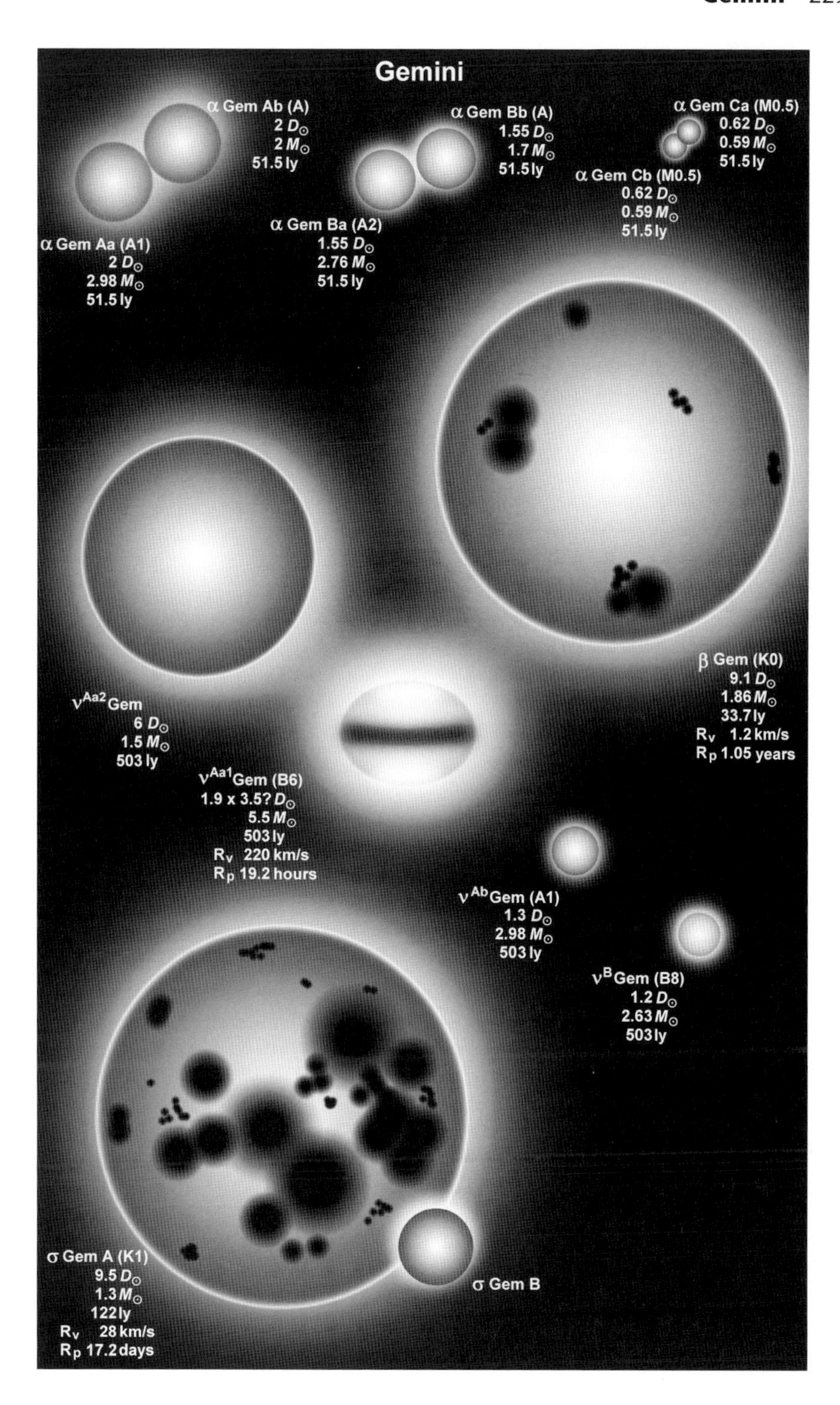
Gemini
α Gem Ab (A)
2 $D_\odot$
2 $M_\odot$
51.5 ly
α Gem Bb (A)
1.55 $D_\odot$
1.7 $M_\odot$
51.5 ly
α Gem Ca (M0.5)
0.62 $D_\odot$
0.59 $M_\odot$
51.5 ly
α Gem Cb (M0.5)
0.62 $D_\odot$
0.59 $M_\odot$
51.5 ly
α Gem Ba (A2)
1.55 $D_\odot$
2.76 $M_\odot$
51.5 ly
α Gem Aa (A1)
2 $D_\odot$
2.98 $M_\odot$
51.5 ly
β Gem (K0)
9.1 $D_\odot$
1.86 $M_\odot$
33.7 ly
R_v 1.2 km/s
R_p 1.05 years
ν^{Aa2}Gem
6 $D_\odot$
1.5 $M_\odot$
503 ly
ν^{Aa1}Gem (B6)
1.9 x 3.5? $D_\odot$
5.5 $M_\odot$
503 ly
R_v 220 km/s
R_p 19.2 hours
ν^{Ab}Gem (A1)
1.3 $D_\odot$
2.98 $M_\odot$
503 ly
ν^{B}Gem (B8)
1.2 $D_\odot$
2.63 $M_\odot$
503 ly
σ Gem A (K1)
9.5 $D_\odot$
1.3 $M_\odot$
122 ly
R_v 28 km/s
R_p 17.2 days
σ Gem B

heading out to 20 AU is the third component which takes 13 years to complete an orbit. It is an A1 with a m_v of +9.0 and mass of 2.98 $M_\odot$. The fourth star, an m_v +8.01, is a B8 with diameter of 1.2 $D_\odot$ and a mass of 2.63 $M_\odot$.

Gemini contains a number of naked eye stars that are similar in size to the Sun including the 4th magnitude **ρ Geminorum.** ρ Gem is 1.3 $D_\odot$ F0 dwarf lying at a distance of 60.3 ly. It is another complex quadruple system. One of its companions is a G8 dwarf in an 897,000 year long orbit. Another is an 8th magnitude G8 which also has as 11th magnitude orbiting companion.

38 Geminorum consists of a pair of identical 1.4 $D_\odot$ F0 stars, both with surface temperatures of 7,200 K, in an orbit that takes 1,944 years to complete. They are 91.1 ly from Earth. Slightly larger in size, **64 Geminorum** is a much hotter A4 at 8,400 K and farther away at 163 ly.

σ Geminorum belongs to the RS Canum Venaticorum class of eclipsing variable stars. Its magnitude brightens from m_v +4.29 to +4.13 before settling again with a clockwork period of $19^d\ 10^h\ 9^m$. A K1 giant of 9.5 $D_\odot$ its companion orbits at just 0.2 AU (30 million km), so close that each disrupts the other's magnetic field causing brilliant flares. About one-third of the primary star can be covered in dark starspots which add to the changes in magnitude.

Giant **ω Geminorum** is the outmarker for the constellation at 1,656 ly, though it could be as close as 1,153 ly or as far as 2,160 ly such is the uncertainty in our measurements. It is a bright yellow G5 with a luminosity of 1,753 $L_\odot$. Rotating at 8 km/s it takes slightly more than 221 days – two-thirds of a year – to complete a single rotation. Like ζ Gem it is a Cepheid variable with a period of $17^h\ 28.5^m$, making it the shortest period naked eye Cepheid, and has a magnitude range of m_v +5.14 to +5.23.

37 Geminorum is one of the 30 stars most likely to have planetary systems that harbor life, according to some researchers. A G0 class with a diameter slightly larger than the Sun (1.03 $D_\odot$) it is more luminous by about 23% and about 1,000 million years older. A Message to Extra-Terrestrial Intelligence (METI) was sent to the star on 3 September 2001 and should reach it in 2057.

With a radial velocity of 81 km/s **81 Geminorum** is one of the 1% highest velocity stars. Moving away from us its velocity actually varies between 76 and 90 km/s suggesting the presence of another star in orbit around it or possibly a planetary system but so far nothing has been detected. It is a K4 giant, 27 $D_\odot$ across, and about 100 times more luminous than the Sun.

M35 or **NGC 2168** is a cluster of more than 500 post-Main Sequence stars that are at least 95 million years old. Lying at a distance of 2,700 ly the cluster core is 24 ly in diameter and has a density of just over 6 stars per cubic parsec, but the entire cluster is probably more than twice this size. The cluster can be found slightly to the north west of η Gem just above the Ecliptic. A small telescope or binocular will reveal many of the brighter stars.

Open clusters in Gemini

Name	Size arc min	Size ly	Distance ly	Age million yrs	Brightest star in region*	No. stars m_v >+12*	Apparent magnitude m_v
Collinder 89	50′	38	2,600	32	HD 43740 m_v +6.57	95	+5.7
M35	67′	52	2,700	95	HD 41996 m_v +7.41	209	+5.0
NGC 2129	25′	36	5,000	21	HD 250290 m_v +7.38	22	+6.7
NGC 2169	3′	3	3,400	12	HD 41943 m_v +6.92	18	+5.9
NGC 2395	27′	13	1,700	1,180	TYC 776-1313-1 m_v +9.75	17	+8.0

*May not be a cluster member.

Grus

Constellation:	Grus	**Hemisphere:**	Southern
Translation:	The Crane	**Area:**	366 deg^2
Genitive:	Gruis	**% of sky:**	0.887%
Abbreviation:	Gru	**Size ranking:**	45th

Once called Phoenicopterus, meaning flamingo, Grus was created in the 16th Century by Petrus Plancius using observations made by the Dutch explorers Pieter Keyser and Frederick de Houtman.

The brightest star in the constellation, α **Gruis** or Alnair, is also the second closest at 101.4 ly. It is a textbook B-class star: a B6 with a temperature of 13,000 K, a diameter of 3.6 $D_\odot$, a mass of 4.5 $M_\odot$, an absolute magnitude of M_v -0.7, a luminosity of 380 $L_\odot$ and a rotational velocity of 236 km/s. All very normal and a useful standard against which other stars can be compared.

There are little more than a dozen naked eye M5 red giants – less than half a percent of all naked eye stars – and two of them lie in Grus. β **Gruis** is the smaller of the two at 82 $D_\odot$ (about the size of Mercury's orbit), the closer at 170 ly and the most luminous, pouring out 3900 $L_\odot$. The other M5 is δ^2 **Gruis**: 135 $D_\odot$ (86% the size of Venus' orbit), 325 ly and 2,200 $L_\odot$. They are both irregular variables, β Gru pulsating between m_v +2.00 and +2.30 while δ^2 Gru has a slightly smaller range of +3.99 to +4.20. Right next to δ^2 Gru on the celestial sphere is δ^1 **Gruis**. In reality it is 29 ly closer to us and has nothing to do with its apparent companion, being a 22 $D_\odot$ giant yellow G7.

Perhaps too faint for most town and city dwellers to see, the yellow dwarf **HD 211415** shines at m_v +5.40 from a distance of 44.4 ly. A G1 it has a diameter and mass very similar to the Sun's, 0.96 $D_\odot$, 1.01 $M_\odot$, and is slightly more luminous at 1.04 $L_\odot$. It has a rather greater metal content than the Sun, enriched by about 50%, and is somewhat younger at 3,300 million years (the Sun's age is 4,560 million years). The real difference, however, is that HD 211415 is a binary. Its companion is a red dwarf believed to be about half the size and mass of the Sun and just 0.018 $L_\odot$, its temperature being a cool 3,800 K. Its orbit is not particularly well known but it is thought that the average separation is 46 AU. Statistically stars with high metal contents are more likely to have planetary systems. Although no planet has yet been detected around HD 211415 it is regarded as being among the top 30 stars to have planetary systems that may harbor life. Several stars in Grus do, in fact, have planets, including τ^1 **Gruis** which is listed as **HD 216435 b**.

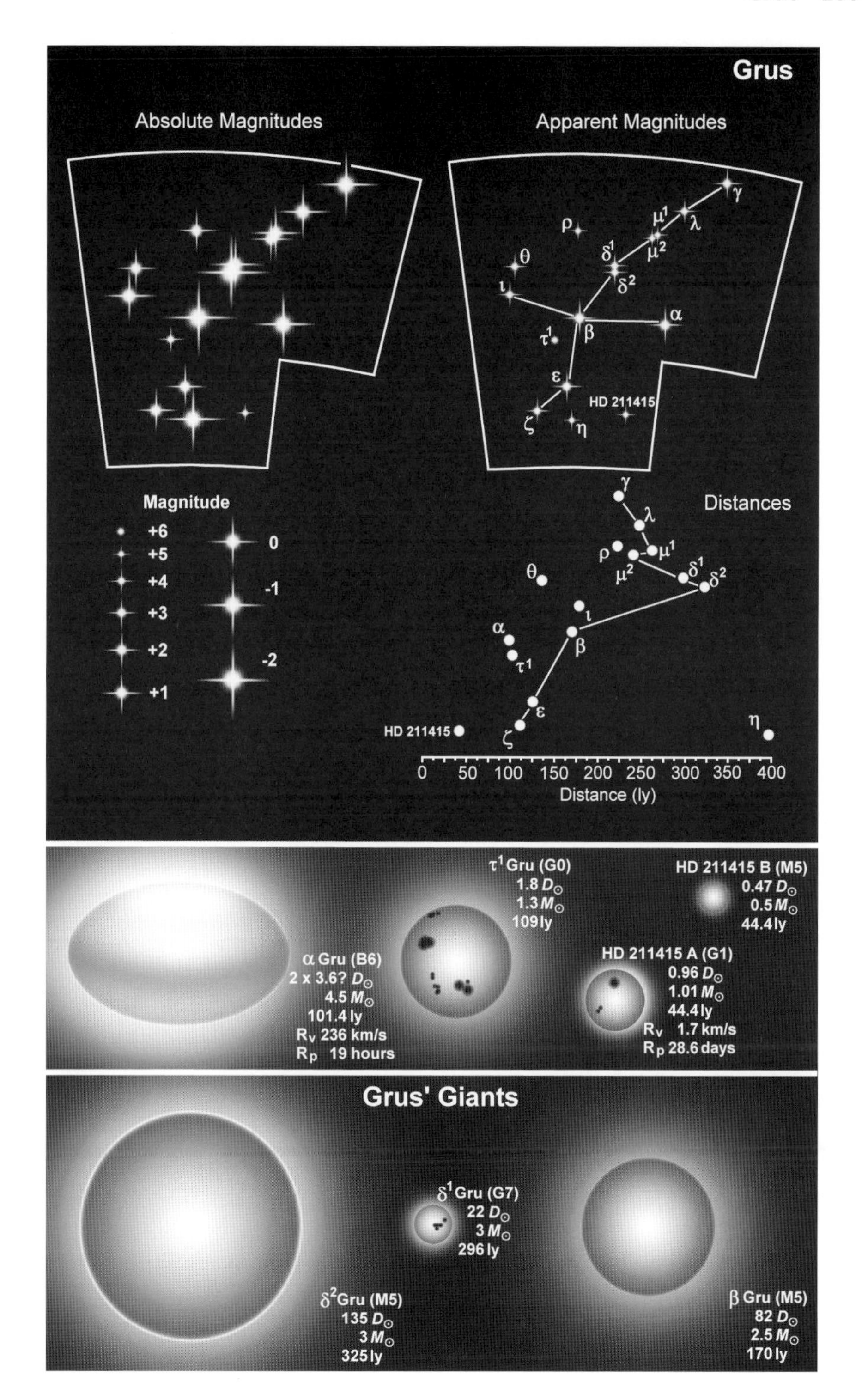

Grus
Absolute Magnitudes
Apparent Magnitudes
γ
λ
μ1
μ2
ρ
θ
δ1
δ2
ι
α
β
τ1
ε
HD 211415
ζ
η
Magnitude
+6
+5
+4
+3
+2
+1
0
-1
-2
Distances
0 50 100 150 200 250 300 350 400
Distance (ly)
τ1 Gru (G0)
1.8 D☉
1.3 M☉
109 ly
HD 211415 B (M5)
0.47 D☉
0.5 M☉
44.4 ly
α Gru (B6)
2 x 3.6? D☉
4.5 M☉
101.4 ly
Rv 236 km/s
Rp 19 hours
HD 211415 A (G1)
0.96 D☉
1.01 M☉
44.4 ly
Rv 1.7 km/s
Rp 28.6 days
Grus' Giants
δ1 Gru (G7)
22 D☉
3 M☉
296 ly
δ2 Gru (M5)
135 D☉
3 M☉
325 ly
β Gru (M5)
82 D☉
2.5 M☉
170 ly

Planetary systems in Grus

Star	$D_{\odot}$	Spectral class	ly	m_v	Planet	Minimum mass	q	Q	P
τ^1 Gru	2.00	G0	109	+6.03	HD 216435 b	1.25 M_J	2.38	2.74	1.311 d
HD 204961	0.50	?	16.1	+8.67	GJ 832 b	0.64 M_J	2.99	3.81	9.35 y
HD 208487	1.15	G2	147	+7.48	HD 208487b	0.45 M_J	0.33	0.65	123 d
c unconfirmed					HD 208487c?	0.46 M_J	1.46	2.14	2.49 y
HD 213240	1.50	G4	133	+6.08	HD 213240 b	4.50 M_J	1.12	2.94	2.60 y

Hercules

Constellation:	Hercules	**Hemisphere:**	Northern
Translation:	Hercules	**Area:**	1,225 deg^2
Genitive:	Herculis	**% of sky:**	2.969%
Abbreviation:	Her	**Size ranking:**	5th

The constellation represents the mythical Greek hero who was the son of Zeus and a mortal, Alkmene. He achieved 12 seemingly impossible tasks – the Labors of Hercules – and was placed in the heavens at the end of his Earthly life.

α **Herculis** or Rasalgethi could be one of the largest stars visible in the night sky. The current consensus is that it is about 400 $D_{\odot}$ across, although a dozen or so estimates over the past century average out at about 320 $D_{\odot}$. If it truly is 400 times as big as the Sun then, in Solar System terms, it would have the same diameter as the inner edge of the Asteroid Belt. It is a semi-regular, SRc, red supergiant of spectral class M5, some 382 ly distant and is an impressive 17,000 $L_{\odot}$, mostly in the infrared. Its magnitude dips from m_v +2.74 to +4.00 with a main period of about 90 days. However, superimposed on this period are several others lasting from a few weeks to several years. Just 5″ away is a 5th magnitude yellow star trapped in a 550 AU, 3,600 year orbit. Known as **α^2 Herculis** – the main star, of course, is α^1 – it is a G5 giant of 15 $D_{\odot}$ and 4.1 $M_{\odot}$. It also has an orbiting companion, a 2.5 $M_{\odot}$ F2 dwarf in an orbit of less than half an AU leading to a 51.6 day orbital period. The supergiant has blown off at least one shell of material which now has a radius of 522 AU, more than 17 times the size of Neptune's orbit. Despite the star's huge diameter its mass is only about 15 $M_{\odot}$.

β **Herculis** or Kornephoros is also a G-class giant. At 12 $D_{\odot}$ it is not quite as big as α Her's companion and is certainly closer at 148 ly from Earth. A G7 it is a few hundred degrees cooler, 4,900 K, and rotates at 4.8 km/s taking a third of a year – 127 days – to complete one rotation. Again, it is not alone but has a spectroscopic companion in a 410.6 day orbit.

ζ **Herculis** is an interesting star, or rather collection of stars. ζ Her is a binary system. The primary is a m_v +2.90 G0 class star of 2.5 $D_{\odot}$, 1.5 $M_{\odot}$ and 6 $L_{\odot}$. Separated by 1″ on the celestial sphere is its companion, ζ^B Herculis. It is a faint m_v +5.23 at the other end of the spectral class, G7, and weighs in at a much smaller 0.85 $M_{\odot}$ and 0.65 $L_{\odot}$. The pair are in a 34.5 year long orbit during which they approach one another to within 8 AU and are then separated by 21 AU. The most intriguing thing about ζ Herculis is that it is the leader of a group of 10 stars (originally thought to be 22) that seem to be moving in unison. The group includes β Hydri, ζ^1 Reticuli, ζ^2 Reticuli, ϕ^2 Pavonis, 1 Hydrae, HD 14680 in Fornax, HD 158614 in Ophiuchus and Gliese 456 in Virgo.

ζ Her marks the south west corner of what is commonly called the 'Keystone'. Diagonally opposite is π **Herculis,** slightly dimmer than ζ Her at m_v +3.14 but much, much larger being a 67 $D_{\odot}$ K3 supergiant. To the south we find ε **Herculis,** a sparkling white A0 dwarf, twice as large as the Sun and 163 ly away.

The north west corner is the home of η **Herculis,** another G-class star, this time a G8, with a diameter of 5.6 $D_\odot$ and lying 112 ly from Earth. Just below η Her is one of the most magnificent globular clusters in the northern skies, **Messier 13.** Often just referred to as M13 aficionados of deep sky objects know it as 'The Great Globular Cluster', and for good reason. Although it is 25,100 ly away it can be seen as a grayish smudge on dark, Moonless nights with an apparent magnitude of m_v +5.8 and a diameter of 20'. It is, in fact, 145 ly across and contains anywhere between 100,000 and 1 million stars. Just as uncertain is its age: between 14,000 and 24,000 million years (compare this to the Sun which is 4,560 million years old and which will expire when it reaches 10,000 million years). This is a cluster of very old stars, except for one: Barnard 29, a bright, young B2 which seems to have been captured from interstellar space by the cluster.

θ **Herculis** is one of those stars that illustrate the uncertainties in stellar measurement. Its spectrum suggests a class K1 but direct measurements of its temperature reveal a somewhat cool 4,320 K which is more consistent with a K3. It could lie somewhere between 600 and 750 ly or, just as equally, between 730 and 780 ly. Its luminosity has been estimated at a minimum 784 $L_\odot$ and a maximum of around 2,500 $L_\odot$. Attempts at measuring its diameter do not agree with stellar theory so could be anywhere within the range of 59 to 87 $D_\odot$. Its mass comes in at 5.5 to 6.5 $M_\odot$ and its age is somewhere between 55 and 78 million years. So is there anything we do know about it for certain? Well, it's a star. We think.

ι **Herculis** belongs to one of the more common variable groups, the β Cepheids. Its magnitude ranges between m_v +2.93 and +2.95, an amplitude that is found in 14% of naked eye β Cepheids. It has a diameter of 2.6 $D_\odot$, typical of a third of the class and it is a B3 (B2s and B1s together make up 75% of β Cepheids, B3s come in at 14%). Unlike most of its class it has no obvious period although there is an underlying 4.48 day cycle which is superimposed on several others. It is also a quadruple star system. At about 1 AU is a companion that takes 113.8 days to complete a single orbit. It is impossible to see this particular star through a telescope but it can be detected spectroscopically. At an average distance of 30 AU is a second companion in a 60 year orbit while the fourth member of the system takes about 1 million years to orbit the primary in a path that never brings them closer than 18,000 AU.

The solar analog μ **Herculis** is a well known quadruple star system. Being a G5 it is a couple of hundred degrees cooler than the Sun but larger at 1.76 $D_\odot$ and more luminous at 2.47 $L_\odot$. Despite this it is just 10% more massive and, lying at a distance of 27.4 ly, is the closest star in the constellation. Its companions are a trio of red dwarfs, designated μ^{Ab}, μ^{B} and μ^{C}. μ^{Ab} orbits the primary μ^{Aa} with a period of 65 years. μ^{B} and μ^{C} Herculis orbit one another with a period of 43.2 years, swinging between 1.5 AU and 3.6 AU. The pair maintain an average distance of 300 AU from the primary star with the result that they take more than 3,680 years to complete an orbit.

Hercules contains two α CV rotating variables. ω **Herculis** has a period of 2^d 22^h 49.5^m during which it cycles through m_v +4.57 to +4.65 and back again. A

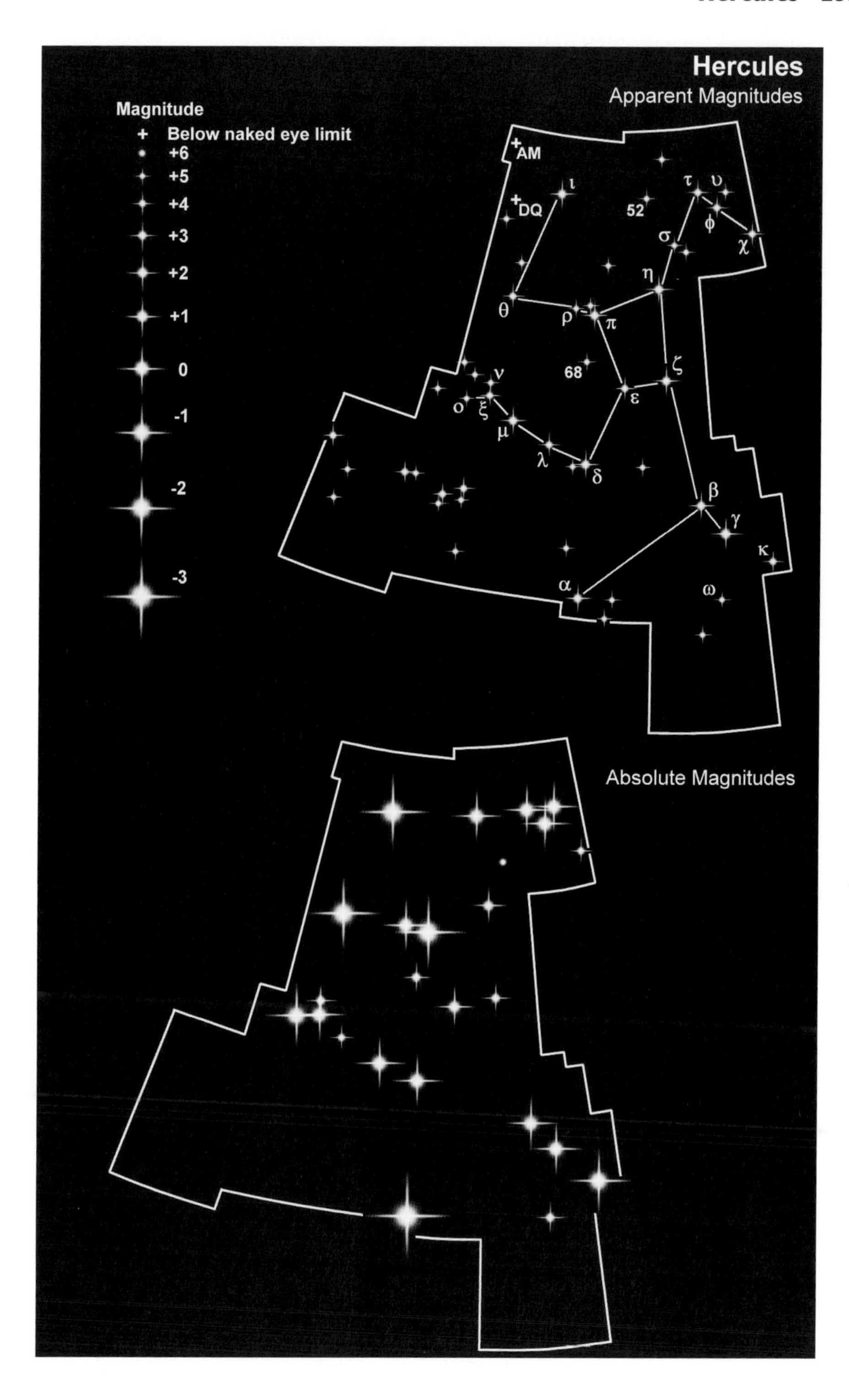
Hercules
Apparent Magnitudes
Magnitude
Below naked eye limit
+6
+5
+4
+3
+2
+1
0
-1
-2
-3
AM
DQ
ι
52
τ
υ
φ
χ
σ
η
θ
ρ
π
68
ζ
ν
ο
ξ
ε
μ
λ
δ
β
γ
κ
α
ω
Absolute Magnitudes

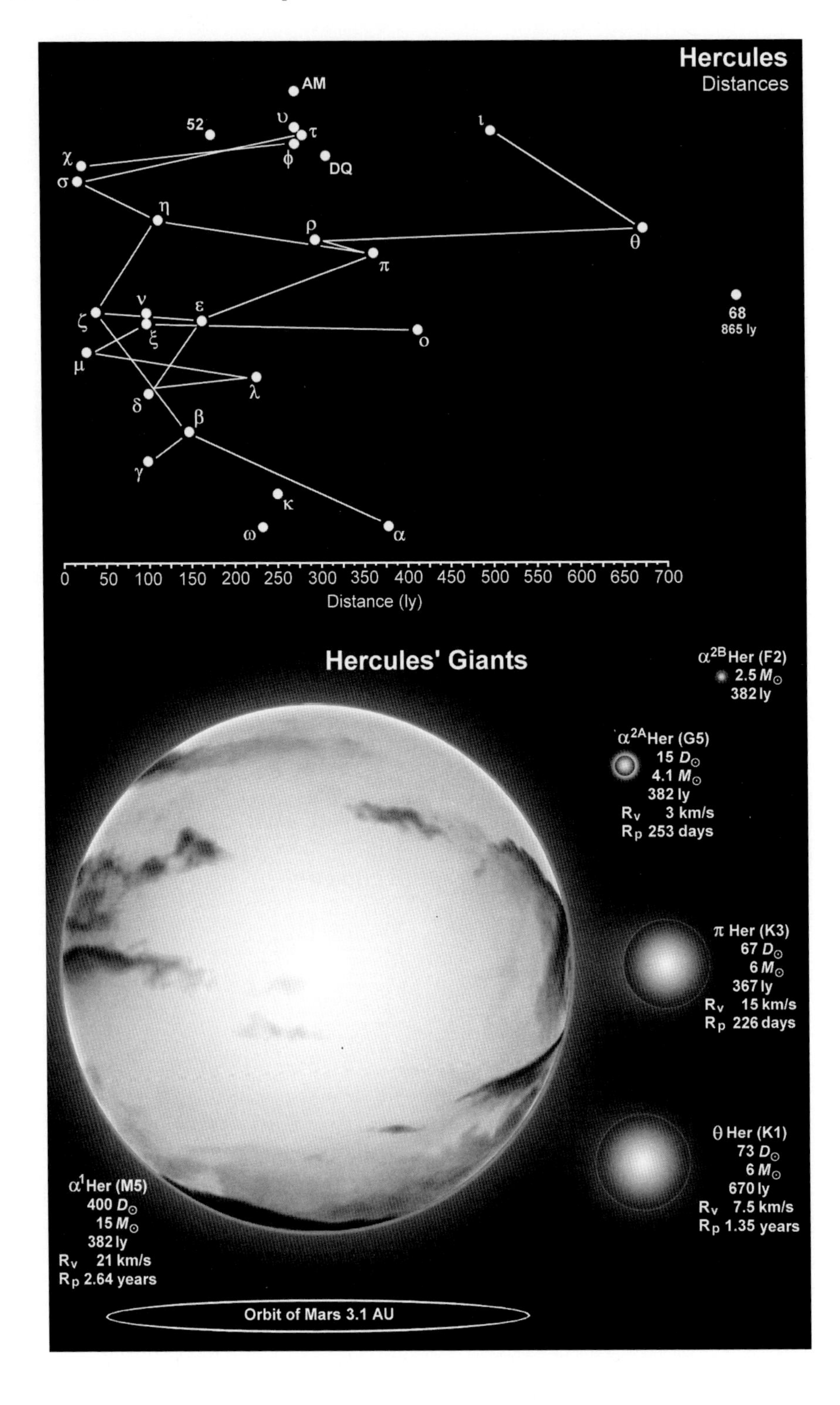

Hercules
Distances
AM
52
υ
τ
ι
χ
φ
DQ
σ
η
ρ
θ
π
ν
ε
ζ
ξ
ο
68
865 ly
μ
λ
δ
β
γ
κ
ω
α
0 50 100 150 200 250 300 350 400 450 500 550 600 650 700
Distance (ly)
Hercules' Giants
α2B Her (F2)
2.5 M⊙
382 ly
α2A Her (G5)
15 D⊙
4.1 M⊙
382 ly
Rv 3 km/s
Rp 253 days
π Her (K3)
67 D⊙
6 M⊙
367 ly
Rv 15 km/s
Rp 226 days
θ Her (K1)
73 D⊙
6 M⊙
670 ly
Rv 7.5 km/s
Rp 1.35 years
α1 Her (M5)
400 D⊙
15 M⊙
382 ly
Rv 21 km/s
Rp 2.64 years
Orbit of Mars 3.1 AU

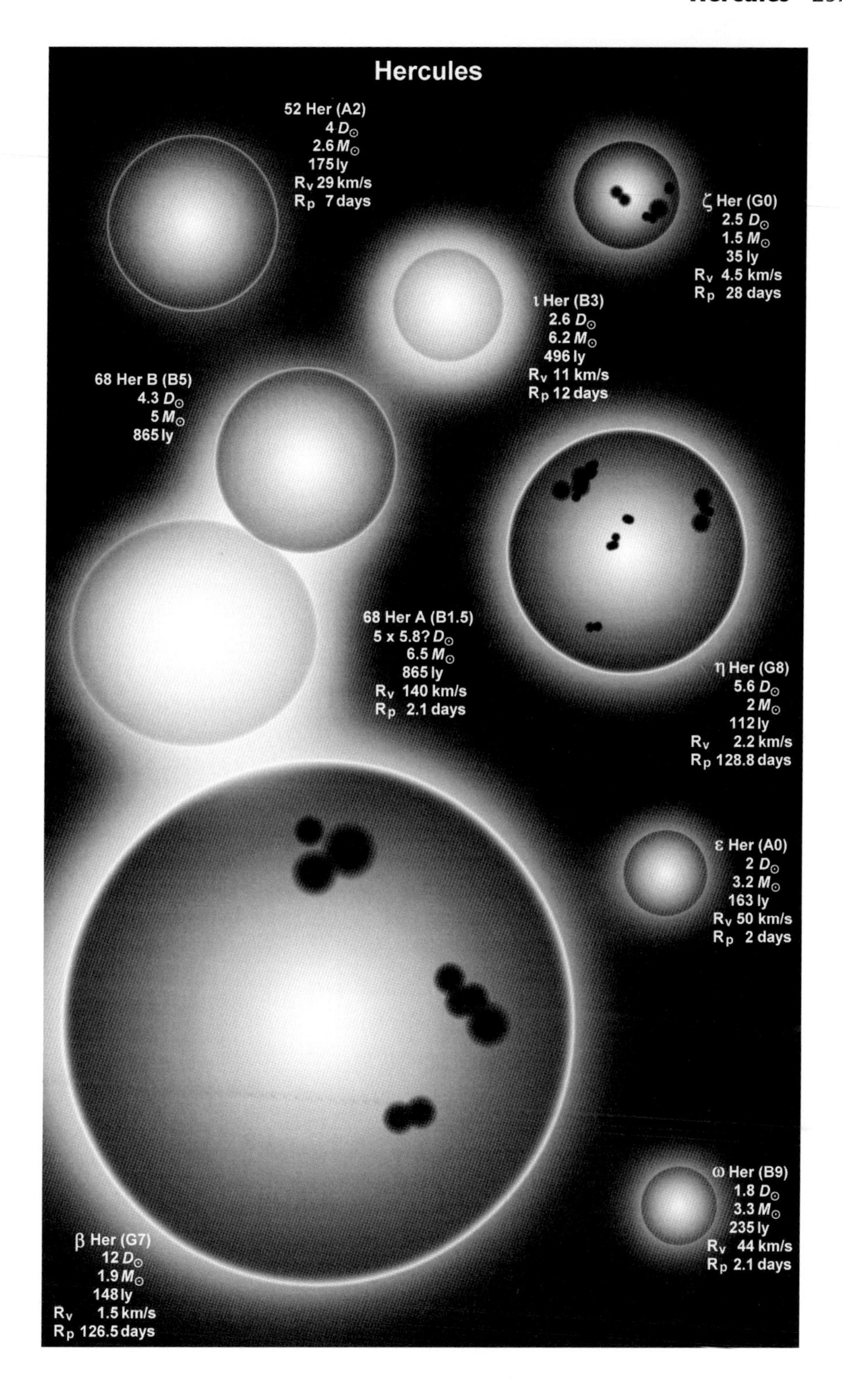

Hercules
52 Her (A2)
4 $D_\odot$
2.6 $M_\odot$
175 ly
R_V 29 km/s
R_P 7 days
ζ Her (G0)
2.5 $D_\odot$
1.5 $M_\odot$
35 ly
R_V 4.5 km/s
R_P 28 days
ι Her (B3)
2.6 $D_\odot$
6.2 $M_\odot$
496 ly
R_V 11 km/s
R_P 12 days
68 Her B (B5)
4.3 $D_\odot$
5 $M_\odot$
865 ly
68 Her A (B1.5)
5 x 5.8? $D_\odot$
6.5 $M_\odot$
865 ly
R_V 140 km/s
R_P 2.1 days
η Her (G8)
5.6 $D_\odot$
2 $M_\odot$
112 ly
R_V 2.2 km/s
R_P 128.8 days
ε Her (A0)
2 $D_\odot$
3.2 $M_\odot$
163 ly
R_V 50 km/s
R_P 2 days
ω Her (B9)
1.8 $D_\odot$
3.3 $M_\odot$
235 ly
R_V 44 km/s
R_P 2.1 days
β Her (G7)
12 $D_\odot$
1.9 $M_\odot$
148 ly
R_V 1.5 km/s
R_P 126.5 days

1.8 $D_\odot$ B9 it spins at 44 km/s. The other α CV is **52 Herculis** or **V637**, a 4 $D_\odot$ A2 spinning at 29 km/s. It switches between m_v +4.78 to +4.85 with a period of 3^d 20^h 33.6^m. A number of other variables can be found in Hercules including **68 Herculis** (sometimes listed as **u Herculis**) which is an Eclipsing Type A or 'EA' binary. Its brightness drops by two-thirds of a magnitude from m_v +4.69 to +5.37 as the two components orbit one another, the period measured as 2^d 1^h 13.5^m. The primary star is a B1.5 with a diameter of 5.8 $D_\odot$. Its companion is a cooler B5 and somewhat smaller at 4.3 $D_\odot$.

AM Herculis loans its name to a particular type of cataclysmic variable. In close binary systems material from one star may flow towards the other and form an accretion disk around it. In AM Her type systems the binary pair consist of a red and a white dwarf. It is the white dwarf that is the most important component. If its magnetic field is sufficiently strong it will disrupt the flow of particles and prevent the accretion disk forming. Instead, the particles follow the lines of magnetic force and flow in towards the white dwarf's magnetic poles at up to 3,000 km/s. The impact point is tiny: just 100th of the diameter of the star. In AM Her's case the impact point is only 10,500 km across – 2,250 km smaller than the Earth. This concentration of particles into such a small funnel at such high speeds superheats the flow resulting in the emission of various types of radiation including light and X-rays. We see a similar though thankfully very weak effect on Earth when particles released from the Sun are channeled towards the Earth's magnetic poles and produce aurorae as they plunge through the atmosphere. In an AM Her system it is believed that the magnetic pole of the white dwarf tilts towards the flow of particles so that one pole points towards its stellar companion with the other pole pointing out into space. The light from such system has a particular polarized signature and so the stars are often referred to as 'polars'. The magnetic field of AM Her type systems is in the order of 10 million to 100 million gauss (the Earth's magnetic field is typically just 0.45 gauss and a small iron magnet is usually 50 to 100 gauss). AM Herculis itself requires a fairly substantial telescope to see it, its magnitude varying between m_v +12.3 and +15.7. It lies at 272 ly from Earth in the most north eastern corner of the constellation.

Closely related to AM Her systems are **DQ Herculis** stars. The main differences are that the magnetic field is weaker – 'only' 1 million to 10 million gauss – the white dwarf is surrounded by an accretion disk but very close to the star the disk is disrupted and the particles flow towards the star's poles, and there is less likelihood that the two stars will synchronize their rotational periods (e.g. in the way the Earth and Moon have so that the same lunar hemisphere always points towards the Earth). Synchronized rotation is believed to occur in at least 90% of AM Her systems. DQ Her is a m_v +14.2 object which, strangely enough, is close to AM Her on the celestial sphere but is 44 ly farther away. The white dwarf is 0.012 $D_\odot$, 0.7 $M_\odot$ and has a luminosity of 0.0064 $L_\odot$, its temperature standing at 14,500 K. The red dwarf companion is 0.44 $D_\odot$, 0.4 $M_\odot$ and has a luminosity of 0.03 $L_\odot$, it being a cool 3,500 K. DQ Her flared to m_v +1.5 on 21 December 1934 and received the designation Nova Her 1934.

Perhaps not surprising in a constellation as large as this there are quite a number of stars that have planetary systems (see table). One of the more interesting is **HAT-P-2 b** discovered by the HATNet global network of automated telescopes that scan the Northern Hemisphere each night in search of exoplanets. Although it is only 1.16 times larger than Jupiter it is 9 times more massive. Its eccentric orbit takes it from 4.5 million kilometers of the star during which its surface temperature reaches a scorching 1,700 K, then out to 15 million kilometers when it cools to a balmy 725 K. **HD 149026 b** is another superdense planet that may have a core of heavy elements equivalent to 65-70 Earth masses – more than all the heavy elements found in the planets and asteroids of the Solar System. It has a diameter of 0.72 D_J but a mass of 0.359 M_J. One model of the planet's structure suggests that the dense core may be enveloped by a layer of highly compressed liquid water, then a layer of metallic hydrogen and helium and finally a deep atmosphere of primarily hydrogen and helium gas. Being in what appears to be a nearly circular orbit its temperature remains at a constant 1,000 K. **HD 156668 b** seems to be a super-Earth with a mass of 4.13 $M_\oplus$ or 0.013 M_J making it one of the smaller exoplanets to be discovered. The star itself is slightly less massive than the Sun at 0.72 $M_\odot$ and considerably less luminous: a mere 0.27 $L_\odot$. **TrES-3 b** – TrES is the Trans-Atlantic Exoplanet Survey – is nearly two Jovian masses but just 1.3 D_J. Its orbital period is extremely short at $31^h\ 21^m$ with the result that the orbit is decaying. Eventually the planet will fall into the star.

Planetary systems in Hercules

Star	$D_\odot$	Spectral class	ly	m_v	Planet	Minimum mass	q	Q	P
14 Her	0.71	K0	59	+6.67	14 Her b	4.64 M_J	1.75	3.79	4.85 y
c unconfirmed					14 Her c?	2.1 M_J	6.90	6.90	18.91 y
Gliese 649	0.52	M1.5	34	+9.7	Gl 649 b	0.328 M_J	0.80	1.48	1.64 y
HD147506	1.64	F8	385	+8.71	HAT-P-2 b	9.09 M_J	0.03	0.10	5.63 d
HD 149026	1.50	G0	257	+8.15	HD 149026 b	0.359 M_J	0.04	0.04	2.88 d
HD 154345	0.88	G8	59	+6.74	HD 154345 b	0.947 M_J	4.01	4.37	9.14 y
HD 155358	0.90	G0	139	+7.5	HD 155358 b	0.89 M_J	0.56	0.70	195 d
					HD 155358 c	0.504 M_J	1.01	1.44	1.45 y
HD 156668	0.75	K2	79	+8.42	HD 156668 b	4.13 $M_\oplus$	0.05	0.05	4 d
HD 164922	0.90	K0	72	+7.01	HD 164922 b	0.36 M_J	2.01	2.22	3.16 y
TrES-3	0.81	G	?	+12.4	TrES-3 b	1.92 M_J	0.02	0.02	1.31 d

Globular clusters in Hercules

Name	Size arc min	Size ly	Distance ly	Age million yrs	Apparent magnitude m_v
M13 (NGC 6205)	20′	145	25,100	14,000-24,000	+5.8
M92 (NGC 6341)	14′	109	26,700	16,000	+6.4

Horologium

Constellation:	Horologium	**Hemisphere:**	Southern
Translation:	The Pendulum Clock	**Area:**	249 deg^2
Genitive:	Horologii	**% of sky:**	0.604%
Abbreviation:	Hor	**Size ranking:**	58th

Another inconspicuous constellation introduced by Abbé La Caille.

α **Horologii** is an unimpressive m_v +3.85 and often impossible to see from the towns and cities of the Southern Hemisphere. It began life as a hot A-class but that was more than a thousand million years ago. Today it is 16 $D_{\odot}$ K2 about 30 times more luminous than the Sun. It lies at a distance of 117 ly, give or take 2 ly, and is drifting away from us at 21.7 km/s. K-class stars make up a quarter of all the naked eye stars. Apart from a handful of supergiants the rest are all under 30 $D_{\odot}$ and nearly one-half are 15 to 17 $D_{\odot}$ across so α Hor falls right in the middle.

β **Horologii** is what α Hor used to be like. Much smaller at 4.4 $D_{\odot}$ and three times as hot at 9,000 K β Hor is an A3 and is between 300 and 328 ly from Earth. It is somewhat larger than average for the spectral group, which comes in at about 2 $D_{\odot}$ and, at 77 $L_{\odot}$, is a bit above the mean of 60 $L_{\odot}$.

The rest of the stars can be divided into A and F-class. At the early end of the A-class is the A2 ν **Horologii**, a 1.8 $D_{\odot}$, 1.9 $M_{\odot}$ dwarf 165 ly away. Moving up the scale to A6 is η **Horologii**, very slightly larger at 1.9 $D_{\odot}$ and very slightly less massive at 1.8 $M_{\odot}$. η takes the prize for the fastest spinning star: a breakneck 315 km/s. Almost spilling over into the B-class is the A9 **δ Horologii**, the largest of the trio at 3.2 $D_{\odot}$ and 1.9 $M_{\odot}$. Not nearly as fast a spinner as η, δ still puts in an impressive 193 km/s.

Of the F-class stars the largest is the F6 2.9 $D_{\odot}$ ζ **Horologii** followed by μ **Horologii,** an FO at 2.4 $D_{\odot}$. Just 56.2 ly from the Sun is the F8 dwarf ι **Horologii.** Slightly larger than the Sun at 1.13 $D_{\odot}$, slightly more luminous at 1.67 $L_{\odot}$ and slightly more massive at 1.2 $M_{\odot}$ this star is considerably younger – just 625 million years old, compared to the Sun's 4,567 million years – and appears to have a planetary system. So far only one planet has been discovered: a two Jovian mass giant averaging 1 AU from the star. Those who name such planets could have called it ι Horologii b, or HD 17051 b or even GJ 108 b but instead decided to use HR 810 b from the *Harvard Revised Photometry Catalog*. Only one other planetary system carries the HR name. Meanwhile, the star ι Hor is believed to have once been a member of the Hyades Moving Group but was ejected, possibly by gravitational perturbations, and is now 130 ly adrift of the Group.

Keep an eye out for **R Horologii**. A rare M7 red giant some 50 times larger than the Sun it is a Mira variable and is usually too faint to be seen without optical aid. It can reach m_v +4.7 but for most of its 407.6 day period it is either plunging towards or just recovering from its minima of m_v +14.3.

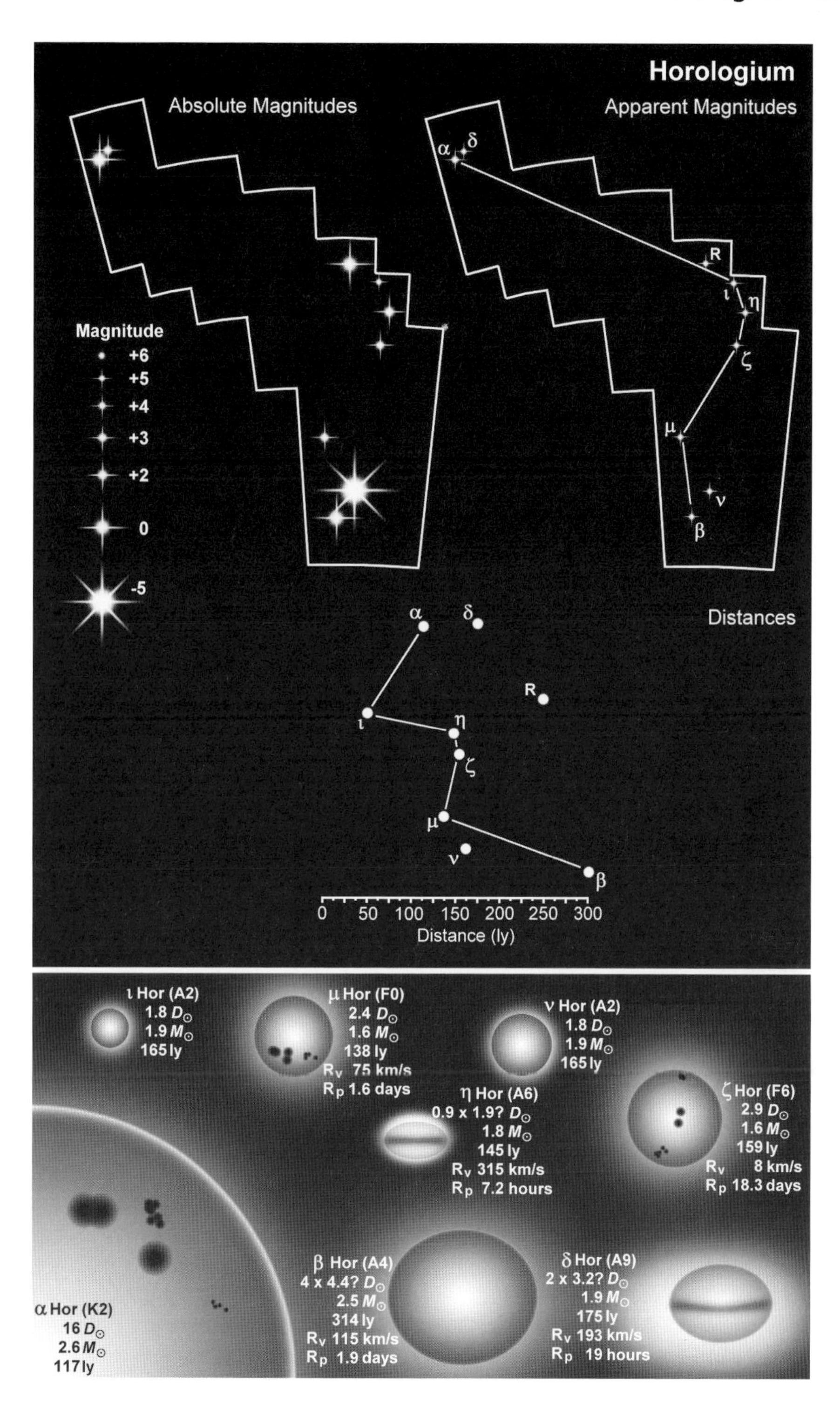

Horologium
Absolute Magnitudes
Apparent Magnitudes
Magnitude
+6
+5
+4
+3
+2
0
-5
α
δ
R
ι
η
ζ
μ
ν
β
Distances
0 50 100 150 200 250 300
Distance (ly)
ι Hor (A2)
1.8 $D_\odot$
1.9 $M_\odot$
165 ly
μ Hor (F0)
2.4 $D_\odot$
1.6 $M_\odot$
138 ly
R_V 75 km/s
R_P 1.6 days
ν Hor (A2)
1.8 $D_\odot$
1.9 $M_\odot$
165 ly
η Hor (A6)
0.9 x 1.9? $D_\odot$
1.8 $M_\odot$
145 ly
R_V 315 km/s
R_P 7.2 hours
ζ Hor (F6)
2.9 $D_\odot$
1.6 $M_\odot$
159 ly
R_V 8 km/s
R_P 18.3 days
α Hor (K2)
16 $D_\odot$
2.6 $M_\odot$
117 ly
β Hor (A4)
4 x 4.4? $D_\odot$
2.5 $M_\odot$
314 ly
R_V 115 km/s
R_P 1.9 days
δ Hor (A9)
2 x 3.2? $D_\odot$
1.9 $M_\odot$
175 ly
R_V 193 km/s
R_P 19 hours

Planetary system in Horologium

Star	$D_\odot$	Spectral class	ly	m_v	Planet	Minimum mass	q	Q	P
ι Hor	1.13	G0	56	+5.40	HR 810 b	1.94 M_J	0.69	1.12	311.3 d

Hydra

Constellation:	Hydra	**Hemisphere:**	Equatorial
Translation:	The Water Snake	**Area:**	1,303 deg^2
Genitive:	Hydrae	**% of sky:**	3.159%
Abbreviation:	Hya	**Size ranking:**	1st

The largest, but far from the most prominent of all the constellations, Hydra represents the monster slain by Hercules as one of his 12 labors. It consists of 69 stars brighter than m_v +5.50 but they are generally faint and it is a difficult constellation to trace across the night sky.

α **Hydrae** is a giant K3 but how big a giant is debatable. Estimates put its size at 23, 26, 37, 40 and 93 $D_\odot$. If we take the average, about 40 $D_\odot$, then from its home 177 ly away its luminosity is in the order of 387 $L_\odot$. It's a reasonably bright star, m_v +1.98, so at 10 pc it would shine at M_v -0.2. Alphard, to give the star its proper name, which means 'the solitary one', was once the companion to a much more massive star that has since died and is now a white dwarf. As it perished it contaminated Alphard with neutrons resulting in enhanced levels of barium which show up in the star's spectrum.

β **Hydrae** is by no means the second brightest star in the constellation - there are 16 others that are brighter! - but it is a well known double and variable. To the naked eye it looks like a single m_v +4.27 star but optical aid reveals a m_v +5.47 companion. Some 365 ly from Earth the primary is a bluish-white B9 of about 5.8 $D_\odot$ and 193 $L_\odot$. Its orbital companion is somewhat smaller, probably about 2.9 $D_\odot$, and a third as luminous. The system is very young, an estimated 200 million years, and the two stars appear to be separated by about 200 AU which leads to an orbital period in excess of 1,000 years. β Hydrae's variability, which dims the star to m_v +4.31 before returning to m_v +4.27 with a period of $2^d\ 8^h\ 15.4^m$ is caused by dark islands of strong magnetic activity where strontium, silicon and chromium have pooled near the star's surface and rotate with it at 72 km/s. It is classed as an α CV rotating variable

Near the end of the Hydra's tail is γ **Hydrae**, a 13 $D_\odot$ G8 lying at a distance of 132 ly. Perhaps more interesting is **R Hydrae**, 2.6° to the East. An M7 red supergiant of 440 $D_\odot$ it resides in the outer limits of the constellation at 1,400 ly. If the Sun was replaced by R Hya it would reach out to midway between Mars and the Asteroid Belt. It is thought that red giants rotate at about half the velocity of the Sun: about 1 km/s. If this is correct then R Hya takes more than 60 years to turn once on its axis. As with many red giants it is a Mira variable, its rollercoaster magnitude ride reaching a peak m_v +3.50 before plunging to m_v +10.90. In 1932 Aitken noted the period as being 425 days, Robert Burnham in 1978 said it had decreased to 400 days and now it is 388.87 days. A couple of other stars, R Aquilae and R Centauri, show similar phenomena, which may be due to internal structural changes.

The head of the Hydra is picked out by five or six stars depending on who

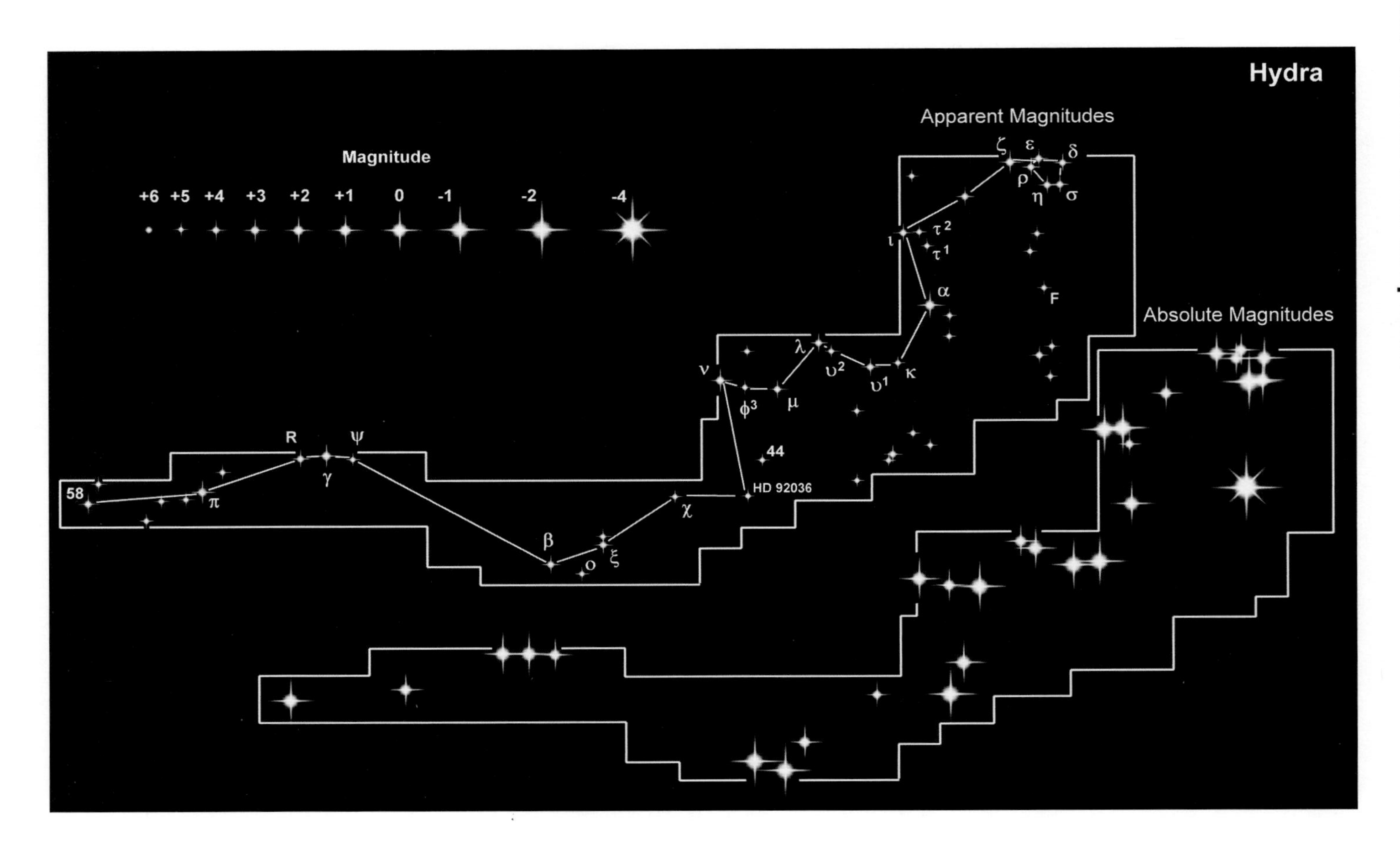
Hydra
Apparent Magnitudes
Absolute Magnitudes
Magnitude
+6 +5 +4 +3 +2 +1 0 -1 -2 -4
ζ ε δ ρ η σ
τ2 τ1 ι α F
λ υ2 υ1 κ ν φ3 μ
44
HD 92036
R ψ γ 58 π
β ο ξ χ

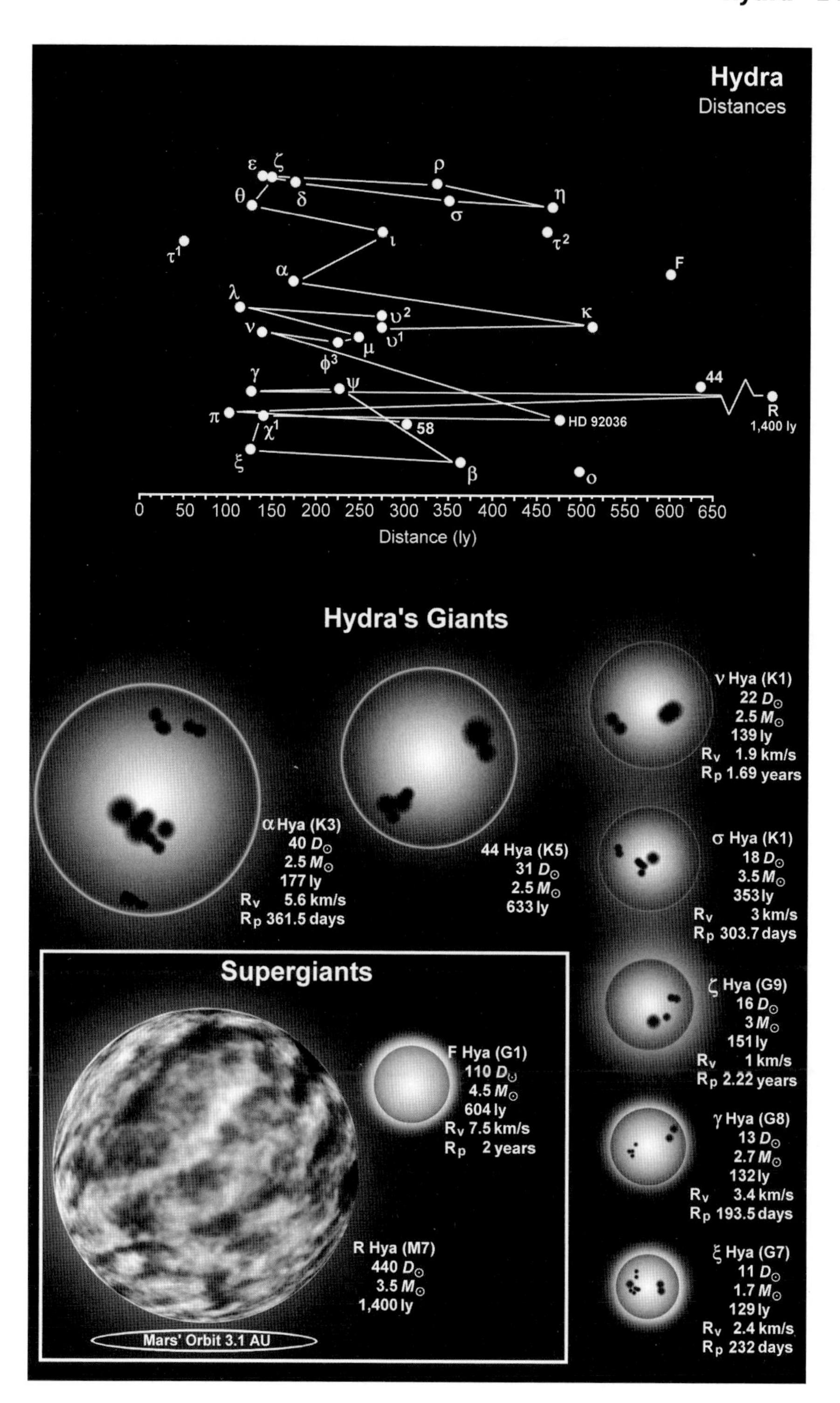

Hydra
Distances
ε ζ ρ
θ δ η
σ
ι τ2
τ1
α F
λ
υ2 κ
ν μ υ1
φ3
γ ψ 44
π R
χ1 58 HD 92036 1,400 ly
ξ β o
0 50 100 150 200 250 300 350 400 450 500 550 600 650
Distance (ly)
Hydra's Giants
α Hya (K3)
40 D⊙
2.5 M⊙
177 ly
Rv 5.6 km/s
Rp 361.5 days
44 Hya (K5)
31 D⊙
2.5 M⊙
633 ly
ν Hya (K1)
22 D⊙
2.5 M⊙
139 ly
Rv 1.9 km/s
Rp 1.69 years
σ Hya (K1)
18 D⊙
3.5 M⊙
353 ly
Rv 3 km/s
Rp 303.7 days
Supergiants
ζ Hya (G9)
16 D⊙
3 M⊙
151 ly
Rv 1 km/s
Rp 2.22 years
F Hya (G1)
110 D⊙
4.5 M⊙
604 ly
Rv 7.5 km/s
Rp 2 years
γ Hya (G8)
13 D⊙
2.7 M⊙
132 ly
Rv 3.4 km/s
Rp 193.5 days
R Hya (M7)
440 D⊙
3.5 M⊙
1,400 ly
ξ Hya (G7)
11 D⊙
1.7 M⊙
129 ly
Rv 2.4 km/s
Rp 232 days
Mars' Orbit 3.1 AU

draws the connecting lines. δ **Hydrae**, a 2.2 $D_\odot$ A1, marks the creature's forehead. Below that, its nose is marked by σ **Hydrae**, an 18 $D_\odot$ K1. Its lower jaw is pinpointed by η **Hydrae**, a hot B3 with a diameter of 3.9 Suns. It is a β Cepheid variable with a period of 4^h 4.8^m during which its magnitude flickers between m_v +4.27 and +4.33. ρ **Hydrae** is often given as the position of its upper jaw. This is an A0 dwarf of 2.0 $D_\odot$. Some charts miss out this star altogether and connect η Hya with ζ **Hydrae** where the head joins the long curving body. ζ Hya is a giant yellowish-orange G9 about 16 $D_\odot$ across and some 400 million years old. Rotating at just 1 km/s, half the speed of the Sun, ζ Hya takes 810 days – 2.2 years – to make a single rotation. ε **Hydrae** is the back of the creature's head. This is another G-class star but right in the middle of the scale, a G5 of 7 $D_\odot$. It belongs to the rare BY Draconis variable class – only one other naked eye BY Dra exists (ξ Boo) – and varies between m_v +3.35 and +3.39. It is believed the variability is due to large spots crossing the star's visible surface. The star is also a quintuplet system. In a 15.06 year orbit around the primary is an A8 dwarf which can come as close as 3.6 AU at periastron and fly out as far as 17.4 AU at apastron. A second binary pair orbit one another in just 9.9 days, separated by 0.09 AU. This pair orbits the inner pair at an average distance of 190 AU, taking 990 years to complete an orbit. Farther out again at 800 AU is a red dwarf of about 0.3 $M_\odot$. It takes more than 8,360 years to complete a single orbit around the other four stars. Drawing a line from ε to δ Hya completes the monster's head.

χ^1 and **44 Hydrae** are the forgotten stars of the constellation. In their eagerness to join the dots some celestial cartographers draw a line from ν to ξ **Hydrae** cutting across the star group of Crater as though it was not there. One or two others take a different route going from ν Hya to 44 Hya to χ^1 Hya and then linking up to ξ Hya, keeping the line of the water snake within the boundaries set by the International Astronomical Union.

We naturally tend to think of yellow stars as being like the Sun. **F Hydrae** is spectroscopically similar – it's a G1 whereas the Sun is a G2 – but there the similarity ends. F Hya is 110 times larger than the Sun and 391 times as luminous. The lifespan of such supergiants can be measured in millions of years compared to the 10,000 million year solar longevity.

Hydra is home to one of the more elusive celestial bodies – a brown dwarf. It was discovered in 1987 by Chilean astronomer Maria Teresa Ruiz who was looking for white dwarfs on two photographs of the same part of the sky but taken 11 years apart. She decided to name it **KELU-1** which means 'red' in the language of the Mapuche people. It is about 33 ly from Earth and has a temperature of less than 1,400 K.

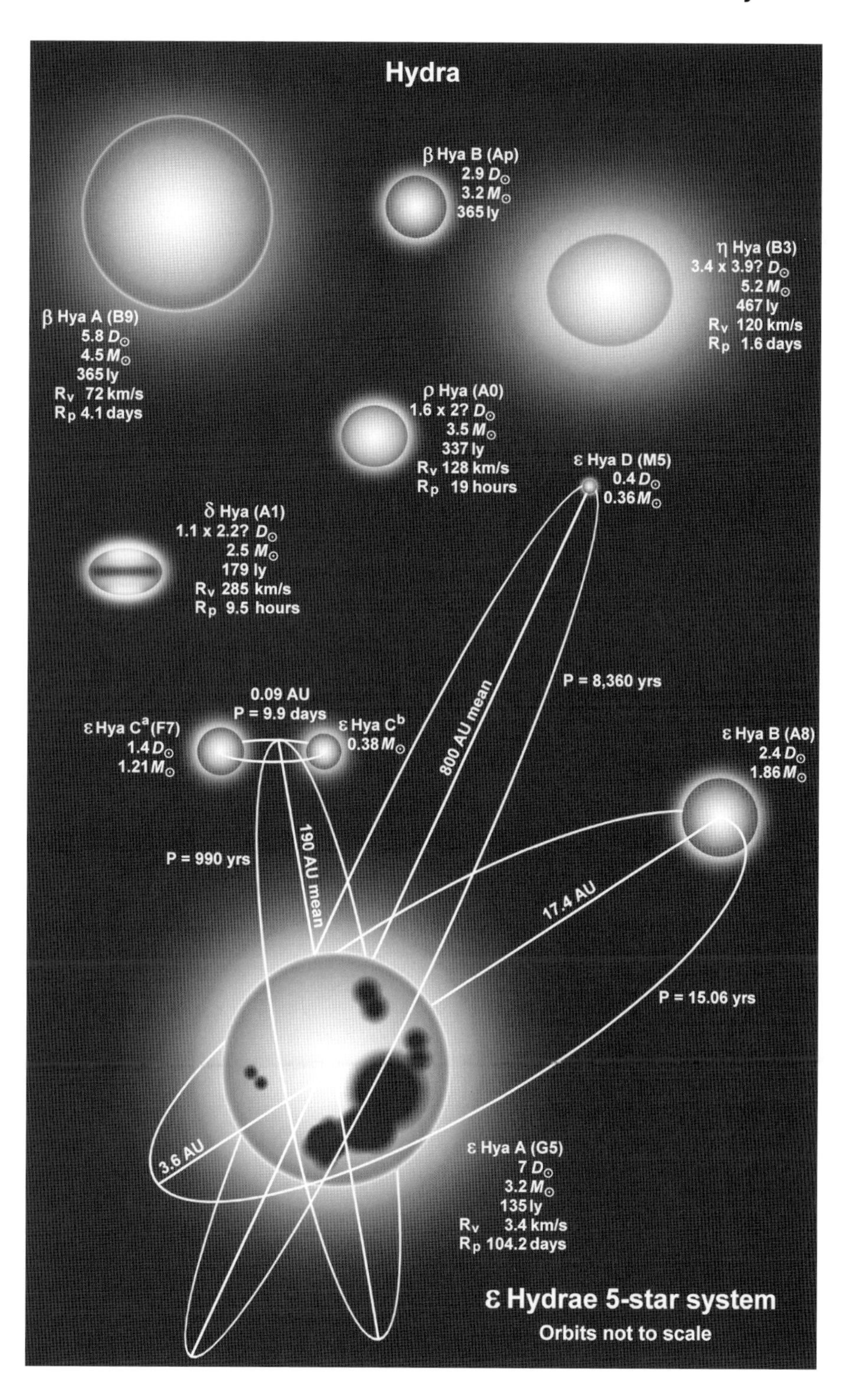

Hydra
β Hya B (Ap)
2.9 D☉
3.2 M☉
365 ly
η Hya (B3)
3.4 x 3.9? D☉
5.2 M☉
467 ly
Rv 120 km/s
Rp 1.6 days
β Hya A (B9)
5.8 D☉
4.5 M☉
365 ly
Rv 72 km/s
Rp 4.1 days
ρ Hya (A0)
1.6 x 2? D☉
3.5 M☉
337 ly
Rv 128 km/s
Rp 19 hours
ε Hya D (M5)
0.4 D☉
0.36 M☉
δ Hya (A1)
1.1 x 2.2? D☉
2.5 M☉
179 ly
Rv 285 km/s
Rp 9.5 hours
P = 8,360 yrs
800 AU mean
0.09 AU
P = 9.9 days
ε Hya Ca (F7)
1.4 D☉
1.21 M☉
ε Hya Cb
0.38 M☉
ε Hya B (A8)
2.4 D☉
1.86 M☉
P = 990 yrs
190 AU mean
17.4 AU
P = 15.06 yrs
3.6 AU
ε Hya A (G5)
7 D☉
3.2 M☉
135 ly
Rv 3.4 km/s
Rp 104.2 days
ε Hydrae 5-star system
Orbits not to scale

Open and globular clusters in Hydra

Name	Size arc min	Size ly	Distance ly	Age million yrs	Brightest star in region*	No. stars m_v >+12*	Apparent magnitude m_v
M48 (NGC 2548)	75′	55	2,500	360	HD 68105 m_v +7.50	460	+5.8
M68 (NGC 4590)	11′	106	33,000	11,200	Globular cluster		+7.8

*May not be a cluster member.

Hydrus

Constellation:	Hydrus	**Hemisphere:**	Southern
Translation:	The Lesser Water Snake	**Area:**	243 deg^2
Genitive:	Hydri	**% of sky:**	0.589%
Abbreviation:	Hyi	**Size ranking:**	61st

A 'modern' constellation created from the observations of the 16th Century Dutch explorers Pieter Keyser and Frederick de Houtman.

In 2600 BC **α Hydri** marked the position of the South Celestial Pole being just 2° away. Today it is 29° adrift of the pole, the result of the Earth's precession. It is an F0 sub-giant 13.3 $D_{\odot}$ across and with a luminosity of 28 $L_{\odot}$. Its rotational period is just under 1.5 days, the star revolving at 118 km/s. When Burnham published his *Celestial Handbook* in the mid 1960s it was generally believed that α Hyi was just 30 ly from Earth. By the late 60s it had drifted out to 41 ly according to Josef Klepešta and Antonín Rükl's *Constellations*. We are now pretty certain it is 71.3 ly.

β Hydri was a slightly better South Pole star than α Hyi, being just under 2° short of the pole. It is not vastly different to our own Sun: 1.9 $D_{\odot}$ and 1.1 $M_{\odot}$, but somewhat older by about 1,000 million years (5,500 million years compared to the Sun's 4,567 million years) and a bit more luminous at 3.5 $L_{\odot}$. Spinning on its axis at 3.3 km/s it takes 29 days to complete one rotation (the Sun takes 25 days). The star sits close to the Small Magellanic Cloud (SMC) and is sometimes used as a finder star for its location. The big difference is in the distance. β Hyi is just 24.4 ly away whereas the SMC is more than 200,000 ly. The star's space velocity relative to the Sun is very high – more than 80 km/s – which would tend to suggest it is a visitor from another part of the Galaxy.

γ Hydri is the only red giant in the constellation. Some 214 ly from Earth it is 60 $D_{\odot}$ across and 655 times more luminous than the Sun. Despite its size its mass is only about 1.7 $M_{\odot}$.

η^2 Hydri (or **HD 11977**) has a confirmed planetary system. The star itself is a larger but cooler version of the Sun: a 12 $D_{\odot}$ G8.5 lying at a distance of 217 ly. Its planet is equally more impressive than our own Jupiter being 6.5 times more massive and in an orbit that takes nearly two years to complete its path around the star. Giant stars with planets are a relative rarity. **GJ 3021** (**HD 1237**) is another Sun-like star. Its planet appears to have high levels of sulfur in its atmosphere.

IC 1717 is the celestial object that does not exist. It was originally recorded by John Dreyer in the late 19th Century as being located at RA 1^h 32.5^m, Dec. -67° 32′ and described as *excessively faint, excessively small, much extended 25°* and a stellar nucleus. Look in that location today and there is nothing that matches Dreyer's description. It was almost certainly a faint comet. Indeed, the renowned comet hunter Charles Messier confused comets with nebulae so often that he listed the latter as his now famous *Messier Catalog.*

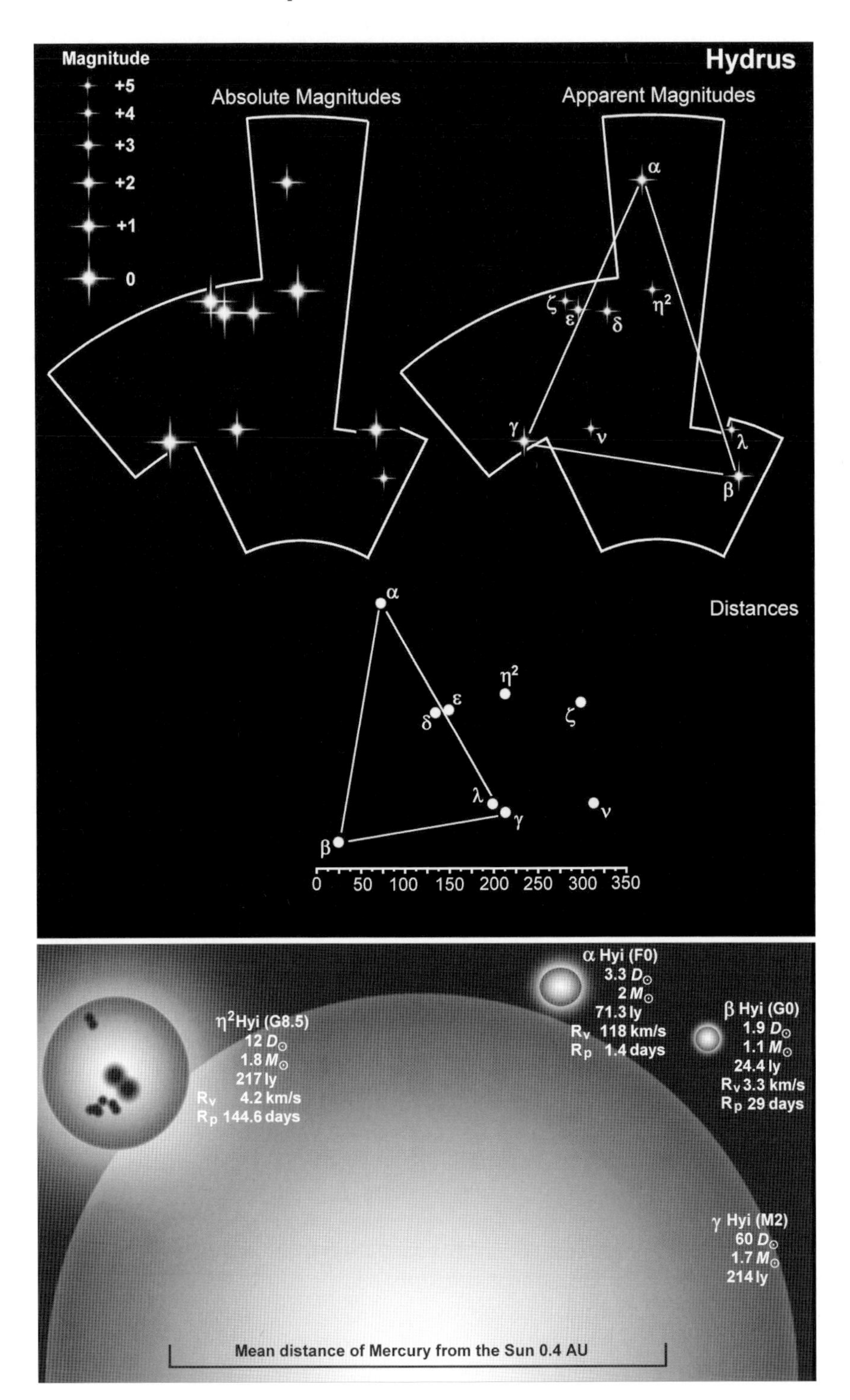

Magnitude
+5
+4
+3
+2
+1
0
Hydrus
Absolute Magnitudes
Apparent Magnitudes
α
ζ
ε
δ
η2
γ
ν
λ
β
Distances
α
η2
ε
δ
ζ
λ
γ
ν
β
0 50 100 150 200 250 300 350
α Hyi (F0)
3.3 D⊙
2 M⊙
71.3 ly
Rv 118 km/s
Rp 1.4 days
η2Hyi (G8.5)
12 D⊙
1.8 M⊙
217 ly
Rv 4.2 km/s
Rp 144.6 days
β Hyi (G0)
1.9 D⊙
1.1 M⊙
24.4 ly
Rv 3.3 km/s
Rp 29 days
γ Hyi (M2)
60 D⊙
1.7 M⊙
214 ly
Mean distance of Mercury from the Sun 0.4 AU

Planetary systems in Hydrus

Star	$D_{\odot}$	Spectral class	ly	m_v	Planet	Minimum mass	q	Q	Period (days)
η^2 Hyi	13.0	G8.5	217	+4.7	HD 11977 b	6.54 M_J	1.16	2.70	1.95 y
HD 1237	1.0	G6	57.5	+6.59	GJ 3021 b	3.37 M_J	0.24	0.74	133.7 d

Indus

Constellation:	Indus	**Hemisphere:**	Southern
Translation:	The Indian	**Area:**	294 deg^2
Genitive:	Indi	**% of sky:**	0.713%
Abbreviation:	Ind	**Size ranking:**	49th

Indus first appeared in Johann Bayer's *Uranometria* in 1603 but it is not clear which part of the Americas Indus is supposed to represent. The Latin term is more correctly applied to a native of India.

Of the ten naked eye stars in Indus, half belong to spectral class K including **α Indi**, a K0 which, for reasons best known to themselves, was named 'the Persian' by Jesuit missionaries. Its diameter is estimated as being 11 $D_{\odot}$. At a distance of 101.3 ly its luminosity results in a m_v +3.10 luminary which at 10 pc would brighten to M_v +1.1. It is a super-rich metal star with twice the abundance of iron as that found in the Sun along with high levels of other metals, all of which suggests it was born on the inner side of our galactic arm, towards the center of the Galaxy.

β Indi is the second K-class, a K1, and a solar diameter larger than α Ind at 12 $D_{\odot}$. It is also the farthest: 603 ly, maybe 675 ly, such is the uncertainty in our knowledge. At 17.3″ separation from β Ind is a m_v +12.5 dwarf (PA 104°) which may or may not be a binary companion.

Rather more certainty surrounds δ **Indi,** a 4 $D_{\odot}$ F0 which is between 177 and 193 ly. It has a doppelgänger companion in either a 6.09 or 12.24 year long orbit. They are both F0 and both m_v +5.2, their combined magnitude coming in at m_v +4.40.

ε **Indi** is an intriguing triple star system. Lying at a distance of just 11.83 ly the main component, ε Indi A, is a K5 dwarf around 0.76 $D_{\odot}$ and a corresponding 0.70 $M_{\odot}$ but with a very low luminosity of just 0.14 $L_{\odot}$. This puts it close to the edge of visibility. If it were just a little farther away it would be below the naked eye limit. At 10 pc it would have an absolute magnitude of M_v +7.0. It rotates at 1.46 km/s taking 26 days to spin once on its axis. Both its components are methane-rich brown dwarfs in an orbit around one another that averages 2.65 AU and which takes 16 years to run full circle. One of the dwarfs, ε Indi B^a belongs to a relatively new spectral class, T1. Its temperature is 1,280 K, its diameter just 0.091 $D_{\odot}$ and its luminosity an almost non-existent 0.000019 $L_{\odot}$. It has a mass of 0.045 $M_{\odot}$ which equates to 47 M_J. Its partner, ε Indi B^b, is an even cooler 854 K and regarded as a T6. Its diameter is 0.096 $D_{\odot}$ and its mass is estimated to be 0.027 $M_{\odot}$ or 28 M_J. It is even fainter at 0.0000045 $L_{\odot}$. This binary pair of brown dwarfs lie at least 1,460 AU from ε Indi A and take more than 63,000 years to complete a single orbit. The whole system is believed to be no more than 1,300 million years old (about 29% of the age of the Solar System) and has a space velocity of 86 km/s indicating the stars are visitors from a different part of the Galaxy.

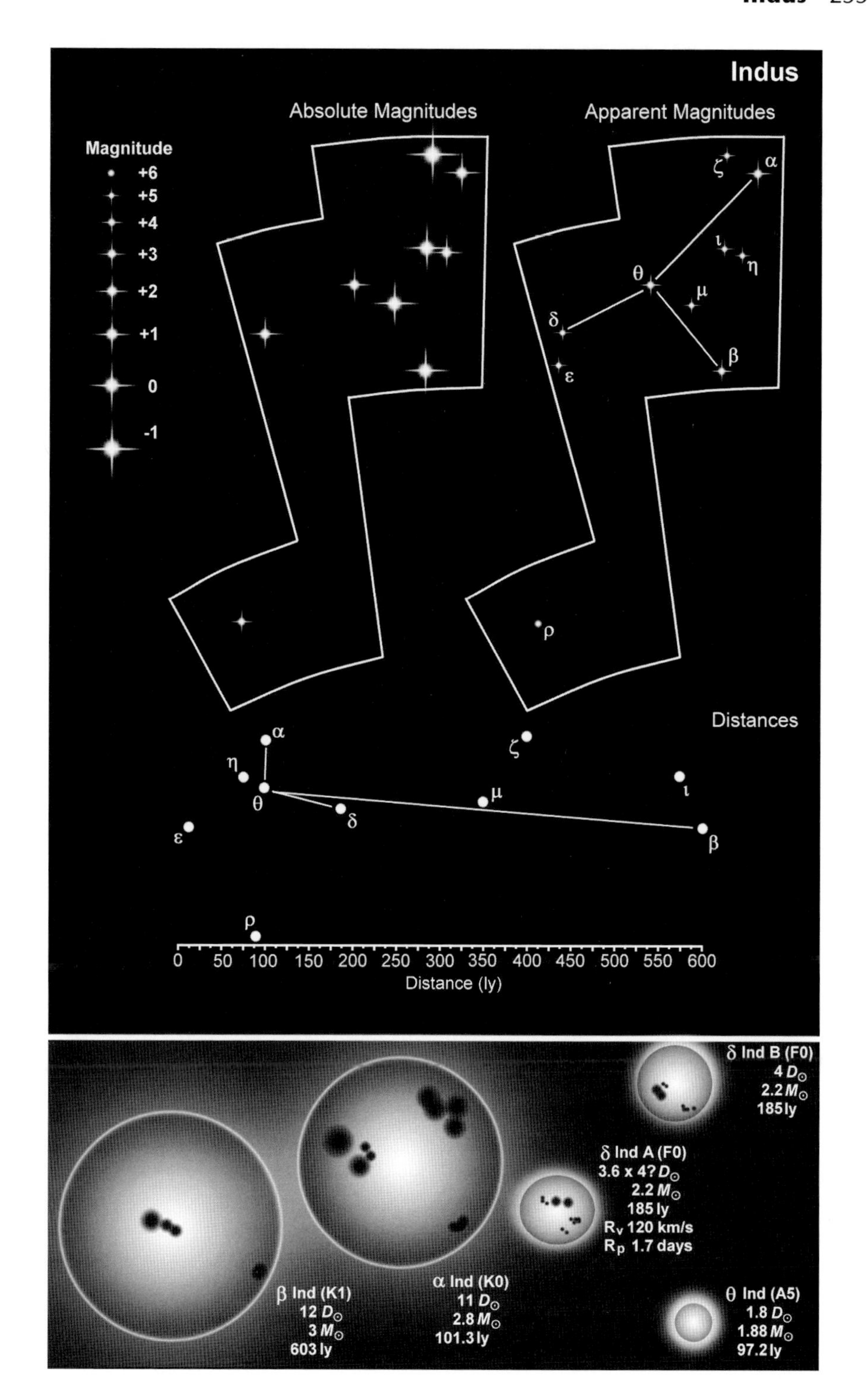

Indus
Absolute Magnitudes
Apparent Magnitudes
Magnitude
+6
+5
+4
+3
+2
+1
0
-1
ζ
α
ι
η
θ
μ
δ
β
ε
ρ
Distances
0 50 100 150 200 250 300 350 400 450 500 550 600
Distance (ly)
δ Ind B (F0)
4 D⊙
2.2 M⊙
185 ly
δ Ind A (F0)
3.6 x 4? D⊙
2.2 M⊙
185 ly
Rv 120 km/s
Rp 1.7 days
β Ind (K1)
12 D⊙
3 M⊙
603 ly
α Ind (K0)
11 D⊙
2.8 M⊙
101.3 ly
θ Ind (A5)
1.8 D⊙
1.88 M⊙
97.2 ly

One of the smaller stars in the constellation is the 1.5 $D_\odot$ η **Indi** which is 78.8 ly away. An A9 it has a luminosity of 7.5 $L_\odot$ and rotates at 122 km/s, completing a full rotation in just under 15 hours.

The two remaining K-class stars in the constellation are ι **Indi**, a K1 of 15 $D_\odot$, and μ **Indi** which is a little larger at 17 $D_\odot$ and which is tagged as a K2.

At 12,960 million years the 6th magnitude ρ **Indi** is about three times as old as the Sun but in many other respects is very similar. It's about 10% larger than the Sun, is a G1 and has a very similar mass at 1.07 $M_\odot$ though is twice as luminous. It also has at least one planet which has twice the mass of Jupiter (see table). While it would have been so easy to call the planet ρ Indi b instead it is **HD 216437 b**.

Planetary system in Indus

Star	$D_\odot$	Spectral class	ly	m_v	Planet	Minimum mass	q	Q	P
ρ Ind	1.1	G4	86.5	+6.06	HD 216437 b	2.1 M_J	1.78	3.62	3.54 y

Lacerta

Constellation:	Lacerta	**Hemisphere:**	Northern
Translation:	The Lizard	**Area:**	201 deg^2
Genitive:	Lacertae	**% of sky:**	0.487%
Abbreviation:	Lac	**Size ranking:**	68th

A constellation created by the 17th Century Polish astronomer Johannes Hevelius. Only two stars have Greek letter designations.

The brightest star in Lacerta is also the closest. **α Lacertae** lies 102.4 ly away and twinkles at a dim m_v +3.76. Belonging to the A1 spectral class this white hot dwarf, with a temperature of around 9,000 K, is 1.5 $D_\odot$ across and takes just under 16 hours to turn once on its axis, its rotational velocity being 115 km/s. It is 25 times more luminous than the Sun.

β Lacertae is the only G-class star in the constellation, but only just. It's a G8.5 and a giant at 13 $D_\odot$ but around 1,000 K cooler than the Sun. Rotating at 17 km/s it takes 38.7 days to revolve once.

DD Lacertae is a β Cepheid or, if you prefer, a β CMa pulsating variable. A blue B2 of 5.2 $D_\odot$ it has a period between successive maxima (or minima) of 4^h 38^m reaching m_v +5.16 at its brightest and m_v +5.28 at its dimmest.

You could be forgiven for thinking that **BL Lacertae** is also a variable. That's exactly what Cuno Hoffmeister thought when he noticed the 14th magnitude star in 1929, hence its 2-letter designation. Forty years later it was found to be the source of strong radio emissions and a faint trace of a galaxy was detected. A few years passed and then news broke that its distance had been measured: it was put at 900 million light years. Several dozen BL Lac objects are now known. They seem to be similar to quasars but missing both absorption and emission lines from their spectra. They are most luminous in the infrared. So what is BL Lac? No one really knows for certain but the wise money is on a rotating black hole that is surrounded by an accretion disk.

Lacerta contains three very different supergiants. The smallest of these is **4 Lacertae,** a 75 $D_\odot$ bluish-white B9. Blue and bluish-white supergiants are quite rare: they make up only 4% of all the naked eye B-class stars in the sky. B9 supergiants, not surprisingly, are even more rare – there are just four examples. 4 Lac is thought to be 2,118 ly away and has a luminosity approaching 5,000 Suns. Its magnitude as seen from Earth is a paltry m_v +4.59 but get close to it – well, the standard distance of 10 pc – and it would blaze at M_v -6.5. The second largest of the supergiants is the 88 $D_\odot$ **5 Lacertae.** A reddish-orange it is 1,847 $L_\odot$ and resides 1,165 ly from us. It has a B8 dwarf companion in a 41.95 year long orbit. The fifth magnitude **HD 216946 (V424 Lacertae)** takes the prize of being the largest supergiant. Being right on the border of the K-M class this reddish-orange M0 is 1,874 ly from Earth. Which is just as well. If it suddenly found itself in the center of our Solar System it would engulf the Earth and reach out just beyond Mars. A huge 350 $D_\odot$ across its radius is a full 1.6 AU, a reminder of what

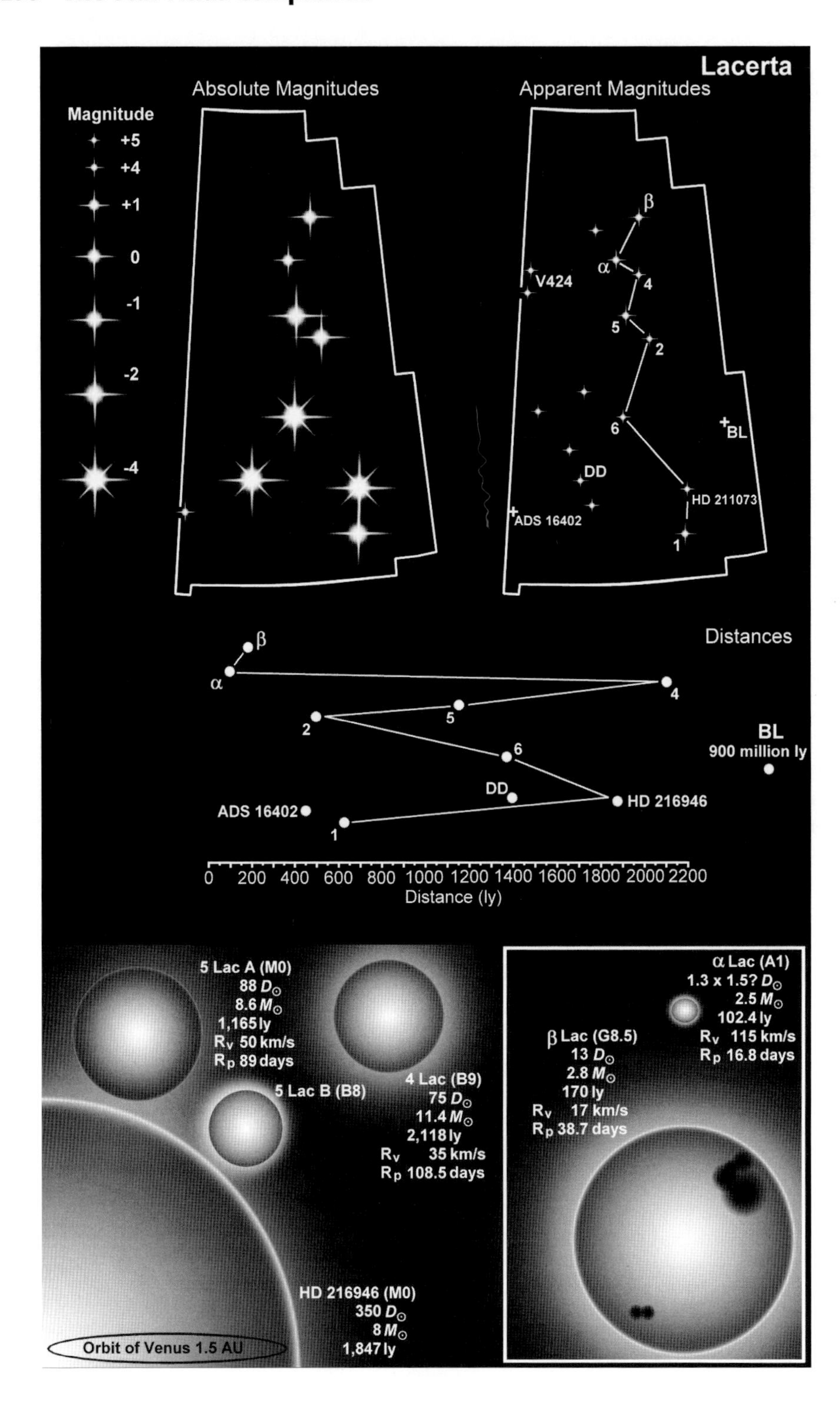

Lacerta
Absolute Magnitudes
Apparent Magnitudes
Magnitude
+5
+4
+1
0
-1
-2
-4
β
α
V424
4
5
2
6
BL
DD
HD 211073
ADS 16402
1
Distances
BL
900 million ly
HD 216946
0 200 400 600 800 1000 1200 1400 1600 1800 2000 2200
Distance (ly)
5 Lac A (M0)
88 $D_\odot$
8.6 $M_\odot$
1,165 ly
R_V 50 km/s
R_P 89 days
5 Lac B (B8)
4 Lac (B9)
75 $D_\odot$
11.4 $M_\odot$
2,118 ly
R_V 35 km/s
R_P 108.5 days
HD 216946 (M0)
350 $D_\odot$
8 $M_\odot$
1,847 ly
Orbit of Venus 1.5 AU
α Lac (A1)
1.3 x 1.5? $D_\odot$
2.5 $M_\odot$
102.4 ly
R_V 115 km/s
R_P 16.8 days
β Lac (G8.5)
13 $D_\odot$
2.8 $M_\odot$
170 ly
R_V 17 km/s
R_P 38.7 days

Leo

Constellation:	Leo	**Hemisphere:**	Equatorial
Translation:	The Lion	**Area:**	947 deg^2
Genitive:	Leonis	**% of sky:**	2.296%
Abbreviation:	Leo	**Size ranking:**	12th

One of the few constellations that bears some resemblance to what it is supposed to be Leo represents the Nemean lion which Hercules slew as one of his labors. It is one of the Zodiacal constellations, the Sun entering its boundaries on 10 August and leaving on 16 September.

α **Leonis**, better known as Regulus, 'the Little King', is a complex arrangement of four stars about 77.5 ly away. Sitting almost on the Ecliptic the m_v +1.35 star is sometimes occulted by the Moon and often grazes the lunar limb. The primary is a hot B7 which has a rotational velocity of 285 km/s. One of the consequences of such a high velocity is that the star is flattened so that pole to pole it measures 3.3 $D_\odot$ while the equatorial diameter is a third larger at 4.3 $D_\odot$. Its equator therefore completes one rotation in 18.3 hours. Another effect of spin-induced distortion is that the star is hottest at the poles where the temperature reaches 15,400 K, some 5,000 K greater than at the equator. In total the star is 150 times more luminous than the Sun and rather more massive at 3.4 $M_\odot$. In a very close 0.35 AU orbit is a white dwarf of 0.3 $M_\odot$ which takes 40.1 days to circle the primary. It is impossible to see through a telescope but its presence has been detected spectroscopically. At the rather greater distance of 4,200 AU is an 8th magnitude orange dwarf of 0.5 $L_\odot$ which takes at least 109,900 years to complete a full orbit. Around this star is an M4 red dwarf of 0.025 $L_\odot$ and which attains only m_v +13.5. It takes 436 years to complete an orbit.

At 36.2 ly β **Leonis** is technically the second closest star in the constellation. The closest is the red dwarf Wolf 359 (see below). At m_v +2.14 however it far outshines 13th magnitude Wolf 359. It is an almost 2 $D_\odot$ A3, about 14 times as luminous as the Sun and 2,000 K hotter. Rotating at about 122 km/s it spins once on its axis every 18 hours. Infrared studies indicate the star, which is often called Denebola (meaning the 'Tail of the Lion'), is surrounded by a dusty cloud of the sort that may produce planets, though none has yet been found. In 1985 it was realized that the star is a pulsating variable of the δ Scuti variety. Its magnitude changes are very subtle though – just three hundredths of a magnitude.

γ **Leonis** or Algieba is a giant star, a binary and hosts a planet. The primary, γ^1 **Leonis**, is a K0 giant of 32 $D_\odot$, 1.23 $M_\odot$ and 180 $L_\odot$. Its magnitude is m_v +2.01. In an orbit that takes in excess of 620 years to complete, γ^2 **Leonis** can come as close as 15 AU and can separate by as much as 180 AU. Its is a much smaller star than γ^1 though still a giant at 10 $D_\odot$, has a similar mass but is only 50 $L_\odot$. At 4,980 K it is about 600 K hotter and falls into the G7 spectral group. It consequently shines at m_v +3.51. A nine Jovian mass planet orbits γ^1 Leo at an average distance of 1.2 AU (see table). The whole system is 126 ly away.

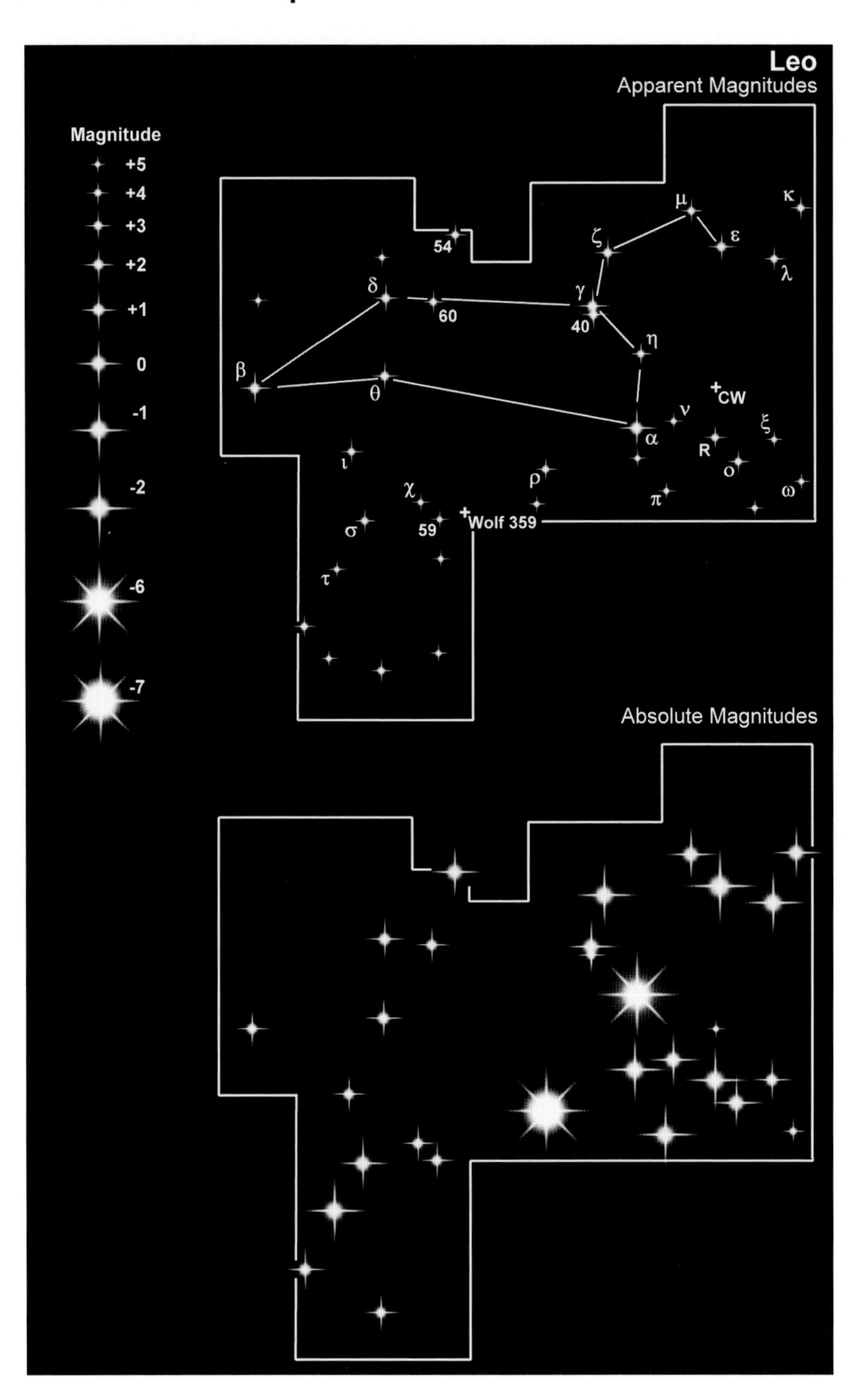
Leo
Apparent Magnitudes
Magnitude
+5
+4
+3
+2
+1
0
-1
-2
-6
-7
μ
κ
54
ζ
ε
λ
δ
γ
60
40
η
β
θ
CW
ν
ξ
α
R
ι
ο
ρ
χ
π
ω
Wolf 359
σ
59
τ
Absolute Magnitudes

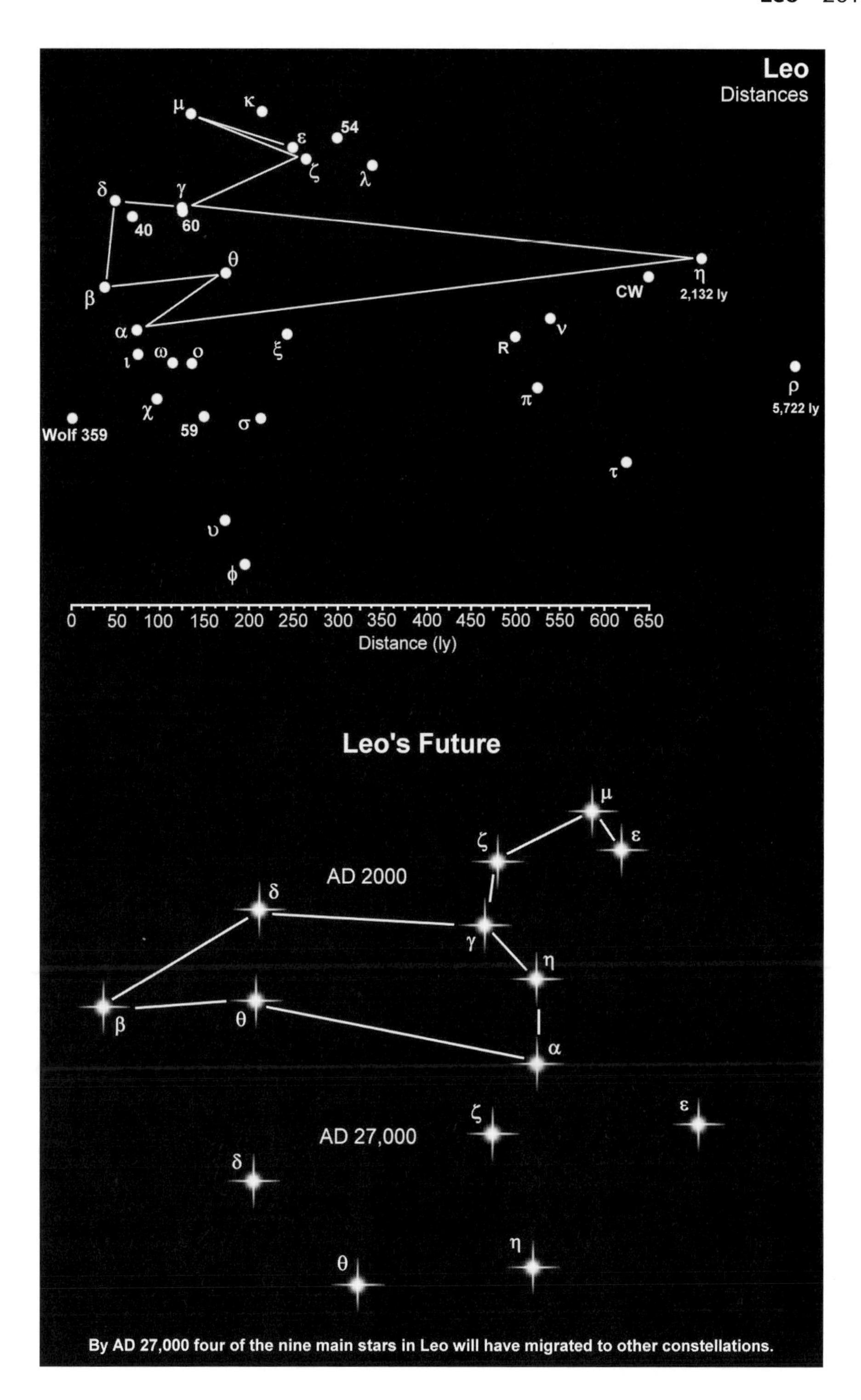

By AD 27,000 four of the nine main stars in Leo will have migrated to other constellations.

The third magnitude ε **Leonis** has a definite yellowish tincture, which is not entirely surprising considering it is a giant G1 some 29 $D_{\odot}$ across and 314, perhaps even 355 times as luminous as the Sun. Currently 251 ly away and creeping away from us at 4.8 km/s this 4 $M_{\odot}$ star has a surface temperature of around 6,000 K. At least five attempts have been made at measuring its rotational velocity. The results range from 4.2 km/s to 10 km/s. The latest reliable data put the speed at 5.7 to 6.0 km/s. This means that at the low end of the scale it takes the star 257.5 days to revolve once while at the upper limit it could take as little as 244.6 days. On the balance of probabilities the most likely rotational period will be around 260 days.

There are a number of sub-2 $D_{\odot}$ stars in Leo. β Leo at 1.8 $D_{\odot}$ has already been mentioned. On par with β is **ϕ Leonis**, an A2, and the slightly warmer **54 Leonis,** an A1 which has a 0.9 $D_{\odot}$ A2 companion. A little smaller is ν **Leonis,** a B9 of 1.7 $D_{\odot}$ while **59 Leonis,** an A5, is a smidgen larger at 1.9 $D_{\odot}$. The B9.5 σ **Leonis** is a 1.5 $D_{\odot}$ star while **θ Leonis** is a 1.4 $D_{\odot}$ A2. The smallest is **ω Leonis** at 1.3 $D_{\odot}$. An F8 it is 6.35 $L_{\odot}$ and resides 112 ly from Earth, glowing at a dim m_v +5.45. It is actually a binary, its partner an A5 dwarf in an orbit which it takes 117 years to complete.

Wolf 359 (aka **CN Leonis**) is an M6 red dwarf and at 7.78 ly is the closest star to us in the constellation. In fact Wolf 359 is the 5th closest star to the Sun over all beaten only by Barnard's Star in Ophiuchus at 5.96 ly, α Centauri A and B at 4.37 ly and Proxima Centauri at 4.24 ly. Named in honor of the German astronomer Max Wolf who, in 1919, issued a catalog or more than 1,000 stars with high proper motions, Wolf 359 is m_v +13.54 and has a luminosity of only 0.0009 $L_{\odot}$. Its diameter is 0.16 $D_{\odot}$ and mass 0.09 $M_{\odot}$. it is believed to be somewhere between 100 and 350 million years old.

CW Leonis at 11th magnitude would normally be of little interest and outside the scope of this book were it not for the fact that the Submillimeter Wave Astronomy Satellite appears to have detected water around the star. It is believed that the water is locked in thousands, perhaps even millions of comets that are sublimating (i.e. turning directly from ice to gas without going through a liquid stage) as they orbit the aging red giant. Classed as a carbon-rich C9.5 supergiant CW Leo is 250 $D_{\odot}$ but has a mass of just 1.5 to 3.5 $M_{\odot}$. It is 650 ly from Earth.

The cataclysmic variable **DP Leonis** is about 1,300 ly from Earth and consists of a white and a red dwarf star of 0.0114 and 0.12 $D_{\odot}$ with masses of 0.6 and 0.09 $M_{\odot}$ respectively. They orbit one another in just 1.5 hours. In this type of system the magnetic field of the white dwarf is so great that matter falling into it from the red dwarf does not form a normal accretion disk but instead follows the lines of magnetic force and flows into the white dwarf's polar regions. Such systems are called 'polars'. DP Leonis is of particular interest in that it also has a least one planet. This is the first case of a planet orbiting a polar and demonstrates that planetary systems are likely to exist around a wider variety of stars than previously speculated.

Leo seems to have more than its fair share of stars with planetary systems (see

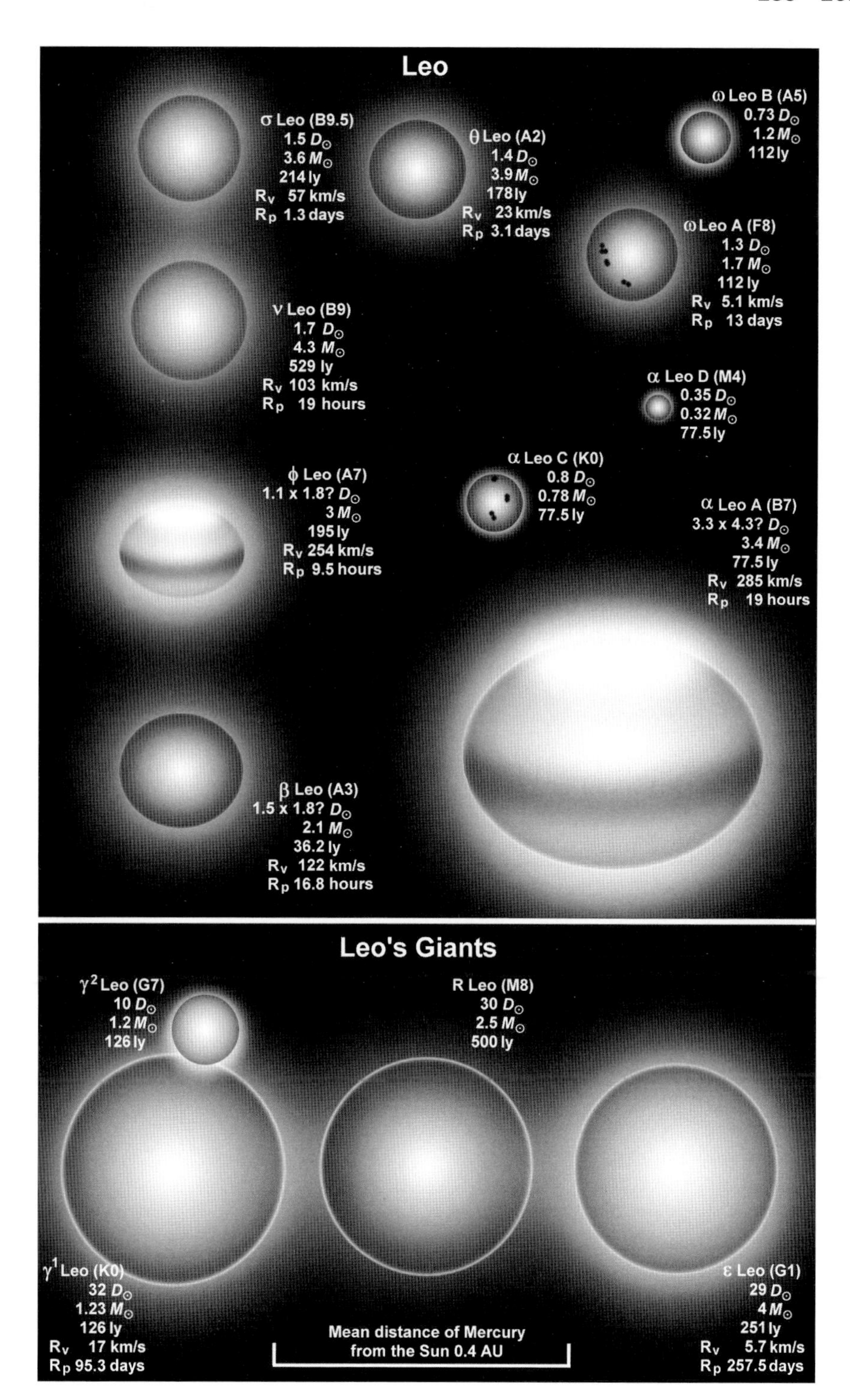
Leo
σ Leo (B9.5)
1.5 D☉
3.6 M☉
214 ly
Rv 57 km/s
Rp 1.3 days
θ Leo (A2)
1.4 D☉
3.9 M☉
178 ly
Rv 23 km/s
Rp 3.1 days
ω Leo B (A5)
0.73 D☉
1.2 M☉
112 ly
ω Leo A (F8)
1.3 D☉
1.7 M☉
112 ly
Rv 5.1 km/s
Rp 13 days
ν Leo (B9)
1.7 D☉
4.3 M☉
529 ly
Rv 103 km/s
Rp 19 hours
α Leo D (M4)
0.35 D☉
0.32 M☉
77.5 ly
α Leo C (K0)
0.8 D☉
0.78 M☉
77.5 ly
ϕ Leo (A7)
1.1 x 1.8? D☉
3 M☉
195 ly
Rv 254 km/s
Rp 9.5 hours
α Leo A (B7)
3.3 x 4.3? D☉
3.4 M☉
77.5 ly
Rv 285 km/s
Rp 19 hours
β Leo (A3)
1.5 x 1.8? D☉
2.1 M☉
36.2 ly
Rv 122 km/s
Rp 16.8 hours
Leo's Giants
γ2 Leo (G7)
10 D☉
1.2 M☉
126 ly
R Leo (M8)
30 D☉
2.5 M☉
500 ly
γ1 Leo (K0)
32 D☉
1.23 M☉
126 ly
Rv 17 km/s
Rp 95.3 days
Mean distance of Mercury
from the Sun 0.4 AU
ε Leo (G1)
29 D☉
4 M☉
251 ly
Rv 5.7 km/s
Rp 257.5 days

table). Those of particular note include **BD+20 2457** (the name comes from the *Bonner Durchmusterung* stellar catalog) which appears to have at least one and possible two brown dwarfs. DP Leo, γ^1 Leo mentioned earlier, **HD 81040** and **HD 102272** all have fairly substantial planets in the range of about six to nine Jovian masses, and **GJ 436,** a 0.46 $D_\odot$ star with a 0.072 M_J planet in a 2.64 day orbit during which its distance from the star varies between 3.6 million km to 4.9 million km – a tenth of the distance that Mercury lies from the Sun.

Planetary systems in Leo

Star	$D_\odot$	Spectral class	ly	m_v	Planet	Minimum mass	q	Q	P
γ^1 Leo	31.9	K0	126	2.01	γ^1 Leo b	8.78 M_J	1.02	1.36	1.18 y
BD+20 2457	?	K2	652	9.57	BD20 2457 b	21.42 M_J	1.23	1.67	1.04 y
					BD20 2457 c	12.47 M_J	1.65	2.37	1.70 y
DP Leo	?	AM Her	?	17.5	DP Leo b	6.28 M_J	8.60	8.60	13.8 y
GJ 436	0.46	M2.5	33.3	10.68	GJ 436 b	0.072 M_J	0.024	0.033	2.64 d
HD 81040	1.00	G2/3	106	7.72	HD 81040 b	6.86 M_J	0.92	2.96	2.74 y
HD 88133	1.93	G5	243	8.01	HD 88133 b	0.22 M_J	0.041	0.053	3.416 d
HD 89307	1.05	G0	101	7.06	HD 89307 b	1.78 M_J	2.46	4.02	5.91 y
HD 99492	0.81	K2	111	7.57	HD 99492 b	0.109 M_J	0.09	0.16	17.04 d
HD 100777	1.00	K0	172	8.42	HD 100777 b	1.16 M_J	0.66	1.40	1.05 y
HD 102272	10.1	K0	1,174	8.71	HD 102272 b	5.9 M_J	0.58	0.65	127.6 d
					HD 102272 c	2.6 M_J	0.50	2.64	1.42 y

Leo Minor

Constellation:	Leo Minor	**Hemisphere:**	Northern
Translation:	The Lesser Lion	**Area:**	232 deg^2
Genitive:	Leonis Minoris	**% of sky:**	0.562%
Abbreviation:	LMi	**Size ranking:**	64th

A faint, inconspicuous constellation introduced in 1690 by Hevelius to mop up the stars that lie between Leo and Ursa Major.

If you are looking for α Leonis Minoris then you are out of luck: there is no such star and there never has been. The title of brightest star goes to – wait for it – **46 Leonis Minoris** at a dim m_v +3.83 and at a distance of 97.7 ly. Its diameter is 8.3 $D_\odot$ although its mass is 1.2 $M_\odot$, typical of K0 class stars. It lies on the border with Ursa Major: another 40″ to the east and it would a member of the Great Bear.

β Leonis Minoris is indeed the second brightest star – it comes almost as a shock – and is a binary. The main component is a yellowish-orange G9 with a diameter of 8.5 $D_\odot$ and a mass of 1.9 $M_\odot$. With a luminosity of 33.6 $L_\odot$ it glows at m_v +4.21 but would brighten to 1st magnitude at the standard 10 pc used to determine absolute magnitude. It spins at 5.6 km/s taking 107.6 days to complete one rotation on it axis. Its companion is an F8 and much smaller, less massive and luminous at 2 $D_\odot$, 1.35 $M_\odot$ and 5.8 $L_\odot$. With an apparent magnitude of m_v +6.12 the star can approach the primary to within 5.4 AU before shooting out to 27 AU, the orbital period being 28.62 years. The system lies 146 ly away.

Sometimes shown connected to 21 LMi, sometimes not, **10 Leonis Minoris** is an 11 $D_\odot$ G8 at a distance of 176 ly, give or take 8 ly. Its magnitude switches between m_v +4.54 and +4.56, the star often being listed as **SU Leonis Minoris.** Its variability has a period of 40.4 days and is most likely due to active regions on the star's surface affecting its brightness.

11 Leonis Minoris is another binary. The primary is a G8 dwarf with a mass of around 0.9 $M_\odot$ and a luminosity of 0.71 $L_\odot$. Its diameter is not well determined but it is likely to be around 0.97 $D_\odot$. Its partner is a 0.2 $M_\odot$ red dwarf of the M5 spectral group with a diameter of 0.24 $D_\odot$ and luminosity of only 0.000687 $L_\odot$. The pair are in a 201 year long orbit which varies between 5.2 and 80.8 AU. The primary is also known as **SV Leonis Minoris** its magnitude varies over an 18 day timescale during which it fluctuates by 0.03 magnitudes between m_v +5.40 and +5.43. Like 10 LMi the variability may be due to large dark spots crossing the star's visible hemisphere.

About 96% of the stars in our corner of the Galaxy have radial velocities of less than 40 km/s and half have velocities of less than 10 km/s. **20 Leonis Minoris** is one of the Galaxy's natural speeders, tearing through the galactic arms at 54.7 km/s. It is not physically dissimilar to the Sun. It is about the same size and a G3 with a luminosity of around 1.27 $L_\odot$ but somewhat older at 6,200 to 7,700 million years (Sun = 4,567 million years). It is currently 48.6 ly away.

When Dutch schoolteacher, Hanny van Arkel, volunteered to help classify

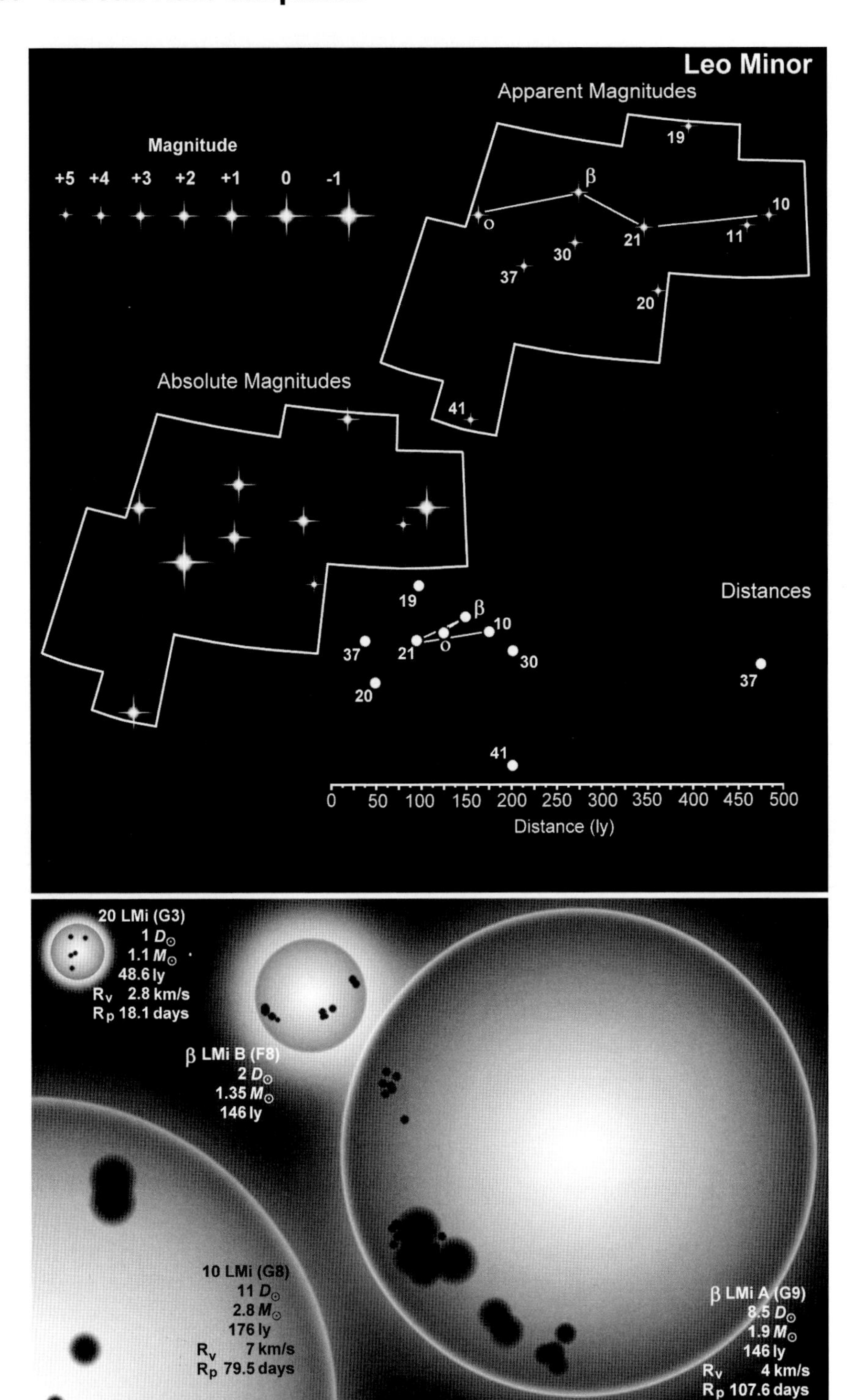

Leo Minor
Apparent Magnitudes
Magnitude
+5 +4 +3 +2 +1 0 -1
19
β
10
o
21
11
30
37
20
41
Absolute Magnitudes
Distances
19
β
10
37
21
o
30
37
20
41
0 50 100 150 200 250 300 350 400 450 500
Distance (ly)
20 LMi (G3)
1 $D_\odot$
1.1 $M_\odot$
48.6 ly
R_v 2.8 km/s
R_p 18.1 days
β LMi B (F8)
2 $D_\odot$
1.35 $M_\odot$
146 ly
10 LMi (G8)
11 $D_\odot$
2.8 $M_\odot$
176 ly
R_v 7 km/s
R_p 79.5 days
β LMi A (G9)
8.5 $D_\odot$
1.9 $M_\odot$
146 ly
R_v 4 km/s
R_p 107.6 days

galaxies as part of the Galaxy Zoo Project she did not expect to make history. But one day in 2007 she came across a strange object on one of the photographs. The object looks like a man carrying something, taken from slightly above. Now universally known as **Hanny's Voorwerp** – Dutch for 'object' – it appears blue or green depending on who has taken the photograph. No one is really sure what it is but it appears to be close to the spiral galaxy IC 2497 at a distance of 700 million ly. Current speculation is that it is the remnant of a small galaxy reflecting a quasar event that occurred in IC 2497 about 100,000 years ago.

Lepus

Constellation:	Lepus	**Hemisphere:**	Southern
Translation:	The Hare	**Area:**	290 deg^2
Genitive:	Leporis	**% of sky:**	0.703%
Abbreviation:	Lep	**Size ranking:**	51st

A relatively easy constellation to find as it lies at the feet of Orion and is being chased across the sky by his dogs. However, it is often ignored by observers who are naturally drawn to more dominant Orion and Canis Major.

Lepus contains relatively few big stars: α **Leporis** or Arneb is the largest. Hanging in space at a distance of 1,284 ly α Lep is a full 75 $D_\odot$ (0.7 AU). Its 12,000 $L_\odot$ causes this yellowish-white to glow at m_v +2.56 but its absolute magnitude is on par with Venus at her most brilliant, M_v -4.7. But supergiant luminaries don't last for long, their lifespans measured in just millions of years.

Belonging to spectral group G5 and with a temperature of 5,200K β **Leporis** or Nihal is not unlike the Sun except that it is 16 times larger and 144 times as luminous. It is one of just four G-class stars in Lepus. Moving to the top of the scale at G9 the 5 $D_\odot$ **HD 35162** can be a difficult binary to locate at m_v +5.35, its companion a m_v +6.71. δ **Leporis** is a G8 about a dozen times as big as the Sun. It is tearing away from us at 99.3 km/s putting it in the top 10 fastest stars. Another swift star is the G8 **HD 34538** which is at the edge of naked eye visibility for most urban dwellers at m_v +5.50. It is also receding from us at 75.3 km/s. It is currently 158 ly from Earth. **HD 41312** holds the galactic record for stars that are moving away from the Sun at 182.2 km/s. A K3 giant of 23 $D_\odot$ it is 109 times more luminous than the Sun. HD 41312 is a visitor to our neck of the galactic woods having probably originated in the halo of the Galaxy. Only one other star is traveling faster than HD 41312 and that is τ^1 Lupi which is closing in on us at 215 km/s.

You have to have very keen eyesight and pitch black skies to separate γ **Leporis** into its individual components with the naked eye, though some people claim to have done it. The primary, γ^A Lep, is a m_v +3.59 F6 with a diameter just about equal to that of the Sun and a bit warmer at 6,400K. It has slightly more mass, weighing in at 1.25 $D_\odot$, and has a luminosity of 2.39 $D_\odot$. Its companion, γ^B Lep, is a K2 of 0.80 $D_\odot$ and 0.83 $M_\odot$ but less than a third as luminous as the Sun at 0.29 $L_\odot$. Their separation on the celestial sphere is 96.3″ (PA 350°) which converts to 864 AU in real space terms. The pair are 29.3 ly from Earth and take more than 18,000 years to complete their mutual orbit. They are believed to be fairly young stars, perhaps less than 1,000 million years old. Astronomers have searched for planets over a number of years but without success. Observers have described the binary as having a variety of colors. In theory they should appear as a very pale yellow and a reddish-orange.

ζ **Leporis** is another young star, no more than a couple of hundred million years old, which is embedded in a series of concentric dust rings out to about 3

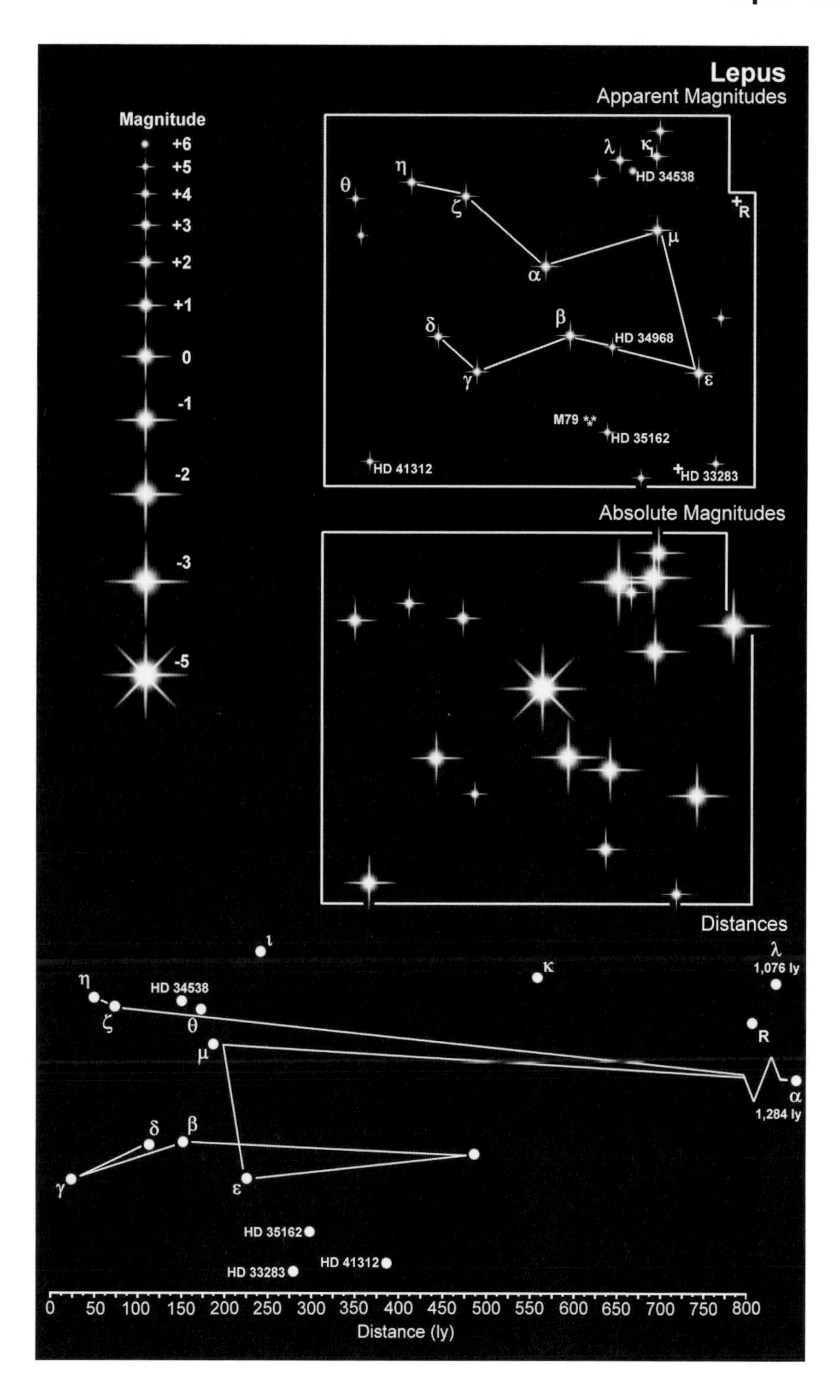
Lepus
Apparent Magnitudes
Magnitude
+6
+5
+4
+3
+2
+1
0
-1
-2
-3
-5
θ
η
ζ
λ
κ
HD 34538
R
μ
α
δ
β
HD 34968
γ
ε
M79
HD 35162
HD 41312
HD 33283
Absolute Magnitudes
Distances
ι
κ
λ
1,076 ly
η
HD 34538
ζ
θ
R
μ
α
1,284 ly
δ
β
γ
ε
HD 35162
HD 41312
HD 33283
0 50 100 150 200 250 300 350 400 450 500 550 600 650 700 750 800
Distance (ly)

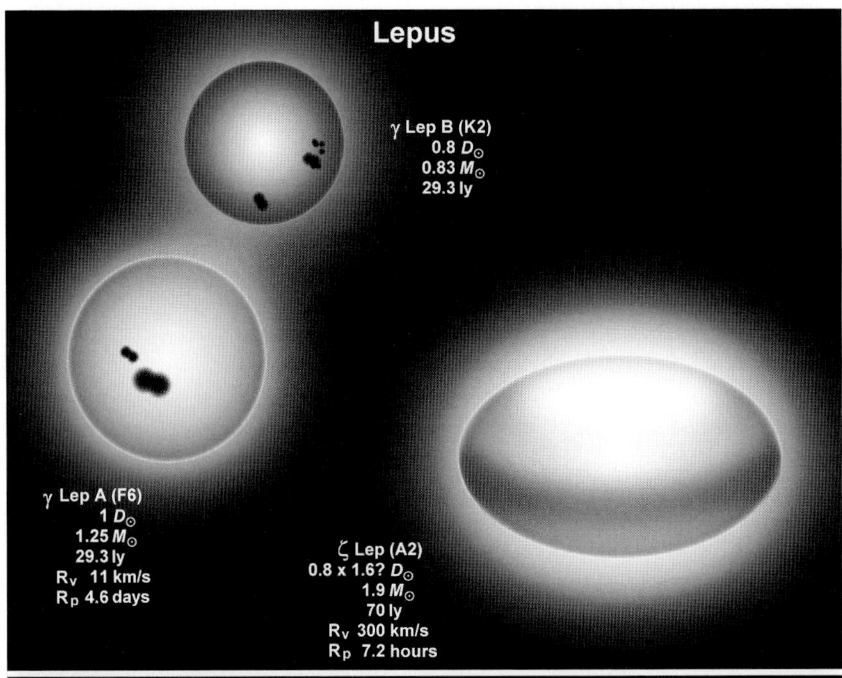
Lepus
γ Lep B (K2)
0.8 D⊙
0.83 M⊙
29.3 ly
γ Lep A (F6)
1 D⊙
1.25 M⊙
29.3 ly
Rv 11 km/s
Rp 4.6 days
ζ Lep (A2)
0.8 x 1.6? D⊙
1.9 M⊙
70 ly
Rv 300 km/s
Rp 7.2 hours

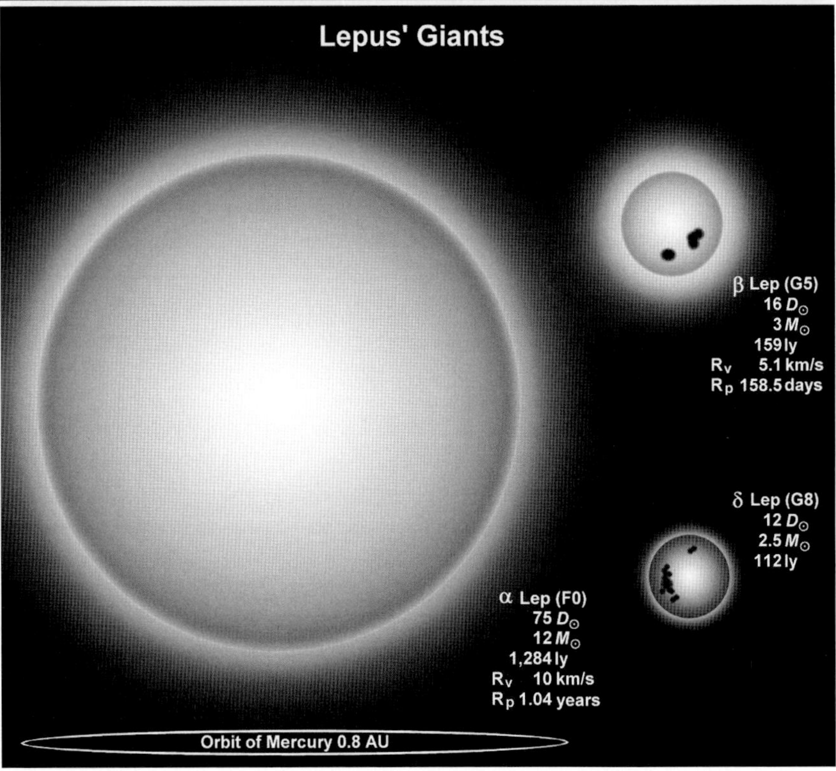
Lepus' Giants
β Lep (G5)
16 D⊙
3 M⊙
159 ly
Rv 5.1 km/s
Rp 158.5 days
δ Lep (G8)
12 D⊙
2.5 M⊙
112 ly
α Lep (F0)
75 D⊙
12 M⊙
1,284 ly
Rv 10 km/s
Rp 1.04 years
Orbit of Mercury 0.8 AU

AU. It is 1.6 $D_{\odot}$ across its equator and spins at a breakneck speed of 300 km/s, teetering on the edge of stability, and revolving once every 5.3 hours. It is a very oblate spheroid with an estimated polar diameter of just 0.8 $D_{\odot}$. Such a high rotational velocity is rare for an A-class (A2) star: just 1.1% of all known A-class stars attain such a high velocity. The average is 104 km/s. It may be a member of the Castor Moving Group.

"... like a drop of blood on a black field", was how John Russell Hind described **R Leporis**. The 19th Century British astronomer discovered that its magnitude varied between m_v +5.5 and +11.7 turning increasingly red as it faded. Often called Hind's Crimson Star, R Lep is a Mira variable with a period of 427.07 days although in Hind's day the period could be anywhere between 418 and 441 days. The star belongs to the rare spectral C-class carbon-rich red supergiants of which there are only a few others visible to the naked eye.

Close to its southern border with Columba **HD 33283** sits quietly 281 ly away. It's another solar analog. A G4 carrying a little more mass than the Sun at 1.24 $M_{\odot}$ and with a diameter to match, 1.20 $D_{\odot}$, it is perhaps as young as 900 million years or could be as old as 5,500 million years. It also has a planet of about a third of a Jovian mass in a very small orbit with a period of just 18 days (see table).

The globular cluster **Messier 79** may be a visitor to our Galaxy from the Canis Major dwarf galaxy. M79 is in the wrong place in the sky. Most clusters are grouped around the galactic center on our inward side but M79 is on the outward side. We are about 26,000 ly from the center of the Galaxy but M79 is a further 42,000 ly. It is easily resolved with a binocular or small telescope being 9.6′ across and m_v +7.7. M79 is about 118 ly in diameter and is traveling away from us at about 200 km/s.

Planetary system in Lepus

Star	$D_{\odot}$	Spectral class	ly	m_v	Planet	Minimum mass	q	Q	P
HD 33283	1.20	G3	281	+8.05	HD 33283 b	0.33 M_J	0.09	0.25	18.2 d

*q = periastron, Q = apastron.

Globular cluster in Lepus

Name	Size arc min	Size ly	Distance ly	Age million yrs	Apparent magnitude m_v
M79 (NGC 1904)	9.6′	+7.7	42,000	12,000?	+7.7

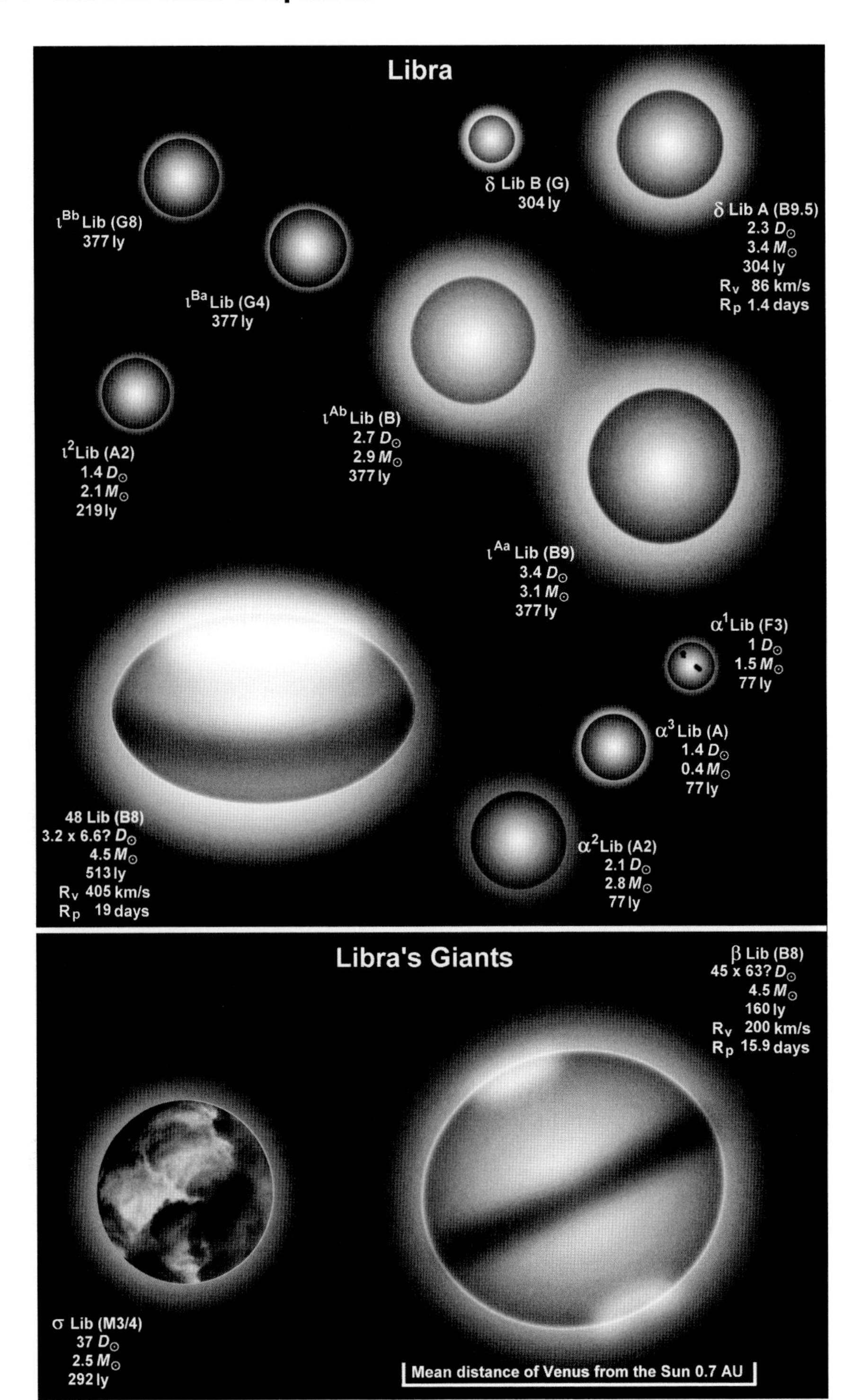
Libra
δ Lib B (G)
304 ly
δ Lib A (B9.5)
2.3 $D_\odot$
3.4 $M_\odot$
304 ly
R_V 86 km/s
R_P 1.4 days
ι^{Bb} Lib (G8)
377 ly
ι^{Ba} Lib (G4)
377 ly
ι^{2} Lib (A2)
1.4 $D_\odot$
2.1 $M_\odot$
219 ly
ι^{Ab} Lib (B)
2.7 $D_\odot$
2.9 $M_\odot$
377 ly
ι^{Aa} Lib (B9)
3.4 $D_\odot$
3.1 $M_\odot$
377 ly
α^1 Lib (F3)
1 $D_\odot$
1.5 $M_\odot$
77 ly
α^3 Lib (A)
1.4 $D_\odot$
0.4 $M_\odot$
77 ly
48 Lib (B8)
3.2 x 6.6? $D_\odot$
4.5 $M_\odot$
513 ly
R_V 405 km/s
R_P 19 days
α^2 Lib (A2)
2.1 $D_\odot$
2.8 $M_\odot$
77 ly
Libra's Giants
β Lib (B8)
45 x 63? $D_\odot$
4.5 $M_\odot$
160 ly
R_V 200 km/s
R_P 15.9 days
σ Lib (M3/4)
37 $D_\odot$
2.5 $M_\odot$
292 ly
Mean distance of Venus from the Sun 0.7 AU

primary the light is predominantly from one side of the brighter star and then the other. Schlesinger measured the differences in radial velocity and from this he was able to obtain the rotational velocity of the primary.

ι Librae is a complex quadruple system lying 377 ly away. **ι^{1Aa} Librae** is an estimated 3.4 $D_\odot$ with a mass to match: 3.1 $M_\odot$. It is possibly a B9, possibly an A0, possibly something in between, and it has a luminosity of 149 $L_\odot$. In a 23.5 year orbit around ι^{1Aa} Lib is **ι^{1Ab} Librae**. It is only 80% the size of its companion at 2.7 $D_\odot$ and 2.9 $M_\odot$ with a luminosity of 94 $L_\odot$. The orbit brings the two stars as close as 11.3 AU and separates them by as much as 18.6 AU. Some 6,600 AU away is another binary pair: the 10th magnitude **ι^{1Ba} Lib**, a G4, and **ι^{1Bb} Lib**, an 11th magnitude G8. They are separated by 230 AU and take 2,700 years to orbit one another. This pair – ι^{1Ba} and ι^{1Bb} – orbit the other pair – ι^{1Aa} and ι^{1Ab} – over a period of 195,000 years. Now, you may be wondering where ι^2 Librae is. Well, it is there. It is an A3 but at a distance of 240 ly it has absolutely nothing to do with the others.

Originally designated γ Sagittarii the red giant **σ Librae**, or Zuben Algubi, is a semi-regular SRb pulsating variable. About 292 ly away it pulsates with a period of 20 days, its magnitude varying between m_v +3.28 and +3.46.

11 Librae is one of the older stars in the Galaxy at about 6,000 million years. Its age, and the fact that it is tearing through space at 83.1 km/s, indicates it a 'thick-disk' star. The galactic disk is about 1,000 ly thick with its outer edges populated by older fast moving stars. The center of the disk – the 'thin-disk' – is only about 400 ly thick and contains both young stars and copious amounts of dust and gas. 11 Lib is a yellowish-orange G8 of 7.8 $D_\odot$ and 1.5 $M_\odot$ with a luminosity of 37.6 $L_\odot$.

48 Librae – sometimes referred to as **FX Librae** – is one of the fastest rotating stars in the sky at 405 km/s. A B8 its diameter is difficult to measure accurately due to equatorial bulging. It could be as wide as 6.6 $D_\odot$ at the equator and just 3.2 $D_\odot$ through the poles. A Be emission star it is surrounded by a disk of material probably thrown off from the star by its rapid rotation. Its 212 $L_\odot$ from a distance of 513 ly results in a m_v +4.94 star that would brighten to M_v -2.0 at 10 pc. As its alternative name suggests it is also a γ Cas eruptive variable, its magnitude fluctuating between m_v +4.79 and m_v +4.96.

Libra contains an interesting crop of planets. **GJ 581**, a red dwarf just 20.4 ly away and sometimes known as **HO Librac**, has at least three super-Earths. Gliese 581c has a mass of 0.01686 M_J or 5.36 $M_\oplus$ which could equate to a physical size of 1.5 $D_\oplus$. Gliese 581d is somewhat more massive at about 7.09 $M_\oplus$ while Gliese 581e is the smallest planet yet detected at 1.94 $M_\oplus$ probably making it about 10% to 15% larger than Earth. **HD 134987** – aka **23 Librae** – has two planets including a Jupiter-size planet in a 13.7 year orbit (see table).

Planetary systems in Libra

Star	$D_\odot$	Spectral class	ly	m_v	Planet	Minimum mass	q	Q	P
HO Lib	0.38	M3	20.4	+10.55	GJ 581 b	0.0492 M_J	0.041	0.041	5.37 d
					GJ 581 c	0.01686 M_J	0.058	0.082	12.93 d
					GJ 581 d	0.02231 M_J	0.14	0.31	66.8 d
					GJ 581 e	0.006104 M_J	0.03	0.03	3.1419 d
HD 134987	1.20	G5	81.6	+6.45	HD 134987 b	1.59 M_J	0.621	0.999	258.2 d
					HD 134987 c	0.82 M_J	5.10	6.49	13.69 y
HD 141937	1.06	G2/G3	109	+7.25	HD 141937 b	9.7 M_J	0.897	2.143	1.79 y

Lupus

Constellation:	Lupus	**Hemisphere:**	Southern
Translation:	The Wolf	**Area:**	334 deg^2
Genitive:	Lupi	**% of sky:**	0.810%
Abbreviation:	Lup	**Size ranking:**	46th

There are several myths associated with the wolf. The most likely as far as this constellation is concerned relates to King Lycaon of Arcadia who killed one of his own sons, Nyctimus – he had 50 sons – and fed him to Zeus. The outraged god brought Nyctimus back to life, killed the other 49 sons with lightning bolts and turned the king into a wolf. Most of the constellation's stars are third magnitude or less, although there are plenty of them and over half are B-class.

The brightest star in the constellation, **α Lupi** or Men, attains only m_v +2.29 at best. One of many B-class stars it is a hot, 21,600 K, B1.5 with a luminosity of 2,804 $D_\odot$. Not particularly big at 5.7 $D_\odot$ it lies at a distance of 548 ly and, unlike many B-class stars, is a slow rotator at about 20 km/s taking 14.4 days to spin once on its polar axis. It is one of three stars in the constellation that are β Cepheids. At minimum its magnitude dips to m_v +2.34, its period being 6^h 14.2^m. Such an amplitude (0.05 magnitude) is the most common among β Cepheids as is its spectral class, B1 and B2 accounting for 75% of such variables. The other two β Cepheids are δ Lupi and τ^1 Lupi. **δ Lupi** is not much larger at about 7 $D_\odot$ and lying at 510 ly distance. It is also a B1.5 with a slightly smaller amplitude of 0.04 magnitude fluctuating between m_v +3.20 and +3.24 and back with a period of 3^h 58.3^m. Unlike α Lup it is a fast spinner with estimates averaging 207 km/s. Identical in size and amplitude is **τ^1 Lupi** but at 1,035 ly it shines at just m_v +4.54 to +4.58, its period being 4^h 15.4^m. It is also a slow rotator at 30 km/s. It is particularly noteworthy as it has the highest radial velocity of any star: 215 km/s heading roughly in our direction. **τ^2 Lupi** is not related to τ^1 being 720 ly closer to us. It is the same size as the Sun and, being an F7, should appear as yellow in a binocular or small telescope, contrasting with the blue of τ^1.

χ Lupi is a spectroscopic binary consisting of a luminous B9 and an A2. The primary is 3.7 $D_\odot$ and has a luminosity of about 90 $L_\odot$. Weighing in at an estimated 3 $M_\odot$ it is separated from its A-class companion by just 0.21 AU – half the Mercury-Sun distance – which is a much less luminous 25 $L_\odot$ and slightly less massive at 2.2 $M_\odot$.

Another unconnected pair is **ϕ^1** and **ϕ^2 Lupi.** At m_v +3.56 ϕ^1 Lup is almost a full magnitude brighter than its neighbor. Hardly surprising really, ϕ^1 Lup is a K5 giant of 29 $D_\odot$ and 306 $L_\odot$ lying some 326 ly away. Although it is more luminous at 440 $L_\odot$, ϕ^2 Lup is just 8.6 $D_\odot$ and resides at about twice the distance: 606 $\pm$76 ly. It is a bluish B4.

Lupus has three supergiants: the 110 $D_\odot$ **1 Lupi**, an F1, 1,140 ly away. From Earth it appears as modest m_v +4.93 but at 10 pc it would be an impressive M_v -6.7.

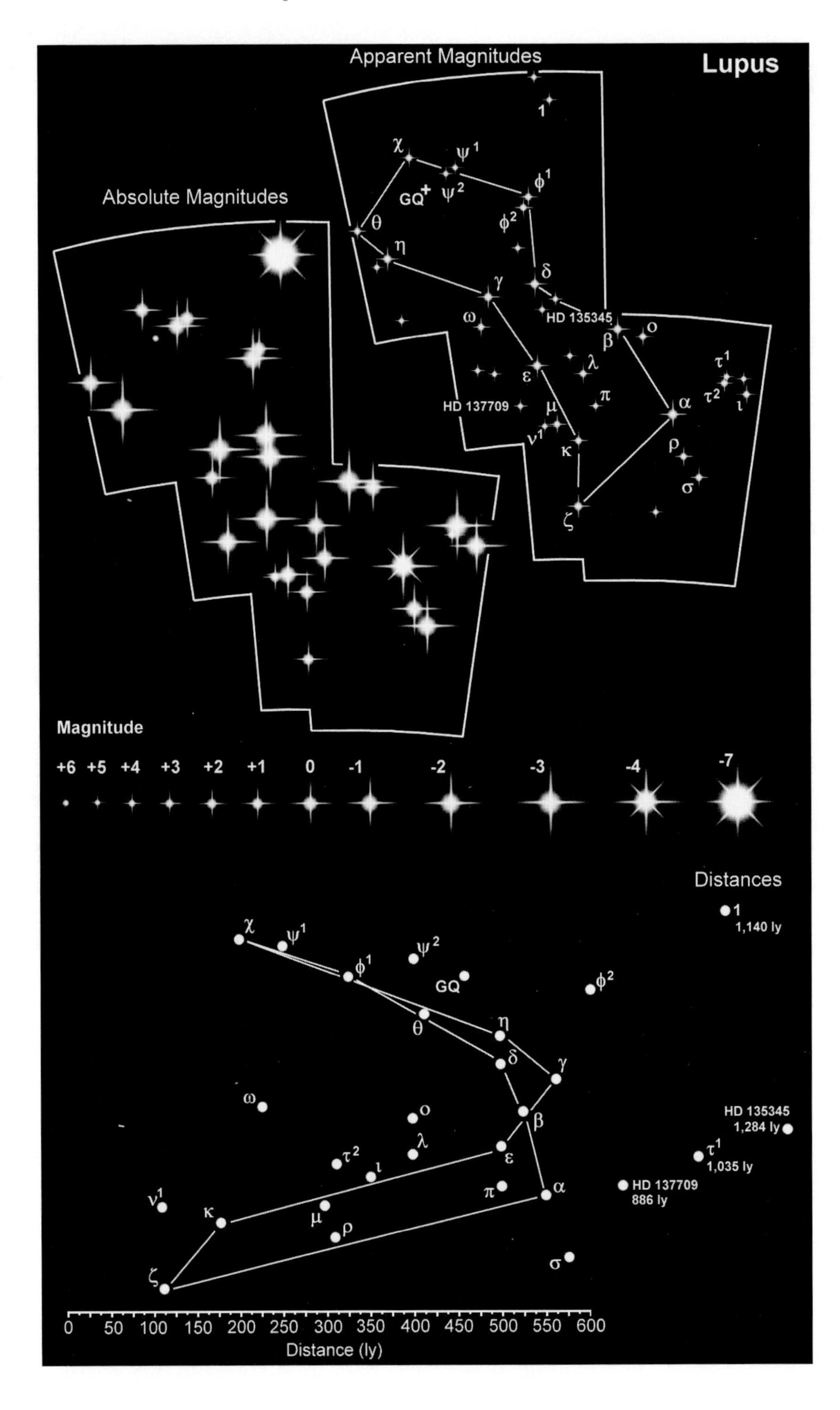

Lupus
Apparent Magnitudes
Absolute Magnitudes
Magnitude
+6 +5 +4 +3 +2 +1 0 -1 -2 -3 -4 -7
Distances
1
1,140 ly
HD 135345
1,284 ly
1,035 ly
HD 137709
886 ly
0 50 100 150 200 250 300 350 400 450 500 550 600
Distance (ly)

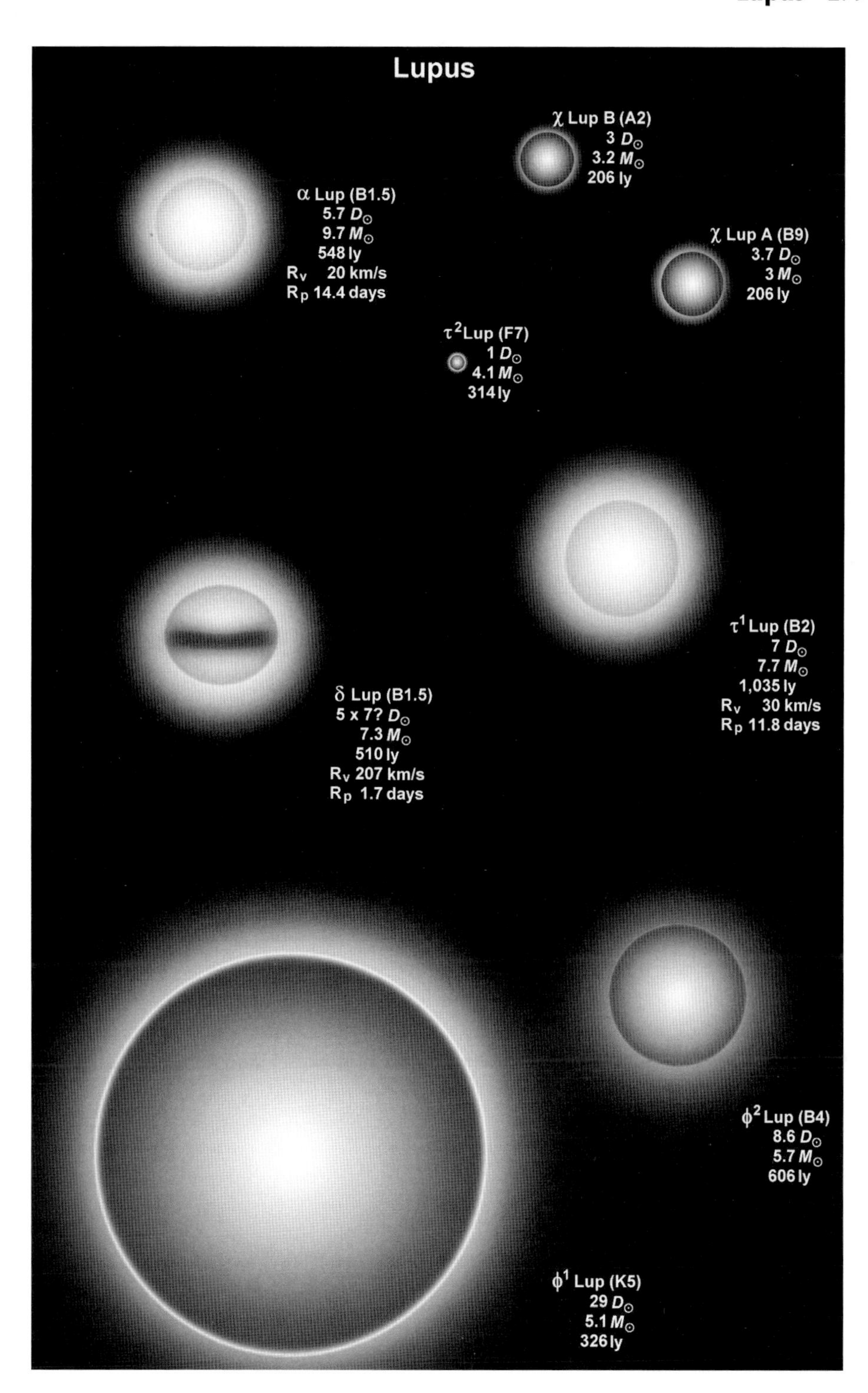
Lupus
χ Lup B (A2)
3 D⊙
3.2 M⊙
206 ly
α Lup (B1.5)
5.7 D⊙
9.7 M⊙
548 ly
Rv 20 km/s
Rp 14.4 days
χ Lup A (B9)
3.7 D⊙
3 M⊙
206 ly
τ2 Lup (F7)
1 D⊙
4.1 M⊙
314 ly
τ1 Lup (B2)
7 D⊙
7.7 M⊙
1,035 ly
Rv 30 km/s
Rp 11.8 days
δ Lup (B1.5)
5 x 7? D⊙
7.3 M⊙
510 ly
Rv 207 km/s
Rp 1.7 days
φ2 Lup (B4)
8.6 D⊙
5.7 M⊙
606 ly
φ1 Lup (K5)
29 D⊙
5.1 M⊙
326 ly

A K4, **HD 137709**, rather closer at 886 ly and about 235 $D_{\odot}$, and a very yellow G5, **HD 135345**, which is 310 $D_{\odot}$ (2.9 AU) and 1,284 ly distant.

GQ Lupi is a very young T-Tauri star in the region of 100,000 to 2 million years old and about 457 ly away. It has a mass of 0.7 $M_{\odot}$. In orbit around the star is an object about 1.8 D_J and 21.5 M_J which probably means it is a brown dwarf. The planet orbiting **Lupus-TR-3** is a much smaller body altogether being just 89% the diameter of Jupiter. Its circular orbit keeps it 6.94 million km from the star. A 5 Jovian mass planet orbits **HIP 70849** but relatively little is known about it.

Planetary systems in Lupus

Star	$D_{\odot}$	Spectral class	ly	m_v	Planet	Minimum mass	q	Q	P
GQ Lup	?	K7	457	+11.4	GQ Lup b	21.5 M_J	103	103	1,000 y
HIP 70849	0.59	K7	78.3	+10.4	HIP 70849 b	5 M_J	?	?	8.2+ y
Lupus-TR-3	0.82	K1	6,500	+17.4	Lupus-TR-3 b	0.81 M_J	0.046	0.046	3.914 d

Lynx

Constellation:	Lynx	**Hemisphere:**	Northern
Translation:	The Lynx	**Area:**	545 deg^2
Genitive:	Lyncis	**% of sky:**	1.321%
Abbreviation:	Lyn	**Size ranking:**	28th

Introduced in 1690 by Johannes Hevelius, astronomer, selenographer and Mayor of Danzig, this large constellation is populated with faint stars.

α Lyncis, the only star in the constellation to be assigned a Greek letter, is a K7 supergiant of 65 $D_{\odot}$. Lying somewhere between 210 and 234 ly its luminosity is around 212 $L_{\odot}$ so it appears as a m_v +3.12 star. At 10 pc it would just creep into the brighter side of zero at M_v -0.4.

The constellation has two variable stars. **1 Lyncis** (aka **UW Lyncis**) is a 42 $D_{\odot}$ red giant of the Lb pulsating variety. Its magnitude slides between m_v +4.95 and +5.06 with no obvious period. An M3 it is about 590 ly distant. The other is **2 Lyncis** (or **UZ Lyncis** if you prefer). This is a much more complex star. Not much larger than the Sun at 1.6 $D_{\odot}$ it is both an eclipsing binary and a δ Scuti variable. The main star belongs to the A2 spectral group and is about 29 times as luminous as the Sun. Relatively little is known about its spectroscopic companion except that it is in a 20.82 day orbit (there is also the possibility that the orbital period is 33 days, or perhaps 87 days). The magnitude varies between m_v +4.43 and +4.73.

To the naked eye **12 Lyncis** appears as a single star of m_v +5.44, just on the edge of visibility for most urban dwellers. A binocular or small telescope however will reveal it to be a triple star system. The primary is an A3 of 2.5 $D_{\odot}$. Just 1.8″ away is a m_v +6.0 A2 separated by 120 AU in real space. It is smaller than the primary: only 1.7 $D_{\odot}$. Somewhat farther away at 610 AU is the third member. Its spectral class and other characteristics have yet to be determined but its magnitude comes in at m_v +7.03.

HD 76943, an F4 star about 53 ly away, is sometimes listed as 10 Ursae Majoris. When the IAU decided to redraw the constellation boundaries it adopted nice straight lines: 10 UMa just happened to be in the wrong place and fell – or should that be was pushed? – into Lynx (similarly, 41 Lyncis ended up in Ursa Major). At m_v +3.96 HD 76943 is actually the third brightest star in the constellation. It is rather more massive than the Sun, 1.44 $M_{\odot}$ and more luminous at 4.8 $L_{\odot}$. It is also a binary. Its m_v +6.18 partner is a solar analog, a G5 of 1.1 $M_{\odot}$ and 0.8 $L_{\odot}$. The orbit varies between 9 and 12.2 AU with a period of 21.78 years.

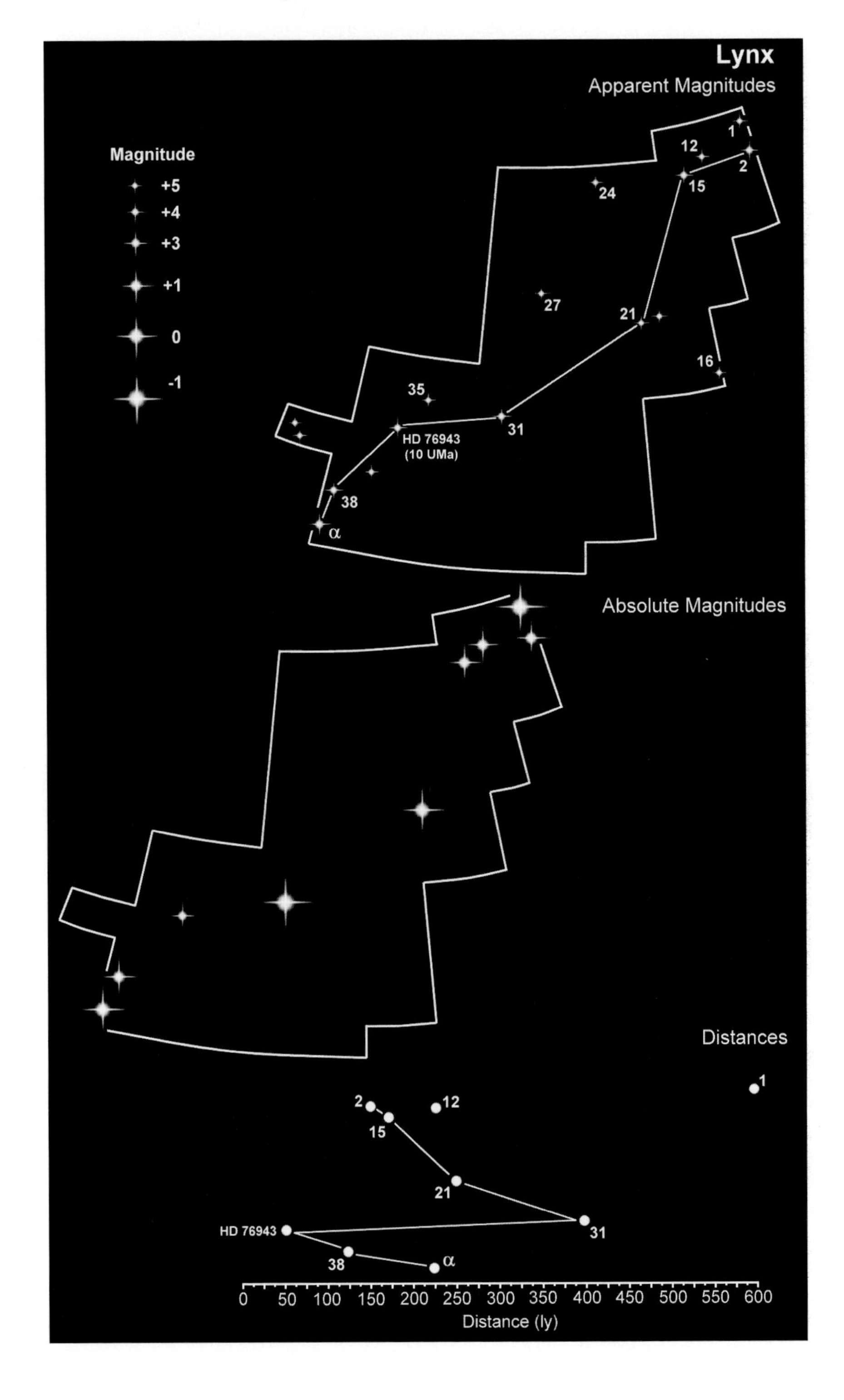

Lynx
Apparent Magnitudes
Magnitude
+5
+4
+3
+1
0
-1
1
12
2
15
24
27
21
16
35
31
HD 76943
(10 UMa)
38
α
Absolute Magnitudes
Distances
1
2
12
15
21
HD 76943
31
38
α
0 50 100 150 200 250 300 350 400 450 500 550 600
Distance (ly)

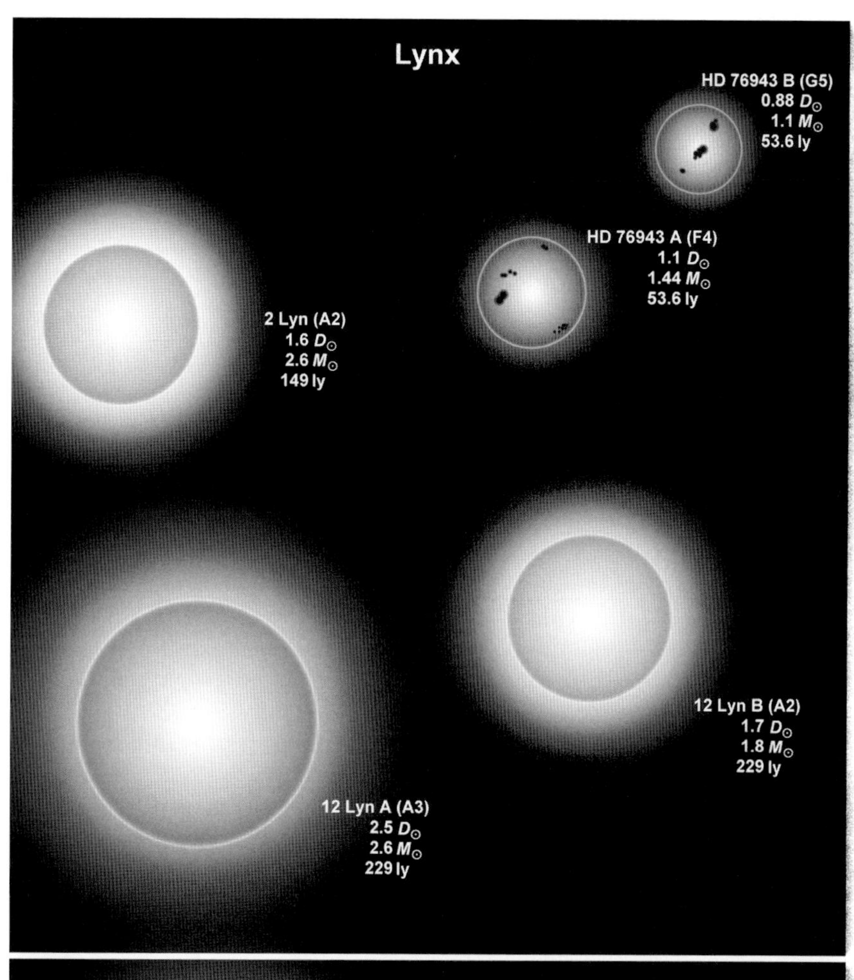
Lynx
HD 76943 B (G5)
0.88 $D_\odot$
1.1 $M_\odot$
53.6 ly
HD 76943 A (F4)
1.1 $D_\odot$
1.44 $M_\odot$
53.6 ly
2 Lyn (A2)
1.6 $D_\odot$
2.6 $M_\odot$
149 ly
12 Lyn B (A2)
1.7 $D_\odot$
1.8 $M_\odot$
229 ly
12 Lyn A (A3)
2.5 $D_\odot$
2.6 $M_\odot$
229 ly

Lynx's Giant Stars
α Lyn (K7)
65 $D_\odot$
4.6 $M_\odot$
222 ly
1 Lyn (M3)
42 $D_\odot$
4.9 $M_\odot$
590 ly

Lyra

Constellation:	Lyra	**Hemisphere:**	Northern
Translation:	The Lyre	**Area:**	286 deg^2
Genitive:	Lyrae	**% of sky:**	0.693%
Abbreviation:	Lyr	**Size ranking:**	52nd

A small but prominent constellation depicting the lyre played by Orpheus.

α **Lyrae**, better known as Vega, is the second brightest star in the Northern Celestial Hemisphere – surpassed only by Arcturus, α Boo – and the 5th brightest in the entire night sky. Its average magnitude of m_v +0.03 is a combination of its proximity, just 25.3 ly from Earth, and its luminosity, some 45 $L_\odot$. Plus we are looking almost directly at one of Vega's hot, highly luminous poles. Vega is a fast spinner. It rotates at 274 km/s, nearly three times the average speed of 104 km/s for A-class stars. At this velocity the poles are drawn towards the center of the star while the equator bulges outwards and the star turns into an oblate spheroid. In Vega's case its polar diameter is 2.26 $D_\odot$ while its equatorial diameter is 22% larger at 2.75 $D_\odot$. Strange things start to happen when a star is distorted in this way. The equatorial region cools to 7,950 K and darkens while the polar temperature soars to 10,150 K and lightens, often developing massive bright patches. Vega is a Main Sequence A0 of 2 $M_\odot$ and is between 385 and 570 million years old – just a tenth of the age of the Sun. Even so it is now middle-aged and in another 400 to 500 million years will rapidly evolve into a red giant before shrinking to a tiny white dwarf. Some 12,000 years ago it was the Pole Star and will be so again in about 13,700 years when it will be at Dec. +86° 14′. It is also heading in our direction at 13.9 km/s and in about 300,000 years is likely to be the brightest star in the night sky. Like many stars it pulsates, the pulsations causing its magnitude to fluctuate between m_v -0.02 and +0.07 with a period of 4^h 33.5^m. As a result it is classed as a δ Scuti variable: the brightest of the class. Vega occupies a special place in the annals of astronomy. On the night of 16/17 July 1850 astronomers at the Harvard Observatory rigged a daguerreotype camera to the 15-inch (38 cm) refractor and, with a 100 second exposure, captured the first photographic image of a night-time star.

The bottom right hand (south west) corner of the famed parallelogram of stars near Vega is marked by **β Lyrae** or Sheliak, a fast spinning, 131 km/s, bluish B7 with a diameter of 3.9 $D_\odot$. Technically the third brightest star at m_v +3.25 – γ Lyr just beats it by 3/100th of a magnitude – β Lyr is an EB eclipsing binary with a period of 12^d 22.5^h, which has increased by 42 minutes over the past century, its magnitude reaching a low of +4.36. The secondary is an A8 which orbits so close to the primary that both stars are distorted with matter flowing towards the A8 component forming a disk around it. A modest telescope will split β Lyr – but don't be fooled. The eclipsing component is spectroscopic. The visual star revealed through a telescope is **HD 174664**, a m_v +6.7 interloper which is a B5. While the β Lyr system is about 882 ly away, HD 174664's distance is believed to

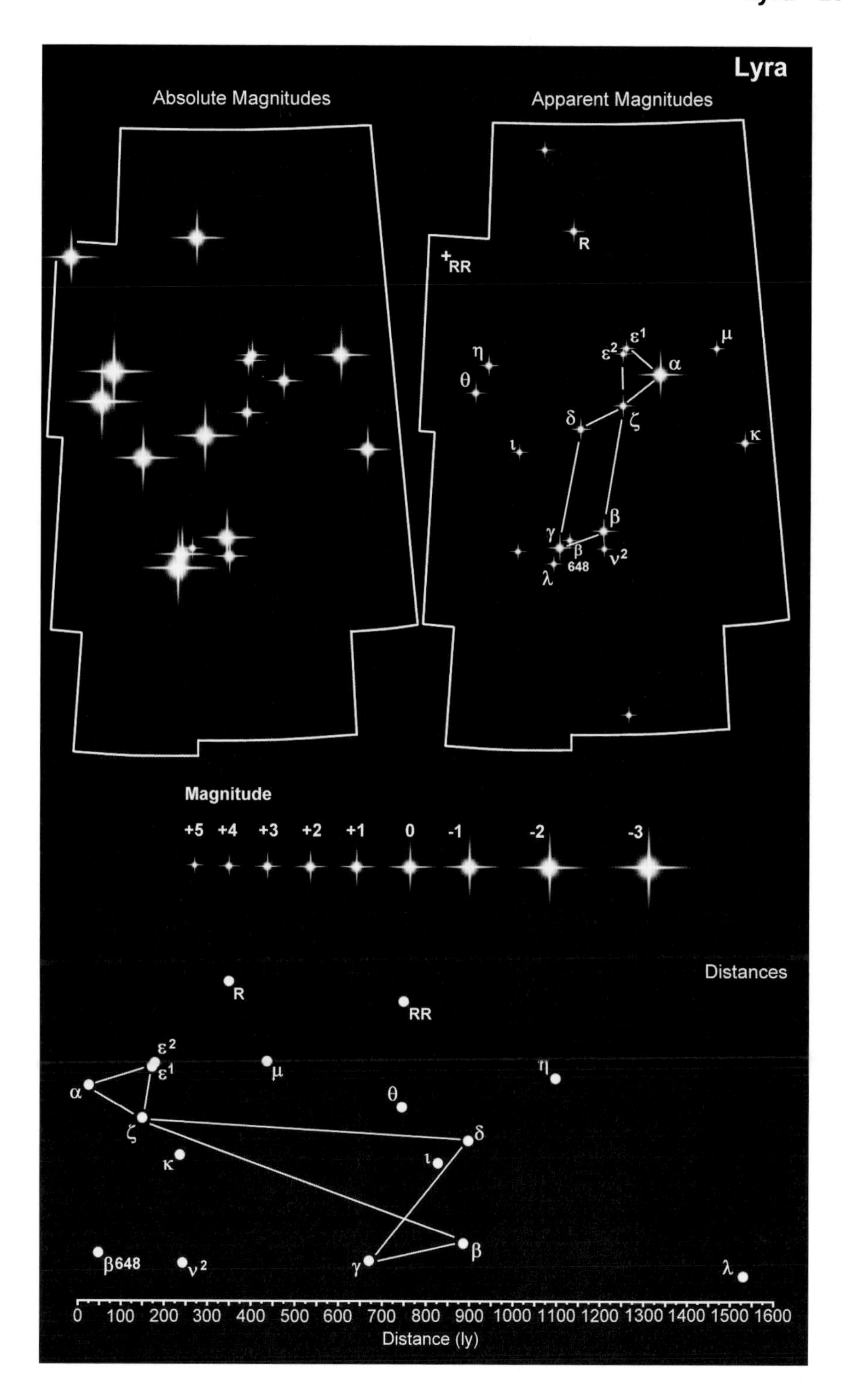
Lyra
Absolute Magnitudes
Apparent Magnitudes
R
RR
ε1
ε2
μ
η
α
θ
δ
ζ
ι
κ
β
γ
β 648
ν2
λ
Magnitude
+5 +4 +3 +2 +1 0 -1 -2 -3
Distances
R
RR
ε2
ε1
α
μ
η
θ
ζ
δ
κ
ι
β
β648
ν2
γ
λ
0 100 200 300 400 500 600 700 800 900 1000 1100 1200 1300 1400 1500 1600
Distance (ly)

Lyra
α Lyrae - Vega
From Earth we are looking at one of Vega's rotational poles, so the star looks like a sphere 2.75 $D_{\odot}$ across (image left).
Vega rotates at 274 km/s, spinning once on its axis every 12 hours. This high rotational velocity causes the star to flatten into an oblate spheroid (image below). As the material between the poles is highly compressed its temperature increases and reaches 10,200 K. Across the equator the material is more thinly spread and cooler - just 7,900 K.
The temperature gradient between the poles and the equator results in a color change from white to a dark blue.
Sun
α Lyr (A0)
2.26 x 2.75? $D_{\odot}$
2 $M_{\odot}$
25.3 ly
R_v 274 km/s
R_p 12 days

Lyra's Giants
R Lyr (M5)
180 $D_{\odot}$
4.5 $M_{\odot}$
350 ly
δ^{2A} Lyr (M4)
55 $D_{\odot}$
7.5 $M_{\odot}$
899 ly

be closer at 652 ly. β Lyrae is the prototype for eclipsing binaries that are ellipsoidal in shape. The closeness of the two components present an ever varying shape to an observer on Earth as the two stars dance around their common center of mass, distorting one another as they do so. This produces the observed changes in magnitude but it is impossible to predict the start or end of the eclipse. There is always a secondary minimum, but this is usually just a fraction of the primary minimum.

Moving around the parallelogram in a clockwise direction the south eastern corner is home to γ **Lyrae** or Sulaphat, another B-class star, this time a B9. Lying 635 ly away it has a luminosity of 1,580 $L_{\odot}$ and a diameter of 2.7 $D_{\odot}$. Between it and β Lyr, though slightly closer to β, is perhaps the best known of all planetary nebulae, **M57** – the Ring Nebula – discovered in 1779 by Antoine Darquier. Early observers thought the 9th magnitude ring was composed of a multitude of individual stars but astronomers today believe planetary nebulae are the outer layers of stars puffed off as the stars rapidly begin to die. M57 is estimated to be 2,280 ly from Earth.

δ Lyrae occupies the north eastern corner and is a complex double system. The brightest star is **δ^2 Lyrae** at m_v +4.22 but variable down to +4.33. It is an M4 red giant of 55 $D_{\odot}$. It may or may not be a triple star system. At 86″, corresponding to 24,000 AU, a pair of 11th magnitude G-class subdwarfs may be in a 1.3 million year long orbit. The subdwarfs are separated from each other by 2.2″, equivalent to 600 AU, and have their own mutual orbital period in excess of 10,000 years. The system is 899 ly away. Too faint at m_v +5.57 to be visible from most urban areas, **δ^1 Lyrae** is not part of the δ^2 Lyr system being a further 181 ly from Earth. It is a B2.5 of 5.6 $D_{\odot}$ and 7 $M_{\odot}$. There is a spectroscopic companion of which nothing is known, other than its orbital period of 88.35 days. A third component keeps its distance of 58,000 AU taking 4.5 million years to complete a single orbit. However, the δ^1 Lyrae family is actually part of a small cluster of stars called, unsurprisingly, the δ Lyrae cluster (aka Stephenson 1 cluster) containing a dozen, perhaps two dozen members. The cluster has a common origin with the Pleiades and the α Persei cluster about 180 million years ago. In the intervening years the clusters have become gravitationally unbound but still move through space in the same general direction.

ζ Lyrae completes the parallelogram in spectacular fashion consisting of six and possibly seven stars in a complex and inherently unstable orbital array. **ζ^{1Aa} Lyrae** (often just identified as ζ^1 Lyrae) is the star that is visible to the naked eye as a m_v +4.32 white Am. It is rather larger than the Sun at 2.8 $D_{\odot}$ with a mass and luminosity of 2.2 $M_{\odot}$ and 30 $L_{\odot}$. It has a close spectroscopic companion, **ζ^{1Ab} Lyrae**, just 0.07 AU away, in a 4.3 day orbit. At a distance on the celestial sphere of 25″ is **ζ^B Lyrae** at m_v +15.8. Next comes the m_v +5.73 **ζ^2 Lyrae** (sometimes called ζ^D) an F0 star of 2.0 $D_{\odot}$ and 16 $L_{\odot}$. It is 2,000 AU from ζ^{1A} Lyrae (44″) and takes at least 47,000 years to orbit the primary. Slightly farther out at 46″ is the 13th magnitude **ζ^C Lyrae** with the most distant member being **ζ^E Lyrae** at 62″ and m_v +11.5. There is the possibility that ζ^2 also has a spectroscopic companion making seven stars in all. Such orbital spaghetti is fundamentally unstable and some of the stars will either coalesce or be ejected from the system or both. Some writers will refer to all the stars by letter rather than number (e.g. ζ^A Lyrae) or put the letter at the end: ζ Lyrae A, ζ Lyrae B, etc, but as planets are now designated in just about the same way the scope for confusion is obvious.

Having got to grips with the complexities of ζ Lyrae, welcome to **ε Lyrae**, the renowned 'double double' containing four stars and a completely different designation system. **ε^{1A} Lyrae** is a m_v +5.06 A3 of 2 $D_\odot$ and 16 $L_\odot$. In orbit around ε^{1A} Lyr is **ε^{1B} Lyrae** a m_v +6.02 F1 of 0.9 $D_\odot$ and 7 $L_\odot$. They are about 140 AU apart and take more than 1,000 years to complete an orbit. Of the second pair **ε^{2C} Lyrae** is a m_v +5.14 A5 of 1.3 $D_\odot$ and 11 $L_\odot$. In orbit around ε^{2C} Lyr is **ε^{2D} Lyrae** a m_v +5.37 F0 of 1.2 $D_\odot$ and 13 $L_\odot$. They are similarly about 140 AU apart. The ε^1 pair and the ε^2 pair may orbit their common center of mass though no one really knows for sure. If they do then the orbital period must be in excess of 500,000 years, the two binary systems being at least 10,000 AU apart. A binocular will separate the components into two, while a good 75 cm (3″) refractor or 100cm (4″) reflector will show all four stars. Most observers see all the components as white although various color combinations have been reported.

B-class stars tend to be fast rotators. They average 153 km/s putting them a clear 20 km/s ahead of their closest rivals, the O-class. Only 2.5% of all B-class stars rotate at 10 km/s or less. **η Lyrae** is one of them. A 12 $D_\odot$ bluish B2.5 it takes almost 61 days to spin once on its axis. It appears as a faint m_v +4.39 star but, if all the stars in Lyra were lined up 10 pc from Earth, η Lyr would be the brightest glowing at a full M_v -3.

R Lyrae is an M5 supergiant and, like many such stars, is a semi-regular SRb pulsating variable, its magnitude varying between m_v +3.88 and +5.00 with a period of 46 days, although there are also secondary 53 and 64 day cycles. It is 350 ly away, which is just as well. Placed in the center of the Solar System it would engulf Venus and possibly even the Earth (there is some uncertainty about its diameter which could be anywhere between 150 and 212 $D_\odot$).

A star too faint to be seen by most urban dwellers is one of the most important in the entire sky. **RR Lyrae** varies between m_v +7.06 and +8.12 with a period of 13^h 36.3^m. Because all RR Lyr variables have roughly the same absolute magnitude of M_v -0.7 they can be used as a 'standard candle' to accurately determine the distances to globular clusters, in which they are principally found, up to 200 kpc (652,300 ly). As RR Lyr pulsates, its temperature and spectral class change from an F7 of 6,200 K to an A8 of 7,000 K. Its diameter has a fair degree of uncertainty – it could be as small as 4 $D_\odot$ or as large as 8 $D_\odot$ – and its luminosity could be as low as 30 $L_\odot$ or could be three times that estimate. RR Lyr is a visitor from the galactic halo, its very high velocity of 285 km/s giving the game away (true members of the solar neighborhood tend to amble along at about 14 km/s).

There is a second β in Lyra: a binary system sometimes referred to as **β 648 (HD 176051)**. The primary is a m_v +5.28 not dissimilar to the Sun. Its an F9, a few hundred degrees hotter than the Sun's G2, but not much larger at 1.3 $D_\odot$, massive, 1.07 $M_\odot$ or luminous at 1.39 $L_\odot$. Its companion, which is probably a K1 of lower mass, 0.71 $M_\odot$, and less luminous, is m_v +7.82. Their orbital period is

61.2 years. In orbit around the primary is a 1.5 M_J planet in an almost circular orbit of 1.76 AU which it completes in 976 to 1,056 days.

Open and globular clusters in Lyra

Name	Size arc min	Size ly	Distance ly	Age million yrs	Brightest star in region*	No. stars m_v >+12*	Apparent magnitude m_v
Stephenson 1	150′	56	1,300	54	m_v +6.12	82	+3.8
M56 (NGC 6779)	8.8′	84	32,900	1,300	Globular cluster		+8.3

*May not be a cluster member.

Mensa

Constellation:	Mensa	**Hemisphere:**	Southern
Translation:	The Table Mountain	**Area:**	153 deg^2
Genitive:	Mensae	**% of sky:**	0.693%
Abbreviation:	Men	**Size ranking:**	75th

Introduced in the 18th Century by Abbé de la Caille to celebrate the site of his observatory near modern day Cape Town, Mensa is a difficult constellation to find, the brightest star being just m_v +5.08.

α **Mensae** is just m_v +5.08 but it wasn't always that way. About 250,000 years ago it passed within 11 ly of Earth and was a star of the second magnitude. It is Sun-like, a G6 possibly a G7, so a bit cooler than the Sun, but with a comparable diameter, 0.91 $D_\odot$, mass 0.87 $M_\odot$, and luminosity, 0.78 $L_\odot$. It even revolves at about the same speed as the Sun, 1.8 km/s compared to the Sun's 2 km/s, so it rotates once in almost exactly the same time: 25.6 days. And it is just 33.1 ly away. Needless to say it has attracted the interests of planet and alien hunters although, so far, it appears to have neither. Time is running out for α Men. It seems to be about 10,000 million years old and will soon begin to swell into a red giant. α Men is at the stage our Sun will reach in 5,000 million years from now.

β **Mensae** is also a G-class star, this time a G8, but rather larger than the Sun at 9.6 $D_\odot$ and radiating 238 $L_\odot$. It is actually the third brightest star in the constellation – γ Men is slightly brighter – and is set against the backdrop of the Large Magellanic Cloud which lies at a distance of 160,000 ly. β Men is much closer at 642 ly and has an apparent magnitude of m_v +5.30.

γ **Mensae** is a K2 giant of 19 $D_\odot$ and 6.5 $L_\odot$. It is 100.6 ly away and shines as a dim m_v +5.19 but would brighten to M_v -0.3 at 10 pc. It is also a high velocity star, heading away from us at 56.7 km/s suggesting it is just a visitor to our corner of the Galaxy.

Just 10° from the South Celestial Pole, π **Mensae** is the only star in the constellation known to have a planetary system. The star itself is similar to the Sun: a G0 of about the same mass, 1.1 $M_\odot$, but more than twice its diameter, 2.1 $D_\odot$, and younger at 3,830 million years (the Sun is 4,560 million years old). Its lone planet at 10.35 M_J is verging on being a brown dwarf. It averages 3.29 AU from the star but gets as close as 1.25 AU at periastron before journeying out to 5.33 AU at apastron (about the same distance as Jupiter is from the Sun). It has an orbital period of 5.7 years. While it would have been so easy to name the planet π Mensae b instead it goes by the instantly forgettable name of HD 39091 b.

Planetary system in Mensa

Star	$D_\odot$	Spectral class	ly	m_v	Planet	Minimum mass	q	Q	P
π Men	2.1	G1	67.1	+5.67	HD 39091 b	10.35 M_J	1.25	5.33	5.65 y

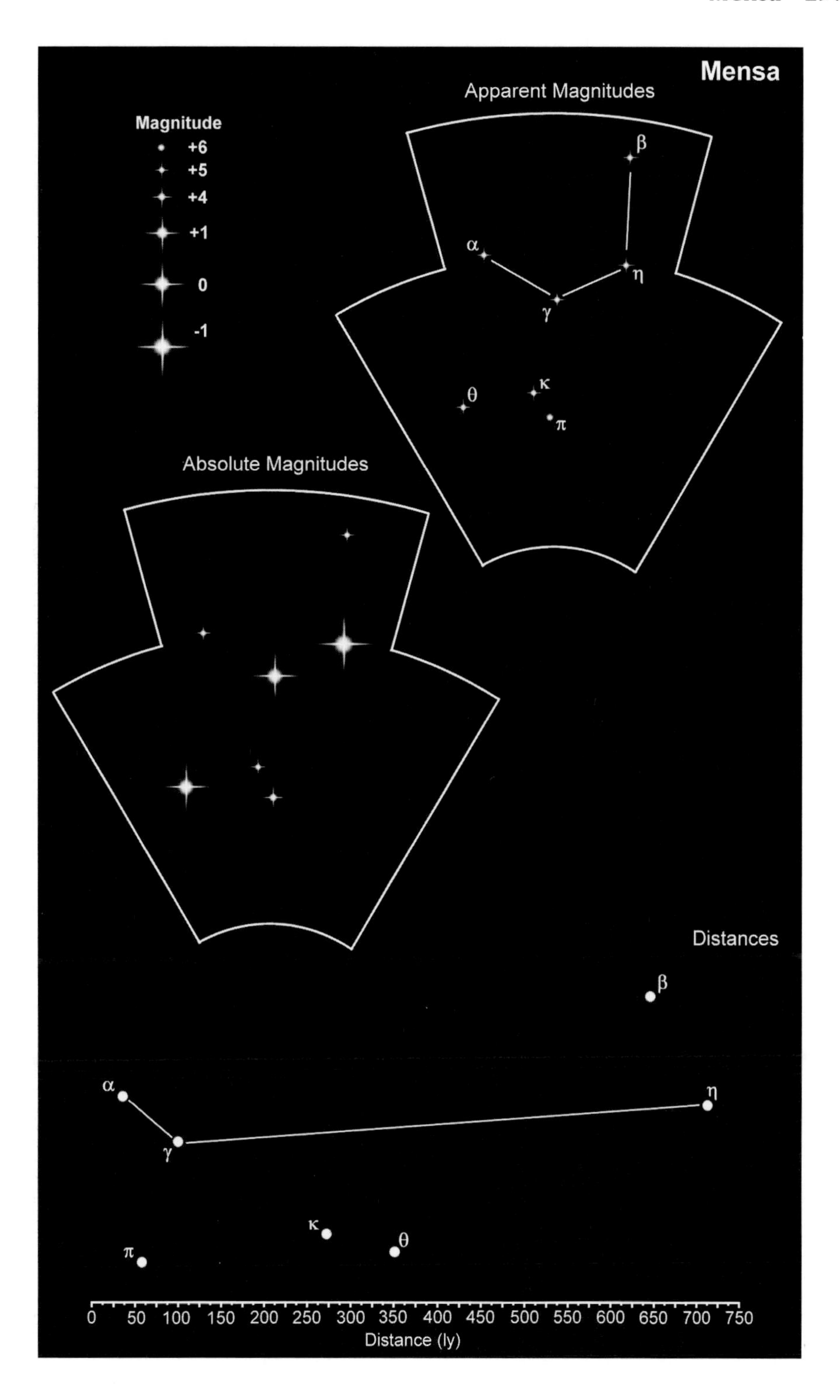
Mensa
Apparent Magnitudes
Magnitude
+6
+5
+4
+1
0
-1
β
α
η
γ
θ
κ
π
Absolute Magnitudes
Distances
β
α
η
γ
κ
θ
π
0 50 100 150 200 250 300 350 400 450 500 550 600 650 700 750
Distance (ly)

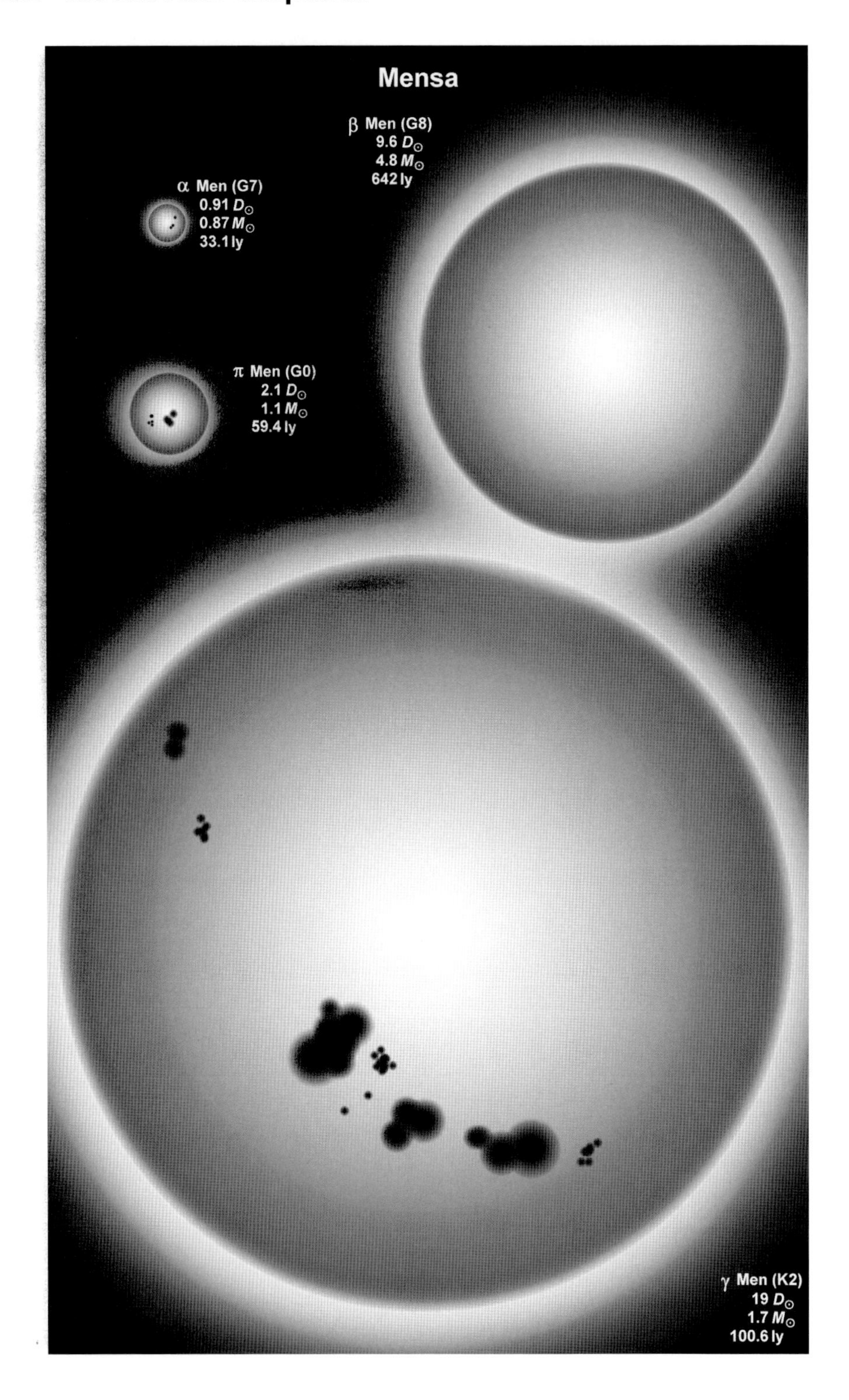
Mensa
β Men (G8)
9.6 $D_{\odot}$
4.8 $M_{\odot}$
642 ly
α Men (G7)
0.91 $D_{\odot}$
0.87 $M_{\odot}$
33.1 ly
π Men (G0)
2.1 $D_{\odot}$
1.1 $M_{\odot}$
59.4 ly
γ Men (K2)
19 $D_{\odot}$
1.7 $M_{\odot}$
100.6 ly

Microscopium

Constellation:	Microscopium	**Hemisphere:**	Southern
Translation:	The Microscope	**Area:**	210 deg^2
Genitive:	Microscopii	**% of sky:**	0.509%
Abbreviation:	Mic	**Size ranking:**	66th

Another faint 18th Century constellation created by Abbé de la Caille in recognition of the importance of the microscope in scientific investigation.

There are two G-class stars in Microscopium, both mid-G, both about the same size but certainly not related. The first of these is α **Microscopii**, a G7, the fourth brightest star in the constellation at m_v +4.90. It is 12 $D_\odot$ across and was once thought to have a 10th magnitude companion, but this turned out to be a line of sight coincidence. The second G6 is γ **Microscopii** which is just slightly smaller at 11 $D_\odot$ but less than half as luminous, coming in at 52.1 $L_\odot$ compared to α Mic's 121 $L_\odot$. At 224 ly γ Mic is the closer of the two by 157 ly. α Mic is moving towards us at -14.5 km/s whereas γ Mic is heading in the opposite direction at 17.6 km/s. Both stars will have started their lives as late B-class objects, α Mic about 450 million years ago while γ Mic is the older of the two at about 600 million years.

β **Microscopii** is an A2 dwarf of 2.2 $D_\odot$ but at m_v +6.02 is too faint to be seen by most urbanites. It is 483 ly away but is heading in our direction at 12 km/s.

θ^1 **Microscopii's** brightness fluctuates by 1/10th of a magnitude between m_v +4.77 and +4.87. The star belongs to the α CV class of rotating variables and has a period of $2^d\ 2^h\ 55^m$. Some 2.8 $D_\odot$ across it has a peculiar spectral signature of ApCrEuSr indicating its outer layer is enriched in chromium, europium and strontium.

At just 10% larger than the Sun, ι **Microscopii** is probably the smallest naked eye star in the constellation. An F1 it is 134 ly from Earth.

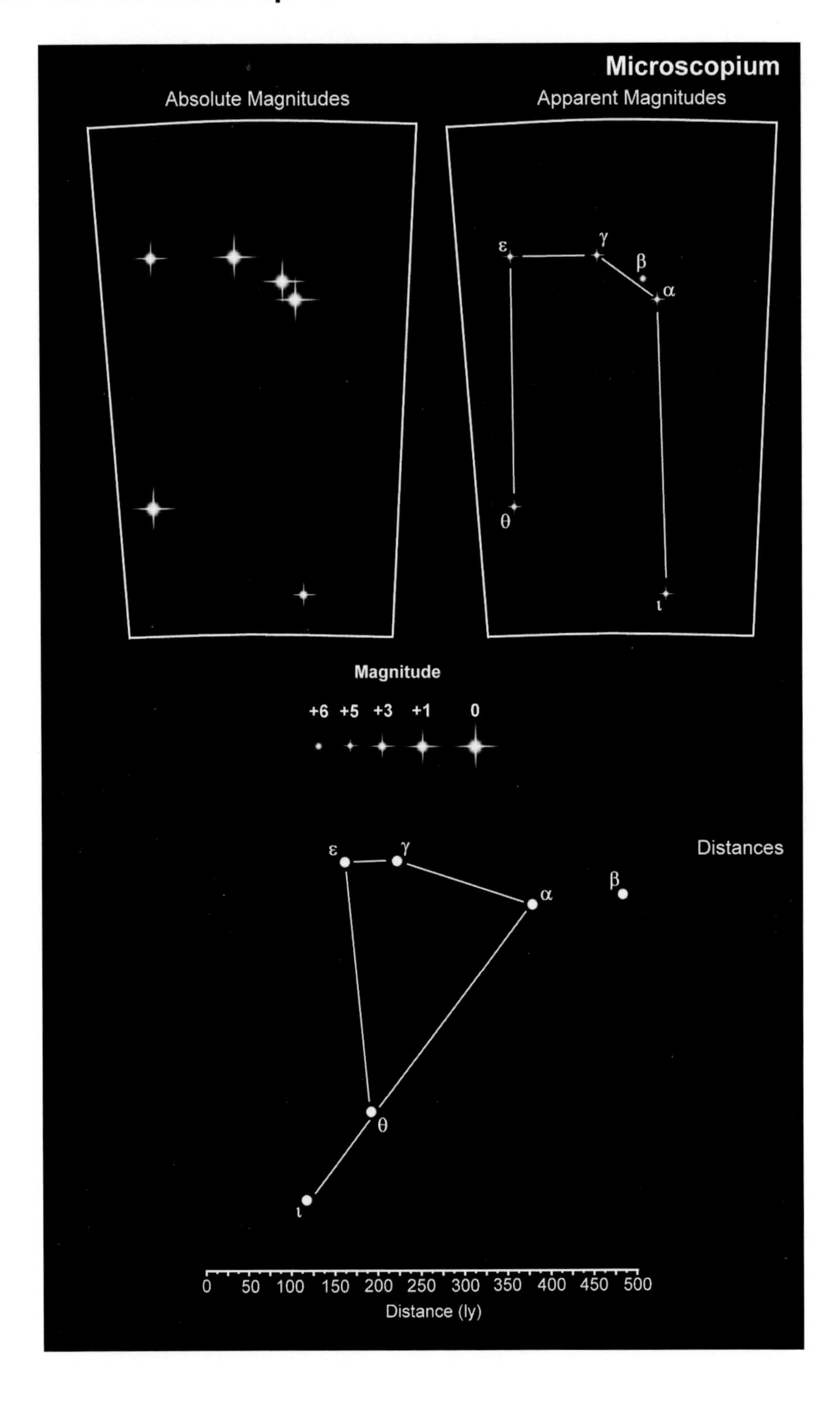
Microscopium
Absolute Magnitudes
Apparent Magnitudes
ε
γ
β
α
θ
ι
Magnitude
+6 +5 +3 +1 0
Distances
ε
γ
α
β
θ
ι
0 50 100 150 200 250 300 350 400 450 500
Distance (ly)

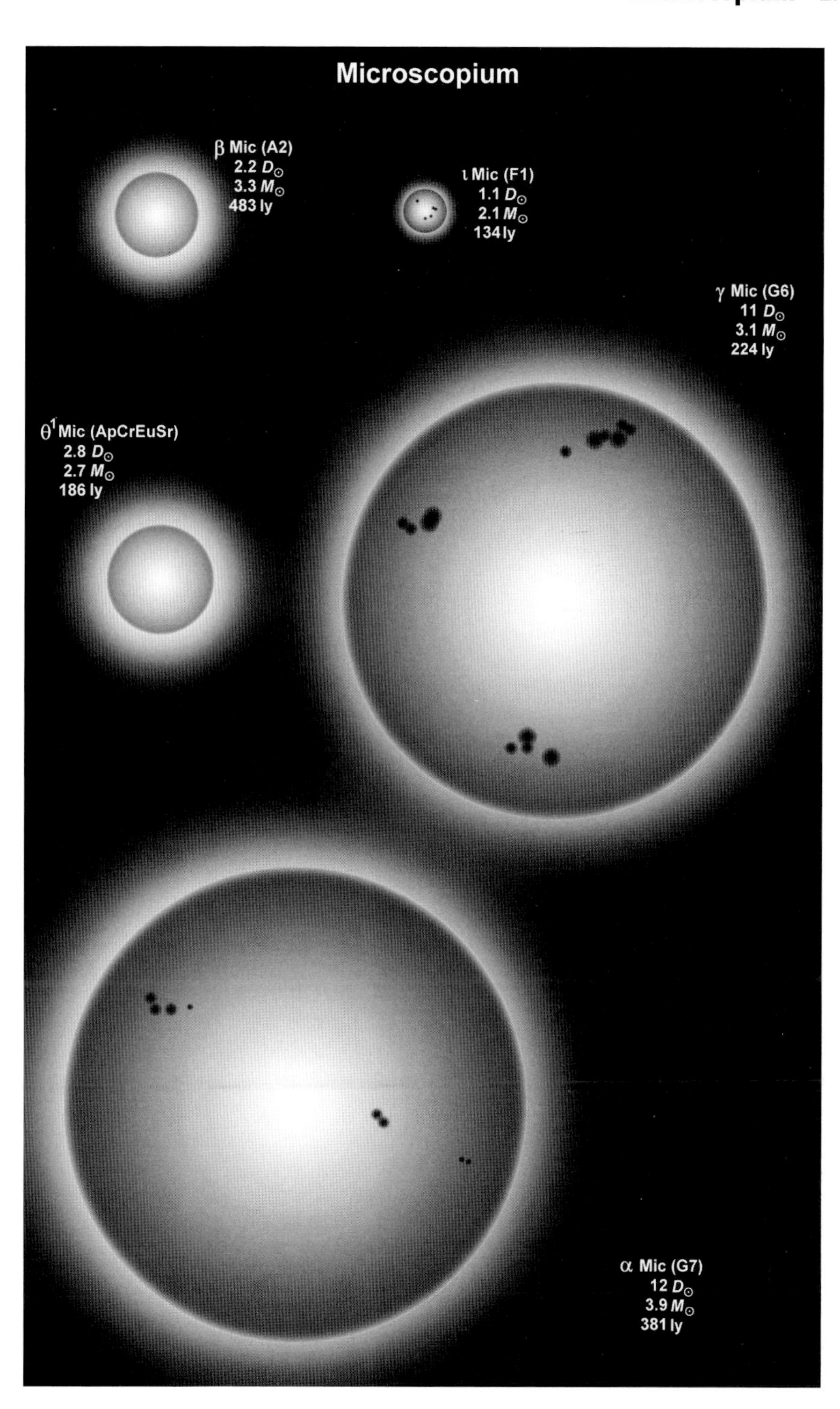
Microscopium
β Mic (A2)
2.2 $D_{\odot}$
3.3 $M_{\odot}$
483 ly
ι Mic (F1)
1.1 $D_{\odot}$
2.1 $M_{\odot}$
134 ly
γ Mic (G6)
11 $D_{\odot}$
3.1 $M_{\odot}$
224 ly
θ1 Mic (ApCrEuSr)
2.8 $D_{\odot}$
2.7 $M_{\odot}$
186 ly
α Mic (G7)
12 $D_{\odot}$
3.9 $M_{\odot}$
381 ly

Monoceros

Constellation:	Monoceros	**Hemisphere:**	Equatorial
Translation:	The Unicorn	**Area:**	482 deg^2
Genitive:	Monocerotis	**% of sky:**	1.168%
Abbreviation:	Mon	**Size ranking:**	35th

Surrounded by the great constellations of Orion, Gemini and Canis Major this faint collection of stars is often overlooked. It is the invention of Petrus Plancius, a Dutch explorer of the 17th Century.

α **Monocerotis** is what is often referred to as a 'clump star'. Plot such helium-fusing giants on a Hertzsprung-Russell diagram of temperature vs. luminosity and they tend to clump together in one place. α Mon is 12 $D_\odot$ across and has a luminosity of 43 $L_\odot$ with a mass of around 2.5 $M_\odot$. Like all the stars in Monoceros it is a fair distance away: 139 to 149 ly. Its apparent magnitude is just m_v +3.93. There are web pages that claim α Mon goes by the name of Ctesias. This is actually the name of a Greek physician of the 5th Century BC and seems to have been added to the star only recently. Similarly, β Mon is not called Cerastes, γ Mon is not Tempestris and δ Mon is not Kartajan. HD 47129 is, however, called Plaskett's Star but we'll come to that in due course.

β **Monocerotis** is a triple star system of almost identical B-class stars. β^B and β^C **Monocerotis** are in a 590 AU orbit around one another that takes 4,200 years to complete. β^B is the more luminous and brighter of the two at 1,600 $L_\odot$ and m_v +4.63. β^C is 2.8″ away and shines at m_v +5.33, its luminosity being 1,300 $L_\odot$. The pair are estimated to be about 34 million years old. Orbiting these two stars at a distance of 1,570 AU – 7.4″ on the celestial sphere – is β^A, a somewhat older star at 43 million years and the most luminous at 3,200 $L_\odot$. Its orbital period is in excess of 14,000 years. There is not much difference in their masses: β^A is 7 $M_\odot$, β^B 6.2 $M_\odot$ and β^C is 6 $M_\odot$, and all appear to be Be emission stars with surrounding dust rings. The system lies 691 ly from us.

A binary star system can be found in ε **Monocerotis.** ε^A **Monocerotis** is an A5 of 4.0 $D_\odot$ and with a mass of 3.2 $M_\odot$ and luminosity of 60 $L_\odot$. Its magnitude is m_v +4.41. Separated by 12.1″, 500 AU in real space, ε^B **Monocerotis** is an F5 of 1.4 $D_\odot$ and with a mass of 1.4 $M_\odot$ and luminosity of 3.8 $L_\odot$. Its magnitude is m_v +6.72. Together they give the appearance of a single naked eye m_v +4.44 star. The pair has an orbital period in excess of 6,000 years. There is however a third member, ε^{Ab} **Monocerotis,** a spectroscopic component in a 331 day long orbit with ε^A Mon.

ε Mon is a good marker for finding the **Rosette Nebula** which is just 2° almost due east. The nebula is also known as **NGC 2237, 2238, 2239** and **2246** depending on which part you are looking at. It is more than 1° across corresponding to 130 ly and lies at a distance of 5,500 ly. This is a region of star formation, the redness of the nebula being caused by radiation from young stars exciting the hydrogen atoms of the nebulous cloud.

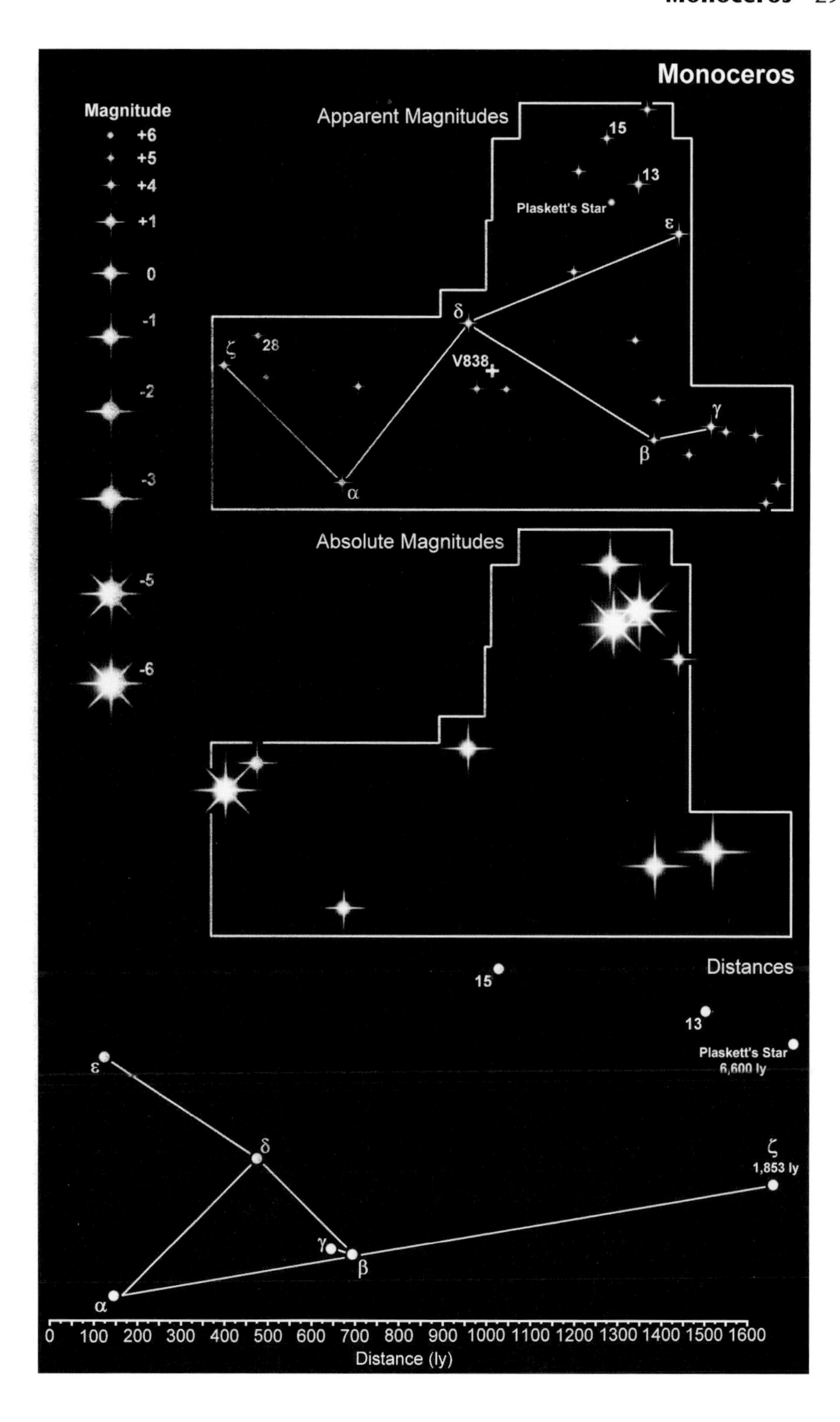
Monoceros
Magnitude
+6
+5
+4
+1
0
-1
-2
-3
-5
-6
Apparent Magnitudes
15
13
Plaskett's Star
ε
δ
28
ζ
V838
γ
β
α
Absolute Magnitudes
Distances
15
13
Plaskett's Star
6,600 ly
ε
δ
ζ
1,853 ly
γ
β
α
0 100 200 300 400 500 600 700 800 900 1000 1100 1200 1300 1400 1500 1600
Distance (ly)

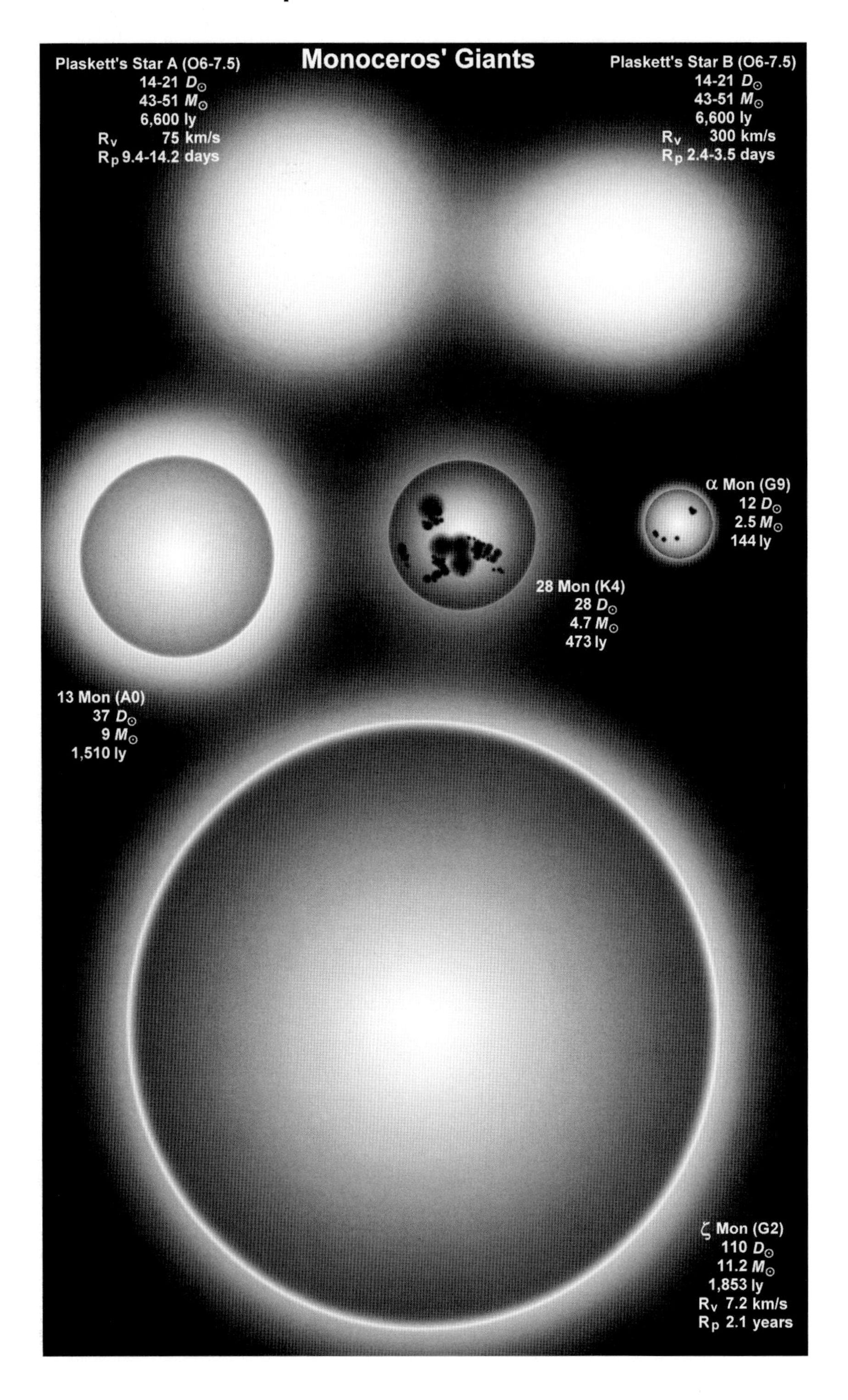
Monoceros' Giants
Plaskett's Star A (O6-7.5)
14-21 D☉
43-51 M☉
6,600 ly
Rv 75 km/s
Rp 9.4-14.2 days
Plaskett's Star B (O6-7.5)
14-21 D☉
43-51 M☉
6,600 ly
Rv 300 km/s
Rp 2.4-3.5 days
α Mon (G9)
12 D☉
2.5 M☉
144 ly
28 Mon (K4)
28 D☉
4.7 M☉
473 ly
13 Mon (A0)
37 D☉
9 M☉
1,510 ly
ζ Mon (G2)
110 D☉
11.2 M☉
1,853 ly
Rv 7.2 km/s
Rp 2.1 years

At 110 $D_{\odot}$ and 1,853 ly away ζ **Monocerotis** is probably the largest and one of the most distant stars in the constellation. This G2 supergiant is 4,718 times more luminous than the Sun so shines at a respectable m_v +4.36. It rotates at 7.2 km/s which means it must take more than 2 years – 773 days – to rotate just once. At 10 pc it would appear as bright as Venus at her most brilliant, M_v -4.5.

One star whose absolute magnitude would surpass Venus is **13 Monocerotis.** It would make M_v -5.2. 13 Mon has a diameter of 37 $D_{\odot}$ but a mass of only 9 $M_{\odot}$. Its distance is believed to be 1,510 ly and it is surrounded by a thin nebula which it helps to illuminate.

Monoceros is host to a couple of rare variable stars. **28 Monocerotis** or **V645 Monocerotis** is a 28 $D_{\odot}$ K4 about 473 ly away. Its magnitude varies between m_v +4.68 and +4.70 with a period of 4^h 59.3^m due to the presence of large star spots. It is classed as an FK Comae Berenicis variable. **15 Monocerotis** or **S Monocerotis** is a member of the unusual IA eruptive variable class that are so rare they have not been fully investigated. 15 Mon itself is a 1 $D_{\odot}$ O7 with a luminosity of 1,079 $L_{\odot}$ which fluctuates by 6/100th of a magnitude from m_v +4.62 to +4.68 with no particular period. It is set against a backdrop of **NGC 2264**, the Christmas Tree Cluster.

HD 47129 is more often referred to as **Plaskett's Star** after the Canadian astronomer who undertook a study of the object in 1922. At m_v +6.06 it is too faint to be seen by many people but it is nonetheless worth a mention as it appears to be a huge binary system. The pair is estimated to be 6,600 ly away and consist of two similar, perhaps even identical O-class stars. They are probably O6 to O7.5, have masses in the range of 43 to 51 $M_{\odot}$, temperatures of 35,000 to 40,000 K, luminosities of between 372,000 to 870,000 $L_{\odot}$ and diameters of 14 to 21 $D_{\odot}$. Of greater certainty are their rotational velocities: 75 km/s for one star and 300 km/s for the other. They are separated by just 0.5 AU and have an orbital period of 14.4 days. As far as anyone is aware HD 47129 could be the most massive binary system visible from Earth.

If you were expecting to find Luyten's Star at this point then go to Canis Minor. Despite countless web sites stating that the star is in Monoceros, it is simply not true!

Amateur astronomers often dream about making a discovery – a new comet, an asteroid, a supernova, a meteorite find – but for Australian amateur Nicholas J. Brown his dream came true in a big way. On 6 January 2002 Nicholas stumbled across a stellar explosion that is so rare only three others have previously been recorded. The star in question, **V838 Monocerotis,** brightened from a 12th magnitude object to m_v +6.5 in just a day. It then faded over the following two weeks until it reached m_v +9.0 at which point it brightened again to m_v +7.5 for a couple of days before subsiding to m_v +16. The star quickly attracted the attention of professional astronomers around the world with the result that the Hubble Space Telescope was pointed in its direction. It revealed that the star was surrounded by 'light echoes' caused by light being deflected by dust shells. Light echoes have been witnessed only on three previous occasions: in 1901, in 1936 and in supernova 1987A. V838 – will it ever become known as Brown's Star? – is a

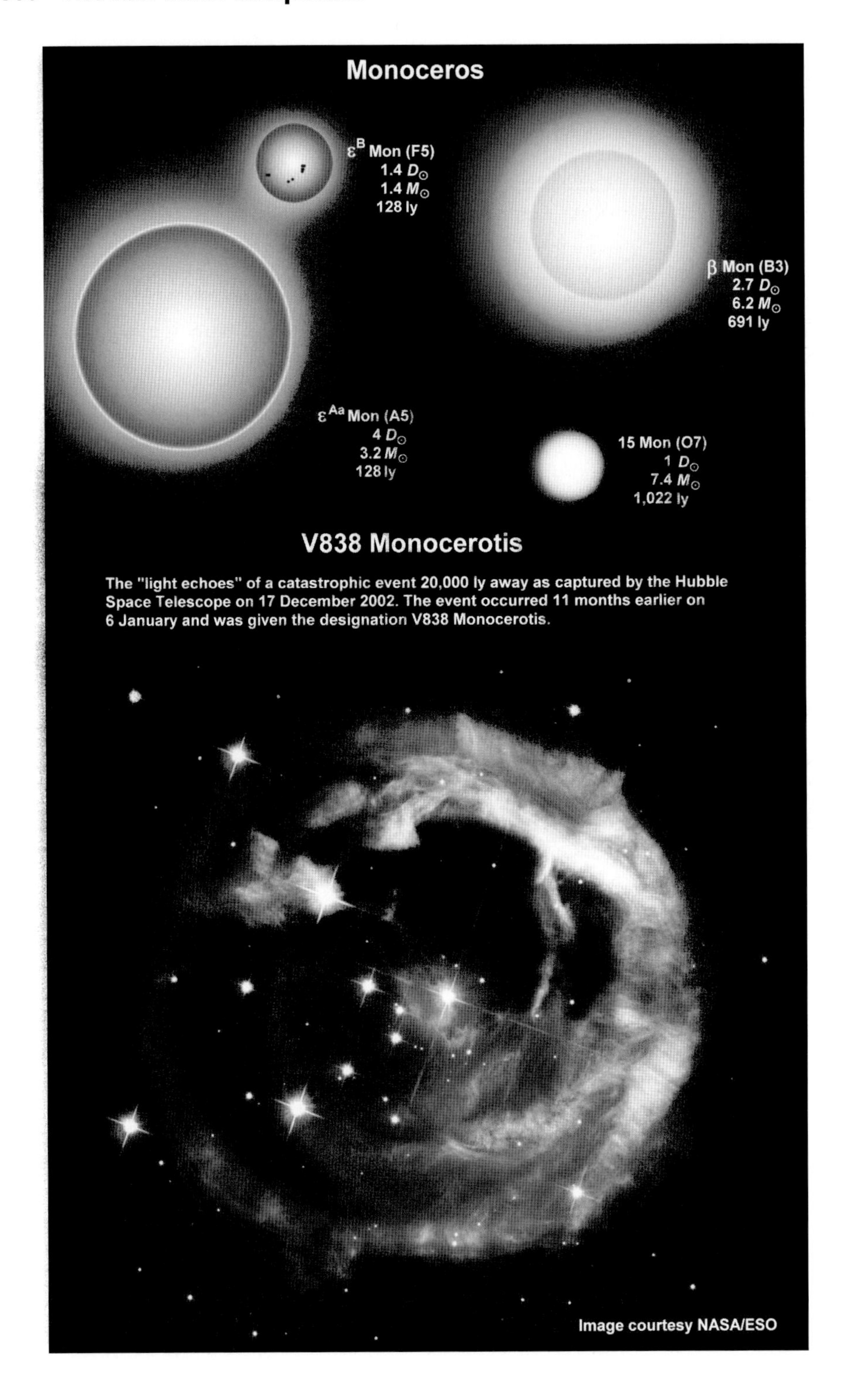
Monoceros
ε^B Mon (F5)
1.4 D☉
1.4 M☉
128 ly
β Mon (B3)
2.7 D☉
6.2 M☉
691 ly
ε^Aa Mon (A5)
4 D☉
3.2 M☉
128 ly
15 Mon (O7)
1 D☉
7.4 M☉
1,022 ly
V838 Monocerotis
The "light echoes" of a catastrophic event 20,000 ly away as captured by the Hubble Space Telescope on 17 December 2002. The event occurred 11 months earlier on 6 January and was given the designation V838 Monocerotis.
Image courtesy NASA/ESO

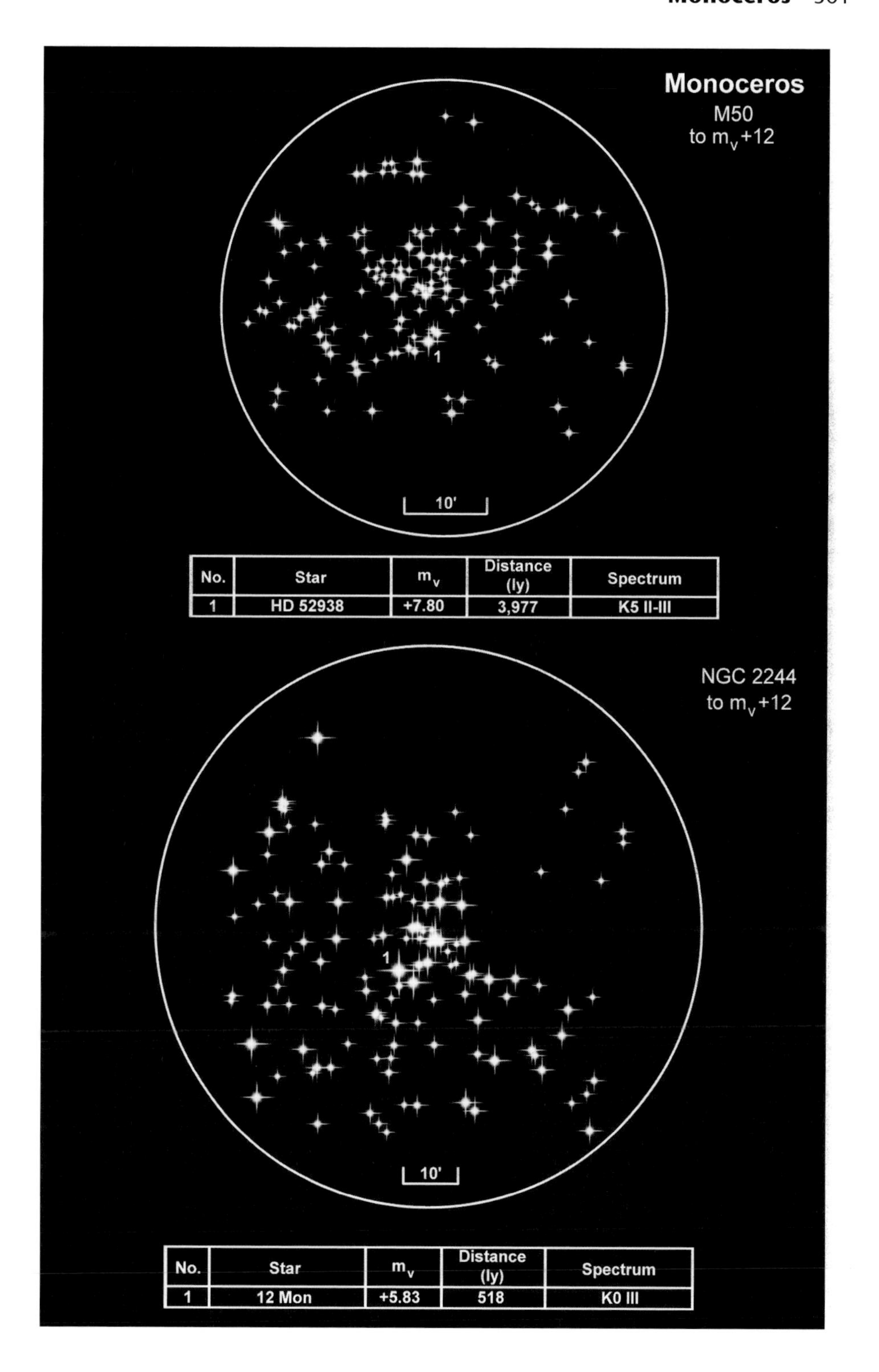

No.	Star	m_v	Distance (ly)	Spectrum
1	HD 52938	+7.80	3,977	K5 II-III

No.	Star	m_v	Distance (ly)	Spectrum
1	12 Mon	+5.83	518	K0 III

mystery and several hypotheses have been advanced. Perhaps we have observed a star rapidly evolving from the Main Sequence to become a red supergiant, and becoming unstable in the process. Normally, this process would take hundreds to thousands of years. In this case it appears to have happened over just a few months. Another strong possibility is that it is the result of two stars colliding.

Several stars within Monoceros have planetary systems. One of the most interesting is **CoRoT-7 b** which orbits its parent star once every 20.4 hours at a distance of just 2.6 million km. Mercury's orbit, for comparison, is 59.9 million km. At such close proximity the planet could reach a temperature of 2,200 K, melting and vaporizing its surface which would them be blown off into space. If the planet's orbit is not truly circular then tidal forces will distort the shape of the planet perhaps inducing widespread volcanic activity. The planet may also be locked in a gravitational embrace such that it always shows the same face to the star (in the same way that out Moon always displays the same face towards Earth). Should that be the case then temperatures on the dark side of the planet could be as cold as -210 K. The planet has a mass of about 4.8 $M_{\oplus}$ but is only slightly larger than Earth. What we could be witnessing is a Jupiter-size planet that has ventured so close to its parent star that its atmosphere has been stripped away and all that remains is its rocky core which is now boiling away. The planet could literally disappear. The star itself is a K0 dwarf of 0.87 $D_{\odot}$. Just m_v +11.7 it lies at a distance of 489 ly and has at least one other planet, a 0.0264 M_J object in a 3.698 day orbit.

Messier 50 (M50) is a reasonably bright, m_v +5.9 open cluster of about 200 stars. Situated 3,200 ly away and up to 20′ across, which equates to a diameter of about 10 ly, the cluster is believed to be about 78 million years old.

Open clusters in Monoceros

Name	Size arc min	Size ly	Distance ly	Age million yrs	Brightest star in region*	No. stars m_v >+12*	Apparent magnitude m_v
M50 (NGC 2323)	38′	33	3,000	125	HD 52938 m_v +7.80	134	+5.9
NGC 2232	50′	17	1,200	53	10 Mon m_v +5.05	34	+3.9
NGC 2244	58′	80	4,700	8	12 Mon m_v +5.83	137	+4.8
NGC 2264 Christmas Tree	40′	26	2,200	9	S Mon m_v +4.65	50	+3.9
NGC 2301	14′	11	2,800	165	TYC 148-2862-1 m_v +8.02	71	+6.0
NGC 2343	18′	18	3,400	13	HD 54388 m_v +8.42	28	+6.7
NGC 2353	17′	18	3,700	94	HD 55879 m_v +6.04	29	+7.1

Musca

Constellation:	Musca	**Hemisphere:**	Southern
Translation:	The Fly	**Area:**	138 deg^2
Genitive:	Muscae	**% of sky:**	0.335%
Abbreviation:	Mus	**Size ranking:**	77th

This constellation has gone through several transformations. Petrus Plancius introduced the constellation in the 16th Century as Apis, the Bee. Edmond Halley, of Halley's Comet fame, renamed it Musca Apis, the Fly Bee. In the mid-18th Century Abbé de La Caille called it Musca Australis, the Southern Fly, to distinguish it from the Musca in the Northern Hemisphere which was eventually absorbed by Aries. Finally, the International Astronomical Union decided that it should just be called Musca. The constellation resembles Ursa Minor.

α **Muscae** is one of the many bluish-white stars that make up nearly a quarter of all the naked eye stars in the night sky. A B2 it lies about 305 ly away and has a luminosity of 4,520 $L_\odot$. Indirect measurement of its size suggests it is 4.8 $D_\odot$ across, its surface temperature being 21,900 K, more than 3.6 times hotter than the Sun. Like many B-class stars it is a fast spinner – 150 km/s – and so takes only 1.6 days to turn once on its poles. It is also a β Cepheid, fluctuating in brightness by 5/100th of a magnitude between m_v +2.68 and +2.73 with a period of $2^h\ 10^m$.

β **Muscae** is a fine binary of two almost identical 2.6 $D_\odot$ bluish-white dwarfs though observers disagree on their color with some claiming they are white, some blue and some bluish-white. Discovered in 1880 by H.C. Russell, the Government Astronomer at Sydney Observatory in Australia, the brighter of the two, m_v +3.7, is a B2.5 while the slightly fainter m_v +4.0, is a B3. Combined they look like a single +3.54 star. The primary is, like most B-class stars, a fast spinner at 185 km/s and it is likely its companion also has a high rotational velocity. The pair complete a single orbit in 383.12 years and lie 311 ly from Earth.

α and β Muscae are among a handful of stars in the constellation that are of similar distance from Earth. The graphic below shows the quoted distance of each star, but there is uncertainty in their exact distances with the result that all five stars could actually be close neighbors to within just a few light years. While α, β and γ Muscae are all B-class stars, γ **Muscae** is the largest of the three at 5 $D_\odot$: ε **Muscae** is an M5 red giant. Although its mass is only 1.5 to 2 $M_\odot$ its diameter is 130 $D_\odot$, almost as big as the orbit of Venus. Typical of such lumbering giants, its magnitude is unstable, the star belonging to the semi-regular SRb category of pulsating giants, and fluctuates between +3.99 and +4.31 with a main period of 40 days. It's a stranger to our neighborhood, passing through at 100 km/s compared to the local traffic at just 20 km/s or less.

Like β Mus, η **Muscae** is a binary system of two B-class stars but at the opposite end of the spectral class. Both classed as B8 recent studies reveal they are almost indistinguishable. They weigh in at 3.30 and 3.29 $M_\odot$, are 2.14 and 2.13 $D_\odot$, and have temperatures of 12,700 K and 12,500 K. They appear to be in a near

Five stars in Musca are of similar distance from Earth and could be closer together than we imagine, perhaps clustering at around 315 ly. The graphic shows the minimum and maximum distance for each star and their mid-distance values.

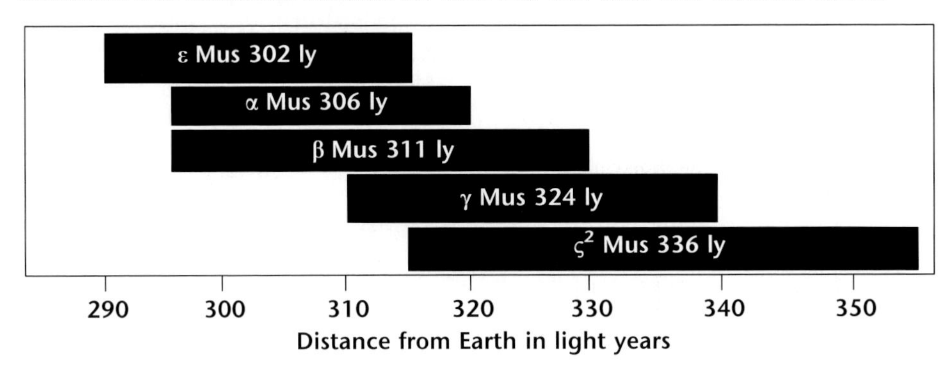

circular orbit of 9.8 million km with an orbital period of just 57.5 hours which means that their orbital speed – how fast they are traveling in their orbit – is in excess of 1 million km/h or about 298 km/s. It was previously thought the orbital period was as long as 20 days. The way the orbit is aligned to us also means that the eclipses are also identical in duration and amplitude, the magnitude fluctuating between m_v +4.76 and +4.81.

Just below naked eye visibility for most people θ **Muscae** is a rare m_v +5.88 Wolf-Rayet star (**WR 48**). It is 10,900 ly distant.

The 5th magnitude variable **GT Muscae** is a quadruple system. The primary component – HD 101379 – is a G2 giant of 9.5 $D_\odot$ across. It has a high level of star spot activity which causes it magnitude to vary over a period of about 64 days. The presence of an A0 class dwarf in a close orbit disrupts the primary's magnetic field resulting in occasional flaring. A further two A-class dwarfs, an A0 and an A2 cataloged as HD 101380, are in a 61.5 day orbit around the primary. They also eclipse one another every 2.7546 days. This complex arrangement means that the system is both an RS CVn and an E class variable with the magnitude varying between m_v +5.08 and +5.21.

A much under-rated globular cluster **NGC 4833** lies at a distance of 21,200 ly and contains thousands of young stars. It was discovered by Abbé de La Caille in 1751-52 while in South Africa and makes a good target for a binocular or small telescope.

About 8,000 ly away and well below naked eye limiting magnitude at m_v +13 **MyCn 18** is a young planetary nebula that has earned the nickname of the Engraved Hourglass Nebula (not to be confused with the Hourglass Nebula in Sagittarius). Such structures provide clues to the ejection of material from dying stars.

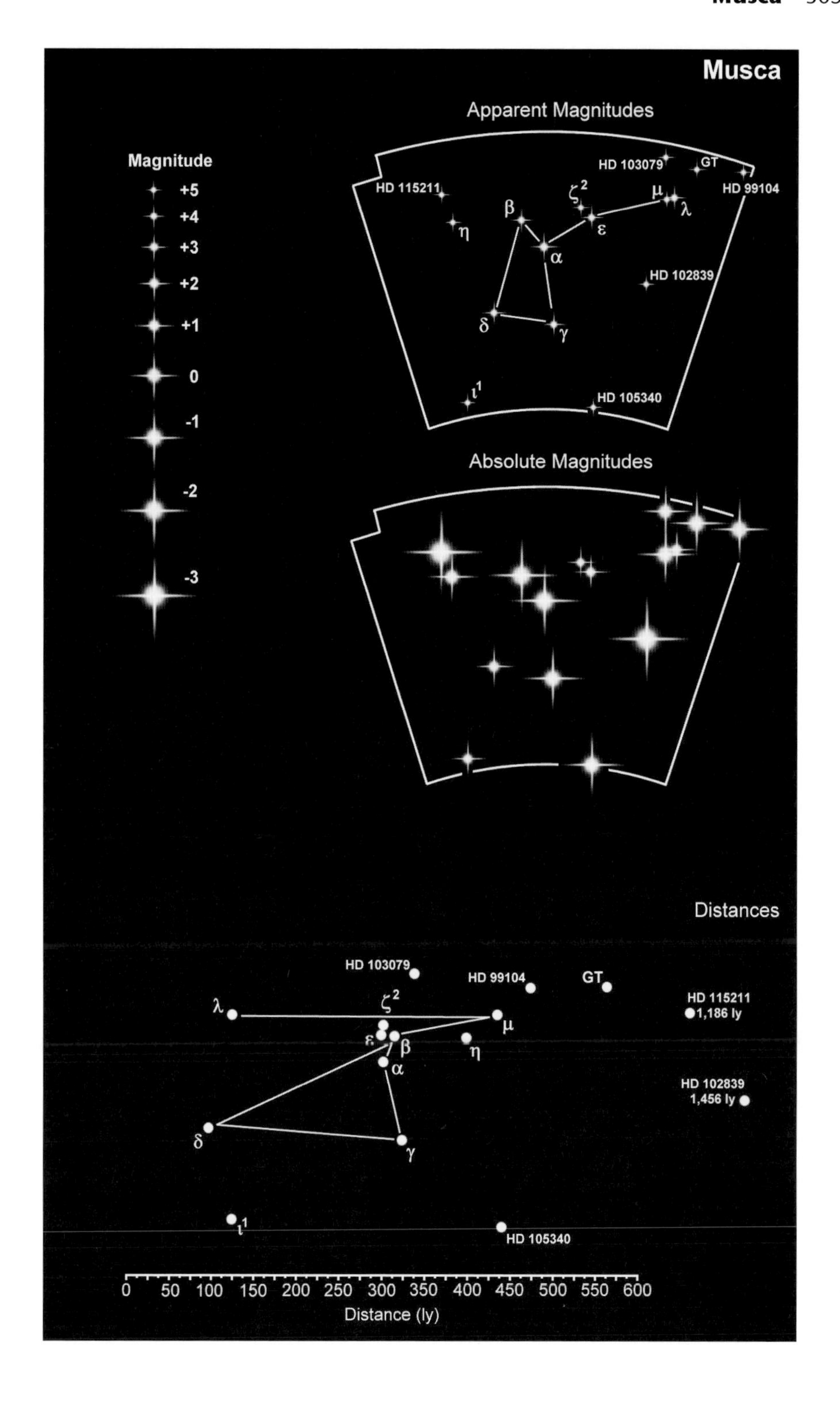

Musca
Apparent Magnitudes
Magnitude
+5
+4
+3
+2
+1
0
-1
-2
-3
HD 103079
GT
HD 115211
HD 99104
β
ζ2
μ
λ
η
ε
α
HD 102839
δ
γ
ι1
HD 105340
Absolute Magnitudes
Distances
HD 103079
HD 99104
GT
HD 115211
1,186 ly
HD 102839
1,456 ly
HD 105340
0 50 100 150 200 250 300 350 400 450 500 550 600
Distance (ly)

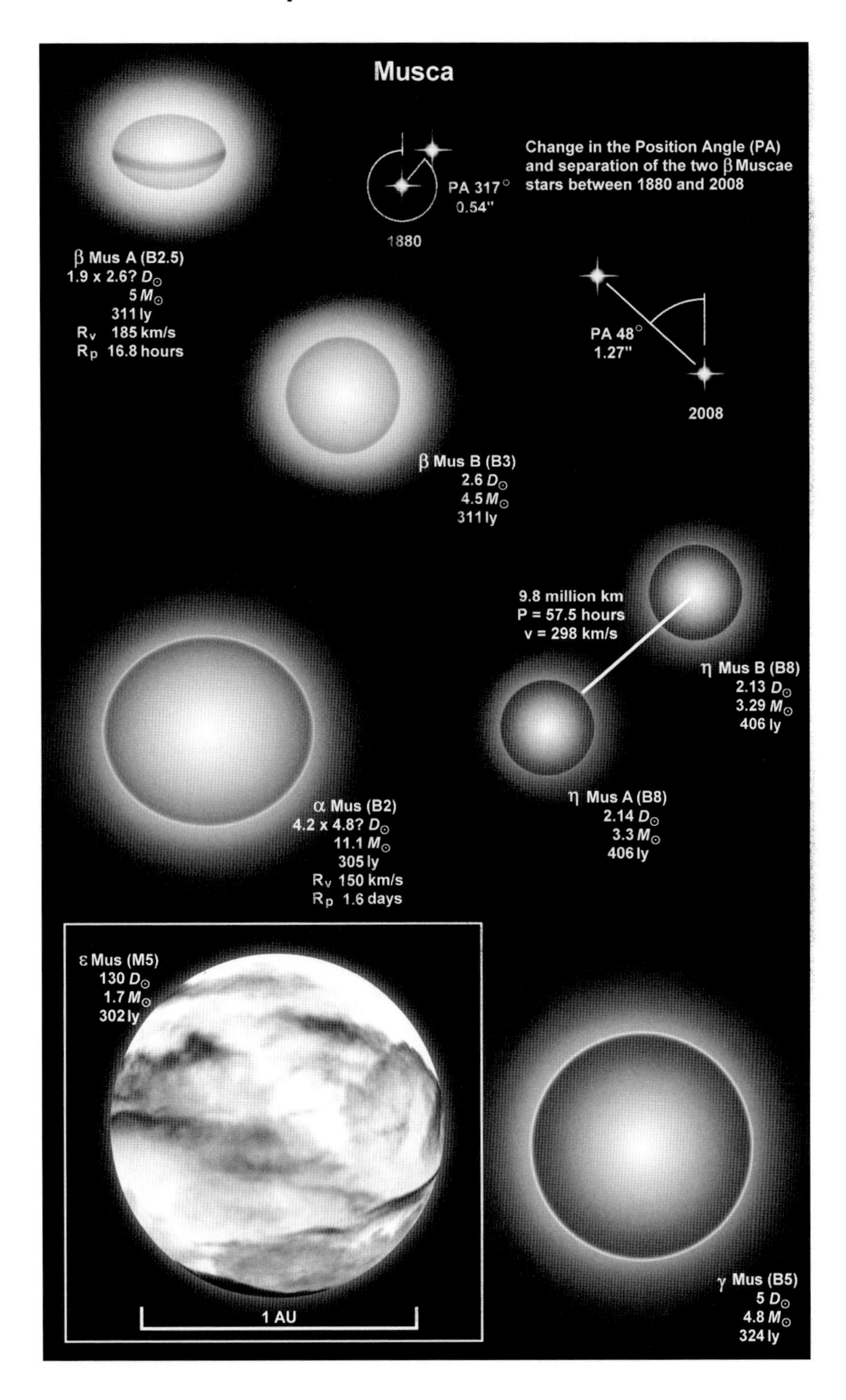

Musca
Change in the Position Angle (PA) and separation of the two β Muscae stars between 1880 and 2008
PA 317°
0.54"
1880
PA 48°
1.27"
2008
β Mus A (B2.5)
1.9 x 2.6? $D_\odot$
5 $M_\odot$
311 ly
R_v 185 km/s
R_p 16.8 hours
β Mus B (B3)
2.6 $D_\odot$
4.5 $M_\odot$
311 ly
9.8 million km
P = 57.5 hours
v = 298 km/s
η Mus B (B8)
2.13 $D_\odot$
3.29 $M_\odot$
406 ly
η Mus A (B8)
2.14 $D_\odot$
3.3 $M_\odot$
406 ly
α Mus (B2)
4.2 x 4.8? $D_\odot$
11.1 $M_\odot$
305 ly
R_v 150 km/s
R_p 1.6 days
ε Mus (M5)
130 $D_\odot$
1.7 $M_\odot$
302 ly
1 AU
γ Mus (B5)
5 $D_\odot$
4.8 $M_\odot$
324 ly

Open and globular clusters in Musca

Name	Size arc min	Size ly	Distance ly	Age million yrs	Brightest star in region*	No. stars m_v >+12*	Apparent magnitude m_v
NGC 4463	5′	5	3,400	32	TYC 8983-2225-1 m_v +8.35	11	+7.2
NGC 4372	18.5′	102	18,900	13,000	Globular cluster		+7.1
NGC 4833	12.7′	78	21,200	13,000	Globular cluster		+7.8

Norma

Constellation:	Norma	**Hemisphere:**	Southern
Translation:	The Level	**Area:**	165 deg^2
Genitive:	Normae	**% of sky:**	0.400%
Abbreviation:	Nor	**Size ranking:**	74th

Another of Abbé La Caille's faint constellations. It depicts a level or set square. When the IAU reorganized the boundaries in 1922, α and β Normae were incorporated into neighboring constellations.

There are two γ Normae stars but they are not related. **γ^1 Normae** is the larger and more distant of the two at 160 $D_\odot$ across and 1,437 ly away. F-class stars of this size are a relative rarity. Of the 350 or so naked-eye F-stars only a dozen are larger than 100 $D_\odot$.The star appears as a feeble m_v +4.98 but with a luminosity of about 1,600 it would brighten to an impressive M_v -6.3 at the standard distance of 10 pc.

γ^2 Normae is almost a whole magnitude brighter than its namesake coming in at m_v +4.01. Astronomers disagree as to its size, some putting it as small as 7 $D_\odot$ while others reckon it could be twice that size: the reality is probably somewhere in between. A G8 it is 128 ly from Earth and rotates at 12 km/s.

γ^2 Nor is one of three G8 stars in the constellation, all of which are very similar in diameter. **η Normae** is 11 $D_\odot$ and 218 ly away while **κ Normae** is 13 $D_\odot$ and 438 ly distant. Neither is particularly unusual although κ Nor is useful for finding a couple of open clusters, one to the north and the other to the south. The northerly one is **NGC 6067**, a group of up to 200 stars at a distance of 4,600 ly. The other is **NGC 6087**, a less compact cluster 3,500 ly away of perhaps a couple of dozen stars centered on **S Normae**, a classic Cepheid that varies between m_v +6.1 and +6.8 with a period of 9.75 days. Various attempts at measuring S Nor's diameter range from 5.3 $D_\odot$ to 8.5 $D_\odot$ with the average being 6.6 $D_\odot$. As the star pulsates it changes between F8 and G0.

ε Normae is a triple system. The primary is a m_v +4.47 B4 of 3.5 $D_\odot$ and has a B9.5 spectroscopic companion in a 3.6 day long orbit. Separated by 22.8″ the secondary is a late B-class, probably a B9, of m_v +7.46 and about half the size and mass. It orbits the primary with a period of about 106,400 years.

ι^1 Normae looks like a m_v +4.68 single star but its magnitude is enhanced by a couple of companions. The two brightest stars are m_v +5.14 and +5.70 with an orbital period of 26.93 years. The main star is an A7 of 1.5 $D_\odot$. In orbit around these two stars is a third dwarf that takes about 4,800 years to complete one orbit.

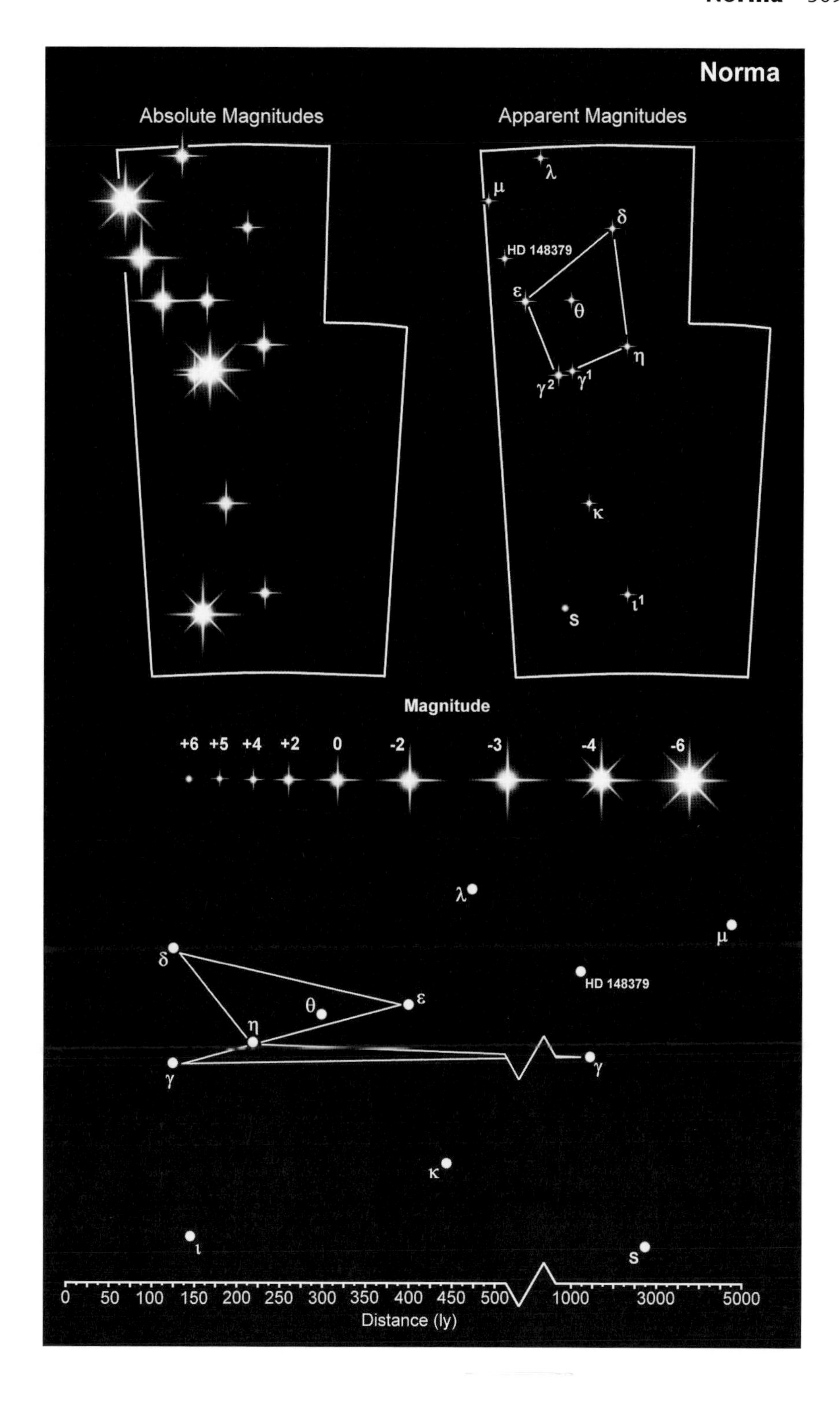

Norma
Absolute Magnitudes
Apparent Magnitudes
λ
μ
δ
HD 148379
ε
θ
η
γ2
γ1
κ
ι1
S
Magnitude
+6 +5 +4 +2 0 -2 -3 -4 -6
λ
μ
δ
HD 148379
θ
ε
η
γ
γ
κ
ι
S
0 50 100 150 200 250 300 350 400 450 500 1000 3000 5000
Distance (ly)

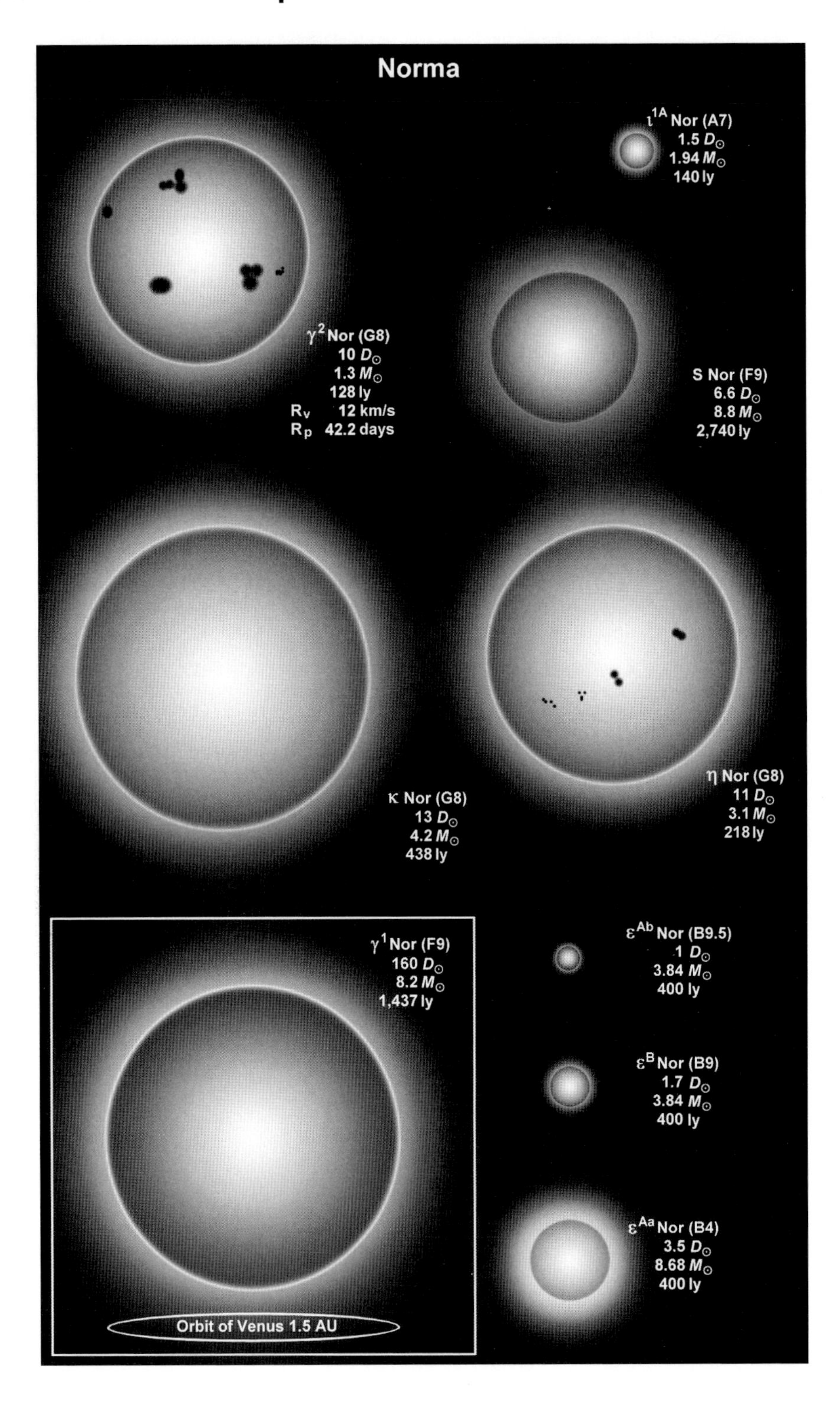
Norma
ι1A Nor (A7)
1.5 D☉
1.94 M☉
140 ly
γ2 Nor (G8)
10 D☉
1.3 M☉
128 ly
Rv 12 km/s
Rp 42.2 days
S Nor (F9)
6.6 D☉
8.8 M☉
2,740 ly
κ Nor (G8)
13 D☉
4.2 M☉
438 ly
η Nor (G8)
11 D☉
3.1 M☉
218 ly
γ1 Nor (F9)
160 D☉
8.2 M☉
1,437 ly
Orbit of Venus 1.5 AU
εAb Nor (B9.5)
1 D☉
3.84 M☉
400 ly
εB Nor (B9)
1.7 D☉
3.84 M☉
400 ly
εAa Nor (B4)
3.5 D☉
8.68 M☉
400 ly

Open clusters in Norma

Name	Size arc min	Size ly	Distance ly	Age million yrs	Brightest star in region*	No. stars m_v >+12*	Apparent magnitude m_v
NGC 6067	13′	17	4,600	119	V340 Nor m_v +8.25	84	+5.6
NGC 6087	108′	91	2,900	95	HD 145782 m_v +5.63	201	+5.4

Octans

Constellation:	Octans	**Hemisphere:**	Southern
Translation:	The Octant	**Area:**	291 deg^2
Genitive:	Octantis	**% of sky:**	0.705%
Abbreviation:	Oct	**Size ranking:**	50th

There seems to have been no end to Abbé de La Caille's talent to create mundane objects out of faint stars. The constellation contains Polaris Australis, the star that is closest to the South Celestial Pole.

Bayer's idea that stars should be designated according to their brightness appears to have been forgotten when celestial cartographers got down to mapping the southern skies. **α Octantis** is the constellation's 5th brightest star, the brightest being....β perhaps? No, of course not. Nor is it γ or δ or ε or....this could take a while. In fact, the brightest star is ν – the 13th letter of the Greek alphabet! α Oct gives the impression it is a single m_v +5.13 star but in fact it is binary of two nearly identical components. The two stars are in a very close orbit that starts out at 0.17 AU (25.4 million km) and then closes in to just 0.08 AU (12 million km). Mercury, for comparison, averages 0.4 AU (59.8 million km) from the Sun. This makes the two stars impossible to separate visually and astronomers have to rely on spectroscopic data. It is likely that the stars are an F6 and an F5 with luminosities of about 7.4 $L_\odot$ each and diameters of 2.4 $D_\odot$. They are heading away from us at some pace, 45 km/s, suggesting they may be strangers to our corner of the Galaxy, and are currently 148 ly from Earth.

β Octantis is indeed the second brightest star at m_v +4.13, a full magnitude brighter than α Oct. It is the same size as the Sun but there the similarity ends. Whereas the Sun has a surface temperature of 6,000 K, β Oct is about 7,600 K and belongs to the A9 spectral group. At 140 ly it is at a similar distance to α Octantis but there is no obvious link between the two.

Octans contains three stars with the designation γ. **γ^1 Octantis** is the brightest at m_v +5.11 and 49 $L_\odot$. A giant of 11 $D_\odot$ it is a G7 and lies somewhere between 257 and 277 ly away. **γ^3 Octantis** is the next brightest at m_v +5.29 and is also a giant, this time a G8 of 13 $D_\odot$ and 34.3 $L_\odot$. It lies between 233 and 251 ly from Earth. The faintest is the m_v +5.73 **γ^2 Octantis**, an 11 $D_\odot$ K0 lying between 300 and 330 ly distant. Although there are clear similarities between the three stars they are not thought to be associated.

The red giant **ε Octantis** is a semi-regular SRb variable with a period of around 50 to 55 days during which its magnitude dips from a maximum of m_v +4.58 to a minimum of +5.30. It is 38 $D_\odot$ across and has a luminosity of 46 $L_\odot$. From a distance of 268 ly it appears as a 5th magnitude star but at 10 pc it would have an absolute magnitude of M_v -1.0.

λ Octantis is a binary star of m_v +5.47 and +7.17 at a distance of 436 ly. The brighter component is a G9 giant of 10 $D_\odot$ accompanied by an A3 dwarf of 1.3 $D_\odot$.

ν Octantis is, strangely enough, the brightest star in the constellation at m_v

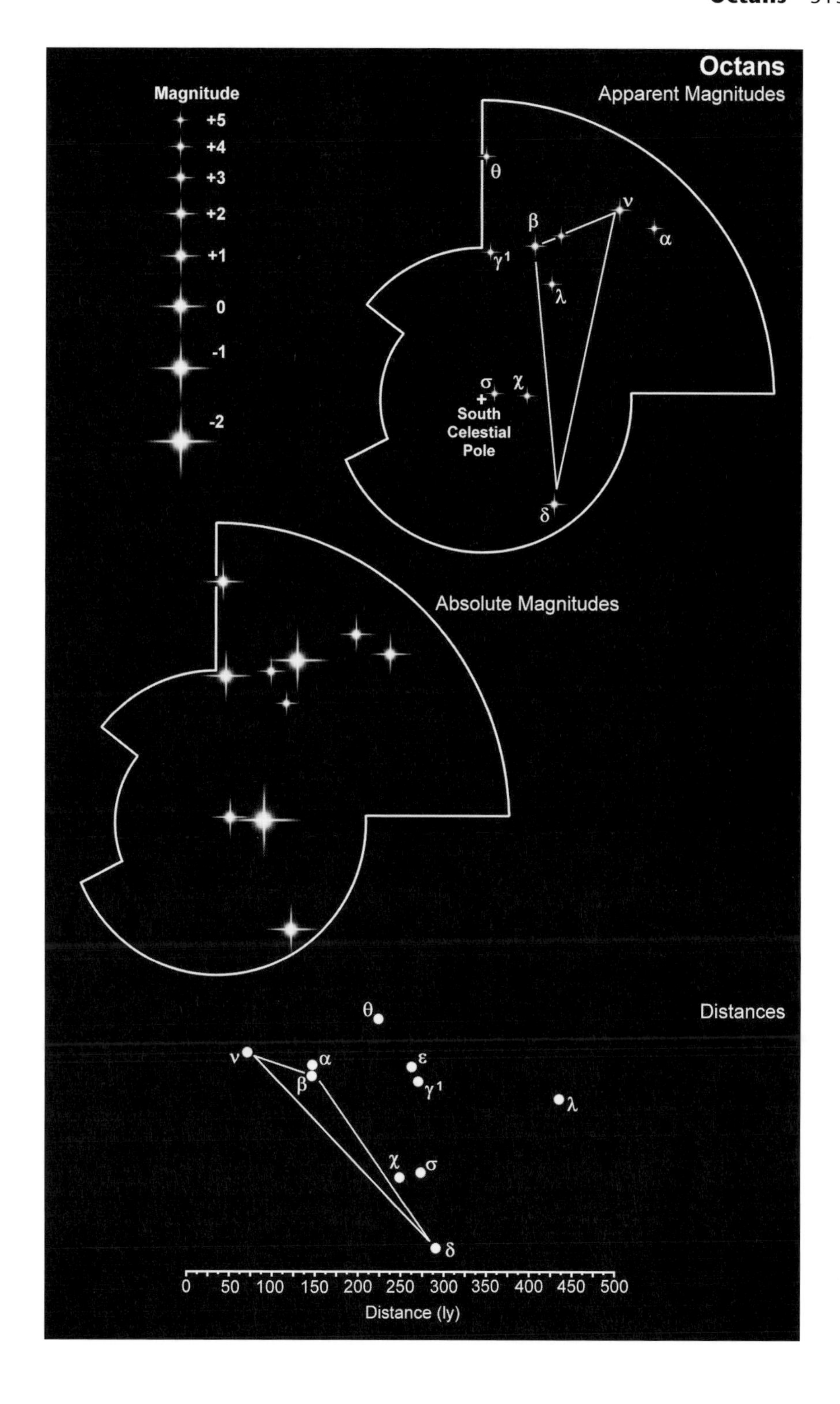
Octans
Apparent Magnitudes
Magnitude
+5
+4
+3
+2
+1
0
-1
-2
θ
ν
β
α
γ1
λ
σ
χ
South
Celestial
Pole
δ
Absolute Magnitudes
Distances
θ
ν
α
β
ε
γ1
λ
χ
σ
δ
0
50
100
150
200
250
300
350
400
450
500
Distance (ly)

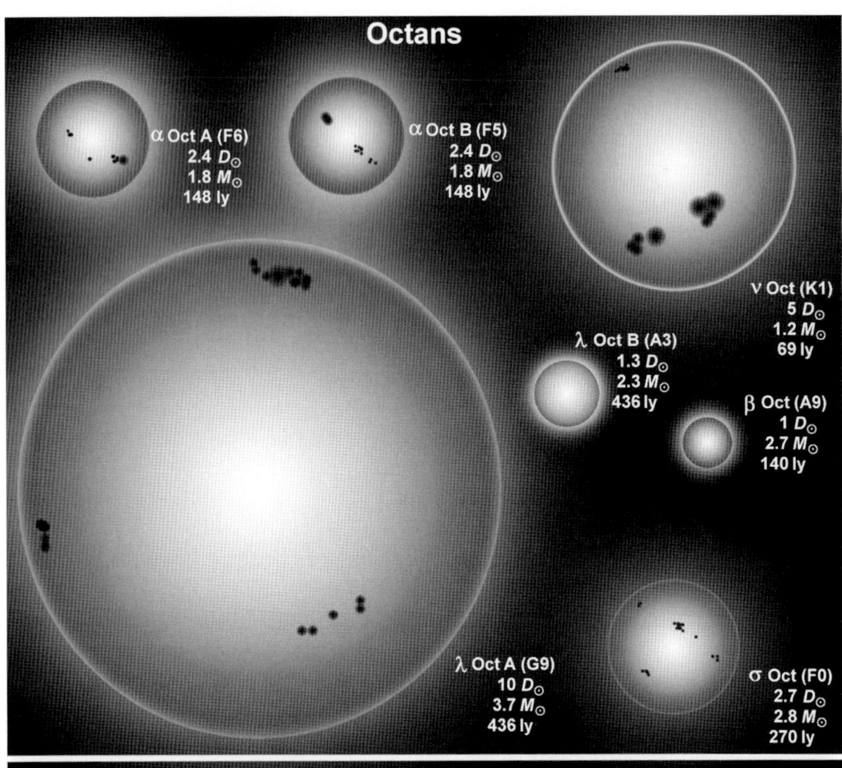
Octans
α Oct A (F6)
2.4 $D_\odot$
1.8 $M_\odot$
148 ly
α Oct B (F5)
2.4 $D_\odot$
1.8 $M_\odot$
148 ly
ν Oct (K1)
5 $D_\odot$
1.2 $M_\odot$
69 ly
λ Oct B (A3)
1.3 $D_\odot$
2.3 $M_\odot$
436 ly
β Oct (A9)
1 $D_\odot$
2.7 $M_\odot$
140 ly
λ Oct A (G9)
10 $D_\odot$
3.7 $M_\odot$
436 ly
σ Oct (F0)
2.7 $D_\odot$
2.8 $M_\odot$
270 ly

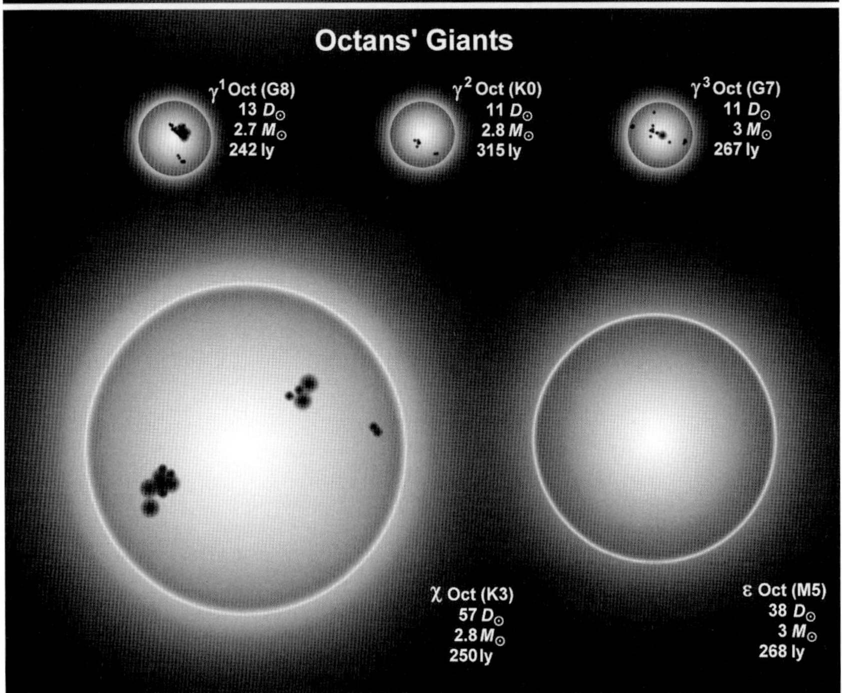
Octans' Giants
γ^1 Oct (G8)
13 $D_\odot$
2.7 $M_\odot$
242 ly
γ^2 Oct (K0)
11 $D_\odot$
2.8 $M_\odot$
315 ly
γ^3 Oct (G7)
11 $D_\odot$
3 $M_\odot$
267 ly
χ Oct (K3)
57 $D_\odot$
2.8 $M_\odot$
250 ly
ε Oct (M5)
38 $D_\odot$
3 $M_\odot$
268 ly

+3.73. One of many K1 class stars it has a diameter of 5 $D_{\odot}$, a mass of about 1.2 $M_{\odot}$ and is 12 times more luminous than the Sun. It is also the closest star in the constellation to us at 69 ly and has a spectroscopic binary about which virtually nothing is known except that the orbital period is 2.84 years and the pair are separated by 1-2 AU. ν **Oct** is a member of the Wolf 630 moving group. This group consists of as many as 200 stars that have similar physical and orbital characteristics.

σ **Octantis** is sometimes referred to as Polaris Australis or the Southern Pole Star. However it is not as bright (m_v +5.45 compared with Polaris' m_v +1.98), as close to the pole (1.05° vs. 0.75°) or as big (2.7 $D_{\odot}$ vs. 67 $D_{\odot}$) as its northern counterpart. At 270 ly it is a lot closer than Polaris, which is 431 ly away, but at 37 $L_{\odot}$ it is just 1.6% as bright at the North Pole Star. Like Polaris though it is an F-type star though at opposite ends of the class (F0, while Polaris is an F7) and it is variable but belongs to the δ Scuti brigade rather than being a Cepheid. Its magnitude varies between m_v +5.45 and +5.50 with a period of 2^h 19.7^m. In fact, σ Oct is *the* Mr Average of naked eye δ Scuti variables as its amplitude, variable period and diameter are all spot on average for the type while its spectral class is the most common among this sort of variable.

The largest star in the constellation is χ **Octantis**, a 57 $D_{\odot}$ K3 which lies at a distance of 250 ly. It also has the brightest absolute magnitude of M_v -2.3.

Ophiuchus

Constellation:	Ophiuchus	**Hemisphere:**	Equatorial
Translation:	The Serpent Holder	**Area:**	948 deg^2
Genitive:	Ophiuchi	**% of sky:**	2.298%
Abbreviation:	Oph	**Size ranking:**	11th

Ophiuchus is sometimes referred to as the 13th Zodiacal sign as the Ecliptic passes through the southernmost part of the constellation. The Sun actually enters Ophiuchus on 29 November and leaves on 18 December. In mythology Ophiuchus is associated with Asclepius, the Greek god of medicine who traditionally holds a snake – the neighboring constellation of Serpens which it divides. The serpent is still used today in medical circles usually occurring as a pair entwined around a herald's staff called a caduceus.

α **Ophiuchi** has the proper name of Ras Alhague meaning 'Head of the Snake Collector'. It is an A5 of 2.8 $D_{\odot}$ and rotates at an average of 219 km/s taking 15.5 hours to complete one full turn. This can vary however between 210 and 228 km/s so a single rotation can last from 14.9 to 16.2 hours (compared to the Sun's 25 days). It is one of the closer stars in the constellation at 46.7 ly – there are a couple of others that are half as close again – and is an astrometric binary. Very little is known about its dwarf companion other than it is in an 8.7 year long orbit and the two are separated by an average of 7 AU. α Oph is suspected of being a δ Scuti type variable although the jury is still out on that one.

At first glance **β Ophiuchi** or Cheleb looks like a common or garden K2 giant traveling through space towards us at 12 km/s and currently 82 ly away, but it seems to have a number of variable periods. The longest lasts 142 days during which its magnitude varies ever so slightly. Current thinking is that the variation is due to a large single or large groups of star spots. If the star is 18 $D_{\odot}$ across then this suggests a rotational velocity of 6.4 km/s. A secondary fluctuation with a period of 13.1 days may be due to the star wobbling like gelatin.

δ and ε Ophiuchi are linked by name – Yed Prior and Yed Posterior – although they have nothing to do with one another other than appearing in the same small segment of the celestial sphere. **δ Ophiuchi** is an M0.5 red giant of 54 $D_{\odot}$ and about 650 $L_{\odot}$ and lies at a distance of 170 ly. In Solar System terms it would swallow up Mercury. **ε Ophiuchi** is much closer at 108 ly and is a G9.5 yellowish-orange giant of 12 $D_{\odot}$.

Despite its designation **ζ Ophiuchi** is the third brightest star in the constellation at m_V +2.56. What cannot be seen with the naked eye is a huge nebulosity that surrounds the star. ζ Oph is a Be emission star that actually belongs to spectral group O9. Some 458 ly away it is 68,000 times more luminous than the Sun. Estimates of its rotational velocity range from 340 to 379 km/s putting it in the top 0.4% of fast spinners and signaling that the star is inherently unstable. It probably measures 8 $D_{\odot}$ pole to pole while the equatorial diameter

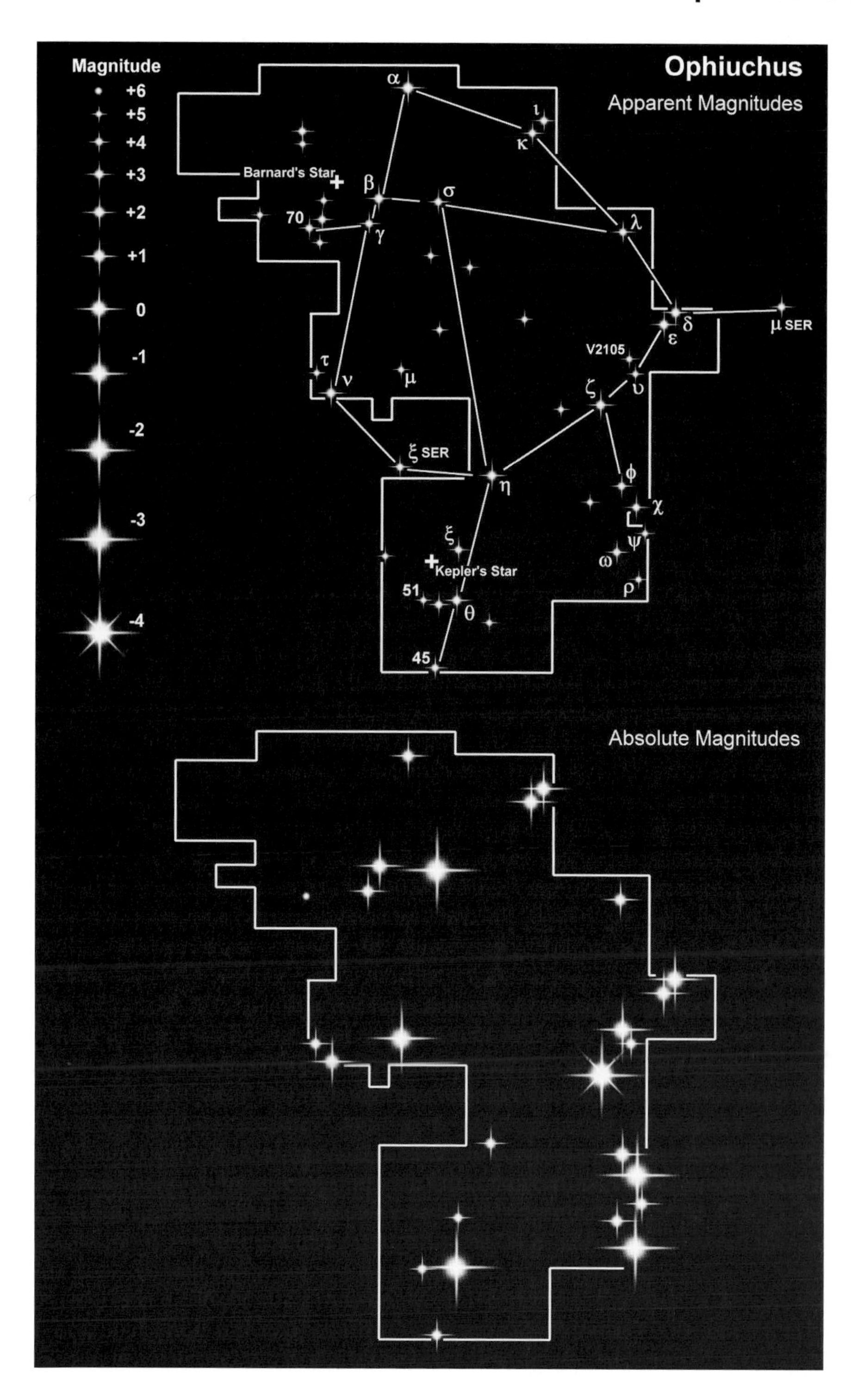
Magnitude
+6
+5
+4
+3
+2
+1
0
-1
-2
-3
-4
Ophiuchus
Apparent Magnitudes
α
ι
κ
Barnard's Star
β
σ
70
γ
λ
δ
ε
μ SER
V2105
τ
ν
μ
ζ
υ
ξ SER
η
φ
χ
ψ
ω
ξ
Kepler's Star
ρ
51
θ
45
Absolute Magnitudes

could be as much as 24 $D_{\odot}$. Its instability manifests itself as slight variations in brightest, the magnitude flickering between m_v +2.56 and +2.58 putting the star in the γ Cas eruptive variable class.

The eccentricity of an orbit – effectively how non circular it is – can go from a value of 0, indicating it is perfectly circular, to almost 1 in which case it is a very long narrow ellipse. Beyond 1 the orbit becomes an open ended curve called a parabolic 'orbit' or, in extreme cases, a hyperbolic 'orbit'. The Solar System's planets have near-circular orbits while comets tend to have highly elliptical orbits, spending most of their lives in the cold outer reaches of the Solar System and only briefly passing close to the Sun. Comets in parabolic or hyperbolic 'orbits' pay only one visit to the Sun before being ejected from the Solar System. The 88 year long orbit of the binary system η **Ophiuchi** is very much comet-like with an eccentricity of e = 0.94. In practice this means that the two stars come as close together as 2 AU but then separate by up to 65 AU (300 million km to 9,724 million km). The primary is a m_v +3.2 A2 with a diameter and mass of a little more than two Suns and a luminosity of 35 $L_{\odot}$. The secondary component is an A3, a bit smaller – about twice the diameter and mass of the Sun – but with a much lower luminosity of 21 $L_{\odot}$. The pair are 84.1 ly from Earth so appear as a single star of m_v +2.47 making it the second brightest star in the constellation.

θ Ophiuchi is interesting not just because it is a pulsating β Cepheid variable but because no one really knows how many stars exist in the system. It is definitely an astrometric binary but there could be at least one more star, and possibly two lurking around. The main component is a B2 of 7.3 $D_{\odot}$ and 11,500 $L_{\odot}$. Its magnitude varies between m_v +3.25 and +3.31 with a period of 3^h 22.4^m. The whole system, if indeed it is a system, lies at a distance of 563 $\pm$ 60 ly.

κ Ophiuchi is a giant pulsating K2 variable, probably an Lb, with a diameter of 18 $D_{\odot}$ and a luminosity of about 30 $L_{\odot}$. It is a high velocity star, closing in on us at 56.5 km/s. The Germany astronomer August Kopff (1882-1960) is credited with the discovery of its variability. He considered it to be irregular with an amplitude of half a magnitude. His observations were doubted by most of the astronomical community who could not verify the variability. The first issue of *The Variable Star Observer* in July 1991 carried an article by Tristram Brelstaff in which he suggested that there was confusion between κ Oph and **χ Ophiuchi**, a 14 $D_{\odot}$ B2 which is known to be variable between m_v +4.18 and +5.00. Although the two stars are at opposite ends of the constellation Brelstaff pointed out that, when written by hand, κ and χ can look very similar. He even cites a documented case of this happening and reported in a 1948 issue of *Popular Astronomy*. The 17th Edition of *Norton's Star Atlas* (1973) did not consider the star to be variable: the current 20th Edition does. Today κ Ophiuchi's variability is taken as read, its brightness fluctuating between m_v +4.18 and +5.00, nearly a full magnitude. It is possible that the star was constant for a while and has since returned to its variability cycle, but is there also a possibility that the star is actually constant and that there is a confusion of identities?

λ Ophiuchi is a binary system of two A-class stars. The brightest at m_v +4.17 is an A0 of 2.5 $D_{\odot}$ and 2.6 $M_{\odot}$. Its companion is a bit smaller at 2.0 $D_{\odot}$ and 2.22

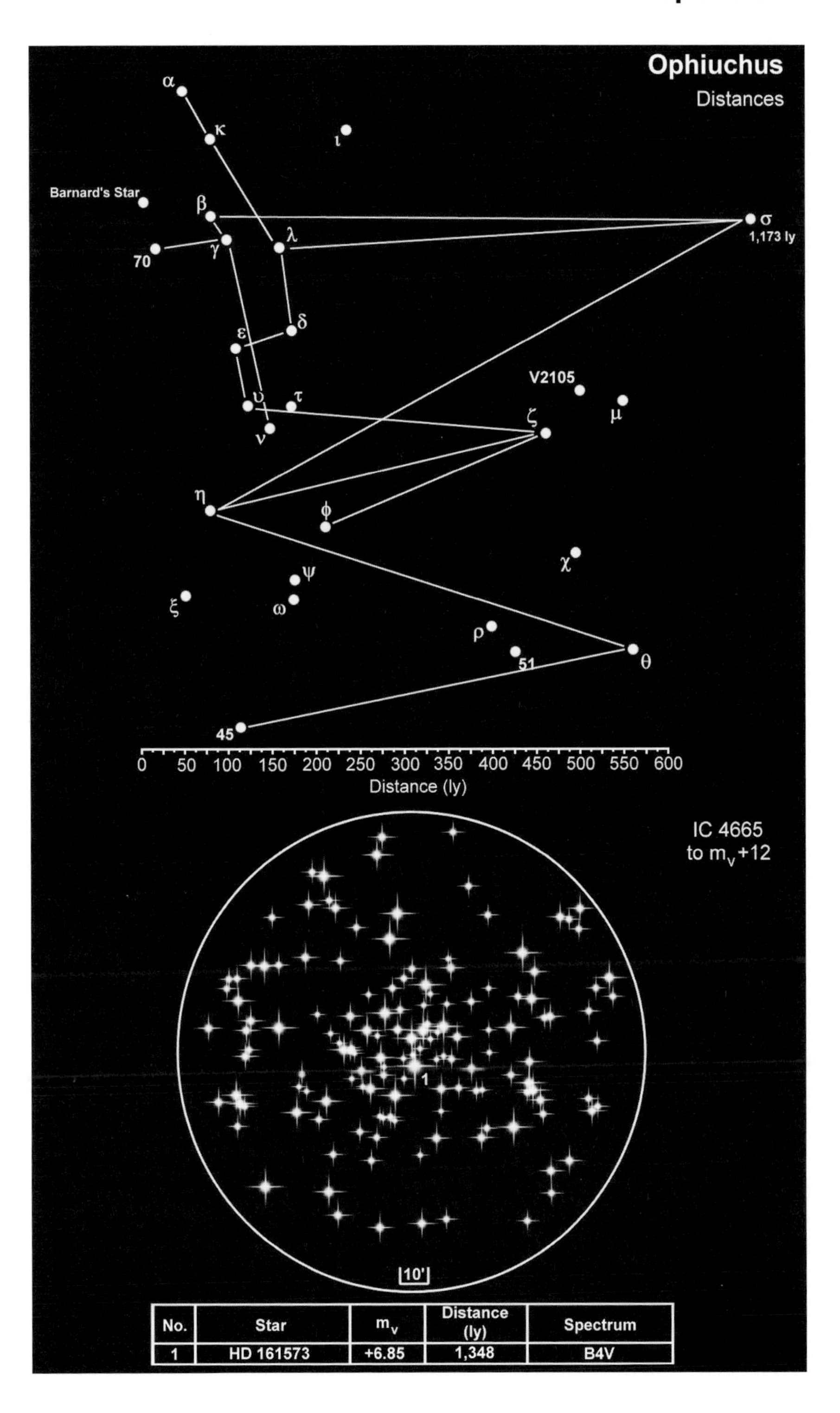

No.	Star	m_v	Distance (ly)	Spectrum
1	HD 161573	+6.85	1,348	B4V

$M_{\odot}$ and is an A4 of m_V +5.21. The pair orbit their common center of mass once every 129.87 years during which their distance varies between 18 and 68 AU. A third star of m_V +10.7 about 120″ south may be part of the system though this is doubtful.

A small telescope will reveal ρ **Ophiuchi** to be a binary: a pair of hot B2-class stars locked in a gravitational embrace that sees them orbit their barycenter once every 2,000 years or so. Just 3″ separates them on the celestial sphere but they are in reality about 400 AU apart and lie at a distance of 394 ly. They are also embedded in a thick dust cloud that is giving birth to new stars: most are just a few hundred thousand years old or less. Without the cloud absorbing much of the star's light, ρ Oph would be a 3rd magnitude object. Instead it shines at just m_V +5.05.

A small, hot A-class star in Ophiuchus has one of the most compact dust shells so far detected, suggesting a young star in the late stages of planet formation. **51 Ophiuchi** is just 0.8 $D_{\odot}$ and lies at a distance of 426 ly. The compact disk around the A0 star extends out to about 4 AU and is 100,000 times as dense as the Solar System's Zodiacal Cloud. It is thought that this indicates the disk is still very young and contains numerous asteroid and comet-like bodies that frequently collide, adding debris to the cloud. Beyond the inner disk is another which billows outwards, extending to about 1,200 AU. This outer disk contains much smaller dust grains, about the size of smoke particles, that are blown away from the star by the relentless pressure of radiation. So far no planets have been discovered around 51 Oph which is rotating at 228 km/s, taking just 4.3 hours to complete one rotation.

70 Ophiuchi is a well known binary consisting of two dwarf K-class stars. **70^A Ophiuchi** (or 70 Ophiuchi A, if you prefer) is the brighter of the two at m_V +4.22. Only 85% the size of the Sun and with 89% of its mass it has a rotational velocity of 16 km/s and completes one turn every 2.7 days. Its companion, **70^B Ophiuchi,** nearly two magnitudes fainter at m_V +5.91, is a K5 and smaller and less massive at 0.70 $D_{\odot}$ and 0.71 $M_{\odot}$. When at periastron the two are 11.6 AU apart, translating to 1.7″ on the celestial sphere but this widens to 6.7″ at apastron when the two stars are 34.8 AU apart. The orbital period is 88.4 years and, because the stars are only 16.6 ly from Earth, it is one of the most thoroughly studied binary systems. Even so, observers cannot agree on what color the stars are with yellow and red, gold and violet, pale topaz and violet, bright yellow and orange and gold and rusty orange all being reported.

The variable star **V2105 Ophiuchi** has one of the greatest radial velocities of all stars, heading away from us at 99.3 km/s and putting it in the top ten of fast naked-eye stars. An M3 red giant of 38 $D_{\odot}$ it is a semi-regular, SRb, variable that oscillates between m_V +5.00 and +5.38. It is currently 507 ly away.

Ophiuchus is home to one of the most famous of all stars, **Barnard's Star (V2500 Ophiuchi).** Discovered in 1916 by Edward E. Barnard (1857-1923), a Nashville, Tennessee born astronomer who worked at the Yerkes Observatory in William's Bay, Wisconsin, it is the second closest star to us lying just 5.98 ly away (that is if you count the entire α Centuari system as the closest). At m_V +9.54 it is

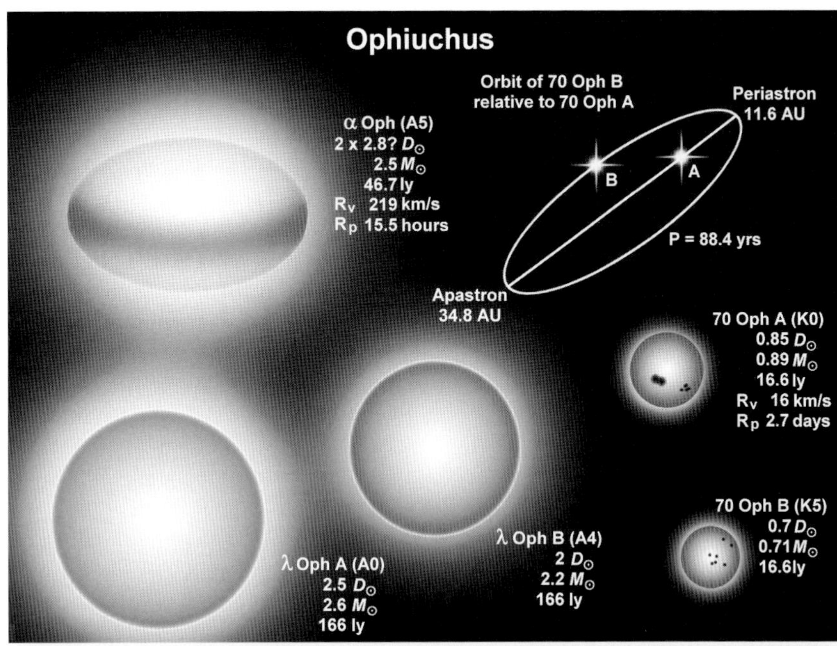
Ophiuchus
Orbit of 70 Oph B
relative to 70 Oph A
Periastron
11.6 AU
α Oph (A5)
2 x 2.8? D☉
2.5 M☉
46.7 ly
Rv 219 km/s
Rp 15.5 hours
B
A
P = 88.4 yrs
Apastron
34.8 AU
70 Oph A (K0)
0.85 D☉
0.89 M☉
16.6 ly
Rv 16 km/s
Rp 2.7 days
70 Oph B (K5)
0.7 D☉
0.71 M☉
16.6ly
λ Oph B (A4)
2 D☉
2.2 M☉
166 ly
λ Oph A (A0)
2.5 D☉
2.6 M☉
166 ly

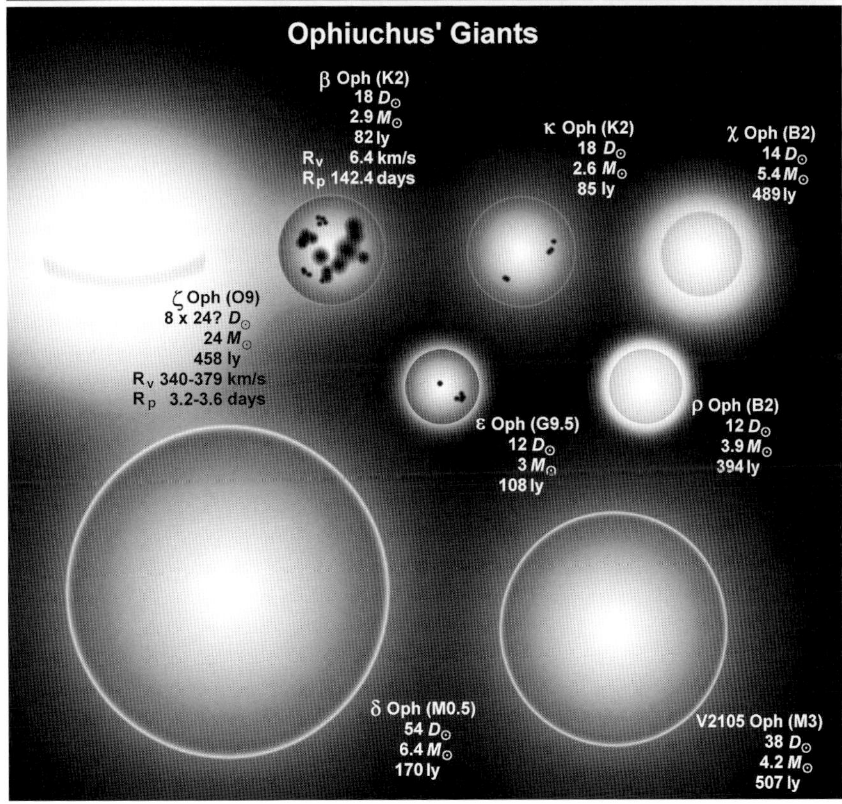
Ophiuchus' Giants
β Oph (K2)
18 D☉
2.9 M☉
82 ly
Rv 6.4 km/s
Rp 142.4 days
κ Oph (K2)
18 D☉
2.6 M☉
85 ly
χ Oph (B2)
14 D☉
5.4 M☉
489 ly
ζ Oph (O9)
8 x 24? D☉
24 M☉
458 ly
Rv 340-379 km/s
Rp 3.2-3.6 days
ε Oph (G9.5)
12 D☉
3 M☉
108 ly
ρ Oph (B2)
12 D☉
3.9 M☉
394 ly
δ Oph (M0.5)
54 D☉
6.4 M☉
170 ly
V2105 Oph (M3)
38 D☉
4.2 M☉
507 ly

well below the naked-eye limit, testament to the fact that this tiny star, just 0.15 to 0.20 $D_{\odot}$ and 0.17 $M_{\odot}$, is a faint red dwarf belonging to spectral group M4. It is also one of the oldest stars in our neighborhood at 10,000 to 12,000 million years; nearly as old as the Galaxy itself (13,200 million years). Formed in an era when the Galaxy as a whole was metal-poor, Barnard's Star has just 10% the metallicity of the Sun. It is just visiting our part of the Galaxy though, its high velocity of 139 km/s sweeping it across the sky by 0.5° in a human lifetime and indicating that it originated in the galactic halo.

A little over 400 years ago, on 9 October 1604, a bright new star was seen just to the east of ξ Ophiuchi. Although first noticed by observers in Italy it came to be called **Kepler's Star (SN 1604)** after Johannes Kepler who described its appearance in detail. This supernova remained visible for 18 months and lies no more than 20,000 ly from the Sun.

Several stars in Ophiuchus harbor planets including a 2.7 $D_{\oplus}$ super-Earth, GJ 1214 b, which is 6.5 times as massive as our own planet, and a planet that is approaching the limit for a brown dwarf, HD 156846 b.

Ophiuchus contains a number of clusters. The open star cluster **IC 4665** was first noticed by the Swiss astronomer Philippe Loys de Chéseaux (1718-51) who was more famous for his discovery of a 7-tailed comet. It's an easy target in a binocular or wide field telescope and consists of at least 35 stars It lies at a distance of 1,400 ly and is estimated to be 36 million years old. de Chéseaux also discovered **NGC 6633**, an open cluster about the size of a full Moon and containing 30 stars set at a distance of 1,040 ly. It is believed to be much older at 660 million years.

Messier 9 is just about visible in very dark skies having an apparent magnitude of m_v +7.7. To the eye this open globular looks as though it is 3′ to 4′ across but photographs reveal it to be at least 12′ across. At a distance of 25,800 ly – only 5,500 ly from the galactic center – its angular size translates into a diameter of 90 ly. M9 is moving away from us at a staggering 224 km/s.

Messier 10 is a brighter, larger and easier globular cluster to find at m_v +6.6. Some 20′ across it has a diameter of 83 ly and is heading away from us at 69 km/s. It is currently 14,300 ly away.

Slightly smaller at 75 ly and farther away at 16,000 ly **Messier 12** could otherwise be the twin of M10. It is about 16′ across and is m_v +6.7.

Messier 14, another globular cluster, is slightly elliptical. It is home to more than 70 variables, an unusually high number, and lies 38,000 ly from the Sun. It is about 100 ly across at its widest point.

Messier 19's proximity to the galactic center, just 1,900 ly away, has distorted the globular cluster into an oblate spheroid, approximately 140 by 70 ly. It lies 31,300 ly from us.

Messier 62 is somewhat farther from the galactic center at 6,200 ly but even so tidal forces have resulted in a very irregular shape about 98 ly across.

An estimated 79 ly across **Messier 107** is 20,900 ly from Earth and appears as a m_v +7.9 object.

Planetary systems in Ophiuchus

Star	$D_{\odot}$	Spectral class	ly	m_v	Planet	Minimum mass	q	Q	P
HD 148427	3.22	K0	193	+6.89	HD 148427 b	0.96 M_J	0.781	1.079	332 d
HD 149143	1.2?	G0	205	+7.9	HD 149143 b	1.33 M_J	0.052	0.054	4.07 d
HD 156846	1.4?	G0	160	+6.51	HD 156846 b	10.45 M_J	0.15	1.83	360 d
HD 170469	1.22	G5	212	+8.21	HD 170469 b	0.67 M_J	1.99	2.49	3.14 y
HD 171028	1.95	G0	294	+8.31	HD 171028 b	1.83 M_J	0.50	2.08	1.47 y
CoRoT-6	1.025	F5	45.3	+13.9	CoRoT-6 b	2.96 M_J	0.078	0.093	8.89 d
GJ 1214	0.211	M4.5	42.4	+14.67	GJ 1214	6.5 $M_{\oplus}$	0.010	0.018	1.58 d

Open and globular clusters in Ophiuchus

Name	Size arc min	Size ly	Distance ly	Age million yrs	Brightest star in region*	No. stars m_v >+12*	Apparent magnitude m_v
IC 4665	117′	39	1,150	45	HD 161573 m_v +6.85	149	+4.2
NGC 6633	71′	25	1,200	425	HD 170200 m_v +5.73	144	+4.6
M9 (NGC 6333)	12′	90	25,800		Globular cluster		+7.7
M10 (NGC 6254)	20′	83	14,300		Globular cluster		+6.6
M12 (NGC 6218)	16′	75	16,000		Globular cluster		+6.7
M14 (NGC 6402)	9′	100	38,000		Globular cluster		+7.6
M19 (NGC 6273)	17′	155	31,300		Globular cluster		+6.8
M62 (NGC 6266)	15′	98	22,500		Globular cluster		+6.5
M107 (NGC 6171)	13′	79	20,900		Globular cluster		++7.9

Orion

Constellation:	Orion	**Hemisphere:**	Equatorial
Translation:	Orion the Hunter	**Area:**	594 deg^2
Genitive:	Orionis	**% of sky:**	1.440%
Abbreviation:	Ori	**Size ranking:**	26th

One of the most magnificent and best known constellations in the entire sky Orion's equatorial position makes it a favorite of astronomers in both hemispheres. Homer mentions Orion in the XIth Book of *The Odyssey* as the lover of Aurora and he is also said to be chasing the Pleiades nymphs across the heavens, but he is also a great hunter, accompanied on his never ending journey by his loyal dogs, Canis Major and Canis Minor, who are often depicted as chasing Lepus the hare.

It is one of the brightest and most widely studied stars and most people - even those who have no interest in astronomy - know it by its proper name of Betelgeuse. Yet we know very little of any substance about α **Orionis**. We do know, or at least we think we know, that it belongs to spectral class M2, that its magnitude is variable and that it has a surface temperature of 3,650 K. We also know that it is big, but exactly how big is almost anyone's guess. Estimates range from a 'mere' 230 $D_\odot$ (2.1 AU) to an enormous 1,500 $D_\odot$ (14 AU - about half the distance between Jupiter's and Saturn's orbits). It depends on what method is used to measure the star with infrared observations resulting in larger estimates. The problem is compounded by the fact that the star is embedded in multiple dust and gas shells stretching out to 20,000 AU, the result of a lifetime of ejecting vast amounts of material into space. In all Betelgeuse has lost an entire solar mass in this way. Estimates of its distance are no better ranging from 427 ly to 640 ly, and its luminosity could be anywhere between 85,000 and 105,000 $L_\odot$. Whatever method is used to measure α Ori it does appear to be shrinking by 0.75% per year, although no one knows if this trend will continue, cease or go into reverse. Betelgeuse appears to be oval and has a number of hot spots on its surface. It belongs to the SRc class of semi-regular variables, changing from m_v +0.50 to +1.30 and back with a period of 2,335 days (6.4 years) although there are other underlying periods of between 200 and 400 days. While our knowledge of Betelgeuse is sketchy, to say the least, we know even less about its two companions. Discovered in 1985 using speckle interferometry the closest companion averages 5 AU from Betelgeuse and takes 2 years to complete an orbit. The other averages 40 to 50 AU and its period is thus far unknown.

Diagonally opposite Betelgeuse is another well known star, Rigel. Despite being given the β **Orionis** label it is usually the brightest star in the constellation although α Orionis' variability may sometimes relegate it to second place. At m_v +0.12 Rigel is therefore generally considered to be the 7th brightest star in the entire sky. Place Rigel and Betelgeuse at the same distance of 10 pc from Earth however and the difference between the two stars would immediately become apparent. Rigel, a hot B8 supergiant of 11,500 K and 35,170 $L_\odot$, would sparkle at

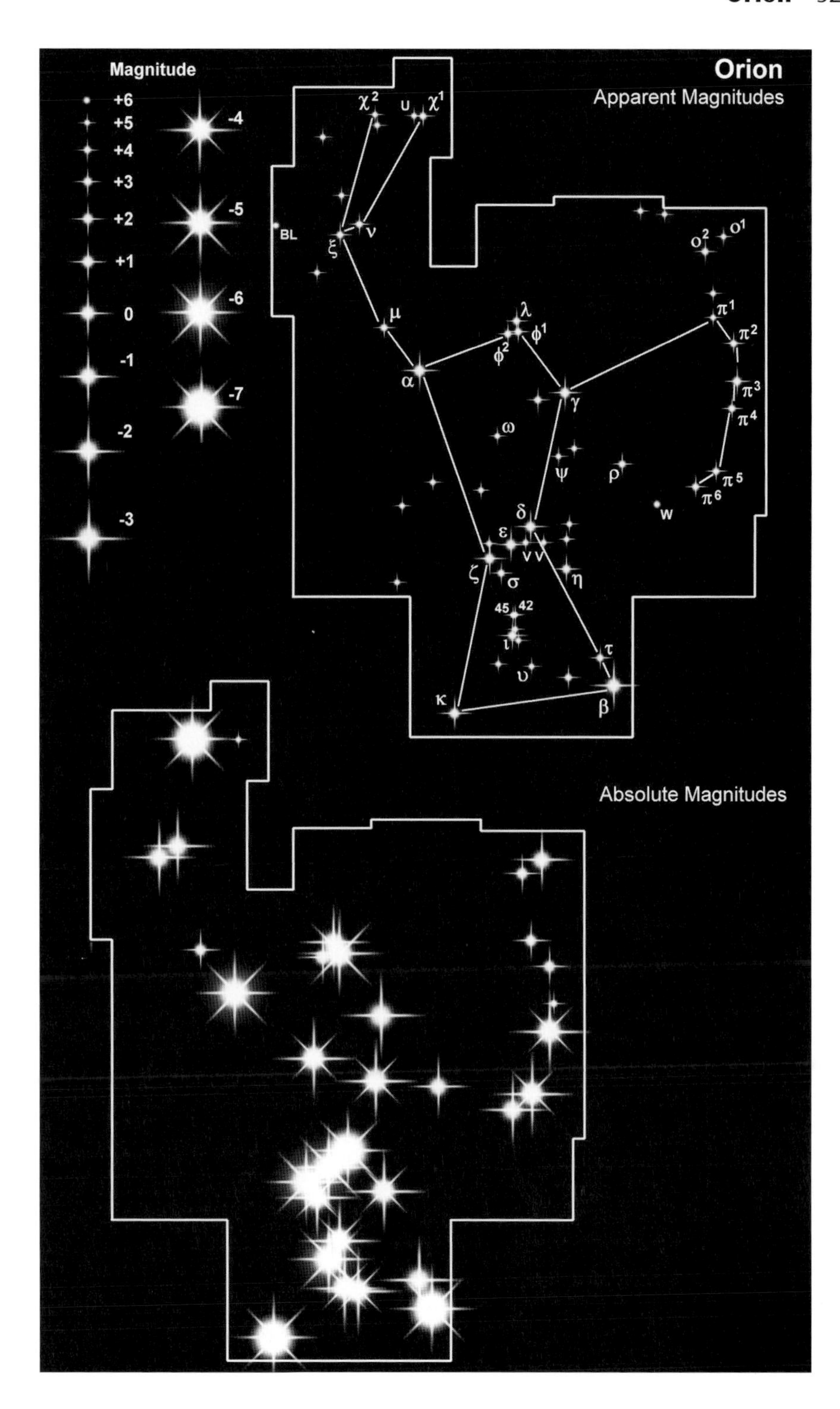
Orion
Apparent Magnitudes
Magnitude
+6
+5
+4
+3
+2
+1
0
-1
-2
-3
-4
-5
-6
-7
χ²
U
χ¹
ν
BL
ξ
μ
λ
ϕ¹
ϕ²
α
γ
ω
ψ
ρ
o²
o¹
π¹
π²
π³
π⁴
π⁵
π⁶
W
δ
ε
V V
ζ
σ
η
45
42
ι
τ
υ
κ
β
Absolute Magnitudes

a brilliant blue-white M_V -7.1 while Betelgeuse would glow a distinct red at M_V -5.6, still outshining Venus (m_V -4.6) but noticeably dimmer than the luminary Rigel. At least 58 $D_\odot$ (0.5 AU) and possibly as big as 74 $D_\odot$ (0.7 AU) across Rigel is much farther away than Betelgeuse at 773 ly. If Rigel is on its way to becoming a red supergiant for the first time then its mass is likely to be in the order of 17 $M_\odot$, but if it has already been there, done that and returned as a blue-white supergiant then its mass will be around 13 to 14 $M_\odot$. Rotating at 36 km/s it will complete one revolution in 81.5 to 104 days depending on its true diameter. Rigel is also a multiple star system. It has a B9 visual companion which is m_V +6.5 and is separated from Rigel by 9.5″, so it should be easy to find in a small telescope. Except that it isn't. Rigel's sheer brilliance hides its faint companion and is a challenge in smaller telescopes. In real space the two are separated by at least 2,600 AU and have an orbital period of around 18,600 years. Completely hidden is the companion's companion: another B9 somewhat less massive – 2.94 $M_\odot$ compared to 3.84 $M_\odot$ – and in a 9.86 day long orbit. A fourth star, again a B9 is in a 46.9 year orbit around the m_V +6.50 star.

About 10′ to the west of Rigel is the bright reflection nebula **IC 2118,** better known as the Witch's Head Nebula.

Compared to Rigel and Betelgeuse, Bellatrix seems positively boring. The proper name of γ **Orionis** means 'female warrior' and it is therefore sometimes called 'The Amazon Star'. Belonging to the B2 spectral group it is a hot, 21,500 K, giant of 8.1 $D_\odot$ and perhaps 8 $M_\odot$. Its 6,400 $L_\odot$ shine at m_V +1.64 from a distance of 243 ly but the star would have an absolute magnitude of M_V -2.7 at 10 pc. It is suspected of being micro-variable by less than 3/100th of a magnitude. Its rotational velocity is 55 km/s, on the slow side for a B2 class which average 142 km/s, so it takes just over a week to complete a single revolution.

The last corner of the great rectangle that, for many, easily identifies Orion is marked by κ **Orionis** or Saiph. Another B-class star, though only just at B0.5, the 65,000 $L_\odot$ pouring out at a distance of 722 ly, most of it ultra-violet, gives rise to a modest m_V +2.05. Once again though its magnitude would jump to -7 at 10 pc. Believed to have a diameter of 38.2 $D_\odot$ it rotates at a modest 65 km/s taking 29.7 days to spin once on its axis.

The three stars that make up Orion's Belt are similar in many respects. δ **Orionis** or Mintaka, the most westerly of the three, is the dimmest fluctuating between m_V +2.14 and +2.26 with a period of 5^d 17^h 34.8^m. The variability is due to a companion, the two stars eclipsing one another and therefore belonging to the EA category of variables. The primary, which is usually referred to as δ^A **Orionis,** is an O9.5 of 13 $D_\odot$ and lies at a distance of 916 ly. If you are wondering where δ^B **Orionis** is, it is hugging δ^A Ori and may be identical to it. They are an astrometric pair with an orbital period of 5.732 days. Almost due north of δ^A Orionis at 52.5″ is δ^C **Orionis**, a B2 of m_V +6.85 and with a diameter of 6.6 $D_\odot$. The separation translates into 16,000 AU – about a quarter of a light year – with an orbital period in excess of 360,000 years. δ Orionis has an important role in the history of astronomy. In 1904 the German astronomer Johannes Hartmann (1865-1936) was studying the star from Potsdam Observatory when he noticed that the spectral lines

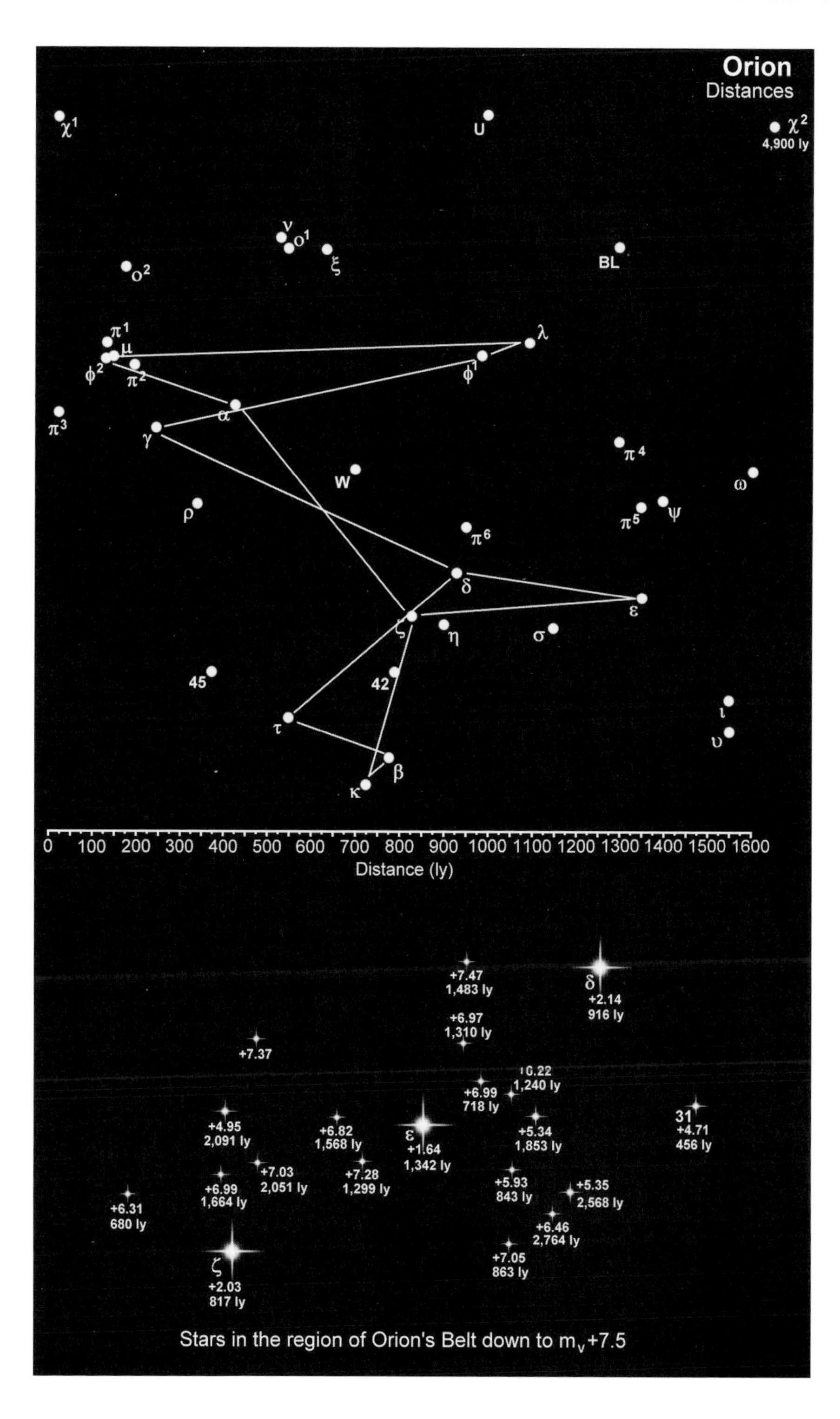
Orion
Distances
4,900 ly
BL
W
U
45
42
31
0 100 200 300 400 500 600 700 800 900 1000 1100 1200 1300 1400 1500 1600
Distance (ly)
+7.47
1,483 ly
+2.14
916 ly
+6.97
1,310 ly
+7.37
+6.22
1,240 ly
+6.99
718 ly
+4.95
2,091 ly
+6.82
1,568 ly
+1.64
1,342 ly
+5.34
1,853 ly
+4.71
456 ly
+7.03
2,051 ly
+7.28
1,299 ly
+6.99
1,664 ly
+5.93
843 ly
+5.35
2,568 ly
+6.31
680 ly
+6.46
2,764 ly
+7.05
863 ly
+2.03
817 ly
Stars in the region of Orion's Belt down to m_v+7.5

of calcium were fixed indicating the star's light was passing through clouds of dust and gas: the interstellar medium had been discovered.

The middle star of Orion's Belt is ε **Orionis** or Alnilam. A B0 supergiant of 32 $D_{\odot}$ and 40 $M_{\odot}$ it burns at 25,000 K and emits 375,000 $L_{\odot}$. It is an α Cygni pulsating variable, peaking at m_v +1.64 before falling by a tenth of a magnitude with no discernable period. This young star, just 4 million years old, rotates at 65 km/s taking 25 days to complete a single rotation. Photographs show it bathed in a huge molecular cloud which, because of the star's spectral class, looks blue. It is the farthest of the three Belt stars at 1,342 ly.

The easternmost star is, of course, the m_v +1.89 ζ **Orionis** or Alnitak. It is definitely a binary and may even be a triple star system. The primary, ζ^A **Orionis,** is a m_v +2.03 O9.7 making it the brightest O-class star though, to be fair, there are only a couple of dozen naked eye O-class stars in the entire sky. It is 20 $D_{\odot}$ across, weighs in at 20 $M_{\odot}$ and is nearly 9,000 times as luminous at the Sun. Spinning at 135 km/s it rotates once in just 7.5 days and although it is only about 6 million years old it is already beginning to die. Its binary companion, ζ^B **Orionis,** is a smaller 8.1 $D_{\odot}$, 14 $M_{\odot}$ B0 separated from the primary by at least 1,300 AU – 2.6″ on the celestial sphere – and taking more than 1,500 years to complete a single circuit of its orbit. Its magnitude is m_v +4.21 but the combined magnitudes of the two stars give the impression of a single star of m_v +1.89. They are the closest of the three Belt stars at 817 ly. The third component, ζ^C **Orionis,** is of the 10th magnitude and separated by almost a degree (57.6″). If it is a genuine member of the system, and not just a line of sight coincidence, then the orbital period will be more than 190,000 years. Just below ζ Ori is the famous **Horsehead Nebula (IC2118)**, a vast dense cloud of dust set against the bright nebula **IC 434**. Discovered in 1888 by the Scottish astronomer Williamina Fleming it requires a substantial telescope to clearly show its shape.

σ **Orionis** is a complex of no fewer than five stars. The primary, σ^A **Orionis,** is a m_v +4.20 O9.5 with a temperature of 32,000 K, a mass of 18 $M_{\odot}$ and a diameter of 7 $D_{\odot}$ although it looks like a m_v +3.78 star. The next brightest star at m_v +5.1 is σ^B **Orionis,** a B0.5 so somewhat cooler at 29,600 K, a less massive 13.5 $M_{\odot}$ but larger at 8.1 $D_{\odot}$. This makes the pair one of the most massive visual binary systems known. σ^A and σ^B Ori have an average separation of 0.25″, equivalent to 90 AU, and an orbital period of 155.3 years. σ^D **Orionis** comes next in the magnitude stakes at m_v +6.62. It is a B2 of 7 $M_{\odot}$. It orbits σ^A and σ^B Ori at an average distance of 4,600 AU taking at least 67,000 years to complete an orbit. Just a smidgen fainter at m_v +6.65 is σ^E **Orionis.** In some respects it resembles σ^D Ori: 7 $M_{\odot}$ and a B2, but it is unusually helium-rich which appears to be concentrated in pools near its surface. It is also much farther away from the primary at 15,000 AU, possibly more, and with an orbital period in excess of 266,000 years. Finally there is σ^C **Orionis**, by far the faintest at m_v +9. It is an A2 in a 40,000 year orbit with an average distance form the primary of 3,900 AU. We say 'finally' but that is not strictly true. These five stars seem to be embedded in a cluster of low mass stars and brown dwarfs and no one is entirely sure of the relationships that exist between them. The whole cluster lies at a distance of 1,148 ly.

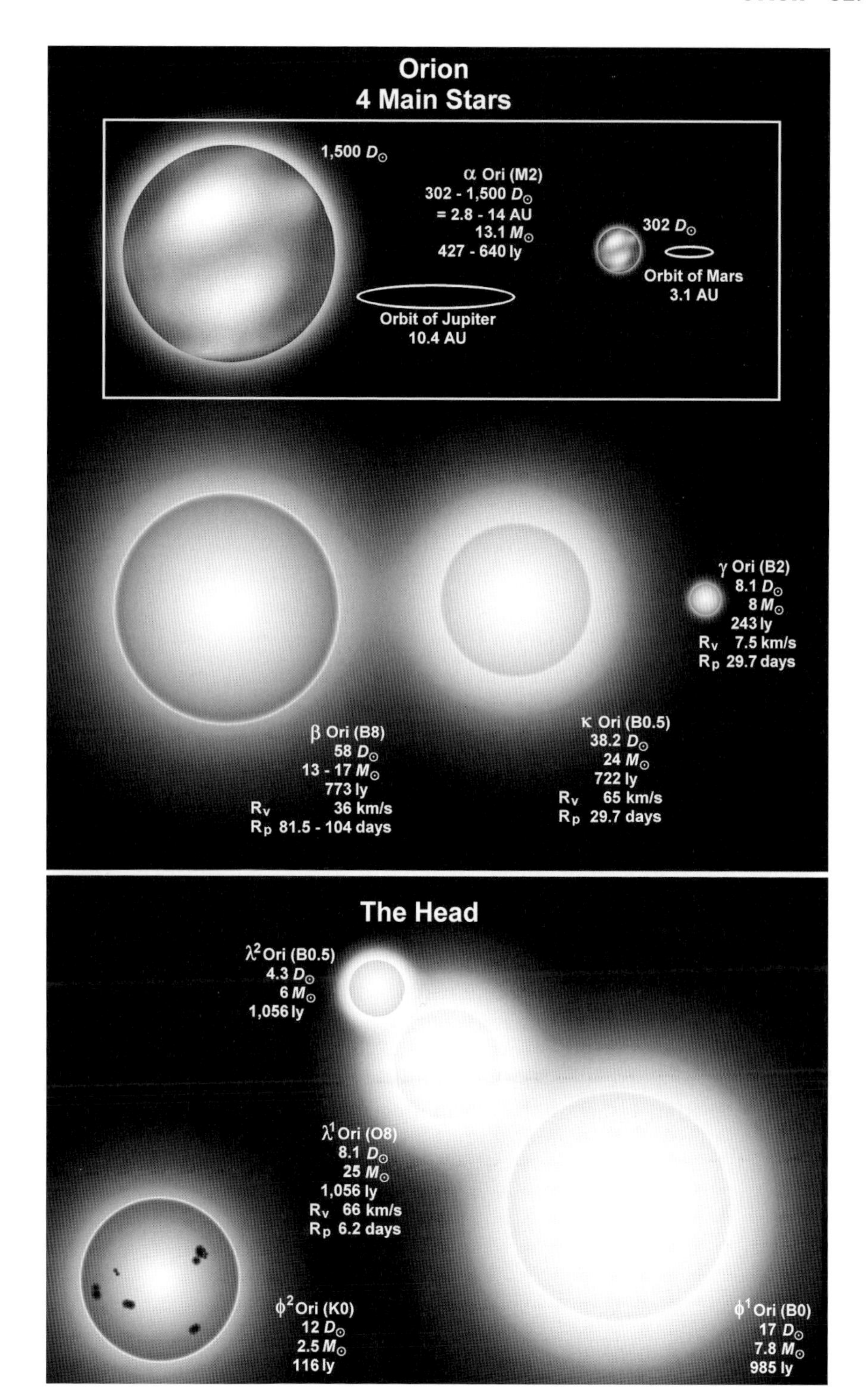

Orion
4 Main Stars
1,500 $D_\odot$
α Ori (M2)
302 - 1,500 $D_\odot$
= 2.8 - 14 AU
13.1 $M_\odot$
427 - 640 ly
302 $D_\odot$
Orbit of Mars
3.1 AU
Orbit of Jupiter
10.4 AU
γ Ori (B2)
8.1 $D_\odot$
8 $M_\odot$
243 ly
R_v 7.5 km/s
R_p 29.7 days
β Ori (B8)
58 $D_\odot$
13 - 17 $M_\odot$
773 ly
R_v 36 km/s
R_p 81.5 - 104 days
κ Ori (B0.5)
38.2 $D_\odot$
24 $M_\odot$
722 ly
R_v 65 km/s
R_p 29.7 days
The Head
λ^2 Ori (B0.5)
4.3 $D_\odot$
6 $M_\odot$
1,056 ly
λ^1 Ori (O8)
8.1 $D_\odot$
25 $M_\odot$
1,056 ly
R_v 66 km/s
R_p 6.2 days
ϕ^2 Ori (K0)
12 $D_\odot$
2.5 $M_\odot$
116 ly
ϕ^1 Ori (B0)
17 $D_\odot$
7.8 $M_\odot$
985 ly

A string of six stars due south of the Belt is often depicted in drawings of Orion to be his sword or scabbard. The most northerly of these is **42 Orionis**, a 4.8 $D_{\odot}$ B1 some 786 ly away. It has an unusually slow rotational velocity. B1-class stars average 150 km/s but 42 Ori manages just 20 km/s – under 3% have velocities of 20 km/s or less – though this may just be because of the angle of the star's rotational pole which could be pointing towards us and therefore giving a false reading. Its apparent visual magnitude is m_v +4.59 but its absolute magnitude is M_v -3.6. It is suspected of being a variable, fluctuating by 1/10th of a magnitude. Just to the east lies **45 Orionis**, an F0 about 371 ly away and shining at m_v +5.26. No one really knows its diameter but a guesstimate would be about 6 $D_{\odot}$. Surrounding 42 and 45 Ori is **NGC 1977** a bright nebula which is peppered with a number of bright stars. **NGC 1981** is a beautiful cluster of about a dozen brightish members just to the north of 42 Ori and centered about 800 ly away. The cluster probably has about 40 members in total.

The next star in line is **θ^1 Orionis**. To the naked eye this is an unremarkable 5th magnitude star but a small telescope or a binocular will reveal three others in a formation that is now widely known as The Trapezium. These four young stars are just the tip of the cosmic iceberg. There are another 1,000 hidden from view by the **M42** gaseous nebula in which they are embedded and from which they were formed. However, the radiation from the brighter stars of the Trapezium, particularly the ferocious 1,000 km/s stellar wind from θ^{1C} Orionis, is slowly but surely destroying the nebula so that in 100,000 years from now it will all but have disappeared. The Trapezium's stars are usually labeled from west to east (or right to left, if you prefer). **θ^{1A} Orionis** is a 5.2 $D_{\odot}$ hot O6 around 30,000 K and with a mass of 5.9 $M_{\odot}$. It is an Algol-type EA variable, with a period of $65^d\ 10^h\ 22.5^m$. At maximum it reaches m_v +6.73 and then dips to +7.7 for 2.3 hours. The cause of its variability is an m_v +8.1 B2 companion just 1 AU from the primary. A bit smaller at 4.8 $D_{\odot}$ the two stars eclipse one another every couple of months. The second of the four bright stars, and the most northerly, is **θ^{1B} Orionis.** This consists of four stars in a complex orbital arrangement. The brightest star is a B0 of m_v +7.96 to +8.60, its variability of 6.471 days due to a 2 $M_{\odot}$ orbiting companion of spectral group B3. **θ^{1C} Orionis** is the most massive of the brighter stars at 40 $M_{\odot}$ and 30 $D_{\odot}$. Being an O5 class it has a surface temperature of 40,000 K and is responsible for emitting 90% of the UV radiation released by the four Trapezium stars. Embedded within the M42 nebula are numerous young T-Tauri stars that, under 'normal' circumstances, would be surrounded by circumstellar gas and dust clouds that could form planets. The massive amounts of ionizing radiation released by θ^{1C} Orionis however is destroying the gas clouds and preventing planetary formation. **θ^{1D} Orionis,** the most easterly of the four, is a m_v +6.71 B0.5 in orbit with θ^{1A} Orionis that takes more than half a million years to complete. Each of the four main stars has other stars in orbit around it, and those orbiting stars are sometimes binary or more complex systems. Such orbital spaghetti is the result of large numbers of stars forming in a relatively small volume of space. The orbits are inherently unstable with some of the stars being gravitationally kicked into deep interstellar space. The Trapezium is about 1,500 ly from Earth.

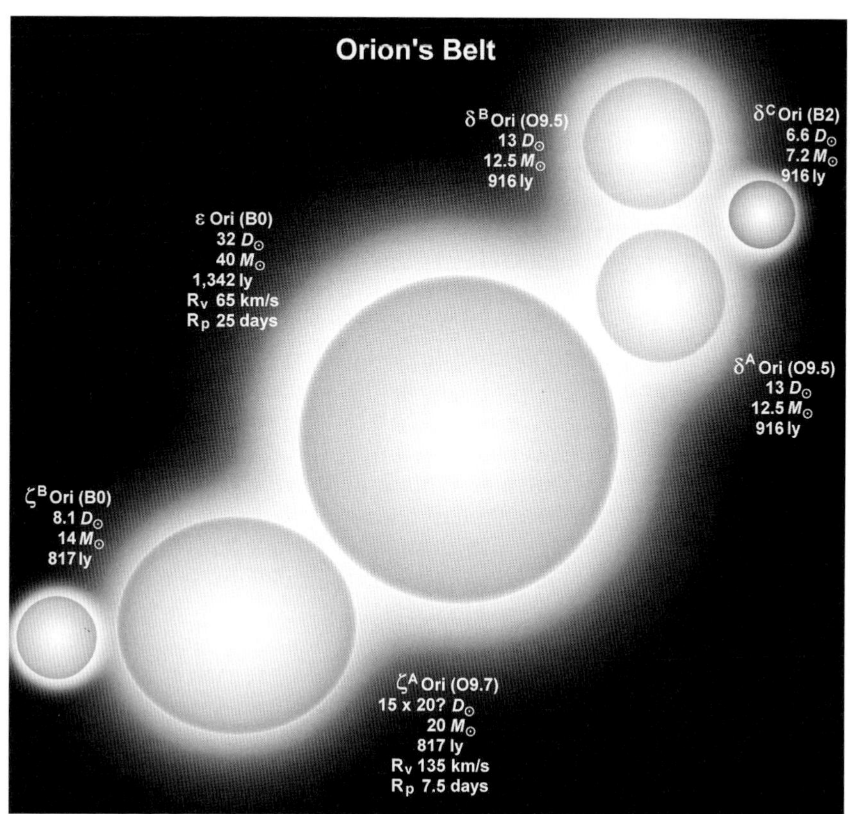
Orion's Belt
δ^B Ori (O9.5)
13 D☉
12.5 M☉
916 ly
δ^C Ori (B2)
6.6 D☉
7.2 M☉
916 ly
ε Ori (B0)
32 D☉
40 M☉
1,342 ly
Rv 65 km/s
Rp 25 days
δ^A Ori (O9.5)
13 D☉
12.5 M☉
916 ly
ζ^B Ori (B0)
8.1 D☉
14 M☉
817 ly
ζ^A Ori (O9.7)
15 x 20? D☉
20 M☉
817 ly
Rv 135 km/s
Rp 7.5 days

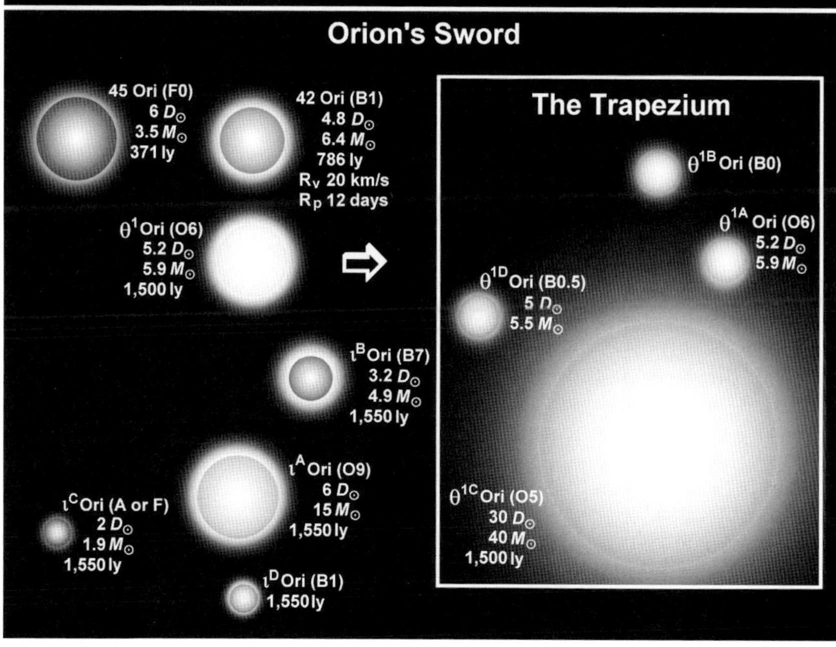
Orion's Sword
45 Ori (F0)
6 D☉
3.5 M☉
371 ly
42 Ori (B1)
4.8 D☉
6.4 M☉
786 ly
Rv 20 km/s
Rp 12 days
θ^1 Ori (O6)
5.2 D☉
5.9 M☉
1,500 ly
The Trapezium
θ^1B Ori (B0)
θ^1A Ori (O6)
5.2 D☉
5.9 M☉
θ^1D Ori (B0.5)
5 D☉
5.5 M☉
ι^B Ori (B7)
3.2 D☉
4.9 M☉
1,550 ly
ι^A Ori (O9)
6 D☉
15 M☉
1,550 ly
ι^C Ori (A or F)
2 D☉
1.9 M☉
1,550 ly
θ^1C Ori (O5)
30 D☉
40 M☉
1,500 ly
ι^D Ori (B1)
1,550 ly

Observers sometimes spend so much time looking at θ^1Orionis that they overlook **θ^2 Orionis** which is also a double. The primary is an 8 $D_\odot$ O9.5 with an apparent magnitude of m_v +5.08 and an absolute magnitude of M_v -4.4. With a rotational velocity of 165 km/s – a bit high for this class of star that averages 135 km/s – it takes just 2.5 days to turn once on its poles. Its companion is a m_v +6.4 B0.5 orbiting at an average distance of 30,300 AU. θ^2 Ori lies 1,896 ly away with a large margin of error: about 700 ly.

The next star in the sword, **ι Orionis**, is also the brightest at m_v +2.74. In fact its Arabic name Na'ir al Saif means 'The Bright One in the Sword' although Antonín Bečvář in his *Skalnate Pleso Atlas of the Heavens* (1951) referred to it as Hatsya and that name is still popular. It is a quadruple system lying at a distance of between 1,000 and 1,650 ly and probably towards the farthest end of the estimate. It is another hot O9, around 31,500 K and with a diameter of 6 $D_\odot$. At 11.3″, some 20,000 AU in space, is a B7 companion **ι^{1B} Orionis** (PA 141°) of m_v +7.1. It is a helium deficient star with enhanced levels of phosphorus and gallium and takes at least 700,000 years to complete an orbit of the primary. **ι^{1C} Orionis** is a much fainter 11th magnitude A or F class star separated by 49.5″ at PA 103°. Its 4,400 AU orbit takes 75,000 years to complete. The third component is an astrometric binary of B1 class in a 29.134 day orbit around the primary ι^{1A} Orionis which sees it come to within 0.11 AU at periastron and then swing out to 0.8 AU at apastron. The high velocity winds from the two stars create copious quantities of X-rays as they collide. About 2.5 million years ago the ι Orionis system was even larger and more complex and part of the Trapezium but it was ultimately less stable. The result was the gravitational ejection of two of the stars which are now designated as AE Aurigae and μ Columbae.

The most southerly star in the sword is **HD 36960.** It is a m_v +4.78 B0.5 which is suspected of being slightly variable by 9/100th of a magnitude. It is another B-type star with an apparently slow rotational velocity of 25 km/s. Couple this with a diameter of 7.7 $D_\odot$ and the rotational period works out at 15.6 days. It is a binary, its companion having very similar characteristics of being a B1, a diameter of 6.97 $D_\odot$ and also rotating at 25 km/s. It is m_v +5.51 and is located at PA 43.7°. The two average 20,600 AU apart. Often mistaken for its companion is **HD 36959** which lies at PA 223°. The 36.2″ distance between the two makes them easy to separate in even the smallest telescopes or binocular. While HD 36960 is an estimated 1,864 ly away HD 36959 is more than three times the distance at 5,930 ly. It is also a B1 with a magnitude that varies between m_v +5.67 and +5.72. Burning at 23,800 K it is 5.24 $D_\odot$ across and is 11.3 $M_\odot$. Just to confuse matters further, it is also a binary, its m_v +8.84 companion considerably less massive at 4.88 $M_\odot$ and located just 0.6″ away at PA 120°. The pair have an orbital period of 1,183 years and average 285 AU distance from one another.

The small triangle of stars that mark Orion's head are designated, from top to bottom, λ, ϕ^1 and ϕ^2 Orionis though, as ever with Orion, nothing is what it seems and ϕ^1 is related to λ and not to ϕ^2. **λ Orionis** is the brighter of the three at m_v +3.54. It is an O8 with a diameter of 8.1 $D_\odot$ and 25 $M_\odot$. Spinning at 66 km/s its 35,000 K surface releases 65,000 $L_\odot$ ionizing the gas and dust cloud in which it

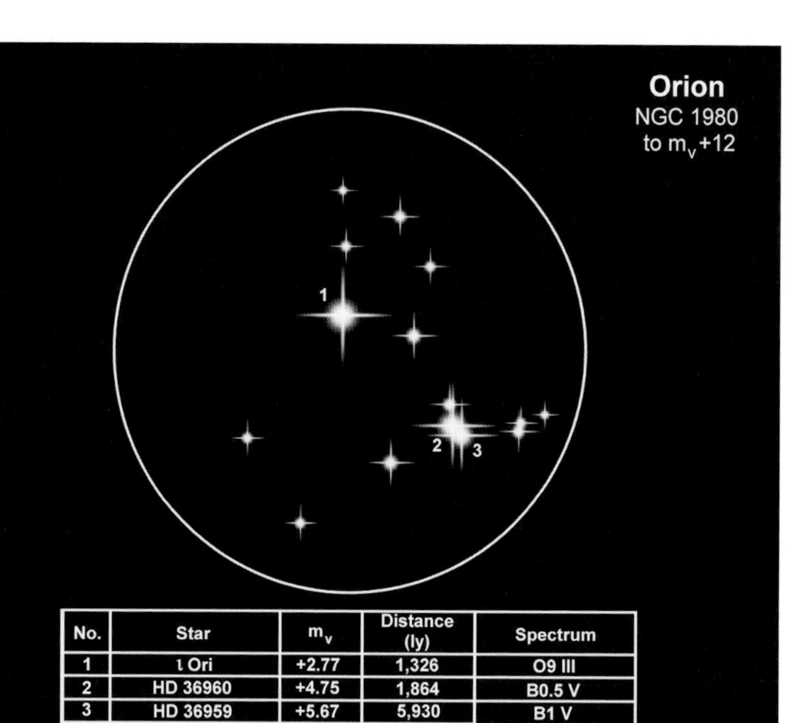

No.	Star	m_v	Distance (ly)	Spectrum
1	ι Ori	+2.77	1,326	O9 III
2	HD 36960	+4.75	1,864	B0.5 V
3	HD 36959	+5.67	5,930	B1 V

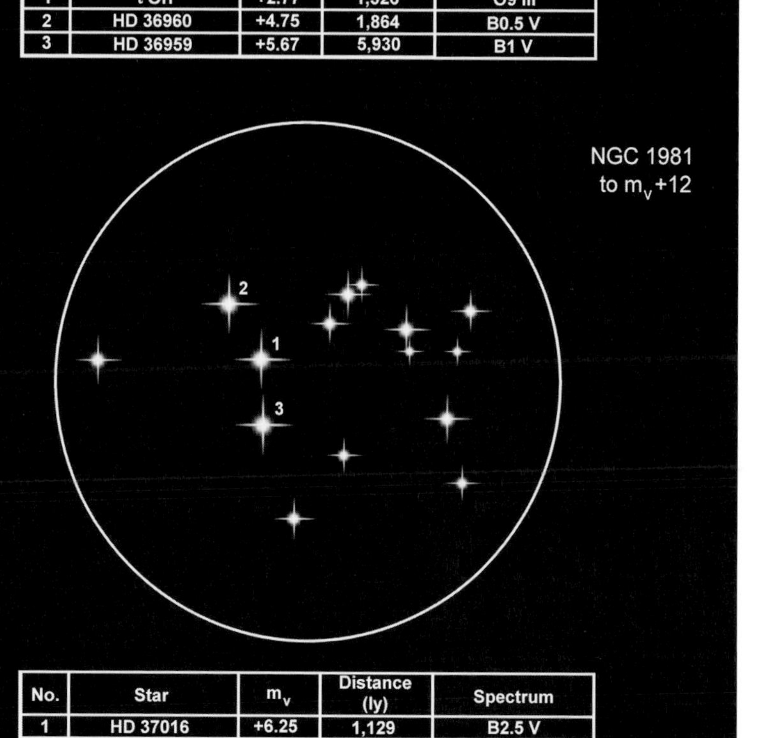

No.	Star	m_v	Distance (ly)	Spectrum
1	HD 37016	+6.25	1,129	B2.5 V
2	HD 37040	+6.48	1,264	B2.5 V
3	HD 37017	+6.56	1,217	B1.5 V

is embedded by up to 150 ly which shows up as concentric red colored rings. It is also a binary system. Its partner is a m_v +5.6 B0.5 separated by 4″ so easily spotted in a binocular. The system is 1,056 ly from Earth. **ϕ^1 Orionis** is a little closer at 985 ly and both it and λ are heading away from us at about 33 km/s. It is a B0 giant of 17 $D_\odot$ and appears as a m_v +4.41 star but would make M_v -4.6 at 10 pc. The unrelated **ϕ^2 Orionis** is a K0 giant of 12 $D_\odot$ and shining at m_v +4.09. It is the closest of the three stars at 116 ly. It is also a runaway star, hurtling through space at 98.7 km/s away from us and a visitor from another part of the Galaxy. Just above the three stars of Orion's head is **HD 36881**, another B-class – this time a B9 – which is 1,463 ly away. It is quite faint at m_v +5.63.

The tip of Orion's club is indicated by χ^1 and χ^2 Orionis but apart from sharing the same name and the fact that both stars are spectroscopic binaries, they have nothing else in common. **χ^1 Orionis** is a solar analog just 28.3 ly from Earth. Slightly larger than the Sun at 1.16 $D_\odot$, more massive, 1.08 $M_\odot$ and more luminous, 1.06 $D_\odot$, it is also cooler at 5,940 K and consequently a paler shade of yellow. It rotates at three times the velocity of the Sun (6 km/s instead of 2 km/s) and so completes a full rotation in just under 10 days. Its apparent magnitude is m_v +4.40 with its absolute magnitude not changing much, M_v +4.71, because its distance is so close to the 10 pc marker used in the calculation. Its companion is a mid-M class red dwarf about 15% the size of the Sun. It is in a 14.25 year long orbit during which the two stars come to within 3.5 AU but then separated at apastron to 9.3 AU. Believed to be about 100 million years old the system is part of the Ursa Major Moving Group. χ^{1A} Ori is barium-rich which would tend to suggest that it has been contaminated by the rapid evolution of a much larger star that has subsequently turned into a white dwarf. There is a suggestion that about 30 million years ago there were two other stars in the system, one of which was the contaminating giant, and because of some cataclysmic event two of the stars were ejected from the system and can now be found in Scorpius as HD147513 and its companion. **χ^2 Orionis** is a different story altogether. Shining at a similar apparent magnitude of m_v +4.64, which is probably why they were given similar designations, if χ^2 Ori suddenly found itself 10 pc from Earth it would have brightened to an impressive M_v -6.8. Part of this is due to its luminosity, a staggering 410,000 $L_\odot$, but it is also due to its great distance. Estimates range from 3,400 to 6,300 ly. No one is really sure, but if χ^2 Ori is part of the Gemini OB1 group, as many astronomers suspect, then it will be 4,900 ly away. It is a 19,000 K B2, 100 $D_\odot$ across and weighing in at 38 $M_\odot$. It is a far younger star, probably just 5 million years old, and spins at 40 km/s taking nearly 126.5 days to rotate once on its axis. As for its companion, apart from being an average of 30 AU from its primary absolutely nothing else is known about it.

η Orionis is a four and possibly five star system about 900 ly away. The primary is a massive B0.5, about 15 $M_\odot$ packed into a sphere of just 7.7 $D_\odot$. It is an eclipsing EA variable with a period of 7^d 23^h 44.5^m during which its magnitude changes from m_v +3.31 to +3.60 due to the presence of an unseen companion in an orbit that averages 0.09 AU (13.5 million km). Spinning once on its axis every 6 days η Ori rotates at 65 km/s. A companion revealed through a small telescope is a m_v +4.91 B2 emission star which is surrounded by a gas cloud. It is smaller than the primary star at 5.6 $D_\odot$, rather less massive, 9 $M_\odot$, and spins more slowly at 50 km/s but takes almost the same time to complete a full rotation, 5.7 days. The two stars are currently 1.7″ apart at PA 78°. In real space their orbital distance varies between 266 and 672 AU and takes more than 1,500 years to complete a single orbit. Rather closer at 12 AU is another member of the system that has an orbital period of 9.219 years. A somewhat more tenuous member is an A-class star of about 1.7 $M_\odot$.

μ Orionis is another complex double binary system. The main star is **μ^{Aa} Orionis.** This is an A2 dwarf, 2.4 $D_\odot$ across and with a mass of 2.1 $M_\odot$. It is by far the most luminous of the quartet at 38.8 $L_\odot$ and looks like a single m_v +4.14 star even through a telescope. In orbit around it is the 1.16 $M_\odot$ G-class **μ^{Ab} Orionis** which stays at 0.077 AU (11.5 million km) and which takes 4.45 days to complete one orbit. Then there is **μ^{Ba} Orionis,** a 1.3 $D_\odot$, 1.46 $M_\odot$ F3 which has a luminosity of 3 $L_\odot$ and a magnitude of m_v +6.24. μ^{Ba} Ori has its own companion, **μ^{Bb} Orionis,** which is almost an identical twin: a m_v +6.91 F5 of 1.3 $D_\odot$ and 1.44 $M_\odot$. This pair are in a very similar orbital arrangement to the Aa-Ab pair: a near-circular orbit of 0.078 AU (11.7 million km) which takes 4.78 days to complete. The A-pair and the B-pair orbit their common center of mass once every 18.85 years, approaching to within 3.3 AU before swinging out to 22 AU. On the celestial sphere the two binaries are separated by just 0.276″. That coupled with the brightness of the main star means they are optically impossible to separate. The system lies 152 ly from Earth.

An arc of six stars near the westernmost boundary of the constellation are given the designation π and in drawings of the Hunter usually depict a lion skin draped over Orion's arm, a shield or a bow. Apart from sharing a common identifier, they have nothing to do with one another. **π^1 Orionis** is the most northerly of the six, and the faintest at m_v +4.65. An A0 class it is 2.9 $D_\odot$, rotates at 110 km/s taking just 1.3 days to complete a spin and lies 121 ly from Earth. Traveling south, **π^2 Orionis** is very similar. Slightly smaller at 2.3 $D_\odot$ it is an A1 but is a much faster spinner at 212 km/s. As a result it is slightly oblate and turns once in just half a day. **π^3 Orionis** is the least luminous, 2.8 $L_\odot$, and brightest, m_v +3.15, of all the π stars due to the fact that it is the closest at a mere 26.2 ly. Belonging to spectral group F6 it is a few hundred degrees warmer than the Sun, 6,400 K, and has a similar diameter of 1.1 $D_\odot$ and mass, 1.2 $M_\odot$. It rotates at 17 km/s and so completes one rotation on its axis every 3.3 days. π^3 Ori is also cataloged as NSV 1731, the code standing for New Suspected Variable. There is some evidence that the magnitude varies between +3.15 and +3.21 in δ Scuti style but no one is yet sure. Another newly suspected variable is, by coincidence, **π^4 Orionis** which gets the catalog number NSV 1742. A good eight solar diameters across and about 10 $M_\odot$ it is a B2 which, from a distance of 1,259 ly and with a luminosity of 4,190 $L_\odot$, shines at us as a m_v +3.65. Its absolute magnitude is easy to remember, just change the sign to a negative M_v -3.6. Its apparent magnitude may vary with a period of 0.62 days but don't try to measure it as the fluctuation is only 3/1,000th of a mag. Also undetectable with the human eye is its companion, another fairly hefty star of 9 $M_{\odot}$ and 6 $D_{\odot}$. Its orbit is just about 0.25 AU leading to a period of 9.52 days. π^5 **Orionis** is another spectroscopic binary but of the EII variety: a pair of stars that are so close they are deformed into egg-shaped objects. No one knows how close the stars are to one another and virtually nothing is known about the invisible companion except that it is a B0-class dwarf. The primary, however, belongs to the B3 group and is 6.1 $D_{\odot}$. Its magnitude varies between m_v +3.66 and +3.73 as the two stars partly eclipse each other. The period is 3^d 16^h 48.5^m, the system lying 1,342 ly from Earth. Finally to π^6 **Orionis,** a rare – well rare for Orion – K2 giant with a diameter of 62 $D_{\odot}$ and lying 954 ly away. It is another NSV, number 1786 in the catalog, the suspected variability being m_v +4.45 to +4.49.

Like some sort of celestial sorcerer the now-you-see-me-now-you-don't **U Orionis** is a clear indication that it is a Mira-type variable. Believed to have a diameter of between 370 $D_{\odot}$ and 485 $D_{\odot}$ – at least the size of the orbit of Mars and possibly as big as the orbits of most of the asteroids – this M8 red supergiant has a mass of less than 1.5 $M_{\odot}$. When visible to the naked eye its magnitude can reach m_v +4.8, but it won't stay that bright for long and plunges to m_v +13 with a period of about 368.3 days. It lies at an uncertain distance of 1,000 ly. If the star is spinning at 1 km/s then it will take between 51 and 67 years to turn once on its axis.

There are always two things that are certain about stars that begin with a double-V: one, they are variable, and two, there is confusion over their name with people reading 'V V' as 'W'. **VV Orionis** – that is V V Orionis not W Orionis – is no exception. A B1 with a diameter of 4.98 $D_{\odot}$ and mass of 10.9 $M_{\odot}$ its magnitude changes rhythmically... m_v +5.34, +5.73, +5.34, +5.55... with a period of 1^d 11^h 39^m. VV Ori is, of course, an eclipsing binary. Its companion is a B4.5, about half the size of the primary at 2.41 $D_{\odot}$ and with a mass of 4.09 $M_{\odot}$. They are separated by a mere 0.063AU (9.4 million km). It has long been believed that a third component, an A4, was in a 119.1 day long orbit but recent studies have cast considerable doubt on this idea.

Like Herschel's Garnet Star in Cepheus, **W Orionis** – that's W not V V – is very red. It is a semi-regular variable, an SRb, which often reaches m_v +5.88 but can be as faint as m_v +12.4. The variability has a period of around 212 days but there is also a long term fluctuation that lasts 6.7 years. About 700 ly away it is estimated to have a diameter of 220 $D_{\odot}$ – a full 1 AU – and a luminosity of 460 $L_{\odot}$. W Ori is a rare, barely naked eye carbon star, a C4.5, which has regurgitated carbon from deep within its interior. Despite its huge size it has a mass of just 2 $M_{\odot}$ and a temperature of 3,200 K. It is not the only carbon star in Orion. **BL Orionis** is sometimes a m_v +6.3 dimming to m_v +9.7 and is a C6.3. About the same size as W Ori it is almost twice as far away at 1,300 ly. It is set in a field of blue and white stars.

Open clusters in Orion

Name	Size arc min	Size ly	Distance ly	Age million yrs	Brightest star in region*	No. stars m_v >+12*	Apparent magnitude m_v
NGC 1977 Trapezium	17′	7.9	1,600	12	42 Ori m_v +4.59	12	+7.0
NGC 1980	16′	8.3	1,800	4.7	ι Ori m_v +2.77	16	+2.5
NGC 1981	25′	9.5	1,300	32	HD 37016 m_v +6.25	20	+4.2

Pavo

Constellation:	Pavo	**Hemisphere:**	Southern
Translation:	The Peacock	**Area:**	378 deg^2
Genitive:	Pavonis	**% of sky:**	0.916%
Abbreviation:	Pav	**Size ranking:**	44th

One of the dozen new constellations created by Petrus Plancius in the 16th Century.

By far the brightest star in this otherwise faint constellation, α **Pavonis** has an apparent magnitude of m_v +1.91 though it is suspected of fading to +1.96. Its absolute magnitude would be M_v -2.3. Lying 183 ly from Earth it is about 4.5 $D_\odot$ across with a mass of around 5.5 $M_\odot$. It is also an astrometric binary, its invisible companion in an 11.8 day orbit. The stars are members of the Pleiades Group. D.H. Sadler in his book, *A Personal History of H.M. Nautical Almanac Office,* states that the name of the star, Peacock, was introduced by HMNAO in the late 1930s. The Nautical Almanac Office was preparing *The Air Almanac,* a book containing details of 57 bright stars that pilots could use for navigational purposes. Only two of the stars did not already have names - α Pavonis and ε Car – so they called them Peacock (the English translation of Pavo) and Avior (from the French *avis* for bird).

β **Pavonis** is an A7 dwarf lying 138 ly from Earth. So is π **Pavonis**. β Pav is a bit larger at 2.5 $D_\odot$ – π Pav comes in at 1.4 $D_\odot$ – and brighter at m_v +3.41 compared to +4.34. But while β Pav is heading away from us at 9.8 km/s π Pav is on its way towards us at 15.6 km/s. β Pav's space velocity is 11 km/s while π Pav is nearly three times larger at 29 km/s. Putting all these facts together suggests that the two stars, despite certain similarities, are unrelated.

The F9 dwarf star γ **Pavonis** is just 30.1 ly from us, making it the second closest star in the constellation. Only slightly larger than the Sun at 1.06 $D_\odot$ it weighs in at 0.8 $M_\odot$ and is 1.5 $L_\odot$. It is believed to be about twice as old as the Sun at around 9,100 million years. Its apparent magnitude of m_v +4.23 is almost the same as its absolute magnitude, M_v +4.5, due to its proximity to the 10 pc limit used in the standard calculation.

The closest star at 19.9 ly is δ **Pavonis**. Some researchers believe that it is the star that is most likely to have an Earth-type planet in the so-called 'habitable zone'. A G8 with a diameter of 1.06 $D_\odot$, a mass of 1.1 $M_\odot$ and luminosity of 1.18 $L_\odot$ it is thought to be at least as old as the Sun, 4,560 million years, and possibly as old as 11,000 million years. Its chromosphere appears to be quiet, so there is less risk of huge levels of radiation inhibiting the formation of life, and it rotates at 3.2 km/s, not much faster than the Sun's 2 km/s. Measurements of its radial velocity indicate that it is closing in on us at 21.7 km/s but, more importantly, its velocity varies by less than 3 m/s. This suggests that there are no giant planets close to the star gravitationally tugging at it. Nor is it a binary. The absence of giant planets and a stellar companion means that planet formation is unlikely to

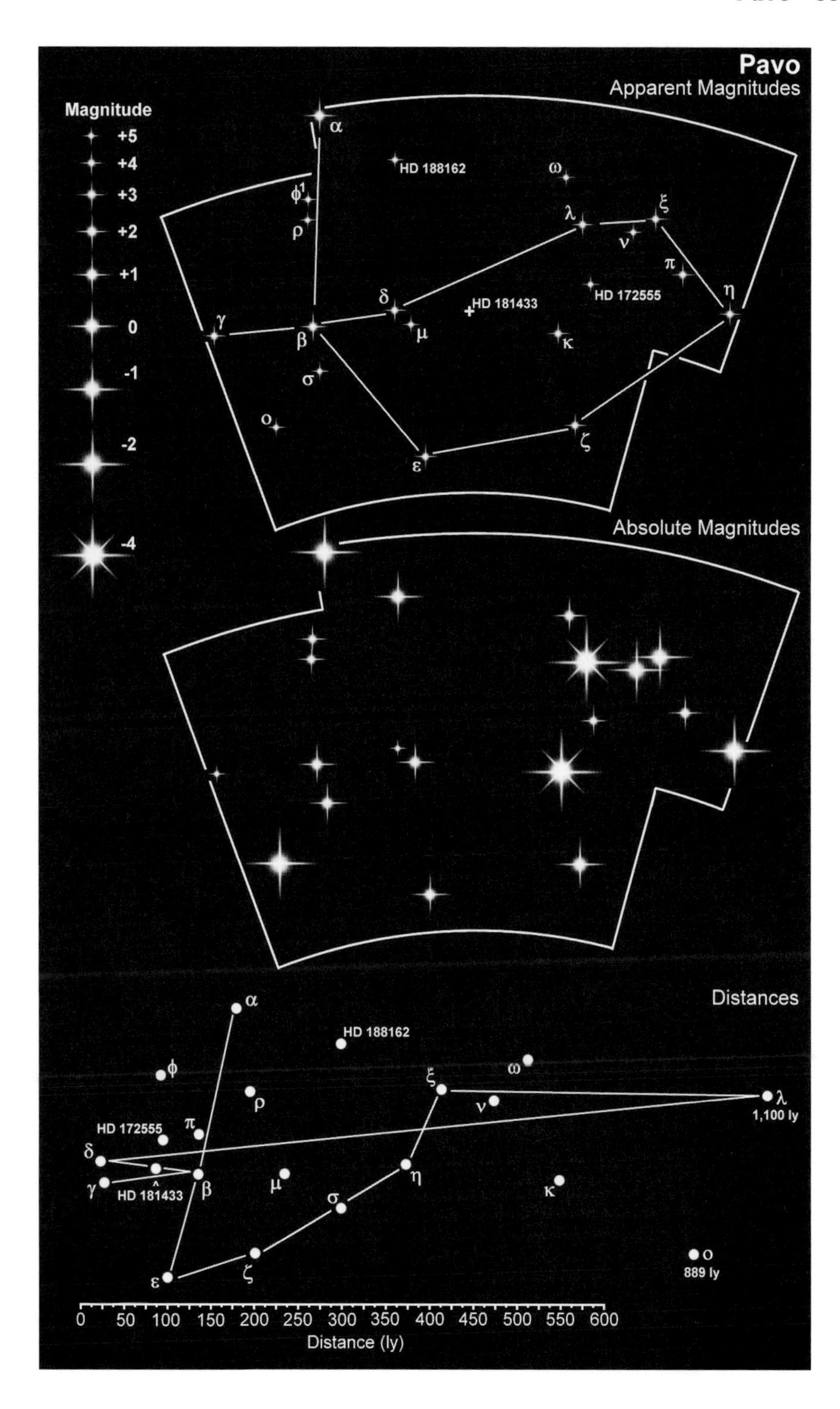
Pavo
Apparent Magnitudes
Magnitude
+5
+4
+3
+2
+1
0
-1
-2
-4
α
HD 188162
ω
ϕ1
ρ
λ
ξ
ν
π
δ
HD 181433
HD 172555
η
γ
β
μ
κ
σ
ο
ζ
ε
Absolute Magnitudes
Distances
α
HD 188162
ϕ
ω
ξ
ρ
ν
λ
1,100 ly
HD 172555
π
δ
η
γ
HD 181433
β
μ
κ
σ
ο
889 ly
ε
ζ
0 50 100 150 200 250 300 350 400 450 500 550 600
Distance (ly)

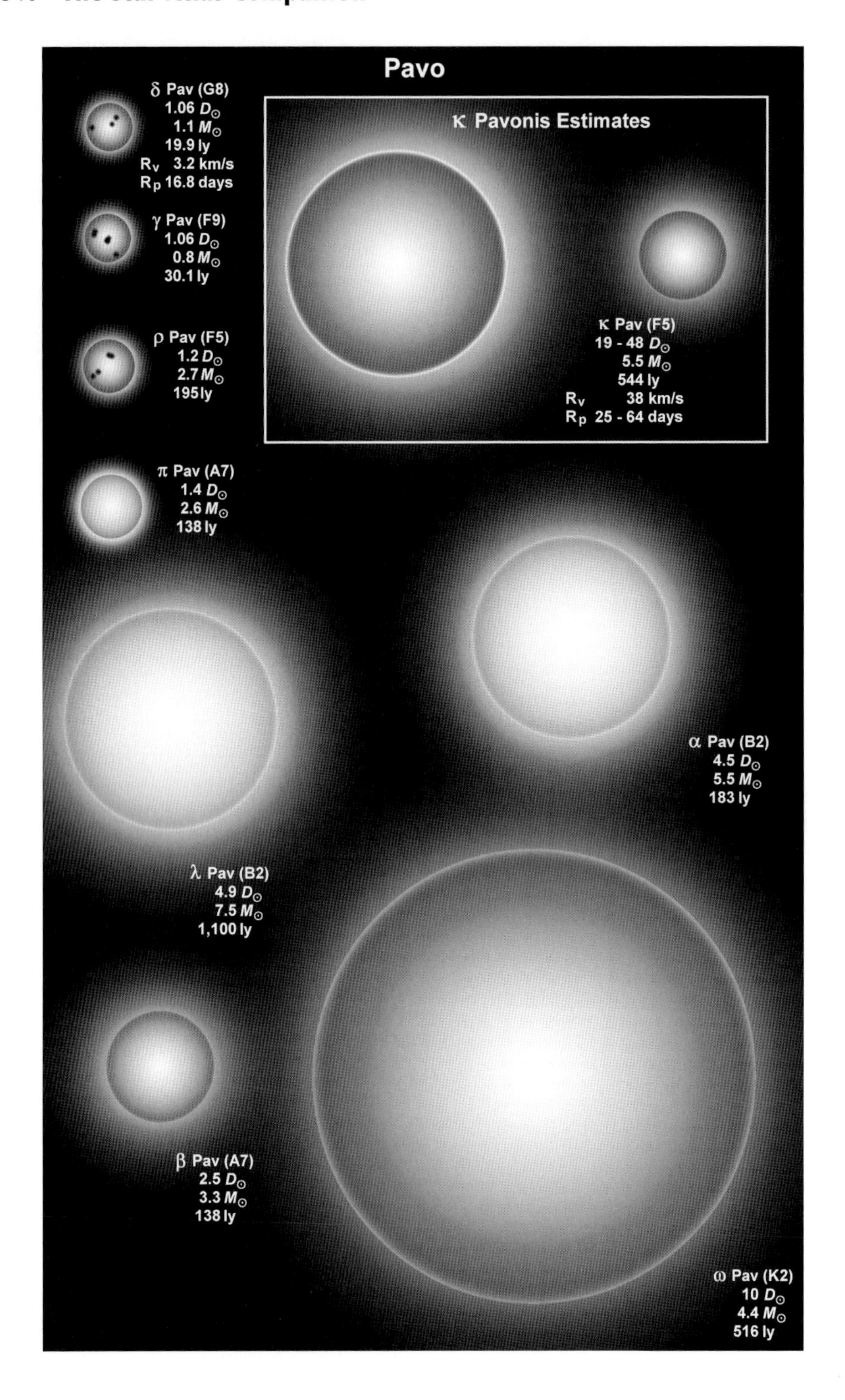

Pavo
δ Pav (G8)
1.06 $D_\odot$
1.1 $M_\odot$
19.9 ly
R_v 3.2 km/s
R_p 16.8 days
κ Pavonis Estimates
γ Pav (F9)
1.06 $D_\odot$
0.8 $M_\odot$
30.1 ly
ρ Pav (F5)
1.2 $D_\odot$
2.7 $M_\odot$
195 ly
κ Pav (F5)
19 - 48 $D_\odot$
5.5 $M_\odot$
544 ly
R_v 38 km/s
R_p 25 - 64 days
π Pav (A7)
1.4 $D_\odot$
2.6 $M_\odot$
138 ly
α Pav (B2)
4.5 $D_\odot$
5.5 $M_\odot$
183 ly
λ Pav (B2)
4.9 $D_\odot$
7.5 $M_\odot$
1,100 ly
β Pav (A7)
2.5 $D_\odot$
3.3 $M_\odot$
138 ly
ω Pav (K2)
10 $D_\odot$
4.4 $M_\odot$
516 ly

be adversely disrupted. As yet, however, there is no evidence of Earth-size planets in either the habitable zone or elsewhere.

Various attempts have been made to estimate the size of κ **Pavonis** but without much success coming in at 19, 26, 37 and 48 $D_\odot$. With a rotational velocity of 38 km/s it could therefore revolve once every 25.3 days if 19 $D_\odot$ across or every 63.9 days if 48 $D_\odot$ across. An F5 some 544 ly away, give or take 10%, it is a Cepheid variable with its magnitude cycling between m_v +3.91 and +4.78 with a period of $9^d\ 2^h\ 6.7^m$. Two other naked eye variables reside in Pavo. λ **Pavonis** is a 4.9 $D_\odot$ B2 emission star with a circumstellar shell. A distant member of the Pleiades Group at 1,100 ly it belongs to the γ Cas type of eruptive variables, its normal brightness of m_v +4.26 increasing to +4.00 at irregular intervals. The third variable, ρ **Pavonis**, represents a third different variable class, this time a δ Scuti. An F5 of 1.2 $D_\odot$ it lies 195 ly from Earth and, with a period of $2^h\ 44.3^m$, changes from m_v +4.850 to +4.795 and back.

Hurtling away from us at 180 km/s is ω **Pavonis**. It has the third highest velocity of all the naked eye stars beaten only by HD 41312 in Lepus, which is also heading away at 182.2 km/s, and τ^1 Lupi, which is coming towards us at 215 km/s. ω Pav is a K2 giant of 10 $D_\odot$ and is currently 516 ly away.

HD 181433 has a system of at least three planets, one of which is a super-Earth of 7.56 $M_\oplus$ (0.0238 M_J). A number of other stars also have planets (see table).

NGC 6752 is a compact globular cluster about 13,000 ly from us and 17,000 ly from the galactic center. It contains more than 100,000 stars and appears as a m_v +5.4 cloud to the naked eye. A small telescope will resolve some of the individual stars.

Planetary systems in Pavo

Star	$D_\odot$	Spectral class	ly	m_v	Planet	Minimum mass	q	Q	P
HD 181433	?	K3	85.3	+8.38	HD 181433 b	7.56 $M_\oplus$	0.048	0.112	9.37 d
					HD 181433 c	0.64 M_J	1.27	2.25	2.52 y
					HD 181433 d	0.54 M_J	1.56	4.44	5.95 y

Globular cluster in Pavo

Name	Size arc min	Size ly	Distance ly	Apparent magnitude m_v
NGC 6752	20.4′	77	13,000	+5.4

Pegasus

Constellation:	Pegasus	**Hemisphere:**	Northern
Translation:	The Winged Horse	**Area:**	1,121 deg^2
Genitive:	Pegasi	**% of sky:**	2.717%
Abbreviation:	Peg	**Size ranking:**	7th

In classical mythology the great white winged horse was the son of Neptune, born from the blood of Medusa the Gorgon when beheaded by Perseus. The constellation represents the upper body of the horse and contains the Great Square of Pegasus, a relatively barren region of the sky.

The third brightest star in the constellation at m_v +2.45, **α Pegasi** or Markab is outshone by β and, oddly, ε. It is a typical bluish-white dwarf: 2 $D_\odot$ with a mass of 3.1 $M_\odot$ and luminosity of 156 $L_\odot$ (or about 200 $L_\odot$ if you take into account its UV radiation). A B9 with a temperature in the order of 10,500 K it spins at 140 km/s – a bit on the high side for B9s which average 115 km/s – and completes a full rotation in 16^h 48^m. The star marks the bottom right hand (south-west) corner of the Great Square and lies at a distance of 140 ly.

Almost directly above α at the north-west corner of the Square is **β Pegasi** or Scheat. A distinctly red star it is not surprising to discover that it is a red giant of 55 $D_\odot$ and spectral class M2.5. Almost 200 ly away its luminosity is a deceptive 308 $L_\odot$ visible rising to over 1,000 $L_\odot$ in the infra-red. Again, not surprisingly, it is an irregular pulsating variable: m_v +2.31 at its brightest falling off to m_v +2.74.

Diagonally opposite β, at the south-east corner, is **γ Pegasi** or Algerib, another variable but this time of the β Cepheid variety. Over a period of 3^h 28.5^m the star goes from m_v +2.78 to m_v +2.89 and back. Situated 333 ly away γ Peg is a hot 21,500 K B2 with a luminosity of 647 $L_\odot$ but rising to a couple of thousand if we were to include UV. Two factors remain a mystery. Its diameter, which has been variously quoted as being 1.8, 4.5 and 6.3 $D_\odot$, and its rotational velocity which appears to be just 3 km/s making it the slowest rotating B2 known (the average is 142 km/s). However, this may just be down to observational error. If we are observing γ Peg pole-on then it will appear to rotate much more slowly than it actually does. The star appears to be an astrometric binary with an orbital period of 6.83 days.

The north-east corner of the Great Square of Pegasus is marked by α Andromedae, making it one of the few constellations that share a star (the others are β Tauri with Auriga and μ and ξ Serpentis with Ophiuchus). It used to be known as **δ Pegasi** but the name is no longer used.

ε Pegasi or Enif may have a common origin with α and β Aquarii. A K2 supergiant of 193 $D_\odot$ – about the size of the Earth's orbit – it has a mass of 10 $M_\odot$ and luminosity of 6,700 $D_\odot$. Classed as an Lc pulsating variable its magnitude normally stays within the range of m_v +2.38 to m_v +3.50 but on 26/27 September 1972 it suddenly brightened to m_v +0.7 and stayed there for about 10 minutes. It is 672 ly away but at 10 pc would brighten to M_v -4.4.

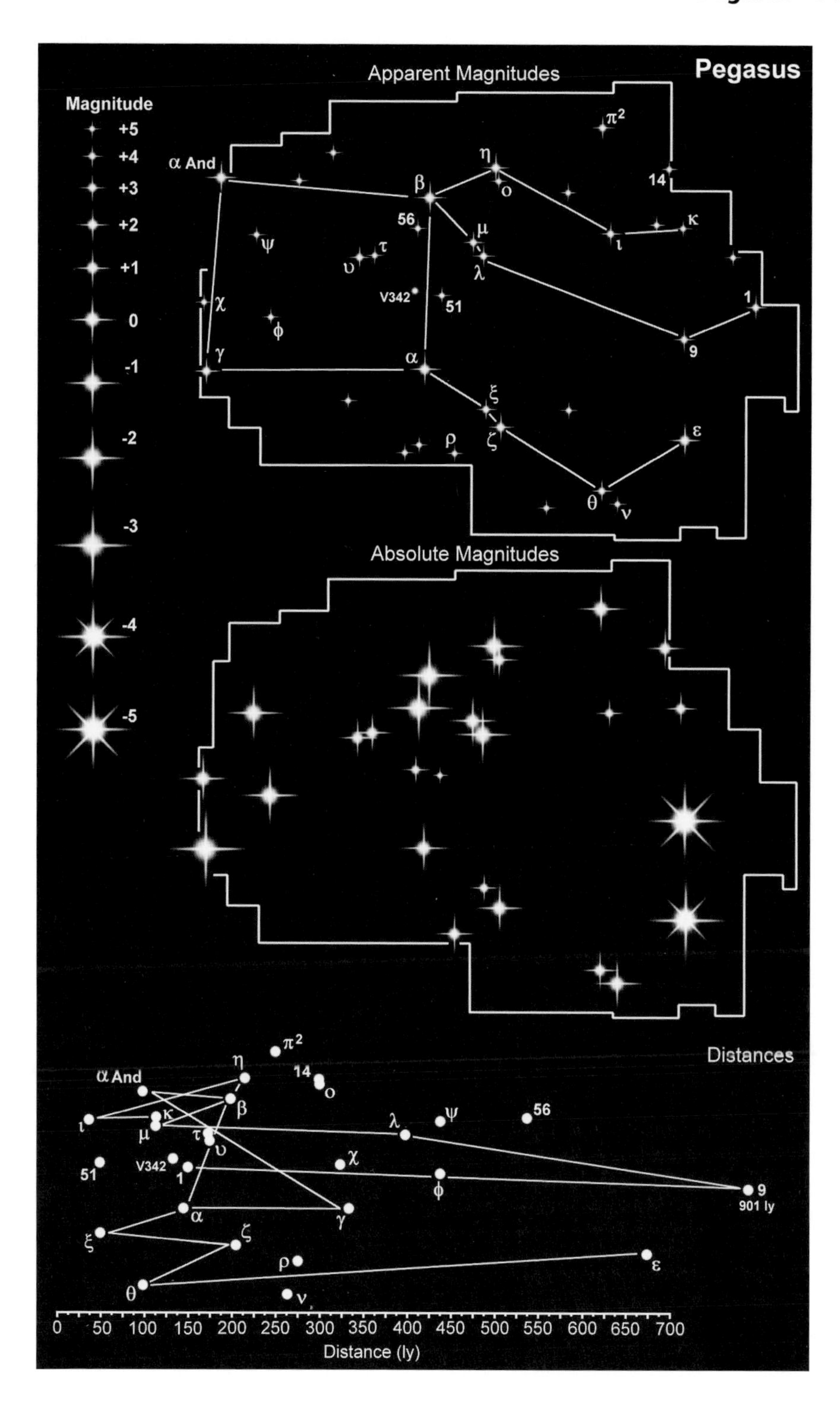

Pegasus
Apparent Magnitudes
Magnitude
+5
+4
+3
+2
+1
0
-1
-2
-3
-4
-5
α And
β
η
ο
π2
14
κ
ι
56
μ
λ
ψ
τ
υ
V342
51
1
χ
ϕ
γ
α
9
ξ
ζ
ρ
ε
θ
ν
Absolute Magnitudes
Distances
901 ly
0 50 100 150 200 250 300 350 400 450 500 550 600 650 700
Distance (ly)

ζ **Pegasi** is another B-class (B8) with a temperature of 11,200 K. Some 209 ly from Earth it has a diameter of 1.4 $D_{\odot}$ and mass of 3.4 $M_{\odot}$. Its 229 $L_{\odot}$ yields a m_V +3.39 star that has microfluctuations of just 0.00049 of a magnitude. It has two companions: ζ^{C} **Pegasi** is an 11th magnitude star in an 11,000 AU orbit that takes 636,000 years to complete. The 12th magnitude ζ^{B} **Pegasi** is just a line-of-sight coincidence.

The jury is still out on whether η **Pegasi** is a double or double-double system. The primary star is a G2 of 18 $D_{\odot}$ and mass of 3.2 $M_{\odot}$. Its spectroscopic companion is likely to be an A5 in an 818 day, 3 AU orbit. At 90.4″ separation (PA 339°) is a second binary pair that carries the designation BD+29° 4740 from the *Bonner Durchmusterung* stellar catalog. Probably a pair of mid-G class stars they orbit one another at an average distance of 13 AU with a period of 34 years. They in turn may orbit the primary pair at a distance of 6,000 AU taking about 170,000 years to complete a single orbit.

ι **Pegasi** is a spectroscopic binary and the closest star in the constellation to us at 38.4 ly. Twinkling at m_V +3.77 its absolute magnitude is not much different at M_V +3.1 due to the star being close to the 10 pc standard distance. An F5 with a diameter of 1.45 $D_{\odot}$ and mass of 1.29 $M_{\odot}$ it has a luminosity output of 3.48 $L_{\odot}$. Its G8 companion is smaller all round: 0.9 $D_{\odot}$, 0.81 $M_{\odot}$ and 0.9 $L_{\odot}$. They are in a near circular orbit of 0.051 AU (7.6 million km) with a period of 10.2 days. Estimates of the age of the system range from 80 million years to more than 2,500 million years.

In the late 1800s κ **Pegasi** was the most rapid binary system then known with an orbital period of 11.53 years. The primary is a m_V +4.8 F5, though combined with the light from its partner it actually looks like a m_V +4.14 star. Some 115 ly away it is 2.3 $D_{\odot}$ across, has a mass of 1.55 $M_{\odot}$ and a luminosity of 22.3 $L_{\odot}$. It rotates at 42.3 km/s, slightly above the average for F5 stars of 35 km/s, and so takes 2.8 days to turn once on its axis. Its binary nature was discovered in 1880 by S.W. Burnham. The secondary component is, in many ways, similar to the primary: an F6 of 2.7 $D_{\odot}$, 1.66 $M_{\odot}$ and 8 $L_{\odot}$ resulting in a m_V +5.2 star. The orbit swings between 5.25 AU and 9.75 AU (similar to the distances of Jupiter and Saturn from the Sun). There is a third spectroscopic component, a K0 of 0.82 $M_{\odot}$ in a 5.9715 day long orbit around the secondary star. The whole system is believed to be about 2,500 million years old.

Pegasus contains three G8 stars all within a stone's throw of one another. λ **Pegasi** is a 24 $D_{\odot}$, 3.8 $M_{\odot}$ giant some 395 ly away. With a luminosity of 312 $L_{\odot}$ it has a visual magnitude of m_V +3.95 and an absolute magnitude of M_V -0.9. Rotating four times faster than the Sun at 8 km/s it takes nearly 152 days to turn once on it axis. About a degree to the north east is the m_V +3.51 μ **Pegasi**. Somewhat smaller at 9 $D_{\odot}$ and less massive at 2.3 $M_{\odot}$ it is only 14% as luminous, 45 $L_{\odot}$, and would attain just M_V +0.2 at 10 pc. Spinning at about the same speed as the Sun, 1.9 km/s, it takes almost 240 days to rotate once. The third G8 star is just to the east of the line that joins α and β Pegasi. Cataloged by Flamsteed as **56 Pegasi** is the third and faintest of the trio at m_V +4.76 but also the brightest at 10 pc: M_V -2.1. Some 55 $D_{\odot}$ and 274 $L_{\odot}$ it resides 537 ly away and spins at 17 km/s,

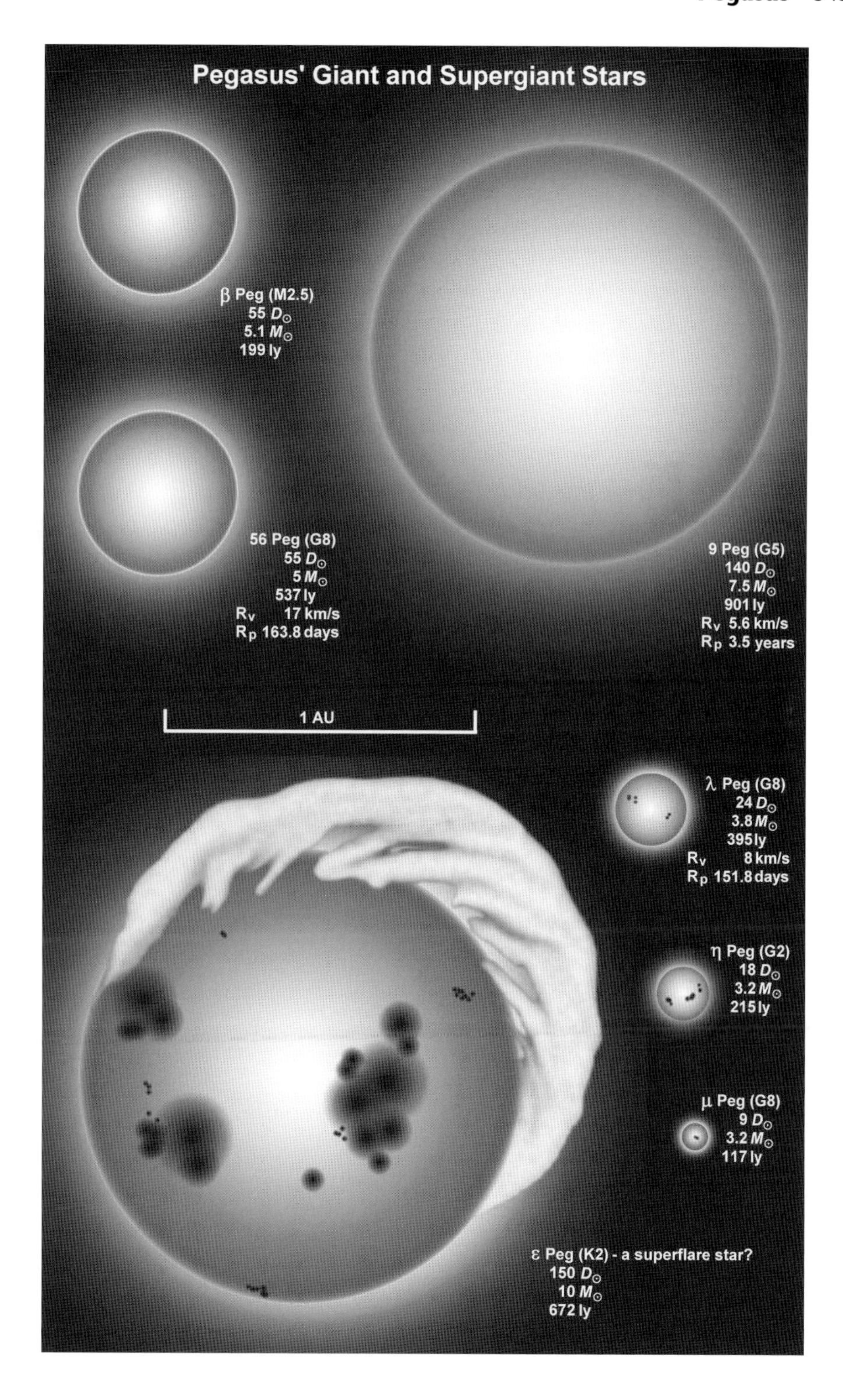

Pegasus' Giant and Supergiant Stars
β Peg (M2.5)
55 $D_\odot$
5.1 $M_\odot$
199 ly
56 Peg (G8)
55 $D_\odot$
5 $M_\odot$
537 ly
R_v 17 km/s
R_p 163.8 days
9 Peg (G5)
140 $D_\odot$
7.5 $M_\odot$
901 ly
R_v 5.6 km/s
R_p 3.5 years
1 AU
λ Peg (G8)
24 $D_\odot$
3.8 $M_\odot$
395 ly
R_v 8 km/s
R_p 151.8 days
η Peg (G2)
18 $D_\odot$
3.2 $M_\odot$
215 ly
μ Peg (G8)
9 $D_\odot$
3.2 $M_\odot$
117 ly
ε Peg (K2) - a superflare star?
150 $D_\odot$
10 $M_\odot$
672 ly

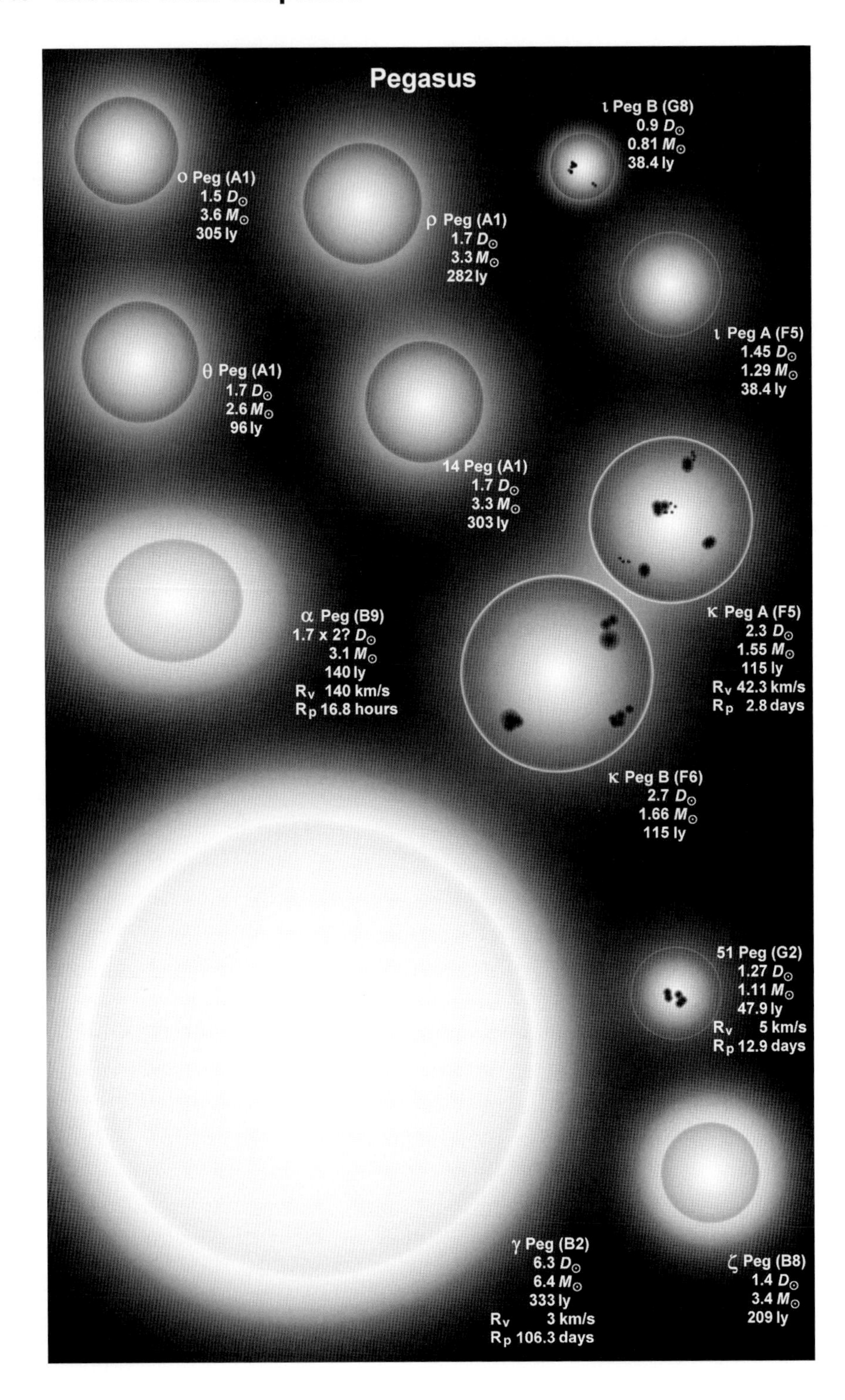
Pegasus
ι Peg B (G8)
0.9 $D_\odot$
0.81 $M_\odot$
38.4 ly
ο Peg (A1)
1.5 $D_\odot$
3.6 $M_\odot$
305 ly
ρ Peg (A1)
1.7 $D_\odot$
3.3 $M_\odot$
282 ly
ι Peg A (F5)
1.45 $D_\odot$
1.29 $M_\odot$
38.4 ly
θ Peg (A1)
1.7 $D_\odot$
2.6 $M_\odot$
96 ly
14 Peg (A1)
1.7 $D_\odot$
3.3 $M_\odot$
303 ly
α Peg (B9)
1.7 x 2? $D_\odot$
3.1 $M_\odot$
140 ly
R_v 140 km/s
R_p 16.8 hours
κ Peg A (F5)
2.3 $D_\odot$
1.55 $M_\odot$
115 ly
R_v 42.3 km/s
R_p 2.8 days
κ Peg B (F6)
2.7 $D_\odot$
1.66 $M_\odot$
115 ly
51 Peg (G2)
1.27 $D_\odot$
1.11 $M_\odot$
47.9 ly
R_v 5 km/s
R_p 12.9 days
γ Peg (B2)
6.3 $D_\odot$
6.4 $M_\odot$
333 ly
R_v 3 km/s
R_p 106.3 days
ζ Peg (B8)
1.4 $D_\odot$
3.4 $M_\odot$
209 ly

its rotational period coming in at 164 days. λ and 56 are heading our way at 3.9 and 26.8 km/s respectively while μ is receding at 13.9 km/s. None of the stars is related.

o **Pegasi** is almost the second smallest star in the constellation at 1.5 $D_{\odot}$ though estimates range from 1.1 to 2.5 $D_{\odot}$. About 305 ly away, give or take 18 ly, it is an A1 of around 86 $L_{\odot}$ and shines at m_v +4.79. Of the six other stars that are less than two solar diameters across, three are A1s. ρ **Pegasi** is a m_v +4.90 (M_v +1.2) lying at a distance of 282 ± 21 ly and is 1.7 $D_{\odot}$. Of the same size is **14 Pegasi** with an overlapping distance of 303 ± 15 ly. It appears as a m_v +5.07 star (M_v +0.6) and has an astrometric companion in a 5.3047 day orbit. **θ Pegasi**, on the other hand, is an A2 just 96 ± 2 ly away. With an output of 28 $L_{\odot}$ it shines at m_v +3.51 (M_v +1.4). It is also a newly suspected variable, believed to fade to m_v +3.56, and carries the catalog number NSV 14057. At the other end of the size scale is, of course, the 150 $D_{\odot}$ ε Pegasi mentioned earlier and the 140 $D_{\odot}$ supergiant **9 Pegasi.** A little over 900 ly from Earth 9 Peg is a m_v +4.33 G5. Erratically variable to m_v +4.2 this monster of a yellow star rotates at just 5.6 km/s taking 1,265 days – 3.5 years – to turn just once on its axis.

Pegasus is home to one of the oldest and probably the densest of all globular clusters. Messier 15 (**M15** or **NGC 7078**) is 33,600 ly from us and appears as a m_v +6.2 gray smudge against the sky. At 10 pc however it would be an impressive M_v -9.3. Containing in excess of 30,000 stars it has no fewer than 9 pulsars, 112 variable stars and one planetary nebula. It has considerable structure. The central core is just 1.4 ly across and is dominated by larger F3 and F4 stars. This process of mass segregation is known as *core collapse* and, in the case of M15, has resulted in a strong X-ray source in the very center of the cluster. Half the mass of the cluster is contained within 10 ly of the center although the cluster is at least 176 ly across. Theoretically the cluster could extend to a diameter of 210 ly beyond which the gravitational tides of the Galaxy would systematically strip the cluster of its member stars. Now 12,000 million years old M15 is closing in on us at 107 km/s.

Pegasus contains several stars that have planetary systems. One of the most notable is WASP-10, a K5 dwarf which has a three Jovian mass planet in an orbit that is only 10% the size of Mercury's orbit. The planet orbits the star in just 3.1 days corresponding to an orbital velocity of 123 km/s or 442,241 km/h – about four times the Earth's orbital velocity. The planet in orbit around the variable star **V342 Peg** (HD 209458 b) was the first to have its atmosphere analyzed. Discovered on 5 November 1999 this 'hot Jupiter' orbits the star in a slightly eccentric path that causes its distance to vary between 6.6 and 7.5 million kilometers. As a result its atmosphere is heated to around 1,400 K and is evaporating at a significant rate. Eventually the atmosphere will disappear completely, possibly leaving behind a rocky terrestrial 'chthonian' planet. As the planet transits the star, spectroscopic studies have revealed an atmosphere rich in hydrogen, oxygen and carbon and possibly water vapor. The first gas to be discovered was sodium, but not as much as expected leading to speculation that clouds high in the planet's atmosphere could be blocking out some of the sodium spectral signature. The planet has earned the unofficial name of Osiris. **51 Pegasi** was the first naked-eye star to show signs of a planet back in 1995 – the sixth extrasolar planet to be discovered.

Planetary systems in Pegasus

Star	$D_{\odot}$	Spectral class	ly	m_v	Planet	Minimum mass	q	Q	P
51 Peg	1.27	G2	47.9	+5.49	51 Peg b	0.468 M_J	0.052	0.052	4.23 d
V342 Peg	?	A5	129	+5.96	HR 8799 b	7 M_J	41.5	94.5	465.4 y
					HR 8799 c	10 M_J	23.2	52.8	188.9 y
					HR 8799 d	10 M_J	14.6	33.4	99.9 y
WASP-10	0.78	K5	293.5	+12.7	WASP-10 b	3.06 M_J	0.035	0.039	3.1 d

Globular cluster in Pegasus

Name	Size arc min	Size ly	Distance ly	Apparent magnitude m_v
M15 (NGC 7078)	18′	176	33,600	+6.2

Perseus

Constellation:	Perseus	**Hemisphere:**	Northern
Translation:	Perseus	**Area:**	615 deg^2
Genitive:	Persei	**% of sky:**	1.491%
Abbreviation:	Per	**Size ranking:**	24th

Perseus, the son of Zeus and Danaë, beheaded Medusa the Gorgon and rescued Andromeda.

α Persei or Mirphak is the unequivocal luminary of the constellation. At m_v +1.79 it outshines its nearest rival, β Persei, by a third of a magnitude and would attain M_v -4.6 at 10 pc: β would just about manage M_v -0.2. It is an F5 supergiant of 55 $D_\odot$ with a mass of 6.65 $M_\odot$. F5 supergiants are a positive rarity among naked eye stars. 35 Cygni is thought to be about 51 $D_\odot$ and the only other supergiant, κ Pavonis, may or may not be as large as 48 $D_\odot$ – no one is really sure – but α Per is the largest of this very small bunch with one estimate suggesting it may even be 63 $D_\odot$ across. It outputs 5,000 $L_\odot$ and may be variable between m_v +1.72 and +1.86, gaining the catalog number NSV 1125 as a newly suspected variable. Spinning at 18 km/s it takes 154.7 days to turn once on its axis. With a temperature of 6,180 K the general consensus is that the star is about 52 million years old.

Looking at α Per without any optical aid will reveal it to be surrounded by about half a dozen stars; the keen sighted will be able to spot a few more. A binocular or small telescope will reveal dozens. α Per is actually a member of **Melotte 20**, an open cluster of 139 stars. The cluster is centered at a distance of about 582 ly – α Per itself lies close to the middle at 592 ly – and is composed of B3 to G3 class stars. The central core of the cluster stretches from about 556 to 610 ly and contains half the entire mass, some 96 mainly B, A and F class stars. Most of the rest of the mass extends from 500 to 680 ly (30 stars) and then there is a gap of about 60 ly in which very few stars exist before coming to a halo of just 13 dwarf stars between 400-460 ly and 710-770 ly. Stars brighter than m_v +7.4 are predominantly B-class, α Per being the exception. Those between m_v +7.4 and m_v +9.0 are mainly A-class while F-class stars dominate between m_v +9.0 and m_v +10.8 leaving the fainter members belonging to the G-spectral class. Melotte 20's naked eye stars include ψ Persei, 29, 31 and 34 Persei, the two variables V396 and V575 and HD 21278. The entire cluster is heading in the direction of β Tauri at 16 km/s. Robert Burham pointed out that it will take 90,000 years for the cluster to move just 1° on the celestial sphere.

No one knows how long **β Persei** has been regarded as an unlucky star. It probably dates back to the time when the myth of Perseus was created. It represents the head of Medusa the Gorgon and is also known as Gorgonea Prima. In ancient Arabia it earned the name al Ghul, from which we get ghoul in modern day English and, by corruption, the common name of the star: Algol the Demon Star (sometimes called the 'Winking Demon'). And demon-like it must have appeared to ancient civilizations, fluctuating between m_v +2.12 and m_v +3.39 with a clockwork regularity of $2^d\ 20^h\ 48^m\ 55^s$, its minima lasting 10 hours. Algol is the

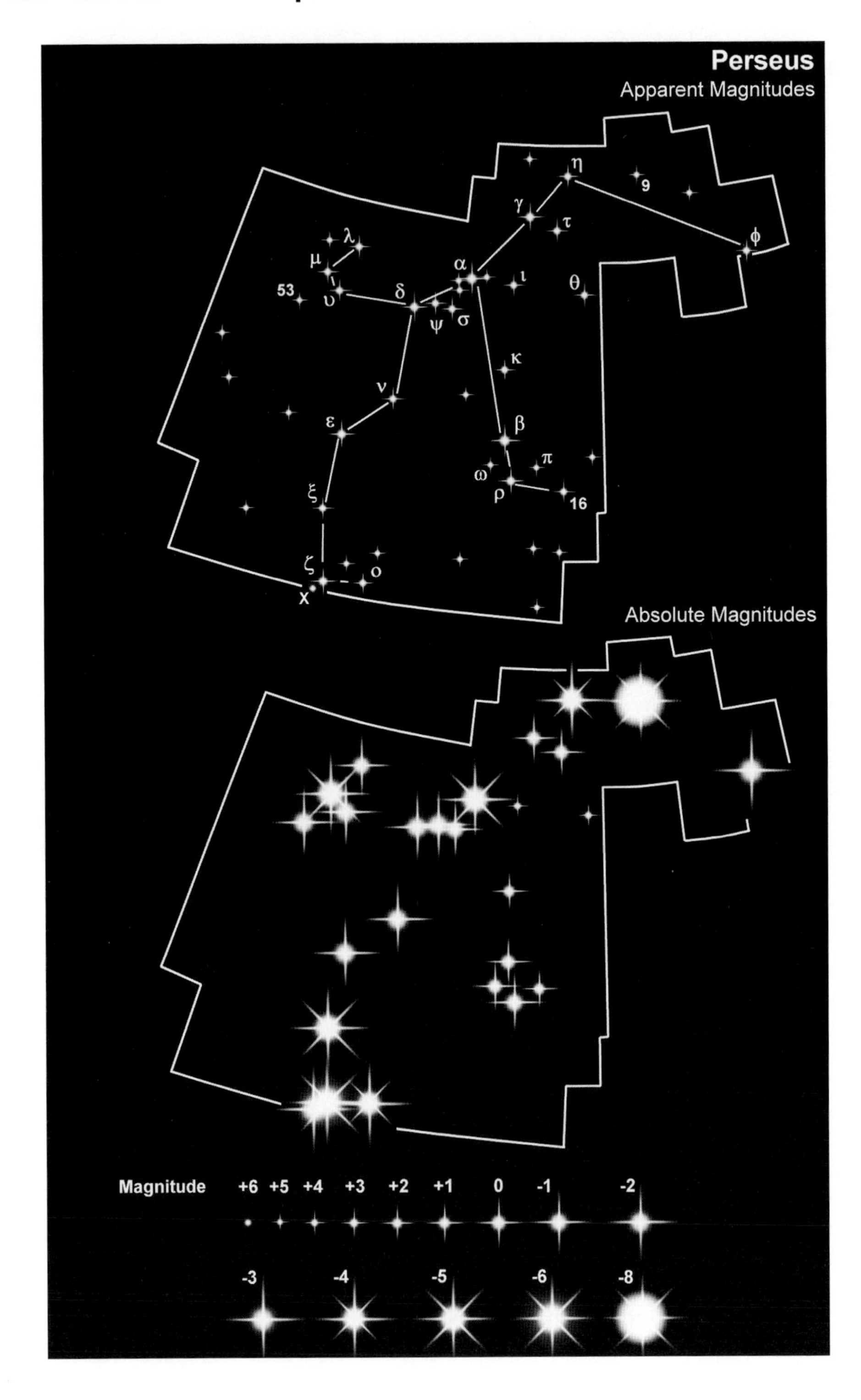
Perseus
Apparent Magnitudes
η
9
γ
τ
ϕ
λ
μ
α
ι
θ
53
υ
δ
ψ
σ
κ
ν
ε
β
π
ω
ρ
16
ξ
ζ
o
X
Absolute Magnitudes
Magnitude
+6 +5 +4 +3 +2 +1 0 -1 -2
-3 -4 -5 -6 -8

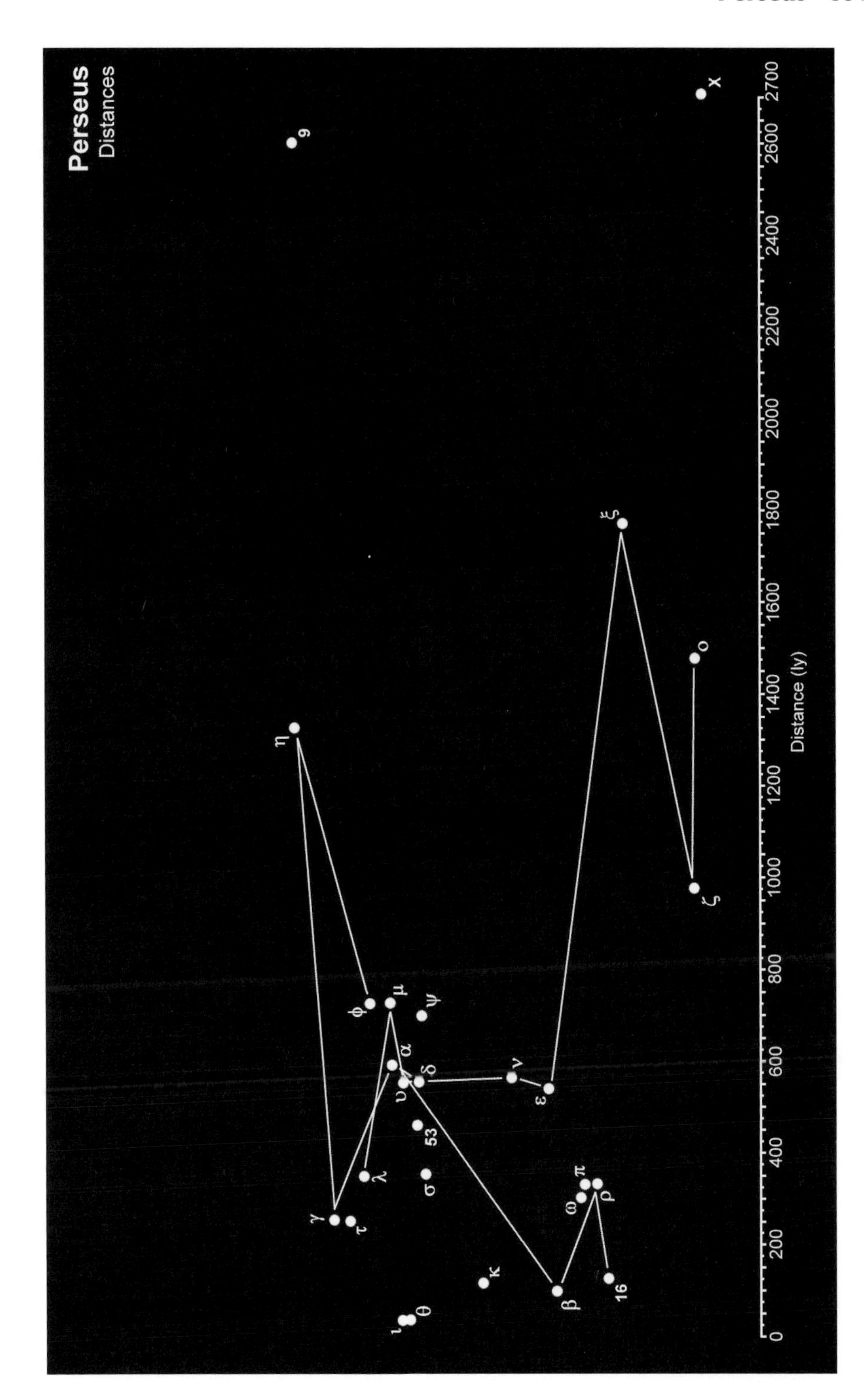
Perseus
Distances
Distance (ly)
0
200
400
600
800
1000
1200
1400
1600
1800
2000
2200
2400
2600
2700
ι
θ
κ
β
16
γ
τ
λ
σ
ω
π
ρ
53
υ
δ
α
ε
ν
ϕ
μ
ψ
ζ
η
o
ξ
9
X

prototype for the EA eclipsing Algol-type binaries of which many are now known. As the name suggests the dip in magnitude is caused by a fainter star passing in front of a brighter star. Periodicity can last from about 2 hours to more than 10,000 days (27.4 years) with amplitudes of between 0.1 to 3 magnitudes. All EA variables are A, B or F spectral class. Algol itself is a 2.8 $D_{\odot}$ B8 with a mass of 3.6 $M_{\odot}$ and a luminosity of 94 $L_{\odot}$ which would yield an M_v -0.2 at 10 pc. Although it rotates at 50 km/s, 25 times faster than the Sun, its rotational velocity is just one-third of the average for B8 stars. This can be attributed to its eclipsing companion causing a drag or rotational braking on Algol. The secondary is physically a larger star at 3.5 $D_{\odot}$ but is much less massive, only 0.81 $M_{\odot}$, and considerably less luminous at 4.5 $L_{\odot}$. Separated by just 0.05 AU (7.5 million km) there is mass transfer from the secondary to the primary which may also be responsible for the X-rays which the system emits. The nature of the secondary is not particularly well understood; it could be anything from a G5 to a K2. The secondary passing in front of the primary causes the noticeable reduction in magnitude. However, when Algol eclipses its companion the dip in brightness is so small it can only be detected photoelectrically. There is also a third component in orbit around the two main stars at an average distance of 3 AU. Possibly an A-class, perhaps an F, its mass is calculated to be 1.8 $M_{\odot}$. Its orbital period is 681 days (1.87 years). The entire system lies 92.8 ly from Earth and is thought to be less than 300 million years old.

γ **Persei** is another EA eclipsing binary, but not as dramatic or as frequent as Algol. Around 6.6 $D_{\odot}$ γ Per is a G8 with a mass of 2.5 $M_{\odot}$. For most of the time it shines at a steady m_v +2.92 but every 5,346 days – 14.64 years – it fades to m_v +3.1. Its spectroscopic companion is a 1.9 $M_{\odot}$ dwarf, an A2, in an orbit that varies between 2 and 18 AU. Around 1,900 million years old, 40% the age of the Sun, the pair are 256 ly away.

δ **Persei** is a slightly variable B5 some 528 ly from Earth. Belonging to the γ Cas class of eruptive variables it changes from m_v +2.99 to +3.04 with no regularity and would brighten to M_v -2.2 at 10 parsecs. Estimates of its diameter range from 2.9 to 10 $D_{\odot}$ with 5 $D_{\odot}$ being about average. Like many B-class stars it is a fast spinner at 190 km/s, about 40 km/s above average for its type. Needless to say, it is probably more of an oblate spheroid than a circular ball, centrifugal forces causing its equator to bulge and darken while its poles are sucked in by gravity: this is known as the Von Zeipel Effect. With a mass of around 6 $M_{\odot}$ and luminosity of about 1,400 $L_{\odot}$ the star is estimated to be just 50 million years old.

Definitely a binary, but possibly a triple star system, ε **Persei** is 538 ly from Earth. A pulsating variable of the β Cephei or β CMa variety its magnitude fluctuates between m_v +2.88 to +3.00. Another rapid spinning B0.5 its rotational velocity has been estimated to be 155 km/s give or take 3 km/s. Its diameter is less certain: published estimates include 3.4, 6.1, 7.0 and 7.7 $D_{\odot}$. At 13.5 $M_{\odot}$ it is considerably more massive than its A2 companion which has a mass of 2.29 $M_{\odot}$ but which spins at almost twice its rotational velocity: 300 km/s (more than three times the average for its class). Separated by around 1,600 AU the pair take more than 11,200 years to circle one another. The third component, if it really exists, is a much smaller 0.94 $M_{\odot}$ star in a 0.3 AU orbit which it takes 14.08 days to complete.

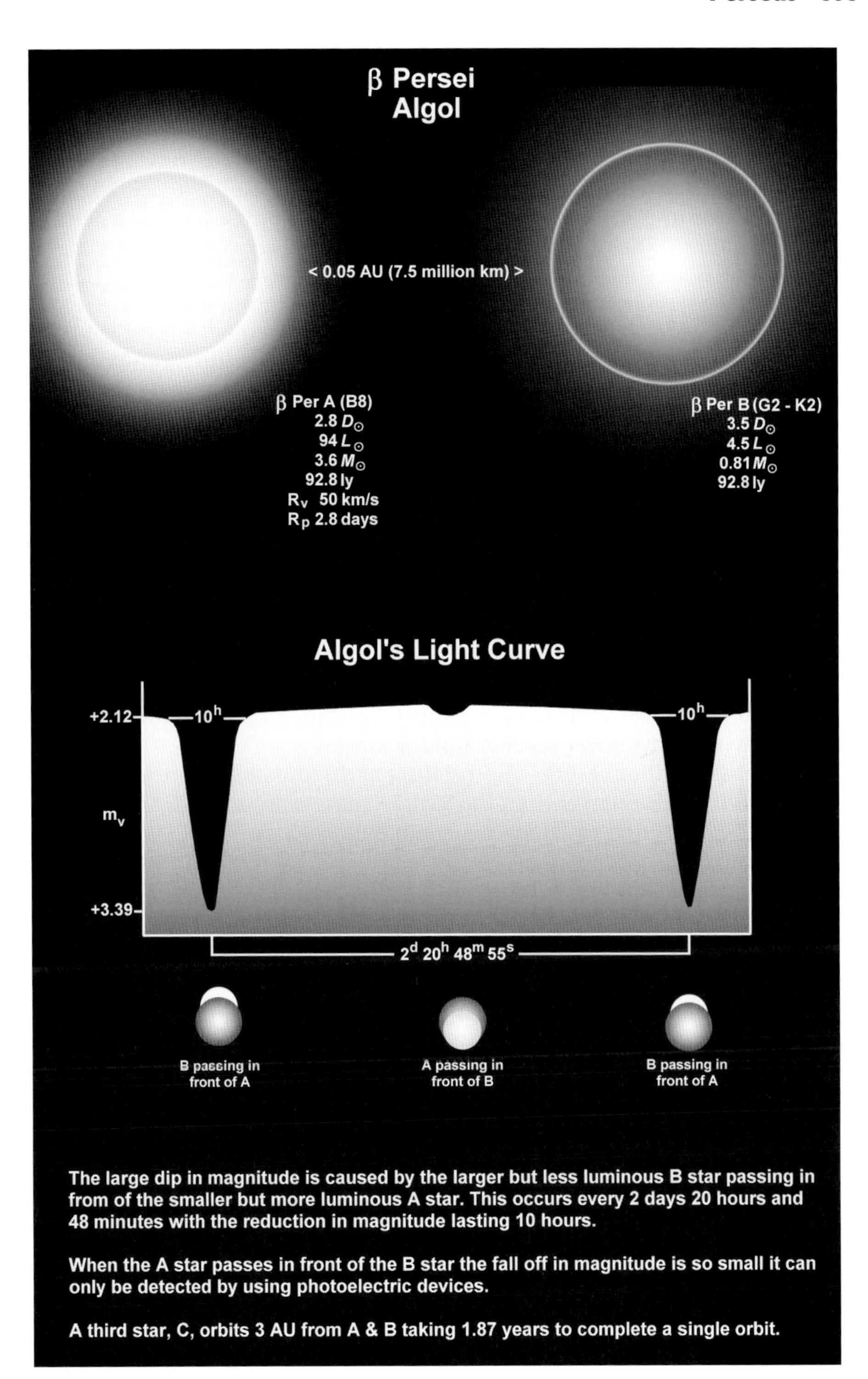
β Persei
Algol
< 0.05 AU (7.5 million km) >
β Per A (B8)
2.8 D☉
94 L☉
3.6 M☉
92.8 ly
Rv 50 km/s
Rp 2.8 days
β Per B (G2 - K2)
3.5 D☉
4.5 L☉
0.81 M☉
92.8 ly
Algol's Light Curve
+2.12
10h
10h
mv
+3.39
2d 20h 48m 55s
B passing in front of A
A passing in front of B
B passing in front of A
The large dip in magnitude is caused by the larger but less luminous B star passing in from of the smaller but more luminous A star. This occurs every 2 days 20 hours and 48 minutes with the reduction in magnitude lasting 10 hours.
When the A star passes in front of the B star the fall off in magnitude is so small it can only be detected by using photoelectric devices.
A third star, C, orbits 3 AU from A & B taking 1.87 years to complete a single orbit.

ζ **Persei** is a quadruple star system some 982 ly from Earth. The primary is a m_V +2.85 B1 of 55 $D_\odot$ and with a mass of 19 $M_\odot$. About 9 million years old it is a spectroscopic binary but virtually nothing is known of its companion. Four visual stars have received the designations B to E but only B and E are truly associated. ζ^B Persei is a m_V +9.16 B8. It has an average distance of at least 3,900 AU (13.3″ at PA 208°) and takes in excess of 50,000 years to orbit the primary. Almost ten times farther away at 36,000 AU (120″ at PA 286°) is ζ^E Persei, an A2 in a highly unstable 1.5 million year long orbit. At 10 pc ζ Per would become an impressive M_V -5.7 star.

Some stars are just plain awkward. η **Persei** is one. Originally classed as an M3 it is now generally regarded as a K3. Its diameter has been estimated as being somewhere between 44 and 221 $D_\odot$ and its rotational velocity could be as low as 6.8 km/s or as high as 155 km/s. Its mass is a more certain 9 to11 $M_\odot$ depending on how far along the evolutionary path the star really is. All things taken into consideration it is likely to be a slow spinning supergiant as big as the Earth's orbit and taking 4.5 years to turn once on its axis. The star's distance from us is 1,331 ly, give or take 312 ly.

Tearing through our neighborhood at 92 km/s ι **Persei** is a very good solar analog. A G0 with a diameter of 1.08 $D_\odot$ and mass of 1.1 $M_\odot$ it is a little over twice as luminous as the Sun, 2.16 $L_\odot$, and is about eight times more metallic in composition. Its age has been estimated to be between 2,089 million and 4,571 million years with one other estimate putting it at 8,100 million years. Currently 34.4 ly away it rotates at 10 km/s. Although it has been searched for planets none has been found. However, this may simply be because our detection methods are not yet able to find smaller terrestrial size planets.

μ **Persei** is another G0 star also rotating at 10 km/s but there the similarity ends. A good 53 times larger than the Sun, its mass is around 6 $M_\odot$ and its age is just 60 to 70 million years. Hanging in our skies as a m_V +4.14 star its absolute magnitude of M_V -4.5 would be on par with Venus. It has an unseen spectroscopic companion; a B9 in a 1.7 AU orbit that takes 283.3 days to complete. The system is 723 ly away.

O-class stars are relatively rare. There are only a couple of dozen naked eye examples – 0.86% of all naked eye stars – and six of those hang out in Orion. They are incredibly hot, up to 60,000 K, and painfully luminous at up to 15 million $L_\odot$, theoretically at least. Their radiation output can destroy nebulae and planetary atmospheres alike, sterilizing any orbiting planets and preventing the establishment of even the most simple forms of life. ξ **Persei** is one such star. An O7.5 with a diameter of 26 $D_\odot$ and temperature of around 40,000 K it rotates at 216 km/s (the average for the class is 138 km/s). From a distance of between 1,800 and 2,260 ly it appears as a 4th magnitude star, slightly variable between m_V +4.00 and +4.06 but has an absolute magnitude of about M_V -5.

As we noted above Algol represents Medusa, the Queen of the Gorgons or the Gorgonea Prima. The legend of Perseus mentions a total of three Gorgons but the constellation of Perseus has four! The second Gorgon, Stheno the Mighty, is marked by π **Persei** – Gorgonea Secunda. π Per is an A2 star lying 326 ly from Earth. Almost twice as large as the Sun at 1.8 $D_\odot$ it has a luminosity of 108 $L_\odot$

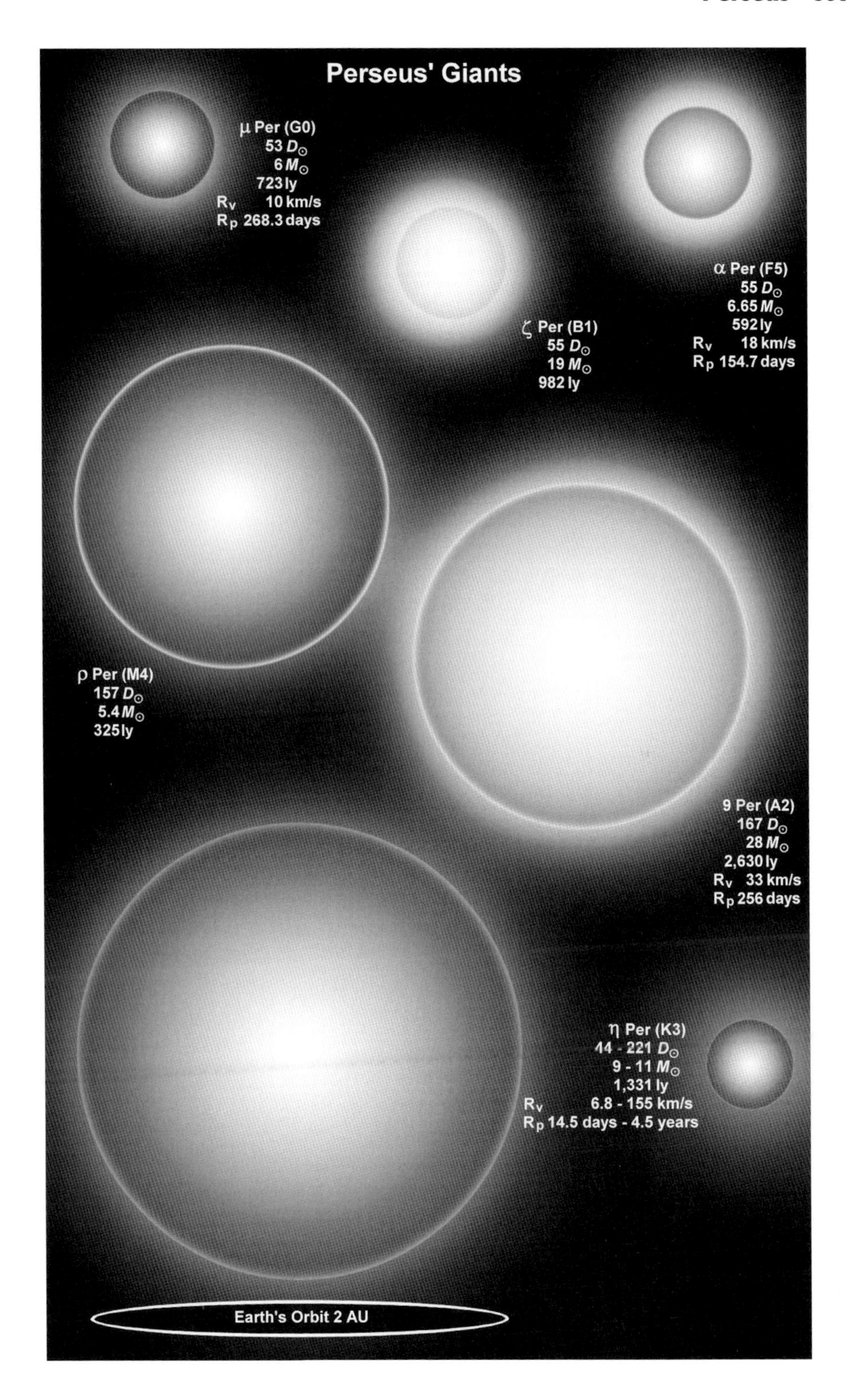
Perseus' Giants
μ Per (G0)
53 D☉
6 M☉
723 ly
Rv 10 km/s
Rp 268.3 days
α Per (F5)
55 D☉
6.65 M☉
592 ly
Rv 18 km/s
Rp 154.7 days
ζ Per (B1)
55 D☉
19 M☉
982 ly
ρ Per (M4)
157 D☉
5.4 M☉
325 ly
9 Per (A2)
167 D☉
28 M☉
2,630 ly
Rv 33 km/s
Rp 256 days
η Per (K3)
44 - 221 D☉
9 - 11 M☉
1,331 ly
Rv 6.8 - 155 km/s
Rp 14.5 days - 4.5 years
Earth's Orbit 2 AU

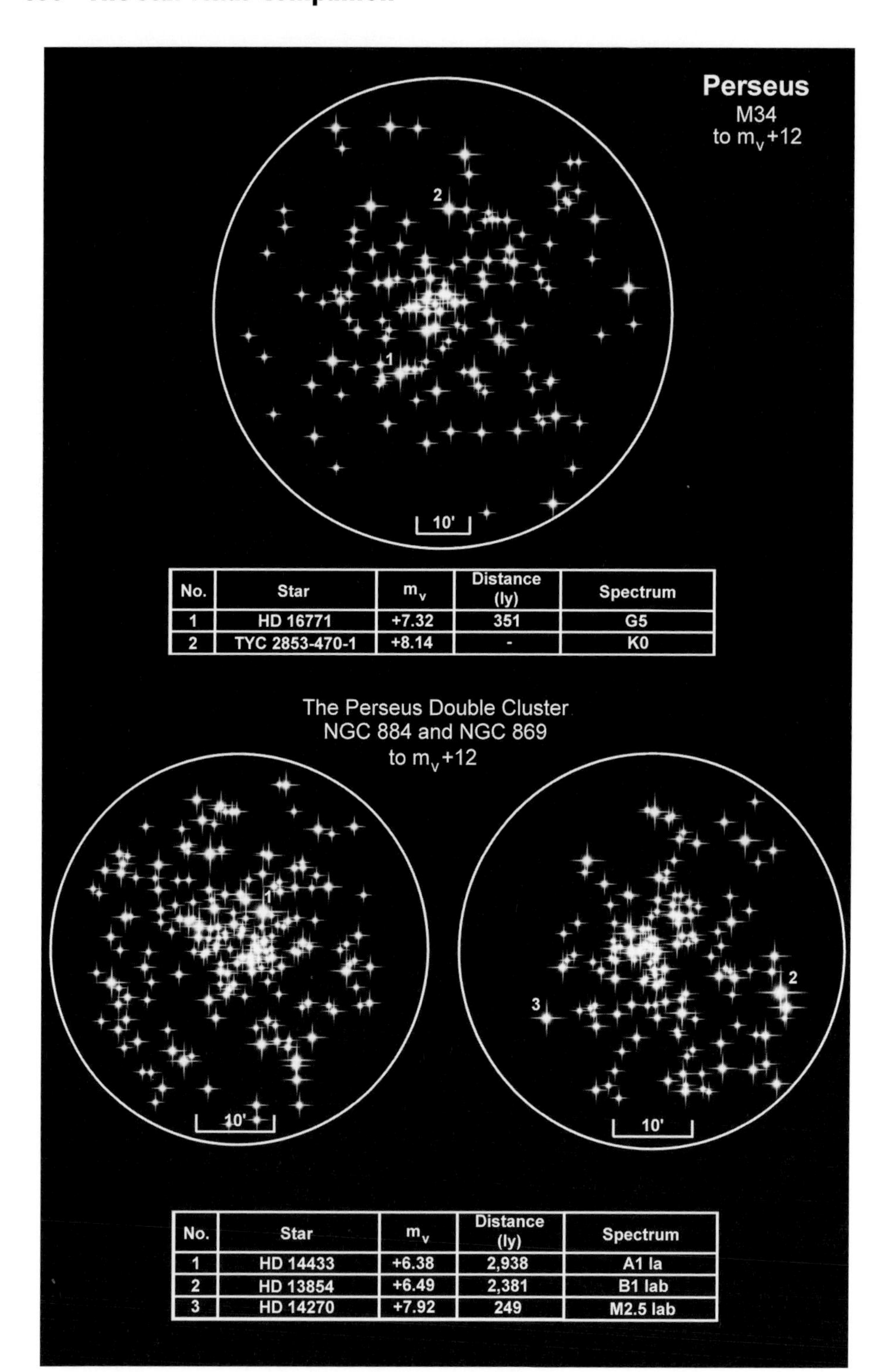

No.	Star	m_v	Distance (ly)	Spectrum
1	HD 16771	+7.32	351	G5
2	TYC 2853-470-1	+8.14	-	K0

No.	Star	m_v	Distance (ly)	Spectrum
1	HD 14433	+6.38	2,938	A1 Ia
2	HD 13854	+6.49	2,381	B1 Iab
3	HD 14270	+7.92	249	M2.5 Iab

and appears as a modest m_v +4.68 but with an absolute magnitude of M_v +1.4. It rotates at 185 km/s, literally twice the average for its class. Of the 600 or so A2 stars whose rotational velocity has been measured or at least estimated only 15% rotate faster than 184 km/s.

ρ **Persei** is Gorgonea Tertia, the third Gorgon whose name was Euryale the Far-springer. Very different to both her sisters ρ Per is a bright, big red giant of 157 $D_\odot$ (0.7 AU) and with a visual luminosity of 362 $L_\odot$, though you would have to multiply that by a factor of ten to get the full luminosity, much of its output being radiated in the infra-red. It will come as no surprise to learn that this is a semi-regular pulsating variable, an SRb, noticeably sliding between m_v +3.30 and +4.00 over a 50 day period. At 10 pc it would brighten to M_v -0.5.

The mysterious fourth Gorgon, Gorgonea Quarta, is marked by **ω Persei**, a m_v +4.61 K1 giant, 16 $D_\odot$ and 305 ly away. Spinning at 20 km/s it takes 40.5 days to rotate once on its axis. Although three of the Gorgons are of similar distance from Earth they are not related.

τ **Persei** is a binary consisting of a giant 14 $D_\odot$ G4 and a 2.2 $D_\odot$ A4 dwarf. The giant primary has a luminosity of 150 $L_\odot$ and mass of 3.3 $M_\odot$. With a rotational velocity of 25 km/s it takes 28.3 days to turn once. Its much smaller companion rotates twice as fast and so turns once every 2.2 days. The two stars are in a 4.42 year long orbit during which they reach a periastron of 1.13 AU and an apastron of 7.2 AU. The alignment of the two stars with the Earth means that they eclipse one another, the magnitude dipping from m_v +3.94 to +4.07.

Perseus is host to an X-ray pulsar called, appropriately enough, **X Persei**. Somewhat below naked eye visibility at m_v +6.1 and with a variability than can take it down to 7th magnitude, X Per is an O9.5 of 13 $D_\odot$ and with a mass of 15 $M_\odot$. Its distance is problematic – estimates range from 2,300 to 4,200 ly – as is its rotational velocity; 162, 199 and 360 km/s have all been quoted. Like all O-class stars it is highly luminous with an output of 24,000 $L_\odot$, and is probably less than 10 million years old. Its companion is a ball of neutrons – a neutron star – just 20 km across and with a mass of 0.0000002 $M_\odot$. That may not seem like much but its density works out at about 100 metric tonnes per cubic centimeter! Just to put that into context, the Sun has a density of 1.4 grams per cubic centimeter, water is 1 g/cm^3 and gold 19.3 g/cm^3. The two stars are in an almost circular orbit with their separation varying between 1.94 and 2.06 AU. The interaction between the two stars causes bursts of X-rays every 13^m 55^s and these pulsations have been used to calculate an orbital period of 250 days.

9 Persei is the brightest member of the **Perseus OB1** association. This huge group of stars straddles an 8° × 6° section of the celestial sphere. In real terms the dimensions are about 100 × 750 ly and includes the famous Double Cluster (see below). 9 Per is an A2 bright supergiant and may be the largest naked eye star in the constellation at 167 $D_\odot$ perhaps only beaten by η Per. It is slightly variable between m_v +5.15 and +5.25 with a period in excess of 200 days and belongs to the α Cygni class of variables. At 10 pc it would be a brilliant M_v -7.5, its luminosity coming in at about 120,000 $L_\odot$. A tentative estimate of its rotational velocity worked out at 33 km/s meaning that it takes more than 256 days to turn once on its axis. 9 Per, which is also known as **V474 Persei**, lies at a distance of 2,630 ly but with a fair amount of error.

At first glance **53 Persei** or **V469 Persei** would appear to be just another β Cephei variable, but a closer examination reveals it differs from most of the type. Fluctuatng between m_v +4.81 and +4.86 with a period of $7^h\ 17^m\ 46^s$ 53 Per is a B4 of 4.9 $D_\odot$ and with a mass of 5.8 $M_\odot$. Its pulses are non-radial, that is to say they travel in all directions across the surface of the star, rather like a wobbling jelly. While β Cepheids normally have periods of 0.1 to 0.5 days several have longer periods of up to 2 days: these belong to the 53 Persei sub-group and are often referred to as Slowly Pulsating B-class or SPB stars. In addition to the main pulses, there are often several other pulsation periods, usually quite minor and unstable.

χ and **h Persei** are not stars at all but open clusters now cataloged as **NGC 884** and **NGC 869**. Together they form the famous Double Cluster. Separated by about 300 ly, NGC 884 is the more easterly cluster and the farthest at about 7,400 ly. It is also the youngest at 3.2 million years. Slightly closer at 7,100 ly is NGC 869. Believed to be 5.6 million years old the two clusters were first recorded by Hipparchus *c.* 130 BC. At m_v +4.4 they are visible to the naked eye but modest optical equipment is required to reveal the individual stars.

Lying midway between Algol and γ Andromedae is **Messier 34 (NGC 1039).** Discovered by Giovanni Hodierna in 1654 it was rediscovered by Charles Messier more than a century later in 1764. Messier cataloged the open cluster to avoid mistaking it for a comet, which is what he was really interested in. M34 is well worth hunting down. Its 23 brightest stars are spread over and area of about 1° which, at an estimated distance of 1,630 ly, corresponds to 30 ly. There is a central cluster about 20′ in diameter containing about a third of the members which is easily resolved in quite a modest 10 × 50 binocular and for which a rich field telescope is ideal. The brightest stars belong mainly to B and A classes while fainter stars tend to be F and G.

Open clusters in Perseus

Name	Size arc min	Size ly	Distance ly	Age million yrs	Brightest star in region*	No. stars m_v >+12*	Apparent magnitude m_v
M34	63′	30	1,630	180	HD 16771 m_v +7.32	23	+5.2
NGC 869	31′	61	6,800	12	HD 14143 m_v +6.55	128	+4.3
NGC 884	49′	110	7,600	11	HD 14433 m_v +6.38	188	+4.4
NGC 1528	40′	29	2,500	370	HD 26603 m_v +8.75	78	+6.4
NGC 1545	2.6′	1.8	2,300	280	HD 27292 m_v +7.13	21	+6.2

*May not be a member of the cluster.

Phoenix

Constellation:	Phoenix	**Hemisphere:**	Southern
Translation:	The Phoenix	**Area:**	467 deg^2
Genitive:	Phoenicis	**% of sky:**	1.137%
Abbreviation:	Phe	**Size ranking:**	37th

Another of the constellations of Keyser and de Houtman to depict the legend of the bird that lived for centuries, died, burst into flames and was reborn.

At m_v +2.37 α **Phoenicis** is a full magnitude brighter than its nearest rival. A 15 $D_\odot$ yellow-orange giant it has a mass of 2.5 $M_\odot$ and luminosity of 51 $L_\odot$. It is not particularly far, just 77.4 ly, but is speeding away from the general direction of the Sun at 74.6 km/s. Its unseen and mysterious companion orbits the primary in 10.5 years at an average distance of 7 AU.

Rather more yellow is the G8 β **Phoenicis**, a 4.6 $D_\odot$ giant with a mass of 1.95 $M_\odot$ and a luminosity of 134 $L_\odot$. Almost 200 ly away it shines at m_v +3.37 and has a surface temperature of around 5,100 K.

With a diameter of 150 $D_\odot$ (1.4 AU) γ **Phoenicis** is the largest star in the constellation though at just 1.75 $M_\odot$ it is by no means the most massive. An Lb pulsating variable its magnitude fluctuates between m_v +3.39 and +3.49 though with no particular period. At 10 pc it would have an absolute magnitude of M_v -4.4 and would only be distinguishable from Venus by its reddish-orange color belonging, as it does, to the M0 spectral group, and the fact that it is a long way from the Ecliptic. Like β Phe it has a spectroscopic companion with an orbital period of 193.9 days. It is 234 ly from Earth.

ζ **Phoenicis** is an Algol-type EA eclipsing binary consisting of two B-class stars. The primary is a 2.85 $D_\odot$ B7 with a mass of 3.93 $M_\odot$. Its companion is a smaller 1.85 $D_\odot$ B8 with a lower mass of 2.55 $M_\odot$. The orbital arrangement is such that the magnitude dips from m_v +3.91 to +4.42 before recovering, then falls again but only slightly to +3.73 as the two stars take it in turns to eclipse one another. The period is $1^d\ 16^h\ 8.5^m$. A third star orbits the eclipsing binaries at an average distance of 600 AU and with an orbital period of 4,245 years.

Apart from γ and ζ the Phoenix contains a few other variables. ι **Phoenicis** changes from m_v +4.70 to +4.75 and back over a period of 12.5 days. About 3.9 $D_\odot$ and with a mass of 2.43 $M_\odot$ its peculiar spectrum, A2VpSrCrEu, reveals it to be an β^2 CVn rotating variable with high concentrations of strontium, chromium and europium. ρ **Phoenicis** is a δ Scuti pulsating variable with a period of $2^h\ 28.5^m$ during which it drops by a tenth of a magnitude from m_v +5.17 to +5.27 before returning to its brighter state. An F2 with a temperature of around 6,900 K various estimates of its diameter have resulted in 1.6, 3.8, 4.07 and 4.4 $D_\odot$. It is 250 ly from Earth. The M4 giant ψ **Phoenicis** is, not surprisingly, a semi-regular variable with a period of about 30 days. At a distance of 321 ly it appears to change from m_v +4.30 to +4.50 as its average 34 $D_\odot$ expands and contracts.

The closest star to us in Phoenix is ν **Phoenicis** at 49.1 ly. An F9 verging on

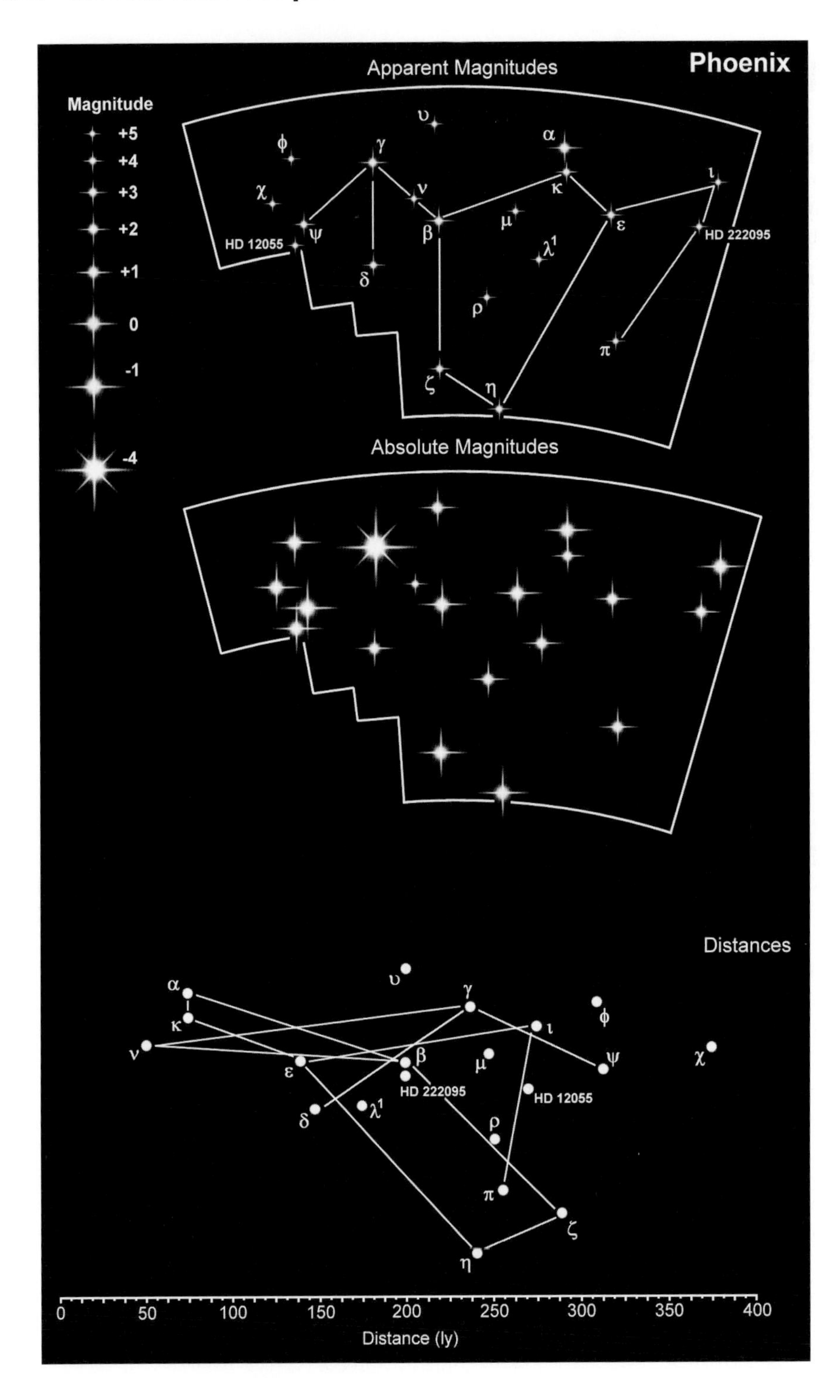

Phoenix
Apparent Magnitudes
Magnitude
+5
+4
+3
+2
+1
0
-1
-4
υ
φ
γ
α
χ
ν
κ
ι
ψ
β
μ
ε
HD 12055
HD 222095
λ[1]
δ
ρ
π
ζ
η
Absolute Magnitudes
Distances
0
50
100
150
200
250
300
350
400
Distance (ly)

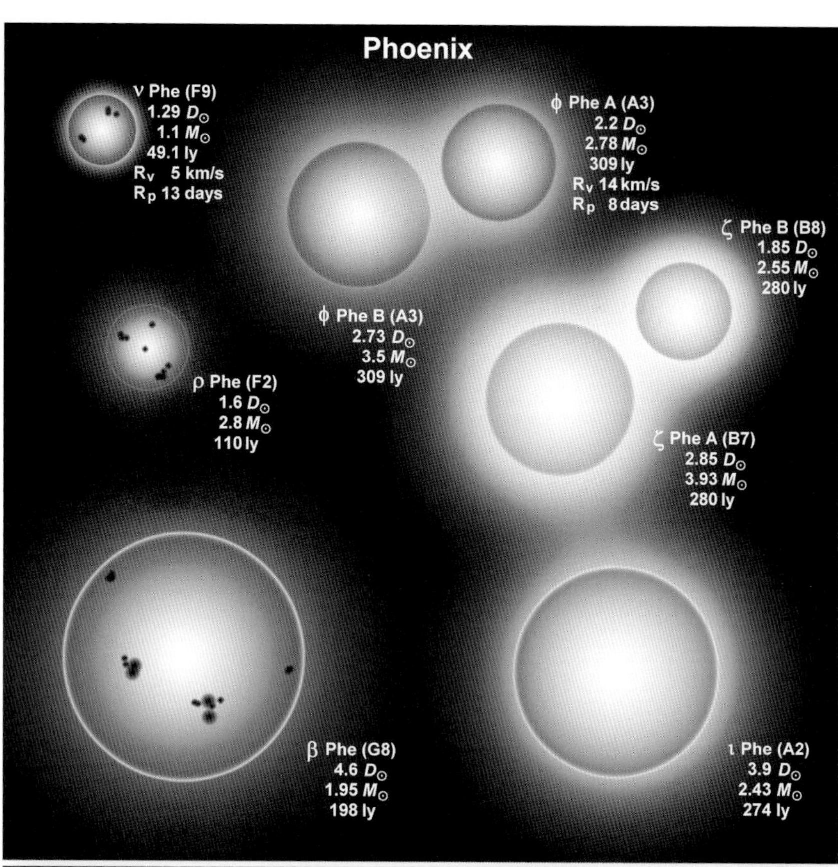
Phoenix
ν Phe (F9)
1.29 $D_{\odot}$
1.1 $M_{\odot}$
49.1 ly
R_V 5 km/s
R_P 13 days
φ Phe A (A3)
2.2 $D_{\odot}$
2.78 $M_{\odot}$
309 ly
R_V 14 km/s
R_P 8 days
ζ Phe B (B8)
1.85 $D_{\odot}$
2.55 $M_{\odot}$
280 ly
φ Phe B (A3)
2.73 $D_{\odot}$
3.5 $M_{\odot}$
309 ly
ρ Phe (F2)
1.6 $D_{\odot}$
2.8 $M_{\odot}$
110 ly
ζ Phe A (B7)
2.85 $D_{\odot}$
3.93 $M_{\odot}$
280 ly
β Phe (G8)
4.6 $D_{\odot}$
1.95 $M_{\odot}$
198 ly
ι Phe (A2)
3.9 $D_{\odot}$
2.43 $M_{\odot}$
274 ly

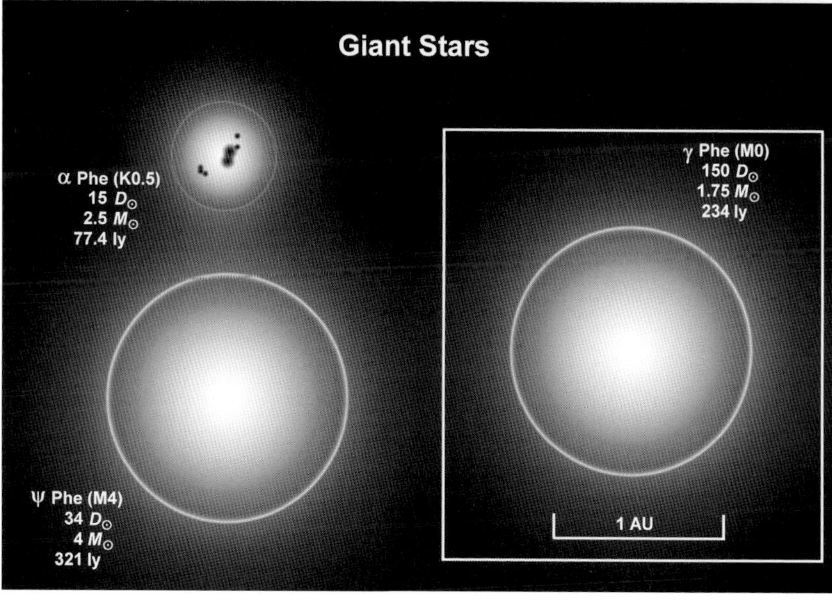
Giant Stars
α Phe (K0.5)
15 $D_{\odot}$
2.5 $M_{\odot}$
77.4 ly
γ Phe (M0)
150 $D_{\odot}$
1.75 $M_{\odot}$
234 ly
ψ Phe (M4)
34 $D_{\odot}$
4 $M_{\odot}$
321 ly
1 AU

6,000 K it is slightly larger than the Sun at 1.29 $D_{\odot}$ but about the same mass. It takes just over 13 days to spin once on its axis, its rotational velocity coming in at 5 km/s, 2½ times the spin speed of the Sun.

Possibly as small as 0.8 $D_{\odot}$ but probably truly 2.2 $D_{\odot}$ **ϕ Phoenicis** weighs in at 2.78 $M_{\odot}$. An A3 dwarf about 309 ly away, it rotates at 14 km/s and so takes between 2.9 and 8 days to complete a full rotation. Estimates put ϕ Phe – it sounds like something out of a pantomime – at a temperature of 10,500 K, rather hotter than usual for its class. It has a spectroscopic B9 companion of 2.73 $D_{\odot}$ which is in a 41.49 day long orbit.

Pictor

Constellation:	Pictor	**Hemisphere:**	Southern
Translation:	The Painter's Easel	**Area:**	247 deg^2
Genitive:	Pictoris	**% of sky:**	0.599%
Abbreviation:	Pic	**Size ranking:**	59th

Another of Abbé de La Caille's constellations it was originally called Le Chevalet et la Palette – the Easel and the Palette. Its name was changed to Equuleus Pictoris – the Painter of the Foal – and changed again to its current name by the IAU.

α **Pictoris** shines as a m_v +3.24 star from its home 98.9 ly away. An A8 about twice as big and as massive as the Sun, 1.9 $D_\odot$, 2.07 $M_\odot$, its 37.7 $L_\odot$ would result in a M_v +2.1 luminary at 10 pc. Not that it will ever get that close, drifting away from us at 20.6 km/s. Estimated to be about 1,000 million years old, and with a surface temperature of 7,600 K, it's a fast spinner at 230 km/s taking just 9.6 hours to complete a single rotation. It is also an X-ray source hinting at an as yet undetected companion.

β **Pictoris** is one of several stars in the constellation that harbors a planet. The star itself is about half a magnitude fainter than α Pic at m_v +3.85 and, at 62.9 ly, is the closest to us of all the stars in Pictor. Various estimates of its diameter average out at 1.6 $D_\odot$ with its mass coming in at 1.54 to 1.8 $M_\odot$ and luminosity at 8.7 $L_\odot$. Belonging to spectral group A6 and with a surface temperature of 8,130 K, infra-red studies several decades ago revealed that it was encircled by a relatively thick dusty disk stretching out to about 400 AU. More recently, in 2008, a planet was discovered in a 6,000 day (16.4 years) orbit about 8 AU from the star. The planet is believed to have a mass of between 6 and 13 Jovian masses, the greater mass pushing it towards being a brown dwarf. This is a young system, somewhere between 8 and 20 million years old, with the star spinning on its axis at around 128 km/s.

δ **Pictoris** is an eclipsing binary, EB, of the β Lyrae variety, the two components being distorted into ellipsoids. The magnitude changes between m_v +4.65 and +4.90 with a period of $1^d\ 16^h\ 8.5^m$. The primary star is a 7.4 $D_\odot$ B3 with a mass of 16.2 $M_\odot$; its partner a 5.1 $D_\odot$ B1 with a mass of 8.6 $M_\odot$. Their orbit varies between 4.18 and 7.95 million km, the two stars exchanging matter and rotating at 180 km/s and 100 km/s respectively. The system is believed to be 1,656 ly away.

When galaxies collide the gravitational forces can scatter the component stars far and wide, and we have an example on our very doorstep. At 12.7 ly lies **Kapteyn's Star**, a m_v +8.87 M1 dwarf in a largely empty part of the constellation. Named posthumously after Jacobus Cornelius Kapteyn (1851-1922) who first noted its large proper motion, the star covers 1° of the celestial sphere in just 414 years making it the second swiftest star, beaten only by Barnard's Star in Ophiuchus. In real terms Kapteyn's Star is traveling through

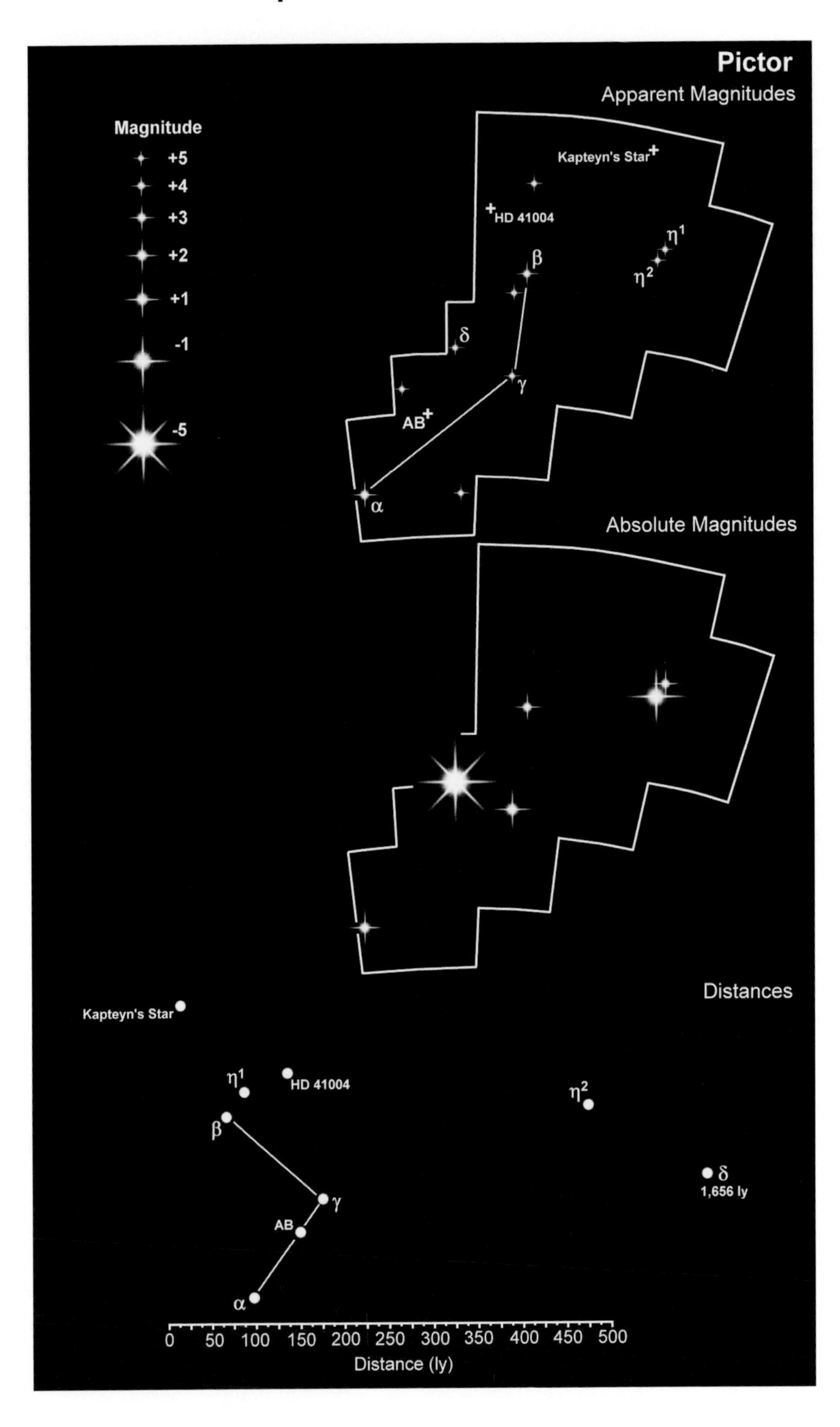
Pictor
Apparent Magnitudes
Magnitude
+5
+4
+3
+2
+1
-1
-5
Kapteyn's Star
HD 41004
η1
η2
β
δ
γ
AB
α
Absolute Magnitudes
Distances
Kapteyn's Star
η1
HD 41004
η2
β
δ
1,656 ly
γ
AB
α
0 50 100 150 200 250 300 350 400 450 500
Distance (ly)

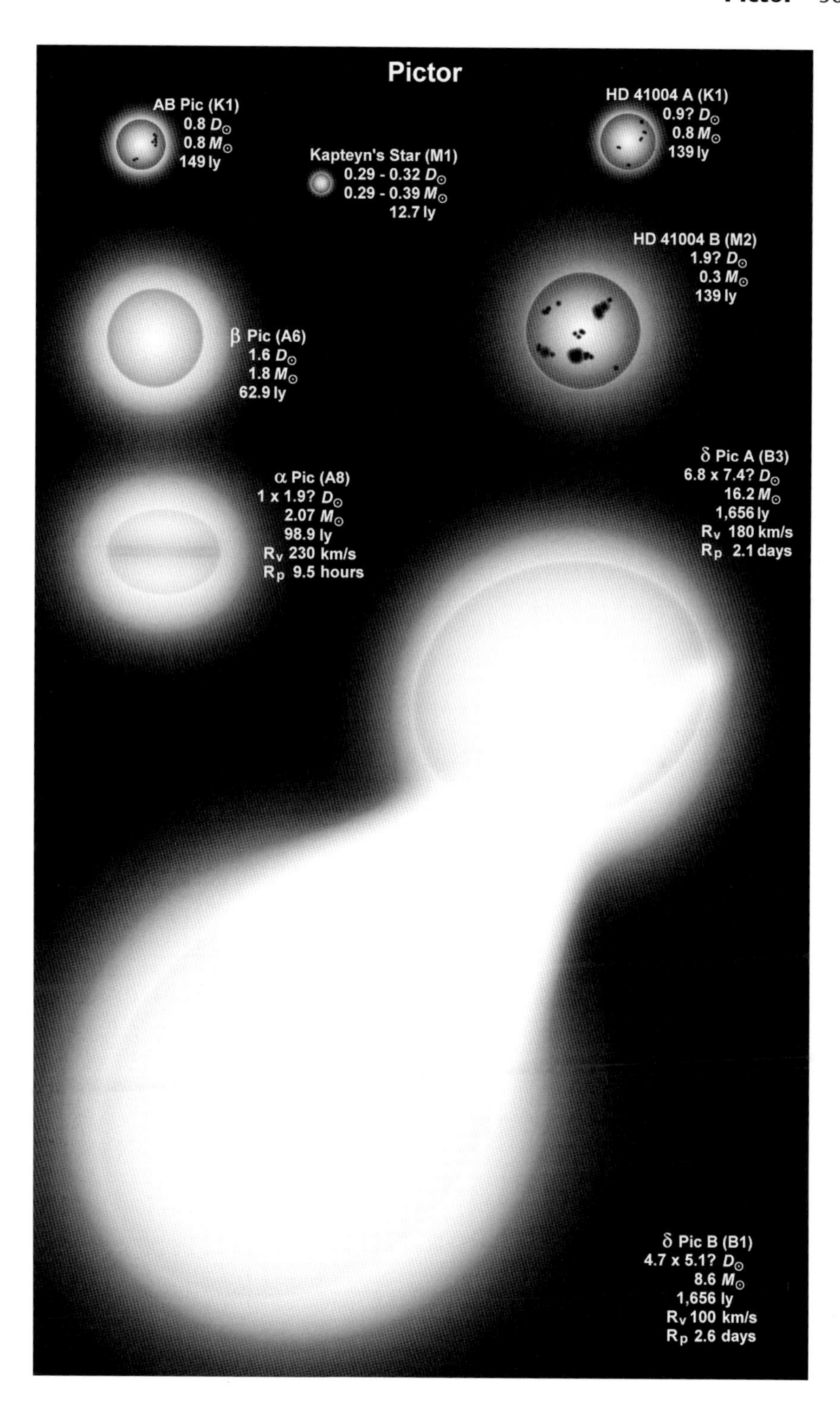
Pictor
AB Pic (K1)
0.8 $D_\odot$
0.8 $M_\odot$
149 ly
Kapteyn's Star (M1)
0.29 - 0.32 $D_\odot$
0.29 - 0.39 $M_\odot$
12.7 ly
HD 41004 A (K1)
0.9? $D_\odot$
0.8 $M_\odot$
139 ly
HD 41004 B (M2)
1.9? $D_\odot$
0.3 $M_\odot$
139 ly
β Pic (A6)
1.6 $D_\odot$
1.8 $M_\odot$
62.9 ly
α Pic (A8)
1 x 1.9? $D_\odot$
2.07 $M_\odot$
98.9 ly
R_V 230 km/s
R_P 9.5 hours
δ Pic A (B3)
6.8 x 7.4? $D_\odot$
16.2 $M_\odot$
1,656 ly
R_V 180 km/s
R_P 2.1 days
δ Pic B (B1)
4.7 x 5.1? $D_\odot$
8.6 $M_\odot$
1,656 ly
R_V 100 km/s
R_P 2.6 days

space at 282 km/s. What makes it particularly interesting is that it is traveling in a retrograde orbit, that is to say in a direction opposite to the vast majority of stars. It has a mass of between 0.29 and 0.39 $M_{\odot}$, a diameter of 0.29 to 0.32 $D_{\odot}$ and a luminosity of just 0.0038 $L_{\odot}$. Its metallicity is only about 10% that of the Sun suggesting it formed early in the history of the Galaxy when supernovae were rare and had not seeded space with heavy metallic elements: perhaps about 12,000 million years ago. It and 16 other stars that form the Kapteyn Star Group have very similar compositions to the stars of the ω Centauri globular cluster – yet that is 17,000 ly away. So what is the connection? It is now believed that the ω Centauri globular cluster is the core of a dwarf galaxy that collided with ours at some point in the past. Many of the outer stars of the globular cluster were stripped away and, in the case of the Kapteyn Star Group, were flung into retrograde orbits around the Galaxy. Kapteyn's Star is also known as **VZ Pictoris.**

HD 41004 is a visually close binary system just 0.5″ apart. The primary, a K1 pre-Main Sequence star, is usually designated **HD 41004 A** while the secondary is **HD 41004 B.** In orbit around the primary is a 2.54 M_J planet designated **HD 41004 Ab** while an 18.4 M_J brown dwarf orbits the secondary and is cataloged as **HD 41004 Bb,** which is all rather messy and easy to confuse for a quadruple star arrangement. The system is considered to be 1,600 million years old. Meanwhile **AB Pictoris b** at 13.5 Jovian masses is a borderline brown dwarf (see table).

Planetary systems in Pictor

Star	$D_{\odot}$	Spectral class	ly	m_v	Planet	Minimum mass	q	Q	P
β Pic	?	A6	63	+3.86	β Pic b	8 M_J	8	8	16.4 y
AB Pic	?	K1	149	+9.16	AB Pic b	13.5 M_J	275	275	?
HD 41004 A	?	K1	139	+8.65	HD 41004 Ab	2.54 M_J	1.00	2.28	2.64 y
HD 41004 B	?	M2	139	+12.33	HD 41004 Bb	18.4 M_J	0.0163	0.0191	1.33 d

Pisces

Constellation:	Pisces	**Hemisphere:**	Equatorial
Translation:	The Fish	**Area:**	889 deg^2
Genitive:	Piscium	**% of sky:**	2.155%
Abbreviation:	Psc	**Size ranking:**	14th

A Zodiacal constellation which the Sun enters on 12 March and leaves on 18 April. It is associated with the Greek myth of Aphrodite and her son Eros (Venus and Cupid in Roman mythology) who dived into the River Euphrates in an attempt to escape the multi-headed monster Typhon.

Something went badly wrong with the ordering of the stars in Pisces. Bayer usually gave the brightest star the designation α, the second β and so on. Occasionally the stars are out of sequence usually because they are of similar magnitude. In the case of Pisces, however, the system is way out. The table below shows how Bayer designated the stars, how they should have been designated and by how many places up or down they have been displaced.

Designation of stars in Pisces

Bayer designation	Should be...	m_v	Movement
η Piscium	α Piscium	+3.61	↑ 6
γ Piscium	β Piscium	+3.70	↑ 1
ω Piscium	γ Piscium	+4.02	↑21
α Piscium	δ Piscium	+4.10	↓ 3
ι Piscium	ε Piscium	+4.13	↑ 4
ε Piscium	ζ Piscium	+4.26	↓ 1
ο Piscium	η Piscium	+4.27	↑ 8
θ Piscium	θ Piscium	+4.28	⇄ 0
δ Piscium	ι Piscium	+4.43	↓ 5
ν Piscium	κ Piscium	+4.44	↑ 3
β Piscium	λ Piscium	+4.47	↓ 9
λ Piscium	μ Piscium	+4.49	↓ 1
τ Piscium	ν Piscium	+4.52	↑ 6
ξ Piscium	ξ Piscium	+4.61	⇄ 0
χ Piscium	ο Piscium	+4.66	↑ 7
φ Piscium	π Piscium	+4.68	↑ 5
υ Piscium	ρ Piscium	+4.75	↑ 3
μ Piscium	σ Piscium	+4.85	↓ 6
κ Piscium	τ Piscium	+4.87	↓ 9
ζ Piscium	υ Piscium	+5.19	↓14
ψ Piscium	φ Piscium	+5.27	↑ 2
ρ Piscium	χ Piscium	+5.35	↓ 5
σ Piscium	ψ Piscium	+5.49	↓ 5
π Piscium	ω Piscium	+5.49	↓ 8

Only two stars, θ and ξ Piscium are in the right order! ω Piscium has climbed 21 places up the table while ζ Piscium has fallen 14 places. Of course, there is more to this than meets the eye. Several of the stars are variable and so could swap places by a few positions but even so it does not explain some of the larger discrepancies. Within today's boundaries there are a further 20 stars that have magnitudes of between m_v +4.31 and +5.44.

And so to **α Piscium**, the 4th brightest star and a quadruple system. The primary is a m_v +4.33 A2, the secondary a m_v +5.23 A3 separated by 1.8″ (PA 271°). They are in a 933 year long orbit that swings between 50 and 190 AU. The primary is estimated to be 2.3 $D_\odot$ across with a mass of 2.43 $M_\odot$. Its temperature is 9,120 K and it spins at 81 km/s taking 1.4 days to rotate once. Its companion is smaller and less massive at 1.5 $D_\odot$ and 1.67 $M_\odot$. It is also cooler, 8,200 K, spins more slowly at 70 km/s and completes a full turn in just 1.1 days. The spectrum of the primary suggests quantities of silicon and strontium and there is some indication that its brightness varies by about 1/100th of a magnitude as the star rotates and the pools come in and out of view. Consequently it has been classed as an α^2 CVn variable. Both stars are spectroscopic binaries though virtually nothing is known about the unseen components.

β Piscium, the 11th brightest star in the Bayer list and at the opposite end of the constellation to α, is m_v +4.47 but is suspected of being variable by 4/100th of a magnitude. A B6 Be emission star surrounded by a disk of dust it has a rotational velocity of at least 121 km/s and may be much faster. Slightly less than two solar diameters across, 1.9 $D_\odot$, it has a mass of 4.8 $M_\odot$ and a luminosity of 300 $L_\odot$ so at 10 pc it would brighten to M_v -1.4. Relatively young, just 60 million years old, it lies at a distance of 493 ly.

To the east of β is a circle of five or six stars, depending on who draws the constellation, known as the Circlet. Starting at 3 o'clock is **γ Piscium**, a yellowish-orange G9, possibly a K0, twinkling at m_v +3.70 from a distance of 131 ly but moving relative to the Sun at 145 km/s, signaling it is an interloper from another part of the Galaxy. Some 11 $D_\odot$ and with a mass of 1.82 $M_\odot$ this 5,000 K giant has a luminosity of 43.2 $L_\odot$ and, rotating at 4.2 km/s, takes 132.6 days to spin once on its axis.

Heading counterclockwise the next star in the Circlet is the m_v +4.28 **θ Piscium**, a K1 giant though how big no one really knows. Estimates range from 11 to 16 $D_\odot$. Its rotational velocity is less that that of the Sun – just 1.8 km/s. If its diameter is the smaller estimate then it will take 30.9 days to complete a single revolution: the larger diameter comes in at more than a year – 450 days. It lies at a distance of 159 ly.

The m_v +4.13 **ι Piscium** at 45 ly is the closest of the Piscean stars. A pale yellow dwarf it has a diameter and mass of 1.5 $D_\odot$ and 1.38 $M_\odot$. Burning at 6,500 K this F7 has a luminosity of 3.44 $L_\odot$ and is thought to be very slightly variable by 3/100th magnitude.

More than twice as far away at 100.7 ly is the A7 dwarf **λ Piscium**. About a third larger than the Sun but with a mass of 1.88 $M_\odot$ and luminosity of 12.4 $L_\odot$ it is a modest m_v +4.49.

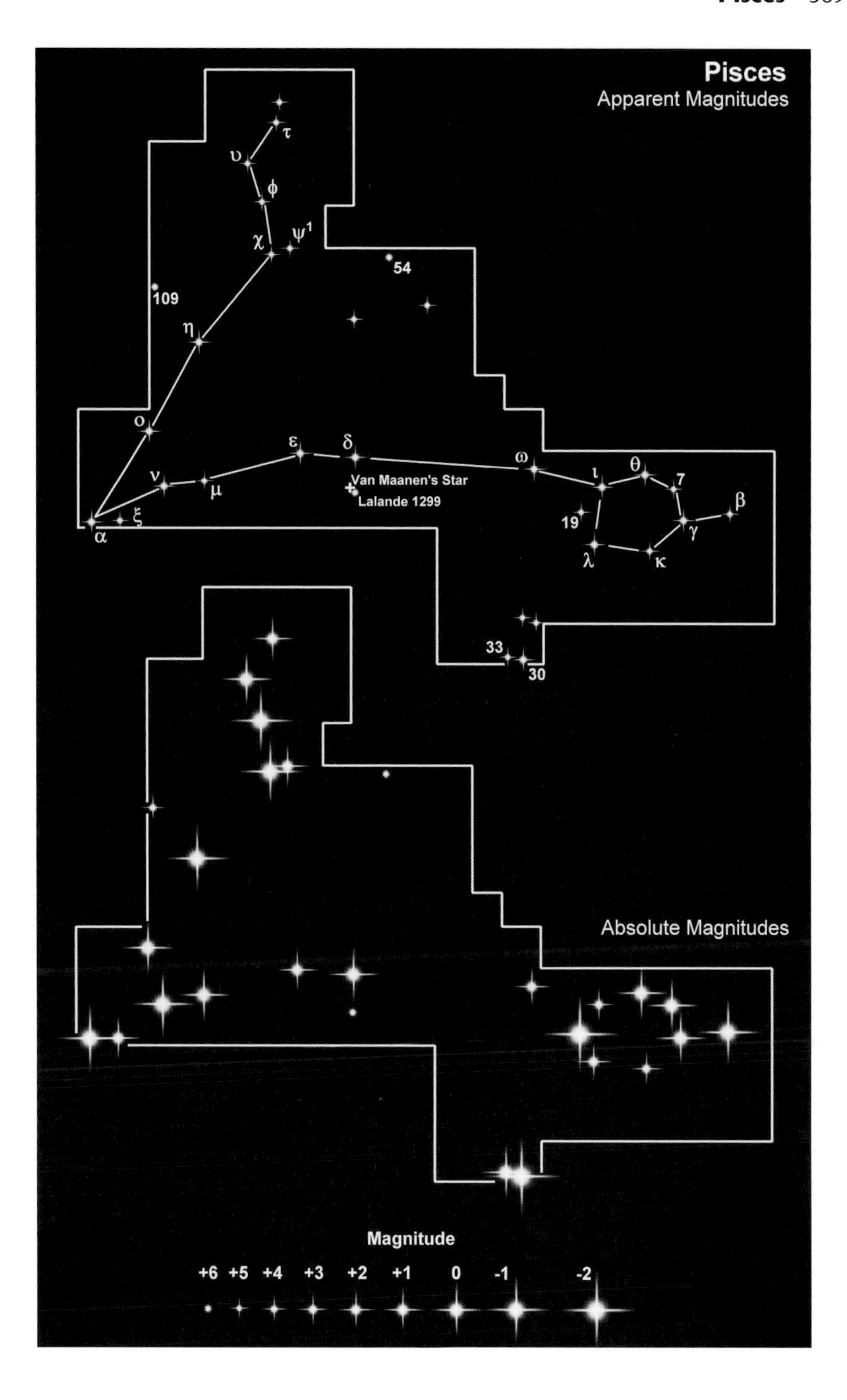
Pisces
Apparent Magnitudes
τ
υ
φ
χ
ψ1
54
109
η
ο
ε
δ
ω
ι
θ
7
β
ν
μ
Van Maanen's Star
Lalande 1299
19
γ
ξ
α
λ
κ
33
30
Absolute Magnitudes
Magnitude
+6 +5 +4 +3 +2 +1 0 -1 -2

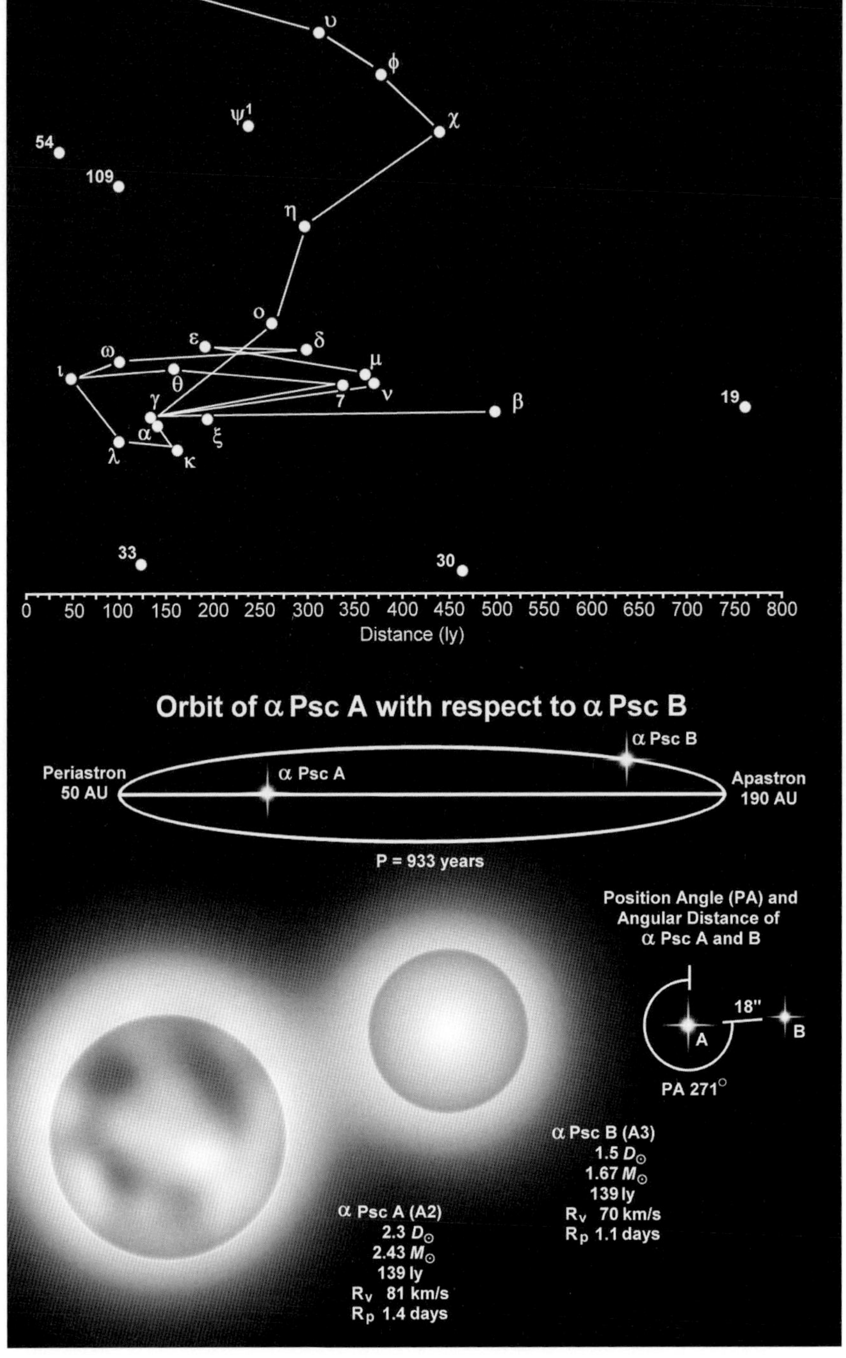

Pisces
Distances
τ
υ
φ
ψ1
χ
54
109
η
ο
ε
δ
ω
μ
ι
θ
7
ν
γ
β
19
α
ξ
λ
κ
33
30
0 50 100 150 200 250 300 350 400 450 500 550 600 650 700 750 800
Distance (ly)
Orbit of α Psc A with respect to α Psc B
α Psc B
Periastron
50 AU
α Psc A
Apastron
190 AU
P = 933 years
Position Angle (PA) and
Angular Distance of
α Psc A and B
18"
A
B
PA 271°
α Psc B (A3)
1.5 D⊙
1.67 M⊙
139 ly
Rv 70 km/s
Rp 1.1 days
α Psc A (A2)
2.3 D⊙
2.43 M⊙
139 ly
Rv 81 km/s
Rp 1.4 days

Next in the Circlet of Pisces is **κ Piscium** which varies between m_v +4.87 and +4.95 with a period of $14^h\ 2^m\ 46^s$, the star being classed as an α^2 CVn rotating variable. With a diameter of 1.9 $D_\odot$ and a mass of 2.15 $M_\odot$ the variability hints at a rotational velocity of 164 km/s. Its A0pCrSiSr peculiar spectrum reveals localized concentrations of chromium, silicon and strontium giving the star a blotchy appearance and causing its brightness to vary. About 370 million years old it has migrated from the Southern Hemisphere, the result of precession, crossing the Celestial Equator in 1771.

Drawing a line from κ to γ Piscium completes the circle – at least for some. Between γ and θ is **7 Piscium** which some include as part of the Circlet. Another K-class giant, this time a K2, it is 20 $D_\odot$ and rotates at half the speed of the Sun: just 1 km/s taking 1,012 days – 2.8 years – to rotate once.

The ruddy colored **19 Piscium** is likely to be the largest star in Pisces at 141 $D_\odot$, more than twice the size of its nearest rival, 30 Piscium and 70% the size of the Earth's orbit. A carbon rich red supergiant its luminosity is 436 $L_\odot$ but take into account the fact that most of its energy is radiated in the invisible infra-red and you would need to increase the luminosity ten fold to get the total output. It is variable between m_v +4.79 and +5.20 but without any particular period and is therefore classed as an Lb variable. It is sometimes given its variable star designation **TX Piscium**. Some 760 ly from Earth and with an absolute magnitude of M_v -2.0 its spectral classification appears to have changed over the years from an N0 to a $C6_2$ to its current C5. However, this is due to a better understanding of the chemical makeup of stars and an improved spectral class system rather than to changes in the star itself.

30 Piscium (or **YY Piscium**) is another Lb variable, 60 $D_\odot$ across and a red supergiant of M3 flavor. Its magnitude varies between m_v +4.31 and +4.41. Long believed to lie at a distance of 415 ly the latest estimates put it 50 ly farther away at 465 ly.

33 Piscium belongs to the RS CVn variable class of which only a dozen naked eye cases are known. Its magnitude falls from m_v +4.61 to +4.69 as it is eclipsed by a secondary component, invisible to the eye but detectable by spectroscope. The primary is a K0 giant of 15 $D_\odot$ but just 1.17 $M_\odot$. The orbital period is 72.93 days.

At m_v +12.36 it is far too faint to be seen in anything less than a 100 mm (4 inch) telescope but **Van Maanen's star** keeps cropping up in the literature and is of historical importance. It was discovered in 1917 by Adriaan Van Maanen who was actually looking for Lalande 1299, a star with a high proper motion (see below). What Van Maanen found was a star with an even greater proper motion, suggesting it was closer to Earth than Lalande 1299. A few years after the discovery Willem Luyten published a catalog of high proper motion stars and referred to the newly discovered object as Van Maanen's Star. Subsequent research revealed it to be a white dwarf. At that time only two other white dwarfs were known, o^2 Eridani B and Sirius B (β CMa), and both of these are part of multiple systems whereas Van Maanen's star was the first lone white dwarf. At 14.1 light years it is the third closest white dwarf after Sirius B (8.6 ly) and

Procyon B (β CMi, 11 ly). Although its mass is between 0.4 to 1.0 $M_\odot$ its diameter is just 0.012 $D_\odot$ and its luminosity is less than 0.0002 $L_\odot$. Belonging to the spectral class DZ7 its temperature is estimated to be 6,750 K and its age around about 3,700 million to 5,000 million years. It crosses the sky at 2.98″ per year, which equates to a degree every 1,208 years, and carries the more familiar catalog designations of Gliese 35 (or GJ 35) and HIP 3829. There is some evidence that Van Maanen's star has a companion but this has been hotly disputed.

The star Van Maanen was looking for, **Lalande 1299** aka **HD 4628**, is a K2.5 dwarf about 24.3 ly away. It has a diameter and mass of about 80% that of the Sun and is about 1,000 degrees cooler at 5,000 K. With a magnitude of m_v +5.75 it is visible to the naked eye under dark skies. Its age is estimated to be between 6,760 and 10,970 million years.

Pisces hosts one of the best examples of a spiral galaxy, **M74 (NGC 628).** Although a difficult 10th magnitude object it is seen face on, looking directly down its rotational axis.

Of the planetary systems in Pisces **HD 10679 b** is worth noting. Originally believed to be 6.38 Jovian masses it is now believed to be somewhere between 25 and 51 M_J putting it very firmly in the realm of the brown dwarfs. Its host star is 109 Piscium, a 1.72 $D_\odot$ G5 of m_v +6.29. The planet orbits the star in 2.95 years varying between 1.94 and 2.38 AU. The system lies at a distance of 106 ly.

Piscis Australis or Piscis Austrinus

Constellation:	Piscis Australis or Piscis Austrinus	**Hemisphere:**	Southern
Translation:	The Southern Fish	**Area:**	245 deg^2
Genitive:	Piscis Australis or Piscis Austrini	**% of sky:**	0.549%
Abbreviation:	PsA	**Size ranking:**	60th

The Egyptian goddess Isis was said to have been saved by a fish and in Greek mythology Piscis Australis is the parent of the northern fish, Pisces. Illustrations of the myth usually depict Aquarius pouring water into its mouth (over it would have made more sense). Note that Piscis is the singular and Pisces is plural.

At m_v +1.16 α **Piscis Australis** stands out in a region of relatively dim stars. Better known as Fomalhaut (pronounced foe-ma-low) this 1.6 $D_{\odot}$ A4 is just 25.1 ly away and has a mass of 1.97 $M_{\odot}$. Its luminosity of 15.4 $L_{\odot}$ is suspected of being slightly variable with the result that the star also carries the designation NSV 14372. About 40% hotter than the Sun at 8,500 K, its rotational velocity is at least 81 km/s but, taking into account its orientation, it could be as high as 102 km/s. A relatively young system, 100 to 300 million years old, studies by the Infra-red Astronomy Satellite IRAS in 1983 revealed it to be surrounded by a torus of dust 133 to 158 AU from the star. This later led to the discovery of a 3 M_J planet in an 876.1 year long orbit that varies between 102.4 and 127.7 AU. The announcement in November 2008 made history as the first extrasolar planet to be viewed in visible light using the Hubble Space Telescope. Fomalhaut is thought to be a member of the Castor Moving Group.

Of the 14 naked eye stars in Piscis Australis 8 are A-class (57%), 2 are B-class, 2 are F and there is one G and one M. β **Piscis Australis** belongs to the A0 group. Shining at m_v +4.28 it is slightly brighter than the average naked eye A-class star (m_v +4.63) but with an absolute magnitude of M_v +0.6 slightly dimmer than the mean of M_v +0.42. It is between 2.14 and 2.3 $D_{\odot}$ across (mean 2.62 $D_{\odot}$) and is 143 to 153 ly away. A0 class stars rotate at up to 425 km/s – there is one whose rotational velocity is 570 km/s but this is likely to be erroneous. The average is 110 km/s but β PsA spins at just 45 km/s, as do 31% of all A0 stars. Although not an average A0 star it's not far off.

Of the non-A-class stars π **Piscis Australis** is a m_v +5.13 spectroscopic binary, although virtually nothing is known of its companion. The primary is a 1.45 $D_{\odot}$ F1 with a mass of 1.53 $M_{\odot}$ and a temperature of around 7,300 K. The secondary, an F3, has an orbital period of 178.32 days, the two coming to within 0.4 AU before being separated by 1.3 AU. The system is 93.2 ly away.

τ **Piscis Australis** is the only other F-class star. An F6 it is the second closest star in the constellation at 61.1 ly. Between 1.2 and 1.4 $D_{\odot}$ across and with a mass of 1.16 $M_{\odot}$ τ Psa has a luminosity of 3 $D_{\odot}$ and a temperature of 6,300 K.

All the naked eye stars in the constellation are less than 4 $D_{\odot}$ with the

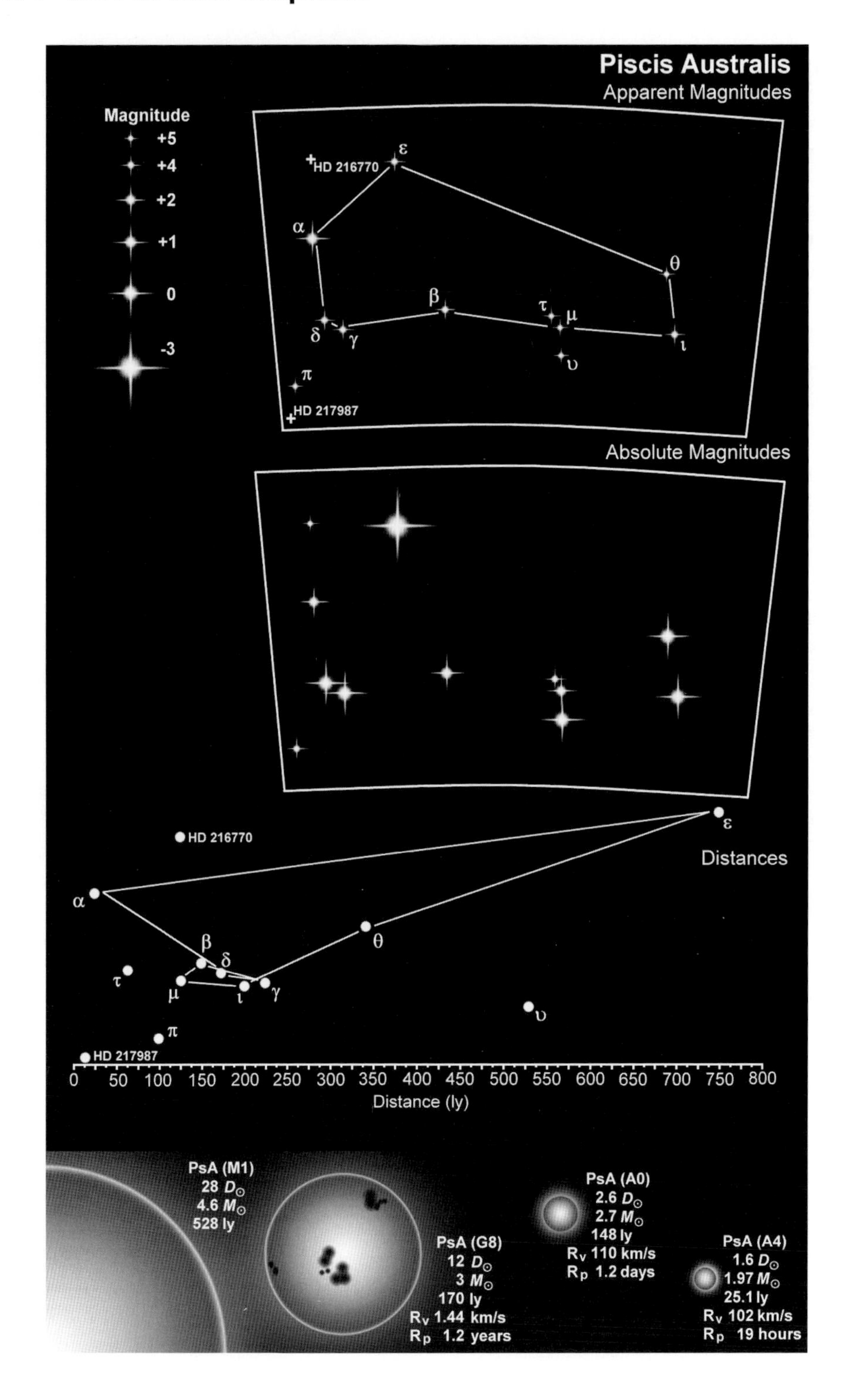

Piscis Australis
Apparent Magnitudes
Magnitude
+5
+4
+2
+1
0
-3
HD 216770
HD 217987
Absolute Magnitudes
HD 216770
Distances
HD 217987
0 50 100 150 200 250 300 350 400 450 500 550 600 650 700 750 800
Distance (ly)
PsA (M1)
28 $D_\odot$
4.6 $M_\odot$
528 ly
PsA (G8)
12 $D_\odot$
3 $M_\odot$
170 ly
R_V 1.44 km/s
R_P 1.2 years
PsA (A0)
2.6 $D_\odot$
2.7 $M_\odot$
148 ly
R_V 110 km/s
R_P 1.2 days
PsA (A4)
1.6 $D_\odot$
1.97 $M_\odot$
25.1 ly
R_V 102 km/s
R_P 19 hours

exception of δ and υ PsA. δ **Piscis Australis** is a 12 $D_\odot$ G8 yellowish-orange giant with a mass of 1.44 $M_\odot$. It lies at a distance of 170 ly. At more than twice the size and three times the distance, υ **Piscis Australis** is an M1 reddish-orange giant glowing at a steady m_v +4.99 from its home 528 ly away.

HD 217987, better known as **Lacaille 9352**, is another M-class star, a slightly cooler M2 but this time at the opposite end of the scale – just 0.56 $D_\odot$ across. A pre-Main Sequence star only 11.68 ly from Earth and traveling parallel with the Sun it has the fourth highest proper motion taking a mere 520 years to cover 1° on the celestial sphere. The only other stars to beat it are Barnard's Star in Ophiuchus, Kapteyn's Star in Pictor and Groombridge 1830 in Ursa Major. It has an apparent magnitude of m_v +7.34 which fades to an absolute magnitude of M_v +9.59 at 10 pc.

Puppis

Constellation:	Puppis	**Hemisphere:**	Southern
Translation:	The Stern	**Area:**	673 deg^2
Genitive:	Puppis	**% of sky:**	1.631%
Abbreviation:	Pup	**Size ranking:**	20th

Originally part of the much larger constellation of Argo Navis, the ship of the Argonauts, Puppis represents the stern while Carina is the keel and Vela the sail. As a result Puppis is missing the stars β to ε which are located in the other two constellations. It is a particularly rich part of the sky.

The brightest star in Puppis is the m_v +2.21 ζ **Puppis** and it's a bit of a rarity. Naked eye O-class stars are thin on the ground – well, in the sky – with only about two dozen of them being visible without optical aid. ζ Pup is the rarest of the rare: one of only two O5 class visible in the entire sky, and there are considerable uncertainties about its physical properties. Estimates of its size range from 6.7 to 26 $D_\odot$ with 17 $D_\odot$ being the most likely size. Its mass is probably 40 $M_\odot$, though it has been put as high as 60 $M_\odot$, and its distance was for many years considered to be 1,400 ly but has now been revised to 970 ly. Its visible luminosity is 22,000 $L_\odot$ but, taking into account that most of its energy is radiated in the ultraviolet, its total luminosity is in the order of 350,000 $L_\odot$. With a surface temperature of 42,400 K it releases a ferocious 2,300 km/s stellar wind that strips the star of vast amounts of mass, equivalent to more than 10 million times the mass lost from the Sun annually. It rotates at 211 km/s. If the star is 6.7 $D_\odot$ across then it turns once in just 1.6 days. If it is as big as the upper estimate of 26 $D_\odot$ then the rotational period is 6.2 days. The more commonly accepted 17 $D_\odot$ means that it rotates once on its axis every 4.1 days. It is slightly variable to m_v +2.17 but there is some disagreement as to the type of variability. Some authorities list it as a BY Draconis, the variation in magnitude due to a large single or large groups of star spots. But BY Draconis variables tend to be dwarf stars. Others regard ζ Pup as a β Cygni pulsating variable but again these tend to be B and A-class stars. Whatever the true nature of its variability ζ Pup is no more than 47 million years old. Originally part of the Trumpler 10 cluster it was ejected about 2.5 million years ago, possibly on the shock wave of an exploding neighbor, and is now hurtling through space at 103 km/s. At 10 pc it would a brilliant M_v -6.8.

Just as there are no stars designated α to ε, neither are there any stars η to μ although there is an **MU Puppis** which is a Mira-type variable that swings between m_v +14.5 and m_v +17.0.

The next bright star is **ν Puppis**, a B8 some 423 ly away. Puppis has a glut of B-class stars – about 38% are so classed – and ν Pup is one of several that are variable, or at least suspected of being so. Cataloged as a newly suspected variable NSV 3062 the star changes between m_v +3.15 and +3.20 with no particular period, the nature of which is suspected of being intrinsic rather than due to, say, an orbiting companion. Its luminosity is 750 $L_\odot$ in the visible rising to 1,340 $L_\odot$

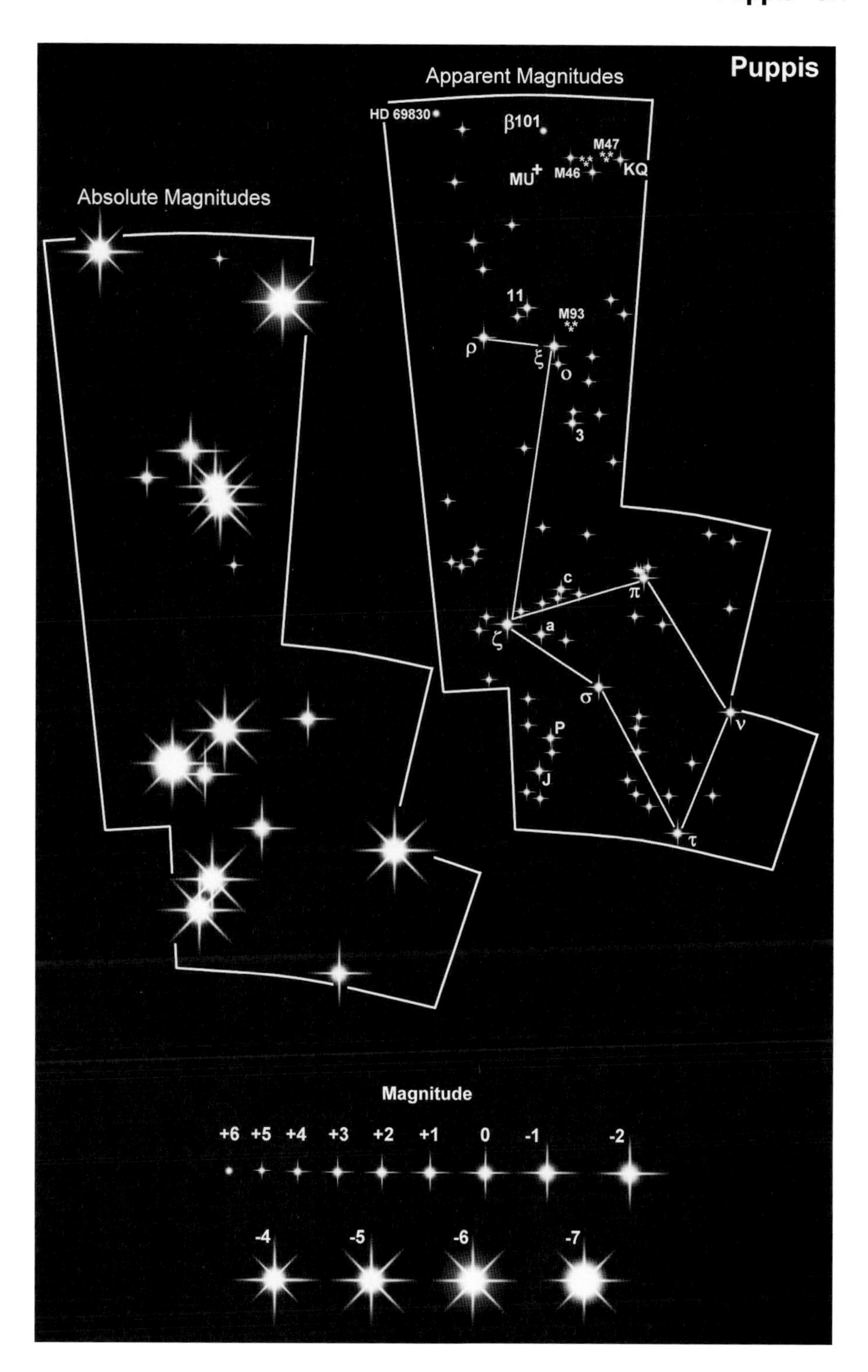
Puppis
Apparent Magnitudes
Absolute Magnitudes
HD 69830
β101
M47
MU
M46
KQ
11
M93
ρ
ξ
o
3
c
π
ζ
a
σ
ν
P
J
τ
Magnitude
+6 +5 +4 +3 +2 +1 0 -1 -2
-4 -5 -6 -7

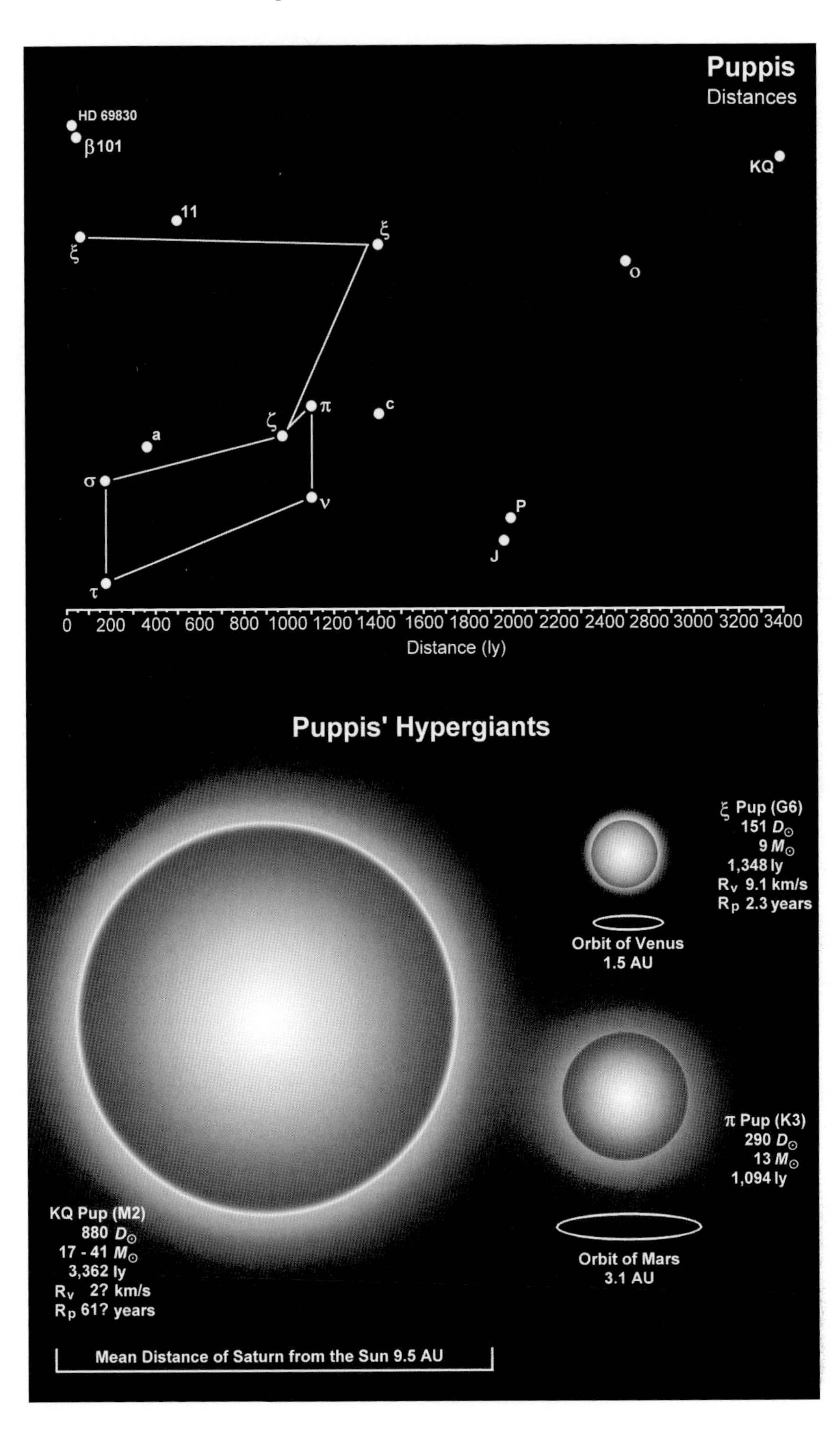
Puppis
Distances
HD 69830
β101
KQ
11
ξ
ξ
o
π
c
ζ
a
σ
ν
P
J
τ
0 200 400 600 800 1000 1200 1400 1600 1800 2000 2200 2400 2800 3000 3200 3400
Distance (ly)
Puppis' Hypergiants
ξ Pup (G6)
151 D⊙
9 M⊙
1,348 ly
Rv 9.1 km/s
Rp 2.3 years
Orbit of Venus
1.5 AU
π Pup (K3)
290 D⊙
13 M⊙
1,094 ly
KQ Pup (M2)
880 D⊙
17 - 41 M⊙
3,362 ly
Rv 2? km/s
Rp 61? years
Orbit of Mars
3.1 AU
Mean Distance of Saturn from the Sun 9.5 AU

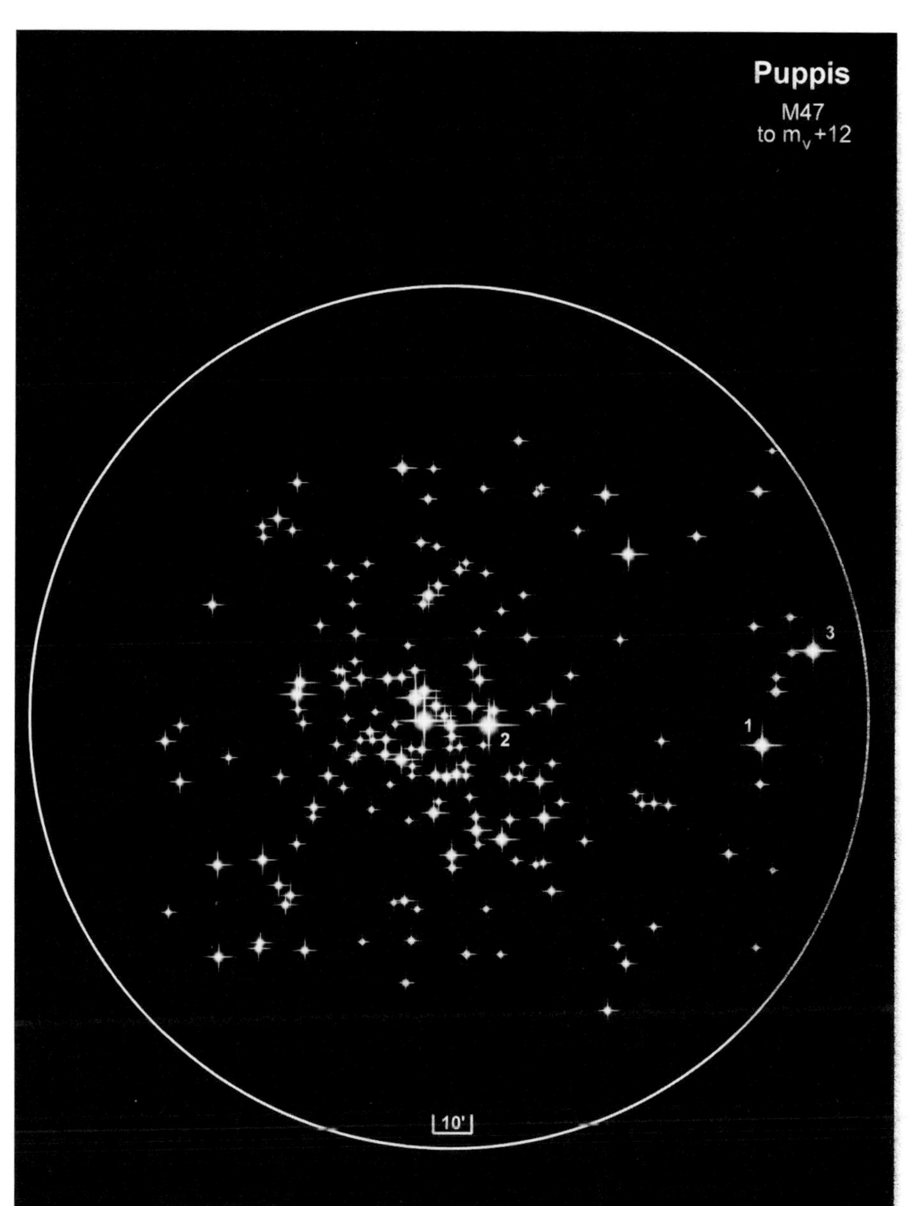

No.	Star	m_v	Distance (ly)	Spectrum
1	KQ Pup	+4.82	1,538	M2 Ia
2	HD 60855	+5.68	1,656	B4 III
3	HD 60325	+6.22	1,510	B1 V

in the ultraviolet. About 95 million years old it rotates at 246 km/s so has distorted into an oblate spheroid of, probably, 8.5 $D_{\odot}$ across the equator and 4.7 $D_{\odot}$ pole to pole. It takes 1.7 days to rotate on its axis.

ξ **Puppis** is a triple star system 1,348 ly from Earth. Twinkling at a modest m_v +3.32 it would shine as brightly as Venus if it were 10 pc away. Puppis contains few big stars but this is one of the larger ones with a diameter of 151 Suns or about as big as the orbit of Venus. Its mass is between 8 and 10 $D_{\odot}$ depending on where it is in the evolutionary cycle. It is a relatively slow rotator at just 9.1 km/s and so takes 840 days (2.3 years) to spin on its axis just once. Often listed as a G3 it is now believed to be a G6 with a surface temperature of 4,700 K. It visible companion is in a 26,000 year long orbit with an average distance of 2,000 AU. A second closer companion can only be detected using a spectroscope and is thought to complete one full orbit in 1 year at an average distance of 2 AU.

The m_v +2.69 π **Puppis** is part of the Collinder 135 loose cluster (see below). A K3 supergiant of 290 $D_{\odot}$ it is 30% larger than the Earth's orbit but has a mass of only 13 to 14 $M_{\odot}$. It lies at a distance of 1,094 ly. Its luminosity of 19,200 $L_{\odot}$ would result in an absolute magnitude of M_v -0.3. It is accompanied by a 2.5 $M_{\odot}$ B9.5 dwarf which has an average orbital distance of 20,000 AU and period in excess of 700,000 years.

Although there is no β Puppis there is a **β101 Puppis**, better known as **9 Puppis**. It is a binary system consisting of a G0 and a G1 dwarf in an 8,467 day (23.18 years) elongated orbit that varies between 2.85 and 15.55 AU. The two stars are similar to our own Sun: 0.8 $D_{\odot}$, 0.64 $M_{\odot}$ and 1.0 $D_{\odot}$, 0.81 $M_{\odot}$. Individually their magnitudes are m_v +5.60 and m_v +6.20 but combined they shine at m_v +5.20. They lie 54.4 ly from Earth.

The jury is still out on whether **Collinder 135** is a true cluster. There are 100 stars brighter than 12th magnitude in the region of Collinder 135. If the physical distance of the cluster is between 700 and 1,400 ly – and that is by no means certain – then only 17 stars belong to the cluster. Extend the distance by 200 ly either way and a further 10 stars can be added. However, take into account the proper motions, ages and composition of the stars and the number could be down to just a handful.

Of greater certainty is **M46 (NGC 2437)**, a 14′ diameter open cluster 4,500 ly away. Although it appears as a m_v +6.1 nebula to the naked eye, only one of its stars is brighter than 9th magnitude (**HD 62000**, m_v +8.67, an A0 at 5,824 ly) and only half a dozen are brighter than 10th magnitude.

About 1½° to the west of M46 is **M47 (NGC 2422)**, a more noticeable and much closer open cluster at 1,600 ly distance and about 17 ly across. Of the 41 stars in the region brighter than 11th magnitude, 19 are above m_v +9.00. Among them is **KQ Puppis**, an M2 hypergiant of 880 $D_{\odot}$ or 8.2 AU, about 80% the size of Jupiter's orbit. At 3,362 ly it is well beyond M47 and not surprisingly it is variable between m_v +4.82 and +5.17. Despite its enormous size it has a relatively modest mass of between 17 and 41 $M_{\odot}$. Its luminosity is 15,400 $L_{\odot}$ but taking into account that nearly all of its energy is radiated in the infra-red its total output could be as high as 1,200,000 $L_{\odot}$.

Lying 9° south of M46 is the open cluster **M93**, smaller but brighter and about 3,400 ly away. It contains 56 stars brighter than 12th magnitude though not all are cluster members. It is about 18′ across which equates, rather conveniently to 18 ly. The brightest star in the area is **HD 62679**, a K2, but there is only a 44% chance that it actually belongs to the cluster.

NGC 2477 is an open cluster with about four dozen stars brighter than 12th magnitude though its combined magnitude is m_v +5.8. Some 4,000 ly away it is thought to be 700 million years old and contains more than 300 stars.

Of the planet bearing stars in Puppis **HD 69830** has two possible terrestrial-type planets with masses of 10.49 $M_{\oplus}$ and 12.08 $M_{\oplus}$. A planet has also been detected around a star in the open cluster **NGC 2423** at a distance of 2,498 ly (see table).

HD 69830 Planetary system in Puppis

Star	$D_{\odot}$	Spectral class	ly	m_v	Planet	Minimum mass	q	Q	Period days
HD 69830	0.895	K0	41.1	+5.95	HD 69830 b	0.033 M_J	0.071	0.086	8.67 d
					HD 69830 c	0.038 M_J	0.162	0.210	31.6 d
					HD 69830 d	0.058 M_J	0.586	0.674	197 d

Open clusters in Puppis

Name	Size arc min	Size ly	Distance ly	Age million yrs	Brightest star in region*	No. stars m_v >+12*	Apparent magnitude m_v
M46 (NGC 2437)	20′	27	4,500	245	HD 62000 m_v +8.68	150	+6.1
M47 (NGC 2422)	80′	37	1,600	75	KQ Pup m_v +4.82	200	+4.4
M93 (NGC 2447)	10′	10	3,400	390	HD 62679 m_v +8.21	56	+6.2
NGC 2396	22′	12	1,900	330	HD 59067 m_v +5.79	37	+7.4
NGC 2439*	9′	31	12,600	30	R Pup m_v +6.61	21	+6.9
NGC 2451	167′	48	1,000	45	HD 63032 m_v +3.61	211	+2.8
NGC 2477	24′	28	4,000	705	DM -38° 3756 m_v +9.65	45	+5.8
NGC 2527	16.5′	9.5	2,000	450	HD 67097 m_v +9.65	46	+6.5
NGC 2571	12′	15	4,400	30	HD 70078 m_v +8.83	30	+7.0

*May not be a true cluster.

Pyxis

Constellation:	Pyxis	**Hemisphere:**	Southern
Translation:	The Mariner's Compass	**Area:**	211 deg^2
Genitive:	Pyxidis	**% of sky:**	0.536%
Abbreviation:	Pyx	**Size ranking:**	65th

Originally called Pyxis Nautica the constellation is the invention of La Caille and is related to Puppis, Carina and Vela, the components of the ship Argo Navis.

Apart from HD 73752 none of the other stars in Pyxis is particularly close to the Sun and the farthest at 845 ly is, as it happens, also the brightest. α **Pyxidis** is a hot, 22,150 K, B1.5 giant of 8.4 $D_{\odot}$ with a mass of 11.4 $M_{\odot}$. Flickering between an almost unnoticeable m_v +3.68 and +3.70 it is probably a β Cepheid pulsating variable (or β Canis Majoris, if you prefer) though no one is exactly sure and the star is thus listed as a newly suspected variable with the catalog number NSV 4220. Probably less than 18 million years old it is surrounded by a circumstellar shell and this, together with clouds of interstellar dust, reduce its brilliance by about half a magnitude. At 10 ly it would be M_v -4.4, about as bright as Venus. Visually it appears to have a luminosity of 1,886 $L_{\odot}$ but it radiates most of its energy in the UV. Take this into account and the output is closer to 16,300 $L_{\odot}$. It rotates at 20 km/s.

Of the dozen stars visible to the naked eye, five are G-class. β **Pyxidis** is a 12 $D_{\odot}$ G7 lying at a distance of 388 ly. It has an apparent magnitude of m_v +3.96 and an absolute magnitude of M_v +0.3. Slightly smaller at 10 or 11 $D_{\odot}$ ζ **Pyxidis** is a G4 and almost a full magnitude fainter at m_v +4.88 but also with an absolute magnitude of M_v +0.3. It is slightly more than two solar masses, 2.13 $M_{\odot}$, and is 236 ly distant. λ **Pyxidis** is also estimated to be 11 $D_{\odot}$ with a mass of 1.69 $M_{\odot}$. A G7 it lies 182 ly away and is m_v +4.72 and M_v +0.3. The closest of the G-stars, in fact the closest star in the constellation, is **HD 73752** at 65 ly. It is also the smallest of the class, 1.4 $D_{\odot}$, its mass coming in at 1.09 $M_{\odot}$. Finally, **HD 75605** is another 10 $D_{\odot}$ giant G5 at 229 ly. It has a mass of 1.98 $M_{\odot}$.

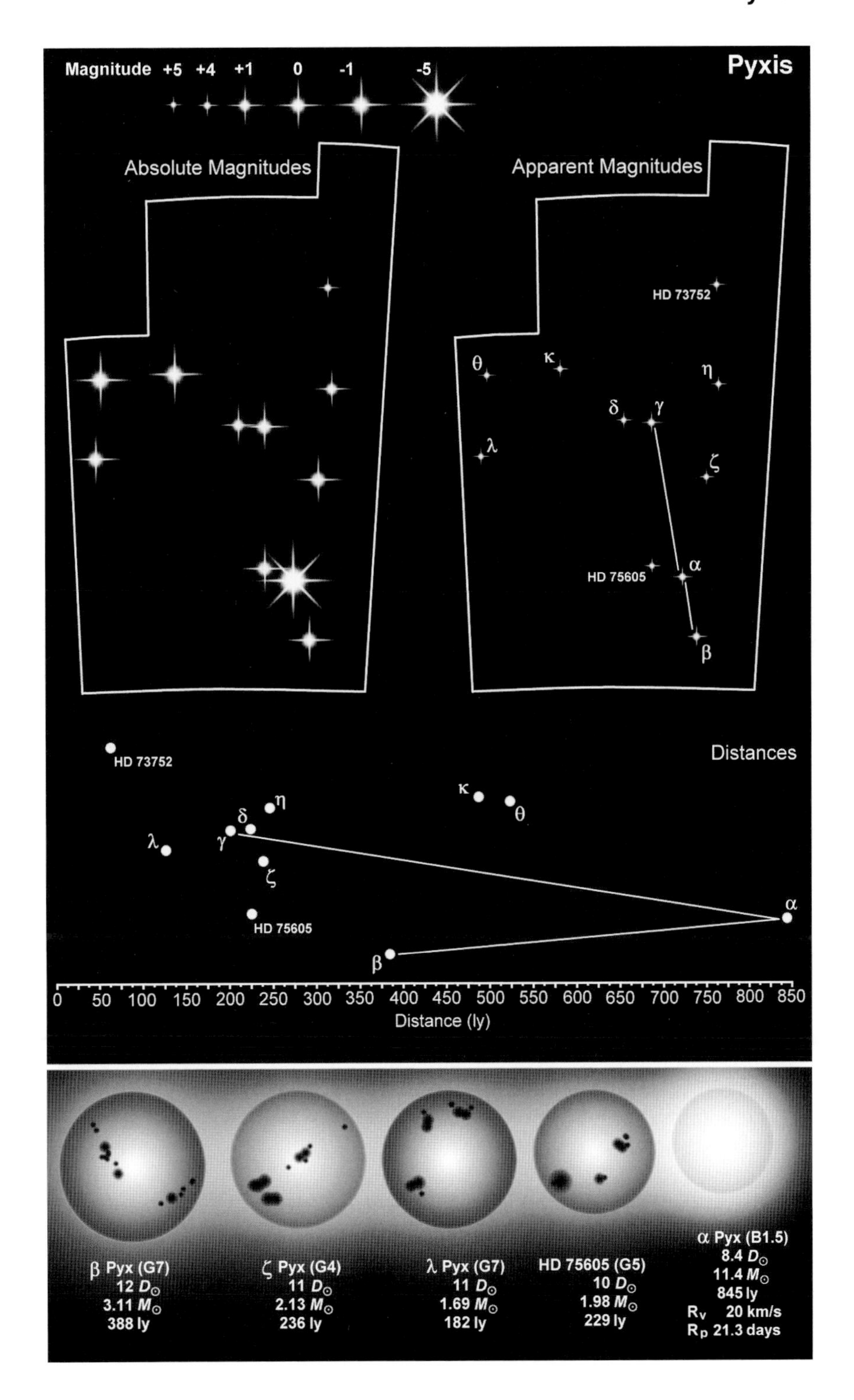

Pyxis
Magnitude +5 +4 +1 0 -1 -5
Absolute Magnitudes
Apparent Magnitudes
HD 73752
θ
κ
η
δ
γ
λ
ζ
HD 75605
α
β
Distances
HD 73752
κ
θ
η
δ
λ
γ
ζ
HD 75605
α
β
0 50 100 150 200 250 300 350 400 450 500 550 600 650 700 750 800 850
Distance (ly)
β Pyx (G7)
12 $D_\odot$
3.11 $M_\odot$
388 ly
ζ Pyx (G4)
11 $D_\odot$
2.13 $M_\odot$
236 ly
λ Pyx (G7)
11 $D_\odot$
1.69 $M_\odot$
182 ly
HD 75605 (G5)
10 $D_\odot$
1.98 $M_\odot$
229 ly
α Pyx (B1.5)
8.4 $D_\odot$
11.4 $M_\odot$
845 ly
R_V 20 km/s
R_P 21.3 days

Reticulum

Constellation:	Reticulum	**Hemisphere:**	Southern
Translation:	The Reticule	**Area:**	114 deg^2
Genitive:	Reticuli	**% of sky:**	0.27%
Abbreviation:	Ret	**Size ranking:**	82nd

Another constellation introduced by La Caille it celebrates the reticule or 'net' used in the eyepiece of a telescope to record the position of a star. Less than a dozen of its stars are bright enough to be seen without optical aid.

α **Reticuli**, the brightest star in the constellation at m_v +3.33, is a double and possibly a triple system. Variously described as being a G5, G6, G7 and G8 it is a 13.8 $D_\odot$ giant with a mass of 2.85 $M_\odot$ and luminosity of 94.5 $L_\odot$. Estimated to be 163 ly away its rotational velocity is 36 km/s which equates to one revolution on its axis every 19.4 days. It has a m_v +12.0 red dwarf companion, an M0 cataloged in the *Cape Photographic Durchmusterung* as CPD-62 332B, which is separated by 49.5″. At 163 ly the separation translates into a real space distance of 2,500 AU. The orbital period is likely to be around 60,000 years. There is some indication that β Ret is also a spectroscopic binary but there is, as yet, no absolute proof.

Reticulum contains two red giants that are visible without optical aid. γ **Reticuli** is the closest at 493 ly and the largest at 45 $D_\odot$. It is a semi-regular SR pulsating variable with a main period of 25 days during which its magnitude changes between m_v +4.42 and +4.64. It is 283 times more luminous than the Sun and belongs to spectral class M4 . The second red giant happens to be the next star that Bayer cataloged, δ **Reticuli**. At 35 $D_\odot$ it is the smaller of the two by 10 solar diameters and farther away at 530 ly. An M2 its luminosity is 318 $L_\odot$ and is suspected of being slightly variable between m_v +4.63 and +4.67.

ε **Reticuli** is a binary system 59.5 ly away consisting of a K2 sub-giant, a white dwarf and at least one planet. The primary is m_v +4.44 and is estimated to be 3.9 $D_\odot$ across with a mass of 0.98 $M_\odot$. It rotates relatively slowly at 2.9 km/s taking just over 68 days to turn once on its axis. In orbit at an average distance of 240 AU – 8 times the distance of Neptune from the Sun – is the white dwarf **WD 0415-594.** The sole detected planet is a 1.28 M_J gaseous giant just over an astronomical unit from the primary star and with an orbital period of 432.8 days. The system is believed to be about 10,000 million years old, a little over twice the age of the Solar System.

Various authors and researchers have speculated that because of the large number of observed binary systems the Sun may also have a companion. This seems highly improbable but, if it does or ever did, then it may look like the ζ **Reticuli** system. ζ Ret consists of two nearly identical solar analogs separated by 3,750 AU and having an orbital period in excess of 170,000 years. Because they lie only 39.5 ly from Earth the pair appear as a wide double in the night sky some 5.2′ apart. ζ^1 **Reticuli** is the fainter of the two at m_v +5.54. It has a diameter of 0.92 $D_\odot$ and mass of 0.96 $M_\odot$. Its spectrum has not been pinned

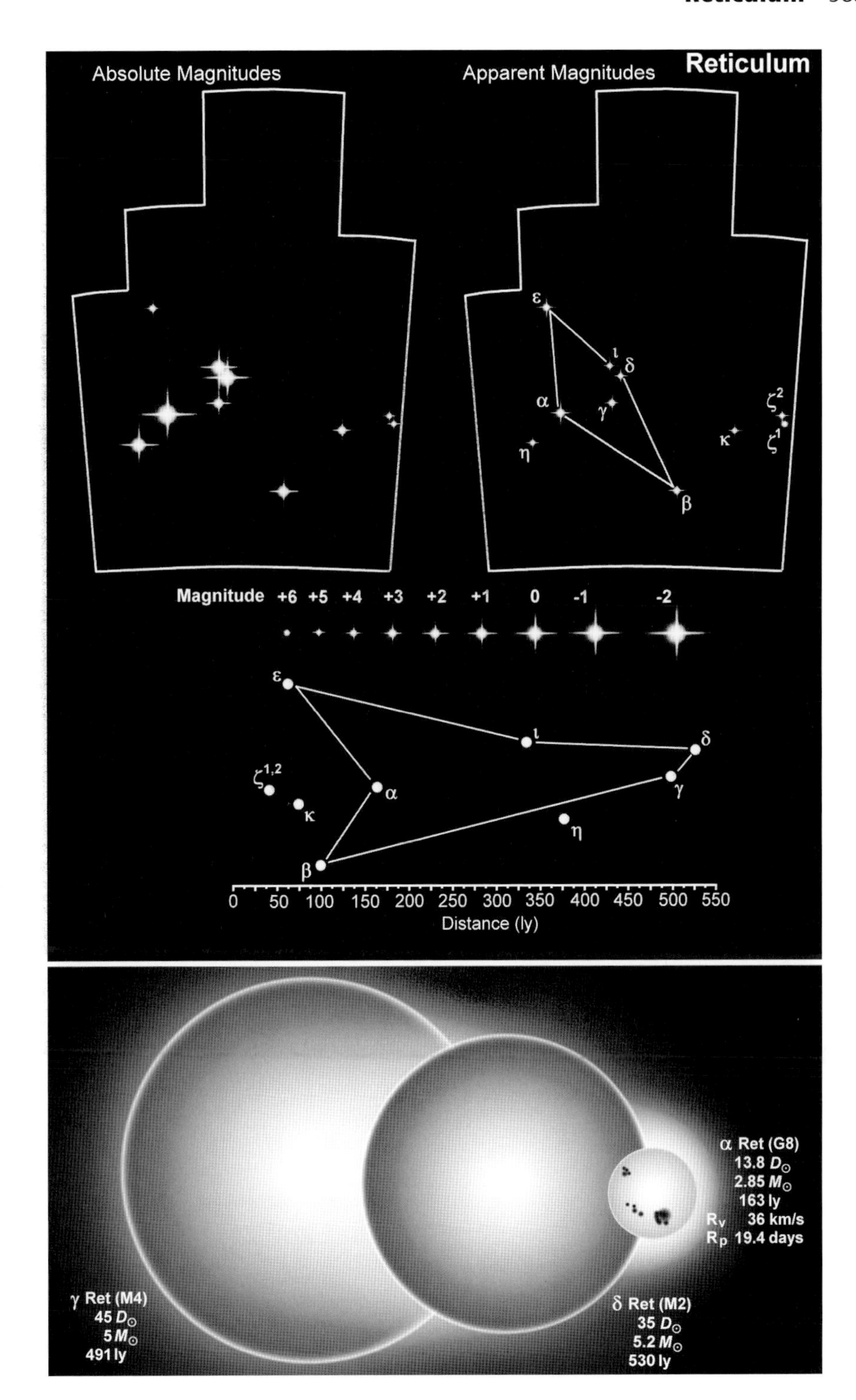
Reticulum
Absolute Magnitudes
Apparent Magnitudes
ε
ι
δ
α
γ
η
β
κ
ζ2
ζ1
Magnitude +6 +5 +4 +3 +2 +1 0 -1 -2
ζ1,2
0 50 100 150 200 250 300 350 400 450 500 550
Distance (ly)
α Ret (G8)
13.8 $D_\odot$
2.85 $M_\odot$
163 ly
R_V 36 km/s
R_P 19.4 days
γ Ret (M4)
45 $D_\odot$
5 $M_\odot$
491 ly
δ Ret (M2)
35 $D_\odot$
5.2 $M_\odot$
530 ly

down exactly but appears to be G3-5. Its partner, ζ^2 **Reticuli**, is an even closer solar analog, being a G2 and with a diameter of 1.0 $D_\odot$ and mass of 0.99 $M_\odot$. It is slightly brighter at m_v +5.24. The luminosity of both stars is 0.95 $L_\odot$ and they are considered to be between 1,560 million and 6,000 million years old (the Sun is 4,567 million years). ζ^2 Ret seems to have a debris disk that reaches out to 4.3 AU. The pair are in a highly elongated orbit that swings between 17,400 and 28,600 ly of the galactic center. ζ Ret shot to fame in the 1960s when Barney and Betty Hill of Portsmouth, New Hampshire claimed to have been abducted by aliens from the star system. The alleged abduction took place on 19 September 1961 but was not made public until 1963 when the couple produced a star map indicating the home of their extraterrestrial abductors. Intrigued by the map Marjorie Fish, a school teacher and amateur astronomer, searched through thousands of stellar combinations and announced in 1968 that the star map was centered on the ζ Reticuli system.

The only F-class star in the constellation is **κ Reticuli**, an F5 dwarf of 1.2 $D_\odot$ and mass of 1.35 $M_\odot$. Fairly close to us at 69.9 ly the star shines at m_v +4.71 and would brighten by 1½ magnitudes to M_v +3.4 at 10 pc. Somewhat hotter than the Sun at 6,750 K its rotational period is exactly 4 days, the star spinning at 15 km/s.

Sagitta

Constellation:	Sagitta	**Hemisphere:**	Northern
Translation:	The Arrow	**Area:**	80 deg^2
Genitive:	Sagittae	**% of sky:**	0.194%
Abbreviation:	Sge	**Size ranking:**	86th

One of Ptolemy's original constellations the arrow is associated with many legends such as Hercules and Eros.

Just seven naked eye stars make up the third smallest constellation and at m_v +4.38 α **Sagittae** is not even the brightest, outshone by both γ and δ and exactly the same magnitude as β with which it may be associated. α Sge has a diameter of 28 $D_\odot$, a mass of 4 $M_\odot$ and a luminosity of 302 $L_\odot$. A G1 it rotates at 6.5 km/s completing a full revolution in 218 days. By comparison **β Sagittae** is somewhat larger at 40 $D_\odot$, has a mass of 4.3 $M_\odot$ and a luminosity of 294 $L_\odot$. It is at the opposite range of the G-class; a G8. α Sge lies at 473 ± 43 ly while β Sge is usually quoted as 467 ± 41 ly, so the two stars could be very close neighbors. β Sge takes 204.5 days to rotate once, spinning at 9.9 km/s.

γ **Sagittae** is one of two M-class giants. Shining at m_v +3.5 its diameter has been estimated at 20, 30 and 48 $D_\odot$, averaging 33 $D_\odot$ with a mass of 2.5 $M_\odot$. A K5 it lies 274 ly from Earth and has a rotational velocity of 5.8 km/s which, assuming a diameter of 33 $D_\odot$, means that it rotates once every 288 days.

The K2, 15 $D_\odot$ η **Sagittae** takes more than two years to rotate once, its rotational velocity being just half that of the Sun, 1 km/s. It is one of the fainter stars at m_v +5.1 but is the closest at 162 ly. Its mass is 0.99 $M_\odot$.

δ Sagittae is a spectroscopic binary. The visible component is an M2 red giant of 152 $D_\odot$ or about as big as the orbit of Venus. Despite its huge size its mass is small, just 3.8 $M_\odot$. It belongs to the Lb class of pulsating variables, its magnitude switching between m_v +3.75 and +3.83. In terms of luminosity it appears 453 times as luminous at the Sun but take into account that 90% of its output is in the infrared and the luminosity rockets to 4,500 $L_\odot$. Its invisible companion is a 2.9 $M_\odot$ A0 white dwarf of 60 $L_\odot$ in a 10.2 year orbit that has an average separation of 8.8 AU. δ Sagittae's distance is 448 ly.

ζ **Sagittae** is a four star system. The primary is a m_v +5.50 A3 dwarf of 1.7 $D_\odot$ which, because of the combined light of the other stars, appears as a m_v +5.04. In a 23.22 year long orbit is the secondary component, a slightly hotter A1. The third component is a m_v +6.02 star which, if it is truly a member of the system, will have an orbital period of at least 8,535 years. Finally there is a 9th magnitude star whose membership of the system is questionable.

The sole remaining naked eye star is **13 Sagittae** which is also known as **VZ Sagittae** on account of the fact that it is an Lb slow irregular pulsating variable, its magnitude fluctuating between m_v +5.27 and +5.57. Apart from the fact that it is an M4 and lies 746 ly from Earth very little is known about this star.

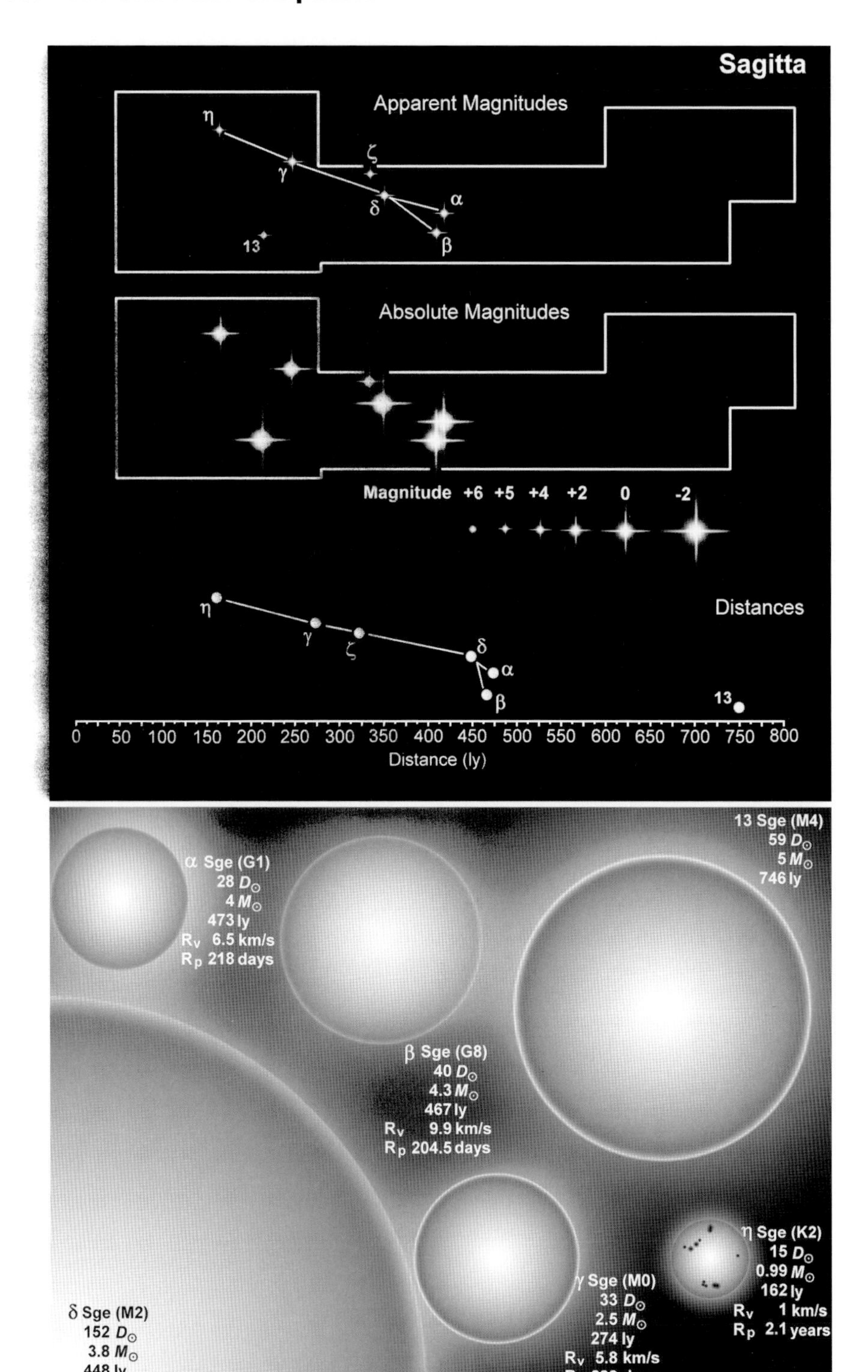

Sagitta
Apparent Magnitudes
η
γ
ζ
δ
α
β
13
Absolute Magnitudes
Magnitude +6 +5 +4 +2 0 -2
Distances
η
γ
ζ
δ
α
β
13
0 50 100 150 200 250 300 350 400 450 500 550 600 650 700 750 800
Distance (ly)
α Sge (G1)
28 D⊙
4 M⊙
473 ly
Rv 6.5 km/s
Rp 218 days
13 Sge (M4)
59 D⊙
5 M⊙
746 ly
β Sge (G8)
40 D⊙
4.3 M⊙
467 ly
Rv 9.9 km/s
Rp 204.5 days
η Sge (K2)
15 D⊙
0.99 M⊙
162 ly
Rv 1 km/s
Rp 2.1 years
γ Sge (M0)
33 D⊙
2.5 M⊙
274 ly
Rv 5.8 km/s
Rp 288 days
δ Sge (M2)
152 D⊙
3.8 M⊙
448 ly

Only one star in Sagitta is currently known to have a planet, the 9th magnitude HD 231701.

Messier 71 was long believed to be a compact open cluster but since the 1970s it has been regarded as a globular. About 27 ly across and 13,000 ly away it has an apparent magnitude of m_v +8.2.

Sagittarius

Constellation:	Sagittarius	**Hemisphere:**	Southern
Translation:	The Archer	**Area:**	867 deg^2
Genitive:	Sagittarii	**% of sky:**	2.102%
Abbreviation:	Sgr	**Size ranking:**	15th

One of the Zodiacal constellations which the Sun enters on 18 December and leaves on 19 January, Sagittarius was a Centaur called Crotus, the son of Pan and the inventor of archery. In traditional celestial cartography Sagittarius is shown firing an arrow at Scorpius, the neighboring constellation. This is a particularly rich region of the sky with the galactic center close to 3 Sagittarii.

Not the brightest, actually the 15th brightest star in the constellation, α **Sagittarii** shines at m_v +3.94 from a distance of 170 ly. A little over twice as hot as the Sun at 12,400 K it is a bluish-white B8 dwarf, 2.3 solar diameters across and with a mass of 3.17 $M_\odot$. It rotates at 85 km/s, a bit on the slow side for its class which average 151 km/s but not unusual, and is consequently slightly oblate. It takes just 1.4 days to complete a single turn on its axis. It is surrounded by a dust cloud, a remnant from its formation, and there is some suggestion that it may be a spectroscopic binary although no one is yet sure.

Sagittarius has not one but two β stars which are totally unrelated. **β^1 Sagittarii** lies at a distance of 378 ly and is a hot, 13,600 K, B9 of 2.7 $D_\odot$ and weighing in at 3 $M_\odot$. Its modest m_v +3.94 would brighten to 1st magnitude at 10 pc, its luminosity being 290 $L_\odot$. It is not alone. In a 3,290 AU orbit that takes 82,000 years to run full circle is a m_v +7.11 A5 dwarf of 1.89 $D_\odot$. While β^1 Sgr spins at a modest 13 km/s taking 10.5 days to turn once, its companion (cataloged as **HD 181484**) whirls around at 140 km/s, its rotational period amounting to just 16.4 hours.

β^2 Sagittarii is a very different star. Much closer at 139 ly but a bit fainter at m_v +4.28 it is a 3.5 $D_\odot$ F2 with a mass of 1.93 $M_\odot$ and surface temperature of 7,100 K. β^2 Sgr is an unusually fast spinner. F2 stars average just 59 km/s. β^2 Sgr, however, rotates at 126 km/s putting it in the fastest 9%. It is thought to be 1,200 million years old.

Like β^1 and β^2 Sgr, γ^1 and γ^2 Sagittarii are also unrelated. **γ^1 Sagittarii** is an F7 supergiant of 58 $D_\odot$ and a mass of around 7 $M_\odot$. Lying 2,077 ly from the Sun it is a Cepheid, varying in magnitude between m_v +4.28 and +5.10 with a period of 7^d 14^h 16.8^m. As a result it sometimes is referred to as **W Sagittarii**. F7 supergiants are relatively rare; only a few are visible to the naked eye. γ^1 Sgr is only expected to last for about 77 million years. It has at least two companions, an F5 in a 4.33 year long orbit and an A0 in a 119 year orbit.

γ^2 Sagittarii is a m_v +2.96 K1 giant of 12, possibly 13 $D_\odot$ and a mass of around 2 $M_\odot$. Just 96.1 ly away it shines at a steady m_v +2.96. With a temperature of 4,600 K its rotational velocity is 2.5 km/s so it takes 263 days – 8 months – to complete a single revolution. It may be a spectroscopic binary.

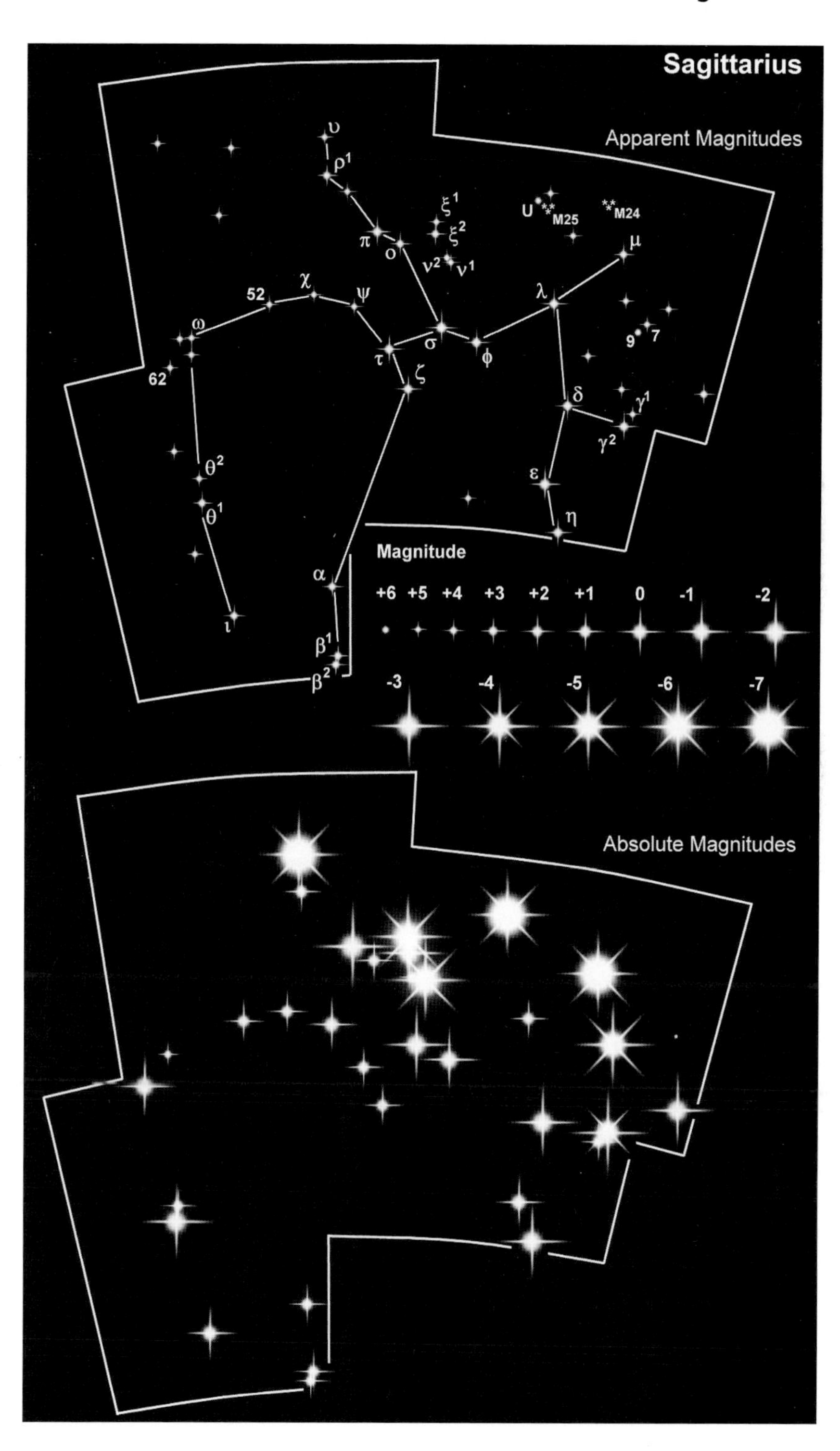
Sagittarius
Apparent Magnitudes
Magnitude
+6 +5 +4 +3 +2 +1 0 -1 -2
-3 -4 -5 -6 -7
M25
M24
Absolute Magnitudes

The brightest star in Sagittarius is ε **Sagittarii** at m_v +1.81. Usually cataloged as a B9.5 it may have just crept into the next spectral class to become an A0. Around 2.6 $D_\odot$ between the poles and perhaps as much as 7 $D_\odot$ equatorially its mass is 3.47 $M_\odot$ and it is 300 times more luminous than the Sun. It lies 145 ly away and its oblate spheroid shape is due to it spinning at 236 km/s. It rotates once every 1.5 days.

ζ **Sagittarii** is a true binary system. It appears to the naked eye as a single star of m_v +2.59 but its magnitude is a combination of a m_v +3.26 A2.5 and a m_v +3.37 A4 separated by just 0.5″. In real space that equates to about 13 AU although the stars separation varies between 10.6 and 16.1 AU with an orbital period of 21.08 years. The primary is a 3.2 $D_\odot$ dwarf of 2.27 $M_\odot$ and a luminosity of 31 $L_\odot$. Its companion has a mass of slightly less, 2.1 $M_\odot$, and is also not as luminous at 26 $L_\odot$ but its diameter is slightly larger at 3.3 $D_\odot$. The system is 89.1 ly from Earth.

Although red giant observers often record η **Sagittarii** as distinctly orange, possibly because of its contrast with a number of blue stars that surround it. An estimated 79 $D_\odot$ across – about as big as the orbit of Mercury – its mass is just 1.5 $M_\odot$. Visually it appears to have a luminosity of 95 $L_\odot$ but most of its output is in the infrared so it total luminosity is around 590 $L_\odot$. Not surprisingly it is a pulsating variable, its magnitude fluctuating between m_v +3.05 and +3.15 without any discernable period. Around 3,000 million years old it is accompanied by an F-class dwarf of unknown diameter but with a mass of 1.3 $M_\odot$. Currently separated by 3.5″ the two orbit at an average 133 AU with a period of around 1,300 years. The companion is m_v +7.8 and the whole system is 149 ly away.

Sagittarius is awash with unrelated 'siblings'. Apart from β and γ mentioned above there is also θ, ν, ξ, ρ and χ. θ^1 **Sagittarii** is a B3 star of 5.6 $D_\odot$ and 6.6 $M_\odot$. A relatively faint m_v +4.34 it would brighten to a respectable M_v -2.3 at 10 pc. Its surface temperature is 20,300 K and it rotates once in just over 4 days, its rotational velocity being 69 km/s. It is a spectroscopic binary, its companion locked in an orbital period of 2.11 days. θ^1 Sgr lies at a distance of 618 ly. θ^2 **Sagittarii** is a much smaller star, estimated to be between 1.8 and 2.1 $D_\odot$ and 1.9 $M_\odot$. It is a cooler A4 of 8,180 K and is significantly closer at 157 ly. Its magnitude is m_v +5.30.

λ Sagittarii, along with ω, are the two closest stars at 77.3 and 77.6 ly respectively. λ **Sagittarii** is a m_v +2.81 K0 giant of 9.5 $D_\odot$ and a mass of 1.78 $M_\odot$. It rotates relatively slowly, just 3.8 km/s, (the average for K1 stars is 12 km/s) and so has a rotational period of 126.2 days. ω **Sagittarii's** rotation has never been measured, which is a pity because it is a solar analog. A G5 its diameter is 1.9 $D_\odot$ with a mass of 1.39 $M_\odot$. It is m_v +4.70.

μ **Sagittarii** is a complex arrangement of half a dozen stars about which nothing is known for sure, except that it is complex! So, let's start with simple observation. First, μ^A **Sagittarii** is variable. At its maximum the main star attains m_v +3.80 but then fades to +3.88. The period is given as 180.45 to 180.55 days. Some of the variability in the period is due, we think, to the star pulsating, causing ripples across its surface and is typical of α Cygni type variables. Most of the dip in magnitude however is due to the presence of a companion that from our vantage point on Earth eclipses the main star, similar to β Persei. In other

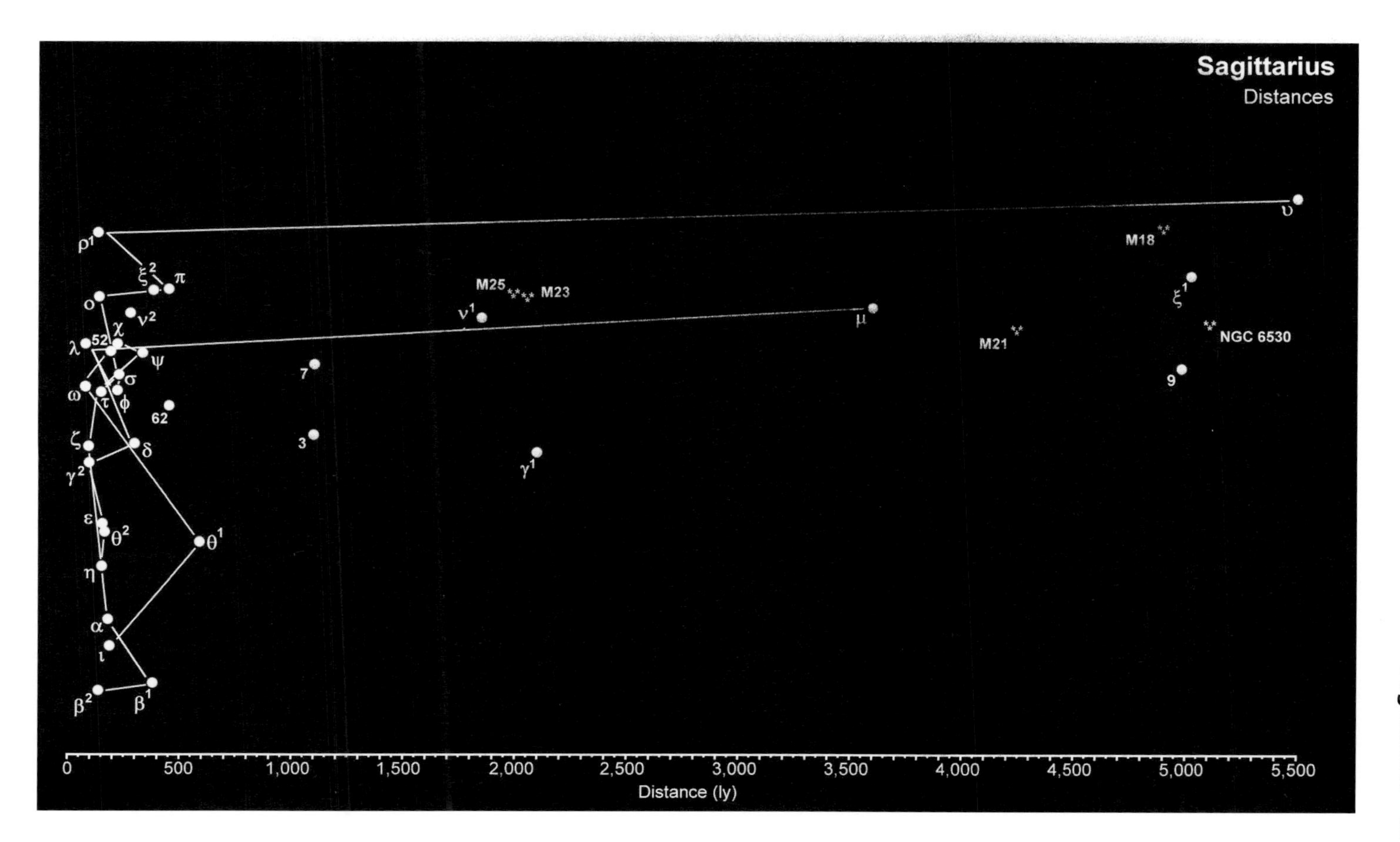
Sagittarius
Distances
M18
M25
M23
M21
NGC 6530
7
3
62
9
52
0
500
1,000
1,500
2,000
2,500
3,000
3,500
4,000
4,500
5,000
5,500
Distance (ly)

words, an Algol (EA) variable. So we already have a hybrid α Cyg-β Per variable. Second, it belongs to the B2 spectral group. Now for the speculation. It is thought to be a bluish-white supergiant. A conservative estimate is that it is 85 $D_\odot$ across. That's as big as the orbit of Mercury (0.8 AU). The upper limit is 115 $D_\odot$ (1.1 AU). Its rotational velocity is about 45 km/s so its rotational period is therefore between 95.6 and 129.3 days. μ^A Sgr's distance is even more difficult to pin down. Published estimates include 3,000, 3,600 and 5,100. The system is thought to belong to the Sagittarius OB1 association in which case 3,000 to 3,600 ly seems the most likely distance. Luminosity has been equally problematic with guesstimates ranging from 55,000 to 180,000 $L_\odot$. Take into account that much of the main star's output will be in the ultraviolet and one published figure suggests a total luminosity of 2 million $L_\odot$. At 10 pc the star could have an absolute magnitude of M_v -7.1. Then there are its five companions. **μ^B Sagittarii** (sometimes referred to as μ Sagittarii B or μ^2) is a 7th magnitude B1.5 separated by 0.011″ which equates to 2 AU in real space. It may have a mass of 8 $M_\odot$. **μ^C Sagittarii** is a B9 dwarf averaging 29,000 AU from the main star, or 17″ on the celestial sphere. It is m_v +10.48. Somewhat farther out at 44,000 AU is **μ^D Sagittarii** which appears 26″ away at m_v +13.5. It is also a B-class dwarf. Next is **μ^E Sagittarii**, a B3 separated by 49″ or 80,000 AU and glowing at m_v +9.69. Finally **μ^F Sagittarii** marks the outer edge of the system at 87,000 AU from the main star. This m_v +9.25 B2 lies 50″ away from μ Sagittarii proper. Components C, D, E and F probably have a combined mass of 10 $M_\odot$. Such a system is technically highly unstable, the gravitational interactions likely to kick some of the stars out of the system. However, the main star is only expected to live for around 5 million years so it is anyone's guess as to whether it will go supernova before losing any of its five companions.

It is easy to understand why the double stars ν^1 and ν^2 Sagittarii were mistaken for a true gravitationally bound binary pair. They are about the same magnitude, are the same spectral class and are close together – at least on the celestial sphere – being separated by about 12′. In truth they are around 1,600 ly apart, a fact that becomes evident when the two stars are 'placed' 10 pc from Earth and there is an absolute magnitude difference of 9 full mags. **ν^1 Sagittarii** is a m_v +4.86 K1 bright supergiant of 360 $D_\odot$ or 3.3 AU. In Solar System terms if the Sun was replaced by ν^1 Sgr it would easily swallow up Mars. Its mass is estimated to be 9.8 $M_\odot$. From its home 1,853 ly away its 3,000 $L_\odot$ displays as a modest star but at 10 pc it would be a sparkling M_v -6.0, far brighter than Venus. It is heading towards us at a reasonable 12 km/s, accompanied by a 3.8 $M_\odot$ B9 in an 11,700 year orbit often referred to as **ν^1 Sagittarii B**. They are currently 2.5″ apart, the secondary star being a faint m_v +10.8. There is also a **ν^1 Sagittarii C** which coincidentally is also m_v +10.8 and 28.2″ from the primary, but it probably is just a coincidence, the stars lying in the same line of sight rather than being gravitationally bound. Much closer to us at 270 ly is **ν^2 Sagittarii.** It is much smaller than ν^1 Sgr – 'just' 85 $D_\odot$ – and significantly less massive at 1.43 $M_\odot$. Its luminosity of 57 $L_\odot$ would have an absolute magnitude of a mere M_v +3.1. Certainly nothing to write home about, were it not for the fact that it is hurtling towards us at 110 km/s on a visit from a more distant part of the Galaxy.

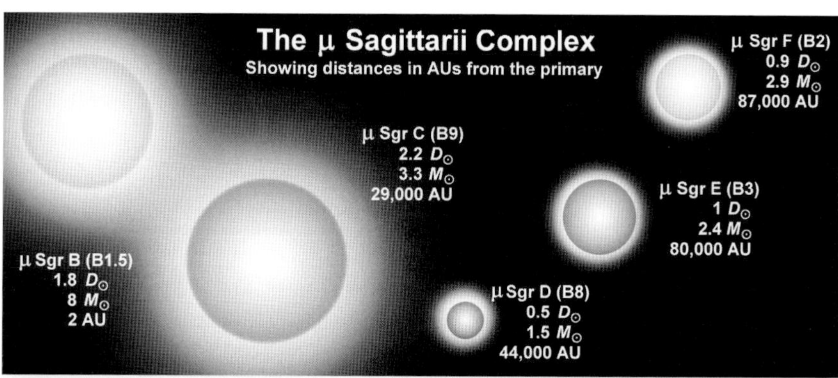
The μ Sagittarii Complex
Showing distances in AUs from the primary
μ Sgr F (B2)
0.9 $D_\odot$
2.9 $M_\odot$
87,000 AU
μ Sgr C (B9)
2.2 $D_\odot$
3.3 $M_\odot$
29,000 AU
μ Sgr E (B3)
1 $D_\odot$
2.4 $M_\odot$
80,000 AU
μ Sgr B (B1.5)
1.8 $D_\odot$
8 $M_\odot$
2 AU
μ Sgr D (B8)
0.5 $D_\odot$
1.5 $M_\odot$
44,000 AU

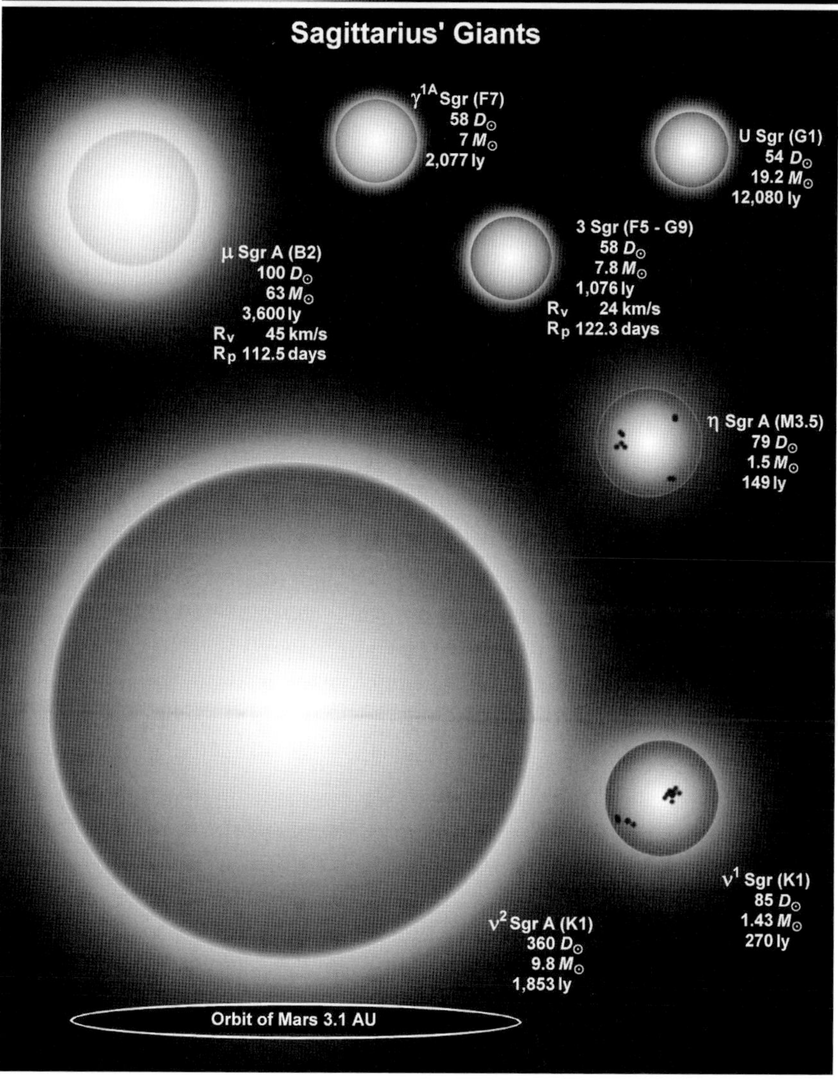
Sagittarius' Giants
γ^{1A} Sgr (F7)
58 $D_\odot$
7 $M_\odot$
2,077 ly
U Sgr (G1)
54 $D_\odot$
19.2 $M_\odot$
12,080 ly
3 Sgr (F5 - G9)
58 $D_\odot$
7.8 $M_\odot$
1,076 ly
R_v 24 km/s
R_p 122.3 days
μ Sgr A (B2)
100 $D_\odot$
63 $M_\odot$
3,600 ly
R_v 45 km/s
R_p 112.5 days
η Sgr A (M3.5)
79 $D_\odot$
1.5 $M_\odot$
149 ly
ν^1 Sgr (K1)
85 $D_\odot$
1.43 $M_\odot$
270 ly
ν^2 Sgr A (K1)
360 $D_\odot$
9.8 $M_\odot$
1,853 ly
Orbit of Mars 3.1 AU

Lying just over a degree from the direction of the galactic center **3 Sagittarii** is better known – and perhaps more appropriately – as **X Sagittarii**, the **X** giving us a clue as to its variable nature. A pulsating δ Cepheid it varies between m_V +4.20 and +4.90 with a period of $7^d\ 14^h\ 16.9^m$ during which its spectrum swings from F5 to G9. Numerous attempts have been made at estimating its diameter with results ranging from 44 to 68 $D_\odot$ and averaging 58 $D_\odot$. Its rotational velocity is known with more certainty coming in at 24 km/s. For the minimum diameter the rotational period then works out at 92.8 days increasing to 122.3 days for the average diameter up to 143.4 for the largest estimate. Believed to be somewhere between 40 and 65 million years old the star lies 1,076 ly from Earth and 25,000 ly from the galactic center. On the opposite side of the sky, 180° away, looking away from the galactic center and out into deep space, is 136 Tauri, m_V +4.55, and TYC 1873-835-1 at m_V +5.99 which we shall come across again when we describe Taurus.

Sagittarius is rich in open and globular clusters (see table). The open cluster **Messier 18 (NGC 6613)** has 29 stars brighter than 12th magnitude and is best seen with a small telescope or binocular. Sixteen of the stars in the region are B-class. It is situated between **M17,** the **Lagoon Nebula,** and the star field **M24.** **Messier 21 (NGC 6531)** has quite a compact core that often leads to an underestimation of its true size which is usually quoted as 13′ but is in fact 22′ across. At an estimated 4.2 million years old it is only one-seventh of the age of M18. By far the oldest open cluster in Sagittarius at 220 million years is **M23 (NGC 6494)**. Most of its stars are late B to early A-class. Of the 148 stars in the area only about two-thirds are actual members of the cluster. **Messier 25 (IC 4725)** is probably the closest open cluster at 2,000 ly but various factors can affect estimations of distances and it could well be that M23 is actually slightly closer. It is a full degree in diameter which, if the distance is correct, works out at nearly 35 ly across. For some strange reason it was never given an NGC number. The δ Cepheid variable **U Sagittarii** lies in the general direction of M25 but at a distance of 12,080 ly sits well beyond the cluster and is not a member. It varies between m_V +6.28 and +7.15 with a period of $6^d\ 17^h\ 52.7^m$.

NGC 6530 formed from the Lagoon Nebula (**M8**) in which it is embedded. The brightest star in the area is **7 Sagittarii** at m_V +5.35 but neither this nor the second brightest star, **9 Sagittarii**, m_V +5.97, are actual members of the cluster, 7 Sgr lying 1,106 ly away and 9 Sgr 4,900 ly. The third brightest, **HD 165052**, m_V +6.87, has only a 39% chance of being a member and we have to look at **HD 164816**, m_V +7.08, before we reach any degree of certainty (98%) that we have a cluster star. The cluster is dominated by early B-class stars with a couple of K-class thrown in for good measure.

Of the seven globular clusters in Sagittarius, **Messier 22** was the first to be discovered and is estimated to contain 70,000 stars. It is possibly the closest globular at 10,400 ly and is certainly one of the brightest at m_V +5.1. If the measurements are correct then the globular cluster **M54** is a huge structure at 305 ly across and relatively easy to find at m_V +7.6. The constellation contains numerous other deep sky objects such as the Omega and Trifid nebulae.

Of the planetary systems so far discovered in Sagittarius MOA-2007-BLG-192-L b is the most intriguing. MOA is the Microlensing Observations in Astrophysics, a collaborative project between New Zealand and Japan. The star has a mass of between 0.039 and 0.088 $M_{\odot}$ putting in it the brown dwarf category (a peculiar name as brown dwarfs are likely to be magenta due to their absorption of elements like potassium and sodium). The planet is a 0.005 to 0.025 Jovian mass (1.59 to 7.95 Earth masses) super-Earth in an orbit that is somewhere between 0.46 and 0.84 AU. A 12.71 $M_{\oplus}$ has also been found in orbit around OGLE-05-169L.

Planetary systems in Sagittarius

Star	$D_{\odot}$	Spectral class	ly	m_v	Planet	Minimum mass	q	Q	P
HD 164604	?	K2	124	+9.7	HD 164604 b	2.7M_J	0.86	1.40	1.66 y
HD 169830	1.84	F8	119	+5.9	HD 169830 b	2.88 M_J	0.48	1.14	226 d
					HD 169830 c	4.04 M_J	2.41	4.79	5.76 y
HD 171238	?	K0	164	+8.66	HD 171238 b	2.6 M_J	1.52	3.56	4.17 y
HD 179949	1.19	F8	88	+6.25	HD 179949 b	0.94 M_J	0.045	0.044	3.09 d
HD 181720	?	G1	182	+7.86	HD 181720 b	0.37 M_J	?	?	2.62 y
HD 187085	?	G0	147	+7.22	HD 187085 b	0.75 M_J	1.09	3.01	2.70 y
HD 190647	?	G5	177	+7.78	HD 190647 b	1.90 M_J	1.78	2.36	2.84 y
MOA-2007-BLG-192-L	?	?	3,262	+19.81	MOA-2007-BLG-192-L b	4.78 $M_{\oplus}$	0.62	0.62?	?
MOA-2007-BLG-400-L	?	?	19,569	?	MOA-2007-BLG-400-L b	0.9 M_J	0.85	0.85	?
OGLE-05-169L	?	?	8,806	?	OGLE-05-169L b	12.71 $M_{\oplus}$	2.8	2.8?	9 y
OGLE-06-109L	?	?	4,925	+17.17	OGLE-06-109L b	0.727 M_J	2.3	2.3?	4.9 y
					OGLE-06-109L c	0.271 M_J	4.5	4.5?	13.5 y
OGLE-235-MOA53	?	?	16,960	+19.7	OGLE-235-MOA53 b	2.60 M_J	5.1	5.1?	?
OGLE-TR-10	1.16	G/K	4,892	+14.93	OGLE-TR-10 b	0.63 M_J	0.041	0.041	3.10 d
OGLE-TR-56	1.32	G	4,892	+16.6	OGLE-TR-56 b	1.29 M_J	0.023	0.023	1.21 d
SWEEPS-04	1.18	?	6,523	?	SWEEPS-04 b	<3.8 M_J	0.055	0.055?	4.2 d
SWEEPS-11	1.45	?	6,523	?	SWEEPS-11 b	9.7 M_J	0.03	0.03?	1.80 d

Open and globular clusters in Sagittarius

Name	Size arc min	Size ly	Distance ly	Age million yrs	Brightest star in region*	No. stars m_v >+12*	Apparent magnitude m_v
M17 (NGC 6618)	11′	40	5,500	1	HD 168302 m_v +9.26	5	+6.0
M18 (NGC 6613)	19′	27.4	4,900	32	HD 168352 m_v +8.70	29	+7.5
M21 (NGC 6531)	22′	27.5	4,250	4.6	HD 164863 m_v +7.28	57	+6.5
M23 (NGC 6494)	43′	27	2,150	220	HD 163245 m_v +6.46	148	+6.9
M25 (IC 4725)	60′	34.9	2,000	89	HD 170764 m_v +6.36	126	+4.6
NGC 6530	58′	88	5,200	2	7 Sagittarii m_v +5.35	182	+4.6
M22 (NGC 6656)	32′	92	10,400		Globular cluster		+5.1
M28 (NGC 6626)	11.2′	60	18,300		Globular cluster		+6.8
M54 (NGC 6715)	12′	305	87,400		Globular cluster		+7.6
M55 (NGC 6809)	19′	97	17,300		Globular cluster		+6.3
M69 (NGC 6637)	9.8′	83	29,700		Globular cluster		+7.6
M70 (NGC 6681)	8′	65	29,300		Globular cluster		+7.9
M75 (NGC 6864)	6.8′	130	67,500		Globular cluster		+8.5

*May not be a cluster member.

Scorpius

Constellation:	Scorpius	**Hemisphere:**	Southern
Translation:	The Scorpion	**Area:**	497 deg^2
Genitive:	Scorpii	**% of sky:**	1.205%
Abbreviation:	Sco	**Size ranking:**	33rd

Like neighboring Sagittarius, Scorpius is set in a rich part of the Galaxy. It is one of the Zodiacal constellations, the Sun entering on 23 November and leaving just six days later on 29 November. The constellation depicts a giant scorpion against which Orion fought and lost, the scorpion inflicting its deadly sting on the great hunter.

The red coloration of α **Scorpii** means that it can easily be mistaken for Mars and has earned it the ancient name Antares. 'Ares' is the Greek for Mars and 'Ant' has the same root as 'anti' so the name is often translated as 'Rival of Mars', 'Against Mars' or 'Opposite Mars', although it could also be translated as 'Not Mars'. Antares is one of the truly great stars; an M1.5 supergiant of 740 $D_{\odot}$ or 6.9 AU which, in Solar System terms, puts it somewhere between the Asteroid Belt (average of 5 AU) and Jupiter's orbit (10.4 AU). Despite its huge size its mass is just 16.5 $\pm$ 1.5 $M_{\odot}$. Its luminosity initially looks like 10,200 $L_{\odot}$ but it should be remembered that most of its radiation is released in the infrared so its absolute luminosity is likely to be 75,000 $\pm$ 15,000 $L_{\odot}$. Lying some 604 ly away its rotational velocity has been estimated at 20 km/s so the star rotates once every 1,873 days or 5.1 years. And therein lies a couple of intriguing mysteries. The first is that angular measurements of the star suggest that it is not spherical and it has been suggested that it may be an oblate spheroid or egg-shaped. The problem is that Antares is not spinning fast enough to deform in such a way. What we could be witnessing is evidence for a star that has a very uneven surface, rather like a dozen golf balls stuck together. Betelgeuse, another red supergiant, also hints at this type of structure. The second mystery is the fact that it is an SRc variable, changing between m_v +0.88 and +1.80 with a period of 1,733 days or 4.75 years. This is very close to its rotational period and, given the margin of error in astronomical observations of such stars and the fact that the equatorial region will rotate faster than the polar regions, it raises the possibility that its variability may be linked to its rotational period rather than the traditional view that variability is caused by the star pulsating. Antares is not alone. At around 550 AU a B3 blue dwarf orbits the supergiant every 2,500 years. At a mere 2.9″ separation this m_v +5.4 star is difficult to see but those who have managed to observe the companion, which has a mass of about half that of the primary, usually say that it looks green, an artifact of its contrast against the much brighter red star. The Antares system is embedded in a nebula about 5 ly across.

β **Scorpii** is a complex of at least six B-class stars in highly unstable orbits. β^{Aa} Scorpii (often referred to as β^1 Scorpii) is a m_v +2.62 B0.5 of 8 $D_{\odot}$ and 13.5 $M_{\odot}$. The double letter superscript gives away the fact that it has an orbiting companion, β^{Ab} Scorpii, which is a B1.5 of 6.5 $M_\odot$ and a magnitude of m_v +3.72. From Earth the two stars appear to be in a vertical orbit with a period of 6.43 days and a separation of 1.42′. Also in orbit around β^{A} Scorpii is β^{B} Scorpii, a m_v +5.98 B-class star of 4.7 $M_\odot$. β^{A} Scorpii and β^{B} Scorpii seem to be in a horizontal orbit separated by 3.9″ and with an orbital period of 610 years. The 13.24 $M_\odot$ β^{C} Scorpii (often referred to as β^{2} Scorpii) is in a 39 year long orbital embrace with the 3.4 $M_\odot$ β^{Ea} Scorpii and has a separation of 0.133″. Needless to say, there is a β^{Eb} Scorpii also in orbit with β^{Ea} Scorpii that has a period of 10.7 days. β^{Eb} Scorpii has a mass of 1 $M_\odot$. They are a spectroscopic pair. The $\beta^{Aa}/\beta^{Ab}/\beta^{B}$ triplet are in a 12,700 year long orbit with the $\beta^{C}/\beta^{Ea}/\beta^{Eb}$ triplet. The two triplets are separated by 13.6″.

The star that was originally designated **γ Scorpii** is now known as **σ Librae** so you won't find it in Scorpius. The next star on the list is **δ Scorpii** or Dschubba, a hot 29,500 K B0.2 some 402 ly away. Spinning at 165 km/s it is an oblate spheroid, its equatorial diameter a bulging 7.4 $D_\odot$ while its polar diameter is probably only about 5 $D_\odot$. Its rapid rotation throws hot gases out into the surrounding space thus making the star a Be emission star. An apparent luminosity of 1,500 $L_\odot$ belies the fact that it radiates most of its energy in the ultraviolet so that its total output is more like 66,000 $L_\odot$. At 10 pc it would have an absolute magnitude of m_v -4.1, about as bright as Venus, while its 'normal' apparent magnitude is m_v +2.28. 'Normal' because in July 2000 it followed γ Cassiopeiae and suddenly brightened, attaining m_v +1.67. It remained high for a few years before fading. A decade later it is still unstable and varying between m_v +1.8 and +2.6. It has a companion in a 0.4 AU orbit that has a period of 20 days and another in a 9.5 AU with a period in excess of 10 years. Both a B-class stars as is a possible third companion.

ζ^{1} Scorpii is a relatively faint m_v +4.78 yellowish star that hides its light under a bushel, or at least behind clouds of interstellar dust. No one really knows how far away it is – 5,700 ly has been suggested – but its light is significantly dimmed and washed out. ζ^{1} Scorpii is actually one of the most massive and most luminous stars in the sky. Without the interstellar dust clouds getting in the way it would be a couple of magnitudes brighter and appear distinctly blue for ζ^{1} Sco belongs to the Luminous Blue Variable (LBV) category, its output a staggering 1.5 million Suns. It is not a particularly large star – 'just' 52 $D_\odot$ across – but it is very massive at 58 $M_\odot$. Such stars live for only a few million years at most, the Galaxy having witnessed their passing many times during its 13,500 million year history. ζ^{1} Sco loses a full solar mass every 100,000 years. Belonging to the B1 spectral group its surface temperature is 21,000 K and its rotational velocity is 57 km/s, so it makes one turn on its axis every 46.2 days.

Seven arc minutes to the east of ζ^{1} Sco is the totally unrelated **ζ^{2} Scorpii,** more than a magnitude brighter at m_v +3.62. A K4 giant of 21 $D_\odot$ its mass is around 1.4 $M_\odot$ and it is 151 ly away. Almost due south is the m_v +5.88 **HD 152293** forming a neat triangle that is easy to spot with a small 'scope or a binocular. Other than the fact that it is an F5 very little else is known about this star.

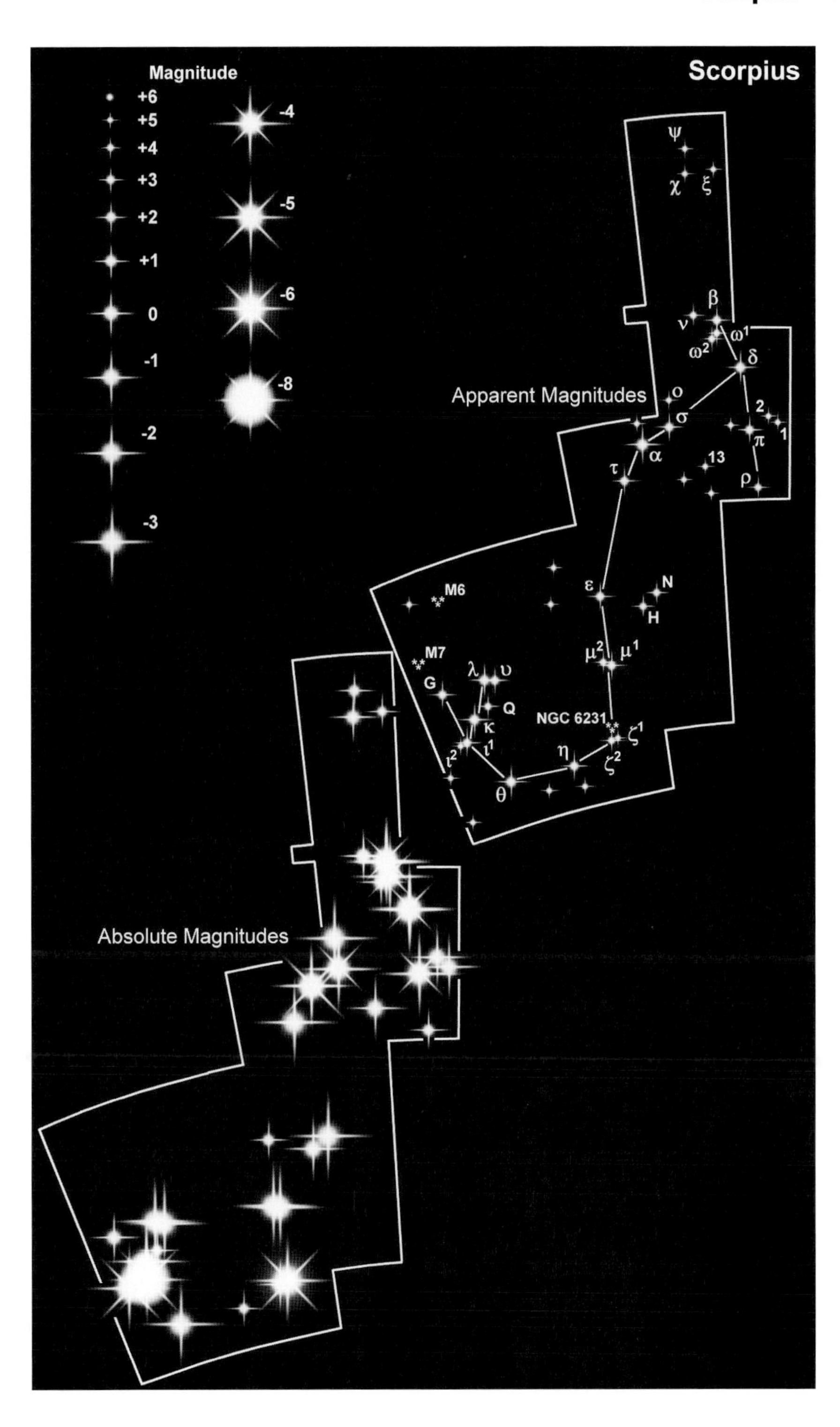
Magnitude
+6
+5
+4
+3
+2
+1
0
-1
-2
-3
-4
-5
-6
-8
Scorpius
Apparent Magnitudes
Absolute Magnitudes
ψ
χ
ξ
β
ν
ω1
ω2
δ
ο
σ
2
π
1
α
τ
13
ρ
ε
N
H
M6
M7
μ2
μ1
λ
υ
G
Q
κ
NGC 6231
ζ1
ι2
ι1
η
ζ2
θ

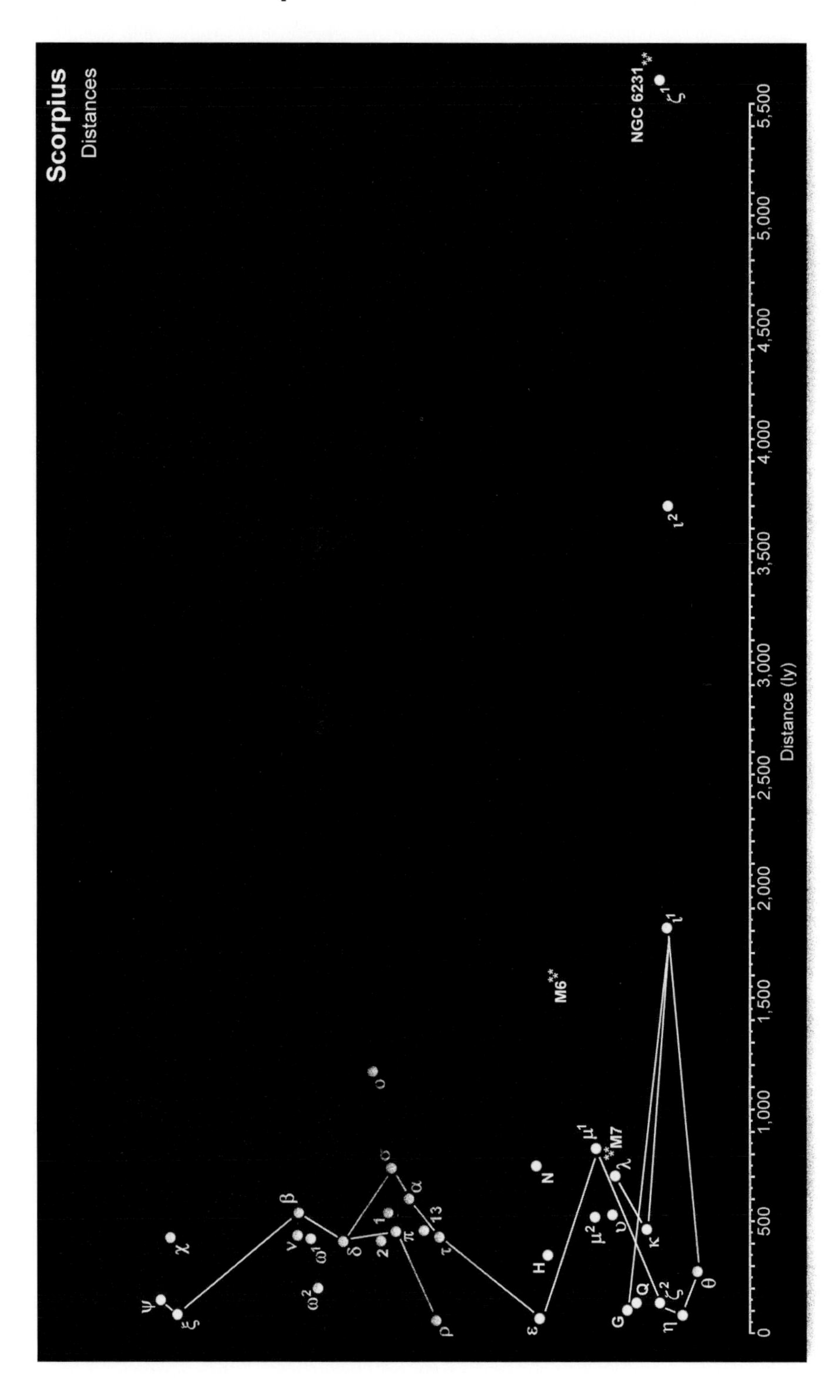
Scorpius
Distances
NGC 6231
M6
M7
Distance (ly)
0
500
1,000
1,500
2,000
2,500
3,000
3,500
4,000
4,500
5,000
5,500

ι **Scorpii** is another double star system which, like ζ^1 and ζ^2 Sco, is not a binary. The brighter of the two is ι^1 Sco at m_v +3.02, a rare F2 supergiant of 400 $D_\odot$ (3.7 AU) with a mass of 12 $M_\odot$ and luminosity of 35,100 $L_\odot$. Long believed to be about 1,790 ly away the latest estimates indicate it is even more distant at 1,930 ly. It rotates at 36 km/s and so takes 1.5 years to complete a single turn on its axis. At 10 pc it would be M_v -8.4. Its namesake, ι^2 Sco, is almost two magnitudes fainter at apparent magnitude m_v +4.81 and with an absolute magnitude of M_v -5.2. While no where near as big its 60 $D_\odot$ is rare for an A2 with only about a dozen naked eye A-class stars known to be larger. It is also much farther away, 3,706 ly, and spins slightly faster at 39 km/s and so completes a single rotation every 80 days.

Scorpius is home to three naked eye β Cepheid pulsating variables. The brightest is λ **Scorpii** or Shaula which is also the second brightest star in the constellation, at m_v +1.63 to +1.65 and with a period of $5^h\ 7^m\ 44^s$. A B2 of 6.6 $D_\odot$ it is just over 700 ly away and rotates once every 2.3 days with a rotational velocity of 145 km/s. The second brightest is κ **Scorpii** or Girtab which over a period of $4^d\ 47^m\ 49^s$ varies between m_v +2.41 and +2.42 and back. A B1.5 similar in size to λ Sco at 7 $D_\odot$ it is much closer at 464 ly and rotates at 130 km/s taking 2.7 days to spin once. It has an estimated mass of 10.5 $M_\odot$ and is accompanied by a 7 $M_\odot$ companion in a 195.7 day orbit which sees them close up 0.87 AU before separating to a maximum 2.5 AU. The faintest of the three β Cepheids is also the most interesting, being a possible quadruple system. The main star, σ **Scorpii** or Alniyat, has a magnitude range of m_v +2.86 to +2.95 with a periodicity of 5^d 55.5^m. Around 8 $D_\odot$ it is potentially the farthest at 735 ly but that depends on how much error there is in the distance estimation. A fairly young system, just 10 million years old, the B1 primary has a rotational velocity of 55 km/s and takes just over a week to complete a single rotation. It has a mass of 15.5 $M_\odot$ while its m_v +5.2 partner is about 3.4 $M_\odot$ and has an orbital period of 33 days. Two other companions are separated from the main pair by 0.4″ and 20″.

The double star μ **Scorpii** consists of a variable B1.5, which fluctuates between m_v +2.94 and +3.22, and a m_v +3.53 B2. μ^1 **Scorpii.** μ^1 Sco is 5.3 by 3.0 $D_\odot$, its shape distorted partly by its rapid rotational velocity of 209 km/s. μ^2 **Scorpii** is a more spherical 7 $D_\odot$, rotating at 52 km/s and turning once every 6.8 days. The two stars, however, are totally unconnected, μ^1 Sco lying some 822 ly away but closing in on us at 25 km/s while μ^2 Sco is 517 ly away and heading deeper into space at a leisurely 1.4 km/s. However, unseen is a B6 in orbit around μ^1 Sco although we can detect its presence by the fact that it eclipses the primary every $1^d\ 10^h\ 42^m\ 38^s$, hence the variability. The μ^1 Sco components weigh in at 13 and 8 $M_\odot$ and are just 10.5 million km (7.5 $D_\odot$) apart, so close that one is losing mass to the other, making it a β Lyrae or EB semi-detached system.

A second β Lyrae pair can be found in π **Scorpii** which is also a quadruple system. The main star is a m_v +2.88 to +2.90 B1 with a diameter of 6.97 $D_\odot$ and mass of 12.4 $M_\odot$. Around 458 ly away the star spins at 90 km/s and takes almost 4 days to turn once on its axis. Its companion is a 5.59 $D_\odot$ B2 with a mass of 7.6 $M_\odot$. The pair are in a roughly circular orbit with a separation of just 3 millionths of an arc second which scales up to 0.071 AU or 10.44 million km – about 7.4 $D_\odot$. The orbital period is a mere 1.57 days. A third companion is in orbit around these two stars at a distance of 8,000 AU with a period in excess of 160,000 years. It also has a partner, probably a K dwarf, in a 90 AU orbit.

To the naked eye ν **Scorpii** or Jabbah looks like a lone 4th magnitude star but a small telescope will reveal a 6th magnitude companion while large apertures show that each of the two components themselves have companions. Add a spectroscope to the set up and a fifth star is evident. There may be more: it is a complicated story. The primary is a m_v +4.35 B2 giant of 12 $D_\odot$ and a mass of 10.96 $M_\odot$. Despite its large size it is also a fast spinner at 200 km/s, completing a turn every 3 days. At these velocities the poles are pulled inwards and the equator bulges so the true physical dimensions are likely to be 12 $D_\odot$ across the equator and 9 $D_\odot$ through the poles. It has an unseen spectroscopic companion of 1.2 $M_\odot$ that orbits the primary every 5.552 days. It is also a B-class star. At a separation of 1.3″ is a m_v +5.37 B9 of 6.5 $M_\odot$. With a more modest rotational velocity of 70 km/s it orbits the primary and its spectroscopic companion at an average distance of 175 AU, taking 452 years to complete a single orbit. Farther out, at 41.1″, is the fourth star, a m_v +6.9 B8 with a mass of 3.88 $M_\odot$. It also has an orbiting companion, a B9 of m_v +7.39 that has a mass of 2.08 $M_\odot$. The two are separated by an average 2″ which corresponds to 320 AU and an orbital period of 1,388 years. The faint pair are in orbit with the brighter pair (and the spectroscopic companion) with a period of at least 68,000 years, the mean orbital distance being 5,500 AU. The system is 437 ly from Earth.

Things do not get any easier with ξ **Scorpii**, another 5 star system at 79.4 ly away. ξ **Scorpii A** is a m_v +4.77 yellowish F6 with a mass of around 1.3 $M_\odot$ and a diameter of 2.3 $D_\odot$. It has an almost identical companion, ξ **Scorpii B,** a slightly smaller 2.0 $D_\odot$ and 1.1 $M_\odot$ but an F5. It has an apparent magnitude of m_v +5.07. On the celestial sphere they are separated by 0.654″ but in real space they are in a 45.648 year long eccentric orbit that brings them as close as 5.5 AU but then separates them by as much as 32.6 AU. ξ **Scorpii C** orbits A and B every 1,500 years at an average distance of 210 AU, the separation currently being 7.8″. It is a yellow G7 of 0.9 $D_\odot$ and about 1.0 $M_\odot$ with a luminosity of about 80% that of the Sun. To this triplet must be added a binary consisting of ξ **Scorpii D** and ξ **Scorpii E** but which are also commonly referred to as Struve 1999A and Struve 1999B (or Σ1999A and Σ1999B). ξ Scorpii D is a yellowish-orange G8 of m_v +7.46 while ξ Scorpii E is a more orange than yellow K0 or K2 of m_v +8.03. Both stars have diameters and masses of about 85% that of the Sun. Their mutual orbit has a period of 3,918 years, the two stars separated by an average 329 AU (11.39″). The binary pair lie about 8,000 AU (282.66″) from the triplet in an orbit that takes more than 282,000 years to complete.

ω **Scorpii** is a fine naked eye double 14.6′ apart and totally unrelated. The brighter of the two is a m_v +3.93 B1 whose diameter has been variously put at 5.6, 6.6 and 10 $D_\odot$. Like many B-class stars it is a fast spinner, rotating at 120 km/s resulting in a rotational period of 2.4, 2.8 or 4.2 days depending on which diameter is correct. It is 424 ly away. ω^2 **Scorpii** is the fainter at m_v +4.32 and its

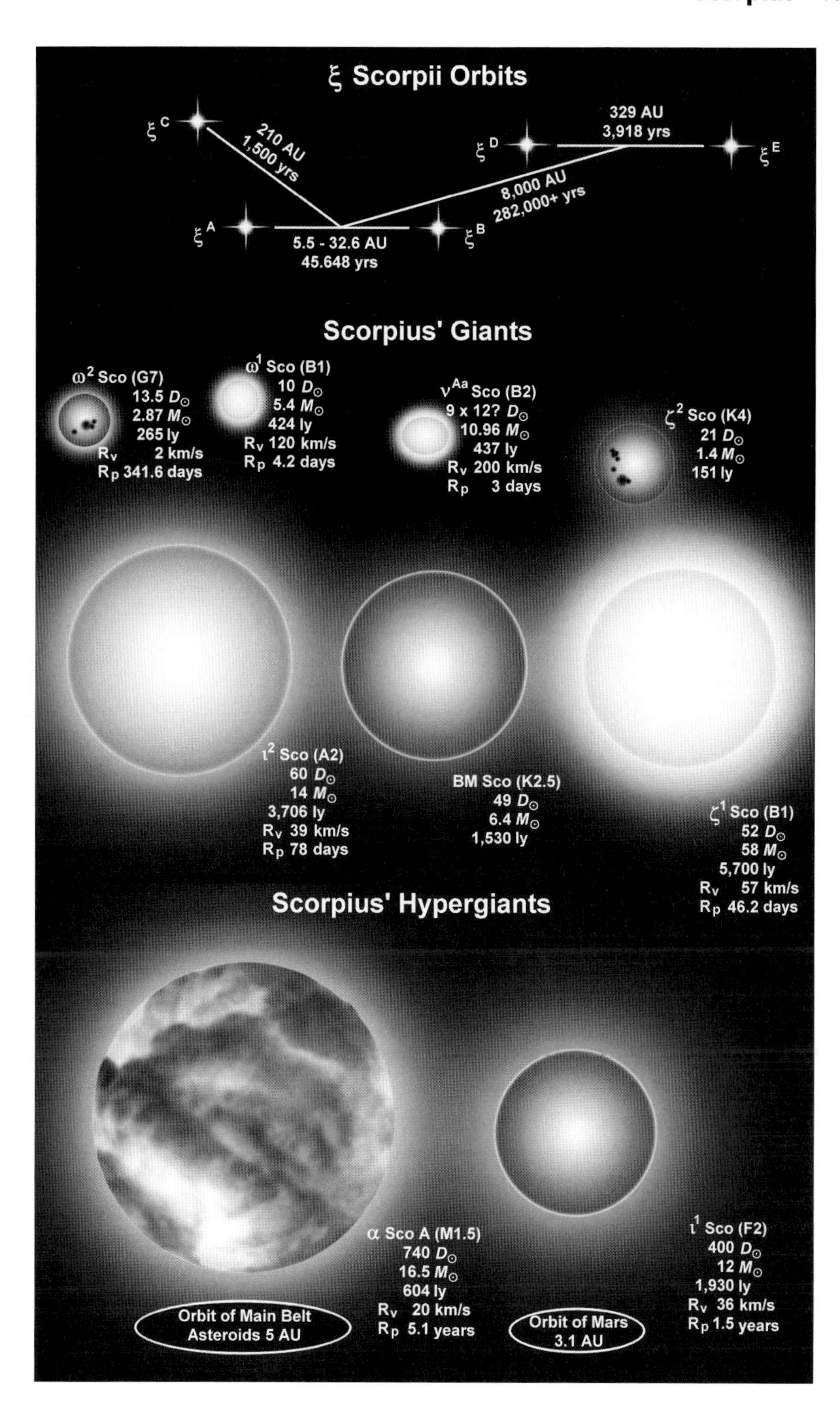
ξ Scorpii Orbits
ξ C
210 AU
1,500 yrs
329 AU
3,918 yrs
ξ D
ξ E
8,000 AU
282,000+ yrs
ξ A
5.5 - 32.6 AU
45.648 yrs
ξ B
Scorpius' Giants
ω2 Sco (G7)
13.5 D☉
2.87 M☉
265 ly
Rv 2 km/s
Rp 341.6 days
ω1 Sco (B1)
10 D☉
5.4 M☉
424 ly
Rv 120 km/s
Rp 4.2 days
νAa Sco (B2)
9 x 12? D☉
10.96 M☉
437 ly
Rv 200 km/s
Rp 3 days
ζ2 Sco (K4)
21 D☉
1.4 M☉
151 ly
ι2 Sco (A2)
60 D☉
14 M☉
3,706 ly
Rv 39 km/s
Rp 78 days
BM Sco (K2.5)
49 D☉
6.4 M☉
1,530 ly
ζ1 Sco (B1)
52 D☉
58 M☉
5,700 ly
Rv 57 km/s
Rp 46.2 days
Scorpius' Hypergiants
α Sco A (M1.5)
740 D☉
16.5 M☉
604 ly
Rv 20 km/s
Rp 5.1 years
ι1 Sco (F2)
400 D☉
12 M☉
1,930 ly
Rv 36 km/s
Rp 1.5 years
Orbit of Main Belt
Asteroids 5 AU
Orbit of Mars
3.1 AU

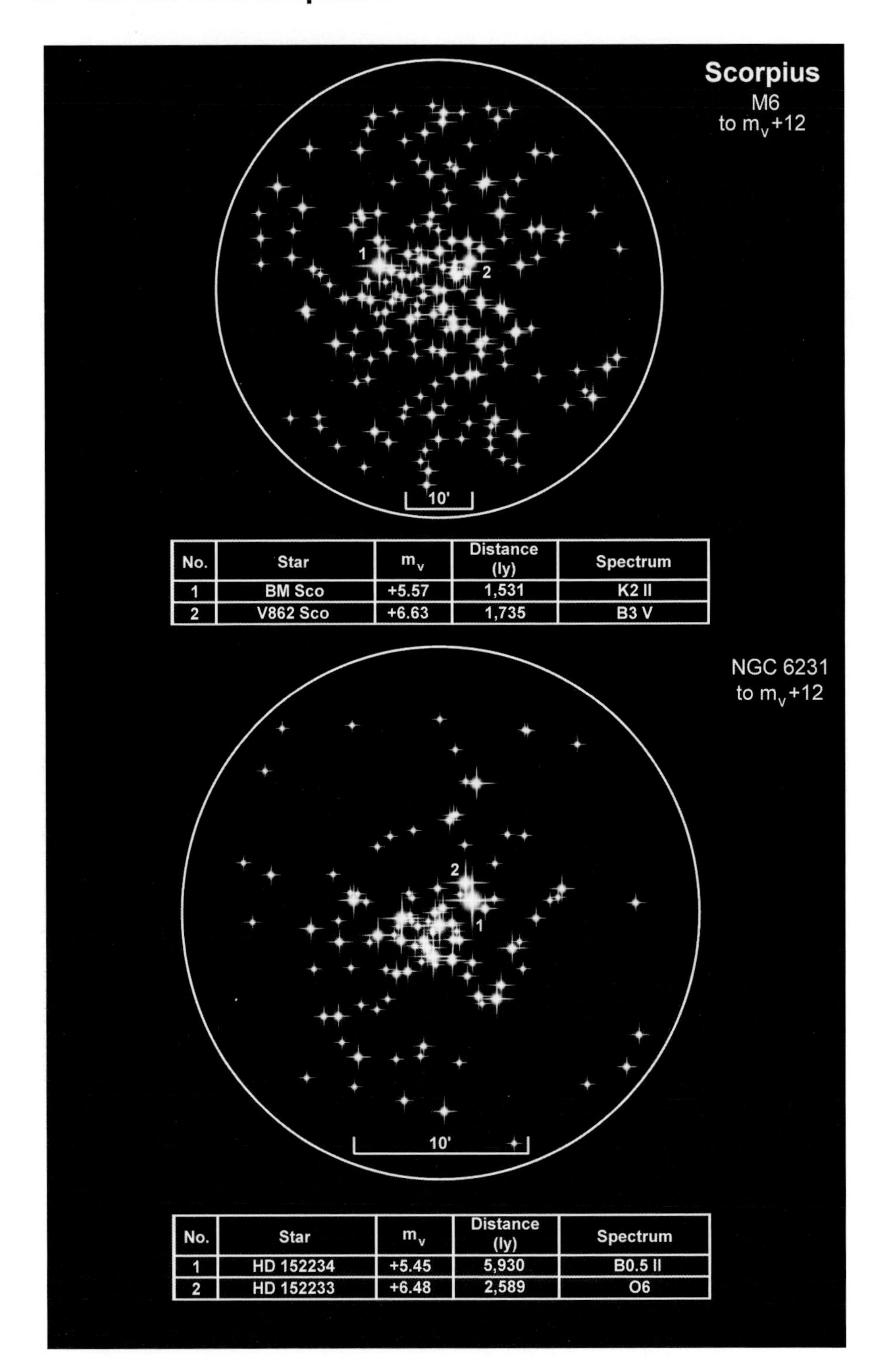

No.	Star	m_v	Distance (ly)	Spectrum
1	BM Sco	+5.57	1,531	K2 II
2	V862 Sco	+6.63	1,735	B3 V

No.	Star	m_v	Distance (ly)	Spectrum
1	HD 152234	+5.45	5,930	B0.5 II
2	HD 152233	+6.48	2,589	O6

spectral class has been described in the literature as being anywhere between a G3 and a G7. Its diameter has been narrowed down to between 10 and 13.5 $D_\odot$ and its mass is reckoned to be 2.87 $M_\odot$. It rotates at the same velocity as the Sun – 2 km/s – which means its rotational period is between 253.1 and 341.6 days; compare this with the Sun's 25 day period. ω^2 Sco is much closer than ω^1 at 265 ly.

Scorpius is home to a number of bright open clusters. **NGC 6231** tops the list at m_v +2.6 and is also the youngest of the three clusters at 3.2 million years. It is dominated by hot O and B class stars and contains 15 variables that attain at least m_v +12. One notable star is **HD 152270** (aka **WR 79**). Its spectral signature, WC7, indicates it is a Wolf-Rayet (WR) star, a relatively rare type of object named after the 19th Century astronomers Charles Wolf and Georges Rayet who discovered the first of the class. WR stars are highly evolved massive stars, in excess of 20 $M_\odot$, with temperatures in the range of 25,000 K to 50,000 K. The stream of particles such stars continually release – the stellar wind – can reach speeds of up to 2,000 km/s compared with the Sun's 400 to 750 km/s. HD 152270 has a mass of 21 $M_\odot$ and lies about 600 ly beyond the center of the open cluster at 6,500 ly. It is a spectroscopic binary, its 6.4 $M_\odot$ O5-8 companion being in an 8.8908 day long orbit. At 10 pc HD 152270 would be a brilliant M_v -6.1. Two other brighter Wolf-Rayet stars exist in Scorpius: **HD 151932 (WR 78)** at m_v +6.61 and the second brightest WR star in the entire sky, **HD 152408 (WR 79a**) at m_v +5.29. Only γ^2 **Velorum** outshines all others at m_v +1.74.

A couple of super-Earths have been detected orbiting stars in Scorpius. OGLE-05-390L b is a 5.4 $M_\oplus$ while GJ 667C b is slightly more massive at 5.7 $M_\oplus$. At the other end of the scale is the 13.75 M_J HD 162020 b which is a brown dwarf in the making (see table).

Messier 7 is often referred to as Ptolemy's Cluster having been mentioned by the Greek astronomer Claudius Ptolemæus in about AD 130. Of the 155 stars in the M7 region that are brighter than m_v +12 about one-third have less than a 30% chance of being cluster members and nearly one-half have more than a 50% chance of being a member. M7 has one of the most densely packed cores of all open clusters.

The faintest though still impressive of the clusters is **Messier 6** with an uncertain age of 51 to 95 million years. Its luminary is sometimes the semi-regular variable **BM Scorpii**, a K2.5 giant which can attain m_v +5.57 but which can often be found in the range m_v +6.8 to +8.7 so can be outshone by ten other stars in the region.

Messier 4 is one of the closest globular clusters at 7,200 ly. Although spread across 75 ly more than half its mass can be found in a compact central core just 16 ly across. M4's claim to fame is that it has a planet in orbit around a white dwarf-pulsar binary system **(PSR B1620-26).** M4 lies just to the west of Antares.

Open clusters in Scorpius

Name	Size arc min	Size ly	Distance ly	Age million yrs	Brightest star in region*	No. stars m_v >+12*	Apparent magnitude m_v
M6 (NGC 6405)	25′	11.6	1,588	51-95	BM Sco m_v +5.57	189	+4.2
M7 (NGC 6475)	80′	18.2	800	220	HD 162587 m_v +5.56	155	+3.3
NGC 6231	20′	30.9	5,900	3.2	HD 152234 m_v +5.45	112	+2.6
M4 (NGC 6121)	36′	75.4	7,200		Globular cluster		+5.6
M80 (NGC 6093)	10′	95.0	32,600		Globular cluster		+7.3
NGC 6388	17′	212.6	44,000		Globular cluster		+6.7
NGC 6453	10′	101.8	34,900		Globular cluster		+9.8

*May not be a cluster member.

Planetary systems in Scorpius

Star	$D_{\odot}$	Spectral class	ly	m_v	Planet	Minimum mass	q	Q	P
GJ 667C	?	M1.5	22.7	+10.22	GJ 667C b	0.018 M_J	?	?	7 d
HD 147513	1.0	G3/G5	42.1	+5.37	HD 147513 b	1 M_J	0.61	1.92	1.48 y
HD 159868	?	G5	172	+7.24	HD 159868 b	1.7 M_J	0.62	3.38	2.70 y
HD 162020	0.71	K2	102	+9.18	HD 162020 b	13.75 M_J	0.052	0.092	8.4 d
OGLE-05-071L	?	?	10,763	+19.5	OGLE-05-071L b	3.5 M_J	3.6?	3.6?	9.86 y
OGLE-05-390L	?	M	21,200	+15.7	OGLE-05-390L b	0.017 M_J	2.1	2.1	9.58 y
PSR B1620-26	?	?	12,400	+24.0	PSR B1620-26 b	2.5 M_J	23?	23?	100 y
TYC 6787-1927-1	1.38	F6	?	+11.6	WASP-17 b	0.49 M_J	0.044	0.058	3.74 d

Sculptor

Constellation:	Sculptor	**Hemisphere:**	Southern
Translation:	The Sculptor	**Area:**	475 deg^2
Genitive:	Sculptoris	**% of sky:**	1.151%
Abbreviation:	Scl	**Size ranking:**	36th

Originally called the Sculptor's Workshop this is another faint constellation invented by de La Caille. The South Galactic Pole – or Polaris Galacticus Australis as Jim Kaler calls it – can be found 2.7° north west of β Sculptoris at RA 00^h 51^m, Dec. -27 08′.

α **Sculptoris** is the brightest star in the constellation but at m_v +4.29 it is not particularly easy to find. It is an SX Arietis or 'helium' variable reaching a minimum m_v +4.35, its variability linked to the rotational period of the star of 21^d 15^h 39^m. Spinning at a relatively slow 15 km/s α Scl is estimated to be 6.4 $D_\odot$ and with a mass of 5.5 $M_\odot$. It is a B7 lying at 672 ly from Earth.

ε **Sculptoris** is definitely a binary, possible a triple but unlikely to be a quadruple system. The second closest star at 89.5 ly the primary ε^A **Sculptoris** (or, if you prefer, ε Sculptoris A) is an F2 of 1.5 $D_\odot$ and mass of 1.37 $M_\odot$. It appears to have three companions: ε^B **Sculptoris** is a m_v +8.50 G9 Main Sequence dwarf which between 1825 and 2002 drifted away from the primary by 1.9″; ε^C **Sculptoris** is a m_v +15.0 star that has shown no movement while ε^D **Sculptoris** closed in on the primary by 1.2″ between 1913 and 1998. A and B appear to be a real binary system with an orbital period of 1,192.2 years. The jury is still out on A and D but it seems at least possible that the star is part of a triple arrangement. A and C are likely to be a line of sight coincidence.

The m_v +5.02 ζ **Sculptoris** marks the position of the dispersed open cluster Blanco 1 although it is not itself a member lying 510 ly from Earth compared to the 877 ly of the cluster. A B4 with a diameter of 5 $D_\odot$ it is another apparently slow spinner at 15 km/s.

The **Blanco 1** cluster was discovered by V.M.Blanco in 1949 and is about 62.5 million years old. It contains about 170 stars brighter than magnitude +12 though no one is quite sure how many are actual cluster members as only about 20% have had their membership probability estimated.

η **Sculptoris** is an M4 red giant 70 $D_\odot$ across; about as big as the orbit of Mercury. Around 548 ly away it is, like many of its class, an Lb pulsating variable, fluctuating between m_v +4.80 and +4.90 with no detectable period. Somewhat smaller at 27 $D_\odot$ is the Mira variable **S Sculptoris** which can attain m_v +5.50 on a good day before plummeting to m_v +13.6. Its period is just about a year long – 362.57 days – its spectrum changing between M3 and M9. S Scl lies at a distance of 570 ly.

The globular cluster **NGC 288** can be found close to the South Galactic Pole though with an apparent magnitude of m_v +8.1 it is not particularly east to find. It is quite widely dispersed with an apparent diameter of about 13′ and a compact core 3′ across. It is 29,000 ly away and has a diameter of about 109.7 ly.

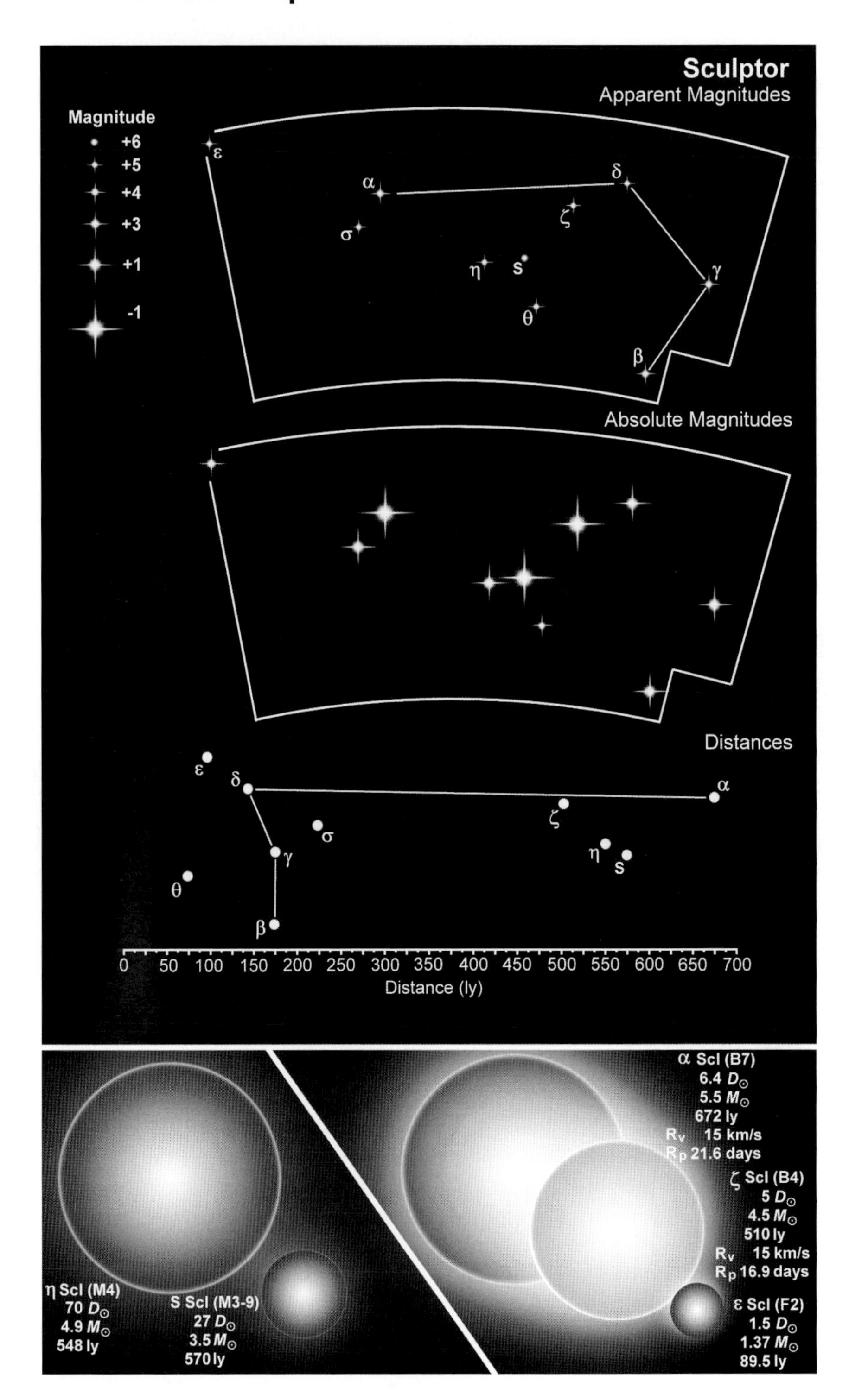
Sculptor
Apparent Magnitudes
Magnitude
+6
+5
+4
+3
+1
-1
ε
α
δ
ζ
σ
η
S
γ
θ
β
Absolute Magnitudes
Distances
0 50 100 150 200 250 300 350 400 450 500 550 600 650 700
Distance (ly)
α Scl (B7)
6.4 D⊙
5.5 M⊙
672 ly
Rv 15 km/s
Rp 21.6 days
ζ Scl (B4)
5 D⊙
4.5 M⊙
510 ly
Rv 15 km/s
Rp 16.9 days
ε Scl (F2)
1.5 D⊙
1.37 M⊙
89.5 ly
η Scl (M4)
70 D⊙
4.9 M⊙
548 ly
S Scl (M3-9)
27 D⊙
3.5 M⊙
570 ly

Scutum

Constellation:	Scutum	**Hemisphere:**	Southern
Translation:	The Shield	**Area:**	109 deg^2
Genitive:	Scuti	**% of sky:**	0.264%
Abbreviation:	Sct	**Size ranking:**	84th

When Johannes Hevelius' observatory burnt down in the late 17th Century, King John III Sobieski of Poland provided the funds for it to be rebuilt. Hevelius named this small constellation Sobieski's Shield in his honor but it is now simply called the Shield. It has the Milky Way as a backdrop and contains a number of open clusters although most of these are optically poor and difficult to see.

α **Scuti** is one of countless K-class stars that populate the Galaxy. In this case it is a K3 giant with a diameter estimated to be somewhere between 20 and 28 $D_\odot$ and with a mass of just 1.3 $M_\odot$. With a temperature of around 4,300 K it shines as a m_v +3.85 star which may be variable between +3.81 and +3.87 but which probably isn't. It rotates more slowly than the Sun – 1.8 km/s compared to 2.0 km/s – and so takes a minimum of 562 days (1.5 years) up to a maximum of 787 days (2.2 years) to complete a single rotation, depending on how big it really is.

β **Scuti** is a spectroscopic double consisting of a 50 $D_\odot$ G4 with a mass of around 4.96 $M_\odot$ and a B9.5 or possibly an A0 class dwarf of about 3 $M_\odot$ in an 834 day (2.28 years) orbit that varies between 2.8 AU at periastron and 4.7 AU at apastron. The primary rotates once every 389 days, spinning at 6.5 km/s.

γ **Scuti** is possibly the smallest star in the constellation that we can see with the naked eye, but with a breakneck rotational velocity of 255 km/s it will be well and truly distorted. An A1 with a temperature of 8,500 K it probably measures 1.6 $D_\odot$ through the poles and 2 $D_\odot$ across the equator. A-class stars average 104 km/s with A1s a bit lower at 100 km/s, so γ Sct is rotating two-and-a-half times as fast as the average. Only about a dozen stars of the class have been observed to rotate faster. One of the effects of such a high rotational velocity is the Von Zeipel Effect in which the equatorial regions darken while the poles appear lighter and may contain bright hot spots.

The most famous star in the constellation is δ **Scuti**, the prototype for a class of short period pulsating variables. Lying some 187 ly away δ Sct has a diameter of 3.8 $D_\odot$ and mass of 2.01 $M_\odot$. An F2 (though sometimes considered to be an F3 or F4) its magnitude fluctuates between m_v +4.60 and +4.79 with a period usually quoted as $4^h\ 39^m\ 2^s$ although chattering away in the background are a further five cycles ranging from $2^h\ 16^m\ 48^s$ to $20^h\ 6^m\ 36^s$. Like 20% of naked eye δ Scuti stars the prototype is not alone. A m_v +12.2 K8 hangs around 870 AU away in a 15,000+ year orbit while a 0.9 $M_\odot$ G7 dwarf seems to be in a 3,000 AU orbit that takes in excess of 85,000 years to complete a single journey around the primary. δ Sct has the largest amplitude of its class, at least among the naked eye cases, at 0.19 magnitude matched only by ρ Puppis. The smallest amplitude, just 1/100th of a magnitude, is

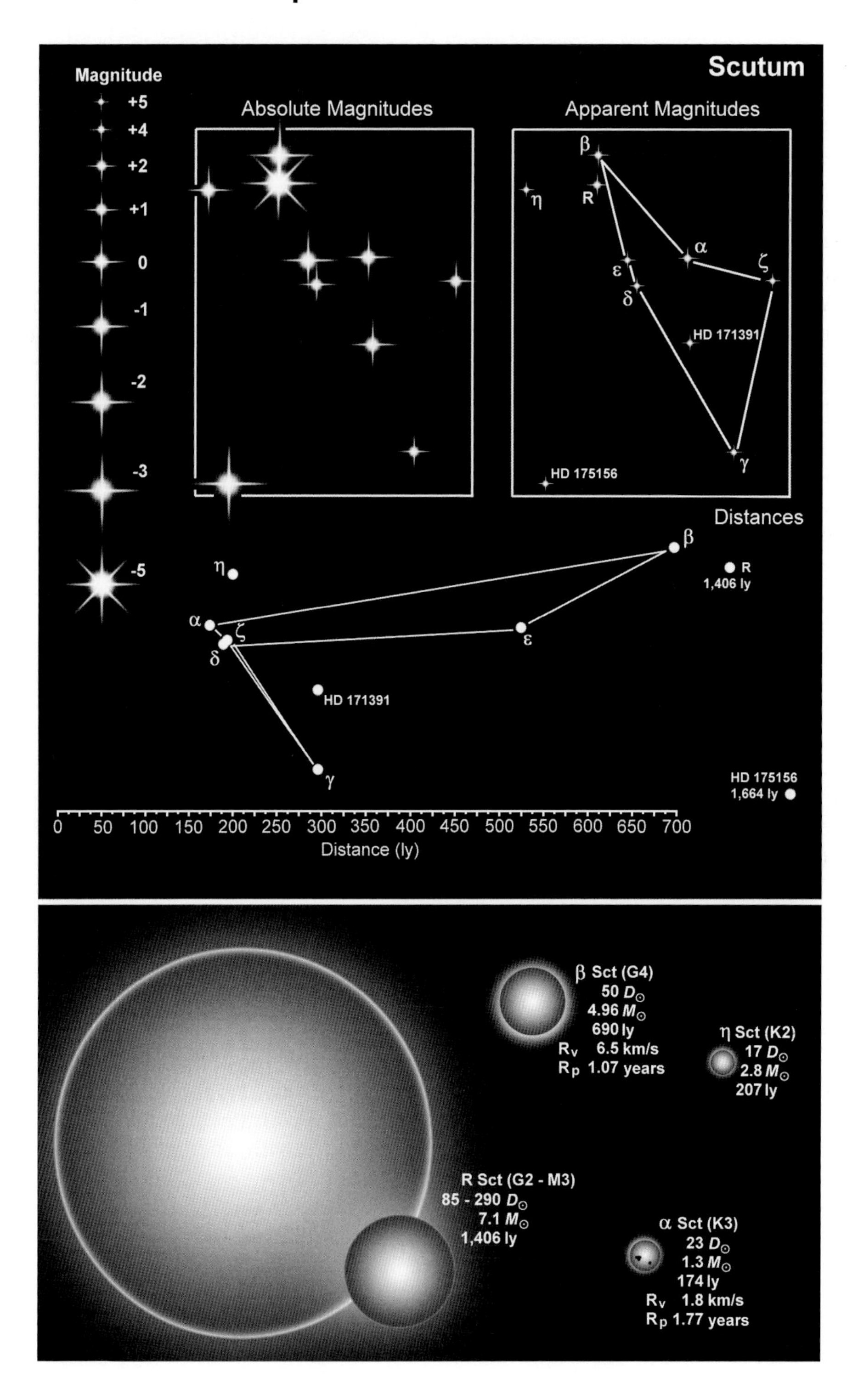

Scutum
Magnitude
+5
+4
+2
+1
0
-1
-2
-3
-5
Absolute Magnitudes
Apparent Magnitudes
β
η
R
α
ζ
ε
δ
HD 171391
HD 175156
γ
Distances
R
1,406 ly
HD 175156
1,664 ly
0 50 100 150 200 250 300 350 400 450 500 550 600 650 700
Distance (ly)
β Sct (G4)
50 $D_\odot$
4.96 $M_\odot$
690 ly
R_v 6.5 km/s
R_p 1.07 years
η Sct (K2)
17 $D_\odot$
2.8 $M_\odot$
207 ly
R Sct (G2 - M3)
85 - 290 $D_\odot$
7.1 $M_\odot$
1,406 ly
α Sct (K3)
23 $D_\odot$
1.3 $M_\odot$
174 ly
R_v 1.8 km/s
R_p 1.77 years

displayed by o Serpentis and ρ Tauri. Although the first to be discovered δ Sct is by no means the brightest of the class; that honor goes to Vega (β Lyrae).

At m_v +4.82 **η Scuti** seems just like a million other K2 giants except that it is hurtling through space at a rate of knots: 92.8 km/s. Currently 207 ly away and heading our way the 17 $D_\odot$ star should be in our neighborhood in about 669 thousand years.

No prizes for guessing that **R Scuti** is a variable star but what type of variable would be a good astropub quiz, if such places exist. It is in fact a rare naked eye RVa Tauri, probably the only example known. It appears to be a low mass Cepheid with a double variable period. The first period sees it change from a sometimes m_v +4.20 to around +6.00 and back over a fairly well defined period of 146.5 days. But then there is a 71 day cycle during which it drops below m_v +8.0 perhaps as a result of different layers expanding and contracting at different rates. Its parameters are difficult to define. It is known to be 1,406 ly away but its diameter has been estimated to be anywhere between 85 and 290 $D_\odot$. Its spectrum changes from a G2 when its temperature is around 5,300 K through a K0 to an M3 at 4,800 K. At 10 pc it would shine as a M_v -5.4 star.

There are a number of open clusters in Scutum although most are faint and difficult to see (see table). **M11**, the Wild Duck Cluster, is one of the richest and most compact with around 3,000 stars and certainly the easiest to spot at m_v +6.3. Not so impressive is **M26** although larger telescopes reveal what appears to be a region of few stars about 3′ from the center. This is likely to be due to interstellar dust obscuring the view.

Open and globular clusters in Scutum

Name	Size arc min	Size ly	Distance ly	Age million yrs	Brightest star in region*	No. stars m_v >+12*	Apparent magnitude m_v
M11 (NGC 6705)	14′	24.4	6,000	250	HD 174512 m_v +8.55	86	+6.3
M26 (NGC 6694)	15′	21.8	5,000	89	HD 173348 m_v +9.17	11	+9.2
NGC 6631	6′	14.8	8,480	?	No ID m_v +10.68	5	+11.1
NGC 6649	11′	16.6	5,200	36.8	DM -10° 4718 m_v +9.52	3	+8.9
NGC 6664	15′	32.7	7,500	14.5	EV Sct m_v +9.17	18	+7.8
NGC 6683	17′	19.3	3,900	10	No ID m_v +10.57	20	+9.4
NGC 6704	7.6′	21.4	9,700	73	No ID m_v +11.35	2	+9.2
NGC 6712	7.4′	48.3	22,500	?	Globular cluster		+8.1

*May not be a cluster member.

Serpens

Constellation:	Serpens	**Hemisphere:**	Equatorial
Translation:	The Serpent	**Area:**	637 deg^2
Genitive:	Serpentis	**% of sky:**	1.544%
Abbreviation:	Ser	**Size ranking:**	23rd

An ancient and unique constellation, Serpens depicts a serpent being held by Ophiuchus. The part of the constellation that contains the serpent's head – Serpens Caput – lies to the west of Ophiuchus while the tail – Serpens Cauda – lies to the east making this the only constellation split in to two parts.

For such an unusual constellation it is a pity that its luminary is so, well, usual **α Serpentis** or Unukalhai is a normal K2 giant: 11 $D_{\odot}$ across, a mass of 0.95 $M_{\odot}$, an apparent magnitude of m_v +2.61 brightening to M_v +1.1 at 10 pc and a temperature of 4,470 K. About 73 ly away it rotates once every 348 days, its rotational velocity being 1.6 km/s, slightly less than the Sun's 2 km/s.

β Serpentis is a binary consisting of an A2 and a K3 dwarf. The primary is m_v +3.64 with a mass of 2.37 $M_{\odot}$ and a high rotational spin of 207 km/s, causing the star to spin once on its axis every 15.3 hours and distorting it into an oblate spheroid of about 1.8 × 2.6 $D_{\odot}$. Its m_v +9.96 companion is separated by 30.6″ at PA 256°. Although a relatively wide double the fainter star is difficult to see against the brightness of the primary. The system lies at a distance of 153 ly and has a 75% chance of being a member of the Sirius Supercluster.

The closest star in the constellation is **γ Serpentis** at 36.3 ly. An F6 it is 40% larger than the Sun at 1.4 $D_{\odot}$ and has a mass of 1.25 $M_{\odot}$. Rotating at 8 km/s it spins once on its axis every 8.9 days. It is estimated to be between 1,600 and 3,400 million years old. Optical aid will reveal a m_v +3.64 companion, TYC 1496-2119-1, but this is simply a line of sight coincidence.

δ Serpentis or Tsin is another binary but this time of nearly identical F0 subgiants. The brighter of the two at about m_v +4.23 has a diameter of 5 $D_{\odot}$ and a mass of 2.8 $M_{\odot}$, its luminosity coming in at 73 $L_{\odot}$. It is also a δ Scuti variable with a period of 3^h 12^m 58^s during which it fades to m_v +4.27 before returning to its higher value. Its m_v +5.2 companion is separated by 3.9″ which, at a distance of 210 ly, translates into 375 AU. A somewhat smaller, 3 $D_{\odot}$, but similar mass star, 2.6 $M_{\odot}$, its luminosity is only about a third that of the primary, 26 $L_{\odot}$, so its magnitude is a fainter m_v +5.20. They have similar rotational velocities of 73 and 75 km/s and so have rotational periods of 3.5 days for the brighter star and 2 days for the fainter component. Their orbital period is about 3,200 years. The system is believed to be about 800 million years old

A third binary, and one that is popular with aficionados of double stars, is **θ Serpentis** the components of which are even more closely matched **$θ^1$ Serpentis** is a 1.7 $D_{\odot}$, 1.9 $M_{\odot}$ A5 with a luminosity of 20 $L_{\odot}$ and an apparent magnitude of m_v +4.58. Its twin, **$θ^2$ Serpentis,** is a 1.4 $D_{\odot}$, 1.8 $M_{\odot}$ A5 with a luminosity of 17 $L_{\odot}$ and an apparent magnitude of m_v +4.91. They both have

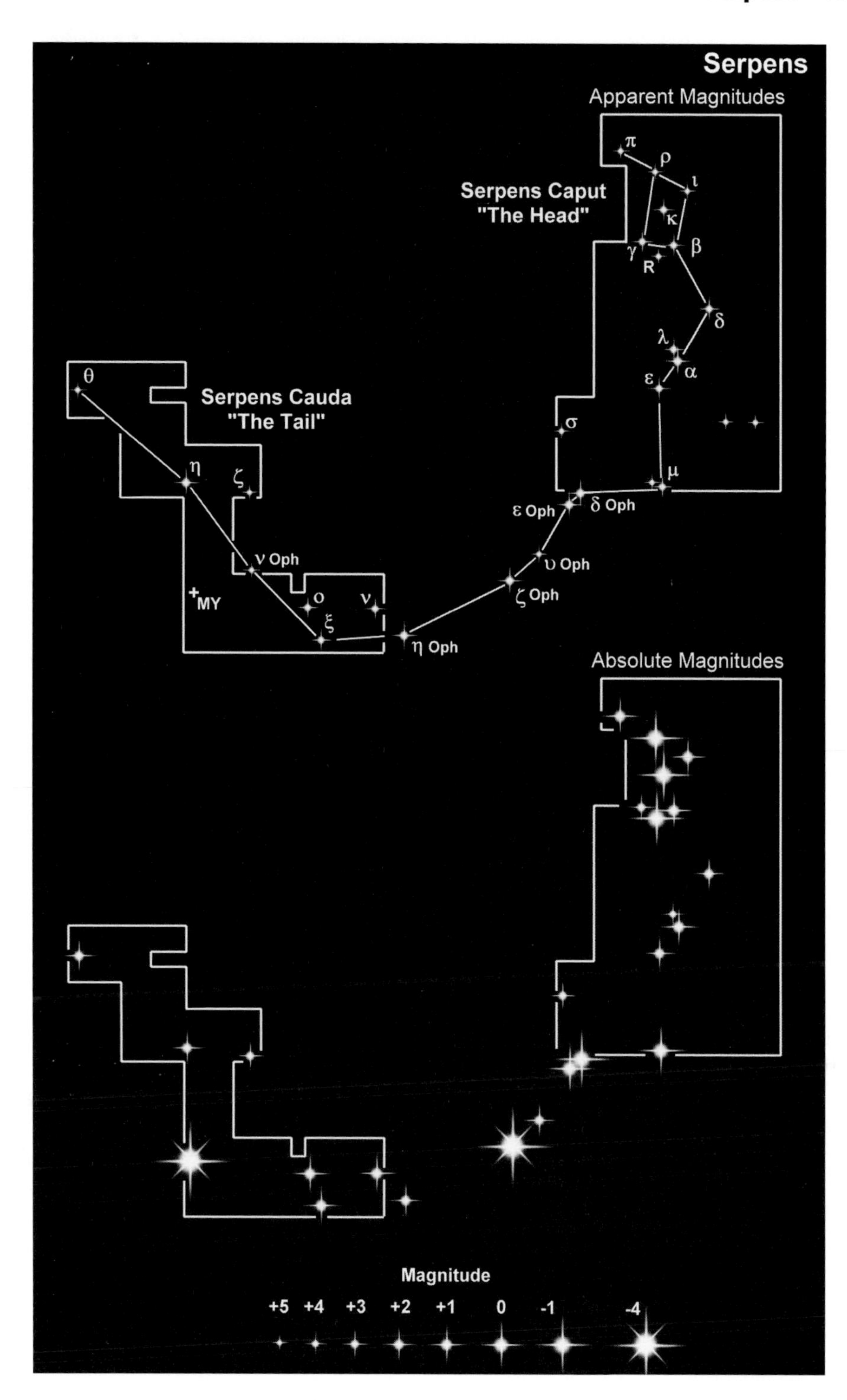
Serpens
Apparent Magnitudes
Serpens Caput
"The Head"
Serpens Cauda
"The Tail"
π
ρ
ι
κ
γ
R
β
δ
λ
α
ε
σ
μ
ε Oph
δ Oph
υ Oph
ζ Oph
η Oph
θ
η
ζ
ν Oph
MY
ο
ν
ξ
Absolute Magnitudes
Magnitude
+5 +4 +3 +2 +1 0 -1 -4

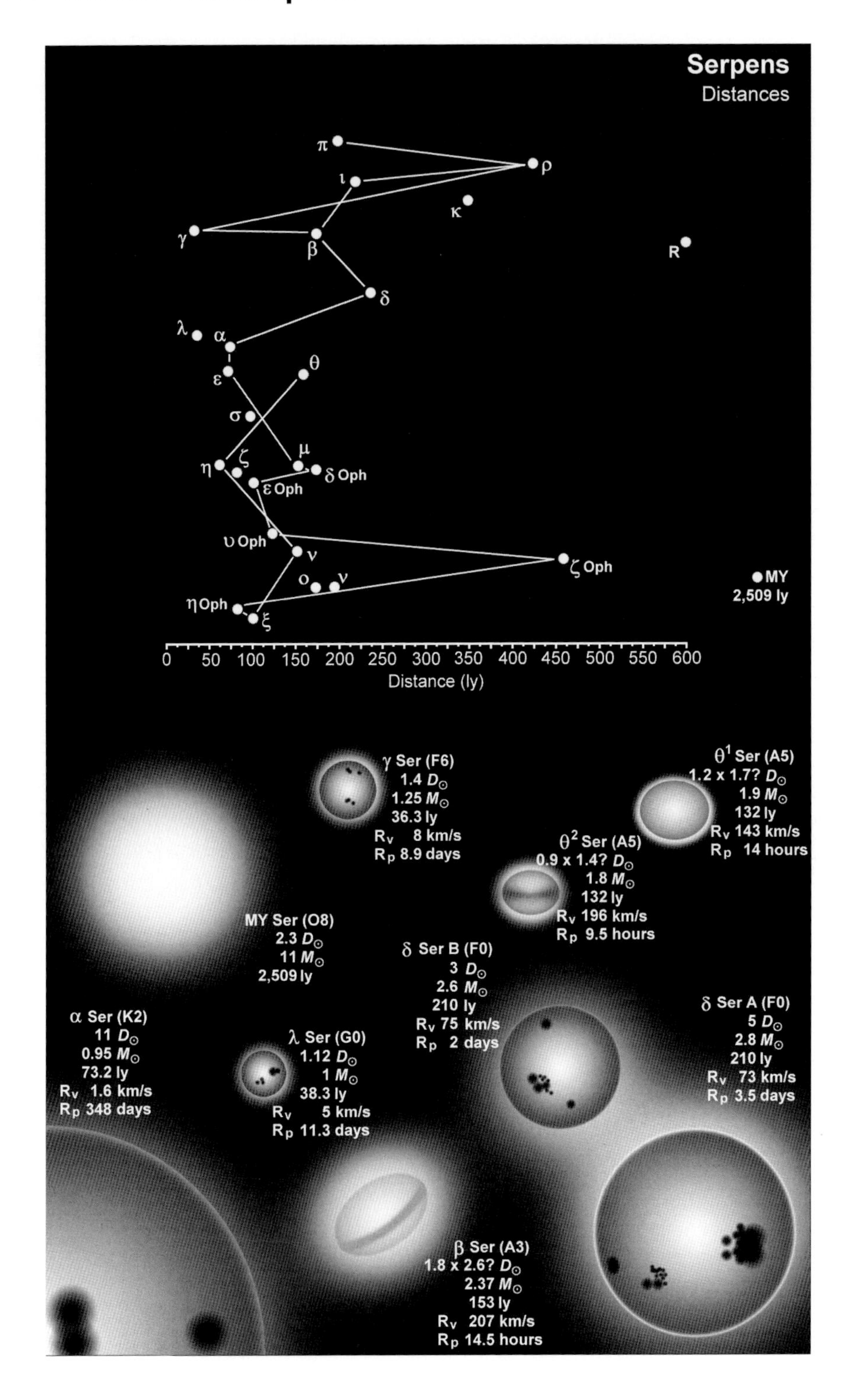

Serpens
Distances
π
ρ
ι
κ
γ
β
R
δ
λ
α
ε
θ
σ
μ
η
ζ
δ Oph
ε Oph
υ Oph
ν
ζ Oph
MY
2,509 ly
ο
ν
η Oph
ξ
0 50 100 150 200 250 300 350 400 450 500 550 600
Distance (ly)
γ Ser (F6)
1.4 D⊙
1.25 M⊙
36.3 ly
Rv 8 km/s
Rp 8.9 days
θ1 Ser (A5)
1.2 x 1.7? D⊙
1.9 M⊙
132 ly
Rv 143 km/s
Rp 14 hours
θ2 Ser (A5)
0.9 x 1.4? D⊙
1.8 M⊙
132 ly
Rv 196 km/s
Rp 9.5 hours
MY Ser (O8)
2.3 D⊙
11 M⊙
2,509 ly
δ Ser B (F0)
3 D⊙
2.6 M⊙
210 ly
Rv 75 km/s
Rp 2 days
δ Ser A (F0)
5 D⊙
2.8 M⊙
210 ly
Rv 73 km/s
Rp 3.5 days
α Ser (K2)
11 D⊙
0.95 M⊙
73.2 ly
Rv 1.6 km/s
Rp 348 days
λ Ser (G0)
1.12 D⊙
1 M⊙
38.3 ly
Rv 5 km/s
Rp 11.3 days
β Ser (A3)
1.8 x 2.6? D⊙
2.37 M⊙
153 ly
Rv 207 km/s
Rp 14.5 hours

temperatures of 8,200 K but while θ^1 Ser rotates at 143 km/s, θ^2 Ser spins at 196 km/s leading to rotational periods of 14.4 hours and 8.7 hours respectively, θ^2 Ser being more oblate than its sister. They are usually described by observers as being yellow or white and are separated by 22.5″ at PA 104°. In real space they average 900 AU apart with an orbital period in excess of 14,000 years.

λ Serpentis is a solar analog that is relatively near at just 38.3 ly. Shining at m_V +4.42 its absolute magnitude would barely change to M_V +4.3, the star being close to the 10 pc standard distance for such calculations. Various attempts have been made at estimating the star's diameter, averaging 1.12 $D_\odot$, its mass reckoned to be about equal to that of the Sun but twice as luminous. A temperature of 6,000 K, indicative of a G0 Main Sequence dwarf, λ Ser rotates at 5 km/s (Sun = 2 km/s) and so turns once every 11.3 days (Sun = 25 days). Its age is uncertain, lying in the range of 4,600 to 7,200 million years (Sun = 4,567 million years). It was believed to have a spectroscopic companion in a 5 year long orbit but this has now largely been discounted.

The Mira variable **R Serpentis** is possibly the largest star in the constellation though exactly how big is anyone's guess: estimates range from 50 to 380 $D_\odot$! At its smallest diameter the star would be half the size of Mercury's orbit while the larger size would put it mid-way between the orbit of Mars and that of the Asteroid Belt. For most of its 356.41 day cycle R Ser is invisible to the naked eye, fading to m_V +14.4 and only remaining above 6th magnitude for about a month. An M7 it is 570 ly away.

A couple of the stars in Serpens appear to have brown dwarfs in orbit around them including an 18.1 M_J and an as yet unconfirmed 42 M_J object (see table).

Of the open and globular clusters in Serpens, **M16** basically has two components. The cluster of young stars is cataloged as NGC 6611: the Eagle Nebula from which they formed is IC 4703. The open cluster was discovered first by Philippe Loys de Chéseaux in 1745 (more famous for his six-tailed comet) but the nebula's discovery had to wait 19 years until another comet hunter, Charles Messier, cataloged it to avoid mistaking it for a comet. The stars in M16 are thought to be no more than 5.5 million years old. **M5,** on the other hand, is one of the oldest globulars at 13,000 million years old. Although there are thousands of stars in the region of **Trumpler 32** few are members of the cluster. Of the 42 that are brighter than m_V +12 none has more than a 29% chance of being a cluster member.

Open and globular clusters in Serpens

Name	Size arc min	Size ly	Distance ly	Age million yrs	Brightest star in region*	No. stars m_v >+12*	Apparent magnitude m_v
IC 4756	92′	42.3	1,580	500	HD 172365 +6.38	181	+4.6
M16 (NGC 6611)	53.3′	88.4	5,700	5.5	No ID +8.01	115	+6.0
NGC 6604	13.3′	23.2	6,000	6.5	MY Ser +7.54	15	+6.5
Tr 32	51.5′	83.9	5,600	300	No ID +8.18	42	+12
M5 (NGC 5904)	23′	164	24,500	13,000	Globular cluster		+5.6
NGC 6535	3.6′	23	22,000	1,000	Globular cluster		+10.5
NGC 6539	3.5′	37.7	27,000	5,000	Globular cluster		+9.3

*May not be a cluster member.

Planetary systems in Serpens

Star	$D_{\odot}$	Spectral class	ly	m_v	Planet	Minimum mass	q	Q	P
HD 168443	1.63	G5	123.5	+6.92	HD 168443 b	8.02 M_J	0.14	0.46	58.1 d
					HD 168443 c	18.1 M_J	2.29	3.53	4.84 y
HD 136118 unconfirmed	1.74	F9	170.6	+6.94	HD 136118 b	42 M_J	0.94	1.96	3.31 y

Sextans

Constellation:	Sextans	**Hemisphere:**	Equatorial
Translation:	The Sextant	**Area:**	314 deg^2
Genitive:	Sextantis	**% of sky:**	0.761%
Abbreviation:	Sex	**Size ranking:**	47th

A 'filler' constellation introduced by Johannes Hevelius in 1690 to mop up a few wayward stars. It consists of just five stars above the naked eye limit.

α **Sextantis** is a barely noticeable m_v +4.48 at 287 ly but would brighten to a very respectable M_v -1.1 at 10 pc. About three and a half times the size of the Sun it is technically a giant, its full spectral classification of A0 III giving the game away. It appears to have a rotational velocity of 21 km/s, well below the average 110 km/s for the A0 class, but this could just be due to the position of its rotational pole. If the pole is pointing towards the Earth then the star will appear to rotate more slowly than it actually does. With a temperature of 9,600 K and a luminosity of 102 $L_\odot$ α Sex made the journey from the northern to the southern hemisphere in December 1923, the result of the Earth wobbling on its axis and causing the co-ordination grid of RA and Declination to appear to drift against the background stars. α Sex, like all the naked eye stars in the constellation, is drifting away from us: its radial velocity is 7.1 km/s.

Not a great deal larger than the Sun at 1.5 $D_\odot$ β **Sextantis** is a hot Main Sequence B6 dwarf with a temperature of 14,000 K and a luminosity of 87 $L_\odot$ Spinning at 80 km/s – half the average velocity for the class – it is an α CV rotating variable changing by a 10th of a magnitude between m_v +5.0 and +5.1, though no one has quite worked out the period. It is the most distant star in the constellation lying 345 ly away, a distance that is increasing by 11.6 kilometers every second.

γ **Sextantis** is a binary and possibly a triple star system. The primary is a m_v +5.58 A2 of 2.8 $D_\odot$ and 2.3 $M_\odot$ Its companion is a slightly dimmer m_v +6.07 A4 in a retrograde (anti-clockwise) orbit with a period of 28,320 days (77.5 years). A third star hangs out with the other two. An A2 of m_v +12.3 it is not clear whether it is a true member of the system. γ Sex is 262 ly away and shines with a combined magnitude of m_v +5.11.

At 300 ly, give or take 19 ly, δ **Sextantis** shines at a constant m_v +5.18, its luminosity estimated to be 58 $L_\odot$ Its diameter is less well known at 1.8, 2.3 or 2.6 $D_\odot$ depending on who you ask. With a rotational velocity of 152 km/s – a bit on the high side for a B9.5 and more in line with a B8 – its rotational period could be anywhere between 14.4 and 20.8 hours. Its mass is reckoned to be 2.68 $M_\odot$.

The faintest naked eye star is also the closest. ε **Sextantis** is a feeble m_v +5.25 and 20 $L_\odot$. A pale yellowish F2 its diameter is 3.5 $D_\odot$ with its mass coming in at 1.73 $M_\odot$. Spinning on its axis with a velocity of 63.5 km/s – just above the average of 59 km/s – its rotational period is about two and a half days. It lies 183 ly away.

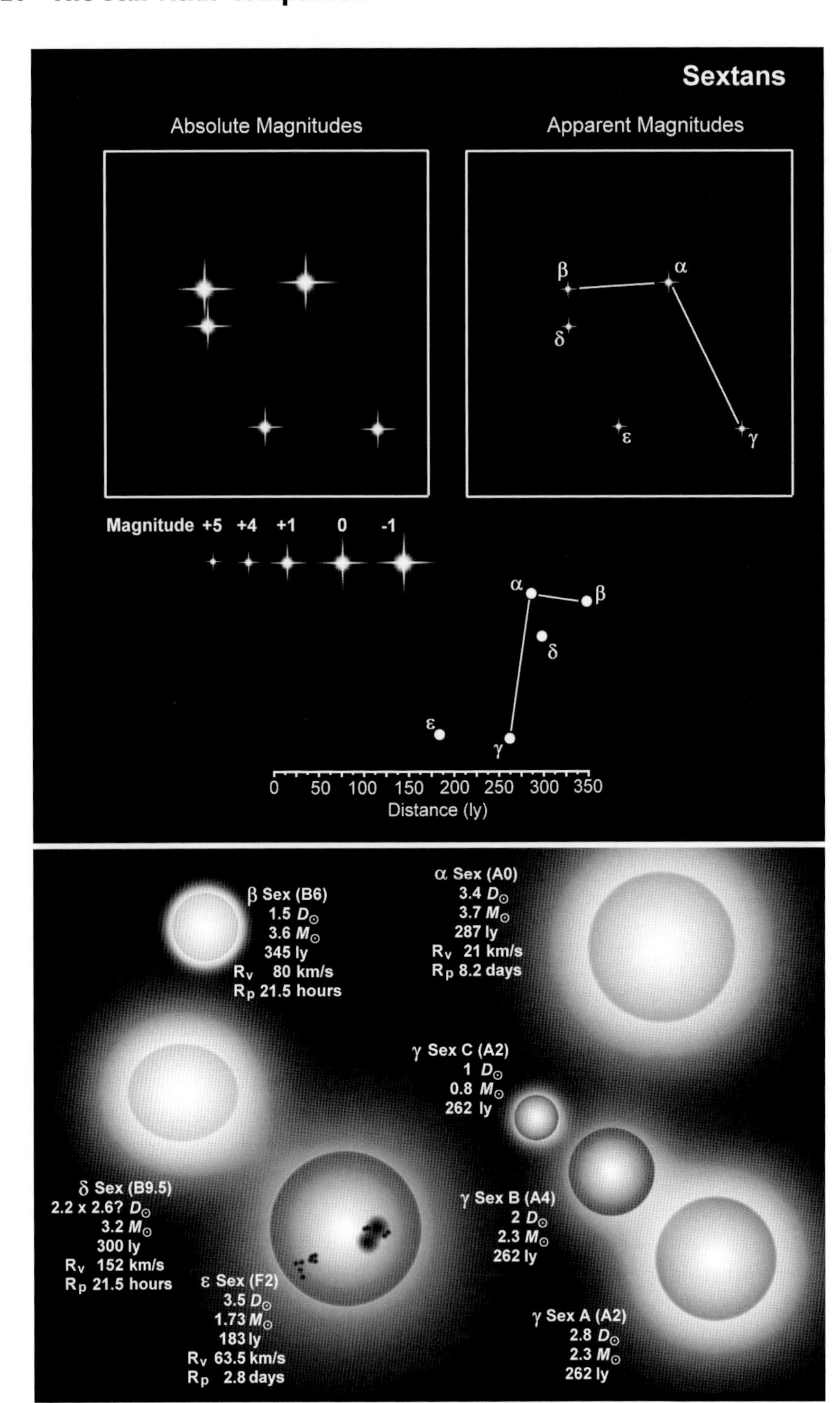

Sextans
Absolute Magnitudes
Apparent Magnitudes
β
α
δ
ε
γ
Magnitude +5 +4 +1 0 -1
α
β
δ
ε
γ
0 50 100 150 200 250 300 350
Distance (ly)
β Sex (B6)
1.5 D⊙
3.6 M⊙
345 ly
Rv 80 km/s
Rp 21.5 hours
α Sex (A0)
3.4 D⊙
3.7 M⊙
287 ly
Rv 21 km/s
Rp 8.2 days
γ Sex C (A2)
1 D⊙
0.8 M⊙
262 ly
δ Sex (B9.5)
2.2 x 2.6? D⊙
3.2 M⊙
300 ly
Rv 152 km/s
Rp 21.5 hours
γ Sex B (A4)
2 D⊙
2.3 M⊙
262 ly
ε Sex (F2)
3.5 D⊙
1.73 M⊙
183 ly
Rv 63.5 km/s
Rp 2.8 days
γ Sex A (A2)
2.8 D⊙
2.3 M⊙
262 ly

Taurus

Constellation:	Taurus	**Hemisphere:**	Equatorial
Translation:	The Bull	**Area:**	797 deg^2
Genitive:	Tauri	**% of sky:**	1.932%
Abbreviation:	Tau	**Size ranking:**	17th

According to legend Zeus changed into a bull and carried off Princess Europa of Phoenicia. Although nearly all the constellation is in the Northern Hemisphere a tiny part of it crosses the Celestial Equator. It is one of the Zodiacal constellations through which the Sun passes between 14 May and 21 June. Taurus is a favorite among many observers as it contains two open and very prominent star clusters: the Pleiades and the Hyades.

α **Tauri** is often simply referred to by its proper name, Aldebaran, which means 'The Follower', presumably following the Pleiades. It marks the beginning of the famous V of the bull's head, the open star cluster Hyades, although it is not part of it lying just 65.7 ly away, less than half the distance of the cluster itself. In illustrations of the animal, Aldebaran is usually depicted as the eye of the bull although it is a somewhat bloodshot eye being orange or even reddish-orange when close to the horizon. The 14th brightest star in the sky at m_v +0.75 this K5 giant is 45 times larger than the Sun and slightly variable, its pulsations causing it to dim to m_v +0.95 without any real period. Despite its large size – half the diameter of Mercury's orbit – Aldebaran has a mass of just 1.7 $M_\odot$. The latest measurements indicate that it rotates at the same velocity as the Sun, 2 km/s, and so takes 1,139 days (3.1 years) to turn just once on its axis. With a temperature of 4,050 K α Tau has a luminosity of 425 $L_\odot$ and would brighten to M_v -0.6 at 10 pc.

The Hyades open cluster also has the catalog designations of Caldwell 41, Collinder 50 and Melotte 25. In mythology they were the daughters of Atlas and Aethra and the half sisters of the Pleiades. The 20 naked eye stars which make up the distinctive V shape (see table) measure about 3.5° across which, at a distance of 151 ly, corresponds to 9.2 ly. However the true cluster is much larger, perhaps up to 42° giving a diameter of 136 ly and containing up to 400 stars. Three of the stars – Aldebaran mentioned above, π Tauri and 75 Tauri – are not cluster members. The cluster is best observed through a good binocular or rich field telescope when, in theory at least, up to about 80 stars down to m_v +12 can be seen in the V region and twice that number over an area of 8.5°. The Hyades appears to have a common origin with The Praesepe or Beehive cluster in Cancer, the two having been formed about 625 million years ago. During the intervening period the Hyades will have lost many of its stars as external gravitational influences tear away some of the outlying members while internal gravitational forces perturb some of the stars into new orbits ejecting them from the cluster. The whole cluster is moving towards a point just to the east of Betelgeuse. If you are into cryogenics then you may like to check out the area in 780,000 years time.

The 20 stars in the V-region of the Hyades to m_v +5.5

Star	Other designation or name	m_v	Star	Other designation or name	m_v
α Tau*	Aldebaran	+0.75	σ^2 Tau		+4.68
θ^2 Tau	Phaesyla	+3.35	π Tau*		+4.69
ε Tau	Ain	+3.55	HD 28527		+4.78
γ Tau	Elizabeth, Hyadum I	+3.61	δ^2 Tau		+4.80
δ^1 Tau	Eudora, Hyadum II	+3.73	75 Tau*		+4.97
θ^1 Tau	Phaeo	+3.84	79 Tau		+5.03
90 Tau		+4.27	σ^1 Tau		+5.07
δ^3 Tau	V776 Tau, Cleeia	+4.29	58 Tau	V696 Tau	+5.26
71 Tau	V777 Tau, Polyxo	+4.73	83 Tau		+5.43
ρ Tau		+4.90	81 Tau		+5.48

*Not a cluster member

γ Tauri, the third brightest of the true Hyades, is also known as Hyadum I, 'Hyadum the First Follower of the Pleiades', and also as Elizabeth. But this is no Thin Lizzie having a diameter of 12.7 $D_\odot$ making it one of the larger stars in the cluster. A K0 with a temperature of 4,900 K it appears to rotate at just 2 km/s, the same as our own Sun, which means that it takes almost a year – 321.4 days – to spin once on its axis. Yet something is amiss. It has a magnetically active outer layer which usually requires a much faster rotational velocity. It could be that we are looking at Elizabeth's pole, which would give the impression of a slow spin, or it could just be that our knowledge of how stars work is not as complete as we sometimes think.

The three δ stars in the Hyades are a bit of a mixed bag. The brightest at m_v +3.76 is **δ^1 Tauri,** a K0 giant of 13 $D_\odot$ weighing in at 2 $M_\odot$. **δ^2 Tauri** is actually the third brightest of the trio at m_v +4.79. About 1½ times as big and twice as massive as the Sun this A7 rotates at 65 km/s and so has a rotational period of 1.1 days. **δ^3 Tauri** is also known as **V776 Tauri,** a clear indication of its variability. Also an A-class star (A2) it belongs to the α CV group of rotating variables switching between an almost imperceptible m_v +4.29 to +4.32 and back over a period of 57.25 days. It is a little over twice the size and mass of the Sun but somewhat hotter at 8,900 K (Sun = 5,778 K).

Another Hyades giant is the 14 $D_\odot$ **ε Tauri**, or Ain, which has a mass of 2 $M_\odot$ is. A little farther away than the middle of the cluster at 155 ly some say it belongs to spectral class G9.5, others put it down as a K0. The difference is not enough to lose sleep over. Of more certainty however is the presence of a planet, the first ever to be discovered in an open cluster. At 7.6 M_J this gaseous giant, known as ε Tau b, takes between 590 and 600 days (1.6 years) to orbit the star, getting as close as 1.64 AU and as far as 2.22 AU (about as far as Mars and the Asteroid Belt from the Sun).

θ^1 and θ^2 Tauri are binary systems, but not with each other. **θ^{1A} Tauri** is the fainter of the two naked eye stars at m_v +3.84. It is a 10.3 $D_\odot$ yellowish-orange K0

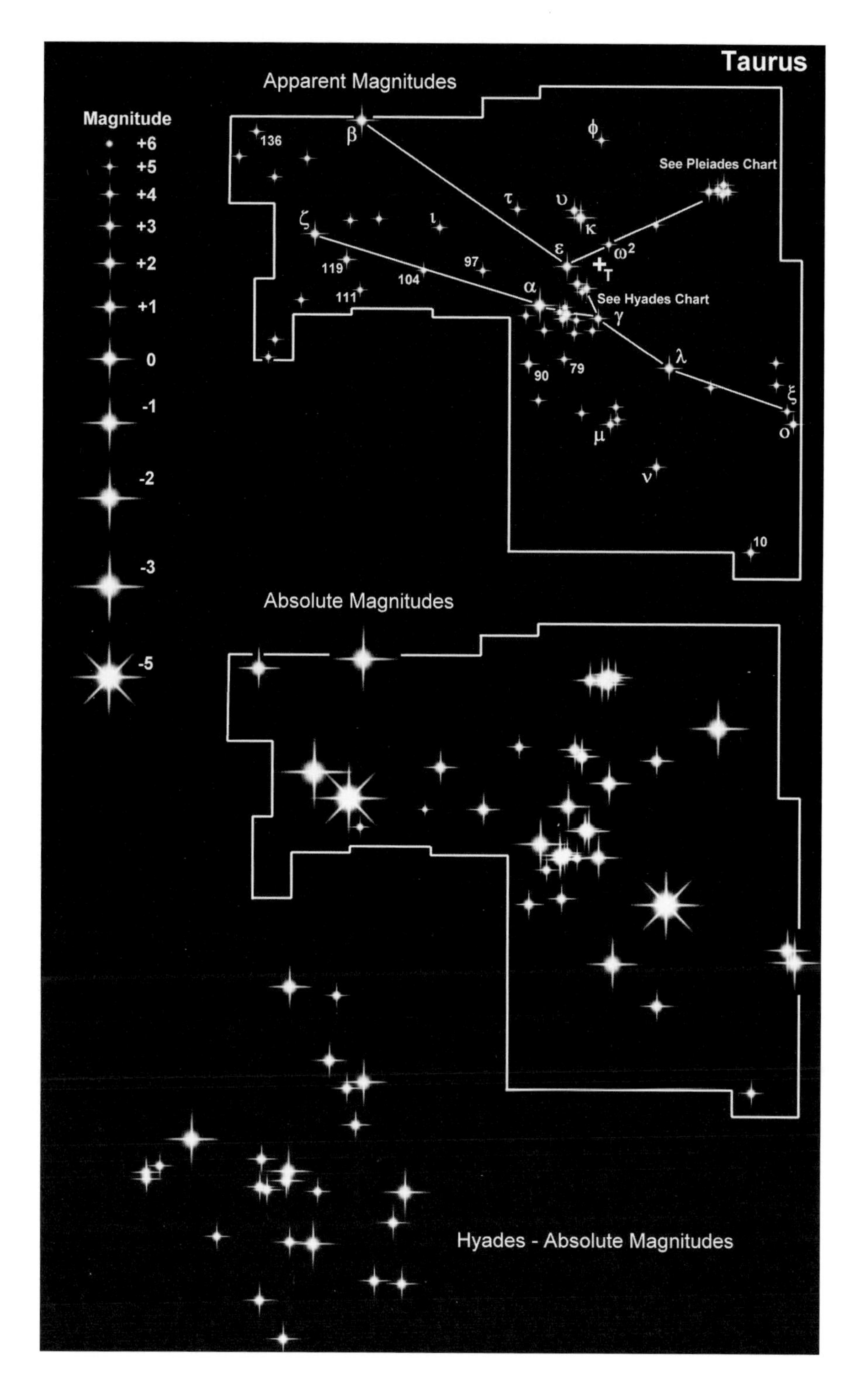
Taurus
Apparent Magnitudes
Magnitude
+6
+5
+4
+3
+2
+1
0
-1
-2
-3
-5
136
β
ϕ
See Pleiades Chart
τ
υ
κ
ζ
ι
ε
ω2
119
104
97
T
111
α
See Hyades Chart
γ
λ
90
79
ξ
ο
μ
ν
10
Absolute Magnitudes
Hyades - Absolute Magnitudes

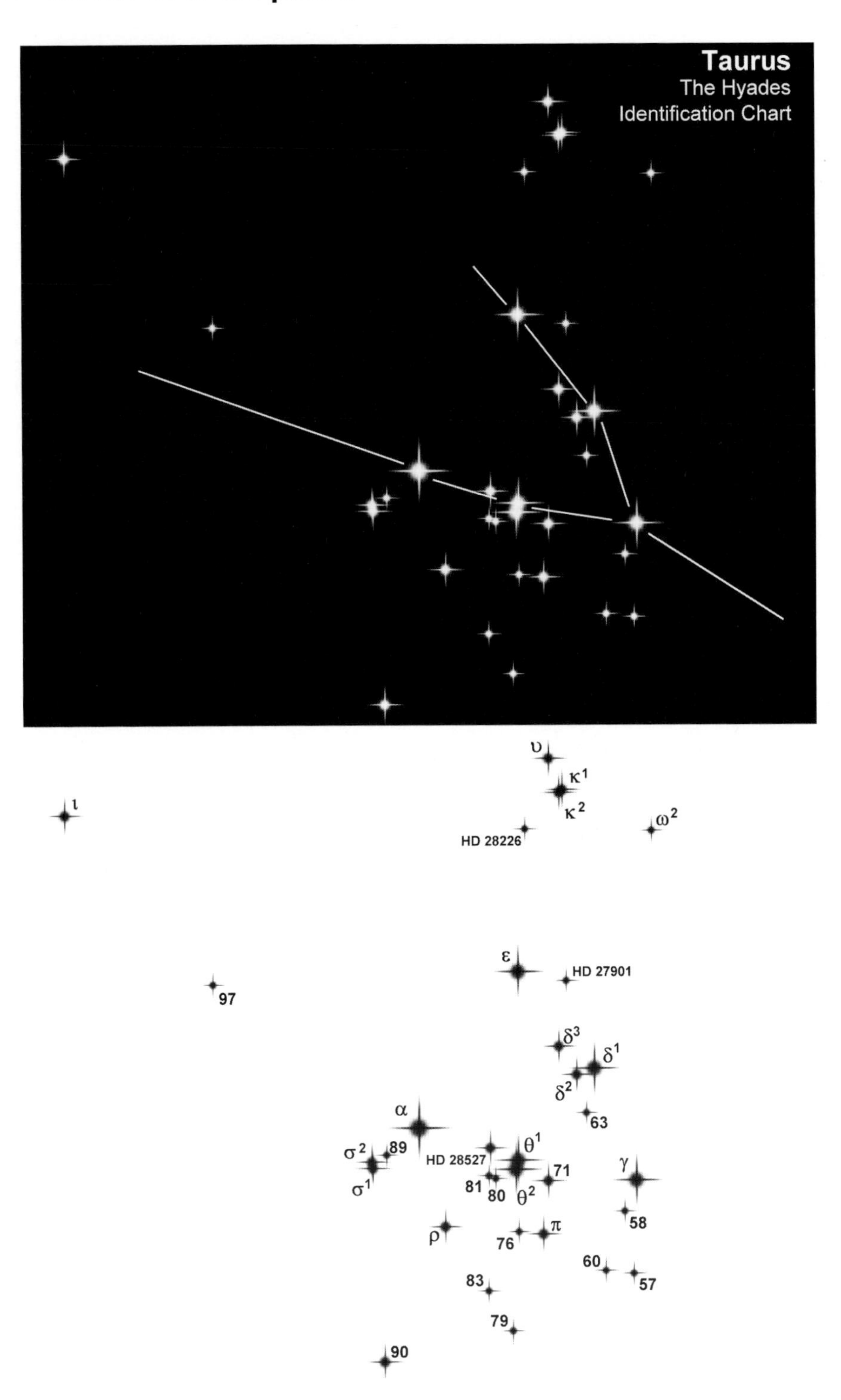

Taurus
The Hyades
Identification Chart
υ
κ1
κ2
ι
HD 28226
ω2
ε
HD 27901
97
δ3
δ1
δ2
63
α
θ1
σ2
89
HD 28527
71
γ
σ1
81
80
θ2
58
ρ
π
76
60
57
83
79
90

with a mass of 1.88 $M_{\odot}$ and luminosity of 55 $L_{\odot}$. Its true companion, **θ^{1B} Tauri,** is probably an F8 dwarf of 1.2 $D_{\odot}$ and about twice as luminous as the Sun. Separated in the sky by 0.2″ in real space they swing between 4.4 and 16 AU with an orbital period of 16.3 years. **θ^{2A} Tauri** is 5.6′ almost due south of θ^{1A} Tau. About half a magnitude brighter it actually varies between m_v +3.35 and +3.42 with a period of 1^h 48^m 55^s, being a δ Scuti-type pulsating variable. It is a 3.9 $D_{\odot}$ A7 burning at 7,760 K and having a luminosity of 75 $L_{\odot}$. It mass works out at 2.4 $M_{\odot}$. Its spectroscopic companion, **θ^{2B} Tauri,** is an F0 of 3 $D_{\odot}$ and 1.8 $M_{\odot}$. It has a much lower luminosity of 19 $L_{\odot}$. Their orbital separation varies between 0.23 and 1.3 AU, the orbital year taking 140.7 days to complete. While the θ^1 pair lie between 152 and 164 ly the θ^2 pair are 144 to 154 ly away so, theoretically, the two binaries could be within 2 ly of one another.

Another δ Scuti variable in the Hyades is **ρ Tauri**. Slightly more rapid than θ^{2A} Tau at 1^h 36^m 29^s it is a 1.5 $D_{\odot}$ A8 with a mass of 1.95 $M_{\odot}$ and luminosity of 25 $L_{\odot}$. It varies by just 1/100th of a magnitude from m_v +4.90 to +4.91. Lying some 153 ly away it is a rapid spinner, its 144 km/s resulting in a 12.7 hour rotational period.

σ^1 and σ^2 Tauri, like θ^1 and θ^2 Tauri, is a non-binary system, as far as we can tell. The brighter of the two, **σ^2 Tauri,** is a m_v +4.67 A5 dwarf, only 1.6 $D_{\odot}$ but with a mass of 2.04 $M_{\odot}$. Rotating at 53 km/s it takes 36.7 hours to spin once on its axis. **σ^1 Tauri,** an A4, is almost the same size at 1.7 $D_{\odot}$ and has a similar mass at 1.97 $M_{\odot}$. It has a spectroscopic companion of which very little is known for certain other than its orbital period of 38.95 days. σ^1 Tau lies between 145.8 and 157.8 ly away while σ^2 Tau is 153 to 165 ly away so the two stars could be very close neighbors near the center of the Hyades cluster.

Of the remaining stars in the Hyades 58 Tau and 71 Tau are both δ Scuti variables. **58 Tauri** (V696 Tau) flickers between m_v +5.22 and +5.28 with a period of 51^m 50^s. An F0 dwarf of 1.6 $D_{\odot}$ and 1.84 $M_{\odot}$ it is approximately 154 ly away. **71 Tauri** (V777 Tau) is a little farther away at 156 ly and has a period of 3^h 50^m 24^s during which it changes from m_v +4.73 to +4.75 and back. It is also an F0 class dwarf of 1.3 $D_{\odot}$ and 2.08 $M_{\odot}$. The m_v +4.98 K2 giant **75 Tauri** is not a member of the Hyades lying behind the cluster at 194 ly, and nor is **π Tauri**, a 13 $D_{\odot}$ G7 at 455 ly. Its magnitude is m_v +4.69.

Arguably the most impressive naked eye cluster in the entire sky, the **Pleiades** (**M45**) consists of 11 stars brighter than 6th magnitude and about 1,000 that are dimmer. Often called the Seven Sisters the cluster actually has nine stars that carry ancient names: the seven daughters Alcyone, Asterope, Celeano, Electra, Maia, Merope and Taygete together with their father Atlas and mother Pleione (see table). These bright stars are spread across slightly more than 1° of sky which equates to 8.4 ly. The closest is Celeano at 384.4 ly while the farthest is Pleione at 437.3 ly, a spread of 53 ly. The entire cluster is thought to be 2.64′ or 19.2 ly across. The distances of the Pleiades have recently been reviewed following the discovery of errors in the Hipparcos data. The cluster contains a number of brown and white dwarfs and is believed to have a total mass of about 800 $M_{\odot}$. Its age is about 100 million years and in 250 million years from now the stars will be

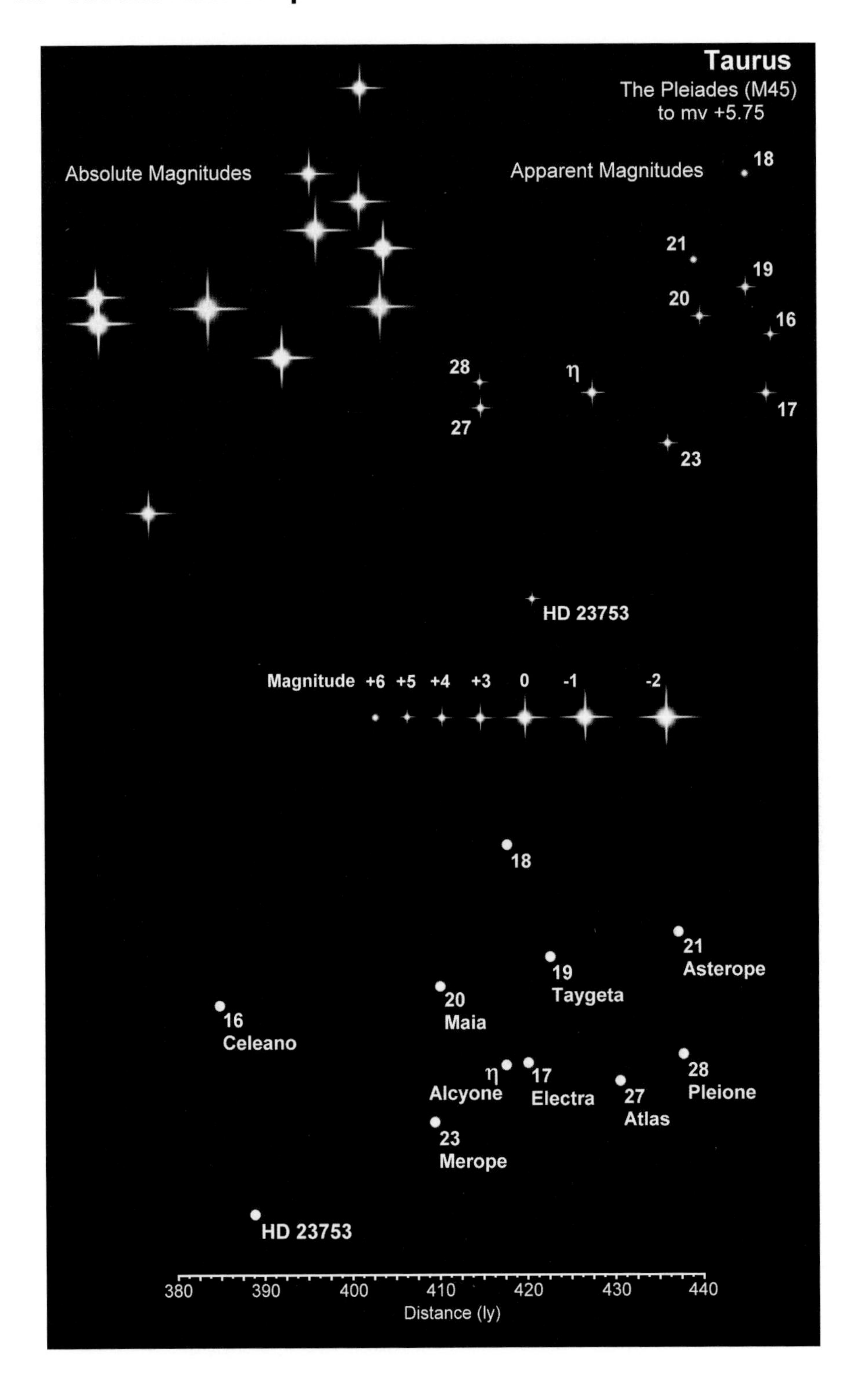
Taurus
The Pleiades (M45)
to mv +5.75
Absolute Magnitudes
Apparent Magnitudes
18
21
19
20
16
28
η
17
27
23
HD 23753
Magnitude +6 +5 +4 +3 0 -1 -2
18
21
Asterope
19
Taygeta
20
Maia
16
Celeano
28
Pleione
η
Alcyone
17
Electra
27
Atlas
23
Merope
HD 23753
380 390 400 410 420 430 440
Distance (ly)

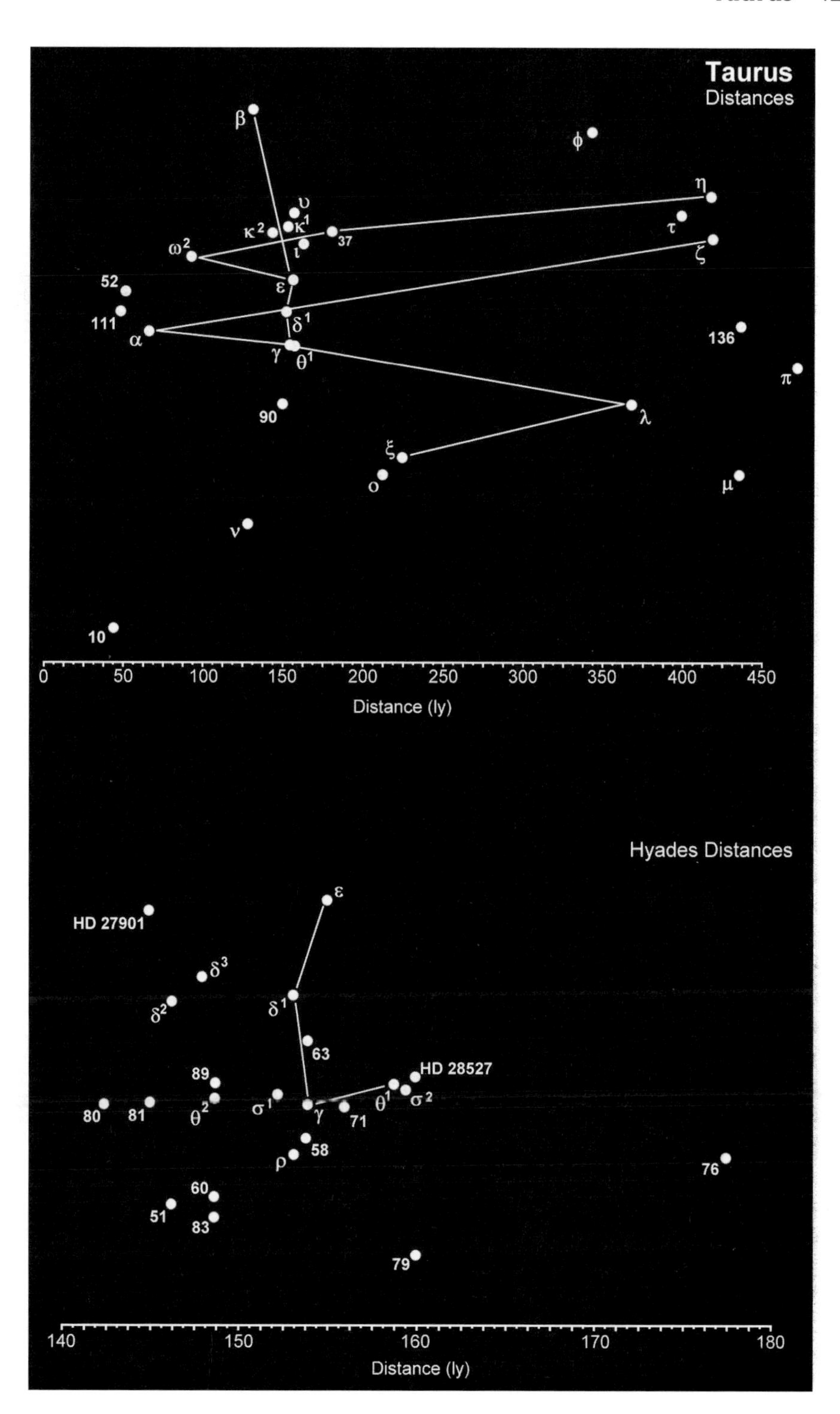
Taurus
Distances
β
φ
η
υ
κ2
κ1
37
τ
ω2
ι
ζ
52
ε
111
α
δ1
136
γ
θ1
π
90
λ
ξ
ο
μ
ν
10
0
50
100
150
200
250
300
350
400
450
Distance (ly)
Hyades Distances
ε
HD 27901
δ3
δ2
δ1
63
89
HD 28527
80
81
θ2
σ1
γ
71
θ1
σ2
58
ρ
76
60
51
83
79
140
150
160
170
180
Distance (ly)

so dispersed that they will not be recognizable as a cluster. Photographs of the Pleiades reveal the stars to be imbedded in a nebula that glows blue because of the color of the stars. In fact the Pleiades are not native to the nebula but are just passing through.

The Pleiades down to m_v +5.75

Star	Other	Mag. mv	Spec. Class	Dist ly	$D_\odot$	$M_\odot$	RotV km/s
η Tau	Alcyone	+2.86	B7	417.6	10	6	215
27 Tau	Atlas	+3.62 to +3.64	B8	430.5	2.7	5	212
17 Tau	Electra	+3.69	B6	420.5	6	5	170
20 Tau	Maia	+3.86	B8	409.9	5.5	4.2	24
23 Tau	Merope, V971 Tau	+4.18 to +4.19	B7	409.1	4.3	4.5	280
19 Tau	Taygete	+4.29	B6	422.6	4.5	4.5	123
28 Tau	Pleione, BU Tau	+4.77 to +5.50	B8	437.3	3.2	3.4	329
HD 23753		+5.44	B9	388.2	3		
16 Tau	Celeano	+5.45	B7	384.4	3	3.7	200
18 Tau	HD 23324	+5.64	B8	417.6	3	3.1	
21 Tau	Asterope	+5.75	B8	436.8	2.2		205

The brightest of the Pleiades is the m_v +2.86 **η Tauri** or Alcyone. At 417.6 ly it is less than 7 ly from the center of the visible cluster (410.9 ly) and is by far the largest of the naked eye Pleiades at 10 $D_\odot$ and 6 $M_\odot$. Alcyone is a binary, the two components separated by about 3.2 AU. The smallest of the Pleiades is Asterope, 2.2 $D_\odot$, which is also one of the farthest at 436.8 ly. Pleione is also known as **BU Tauri** signifying its changing brightness. A γ Cas eruptive variable its magnitude swings between m_v +4.77 and +5.50. It has a companion at an average distance of 28 AU, the orbital period being 35 years. It is not the only naked eye variable in the cluster. Merope, **V971 Tauri**, has an amplitude of just 1/100th of a magnitude, m_v +4.18 to +4.19, and belongs to the β Cepheid class of pulsating variables. Atlas (**27 Tauri**), the daddy of them all, is a rather modest 2.7 $D_\odot$ B8 and is variable between m_v +3.62 and +3.64. Atlas hides a secret (apart from his Hyades daughters): a long period spectroscopic companion which lies an average 52 AU from the primary and which takes 150 years to complete a single orbit. Atlas is also a Be emission star, its 212 km/s rotational velocity throwing matter into space. Electra, **17 Tauri**, also has a companion but in a tight 0.8 AU orbit that takes 100.5 days to complete. Celeano's (**16 Tauri**) binary is also in a small orbit of less than 1 AU. An A3 it has a period of 176 days. Taygete (**19 Tauri**) is in a 3.2 year orbit with a B9 of 3.2 $D_\odot$. They are separated by an average 4.6 AU.

Away from the Hyades and the Pleiades **β Tauri** is another hot bluish star with a temperature of 13,600 K. Relatively close at 131 ly its apparent magnitude is m_v +1.68, its absolute magnitude easy to work out – just change the sign and

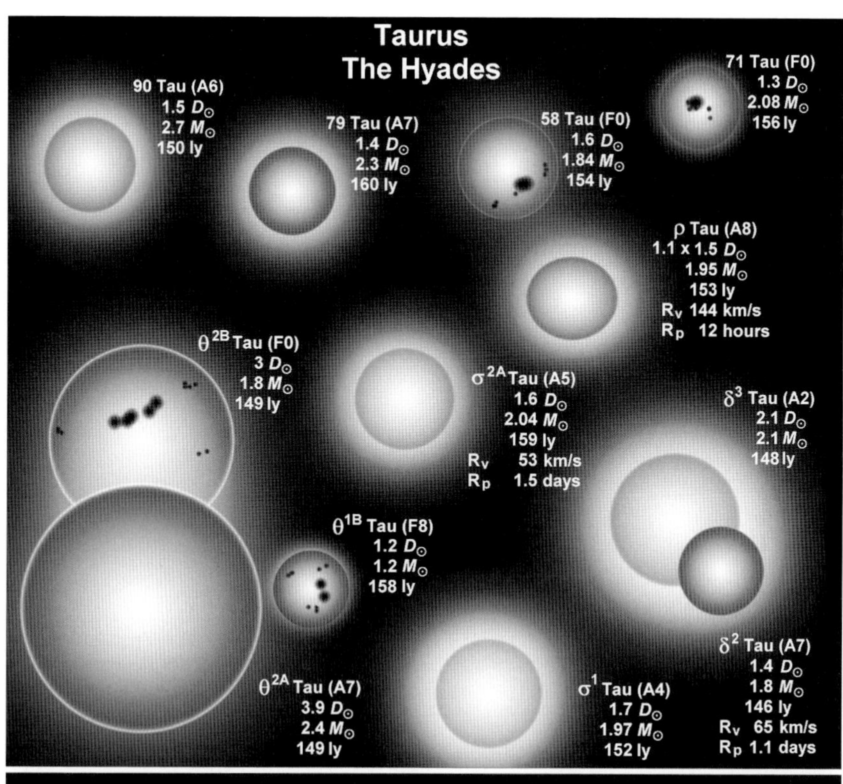
Taurus
The Hyades
90 Tau (A6)
1.5 $D_\odot$
2.7 $M_\odot$
150 ly
79 Tau (A7)
1.4 $D_\odot$
2.3 $M_\odot$
160 ly
58 Tau (F0)
1.6 $D_\odot$
1.84 $M_\odot$
154 ly
71 Tau (F0)
1.3 $D_\odot$
2.08 $M_\odot$
156 ly
ρ Tau (A8)
1.1 x 1.5 $D_\odot$
1.95 $M_\odot$
153 ly
R_v 144 km/s
R_p 12 hours
θ^{2B} Tau (F0)
3 $D_\odot$
1.8 $M_\odot$
149 ly
σ^{2A} Tau (A5)
1.6 $D_\odot$
2.04 $M_\odot$
159 ly
R_v 53 km/s
R_p 1.5 days
δ^3 Tau (A2)
2.1 $D_\odot$
2.1 $M_\odot$
148 ly
θ^{1B} Tau (F8)
1.2 $D_\odot$
1.2 $M_\odot$
158 ly
δ^2 Tau (A7)
1.4 $D_\odot$
1.8 $M_\odot$
146 ly
R_v 65 km/s
R_p 1.1 days
θ^{2A} Tau (A7)
3.9 $D_\odot$
2.4 $M_\odot$
149 ly
σ^1 Tau (A4)
1.7 $D_\odot$
1.97 $M_\odot$
152 ly

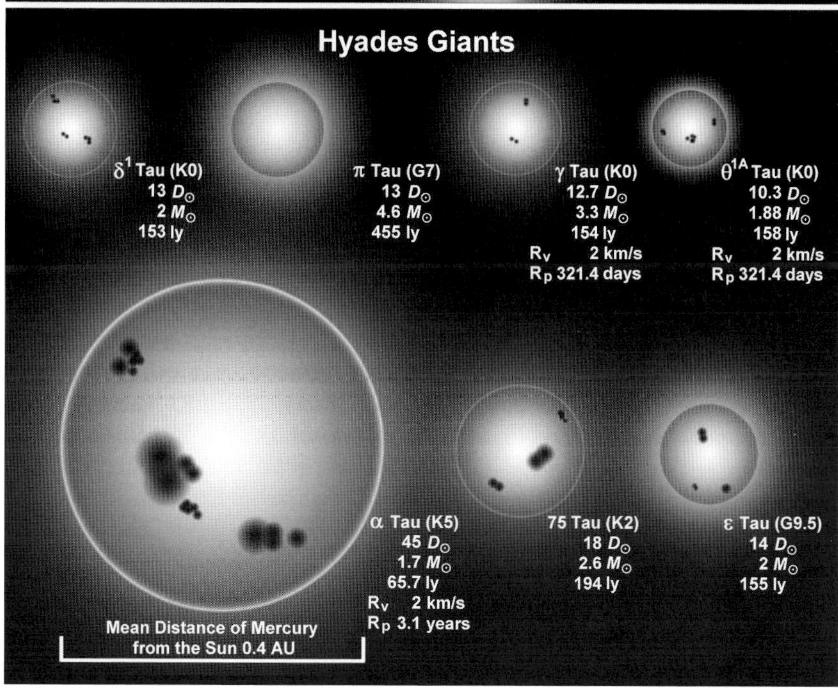
Hyades Giants
δ^1 Tau (K0)
13 $D_\odot$
2 $M_\odot$
153 ly
π Tau (G7)
13 $D_\odot$
4.6 $M_\odot$
455 ly
γ Tau (K0)
12.7 $D_\odot$
3.3 $M_\odot$
154 ly
R_v 2 km/s
R_p 321.4 days
θ^{1A} Tau (K0)
10.3 $D_\odot$
1.88 $M_\odot$
158 ly
R_v 2 km/s
R_p 321.4 days
α Tau (K5)
45 $D_\odot$
1.7 $M_\odot$
65.7 ly
R_v 2 km/s
R_p 3.1 years
75 Tau (K2)
18 $D_\odot$
2.6 $M_\odot$
194 ly
ε Tau (G9.5)
14 $D_\odot$
2 $M_\odot$
155 ly
Mean Distance of Mercury
from the Sun 0.4 AU

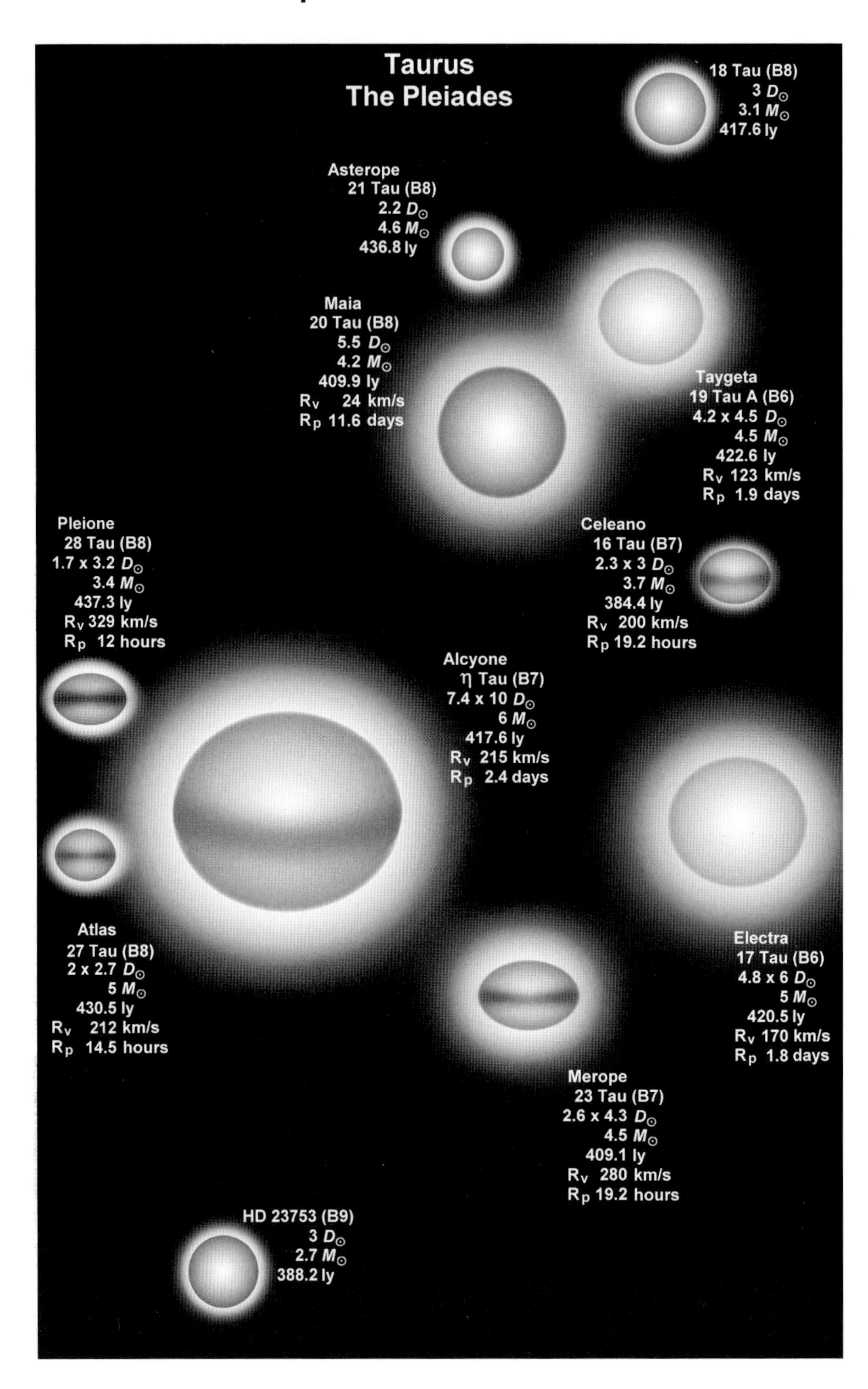

Taurus
The Pleiades
18 Tau (B8)
3 D⊙
3.1 M⊙
417.6 ly
Asterope
21 Tau (B8)
2.2 D⊙
4.6 M⊙
436.8 ly
Maia
20 Tau (B8)
5.5 D⊙
4.2 M⊙
409.9 ly
Rv 24 km/s
Rp 11.6 days
Taygeta
19 Tau A (B6)
4.2 x 4.5 D⊙
4.5 M⊙
422.6 ly
Rv 123 km/s
Rp 1.9 days
Pleione
28 Tau (B8)
1.7 x 3.2 D⊙
3.4 M⊙
437.3 ly
Rv 329 km/s
Rp 12 hours
Celeano
16 Tau (B7)
2.3 x 3 D⊙
3.7 M⊙
384.4 ly
Rv 200 km/s
Rp 19.2 hours
Alcyone
η Tau (B7)
7.4 x 10 D⊙
6 M⊙
417.6 ly
Rv 215 km/s
Rp 2.4 days
Atlas
27 Tau (B8)
2 x 2.7 D⊙
5 M⊙
430.5 ly
Rv 212 km/s
Rp 14.5 hours
Electra
17 Tau (B6)
4.8 x 6 D⊙
5 M⊙
420.5 ly
Rv 170 km/s
Rp 1.8 days
Merope
23 Tau (B7)
2.6 x 4.3 D⊙
4.5 M⊙
409.1 ly
Rv 280 km/s
Rp 19.2 hours
HD 23753 (B9)
3 D⊙
2.7 M⊙
388.2 ly

drop the last digit: M_v -1.6. With a diameter of 4.2 $D_\odot$ and a mass of 4.1 $M_\odot$ β Tau looks like a 278 $L_\odot$ but, taking into account that much of its energy is radiated in the ultraviolet, the figure shoots up to 700 $L_\odot$. Its rotational period is 2.8 days, the rotational velocity being 75 km/s. It is unusual and important in a couple of respects. First, on celestial charts it is traditionally tied to ι Aurigae to the north west and θ Aurigae to the north east, one of the few 'link' stars in the entire sky (others include α Andromedae with Pegasus and ξ and μ Serpentis with Ophiuchus). Second, it is only about 3° from the galactic anticenter; the point on the celestial sphere that lies directly opposite the galactic center. Half as close again to the anticenter is **136 Tauri**, a 2.1 $D_\odot$ A0 that spins at a more leisurely 15 km/s leading to a rotational period of 7.1 days. 136 Tau has a magnitude of m_v +4.54 and is accompanied by a spectroscopic companion in a 5.97 day long orbit.

ζ Tauri is a γ Cas eruptive variable. Over a period of 132^d 23^h 22^m it fades from m_v +2.88 to +3.17 and back. A B2, possibly a B4, it is thought to be 5.5 $D_\odot$ across though it may be as little as 3.7 $D_\odot$ pole to pole due to its rapid rotational velocity of 241 km/s. A second consequence of its high spin rate is the creation of a shell of dust around the star up to a distance of 89 million km (0.6 AU). Now 25 million years old this 8.9 $M_\odot$ Be 'emission' star burns with a surface temperature of 21,150 K and has a luminosity of 12,700 $L_\odot$ although 93% of the output is in the ultraviolet. It is not alone. An invisible companion orbits the star at a distance of 1 AU with a period of 133 days, and no doubt plays a role in its variability. Situated some 417 ly away ζ Tau would brighten to M_v -3 at 10 pc.

λ Tauri is also a variable, this time of the Algol variety. Lying 370 ly from Earth the star is actually a triple system. The primary is a 6.3 $D_\odot$, 7.2 $M_\odot$ B3 which rotates at 50 km/s. The secondary is a 4.8 $D_\odot$, 1.9 $M_\odot$ A4. As the two stars orbit their barycenter λ Tau changes from m_v +3.30 to +3.80 before brightening again over a period of 3^d 22^h 52^m. The B3 star is apparently more than 3½ times more luminous than its A4 mate (456 $L_\odot$ and 128 $L_\odot$) but taking into account that most of its output is in the ultraviolet the total luminosity is about 5,050 $L_\odot$. Somewhere in the system is a K0 of 0.7 $M_\odot$. Never straying more than about 0.4 AU from the primary it takes 33 days to circle the main star. The triplet is about 100 million years old and at 10 pc would be a brilliant M_v -5.

ξ Tauri is a complex quadruple system. To the naked eye the star looks like a single point of light with a magnitude of m_v +3.72 but most of this is due to two nearly identical B9 dwarfs of m_v +4.52 each. With diameters of 1.8 $D_\odot$ and masses of 3.1 $M_\odot$ each the twins are separated by just 0.13 AU (19.5 million km) giving an orbital period of 7.15 days. The third component is a B8, a bit larger at 2.1 $D_\odot$ but with a lot more mass: 5.5 $M_\odot$. It averages 1.1 AU from the twins and has an orbital period of 145 days. The fourth player is a m_v +7.58 F5 dwarf of 1.24 $M_\odot$. It is at least 50 AU from the inner three and takes 212 years to complete a single orbit. ξ Tau is 222 ly away.

τ Tauri is a deceptive binary. Appearing as a 5th magnitude single star to the naked eye optical aid will reveal a m_v +7.15 star 63″ to the south-west. Designated **HD 284659** this is merely a line of sight coincidence, the A0 star lying at a distance of 259 ly. τ Tau – the name is almost poetic – is much farther away at 401 ly. A B3 its diameter has proven devilishly difficult to determine with 1.1, 2.4, 3.2, 3.9 and 4.71 $D_{\odot}$ quoted in the literature. Its rotational velocity is just as elusive ranging from 100 to 187 km/s, so its rotational period could be as short as 7.1 hours or as long as 2.4 days. Its true companion is a spectroscopic binary in a 2.96 day long orbit. τ Tau is a δ Scuti pulsating variable, which may help to explain the confusion over its diameter, its magnitude shifting between m_V +5.09 and +5.13 with a period of $1^h\ 0^m\ 29^s$.

CE Tauri is an extremely large star but no one is really sure how big it is. Estimates range from 'just' 240 $D_{\odot}$ to 800 $D_{\odot}$ although 540 $D_{\odot}$ seems to be closer to the mark. Its distance is just as perplexing: it could be as close as 1,300 ly or as far as 2,530 ly. With a probable mass of around 14 $M_{\odot}$ and luminosity of 47,000 $L_{\odot}$, most of it in the infrared, this cool (3,700 K) M2 supergiant is also a semi-regular pulsating variable with a period of around 165 days during which is magnitude varies between m_V +4.23 and +4.54. The rotational velocities of red supergiants are not particularly well documented but if we assume it rotates at the same velocity as the Sun, 2 km/s, and it is 540 $D_{\odot}$ across then its rotational period is 37.4 years which would increase to 55.4 years at 800 $D_{\odot}$. At 10 pc it would be an impressive M_V -4.8 blood red beacon.

Too faint to be included in this book, but too important to leave out, **T Tauri** is a very young variable star that marks an important early stage in stellar evolution. Varying irregularly between m_V +9.3 and +14 T Tau is perhaps as young as 1 million years old and is a pre-Main Sequence star with a mass of less than 2 $M_{\odot}$. T Tauri-type stars are thought to still be in the process of contracting, accreting material from the surrounding nebula. They tend to be 100 thousand to 100 million years old and have masses of 0.5 to 3.0 $M_{\odot}$. They are surrounded by hot, dense clouds that are heated by ferocious stellar winds pouring away from the star at a typical speed of 100 km/s. It is believed that T Tauri stars have significant starspots that help to drive the wind. The cores of T Tauri stars are too cool to cause hydrogen to fuse into helium so they shine because of the gravitational energy generated as the star collapses. As a result they are over-luminous for their mass and temperature. T Tauri itself illuminates **NGC 1555** which, because it brightens and fades with the star, is known as Hind's Variable Nebula. A second recorded nebula, **NGC 1554**, appears to have disappeared and is often referred to as Struve's Lost Nebula. The star is believed to be a binary and possibly a triplet.

At first glance **104 Tauri** could almost be a mirror image of our own Sun. It is nearly the same size and mass, rotates at the same speed, has a temperature of 5,880 K and is only a couple of notches farther along the spectral sequence at G4 (the Sun's a G2). A closer look, however, reveals a few differences. It is more than twice as luminous, 2.2 $L_{\odot}$, and twice as old at 9,700 million years. And there are twice as many. Twice as many what? Twice as many stars. 104 Tau is a binary system of identical yellow dwarfs. The orbit is not well determined but the period is thought to be 1.19 or 2.38 years, the two stars possibly averaging 2.8 AU apart. Their individual magnitudes are m_V +5.7 each but combined and viewed from 51 ly they appear as a single star of m_V +4.92. At 10 pc they would brighten only marginally to m_V +4.5.

104 Tau is not the closest naked eye star in the constellation. **111 Tauri**, a 1.3 $D_\odot$, 1.15 $M_\odot$ F8 lies at a distance of 47.8 ly. Dark starspots are believed to be the cause of the star's changing magnitude. Classed as a BY Draconis variable its magnitude ranges from m_v +4.98 to +5.02. Closer still at 44.8 ly is **10 Tauri**, a 1.1 $D_\odot$, 1.25 $M_\odot$ F9.

HD 285968 is an M-class dwarf with a diameter a little over half the size of the Sun (738,000 km). Initially astronomers thought they had discovered a 24.5 $M_\oplus$ planet in a 10.24 day long orbit but further research suggests the planet is less massive, 7.8 $M_\oplus$, and in an 8.78 day orbit (see table).

Planetary systems in Taurus

Star	$D_\odot$	Spectral class	ly	m_v	Planet	Minimum mass	q	Q	P
ε Tau	13.7	K0	147	+3.53	ε Tau b	7.6 M_J	1.64	2.22	1.62 y
HD 285968	0.53	M2.5	30.7	+9.97	HD 285968 b	7.8 $M_\oplus$	0.066	0.066	8.78 d

Taurus is the home of the very first object to be cataloged by Charles Messier: **M1** the Crab Nebula, a supernova remnant some 6,300 ly away. Now at m_v +8.4 the supernova reached m_v -6 in July AD 1054. It was visible in daylight for 23 days and was a naked eye object in the night sky for 653 days (1.79 years). It was recorded by sky watchers in north and central America, China, Korea, Japan and the Middle East but there are no surviving accounts from Europe. In the 1970s those who studied the Crab Nebula would often claim that there were two types of astronomy: the astronomy of the Crab and the astronomy of everything else.

Telescopium

Constellation:	Telescopium	**Hemisphere:**	Southern
Translation:	The Telescope	**Area:**	252 deg^2
Genitive:	Telescopii	**% of sky:**	0.611%
Abbreviation:	Tel	**Size ranking:**	57th

Another faint and obscure constellation invented by La Caille.

The brightest star in the constellation is, perhaps not surprisingly, **α Telescopii**, a m_v +3.45 B3 sub-giant of 5.8 $D_\odot$ and with a mass of 5.18 $M_\odot$. It appears to be a slow spinner. The average rotational velocity for stars of this class is 140 km/s but α Tel saunters along at just 35 km/s. But it may be deceiving us. If its pole is pointed in our direction then it will give the impression that the star is turning more slowly than it really is. From a distance of 249 ly its 18,400 K surface has a luminosity of 200 $L_\odot$ which shoots up to 900 $L_\odot$ when we take into consideration that most of its output is in the ultraviolet. At 10 pc it would be a more impressive M_v -3.

There are no β, γ or θ Telescopii. What was called **β Telescopii** is actually η Sagittarii. Similarly **γ Telescopii** is now G Scorpii and **θ Telescopii** is 45 Ophiuchi.

δ^1 and δ^2 Telescopii are a well matched pair. **δ^1 Telescopii** is a m_v +4.91 B6 of 2.2 $D_\odot$ and 525 $L_\odot$. At 10 pc it would be M_v -1.3 and it is moving away from us at 7 km/s. It has a spectroscopic companion in an 18.85 day long orbit. **δ^2 Telescopii** is a m_v +5.06 B3 of 2.1 $D_\odot$ and 900 $L_\odot$. At 10 pc it would be M_v -1.6 and it is receding at 7.6 km/s. It also has a spectroscopic companion in a 21.71 day long orbit. While this may appear to be the perfect arrangement things are not as straightforward as they seem. δ^1 Tel is 672 to 920 ly away while δ^2 Tel is 884 to 1,350 ly. This could mean that the two δ stars are very close neighbors but it could also mean that δ^2 is twice the distance of δ^1. We just don't know.

Of the 15 naked eye stars in the constellation almost a third belong to the K spectral class including **ε Telescopii**, a 13 $D_\odot$ K0 giant some 409 ly from Earth; **ι Telescopii**, 15 $D_\odot$ and 398 ly; **ζ Telescopii**, 13 $D_\odot$ and the closest at 127 ly and **HD 169405**, 9 $D_\odot$ and the farthest at 529 ly.

ξ Telescopii is the largest and most distant of the naked eye stars. Only just a red giant, an M1, it is 35 $D_\odot$ across and lies at a distance of 1,254 ± 272 ly. It is almost spot on 5th magnitude, m_v +4.93, and has an absolute magnitude of M_v -0.5.

NGC 6584 is a globular cluster 43,700 ly away and 22,800 ly from the galactic center. A m_v +8.27 object it appears 9.4′ across which equates to 199.5 ly.

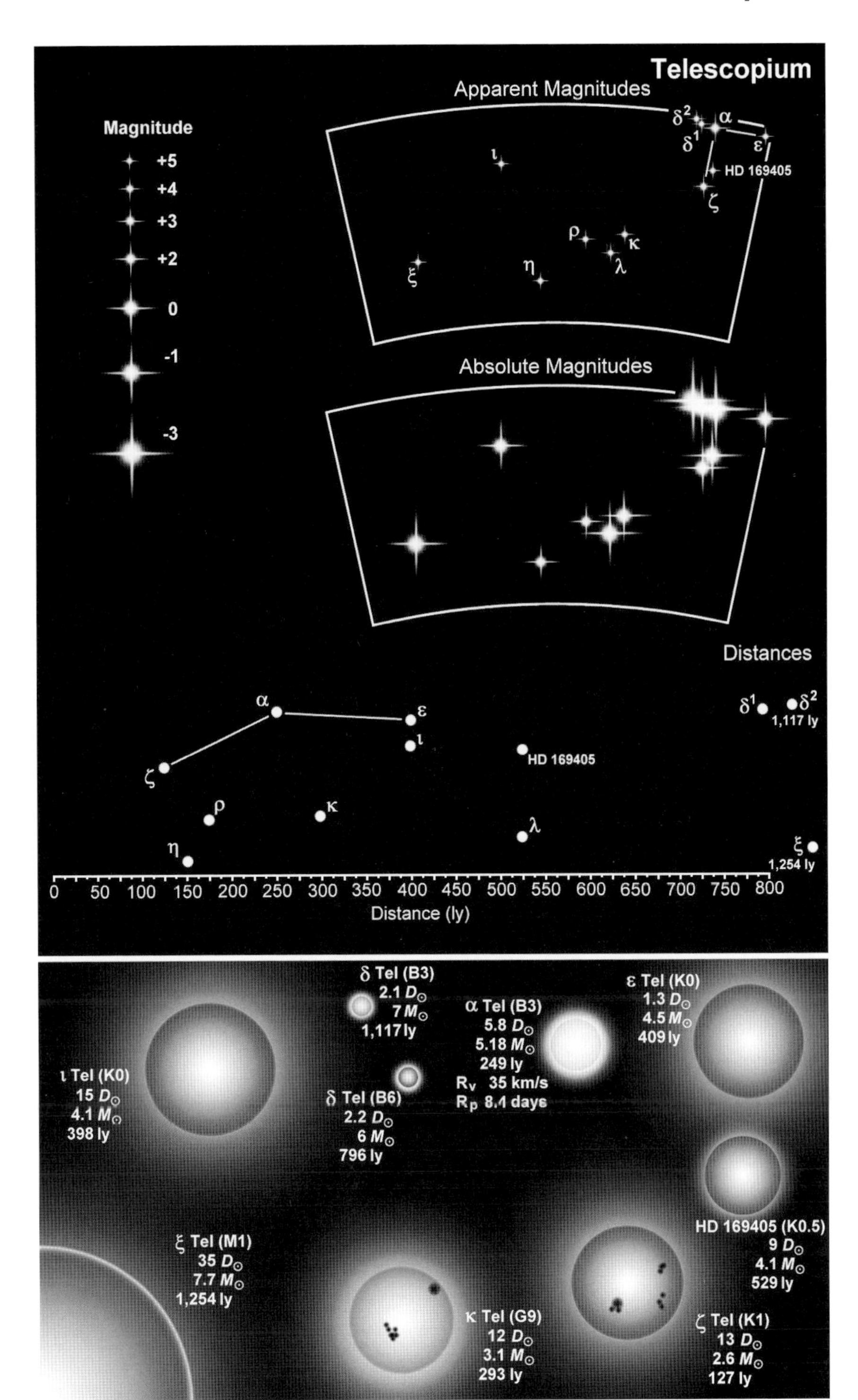
Telescopium
Apparent Magnitudes
Magnitude
+5
+4
+3
+2
0
-1
-3
HD 169405
Absolute Magnitudes
Distances
1,117 ly
HD 169405
1,254 ly
0 50 100 150 200 250 300 350 400 450 500 550 600 650 700 750 800
Distance (ly)
δ Tel (B3)
2.1 D⊙
7 M⊙
1,117 ly
α Tel (B3)
5.8 D⊙
5.18 M⊙
249 ly
Rv 35 km/s
Rp 8.4 days
ε Tel (K0)
1.3 D⊙
4.5 M⊙
409 ly
ι Tel (K0)
15 D⊙
4.1 M⊙
398 ly
δ Tel (B6)
2.2 D⊙
6 M⊙
796 ly
HD 169405 (K0.5)
9 D⊙
4.1 M⊙
529 ly
ξ Tel (M1)
35 D⊙
7.7 M⊙
1,254 ly
κ Tel (G9)
12 D⊙
3.1 M⊙
293 ly
ζ Tel (K1)
13 D⊙
2.6 M⊙
127 ly

Triangulum

Constellation:	Triangulum	**Hemisphere:**	Northern
Translation:	The Triangle	**Area:**	132 deg^2
Genitive:	Trianguli	**% of sky:**	0.320%
Abbreviation:	Tri	**Size ranking:**	78th

An ancient constellation that has been associated with the island of Sicily, the Nile delta and, well, a triangle. It contains only a dozen naked eye stars.

Lying 64.1 ly from Earth α **Trianguli** is a 2.8 $D_\odot$ rotating ellipsoidal binary. An F5 with a mass of 1.57 $M_\odot$ and luminosity of 13.5 $L_\odot$ its temperature is 6,300 K and it rotates at 82 km/s taking 1.74 days to turn on its axis. Its spectroscopic companion is locked in a gravitational embrace that sees it orbit the primary with a period of also 1.74 days, the orbital arrangement causing α Tri's magnitude to change between m_v +3.52 and +3.53 making it the second brightest star in the constellation. The two stars are so close together, probably less than 0.04 AU (5.98 million km), that they distort one another into teardrops. Not particularly old by solar standards, 2,700 million years, α Tri is well past middle age for its mass. The system is heading towards us at 16.2 km/s.

The brightest star, β **Trianguli,** is also a spectroscopic binary. The primary is a 4.6 $D_\odot$, 2.5 $M_\odot$ A5 with a temperature of 7,940 K which rotates once every 3.3 days, its rotational velocity being, as far as we can tell, 70 km/s. Its magnitude is spot on m_v +3.00 and unvarying. Virtually nothing is known about its companion other than its orbit brings the two stars as close as 0.17 AU (25.4 million km) and pushes them as far apart as 0.42 AU (62.8 million km) with an orbital period of 31.39 days. The whole system is enveloped in a dusty disk, testament to the fact that the system is still quite young at 590 million years.

γ Trianguli along with δ Trianguli and 7 Trianguli make a neat little triangle of their own, although they are not related. γ **Trianguli** is a white hot, 9,330 K, A1 of 2.4 $D_\odot$ and 2.32 $M_\odot$ lying 118 ly away. Some 26.5 $L_\odot$ it shines exactly m_v +4.00, brightening to an absolute magnitude of M_v +0.6. It is only about 200 million years old and still retains its youthful speed, spinning at 254 km/s and taking just half a day to complete a single turn. Almost due north is the second star in the trio, the Sun-like δ **Trianguli.** Solar analogs always stir the imagination but this one is particularly intriguing because it is not one but two suns. The primary has an apparent magnitude of m_v +4.87 which, because it is almost 10 pc away (actually 34.5 ly), only brightens very slightly to absolute magnitude of M_v +4.80. In reality its magnitude is m_v +5.2 but the presence of a m_v +6.5 companion makes it appear brighter than it really is. Its diameter is thought to be 0.98 $D_\odot$ and it is believed to have a mass of 1.0 $M_\odot$ and luminosity of 1.07 $L_\odot$. Being a G0.5 it is only 120 K warmer than the Sun at 5,900 K but spins five times faster at 10 km/s, rotating once every 5 days. Its companion lies at the other end of the spectral class, a G9, and is a few hundred degrees cooler at 5,300 K. More yellowish-orange it is much smaller at 0.64 $D_\odot$ and slightly less massive,

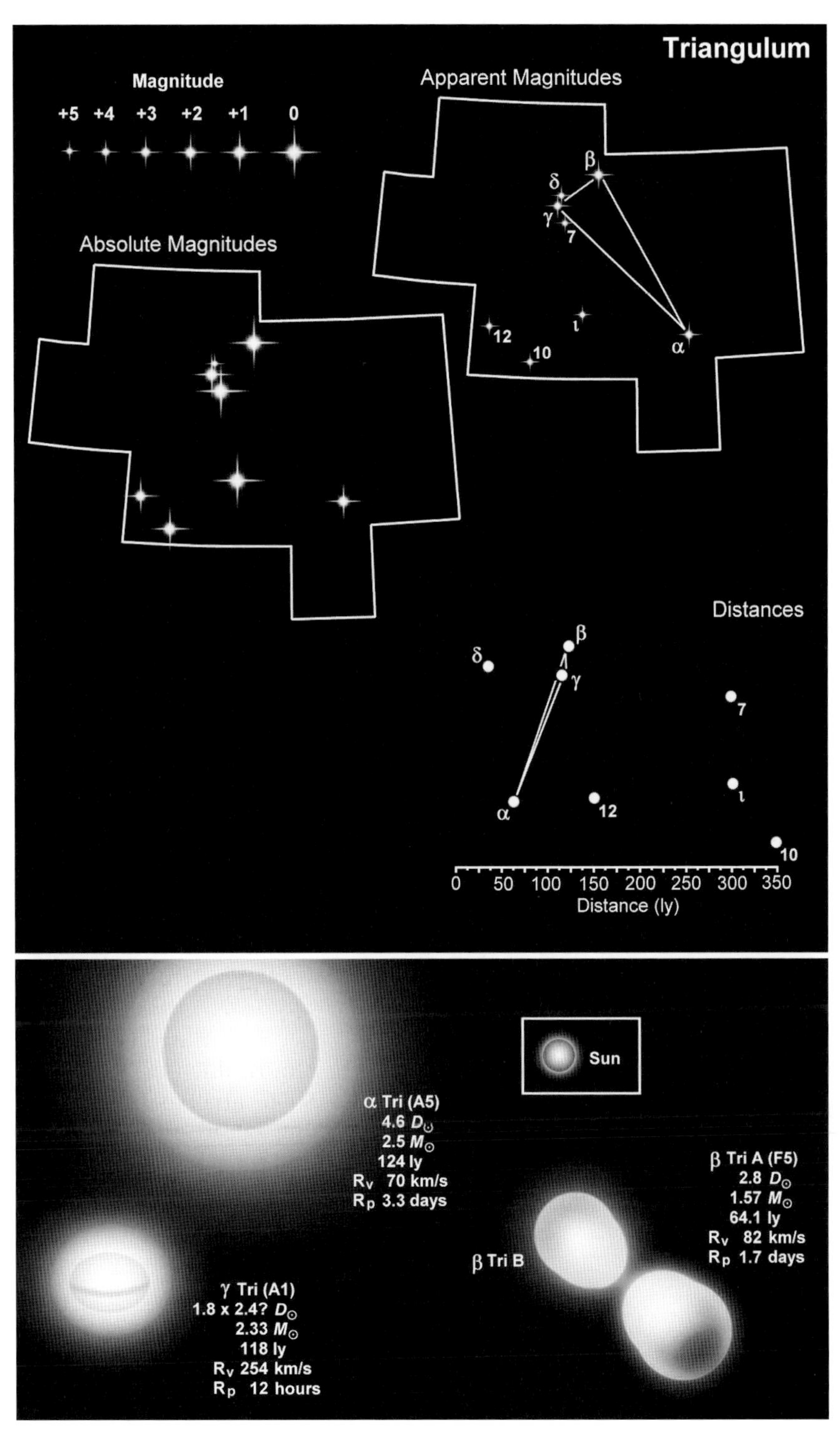
Triangulum
Magnitude
+5 +4 +3 +2 +1 0
Apparent Magnitudes
β
δ
γ
7
ι
12
10
α
Absolute Magnitudes
Distances
β
δ
γ
7
ι
α
12
10
0 50 100 150 200 250 300 350
Distance (ly)
Sun
α Tri (A5)
4.6 D☉
2.5 M☉
124 ly
Rv 70 km/s
Rp 3.3 days
β Tri A (F5)
2.8 D☉
1.57 M☉
64.1 ly
Rv 82 km/s
Rp 1.7 days
β Tri B
γ Tri (A1)
1.8 x 2.4? D☉
2.33 M☉
118 ly
Rv 254 km/s
Rp 12 hours

0.8 $M_\odot$, and considerably less luminous at 0.3 $L_\odot$. The orbit varies between 0.104 and 0.108 AU (15.6 million km to 16.2 million km) with a period of 10.0201 days. The third naked eye star in the arrangement is **7 Trianguli**. In many ways it is like γ Tri. It is an A0, has a diameter of 2.2 $D_\odot$ and 2.48 $M_\odot$ although it is twice as luminous at 52 $L_\odot$ and more than twice the distance, measured at 293 ly. It can be found almost south west of γ Tri twinkling at m_v +5.25.

The 5th magnitude ι **Trianguli** is a complex system of four stars, one pair being an ellipsoidal variable while the other is an eruptive variable. The main star is another G-class, this time a G0 but much larger than the Sun at 13 $D_\odot$ and with a mass of 2.7 $M_\odot$ and luminosity of 61 $L_\odot$. Its companion is an F6 dwarf of 2.3 $M_\odot$ and half the luminosity at 32 $L_\odot$. Separated by 3.8″, equivalent to just 0.2 AU (29.9 million km), the two have an orbital period of 14.732 days. The presence of the second star regularly disrupts the primary's gravitational field leading to eruptions so the system is classed as an RS CVn variable. The second pair consist of mid-F-class stars of 1.7 and 1.5 $D_\odot$ and 18 and 9 $L_\odot$. They have an orbital period of 2.24 days and are so close, a mere 0.05 AU (7.5 million km), that they are distorted into teardrops. As they revolve around one another their magnitude also varies. This pair orbit the other pair at a distance of 355 AU and with a period of 2,465 years. The combined result is that 6 Tri appears to vary between m_v +5.19 and +5.98 with a period of $14^d\ 17^h\ 34.1^m$ and so also has the designation of **TZ Trianguli** (and is also known as **6 Trianguli**). The two sets of binaries are often described as being blue and yellow. Just to make matters even more interesting ι Tri was once part of a smaller triangle called Triangulum Minus which it made up with **10 Trianguli**, a 1.7 $D_\odot$ A2 some 350 ly away, and **12 Trianguli**, a 2.3 $D_\odot$ F0 at a distance of 155 ly. 10 and 12 Tri are both m_v +5.29. The Triangulum Minus was introduced by Johannes Hevelius who renamed the original triangle Triangulum Majus.

Triangulum is also home to the Local Group galaxy **M33**, also known as the Pinwheel Galaxy, which although somewhat smaller than both the Andromeda Galaxy (M31) and our Milky Way is probably near average size for spiral galaxies overall. From our viewpoint the galaxy is seen face-on, enabling one to trace out the loosely wound spiral arms.

Triangulum Australe

Constellation:	Triangulum Australe	**Hemisphere:**	Southern
Translation:	The Southern Triangle	**Area:**	110 deg^2
Genitive:	Trianguli Australis	**% of sky:**	0.267%
Abbreviation:	TrA	**Size ranking:**	83rd

Introduced by Pieter Keyser and Frederick de Houtman, Triangulum Australe is smaller than its northern counterpart but its stars are brighter.

α **Trianguli Australis** also goes by the name of Atria being made up of 'A' for Alpha and 'tria' as in triangle. A m_v +1.92 K2 giant of 37 $D_\odot$ it lies at a distance of 415 ly and has a luminosity of 2,286 $L_\odot$. It is considered to be quite young, about 45 million years old, and would appear at M_v -0.1 at 10 pc.

About 127 ly away β **Trianguli Australis** is a 1.5 $D_\odot$ F1 dwarf weighing in a 1.71 $M_\odot$. An apparent magnitude of m_v +2.82, brightening marginally to an absolute M_v +2.4 this 9.2 $L_\odot$ star burns at 7,400 K and has a rotational velocity of 92 km/s turning once in less than 20 hours.

The third star that makes up the shape of the triangle, γ **Trianguli Australis,** is not a variable despite often being listed as such. It shines at a steady m_v +2.85 from its home 183 ly away. Two and a half times the size of the Sun with a mass of 2.96 $M_\odot$ it is an A1 with a surface temperature of 8,900 K and a luminosity of 220 $L_\odot$. A fast spinner, it completes a single rotation in 13.5 hours rotating at 225 km/s. It appears to be surrounded by a debris disk.

Parked almost midway between β and γ is the m_v +4.11 ε **Trianguli Australis.** A K1.5 giant of 17 $D_\odot$ and 1.5 $M_\odot$ it is 216 ly away and has a temperature of 4,470 K.

With a temperature of 4,800 K, κ **Trianguli Australis** is 1,000 K cooler than our own Sun and is classed as a G5, three grades down from the G2 star around which the Earth orbits. Twinkling away at an almost unnoticed m_v +5.10 it could well be another close solar analog until we start to look at the other measurements. Its distance is the dead giveaway: 3,000 ly. Its luminosity comes in at 6,220 $L_\odot$. And then its absolute magnitude jumps to a stunning M_v -6.2 at 10 pc. We're dealing not with a dwarf yellow but a supergiant some 290 $D_\odot$ across. Placed at the center of our Solar System κ TrA would engulf Mercury, Venus and the Earth, while Mars would just skim its surface (although, in reality, the planet would get sucked into the star). It is big enough to hold more than 24 million Suns. And it is a relatively slow spinner with a rotational velocity of 8.7 km/s which means that it takes 1,687 days – 4.6 years – to rotate just once. Impressive though they are, supergiants burn out in just a few million years.

δ **Trianguli Australis** is another impressive G2. Not quite as big as κ TrA it is still 43 $D_\odot$ which translates to 0.4 AU or 59.8 million km (about half the size of Mercury's orbit). Shining at a steady m_v +3.85 it reaches M_v -2.1 at 10 pc, its luminosity being 850 $L_\odot$. It is somewhere between 555 and 685 ly away.

Although it is an F-class star ζ **Trianguli Australis** is pretty close to our Sun,

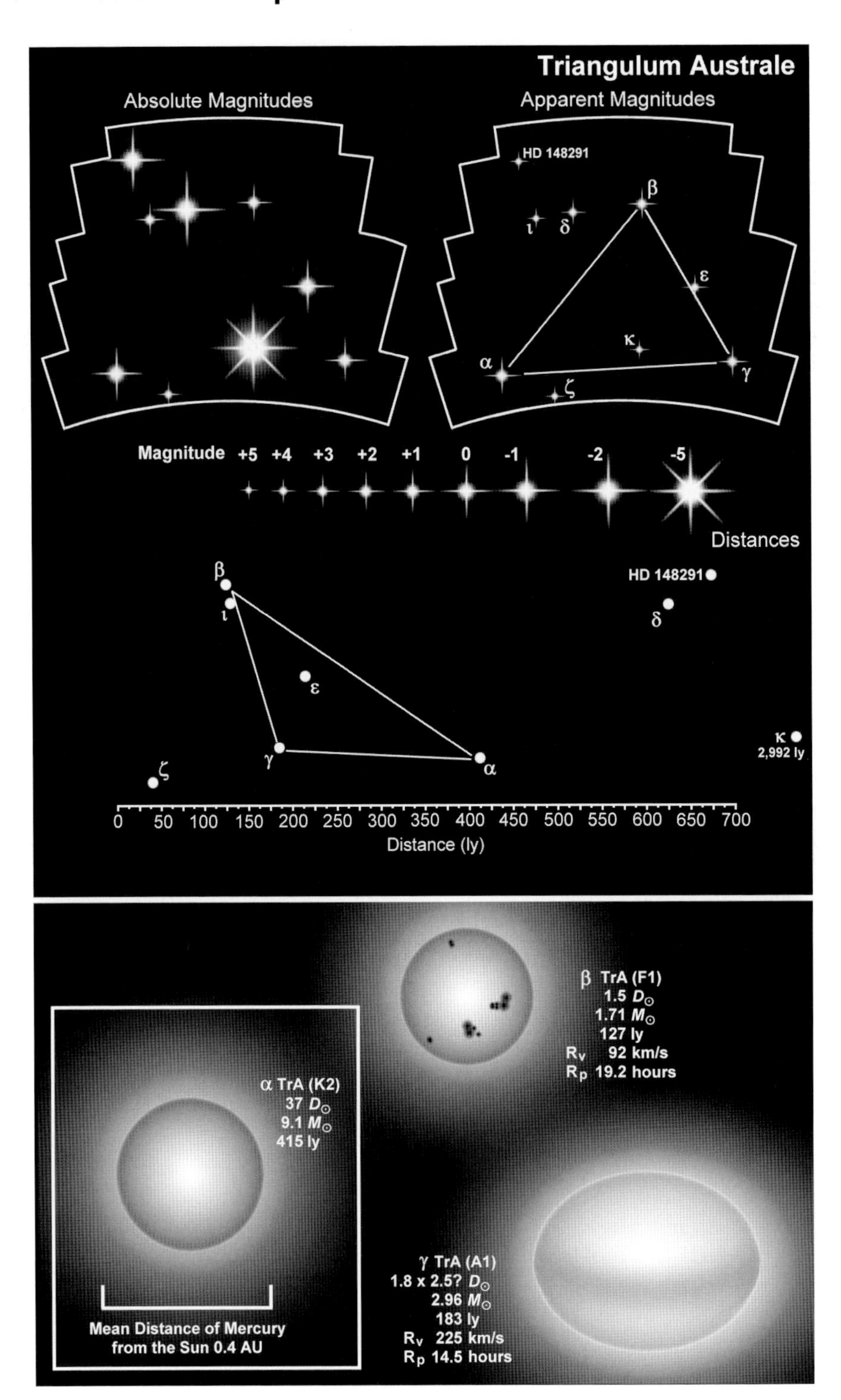

Triangulum Australe
Absolute Magnitudes
Apparent Magnitudes
HD 148291
β
ι
δ
ε
κ
α
γ
ζ
Magnitude +5 +4 +3 +2 +1 0 -1 -2 -5
Distances
HD 148291
δ
β
ι
ε
κ
2,992 ly
γ
α
ζ
0 50 100 150 200 250 300 350 400 450 500 550 600 650 700
Distance (ly)
β TrA (F1)
1.5 D⊙
1.71 M⊙
127 ly
Rv 92 km/s
Rp 19.2 hours
α TrA (K2)
37 D⊙
9.1 M⊙
415 ly
Mean Distance of Mercury from the Sun 0.4 AU
γ TrA (A1)
1.8 x 2.5? D⊙
2.96 M⊙
183 ly
Rv 225 km/s
Rp 14.5 hours

in more ways than one. Just 39.5 ly away it has an apparent magnitude of m_V +4.91 which increases by only half a mag to an absolute M_V +4.40, the standard distance of 10 pc equaling 32.6 ly. Usually listed as an F9 some authorities regard it as a G0, its diameter being, as far as we can tell, the same as the Sun and with only slightly more mass, 1.14 $M_\odot$, and luminosity, 1.29 $L_\odot$. It even rotates at the same speed as the Sun, 2 km/s, and is only a couple of hundred Kelvin hotter at 6,030 K. So almost a perfect match. Almost, but not quite. There is one big difference: it has a companion – an invisible spectroscopic component with a mass of just 0.09 $M_\odot$ in an almost circular 12.98 day orbit of 1.32 million km. The secondary, which is probably a G1, distorts the magnetic field of the primary resulting in high levels of chromospheric activity.

Similar in some ways to ζ TrA is ι **Trianguli Australis**. Just visible to the unaided eye at m_V +5.29 it lies at the other end of the spectral class as an F4. About 1.6 $D_\odot$ and with a mass of 1.72 $M_\odot$ it is ten times more luminous than the Sun and rotates at 13 km/s leading to a rotational period of 6.2 days. It also has a spectroscopic companion which is probably an F5 in a 39.89 day long orbit although little else is known about it. The system is 132 ly distant.

NGC 6025 is an open cluster 2,500 ly away and about 17 ly across. Estimated to be 77.5 million years old it appears to contain about 74 stars brighter than m_V +12 although some are probably not cluster members. The cluster straddles the border with Norma.

Tucana

Constellation:	Tucana	**Hemisphere:**	Southern
Translation:	The Toucan	**Area:**	295 deg^2
Genitive:	Tucanae	**% of sky:**	0.715%
Abbreviation:	Tuc	**Size ranking:**	48th

Another constellation invented by the explorers Pieter Keyser and Frederick de Houtman, Tucana is home to the Small Magellanic Cloud (SMC), a small galaxy 200,000 ly away.

The m_v +2.85 luminary α **Tucanae** is a 28 $D_\odot$ K3 giant with a mass of 2.5 $M_\odot$. Just one light year short of 200 ly, this cool 4,300 K star is 220 times more luminous than the Sun but rotates at just about the same speed: 1.9 km/s (Sun = 2 km/s). Consequently is takes 746 days – almost exactly 2 years – to spin just once on its axis. It has a spectroscopic companion in a 4,197.7 day long orbit (11.5 years).

β Tucanae is a six star complex although there is some uncertainty as to the exact relationships between the components. To the naked eye there are two stars: a m_v +3.70 **β^1 Tucanae** and a m_v +4.49 **β^3 Tucanae** which lies 9′ roughly to the south east (PA 116°). A small telescope or binocular will reveal **β^2 Tucanae** 27″ from β^1 Tuc at PA 169°. All three stars are binary systems with the components identified as β^{1A}, β^{1B}, β^{2A}, β^{2B}, β^{3A} and β^{3B}.

β^{1A} Tuc is a m_v +4.29 B9 of 1.74 $D_\odot$ with a mass of 2.5 $M_\odot$. Its 10,900 K surface has a luminosity of 29 $L_\odot$ and it rotates at 107 km/s making a full turn in 19.8 hours. Just 2.4″ away is its binary companion β^{1B} Tuc: a red dwarf, possibly an M3, with a magnitude of m_v +13.5. The two are separated by an average 110 AU and take 504 years to orbit their common center of gravity.

β^{2A} Tuc is a m_v +4.08 A2 of 1.8 $D_\odot$ with a mass of 2.76 $M_\odot$. Its 9,100 K surface has a luminosity of 17 $L_\odot$. Less than half an arc second away (0.404″) is its binary companion β^{2B} Tuc: a pre-Main Sequence white dwarf, an A7, of 1.5 $D_\odot$, 1.94 $M_\odot$ and 8 $L_\odot$, with a magnitude of m_v +6.0. The two are in a 44.66 year long orbit during which they come as close as 3.3 AU to one another before separating to 30 AU at apastron. The β^2 pair are in orbit with the β^1 pair at an average distance of 1,160 AU and with an orbital period of 13,109 years.

β^3 Tuc is more problematic and opinion is divided as to whether β^3 is a binary or single with the evidence pointing towards a pair of stars. β^{3A} Tuc is a m_v +5.06 A0 of 1.9 $D_\odot$ with a mass of 3.2 $M_\odot$. Its 10,000 K surface has a luminosity of 17 $L_\odot$. Its companion, if it exists, is likely to be identical and at least one authority has calculated the orbital period to be 3.492 years. β^3 is in orbit around the β^1 + β^2 combination with a period of at least 905,000 years. The entire system is about 145 ly away.

ζ **Tucanae** is a Sun-like star just 28 ly away. Almost as big and as massive as the Sun, 0.95 $D_\odot$, 0.91 $M_\odot$, it is about 20% more luminous at 1.22 $L_\odot$. An F9.5, possibly a G0, it has a surface temperature of 5,970 K and appears as a m_v +4.23 point of light which would only brighten to M_v +5.0 at 10 pc, the star lying close

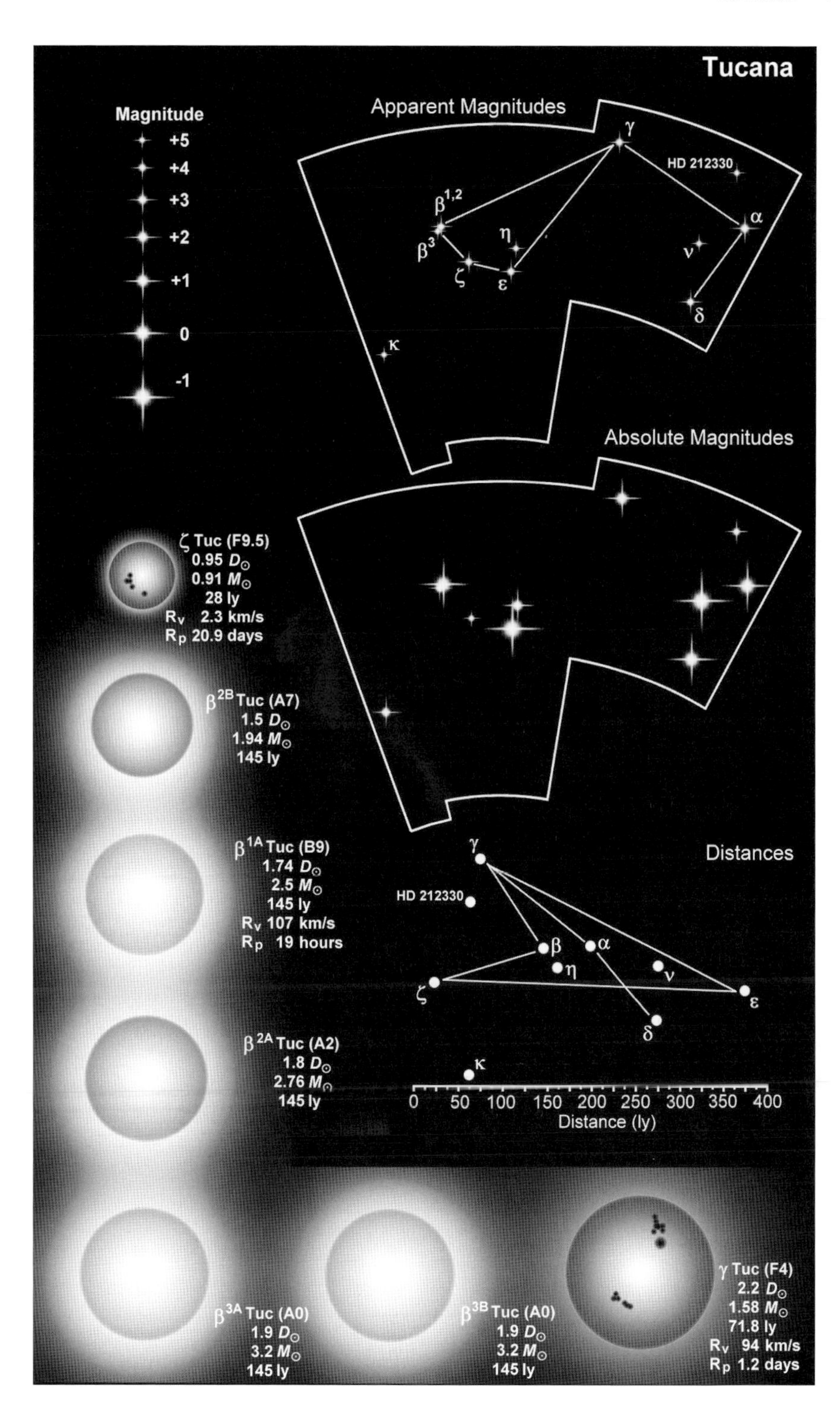
Tucana
Magnitude
+5
+4
+3
+2
+1
0
-1
Apparent Magnitudes
γ
HD 212330
β1,2
β3
η
α
ν
ζ
ε
δ
κ
Absolute Magnitudes
ζ Tuc (F9.5)
0.95 D☉
0.91 M☉
28 ly
Rv 2.3 km/s
Rp 20.9 days
β2B Tuc (A7)
1.5 D☉
1.94 M☉
145 ly
β1A Tuc (B9)
1.74 D☉
2.5 M☉
145 ly
Rv 107 km/s
Rp 19 hours
Distances
β2A Tuc (A2)
1.8 D☉
2.76 M☉
145 ly
0 50 100 150 200 250 300 350 400
Distance (ly)
β3A Tuc (A0)
1.9 D☉
3.2 M☉
145 ly
β3B Tuc (A0)
1.9 D☉
3.2 M☉
145 ly
γ Tuc (F4)
2.2 D☉
1.58 M☉
71.8 ly
Rv 94 km/s
Rp 1.2 days

to the 32.6 ly used to calculate absolute magnitude. Spinning at 2.3 km/s it takes almost 21 days to rotate once, beating the Sun by 4 days. It is somewhat younger than the 4,567 million years of our own star and is estimated to be between 2,100 and 3,000 million years old. There are indications of a debris disk orbiting the ζ Tuc but, as yet, no planets have been detected.

κ Tucanae is a five star system 66.6 ly away. **κ^{Aa} Tucanae** is an F6 of 1.25 $M_\odot$. With a luminosity of 3.54 $L_\odot$ and temperature of 6,600 K it takes only 12 hours to rotate once on its axis, its rotational velocity being an unusually high 111 km/s, almost four times greater than the average for its class. The result is a somewhat oblate star, its equator a bulging 1.1 $D_\odot$ while its polar length is probably about 0.8 $D_\odot$. Its magnitude is m_v +5.10 but is brightened to m_v +4.95 by the presence of a 0.77 $M_\odot$ astrometric companion, **κ^{Ab} Tucanae,** which is probably an F5. Just 4.92″ away is the m_v +7.30 **κ^{B} Tucanae,** a G5 of 0.9 $M_\odot$. The two are locked in an 857 year long orbit. About 5.3′ to the north west in an unstable 252,600 year orbit around κ^{A}- κ^{B} is a pair of orange dwarfs, κ^{Ca} and κ^{Cb}. **κ^{Ca} Tucanae** is a m_v +8.26, 1.07 $D_\odot$ K2 of 0.74 solar mass. Its binary companion, **κ^{Cb} Tucanae,** is separated by 1.12″. A K2 of 0.65 $M_\odot$ it has an orbital period of 86.2 years.

Also at 66.8 ly distant, though at the opposite end of the constellation, is the very close solar analog **HD 212330.** Shining at m_v +5.32 this G2 is 1.0 $D_\odot$ across and 1.1 $M_\odot$. It is younger than the Sun at 5,440 million years and subsequently is more luminous at 2.53 $L_\odot$. It rotates at just about the same speed as the Sun, 1.8 km/s, and has a surface temperature of 5,750 K. As yet no planets have been found.

Not much farther away is γ **Tucanae** at 71.8 ly. An F4 with a diameter of 2.2 $D_\odot$ and a mass of 1.58 $M_\odot$ the 10 $L_\odot$ gives rise to a m_v +3.99 star that would have an absolute magnitude of M_v +0.6. Another high velocity spinner at 94 km/s it turns once every 27.3 hours.

ν **Tucanae** is a red giant variable of 43 $D_\odot$ and lying at a distance of 273 ly. An M4 it belongs to the Lb pulsating category and has a complex series of at least 10 variable periods: 22.3, 24.4, 24.8, 25.1, 25.5, 33.8, 50.6, 80.1, 123.2 and 261.8 days during which its magnitude swings between m_v +4.75 and +4.93. Big enough to swallow more than 79,500 Suns it appears as a 59 $L_\odot$ star but given that most of its energy is radiated in the infrared the true luminosity is in the order of 600 $L_\odot$.

NGC 290 is not a particularly bright object and difficult to see clearly with a small telescope. The globular cluster **NGC 104** is one of the brightest in the sky though at least a 4″ (10 cm) telescope is needed to fully appreciate it. It is also known as **47 Tucanae**.

Open and globular clusters in Tucana

Name	Size arc min	Size ly	Distance ly	Apparent magnitude m_v	Notes
NGC 290	1.1′	65	200,000	+11.7	Open cluster
NGC 104	30.9′	120	13,400	+4.0	Globular cluster
NGC 362	12.9′	104	27,700	+6.6	Globular cluster

Ursa Major

Constellation:	Ursa Major	**Hemisphere:**	Northern
Translation:	The Great Bear	**Area:**	1,280 deg^2
Genitive:	Ursae Majoris	**% of sky:**	3.103%
Abbreviation:	UMa	**Size ranking:**	3rd

One of the best known constellations in the Northern Hemisphere the Great Bear is associated with several legends. Most people are familiar with the seven bright stars that make up the Big Dipper, Plow or Saucepan but the constellation is far more extensive making it the third largest in the sky.

α **Ursae Majoris**, Dubhe, is the farthest of the seven Plow stars at 124 ly, give or take 2 ly, and does not belong to the group of five stars that make up the inner part of the Plow. A distinct orange, indicative of its K0 spectral class, it is 14 $D_\odot$ across and rotates at just 1.6 km/s taking 443 days – 1.2 years – to turn once on its axis. Its mass comes in at 4 $M_\odot$. α UMa appears as a m_v +1.79 star but its real apparent magnitude is m_v +2.02, its brightness being bolstered by the presence of three other companions. The closest of these at 0.58″, 23 AU, is α^B UMa, an F0 class of 1.7 $M_\odot$. It takes 44.4 years to orbit the primary. Somewhat farther away at 4.7′ or 9,500 AU is α^{Ca} UMa, a 1.18 $M_\odot$ F8. Its orbital period is estimated to be 405,000 years around α^A and α^B UMa. α^{Ca} UMa is a close binary, its partner, α^{Cb} UMa, another F-class of 0.37 $M_\odot$ and in an orbit that takes only 6.04 days to go full circle.

Merak, β **Ursae Majoris**, is a much simpler star. Some 2.3 $D_\odot$ across and with a rotational period of 2.5 days – it spins at 46 km/s – β UMa is a typical A1 with a surface temperature of 8,900 K and a luminosity of 55 $L_\odot$ resulting in a m_v +2.35 star twinkling at us from its home 79.4 ly away. Possibly variable by 5/100th of a magnitude it is listed as NSV 5053 in the Newly Suspected Variable catalog. Its mass is about 2.28 $M_\odot$ and it is surrounded by a disk of dust stretching out to about 9 AU. As yet, no planets have been found.

At first glance γ **Ursae Majoris**, Phecda or Phad, is another ordinary, hot 9,500 K A0. Its mass is 2.5 $M_\odot$ and its diameter is 2.3 $D_\odot$. But its rotational velocity is 178 km/s, about 70 km faster than the average A0, and so it takes just 15.7 hours to turn once. More interestingly though is that it is one of only about 100 Ae stars: A-class stars that are encircled by a swirling cloud of gas. With a luminosity of 59 $L_\odot$ and apparent magnitude of m_v +2.39 its distance is typical of that of most of the Plow's stars: 83.7 ly.

Some 450 years ago the Danish astronomer Tycho Brahe recorded δ **Ursae Majoris,** Megrez, as a 2nd magnitude star, considerably brighter than the m_v +3.29 we see today. Lying 81.4 ly away Megrez is an A3 of just over two solar masses. Spinning at 233 km/s, 121 km/s above average, it has deformed into an oblate spheroid, probably 1.5 $D_\odot$ through the poles and 2.1 $D_\odot$ across its equator. Its luminosity is 24 $L_\odot$ but the Von Zeipel effect will have darkened its equator. It is considered to be 50 million years old.

ε **Ursae Majoris** or Alioth is an α^2 CVn rotating variable. About 3.8 $D_\odot$ across and rotating at 38 km/s the star turns once one its axis every $5^d\ 2^h\ 7.7^m$. As it does so light and dark patches appear and reappear causing its magnitude to fluctuate between m_v +1.76 and 1.78, the phenomena cause by concentrations of certain elements in the star's upper layers. Overall, its luminosity is 100 $L_\odot$, the A0 spectrum indicating a temperature of 9,400 K. ε UMa is 80.9 ly away.

To the north east of ε UMa is the 5th magnitude **78 Ursae Majoris** which is also known as **β 1082.** Although it is just half a light year farther away than ε at 81.4 ly their angular separation means that they are at least 1.4 ly apart and so are probably not a real pair. 78 UMa is a 1.6 $D_\odot$ F2 that weighs in at 1.41 $M_\odot$. It has a G6 companion of 0.8 $D_\odot$ and 0.9 $M_\odot$ in a 106.4 year long orbit during which time the stars' separation oscillates between 18 and 49 AU.

Second magnitude Mizar, ζ **Urase Majoris** and 4th magnitude Alcor, **80 Ursae Majoris** to the north east are generally considered to be a test for good eyesight, the two stars separated by 11.8′. A binocular or small telescope will reveal Mizar to have another companion, 4th magnitude Mizar B, a fact discovered by Galileo's student, Benedetto Castelli in 1617. We now know that both Mizar A and B, and 80 UMa, are all binary systems in their own right and may all be part of a gravitationally bound sextuplet arrangement. **Mizar A** (or ζ^A UMa, if you prefer) is a 1.6 $D_\odot$ A2 of 2.5 $M_\odot$ and 60 $L_\odot$. Its apparent magnitude is m_v +2.40, its absolute magnitude works out a full magnitude brighter at M_v +1.40. **Mizar B** (ζ^B UMa) is a 1.4 $D_\odot$ A1 of 1.94 $M_\odot$ and 13 $L_\odot$. Its apparent magnitude is m_v +3.95, its absolute magnitude coming in at M_v +2.3. Mizar A and B are separated by 14.4″ which translates to about 500 AU in real space, the two orbiting their barycenter with a period of at least 2,400 years. Mizar A, however, has a much closer companion: a 4th magnitude star of 2.5 $M_\odot$ that takes just 20.538 days to complete a full orbit. Similarly, Mizar B also has a binary companion: a 0.25 $M_\odot$ A7 just 0.033″ away and in a 175.55 day long orbit. 80 UMa, Alcor, a 1.7 $D_\odot$ A5 of 2.1 $M_\odot$ and 13 $L_\odot$, too has a binary mate, a tiny red dwarf, 0.25 $M_\odot$ and probably an M3. Its magnitude is m_v +9.19. They are parted by 1.11″, around 25 AU, and take at least 88.65 years to orbit one another. But there is a difficulty with this configuration. While the Mizars have been measured as being 78.1 ly away, Alcor appears to be significantly farther away at 81.1 ly, in which case it is not part of the Mizar arrangement. However, instrumental errors in measuring the distances of the stars mean that Mizar and Alcor could actually be as close together as 0.27 ly, in which case their orbital period will be at least 716,000 years, or they could be as far apart as 5 ly, meaning they are not in orbit.

The five inner stars of the Plow, β γ δ ε and ζ together with π^1, 37, 78 and 80 UMa belong to a cluster of stars that have a common origin and which are traveling through space together. Called the Ursa Major Moving Group, or **Collinder 285**, there are perhaps a few dozen other members scattered across the sky, including δ Aqr, γ Lep and ζ Boo, although the membership of some stars such as Sirius has been called into question. Like α UMa, **η Ursae Majoris** does not belong to Collinder 285. Marking the end of the handle of the Plow, Alkaid as it is otherwise known, is a 3.5 $D_\odot$ B3 and a hefty 6 $M_\odot$. Around 101 ly away

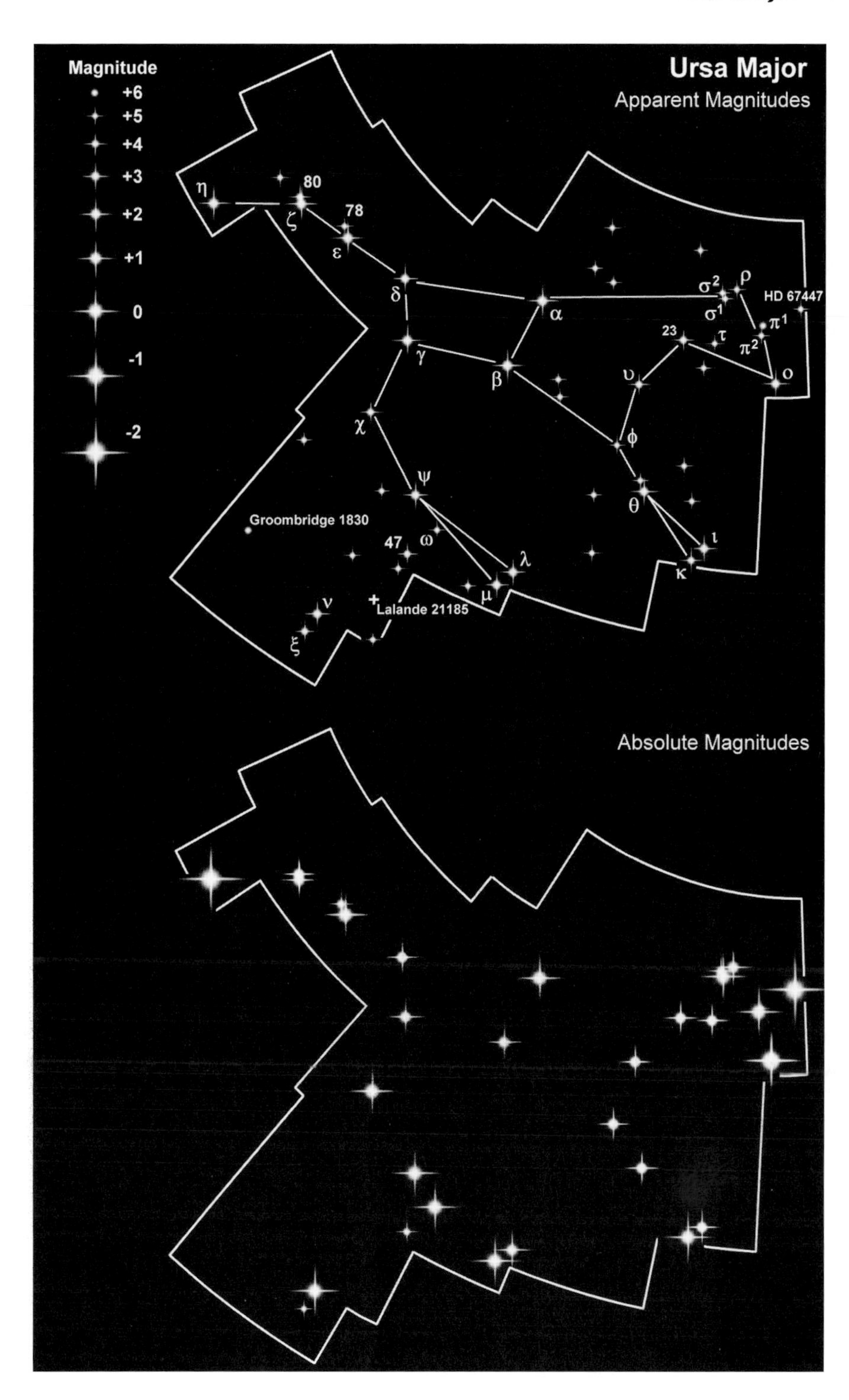
Magnitude
+6
+5
+4
+3
+2
+1
0
-1
-2
Ursa Major
Apparent Magnitudes
η
80
ζ
78
ε
δ
α
γ
β
χ
ψ
ω
47
λ
μ
ν
ξ
Groombridge 1830
Lalande 21185
σ²
ρ
σ¹
HD 67447
π¹
π²
τ
23
υ
ο
φ
θ
ι
κ
Absolute Magnitudes

such massive stars do not last long, perhaps just over 100 million years. With a luminosity of 141 $L_\odot$ and a surface temperature of 20,000 K η UMa appears as a m_v +1.85 star that would have an absolute magnitude of M_v -1.7. It spins as 150 km/s, taking just 28.3 hours to complete a full turn.

θ Ursae Majoris is a spectroscopic binary and possibly a triple star system and one of the closest stars in the constellation at 44 ly. The primary is a 2.5 $D_\odot$ A3 of 1.53 $M_\odot$ and 8 $L_\odot$, shining at m_v +3.17. The spectroscopic companion orbits just over 1 AU away with a period of 371 days. A m_v +13.8 M6 red dwarf appears to be lurking in the background at a distance of perhaps 95 AU. It is less that a third of the size of the Sun, 0.29 $D_\odot$ and with a mass of only 0.15 $M_\odot$. The orbital period, if it is truly a companion, will be in excess of 700 years.

In traditional depictions of the Great Bear, θ along with ι and κ UMa represent one of the front paws. **ι Ursae Majoris** is a quadruple, the primary being a 1.3 $D_\odot$ A7, ten times more luminous than the Sun and almost twice as massive at 1.94 $M_\odot$. About 5 AU away is a 0.82 $M_\odot$ spectroscopic companion about which nothing is known other than it is in an 11.02 year long orbit. Some 127 AU farther away, and with an orbital period of 817.9 years, is a pair of almost identical red dwarves. With apparent magnitudes of m_v +10.8 and +11.0 they are both 0.41 $M_\odot$ and keep an average 10 AU from each other in an orbit that runs full circle in 39.7 years. The entire system is just 47 ly away and has an apparent magnitude of m_v +3.11.

At m_v +3.55 **κ Ursae Majoris** is almost as bright as ι UMA but is exactly nine times farther away, 423 ly. It consists of an identical pair of A1 stars of 3.5 $D_\odot$ and 3.4 $M_\odot$ which are locked into a somewhat eccentric orbit that varies between 11 AU at periastron, their closest point, and 37 AU at apastron, their farthest. The orbital period is 35.6 years. Jim Kaler points out that the pair are a total of 4 $M_\odot$ too light indicating that something is amiss with the measurements.

The hind paw of the Great Bear is traced out by λ, μ and ψ UMa. **λ Ursae Majoris** glows at m_v +3.42 from its home 134 ly away, gradually drifting deeper into space at 18.1 km/s. It is an A2, about 480 million years old, with a diameter of 3.1 $D_\odot$ and a mass of 2.42 $M_\odot$, its luminosity coming in at 59 $L_\odot$. Nearby, at least on the celestial sphere, the red giant **μ Ursae Majoris** is slightly brighter at m_v +3.04. And much larger. Estimates range from 53 to 62 $D_\odot$ (0.5 to 0.6 AU). If the star rotates at half the speed of the Sun, 1 km/s, then it will take between 7.3 and 8.6 years to turn just once on its axis. An M0 class its luminosity is about 850 $L_\odot$ but two-thirds of that is radiated in the infrared. It is not alone having a companion in a 230 day orbit that averages 1.5 AU. Nothing else is known about the unseen consort. **ψ Ursae Majoris** is another giant star but 'only' 20 $D_\odot$ across. A K1 with a mass of 1.84 $M_\odot$ and luminosity of 100 $L_\odot$ (170 $L_\odot$ if you count the infrared glow) it has an apparent magnitude of m_v +2.99. Around 300 million years old it lies at a distance of 147 ly.

Somewhat younger at 100 million years old is the giant **ν Ursae Majoris.** About 66 $D_\odot$ across – the size of Mercury's orbit at perihelion and big enough to swallow up 287,496 Suns – this 5 $M_\odot$ K3 is already well into middle age. Its 1,350 $L_\odot$, only 40% of which is emitted as visible light, produces a m_v +3.48 star that

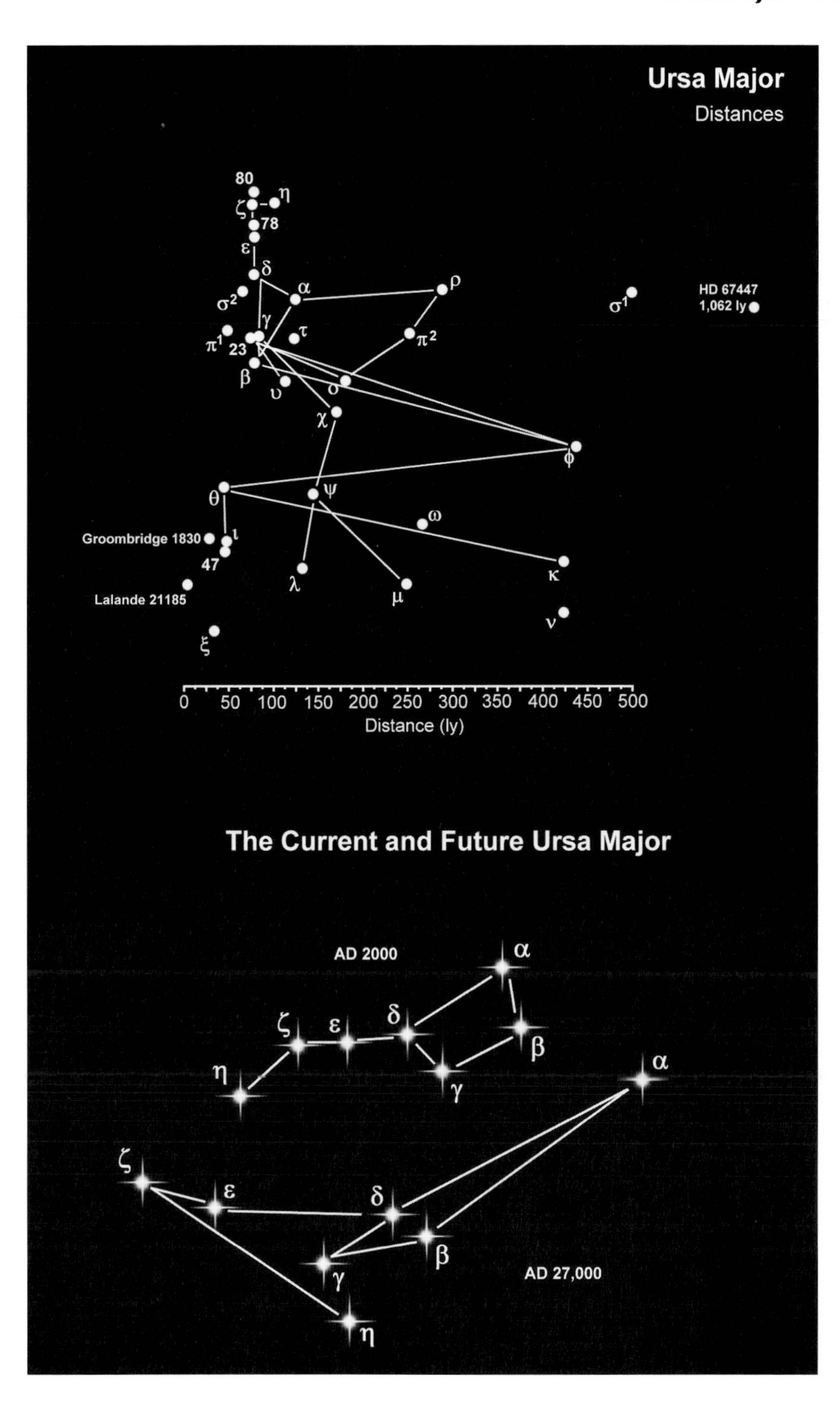
Ursa Major
Distances
80
ζ
η
78
ε
δ
α
ρ
σ2
γ
τ
π1
23
π2
β
υ
o
χ
φ
θ
ψ
ω
Groombridge 1830
ι
47
κ
λ
μ
Lalande 21185
ν
ξ
σ1
HD 67447
1,062 ly
0 50 100 150 200 250 300 350 400 450 500
Distance (ly)
The Current and Future Ursa Major
AD 2000
AD 27,000

would brighten to M_V -0.2 at 10 pc. Rotating once every 1,237 days (2.4 years) it spins slightly faster than the Sun: 2.7 km/s. Around 421 ly away it is accompanied by a m_v +10.1 solar analog, a G1 of 1.1 $D_\odot$ and 1.3 $M_\odot$ which averages 950 AU from the primary - 32 times the Sun-Neptune distance - and which takes more than 12,000 years to complete a single orbit.

ξ Ursae Majoris is a double binary system 34 ly from Earth. **ξ^{Aa} Ursae Majoris** is a 0.9 $D_\odot$ G0, similar to the Sun and with a mass slightly higher mass at 1.1 $M_\odot$ and a similar luminosity. Its M3 companion, **ξ^{Ab} Ursae Majoris,** is only a third of the mass of the Sun, 0.37 $M_\odot$, and is locked in a 1.834 year orbit. **ξ^{Ba} Ursae Majoris** is similarly a G0 of 0.9 $D_\odot$ and 0.9 $M_\odot$. Its binary companion is also an M-class dwarf but much less massive at just 0.15 $M_\odot$ and orbits much closer taking just 3.98 days to complete a full circle. The two binary pairs are in a 59.878 year long orbit during which they close in on each other to 13.4 AU and then separate by 29.6 AU.

A massive Sun *en route* to becoming a red giant probably best describes **o Ursae Majoris**. A very yellow G5 o UMa has a diameter of 14 $D_\odot$ and a mass of 2.96 $M_\odot$. From a distance of 184 ly its 117 $L_\odot$ yield a star of m_v +3.30 fading, without any discernable regularity, to m_v +3.36. Rotating at 3.9 km/s it takes 181.7 days to spin once on its axis. Originally a B-class dwarf o UMa is now a middle aged star around 350 million years old. At 7.1″ (PA 192°), about 400 AU, an M1 red dwarf takes more than 4,000 years to orbit the primary. Virtually nothing else is known about the companion.

Naked-eye rotating variables of the BY Draconis type are quite rare and at m_v +5.64 to +5.72 **π^1 Ursae Majoris** is just about visible for most people. Very similar to the Sun π^1 UMa is a G1.5 of 0.95 $D_\odot$ and 1.1 $M_\odot$. Just 46.5 ly away its variability is believed to be due to huge starspots or chromospheric activity.

A little over half a degree south is the totally unrelated star **π^2 Ursae Majoris.** More than five times farther away at 252 ly π^2 UMa is a m_v +4.60 K1 giant of 16 $D_\odot$ and 1.6 $M_\odot$. Nothing unusual in that except that it has a planet, a 7.1 M_J in a 269 day long orbit which is known by its Flamsteed designation of 4 UMa b.

In classical drawings of the Great Bear one of its ears is marked out by a small triangle of stars - ρ, σ^1 and σ^2 UMa - although they are not physically related **ρ Ursae Majoris** is an M3 red giant of uncertain dimensions; estimates range from 20 through 32 to 59 $D_\odot$. There is similar uncertainty over its stability, revealed in its magnitude which is suspected but not yet proven to vary between m_v +4.75 and +4.82. It is 287 ± 15 ly away. Almost a full degree to the south east is **σ^1 Ursae Majoris** which, despite its designation, is actually the fainter of the two σ stars. Twinkling at a steady m_v +5.15 from 499 ly away it is a K5 rotating slower than the Sun at 1.2 km/s. Its diameter is variously quoted as being 29 or 48 $D_\odot$ which means that its rotational period is between 1,244 and 2,059 days (3.4 to 5.6 years). Slightly brighter at m_v +4.82 is **σ^2 Ursae Majoris** a little more than 19.5′ to the north west. By far the closest of the trio, only 66.7 ly away, this F6 star has a diameter of 1.5 $D_\odot$ and mass of 1.43 $M_\odot$. Rotating at exactly one half of the Sun's velocity, 1 km/s, it spins once every 76 days (Sun 25 days). Keeping it company is a K2 dwarf of m_v +8.16. Currently 3.5″ apart during the course of its

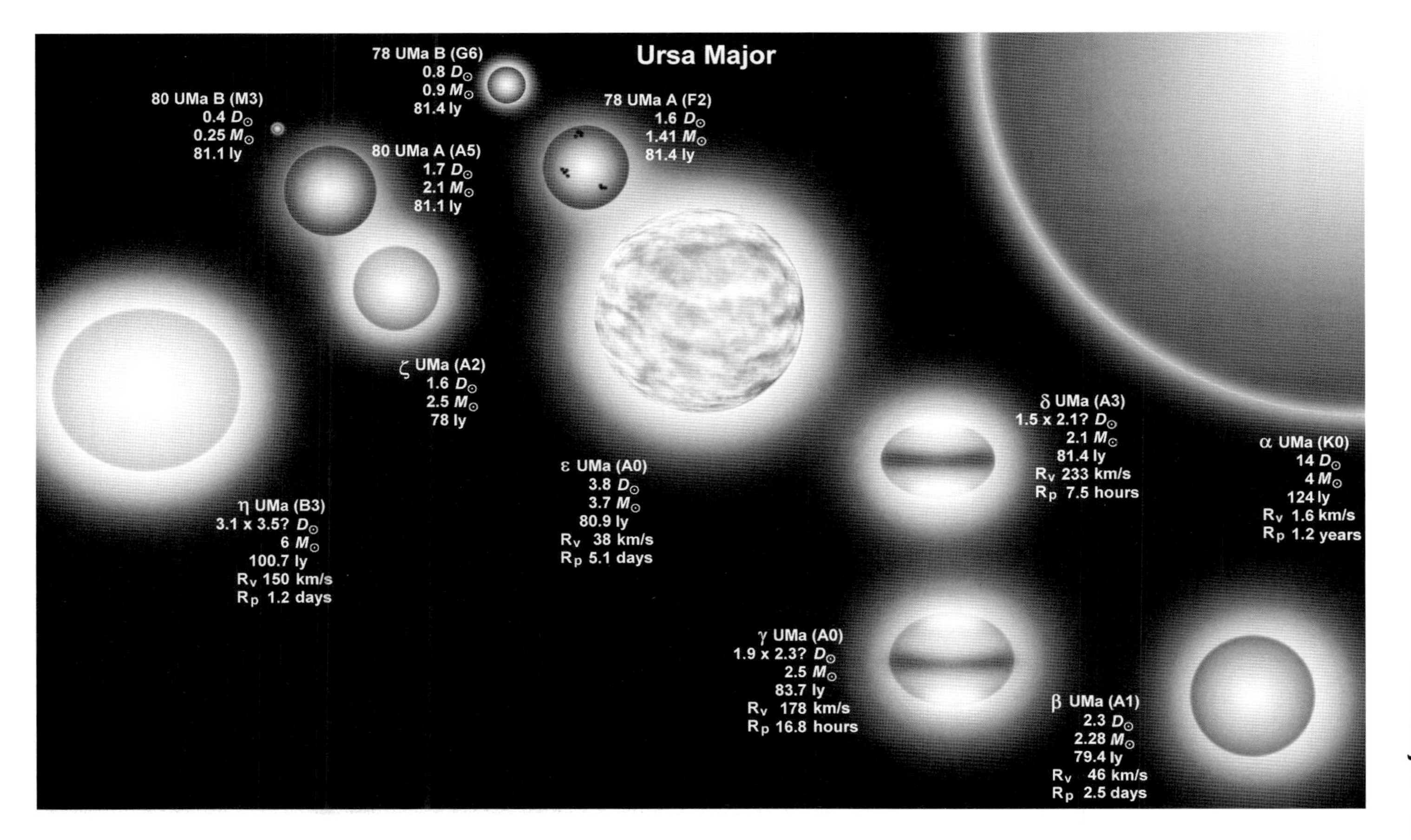

Ursa Major
78 UMa B (G6)
0.8 D☉
0.9 M☉
81.4 ly
80 UMa B (M3)
0.4 D☉
0.25 M☉
81.1 ly
78 UMa A (F2)
1.6 D☉
1.41 M☉
81.4 ly
80 UMa A (A5)
1.7 D☉
2.1 M☉
81.1 ly
ζ UMa (A2)
1.6 D☉
2.5 M☉
78 ly
δ UMa (A3)
1.5 x 2.1? D☉
2.1 M☉
81.4 ly
Rv 233 km/s
Rp 7.5 hours
α UMa (K0)
14 D☉
4 M☉
124 ly
Rv 1.6 km/s
Rp 1.2 years
ε UMa (A0)
3.8 D☉
3.7 M☉
80.9 ly
Rv 38 km/s
Rp 5.1 days
η UMa (B3)
3.1 x 3.5? D☉
6 M☉
100.7 ly
Rv 150 km/s
Rp 1.2 days
γ UMa (A0)
1.9 x 2.3? D☉
2.5 M☉
83.7 ly
Rv 178 km/s
Rp 16.8 hours
β UMa (A1)
2.3 D☉
2.28 M☉
79.4 ly
Rv 46 km/s
Rp 2.5 days

1,067 year long orbit the two stars will come to within 22.5 AU of one another (rather farther apart than Uranus is from the Sun) but will also separate by 157.5 AU (more than 5 times the distance of Neptune from the Sun).

υ **Ursae Majoris** is one of several δ Scuti pulsating variables, flickering between m_v +3.68 and +3.86, putting it towards the top end of the class which has amplitudes of between 0.01 and 0.19 magnitudes. The period is $3^h\ 7^m$. A full 2 $D_\odot$ across and 1.89 $M_\odot$ this F2 has an M0 red dwarf of about 0.5 $M_\odot$ with an orbital period of at least 5,200 years, the two staying on average 410 AU apart.

The m_v +4.56 of ϕ **Ursae Majoris** is deceptive, the magnitude being the combined light from two identical A3 stars of about m_v +5.34 each. Each with a diameter of twice that of the Sun and a mass of three times solar, they orbit each other with a period of 105.5 years reaching 79 AU at apastron and less than half that – 30 AU – at periastron. The system is 437 ly away.

The K0.5 giant χ **Ursae Majoris** is 21.4 $D_\odot$ with a mass of 1.95 $M_\odot$. Some 196 ly away it rotates at almost half the speed of the Sun, just 1.18 km/s, taking 918 days (2.5 years) to complete a single turn. Its luminosity is 98 $L_\odot$ but that can be doubled taking into account that half its output is in the infrared. The result is a m_v +3.69 star that occasionally dims to +3.72.

Try as you might you will not be able to find **10 Ursae Majoris** in Ursa Major. When the IAU redrew the boundaries in the 1920s, 10 UMa became part of Lynx. Just to confuse matters further, 41 Lyncis ended up in Ursa Major. The entry on Lynx has further details. Another star that migrated to Ursa Major is **55 Camelopardalis**, a G8 supergiant 54 $D_\odot$ across and with a luminosity of 630 $L_\odot$. Now usually referred to by its Henry Draper catalog number, **HD 67447,** it rotates at half the speed of the Sun taking 7.5 years to spin once on its axis. It is 1,062 ly away.

23 Ursae Majoris is a binary consisting of an F0 of 2.2 $D_\odot$ and 1.75 $M_\odot$ and a 0.63 $M_\odot$ K7 dwarf. Separated by about 530 AU the two stars, which are m_v +3.65 and +9.19, take at least 8,000 years to orbit one another. The primary is slightly variable between m_v +3.65 and +3.68. The system is about 1,200 million years old.

The solar analog **47 Ursae Majoris** has a family of at least three planets. A G1 it is about 24% larger than the Sun, very slightly more massive, 1.06 $M_\odot$, and more than 1½ times more luminous at 1.55 $L_\odot$. It rotates at 5 km/s, completing a single rotation in 12.6 days (Sun = 25 days). The closest of the three detected planets orbits at a distance comparable to the distance of the inner Asteroid Belt from the Sun while the next farthest planet is about the distance of the outer Asteroid Belt. The third planet has a mean distance just beyond that of Saturn although its eccentric orbit means that it varies between 9.74 and 13.46 AU (Saturn = 9 to 10 AU). Various attempts have been made to estimate the age of the system. Depending on which method is used the star could be as young as 2,512 million years or as old as 11,654 million years. The table below shows some of the estimated age ranges. Because of the possibility of Earth-like planets in orbit around this Sun-like star a Message to Extra-Terrestrial Intelligence (METI) was transmitted on 3 September 2001. It should reach there in July 2047.

Age estimates for 47 UMa (in million years)

Youngest	'Best'	Oldest	Range
2,512	4,677	8,710	6,198
3,890	6,310	8,318	4,428
5,440	6,480	7,920	2,480
6,728	9,191	11,654	4,926

For most people the m_v +6.45 star **Groombridge 1830** (**HD 103095**) is too faint to see without optical aid but it is an interesting object. Discovered by Stephen Groombridge (1755-1834) it was reckoned to have the highest proper motion at 7.04″ per year or 1° in 511 years. The later discoveries of Kapteyn's Star in Pictor (8.70″) and Barnard's Star in Ophiuchus (10.29″) relegated it to third place. What makes Groombridge 1830 particularly interesting is its distance of 29.86 ly. Kapteyn's Star is just 12.7 ly away while Barnard's Star is a mere 5.98 ly so we would expect high proper motions for these two stars, but Groombridge 1830 appears to be traveling at a rate of knots, circling the entire celestial sphere in just 185,000 years. In fact Groombridge 1830 is an exceptionally fast star with a space velocity of 312 km/s, indicating it is one of the Galaxy's Halo stars. A G8 it is not particularly big or massive, 0.66 $D_\odot$ and 0.66 $M_\odot$, and it is just 0.24 $L_\odot$. Reports of it having a companion seem to be spurious. It did, however, brighten to m_v +5.85 in 1939 and it has been suggested that it may be a superflare star, a flare event being mistaken for a companion.

At 8.29 ly **Lalande 21185** (**HD 95735**) is the 4th or 6th closest star to the Sun, depending on whether you class the α Centauri system as being one or three stars. An M2 red dwarf of 0.47 $D_\odot$ and 0.43 $M_\odot$ it is a dim 0.025 $L_\odot$ and shines at m_v +7.49 fading to m_v +10.48 at 10 pc. A cool 3,380 K it rushes across the celestial sphere at 5″ per annum and has a space velocity of 102 km/s. Several claims that it has a planetary system have been discounted. Believed to be between 5,000 and 10,000 million years old it is a suspected flare star and is regarded as a BY Draconis variable.

Planetary systems in Ursa Major

Star	$D_\odot$	Spectral class	ly	m_v	Planet	Minimum mass	q	Q	P
π^2 UMa	16	K1	252	+4.60	4 UMa b	7.1 M_J	0.49	1.25	269.3 d
47 UMa	1.24	G0	45.6	+5.10	47 UMa b	2.53 M_J	2.03	2.17	2.95 y
					47 UMa c	0.54 M_J	3.25	3.95	6.55 y
					47 UMa d	1.64 M_J	9.74	13.46	38.34 y

Ursa Minor

Constellation:	Ursa Minor	**Hemisphere:**	Northern
Translation:	The Little Bear	**Area:**	256 deg^2
Genitive:	Ursae Minoris	**% of sky:**	0.621%
Abbreviation:	UMi	**Size ranking:**	56th

Ursa Minor is one of Ptolemy's original 48 constellations and is associated with the fable of Calisto (or Callisto) in which it is referred to as Arcas.

α **Ursae Minoris** is just 44′ from the North Celestial Pole (NCP) and is known by several names including Polaris, the Pole Star, the North Star, Stella Polaris, Lodestar and Cynosure (which actually means 'dog's tail'). Its altitude above the northern horizon in degrees just about equals the observer's latitude as well as indicating the direction of north and in previous centuries was used as a natural navigation aid. With a diameter 46 times larger than our own Sun, Polaris is a yellowish-white supergiant of 4.57 $M_\odot$ and 2,290 $L_\odot$. Its rotational period is 155.2 days, the star spinning at 15 km/s.

Ask the proverbial man-in-the-street what he can tell you about Polaris and he'll probably say it is the brightest star in the sky. In fact Polaris is a modest m_v +1.92 at best though its absolute magnitude would be M_v -4.6, or about as bright as Venus. It is a Cepheid-type variable, dimming to magnitude +2.07 with a period of $3^d\ 23^h\ 16^m$. Recent research suggests that Polaris may have been a 3rd magnitude object at the time of Ptolemy in the 1st Century AD. In 1779 William Herschel discovered that Polaris has a companion, α^B UMi, a m_v +8.59 F3 star of 1.39 $M_\odot$ which lies at a distance of about 2,400 AU. Seen through a telescope the two stars are 17.8″ apart with the fainter component lying at PA 216°. Their orbital period is estimated to be 48,312 years. There is at least a one other star in the system, α^c UMi, a m_v +7.43 F7 of 1.25 $M_\odot$ which takes 29.59 years to complete an orbit, the two stars varying in distance between 6.7 and 27 AU. There is some indication that Polaris and its companions may actually be part of an open cluster of A and F type stars.

Polaris lies at a distance of 431 ly and appears to be hurtling towards us at about 17 km/s. It will be closest to the NCP in AD 2102 but it is only one of a number of 'pole stars'. As the Earth spins on its axis the North Celestial Pole appears to gyrate around the heavens taking 25,800 years to complete a full circle. Other pole stars include β and γ UMi between 1,500 BC and AD 500, Alderamin in Cepheus, Vega in Lyra (AD 14,000) and Thuban in Draco (2,300 BC).

β Ursae Minoris, or Kochab, is sometimes referred to as the 'Guardian of the Pole'. Of the 16 naked eye stars in the constellation, six are orange giants with β UMi, a K4, being the closest at 126 ly. It is 30 $D_\odot$ across and has a luminosity of 500 $L_\odot$ although only 184 $L_\odot$ is visible to the naked eye, the rest of the output being in the infrared. Suspected of being variable between m_v +2.02 and +2.08 it is also cataloged as NSV 6846. It rotates slightly slower than the Sun, 1.7 km/s, and subsequently takes 893 days – 2.4 years – to turn once.

Magnitude
+6 +5 +4 +3 +2 +1 0 -1 -5
Ursa Minor
Apparent Magnitudes
Absolute Magnitudes
α
North Celestial Pole
δ
ε
ζ
η
θ
4
5
β
γ
11
HD 118904
HD 136064
RR
Distances
α
δ
ε
ζ
η
4
θ
5
β
γ
11
HD 118904
HD 136064
RR
0 50 100 150 200 250 300 350 400 450 500 550 600 650 700 750 800 850
Distance (ly)

Ursa Minor
δ UMi (A1)
2.2 x 2.8? D⊙
2.5 M⊙
183 ly
Rv 180 km/s
Rp 19 hours
α UMi (F7)
46 D⊙
4.57 M⊙
431 ly
Rv 15 km/s
Rp 155.2 days
ε UMi (G5)
9 - 18 D⊙
3.6 M⊙
347 ly
Rv 23 km/s
Rp 9.8 - 39.6 days
ζ UMi (A3)
4.5 x 6? D⊙
3.4 M⊙
376 ly
Rv 210 km/s
Rp 1.4 days
β UMi (K4)
30 D⊙
5.9 M⊙
126 ly
Rv 1.7 km/s
Rp 2.5 years
η UMi (F5)
2 D⊙
1.52 M⊙
100.7 ly
Rv 75 km/s
Rp 1.3 days
γ UMi (A3)
7.9 D⊙
7.4 M⊙
480 ly
Rv 165 km/s
Rp 2.4 days

γ **Ursae Minoris** is another variable, this time of the δ Scuti type. Switching between m_v +3.04 and +3.09 with a period of 3^h 25^m 56^s this A3 is 7.9 $D_\odot$ and 1,110 $L_\odot$. A fast spinner its 165 km/s means it takes 2.4 days to rotate once. It is 480 ly away.

Another star with a high rotational velocity is δ **Ursae Minoris**. An A1 it spins at 180 km/s turning what was once an almost perfect sphere into an oblate mass with a polar length of perhaps as short as 2.2 $D_\odot$ while its equator may bulge to 2.8 $D_\odot$. With a temperature of 9,000 K δ UMi has a mass of 2.5 $M_\odot$ and a luminosity of 47 $L_\odot$. From its home 183 ly away it glows at a rock steady m_v +4.34 but would brighten to M_v +1.2 at 10 pc.

The G5 giant ε **Ursae Minoris** is a triple star system that is both an Algol-type and an RS CVn eclipsing binary. Some 347 ly from Earth the star appears to vary between m_v +4.19 and +4.23 with a period of 39^d 11^h 33^m. Estimates of its size vary from 9 to 18 $D_\odot$ while its mass is estimated to be around 3.6 $M_\odot$. Its rotational velocity is 23 km/s so it takes between 19.8 and 39.6 days to rotate once depending on its diameter. At 76.9″ an m_v +11.2 companion can be seen. Probably a K0 dwarf it lies at an average distance of 8,100 AU – too far away to be the cause of the observed variability. That particular phenomenon is due to a spectroscopic binary component which not only eclipses the primary but is also sufficiently close, just 0.36 AU, as to disrupt its magnetic field resulting in high levels of starspots and other chromospheric activity.

ζ **Ursae Minoris** is another fast spinning A-class star. With a polar diameter of around 4.5 $D_\odot$ and an equatorial diameter of about 6 $D_\odot$ the star takes 1.4 days to turn once on its axis, rotating at 210 km/s. Its mass comes in a 3.4 $M_\odot$ and its luminosity 210 $L_\odot$ producing a m_v +4.27 star twinkling at us from 376 ly away.

Apart from Polaris there are two other naked eye F-class stars in Ursa Minor. η **Ursae Minoris** is a 2 $D_\odot$ F5 of 1.52 $M_\odot$ and with a luminosity of 7.5 $L_\odot$. It is almost dead on 5th magnitude (m_v +4.96) and rotates at 75 km/s – about twice the average velocity for its spectral group – yielding a rotational period of 32.4 hours. It resides 100.7 ly away, somewhat farther than the 82.5 ly of **HD 136064**, the other F-class star. Weighing in at 1.37 $M_\odot$ it has a diameter of 1.4 $D_\odot$ and a luminosity of 4.5 $L_\odot$. About 100 degrees cooler than η UMi at 6,300 K it is classed as an F9. It is closing in on us at 46.8 km/s.

More than a third of Ursa Minor's stars are K-class orange giants. β UMi was mentioned above. θ **Ursae Minoris,** a K5, is the most distant at 832 ly, the largest at 48 $D_\odot$, the most luminous, 528 $L_\odot$, and the fastest spinner, its 3.1 km/s rotational velocity resulting in a rotational period of 784 days, more than 2 years. At 500 ly, the K3 giant **4 Ursae Minoris** is 28 $D_\odot$ across and has a luminosity of 225 $L_\odot$. It rotates at 2.2 km/s and takes 644 days (1.8 years) to rotate once. Drifting away from us at 10.5 km/s it is accompanied by a spectroscopic companion in a 605.8 day long orbit. **5 Ursae Minoris** is 345 ly away and is a K4 of 16 $D_\odot$ and 180 $L_\odot$. It's rotational period is 426 days (1.2 years) spinning at 1.9 km/s. The slowest rotating of the K-class stars is **11 Ursae Minoris.** Its 1.5 km/s coupled with a diameter of 37 $D_\odot$ results in a period of 1,248.5 days or 3.4 years.

It is a K4 at 390 ly. Not much farther is the last of the class, **HD 118904,** a K2 of 17 $D_{\odot}$, 77 $L_{\odot}$ and 401 ly. It is always worth considering whether stars that are similar distances are related. In the case of 11 UMi and HD 118904, the former is heading our way at 16.1 km/s and increasing in both Right Ascension and Declination, while the latter is heading away from us at 14.9 km/s and is decreasing in Right Ascension and Declination. They seem to be just stars 'passing in the night'.

RR Ursae Minoris is a semi-regular variable swinging between m_v +4.53 and m_v +4.73 with a period of 43^d 7^h 12^m. Lying somewhere between 374 and 422 ly it is an M4.5 red supergiant of 51 $D_{\odot}$ (71 million km or about half the size of Mercury's orbit).

Vela

Constellation:	Vela	**Hemisphere:**	Southern
Translation:	The Sail	**Area:**	500 deg^2
Genitive:	Velorum	**% of sky:**	1.212%
Abbreviation:	Vel	**Size ranking:**	32nd

Originally part of the much larger constellation of Argo Navis, the ship of the Argonauts, when the constellation was carved up by the IAU in the 1920s it lost its α and β stars so the luminary is now γ. Similarly missing are ε, ζ, η, θ and ι.

γ Velorum is a double binary system that contains the brightest and possibly closest Wolf-Rayet star known. To the naked eye γ Vel appears as a single star of m_v +1.81 but a small telescope or a good binocular will split the star into two separated by 41.2″ at PA 220°. **γ^1 Velorum** is a B1 with a luminosity of 2,860 $L_\odot$ and a diameter pf 2.7 $D_\odot$. Rotating at 115 km/s it takes 28.5 hours to spin once on its axis. Very little is known about its spectroscopic companion other than the fact that it is in a 1.4826 day long orbit. The two stars appear as a m_v +4.16 star. Of rather more interest is **γ^2 Velorum**, which is another spectroscopic binary. The main component is the brightest naked eye Wolf-Rayet stars in the sky (the others are **HD 152408** in Scorpius at m_v +5.29 and **θ Muscae** at m_v +5.88). With a spectral class of WC8, indicating it is of the carbon-rich variety (those that are nitrogen-rich are classed as WN stars) γ^{2A} Velorum is at least 17 $D_\odot$ across, has a mass of 15 $M_\odot$ and is superluminous, estimated to be 100,000 $L_\odot$. Its temperature is somewhere between 57,000 K and 100,000 K and it emits a ferocious stellar wind that carries mass away from the star at a rate of 100 million times greater than the solar wind. It is slightly variable between m_v +1.81 and +1.87 with a period of just 2^m 34^s. Its binary companion, γ^{2B} Velorum, is an O7.5 which is almost as large, 13 $D_\odot$, twice as massive, 30 $M_\odot$, and almost twice as luminous, 180,000 $L_\odot$, but relatively 'cool' at 32,500 K. Separated by just 1.2 AU the two stars orbit one another with a period of 78.5 days. The distance between the γ^1 pair and the γ^2 pair is thought to be around 15,000 AU. Estimating the distance of the system has proven to be problematic with 840 ly looking favorite although some authorities suggest they could be as far as 1,200 ly. If γ^1 Vel was 10 pc away it would appear as a M_v -3.9 but outshone by γ^2 Vel which would be a very impressive M_v -5.6.

δ Velorum is a triple star system at a more certain 79.7 ly. The primary, **δ^{Aa} Velorum**, is a m_v +2.53 A1 of 2.6 $D_\odot$ and 2.98 $M_\odot$. As the 'Aa' superscript suggests it has a companion, **δ^{Ab} Velorum,** a spectroscopic component thought to be a 1.25 $M_\odot$ F6 which orbits the primary at a distance of about 0.5 AU taking 45.15 days to complete a single orbit. The third star in the system, **δ^B Velorum,** is much farther out, its orbit varying between 26 and 72 AU and with a period of 142 years. It is a m_v +3.04 A5 of 2.54 $M_\odot$. Altogether the three components sparkle as a single m_v +1.99 star which, in about 7,000 years from now, will pinpoint the position of the South Celestial Pole.

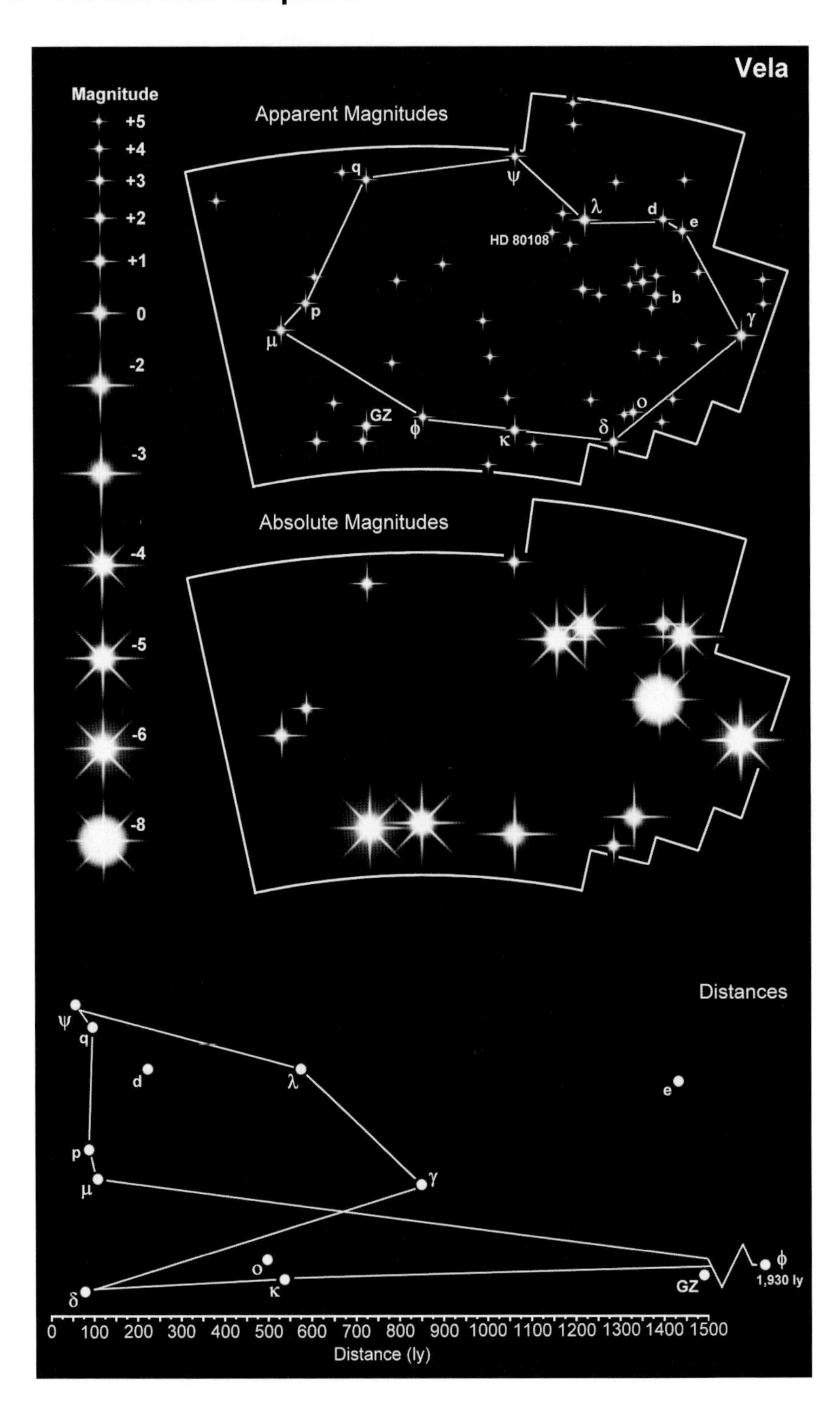
Vela
Magnitude
+5
+4
+3
+2
+1
0
-2
-3
-4
-5
-6
-8
Apparent Magnitudes
q
ψ
λ
d
e
HD 80108
b
p
μ
γ
GZ
φ
κ
δ
o
Absolute Magnitudes
Distances
ψ
q
d
λ
e
p
μ
γ
o
κ
δ
GZ
φ
1,930 ly
0 100 200 300 400 500 600 700 800 900 1000 1100 1200 1300 1400 1500
Distance (ly)

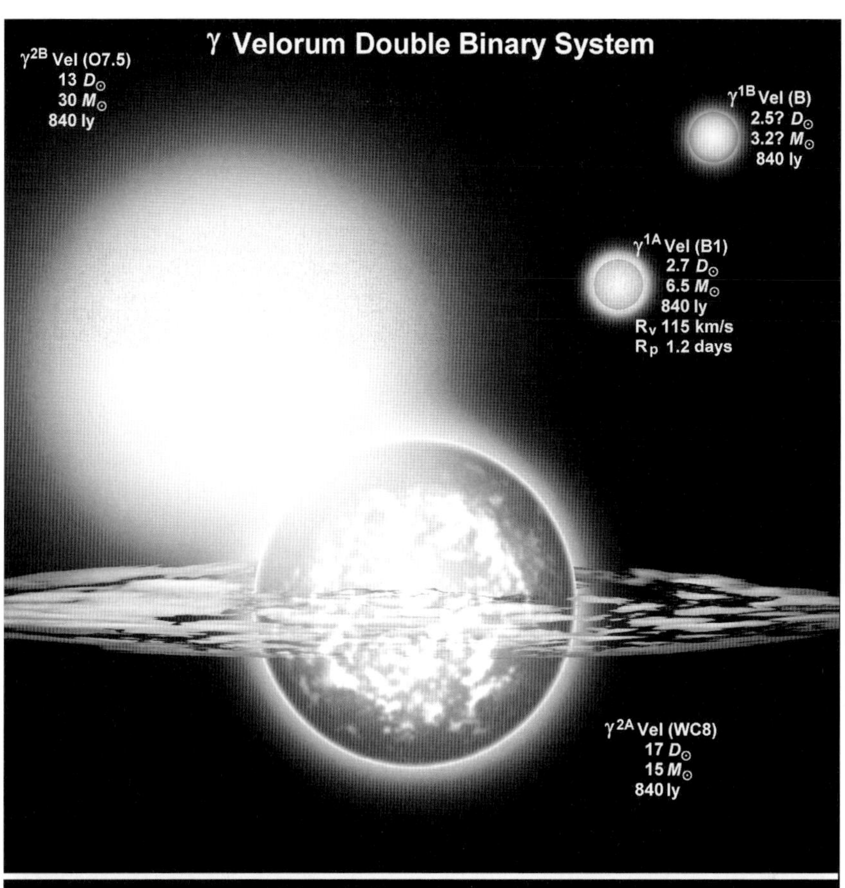
γ Velorum Double Binary System
γ2B Vel (O7.5)
13 D☉
30 M☉
840 ly
γ1B Vel (B)
2.5? D☉
3.2? M☉
840 ly
γ1A Vel (B1)
2.7 D☉
6.5 M☉
840 ly
Rv 115 km/s
Rp 1.2 days
γ2A Vel (WC8)
17 D☉
15 M☉
840 ly

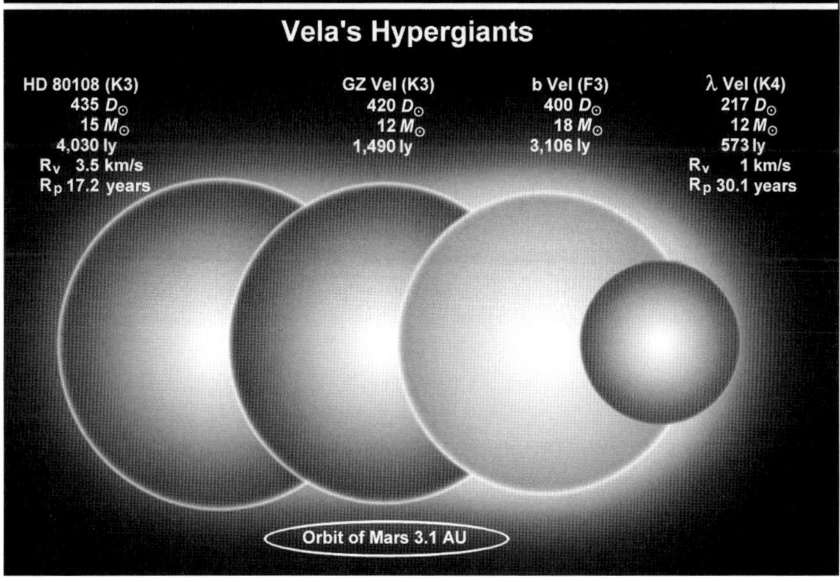
Vela's Hypergiants
HD 80108 (K3)
435 D☉
15 M☉
4,030 ly
Rv 3.5 km/s
Rp 17.2 years
GZ Vel (K3)
420 D☉
12 M☉
1,490 ly
b Vel (F3)
400 D☉
18 M☉
3,106 ly
λ Vel (K4)
217 D☉
12 M☉
573 ly
Rv 1 km/s
Rp 30.1 years
Orbit of Mars 3.1 AU

κ Velorum is a 9.1 $D_\odot$ star some 539 ly away. A B2 with a luminosity of 18,400 $L_\odot$ its mass comes in at about 10.5 $M_\odot$. Taking 8.4 days to turn on its axis, rotating at 55 km/s, κ Vel has a spectroscopic companion in a 116.65 day orbit. The two are separated by an average 0.5 AU.

δ and κ Velorum together with ι and ε Carinae make up the False Cross asterism that is often mistaken for Crux (the Southern Cross).

λ Velorum is as big as the Earth's orbit. Some 217 $D_\odot$ across and with an estimated mass of 12 $M_\odot$ this lumbering K4 supergiant takes 30 years to turn once on its axis, its rotational velocity just one half that of the Sun's: 1 km/s. An Lc pulsating variable its magnitude fluctuates between m_v +2.14 and +2.30 with no discernable period. Distance estimates put it at 573 ly. Its absolute magnitude is M_V -4.4, on par with Venus.

μ Velorum is definitely a binary and could be a triple star. The main star, **μ^A Velorum**, is a m_v +2.72 G6 giant. Exactly how much of a giant no one is absolutely sure but the figures most often quoted are 11, 13.5 and 15 $D_\odot$. It appears to be rotating at 6 km/s so a single spin on its axis could take 92.8, 113.9 or 126.5 days. Just 2.1″ away (PA 51°) is the m_v +5.59 yellow G2 dwarf, **μ^B Velorum.** Their orbit is highly eccentric. Over a period of 138 years they get as close as 8 AU and separate by as much as 93 AU before returning to their start position. But nothing about the system really adds up. The total mass of the stars should be around 6.8 $M_\odot$ which is far too high. And the spectrum is a mess. Enter a spectroscopic companion around μ^A Vel, possibly an F8 dwarf in a 116.24 year orbit, and the masses might work out at 2.92 $M_\odot$ for μ^A Vel, 2.68 $M_\odot$ for its spectroscopic companion and 1.2 $M_\odot$ for μ^B Vel. The system is 116 ly from Earth and, just to muddy the celestial waters even more, μ^A Vel may be a superflare star; the last and only time it was observed to brighten, at least in the ultraviolet, was in 1989.

o Velorum is the brightest member of **IC 2391**, an open cluster whose center is 570 ly away. o Vel is 7.1 $D_\odot$ and lies on the innermost edge of the cluster at 495 ly. A B3 it is apparently rotating at 15 km/s, a long way short of the average for its class of 140 km/s and perhaps suggesting that its pole is pointing towards us. It is a β Cephei type pulsating variable and, over a period of $2^d\ 18^h\ 41^m\ 46^s$ dims from m_v +3.55 to +3.67 and back. The cluster is estimated to be about 45 million years old and is 21 ly across (see table below for details of other clusters).

Vela contains three stars that are about 400 $D_\odot$ across, equivalent to 4 AU or the size of the orbits of those minor planets on the inner edge of the Asteroid Belt. The smallest – if that is the right word! – is the 400 $D_\odot$ **b Velorum**, an F3 that is regarded as a PVSG; a Periodically Variable Supergiant, its magnitude sometimes dimming from m_v +3.80 to +3.91. Lying 3,106 ly away its 22,000 $L_\odot$ would produce a brilliant M_v -8.4 at 10 pc. Larger still by some 20 $D_\odot$ is the 7.9 $M_\odot$ **GZ Velorum,** an Lc variable that oscillates between m_v +3.43 and +3.81. Belonging to the K3 spectral group its absolute magnitude is M_v -6.0. It is the closest of the three at 1,490 ly and has a luminosity of 9,240 $L_\odot$. The farthest and largest of the trio is another K3 cataloged as **HD 80108**. An estimated 4,030 ly away its 11,062 $L_\odot$ results in a m_v +5.08 star that is also variable to +5.14. It is reckoned to be 435 $D_{\odot}$ with a rotational velocity of 3.5 km/s, so it takes 6,291 days – 17.2 years – to spin once on its axis.

Open and globular clusters in Vela

Name	Size arc min	Size ly	Distance ly	Age million yrs	Brightest star in region*	No. stars m_v >+12*	Apparent magnitude m_v
IC 2391	125′	21	570	45	o Vel m_v +3.64	73	+2.5
IC 2395	47′	31.5	2,300	17	HX Vel m_v +5.49	128	+4.6
IC 2488	18′	19.4	3,700	129	HD 30225 m_v +8.87	72	+7.4
NGC 2547	40′	17.5	1,500	36	NN Vel m_v +5.60	94	+4.7
NGC 2645	10′	15.8	5,440	19	HD 73919 m_v +8.85	15	+7.0
NGC 2669	14′	13.9	3,400	84.5	HD 75105 m_v +7.64	19	+6.1
NGC 2910	6′	14.8	8,500	160	HD 82420 m_v +9.24	18	+7.2
NGC 2925	15′	10.9	2,500	71	HD 82812 m_v +8.44	34	+8.3
NGC 3228	28′	14.7	1,800	85.5	HD 89713 m_v +6.94	50	+6.0
NGC 3330	13′	10.9	2,900	169	HD 92348 m_v +8.82	18	+7.4
Pismis 4	18′	10	1,940	34	HD 73075 m_v +7.32	14	+5.9
Trumpler 10	75′	30.5	1,400	35	HD 74772	385	+4.6
NGC 3201	18.2′	77.8	14,700		Globular cluster		+8.2

*May not be a cluster member.

Virgo

Constellation:	Virgo	**Hemisphere:**	Equatorial
Translation:	The Virgin	**Area:**	1,294 deg^2
Genitive:	Virginis	**% of sky:**	3.317%
Abbreviation:	Vir	**Size ranking:**	2nd

The largest Zodiacal constellation and second largest of all the constellations, Virgo has been associated with several mythical females including Astræa or Dike, the Greek Goddess of Justice (the constellation is next to Libra, the Scales of Justice), Ishtar from Babylon, Erigon the daughter of Icarus, Persephone the Goddess of Innocence and Purity, and Demeter the Corn Goddess. The Sun passes through Virgo between 16 September and 31 October.

α **Virginis**, the 15th brightest star in the sky, is probably better known by its common name of Spica. It is a double-variable as well as being a quadruple star. The primary is a 7 $D_\odot$, 9.4 $M_\odot$ B1 with a luminosity of 12,000 $L_\odot$. Its magnitude fluctuates between m_v +0.98 and +1.05 with a period of $4^d\ 0^h\ 21^m$ due to the close proximity of a 3.9 $D_\odot$, 5.9 $M_\odot$ B3. The two stars are separated by an average of just 0.12 AU (18 million km) and physically deform one another into teardrops. As the stars present a continually changing face to us they appear to vary in magnitude. However, apart from this external influence the primary is also intrinsically variable. Belonging to the β Cephei class of variables it pulsates over a period of $4^h\ 5^m$ causing its brightness to dip by 0.015 magnitude. The third star in the system is a m_v +4.50 B-class dwarf of 4.38 $M_\odot$ that has an orbital period of 1.823 years. Farther out, with a period of 55.208 years, is a m_v +7.50 dwarf that has a mass of 1.79 $M_\odot$. It was long believed that the Spica system was 262 ly from Earth but the latest measurements indicate a slightly closer distance of 250 ly.

Not the second but the fifth brightest star in the constellation **β Virginis** has both an apparent and absolute magnitude of about +3.6, a product of it lying at a distance of 35.6 ly, just slightly farther than the 10 pc (32.6 ly) used to calculate absolute magnitude. Sometimes classed as an F8, sometimes an F9, it is not totally dissimilar to our own Sun: a third more massive at 1.36 $M_\odot$, about 1.4 $D_\odot$ across but a rather more luminous 3.5 $L_\odot$ its temperature is around 6,150 K (Sun = 5,780 K) and its rotational period is 1 km/s (Sun = 2km/s) resulting in a rotational period of almost 71 days. β Vir's upper age limit is 4,700 million – comparable to the Sun's 4,567 million years – although the lower age estimate is 2,800 million years. It is drifting away from us at a steady 5 km/s.

Similar in distance to β Vir – just 3 ly farther away – γ **Virginis** or Arich is a pair of identical F0 dwarfs of 1.2 $D_\odot$, 1.4 $M_\odot$ and 4.7 $L_\odot$. In 2007 they came as close as 5 AU, about the distance Jupiter is from the Sun, but over the next 171.4 years the two stars will drift apart until they reach a separation of 81 AU before closing again. A third apparently identical star is a line of sight coincidence.

Although 80% of all stars are M-class they represent less than 7% of those

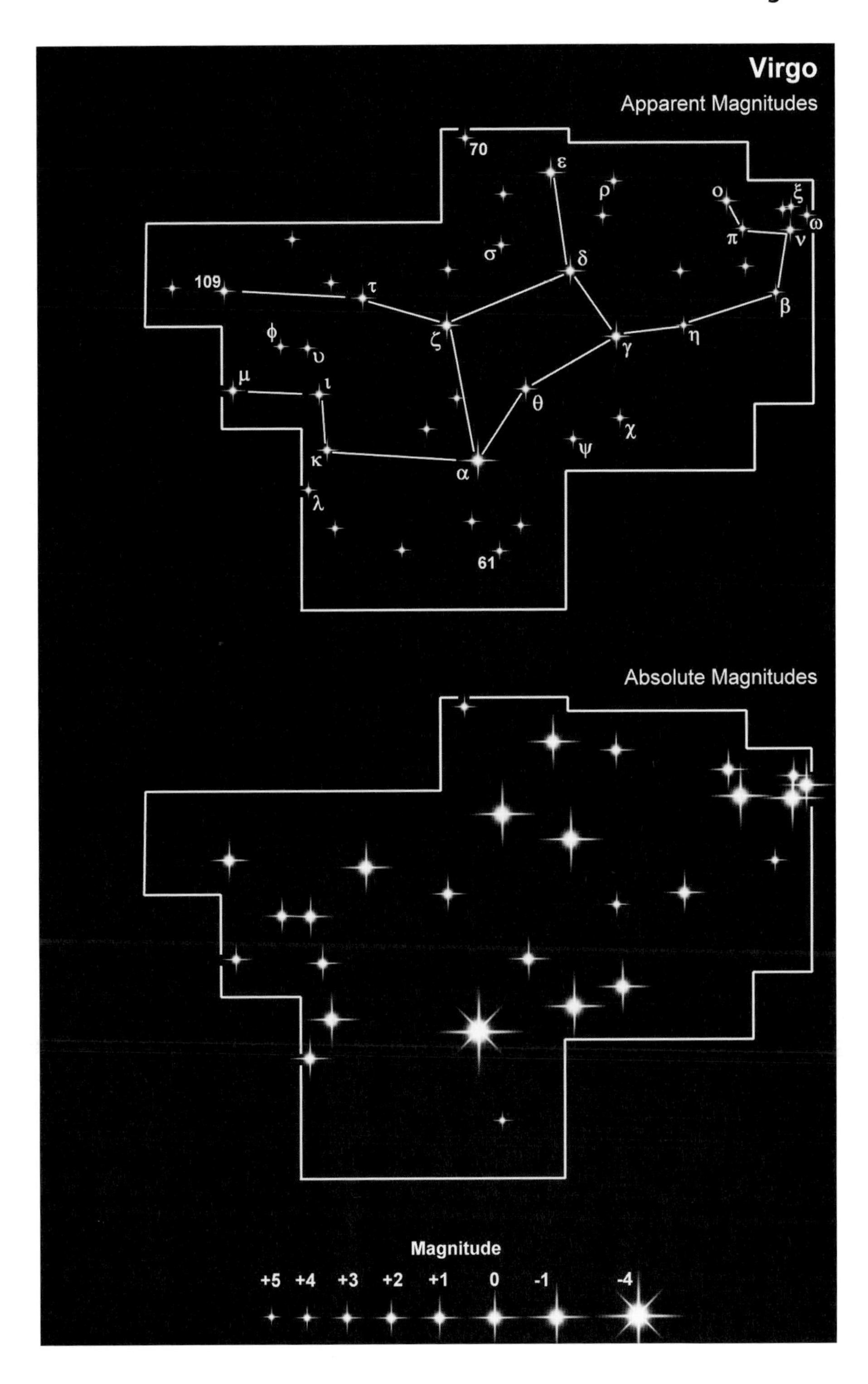
Virgo
Apparent Magnitudes
70
ε
ρ
ο
ξ
ω
π
ν
σ
δ
109
τ
β
ζ
γ
η
ϕ
υ
μ
ι
θ
χ
ψ
κ
α
λ
61
Absolute Magnitudes
Magnitude
+5 +4 +3 +2 +1 0 -1 -4

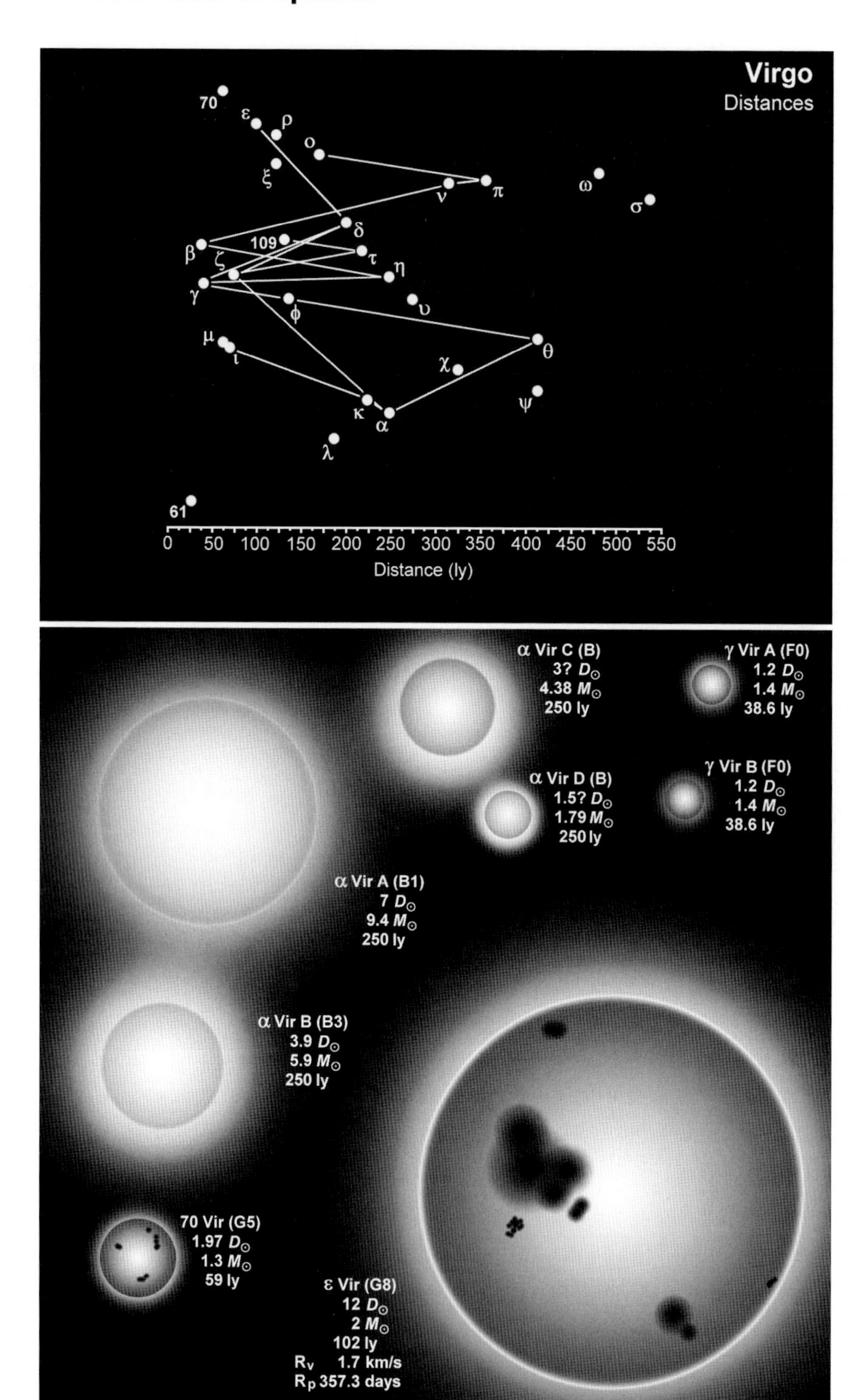

Virgo
Distances
70
ε
ρ
ο
ξ
ν
π
ω
σ
δ
109
β
τ
ζ
η
γ
υ
φ
μ
θ
ι
χ
ψ
κ
α
λ
61
0 50 100 150 200 250 300 350 400 450 500 550
Distance (ly)
α Vir C (B)
3? $D_\odot$
4.38 $M_\odot$
250 ly
γ Vir A (F0)
1.2 $D_\odot$
1.4 $M_\odot$
38.6 ly
α Vir D (B)
1.5? $D_\odot$
1.79 $M_\odot$
250 ly
γ Vir B (F0)
1.2 $D_\odot$
1.4 $M_\odot$
38.6 ly
α Vir A (B1)
7 $D_\odot$
9.4 $M_\odot$
250 ly
α Vir B (B3)
3.9 $D_\odot$
5.9 $M_\odot$
250 ly
70 Vir (G5)
1.97 $D_\odot$
1.3 $M_\odot$
59 ly
ε Vir (G8)
12 $D_\odot$
2 $M_\odot$
102 ly
R_V 1.7 km/s
R_P 357.3 days

visible without optical aid, their relatively low luminosity making them mostly invisible. δ **Virginis** is an M3 and is a giant 65 $D_\odot$ across, its mass coming in at about 1.8 $M_\odot$ and with a luminosity of around 135 $L_\odot$ visible, 630 $L_\odot$ if we take into account that most of its energy is radiated in the infrared. Some 202 ly away it is m_v +3.40 though may be sometimes variable to m_v +3.32; a period of 3^d 1^h 16^m has been suggested.

About 102 ly away lies the G8 giant ε **Virginis.** A dozen times larger than the Sun and twice as massive it takes almost a full year – 357.3 days – to rotate once on its axis spinning at 1.7 km/s. Some 600 degrees cooler than the Sun at 5,100 K its 59 $L_\odot$ result in a m_v +2.82 star that would brighten to M_v +0.2 at 10 pc. ε Vir is useful for locating the Virgo cluster of galaxies that lie between it and β Leonis. The cluster of up to 2,000 galaxies is at the center of the much larger Local Supercluster.

ζ **Virginis** used to be north of the Celestial Equator but in February 1883 it migrated southwards, the result of precession. An A3 lying 73.2 ly away it has a diameter of 1.9 $D_\odot$ and mass of 2.1 $M_\odot$. A3 class stars have an average rotational velocity of 112 km/s but ζ Vir just about reaches double that speed at 222 km/s, spinning once every 10.4 hours. It is thought to be about 500 million years old.

η Virginis is a triple which looks like a single star of m_v +3.87. The primary, η^{Aa} **Virginis**, is a m_v +4.20 A2 of 1.6 $D_\odot$ and mass of 2.68 $M_\odot$. Orbiting just 0.5 AU away is the 1.66 $M_\odot$ η^{Ab} **Virginis** in a 71.792 day long orbit. The third component, η^{B} **Virginis**, is in a 13.07 year orbit averaging 10 AU from the close pair. Its mass is calculated to be 2.04 $M_\odot$. All the stars in the system, which is 250 ly away, are likely to be A-class.

θ Virginis appears to be another triple system but the details remain uncertain. The primary is an A1 of 1.5 $D_\odot$ and carrying a mass of 2.98 $M_\odot$. Its m_v +4.50 is enhanced by the presence of the other components so it appears a 10th of a magnitude brighter. Separated by just 0.219″ is the secondary component, a m_v +6.84 dwarf with an estimated mass of only 0.08 $M_\odot$. Its orbital period is 33.04 years. Farther out at 7.1″ is the m_v +8.03 companion in a 12,197 year orbit. Its spectral class has been variously identified as being an A5, A9 and an F2. A fourth K2 star 71″ away is probably just a line of sight coincidence.

Virgo has an interesting collection of planetary systems. **61 Virginis** is very much a Sun-like star lying just 27.8 ly away. A G7 it is slightly smaller and less massive than the Sun, 0.94 $D_\odot$, 0.90 $M_\odot$, and not as luminous at 0.82 $L_\odot$ though considerably older: 12,000 million years (Sun = 4,567 million years). All three planets so far discovered are less massive than Jupiter by a considerable margin, the least being 61 Vir b which is estimated to be 5.09 $M_\oplus$. Its diameter may be around 1.7 times that of the Earth but any thoughts of it being home to intelligent life would seem optimistic with its orbit taking it to within just 6.6 million kilometers of the star. Mercury, by comparison, stays at least 46 million kilometers from the Sun. The detection of a debris disk around the star hints at the possibility of other terrestrial type planets that are as yet undetectable.

Another close solar analog is **70 Virginis.** A G5, similar to the Sun, but nearly twice the size at 1.97 $D_\odot$ it has just one planet that weighs in at 7.44 M_J. The system is 59 ly away.

PSR 1257+12 is a millisecond pulsar 979 ly distant, which appears to have a planetary system. Pulsars are created when a massive star explodes as a supernova and the core is compressed to a diameter of 20-30 km forming a neutron star (i.e. composed almost entirely of neutrons). The neutron star spins rapidly and emits a high intensity beam of radiation from its magnetic axis which may not be the same as its rotational axis. As the beam points towards the Earth the star appears to flash or pulse. It is not clear how the three terrestrial type planets came into existence but they are thought to have formed after the supernova, the explosion likely to have either destroyed them or ejected them from the system altogether. The masses of the three planets are 0.02 $M_{\oplus}$, 3.81 $M_{\oplus}$ and 4.13 $M_{\oplus}$.

From the tiny to the titanic, the variable star **HW Virginis** has two known planets, one of which has a mass of 19.2 M_J indicating it is a brown dwarf. The star itself is an Algol-type eclipsing binary consisting of a B-class sub-dwarf and an M-class red dwarf.

Planetary systems in Virgo

Star	$D_{\odot}$	Spectral class	ly	m_v	Planet	Minimum mass	q	Q	P
HW Vir	?	sdB+M	290	+10.9	HW Vir b	19.2 M_J	?	?	15.79 y
					HW Vir c	8.5 M_J	?	?	9.09 y
61 Vir	0.94	G7	27.8	+4.74	61 Vir b	5.09 $M_{\oplus}$	0.044	0.056	4.22 d
					61 Vir c	0.0573 M_J	0.19	0.25	38.02 d
					61 Vir d	0.072 M_J	0.31	0.64	123.0 d
70 Vir	1.97	G5	59	+4.97	70 Vir b	7.44 M_J	0.29	0.67	116.7 d
PSR 1257+12	?	?	979	?	PSR 1257+12 b	0.02 $M_{\oplus}$	0.19	0.19	25.26 d
					PSR 1257+12 c	4.13 $M_{\oplus}$	0.35	0.37	66.5 d
					PSR 1257+12 d	3.81 $M_{\oplus}$	0.45	0.47	98.2 d

Volans

Constellation:	Volans	**Hemisphere:**	Southern
Translation:	The Flying Fish	**Area:**	141 deg^2
Genitive:	Volantis	**% of sky:**	0.342%
Abbreviation:	Vol	**Size ranking:**	76th

Originally called Piscis Volans, the constellation is another invention of Keyser and de Houtman.

Lying somewhere between 122 and 126 ly from Earth α **Volantis** is 1.3 or 1.4 or 1.9 or 2.5 or 2.75 $D_\odot$ depending on who you ask (the average works out at 1.97 $D_\odot$). Its mass is better defined – 1.87 to 2.20 $M_\odot$ – but its spectral class seems to lie in the range of A2 to A7. There is widespread agreement that its luminosity is about 30 $L_\odot$ and its apparent magnitude is spot on m_v +4.00 with its absolute magnitude calculated at M_v +2.50. It may or may not be surrounded by a debris disk and if its rotational velocity is anywhere near accurate at 34 km/s – just a third of the average for its class – then it will take between 1.9 and 4.1 days to rotate once. And to cap it all it is actually the 5th brightest star in Volans. Altogether quite a troublesome little star in an otherwise quiet constellation.

β **Volantis**, by comparison, is happily shining away at 4,570 K, 108 ly from Earth. It is 18 $D_\odot$ across, about average for a K3, has a mass of 1.1 $M_\odot$ and 28 $L_\odot$. It is not the only K-class star in the constellation. **HD 70514** is another K1, 296 ly away with a diameter of 13 $D_\odot$ and luminosity of 63 $L_\odot$. **HD 53501,** a 25 $D_\odot$ K3 is slightly closer at 290 ly while ζ **Volantis** is less than half the distance at 134 ly. It is a K0, as is γ^2 **Volantis** which is by far the most interesting of the bunch. A measured 12.7 $D_\odot$ across it has a luminosity of 71 $L_\odot$ and mass of 2.5 $M_\odot$, a fact that has been acquired thanks to the presence of an F2 binary companion called, not surprisingly, γ^1 **Volantis**. Separated by 12″ at PA 297° the 1.9 $D_\odot$ partner is somewhat less massive, 1.6 $M_\odot$, and luminous, 8 $L_\odot$. Their orbital distance must average around 600 AU leading to a 7,500 year long orbital period. The system is between 600 and 700 million years old.

δ **Volantis** is the giant of the constellation at 50 $D_\odot$, although a figure of 106 $D_\odot$ has also been quoted. An F6 of 860 $L_\odot$ it is also the farthest star at 660 ly. At 10 pc its absolute magnitude would be M_v 5.4.

ε **Volantis** is a binary and possible triple star arrangement. The primary is a 2 $D_\odot$ B6 weighing in at a hefty 5.76 $M_\odot$. Its slow apparent rotational period of 20 km/s is probably a matter of perspective with the Earth looking towards the star's pole giving the impression that it is rotating much slower than it actually is. The average velocity for such stars is 147 km/s. Its spectroscopic companion is a B6 of 3.26 $M_\odot$ in an orbit of only 14.168 days. The third component, if it is truly associated with the binary pair, is a B9.5 of 2.86 $M_\odot$. Its orbital period is in excess of 12,000 years. The system is 642 ly away.

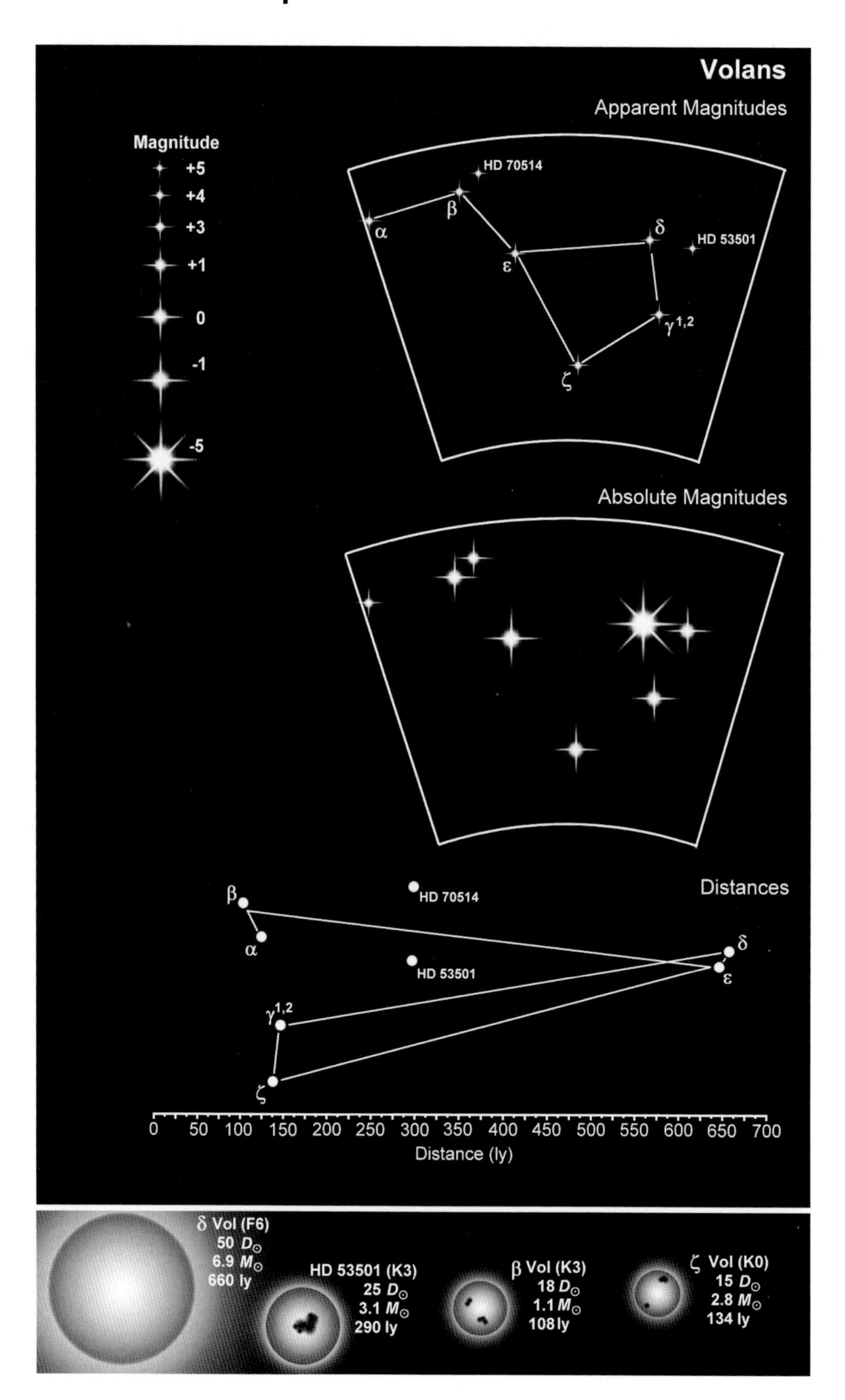

Volans
Apparent Magnitudes
Magnitude
+5
+4
+3
+1
0
-1
-5
HD 70514
β
α
δ
HD 53501
ε
γ 1,2
ζ
Absolute Magnitudes
Distances
β
HD 70514
α
δ
HD 53501
ε
γ1,2
ζ
0 50 100 150 200 250 300 350 400 450 500 550 600 650 700
Distance (ly)
δ Vol (F6)
50 $D_\odot$
6.9 $M_\odot$
660 ly
HD 53501 (K3)
25 $D_\odot$
3.1 $M_\odot$
290 ly
β Vol (K3)
18 $D_\odot$
1.1 $M_\odot$
108 ly
ζ Vol (K0)
15 $D_\odot$
2.8 $M_\odot$
134 ly

Vulpecula

Constellation:	Vulpecula	**Hemisphere:**	Northern
Translation:	The Fox	**Area:**	268 deg^2
Genitive:	Vulpeculae	**% of sky:**	0.650%
Abbreviation:	Vul	**Size ranking:**	55th

One of the most obscure constellations in the Northern Hemisphere, Vulpecula was introduced by Johannes Hevelius who called it Vulpecula cum Ansere meaning 'the little fox with the goose'. The goose has long since flown, courtesy of the IAU, and the fox is encircled by Cygnus, Lyra, Hercules, Sagitta, Delphinus and Pegasus. Only one star has a Bayer designation, the others all carry Flamsteed numbers. The constellation's most notable feature is the Coathanger cluster.

The keen sighted, and those with a binocular, will see **α Vulpeculae** as a double star. The brighter is a m_v +4.45 M0 red giant, one of only five M-class α stars, the others being in Hercules (an M5), Cetus (M1.5), Orion (M2) and Scorpius which is also an M1.5. Unlike the others, however, α Vul does not appear to be variable. At 53 $D_\odot$ and 1.5 $M_\odot$ it is one of the more modest red giants but also one of the speediest hurtling towards us at 85.5 km/s. Its apparent companion, the m_v +5.82 K0 star **8 Vulpeculae**, lies 7′ to the north east at PA 28.5°. Yet the two are unrelated, α Vul some 297 ly away with 8 Vul being much farther at 484 ly.

22 Vulpeculae is a G2 yellow supergiant, 120 $D_\odot$ across; just slightly smaller than the orbit of Venus. With a rotational velocity of 15 km/s it takes more than a year – 405 days – to rotate once. Its mass of 6.25 $M_\odot$ means that it is doomed to a swift existence, burning out in just 100 million years or 1% of the lifetime of our Sun. It is also known by is variable designation of **QS Vulpeculae**. In 1983 it was discovered that 22 Vul is an eclipsing spectroscopic binary. Its unseen companion, a B9 dwarf, orbits the primary in 249.12 days. As it passes in front of the G2 star its magnitude dips from m_v +5.18 to +5.30 for a period of 9.8 days. The stars lie at an uncertain distance of 4,349 ly.

Collinder 399 is not a true cluster but an asterism of 10 stars that resemble a coathanger. Also known as Brocchi's Cluster it includes **4, 5** and **7 Vulpeculae.** The closest star is 5 Vulpeculae at 218 ly with HD 182422 being the farthest at 1,133 ly. Magnitudes range from m_v +5.16 to +7.16.

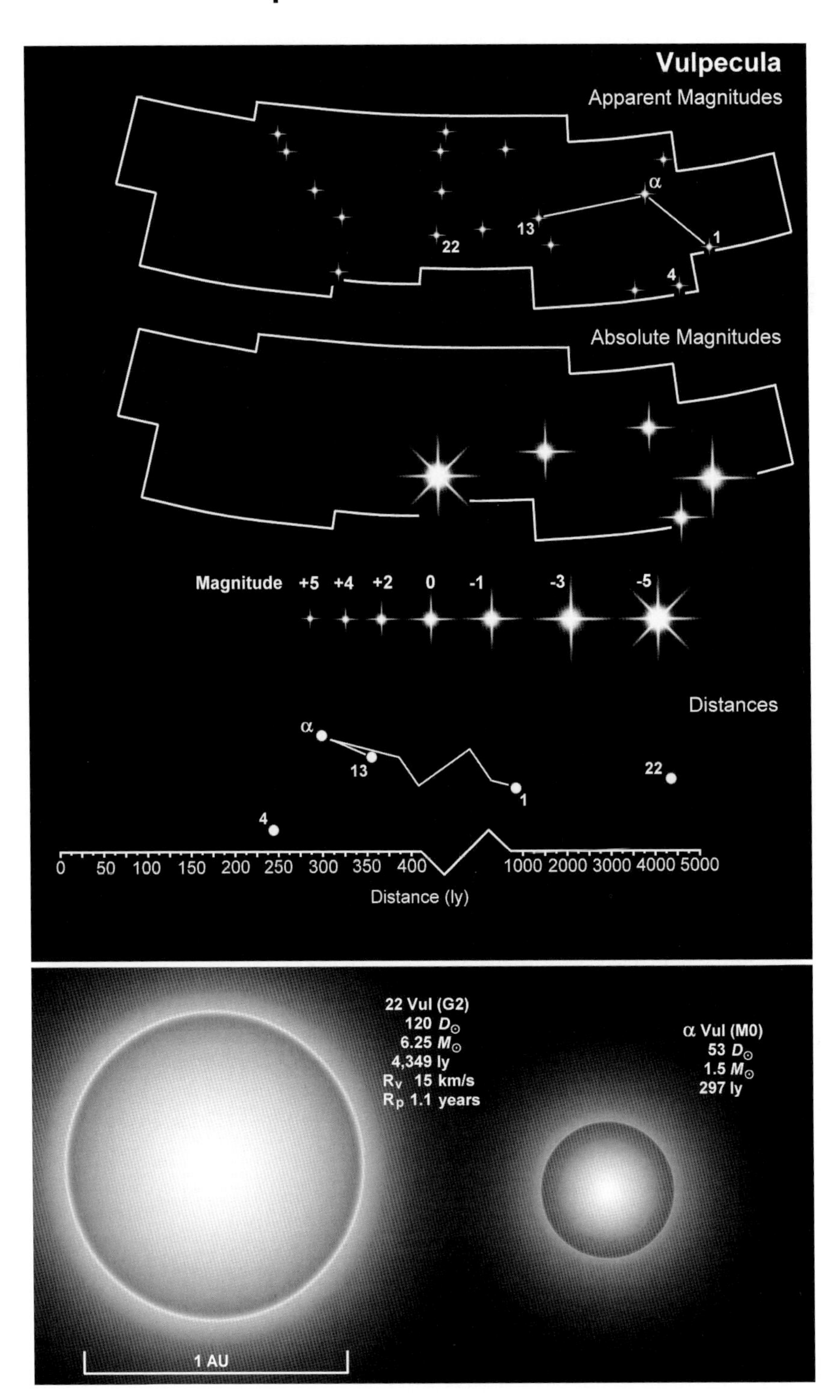

Vulpecula
Apparent Magnitudes
α
13
1
22
4
Absolute Magnitudes
Magnitude +5 +4 +2 0 -1 -3 -5
Distances
α
13
1
22
4
0 50 100 150 200 250 300 350 400 1000 2000 3000 4000 5000
Distance (ly)
22 Vul (G2)
120 $D_\odot$
6.25 $M_\odot$
4,349 ly
R_V 15 km/s
R_P 1.1 years
α Vul (M0)
53 $D_\odot$
1.5 $M_\odot$
297 ly
1 AU

Index

Main Index

Stars – Catalog Numbers

Stars – Named

—

This book is published in conjunction with the exhibition, *Port Cities: Multicultural Emporiums of Asia, 1500–1900*, presented at the Asian Civilisations Museum, Singapore, from 3 November 2016 to 19 Februrary 2017.

Reprint 2018, with minor text corrections

ISBN 978-981-11-1380-2

—

EXHIBITION CURATORS
Guest curator Peter Lee, with ACM curators Clement Onn and Naomi Wang.

Edited by Alan Chong, Richard Lingner, and Clement Onn

Designed by MAKE

Printed by Oxford Graphic Printers Pte Ltd

PRINCIPAL TYPE
Sharp Sans

PAPER
Pacesetter Plus Matt

—

Cataloguing in Publication Data is available from the National Library Board, Singapore, and OCLC.

Frontispiece: Detail of Cat. 107.

PORT CITIES

MULTICULTURAL EMPORIUMS OF ASIA, 1500–1900

Peter Lee, Leonard Y. Andaya, Barbara Watson Andaya,
Gael Newton, Alan Chong

CONTENTS

Foreword

Port cities are liminal spaces. They are in-between forms of human settlement that occupy the threshold 'twixt land and sea, between polities and routes. They have existed for more than a millennium as peripheral places at the edges of the continental land masses and their civilisations. And yet it is through these peripheral dots on the map that the course of world history and culture has changed.

Consider Venice – entrepôt of the western world in the eighth through the fifteenth century – and how it played a role in cultivating a European taste for those odoriferous bits of dried bark, bud, and seed that would become known as "spices". The unprecedented and ravenous demand for spices, and their being worth their weight in gold, would drive Portuguese fleets eastwards to the port cities of Goa, then Malacca, in order to wrestle for a monopoly on the trade.

The Portuguese takeover of Malacca was the prelude to one of the largest movements in people, goods, and cultures in history. In the course of the trade spurred on by colonialism from the 1500s into the 1900s, new port cities emerged to supplant earlier incarnations.

The time frame of our exhibition falls squarely within the four centuries of European colonialism that have shaped the maps and destinies of almost every nation-state in Asia. In the course of the contact, conflict, and negotiation between European and Asian civilisations here on Asian soil, new, hybrid communities and forms of art and architecture evolved. Although much research has been done on port cities in Asia, some of it focusing on links between cities, in our exhibition and in this book we plot a different course. We de-emphasize colonial networks and Eurocentric approaches and instead focus on the people who lived and worked in port cities. The rich mix of clothing styles, modes of living, and tastes in art tell fascinating stories and reveal as yet undiscovered truths. We also encourage a more holistic approach, embracing comparative studies – history of photography, furniture and decorative arts, textile studies, and language, to name a few – to paint a broader picture of how specific kinds of art and artefacts travelled through Asia and evolved.

Singapore, being the latest and possibly the *only* remaining cosmopolitan, multicultural Asian port city in a long line and tradition of them…including Malacca, Manila, Batavia, Nagasaki, Canton, Macau, Calcutta, Goa, Bombay, to name but a few…is ideally placed to explore this topic. Singapore is also a *city-state*, and the story of this particular "red dot" is precisely the story of the people who traded across oceans and made connections across civilisations.

Deeper insight into the nature of Asian port cities can only bring a generation of academics and visitors closer to understanding just what makes Singapore tick, and what it means to have lived and continue to live in this multicultural emporium that is Singapore: past, present, and future.

For this exhibition to have taken place at all, I have to thank our guest curator, local Southeast Asian scholar Mr Peter Lee, and ACM's curators of cross-cultural art, Mr Clement Onn and Ms Naomi Wang. They have laboured tirelessly to assemble the many works of art exhibited and to uncover the various forgotten or unspoken stories behind each of these works. We are extremely privileged and grateful to have loans of exceptional material coming from many partner museums, institutions, and private collectors worldwide; and to have received invaluable help and advice from many individual curators and professionals. We acknowledge every one of them at the conclusion of this catalogue.

Finally, I have to thank my predecessor, Dr Alan Chong, who was responsible for conceptualising this exhibition, and under whose mentorship I have gained the skills and the eye to take my own search for meaning and identity inside the galleries of the museum.

This exhibition is only the beginning of a longer journey of discovery. The core themes of trade, port cities, and hybridity will have fuller expression in ACM's permanent galleries dedicated to trade and the spread of faith and belief systems throughout Asia. We also hope to pursue more exhibitions in this vein as we continue to search for and articulate a unique curatorial voice for ourselves here in tiny, global Singapore.

To the visitor, I invite courage and curiosity. Courage to step out of prescribed boxes into a space that is neither here nor there but somewhere "in-between". And curiosity to probe deeper into what *really* makes us who we are today.

Welcome aboard ship! We are ready to set sail!

–

Kennie Ting
Director, Asian Civilisations Museum
Group Director of Museums, National Heritage Board

1

East Indies market stall, 17th century.
Unknown painter, formerly attributed to Albert Eckhout.
Oil on canvas.
Rijksmuseum, Amsterdam.

Divergence, Convergence, Integration Port cities and the dynamics of multiple networks, 1500-1900

Leonard Y. Andaya and
Barbara Watson Andaya

Trading connections between the Asian region and ports as distant as Rome or the Persian Gulf can be tracked since early times, and by at least the thirteenth century these connections had expanded to become part of an economic network that stretched across much of the world.[1] The bulk of this trade was dominated by land routes, best exemplified by the so-called Silk Road through Central Asia. From the late fifteenth century, however, major advances in ship technology and knowledge of sea routes opened up new lines of communication among Europe, Asia, Africa, and the Americas, and thereby transformed the entire world trading system.

Because these transformations were so extensive, the period between 1500 and about 1800 is increasingly viewed as a new chapter in global history that can be characterized as Early Modern.[2] One of the features of this period is the rapid expansion of maritime trade, and the intensity of cultural encounters in places where trade was concentrated. Movement across the seas was fundamental to this, with interlocking chains of port cities becoming dissemination points for the circulation of people, goods, and ideas. Though European efforts to dominate seaborne trade were unsuccessful, their very presence added to the vitality of Asian ports, which continued to be hubs of cultural exchange where ethnicities were porous and hybrid aesthetics flourished. Even with the extension of European imperialism in the nineteenth century, the defining feature of the Asian port city was its vibrant and interactive culture where divergent peoples and ideas converged to produce a unique synthesis.

Asian port cities and European seaborne empires

When Europeans arrived in the late fifteenth century, Asia's port cities were already integrated into a vast maritime trading web connecting the "single ocean" that stretched from Africa's eastern seaboard to China and Japan.[3] While these early ports reflected the specific linguistic and cultural milieu in which they had developed, they also had much in common. Indeed, in discussing the divergences and convergences that typify Asian port cities, it is useful to keep in mind Michael Pearson's concept of the "littoral", which argues for the shared distinctiveness of coastal and sea-oriented societies.[4] Located along "voyaging corridors", these ports operated as nodes in an interactive sphere where individuals from many different origins collaborated in furthering their common ventures.

In the seventeenth century, Masulipatnam on the Coromandel Coast in India was one such city, its importance due less to its site than to its role as the major trade outlet for the Golconda Sultanate. Here merchants from various parts of India mingled with Persians, Armenians, and other Asians, the multicultural milieu heightened because the city was also a departure point for the pilgrimage to Mecca.[5] Under Mughal rule in the following century, Masulipatnam continued to be known for its diverse population where religious differences mattered little; in the transhipment of elephants, for example, Muslim and Hindu merchants worked closely with their Siamese Buddhist counterparts, often in co-operation with European traders.[6]

A second example of a well-established Asian port city that expanded with global trade was Guangzhou, on the Pearl River estuary in southeast China. By the sixteenth century, Guangzhou had already developed as an important port, fed by an interior producing iron, pottery, and silk. However, its multicultural environment expanded exponentially after 1757, when Guangzhou (or Canton, the European rendering of Guangdong province) became the only port open to Western shipping. In consequence, Guangzhou became one of the busiest harbours in the world, where European traders collaborated with their Asian counterparts in carrying cargoes from Guangdong province and the Yangzi River area to Southeast Asia and onwards to India and Europe.[7] Visiting Calcutta's "China Bazaar" in 1810, the Penang scribe Ahmad Rijaluddin was amazed to see the unlimited supply of Chinese mirrors, silk, teapots, porcelain, and other "unusual" wares that were sold by wealthy Indian merchants.[8]

With the expanding market for unusual wares, the ports of early modern Southeast Asia gained a global reputation as distribution centres for products found only in tropical jungles and seas. The renowned entrepot of Malacca, described as a place "where you find what you want and sometimes more than you are looking for"[9] fell to the Portuguese in 1511. Nonetheless, its location on the straits midway between China and India ensured its importance as an exchange point until it was taken by the Dutch in 1641. Other Southeast Asian ports, notably Ayutthaya in Thailand, Ha Tien in Vietnam, Manila, and Batavia (Jakarta), continued to operate as doorways to the Asian trading world, supplying aromatic woods, spices, and pepper as well as a variety of sea products such as tortoiseshell, pearls, and sea cucumber. Exotic goods like kingfisher feathers, elephant tusks, and rhinoceros horn found a ready market in China, but also in India, the Middle East, and Europe. In Southeast Asia's bustling ports such products were exchanged for silk, brocades, and porcelain from China, for cotton textiles from India,

2

3

4

2

Masulipatnam.
Print from J. Churchill, ed.,
A Collection of Voyages and Travels,
London 1732, vol. 3, pp. 519–877.
[London 1703, pp. 667–829].

3

Canton, around 1820.
Unknown painter. Oil on canvas.
The area of Guangzhou to which Western traders were restricted from 1757–1842.
The flags of Denmark, Spain, the US, Sweden, Britain, and the Netherlands are visible.

4

Bazaar leading to Chitpore Road.
Calcutta, 1826
Chitpore Road divided the native and European parts of the city.

and for frankincense, myrrh, and horses from Arabia and Egypt.

At the beginning of the sixteenth century, this thriving Asian network was forced to contend with a new phenomenon: the European seaborne empires connected by a series of port cities, some taken by force and others newly established. As the first permanent European presence in Asia, the Portuguese developed what was officially termed the Estado da Índia (State of India). Linking the capital at Goa on the west coast of India to Sofala on the east coast of Africa, Malacca on the Malay Peninsula, and Macao in southern China, the Portuguese ports were part of a larger network outside direct Portuguese control: Nagasaki, Siam, Manila, Makassar, and Acapulco.[10]

During the seventeenth century, however, the State of India was overpowered by the financial and military might of the Dutch East India Company (known by its initials in Dutch, the VOC). Either seizing or eliminating Portuguese ports, the Dutch created an Asian maritime empire headquartered at Batavia, which was founded in 1619. In less than fifty years, a web of VOC posts and port cities reached from Cape Town in South Africa, to Persia, India, Sri Lanka, Southeast Asia, Japan, and Korea. Operating under the direction of the VOC government in Batavia, this interconnecting system of communication furnished the Dutch with vital trade and political information. Though more effective than the Portuguese Estado da Índia, the VOC collapsed in 1799 as a result of venality and involvement in costly local wars and territorial acquisitions.

Launched in 1600, the English East India Company was initially unable to challenge the Dutch in Southeast Asia and retreated to the Indian subcontinent, where it became a major power, governing an extensive network of port cities through its capital at Calcutta. In 1905, looking back over this period, an Indian writer remarked that "commerce made Calcutta the home of many nations ... Armenian and Jew, Parsi and Marwari, Frenchman, German, Greek and Chinese are all here for commercial purposes".[11] By the mid-eighteenth century the East India Company's superior naval technology and cartography enabled it to extend its dominance beyond India.

The newly established British settlements of Penang (1786) and Singapore (1819) became major British ports on the all-important route to China. Indian opium became an indispensable trade item in the acquisition of China's tea, and the Company's control of this trade assured its ongoing economic success. Imperial edicts to control opium imports because of escalating addiction among Chinese led to conflict with the East India Company and China's defeat in the First Opium War (1839–1842). Beijing was forced to surrender Hong Kong to the British and open up several Chinese ports to foreign trade, which added to the East India Company's already extensive network.[12] At its dissolution in 1874, the Company had created the greatest seaborne empire Asia had ever known.

5

Portuguese mestizo and wife.
Print. Johan Nieuhof, in *Voyages and Travels, into Brasil, and the East-Indies* (London, 1704), p. 316C.

Private traders and port communities

While trade linked Asian ports, it was the interaction of individuals at all levels of society that generated the eclectic material culture represented in this exhibition. The Portuguese case offers a prime example of the converging communities that emerged after 1500. Though the Estado da Índia was ultimately unsuccessful both as a political and commercial enterprise, open encouragement was given to intermarriage with local women in order to create an Asian-based population of loyal Portuguese-speaking Christians. The resulting informal empire consisted of *moradores* (settlers), who included substantial numbers of *casados* (private Portuguese traders married to local Asian women), their *mestiço* (mixed) descendants, some priests, slaves, drifters, and local Christian converts.

Among the most successful of these private traders were the New Christians (*cristão-novo*), Iberian Jews who had converted to Roman Catholicism to escape persecution. They were especially prominent in the most profitable Portuguese network, which linked Malacca, Macao, Manila, Maluku, and Japan. It has been estimated that the New Christians invested a minimum of 450,000 cruzados in the annual galleons, and another similar amount to acquire goods for the Galleon Trade between Manila and Acapulco. The integration of the Portuguese into the Asian trading world and the elevation of the Portuguese language as a lingua franca in the port cities owe much to the commercial and social influence of these New Christians.[13]

Unlike the Portuguese, Dutch private trade by *vrijburghers*, that is, former VOC employees, was initially encouraged to complement

6

Portuguese in Goa and Street scenes in Goa.
Print. From the *Itinerario* of Jan Huygen van Linschoten. Koninklijk Bibliotheek, The Hague, Netherlands.

the Company's activities. Their impact on Asian trade was limited by their dubious reputation in commercial relations and by restrictions on products and destinations. Certain wealthy, well-placed VOC officials traded surreptitiously, but neither they nor the *vrijburghers* ever reached the levels of integration with the Asian network that occurred among private Portuguese traders.[14] By contrast, the penetration of English "country" traders (so-called because they operated solely in Asian ports) was particularly noticeable in the eighteenth century. They co-operated and prospered alongside the East India Company, and were instrumental in extending British interests throughout Asia.

7
Portuguese in India.
Print. Jan Huygen van Linschoten.

All these private traders spoke the trade lingua franca of Portuguese and Malay, intermarried or took temporary local wives, and understood native customs and the patterns of local commerce. A voyage that entailed buying Gujarati cloth from western India, selling it at a good price in Batavia, using the proceeds to acquire pepper in Sumatra, and then disposing of the cargo to European-bound merchants exemplifies the ease with which they moved through the entire Asian region.[15] In these globally connected operations, English country traders dealing in opium and tea in China could deposit their profits in silver bullion in the East India Company treasury in Guangzhou, and receive bills of exchange that were honoured in India, London, and even New York.[16]

Nonetheless, although usually operating well in the Asian environment, independent European traders were almost always outmatched by their counterparts from the Middle East, China, and India. Indians from Gujarat, Bengal, and the Malabar and Coromandel Coasts were especially active in the western Malay Archipelago. Rich merchants and high-placed Indian officials provided the capital to build ships and invest in cargoes, but much of the shipboard space was allocated to less affluent traders who carried their own packs of cheaper textiles. By the eighteenth century, trade around the Bay of Bengal was dominated by Tamil Muslims (known as Chulias) from the Coromandel Coast, many of whom took up permanent residence overseas. Their value to Southeast Asian rulers is evident in their appointment to important positions as *syahbandar* (official in charge of foreign trade) and *saudagar raja* (royal merchant).[17]

Over the centuries, however, the most effective traders were Chinese, operating through their long-established maritime networks. Defying periodic imperial bans on overseas trade, men from the southern provinces of Guangdong and Fujian continued to sail to the Nanyang (southern ocean), a general reference to Southeast Asia. With capital provided by wealthy entrepreneurs both at home and in the overseas Chinese communities, often supplemented by investments from Europeans based in Asia, Chinese traders extended credit to Southeast Asians who collected the products that could be sold so profitably in the international marketplace. Many Chinese married local women and became part of the community, often adopting Islam or (especially in the Philippines), Christianity.

At the same time, when necessary, they were willing to work closely with other trade networks, such as those of the Bugis from Sulawesi.[18] Chinese willingness to extend credit and operate with small profit margins, combined with their knowledge of local trade conditions and their infinite patience, made them effective traders. The products they bought from local collectors were taken to major port cities like Makassar, Batavia, and Ayutthaya, where they were sold to Chinese merchants from China. Large junks then carried these cargoes back to southern China, primarily to Guangzhou and Quanzhou.[19] Chinese involvement in the Nanyang trade was so marked that in the early seventeenth century, Manila became known as "the second hometown of the Fujianese".[20]

Chinese immigrants – carpenters, bricklayers, smiths, bakers, shopkeepers, tanners, craftspeople, artisans, and numerous others – were also indispensable in the development of the European-controlled port cities. Existing Chinese trade communities were swelled by a new wave of migrants who formed sojourner populations both in the ports and in adjacent towns, mines, and plantations where they found employment. By 1800, Chinese settlers were mining gold and diamonds in southwest Borneo; gold in Sumatra and the northeast coast of the Malay Peninsula; tin in Bangka, Perak, Songkhla, and Phuket; and cultivating pepper in Jambi, Palembang, Riau, Siam and Brunei; and gambier and pepper in Riau and Johor.

In the course of the nineteenth century, European imperialism redirected Asian commerce away from local hands, but the collaboration between individuals of different backgrounds was still a prominent feature of port cities, even under European control. In Singapore, for example, one of the most influential business partnerships involved the wealthy locally born merchant Tan Kim Ching (1829–1892, eldest son of Chinese community leader Tan Tock Seng), and two Scotsmen, W. H. Read (1819–1909) and J. G. Davidson (1838–1891), who had both come to Singapore as young men. Read was a generous supporter of St Andrews Cathedral, but as a masonic Grandmaster he also arranged for a Muslim, Sultan Mahmud of Riau, to be initiated into the Singapore lodge.[21] In this environment, religious and cultural differences meant little as long as co-operation was seen as mutually advantageous.

Cultural interaction in the Asian port city

At first glance, this kind of interaction might seem at variance with the layout of the Asian port cities that arose or expanded between 1500 and 1900, since residential areas were typically allocated or developed from the basis of ethnic similarity. The increasing presence of outsiders intensified this pattern because many Asian rulers regarded incoming merchants with some ambivalence. While generating a new prosperity, these outsiders also had the economic influence and human resources to challenge the status quo. As a result, foreign traders were typically confined to specified ethnic quarters under a chosen leader. Trade was conducted in a controlled environment governed by local regulations and laws administered by a court-appointed official in charge of international commerce.

European-controlled cities were organized in a similar manner, with European residences, administrative offices, religious buildings, cemeteries, and garrisons established in separate areas that were at times walled off from the rest of the population. This pattern is clearly evident in Manila, where the Spaniards occupied the walled city of Intramuros, but assigned the Chinese to an area outside, known as the Parian, or marketplace. In the aftermath of an uprising in 1740, Batavia's Chinese were also moved to a *pecinaan*, a Chinatown, outside the city walls, where their activities could be more easily monitored. But in all port cities, the urban division between European-dominated residential areas and those regarded as native were increasingly apparent.

8
The King Rides on his Elephant.
In Guy Tachard, *Voyage de Siam des pères Jesuites, envoyés par le roy,aux Indes & à la Chine :avec leurs observations astronomiques, & leurs remarques de physique, de géographie, d'hydrographie, & d'histoire* (Amsterdam, 1687). Subjects prostrate in front of the King of Ayutthaya on his elephant.

In Calcutta, for example, Company offices and the European settlement clustered around Fort William, forming what Indians called the *sahib para* (white locality), clearly set apart from the "black town", where sub-castes and occupation groups were concentrated along specific streets or in certain districts.[22] Nevertheless, the idea that city government operated best when ethnic communities lived in assigned quarters did not prevent the mobility of goods and ideas as a medley of servants, artisans, soldiers, midwives, hawkers, and officials moved through the interstices of urban life. In the process the marketplace became not just a site of commercial negotiation, but a domain of aesthetic exchange and creativity. It was in the Parian, for instance, that the Spanish placed their orders for all manner of consumer goods and religious items, which Chinese artisans crafted according to Spanish tastes and specifications. The ivory images of the child Jesus which they carved, "as perfect as can be", would therefore end the need for imports from Europe.[23]

The most extreme case of physical separation was Nagasaki, where the Dutch (the only Europeans permitted to trade in Japan) were confined to the man-made island of Deshima, linked to the mainland by a narrow bridge. Though Chinese traders maintained a key role as interpreters and mediators in Japanese-Dutch relations, they too could only operate within Nagasaki. Yet ukiyo-e artists studied European techniques such as perspective, and from the 1720s, many samurai travelled to Deshima to acquire Western knowledge (*rangaku*).[24] The children born of liaisons between Dutch fathers and Japanese "women of pleasure" are also a reminder of the new identities that arose in the port environment.

The cohabitation or marriage of local women with foreign traders, officials, and even Catholic friars resulted in a growing number of mixed communities such as the Peranakan Chinese (Malay-Chinese), Jawi Peranakan (Malay-Indian), Black Portuguese (mixed Portuguese-local, or Asian Christian converts), and Portuguese/Spanish mestizo (Eurasians). The "Betawi", who were specifically associated with Batavia, were descendants of slaves who had originated from India and more distant areas of the Malay Archipelago.[25] Indeed, the miscegenation that characterized port cities was often a direct result of sexual relations between owners and female slaves, whose children might well be legally acknowledged by their father. Speaking several languages and knowledgeable in the ways of two or more cultures, members of these mixed communities were indispensable intermediaries both in trade and politics. Mutual borrowings of language, food, religious ideas, architecture, and the arts signalled the place of port cities as gateways to ideas, experimentation, and aesthetic expression.

Some of this interaction can be seen in the protocol associated with official occasions. In Batavia, for instance, envoys from Asian kingdoms were taken in European-style carriages to meet the governor-general in a reception hall. In keeping with native tradition, they were welcomed with betel nut and the letters they brought were carried on silver or golden salvers shaded by a yellow parasol, a symbol of royalty. A triumphal arch erected by Manila's Chinese community to honour Ferdinand VII displayed the Spanish king's portrait in a temple setting guarded by lions and dragons, with five vessels arranged like a Buddhist offering.[26] This cultural amalgam flowed over into the construction and decoration of elite buildings, especially those of rulers and high-ranking officials. A number of Hindu temples in Goa incorporated elements of Iberian baroque architecture, and some of Java's early mausoleums are richly carved and guarded by feline-like animals reminiscent of the iconography on Chinese tombs.

In the lives of ordinary people, the cultural boundaries associated with ethnicities were particularly blurred. Boria performances arrived in nineteenth-century Penang with Muslim Indians, but they became a localized genre of Malay theatre that welcomed the participation of Europeans, Portuguese, Indians, and Chinese. In Shanghai, one of the Chinese ports opened to foreign trade after 1842, fascinated spectators watched *moro-moro* plays, introduced by Spanish missionaries, that celebrated Christian victories over Muslims, or gambled at cockfights popularized by Filipino sailors.[27] In every city, certain areas became famous for entertainment, like a Calcutta bazaar where "English, Portuguese, French, Dutch, Chinese, Bengalis, Burmese, Tamils, and Malays visit morning noon and night and so it is always terribly crowded and noisy, as if they were celebrating the end of a war".[28] Even in the more sedate environment of the residence of the governor of Malacca, a group of slaves from Ambon led by "a native of New Guinea, his hair frizzled out 6–8 inches from his head", played European tunes "remarkably well, with very little practice."[29]

The multiple dimensions of cultural exchange become tangible in the material goods purchased and displayed in European, Chinese, and Muslim households. A watercolour by Jan Brandes, *A tea visit in Batavia* (fig. 9), provides a visual depiction of this heterogeneous aesthetic. Two ladies of European and Eurasian origin, attended by slaves, are shown in an ornate reception room where various objects are displayed. On the wall are Chinese paintings and gilded mirrors from Europe, but we also see Chinese porcelain, a Vietnamese vase, large Martaban jars, and a copper spittoon, should a visitor be chewing betel nut.[30]

While documents such as wills and estate sales record the possessions of elite households, local people like the VOC translators in Sri Lanka also became "tastemakers", setting new standards for cultural consumption in their display of Japanese tea kettles, playing cards, or European prints.[31] Asian port cities thus served as dissemination points for the paintings, textiles, ornaments, clocks, weapons, and other imported items desired by Asian consumers of all levels, so that even the wearing of a European hat could become a status symbol for a mestizo man. What began as divergent cultural traditions converged in the Asian port cities to create a new and integrated cosmopolitanism as European designs were incorporated into Indian textiles, or Chinese porcelain was decorated with Arabic-style lettering in order to reach out to Muslim consumers.

9

10

11

9

A tea visit in a house in Batavia, 1779-85.
Jan Brandes, Watercolour over pencil sketch on paper.
Rijksmuseum, Amsterdam [NG-1985-7-2-15].

10

View of the Plaza in front of the Binondo Church, Manila, 1868.
José de la Gándara.
In Richard Chu, *Chinese and Chinese Mestizos of Manila Family, Identity, and Culture, 1860s-1930s* (Leiden, 2010), p. 181, fig. 6.

11

Dutchman at home in Batavia.
From W. L. Ritter, *De Laatste der Oudgasten* (*Indische jaarbookje* 1854, p. 97.).

next to a joss house (and later a temple) that housed Mazu's statue, and Chinese sailors commonly made offerings to images of the Virgin Mary, like the one in an old Chinese temple at Batangas, south of Manila.[35]

Although the arrival of Christianity in the early sixteenth century added another religious dimension to an existing cosmopolitan mixture, it also introduced tensions, especially in China, Vietnam, and Japan. The Christian-Muslim animosities resulting from the long Moorish occupation of the Iberian Peninsula were also transported into Asia. Yet in time most port-city authorities adopted pragmatic measures in order to guarantee a peaceful environment that would attract traders. Indeed, their long experience in accepting compromises both personally and in commerce meant that Europeans in early modern Asia were generally more tolerant of non-Christian beliefs and practices than were their countrymen at home.[36]

During the nineteenth century, however, Islamic reformism from the Middle East and more assertive Christian missionizing associated with European colonial regimes tended to strengthen religious boundaries. The increased presence of European and Chinese women meant a decline in interethnic marriages, with a consequent weakening of the cross-cultural ties that had promoted the adaptation of outside ideas and practices, including religion, to the local environment. Yet enduring features remained. During her visit to Singapore in 1880, the intrepid Isabella Bird remarked that "the English, though the ruling race, make no impression"; rather, she depicted a city "ablaze with color... a fascinating medley... of mingled nationalities."[37] From this perspective, Singapore is emblematic of the unique environment that characterized the Asian port city.

16

Jesuit Convent, Macao, 1856.
Lithograph, based on an 1853 sketch by William Heine (1827–1885), from *Narrative of the Expedition of an American Squadron to the China Seas and Japan performed in the years 1852, 1853 and 1854, under the command of Commodore M. C. Perry...* (Washington, 1856), after p. 300.

Singapore: A new Asian port city

The networks of early modern port cities linked together by the Portuguese Estado da Índia, the Dutch VOC, the English East India Company, and the private traders underwent a significant change in the nineteenth century. With the decisive victory of the British over the Nawab of Bengal and his French allies in 1757, the East India Company was able to consolidate its position in India and expand to other areas of Asia. The English-dominated ports of Bombay, Madras, Calcutta, Singapore, and Hong Kong came to be the principal bases for the rapid expansion of the British colonial empire and hubs for the new influx of people (particularly women), advanced technology, and racist ideology in the guise of science. From an imperial standpoint, the self-confidence and power of Britain's empire in Asia was nowhere more evident than in Singapore.

Located at the southern end of the Straits of Malacca, Singapore was easily accessible to maritime traffic moving both east and west. Archaeological evidence demonstrates that there was a trade-focused settlement on the island as early as the fourteenth century, but it was abandoned in later centuries and became a marginal outpost of Malay kingdoms and an appanage of the Temenggung family. Except for the temporary residents of local sea-peoples

17
The Armenian Church, Singapore.
Photograph, mid-20th century.
National Museum of Singapore
[XXXX-13223].

(*orang laut*) and a few Malay fishing villages, Singapore was not prominent in the Asian trade network. This changed dramatically in 1819, when Stamford Raffles on behalf of the East India Company signed a treaty acquiring the island from the Temenggung of Johor. It was a move that was part of a British plan to establish a series of port cities that would service the highly profitable China tea trade.

Characterized by free trade and a laissez-faire government, and with a large and conveniently located harbour, Singapore's commercial success was rapid. By mid-century it had become a favoured entrepot and destination of Chinese junks and European ships, including steamers from the 1870s, bringing in both goods and immigrant labour. Between the 1880s and the outbreak of the First World War, some five million Chinese, mostly male, passed through Singapore to various regional ports to become labourers in the expansion of Chinese commercial interests.[38] Malacca and Penang, which had been grouped together with Singapore to form the Straits Settlements in 1826, absorbed these new migrants into the traditional ethnic quarters or kampong. Singapore authorities, however, envisaged a simplistic division of neighbourhoods based on the essentialized categories Chinese, Indian, and Malay, ignoring the complexity of caste, religion, clan, and dialect differences. Asian communities conducted their affairs through their own religious and cultural institutions, but the civic space catered primarily to European concerns. Singapore's dependence upon trade as its lifeblood is reflected in the presence of the commercial establishments outside the European town, unlike other European-controlled Asian port cities where military, commercial, and administrative centres were adjacent to each other.[39]

As was the case in earlier entrepots in the region, Singapore became the staple port for Southeast Asia's sea and forest products, as well as for spices, pepper, and rice. It also became the primary redistribution centre for tin and cash crops from the hinterland in the Malay Peninsula, including rubber at the turn of the nineteenth century. Within five years of its founding, Singapore boasted a population that included Chinese, Malays, Indian, numerous traders from the archipelago and mainland Southeast Asia, Arabs, and other traders from central Asia.

Begun as a typical Asian port city, it developed into a global player because of the commercial networks that linked wealthy Chinese and Indian merchants with European representatives of agency houses. The vast differences in wealth in the population produced a society of contrasts, evident in the material culture that became part of Singapore's varied heritage.[40] Yet, although a Malay ruler like Abu Bakar of Johor was regarded in Singapore's elite circles as an English gentleman, he considered an understanding of Chinese culture to be essential to good government.[41] The connections that underwrote this legacy of tolerance are tangibly represented by Singapore's Armenian Church (fig. 17). Dating from 1835, it testifies to a long history of global interaction that linked the port cities of Asia not only to regional exchange points but to the great commercial centres of Europe.[42]

Asian port cities survived the vagaries of politics, conquerors, and empires, with the most successful adjusting to ongoing changes in the political and economic environment. Throughout the period from 1500 to 1900, the enduring element was the human mobility by which they were connected. Asian port cities flourished as unique sites for cultural and material exchanges and as gateways for ideas. The vibrant and hybrid communities that developed here foreshadowed the future of a shrinking globe, where the multicultural societies of contemporary port cities are still exploring new modes of intellectual interaction and artistic expression.

1 Abu-Lughod 1989.
2 Andaya and Andaya 2015, pp. 5–9.
3 Wolters 1999, pp. 42, 44.
4 Pearson 2006.
5 Seshan 2012, p. 10.
6 Dhiravat 2016, pp. 200–201.
7 Faure 1996, pp. 6–9.
8 Skinner 1982, p. 47.
9 Cortesão 1990, vol. 1, p. 228.
10 Gipouloux 2011, pp. 118, 121, 139.
11 Deb 1905, p. 3.
12 Gipouloux 2011, p. 11.
13 Boyajian 1993, pp. 78, 81; Barros 2013, pp. 160–61; Disney 2009, pp. 192–98.
14 Meilink-Roelofsz 1962, pp. 227–38; Boxer 1966, pp. 201–6.
15 Gipoutoux 2011, p. 34.
16 Trocki 2011, p. 205.
17 Cortesão 1944, vol. 2, pp. 265, 273–74; B. Andaya 2012.
18 Blussé 1999, pp. 122–23; Trocki 2011, pp. 200, 205.
19 Chin 1998, pp. 140–41, 147–48, 156–58, 303.
20 Blussé 1999, p. 109, 119.
21 Matheson 1972, p. 138, note 52.
22 Chattopdhyay 2005, p. 77.
23 Felix 1966, p. 126.
24 Earns 2012, p. 43.
25 Raben 1996, pp. 246–47.
26 Negrón 2015.
27 Carpio 2005, p. 103.
28 Skinner 1982, p. 61.
29 Bulley 1992, p. 90.
30 De Bruijn and Raben 2004, pp. 193–96.
31 Kaufmann and North 2014, p. 14; North 2010.
32 Pearson, 2006, pp. 365–66; B. Andaya 2016.
33 Newbold 1971, p. 147,
34 Tan 2014, p. 217.
35 Ang See 1990, pp. 64–67; Dy 2014.
36 For the Portuguese see Disney 2009, pp. 00–201.
37 Isabella Bird, *The Golden Chersonese and the Way Thither* (London, 1883), pp. 144–45.
38 Trocki 2011, pp. 210–11.
39 Pieris 2009, p. 49.
40 Trocki 2006, pp. 40–66.
41 Andaya and Andaya 2016, pp. 160, 181.
42 Aslanian 2011, pp. 57, 61, 140–42.

1

Interior of a horse-drawn tram.
Batavia, 1888. Engraving by J. C. Rappard in Perelaer 1888, vol. 1, pl. 8.

Peter Lee and
Alan Chong

Mixing up things and people in Asia's port cities

Although port cities share many cultural characteristics and dynamics with urban centres generally, they are defined by traffic across maritime routes – a flow of commerce, goods, and people seemingly more intense and less governable. A consideration of the connections among Asia's port cities presents new and intriguing dimensions to understanding this vast region. The diverse networks created by the movement of people and cultures from port to port evince a much more complex, and ultimately more realistic, concept of Asia than a conventional geopolitical map, traditionally defined by nations, empires, and shipping routes. The ceaseless, varied, and multidirectional nature of human encounters of all these networks challenge the traditionally fixed boundaries of race, culture, nationality, and civilization.

This paradigm has long been established in the field of Asian history. The work of Wang Gungwu, Anthony Reid, Victor Lieberman, Leonard and Barbara Andaya, David Ludden, Finbarr Barry Flood, among others, many inspired by Fernand Braudel, have established new ways of thinking about Asian history beyond national and racial narratives. Anthony Reid in particular promoted the idea of intra-Asian networks and of the cultural and racial diversity, and consequent hybridity, in Asia's *cosmopoleis*, or cosmopolitan centres where the entire globe seemed to gather.[1] Leonard Blussé, Heather Sutherland, Frank Broeze, and others furthered this approach with case studies of ports, individuals, and groups, thereby uncovering hidden histories and providing a critical human dimension to the study of networks.[2]

Yet Asian art history, especially when presented in museums, remains entrenched in old historiographies by favouring traditional ethnocentric and nationalistic perspectives. The fact that many museums in Asia are nationally funded institutions further politicizes their displays. The culture of port cities has generally been subsumed into the wider histories of nation-states. For example, the material culture of Batavia (present-day Jakarta) has usually been presented in the context of the Dutch colonial empire or in the development of the Indonesian republic, while its intimate links to other Asian port cities, especially those belonging to other political spheres, have barely been explored. Batavia is, for example, perhaps the most important precursor to Singapore, and not only because Thomas Stamford Raffles had been governor of Java from 1811 to 1815. The multicultural trading port, connecting Asia with Europe and open to Chinese immigration, was a model for Singapore. At the same time, the mixed-race communities of Batavia were also duplicated in Singapore.

The presentation of the art of Asian port cities in the light of new academic developments is a challenging task, and this exhibition cannot hope to be a comprehensive analysis of the material cultures of these networks. Rather, the project attempts to highlight a few key themes. These are based on three basic dynamics of transcultural movement: divergence, convergence, and integration. These forces create a high degree of commonality across large distances. What we understand as globalization and contemporary popular culture are therefore rooted in these very old networks.

A resonant starting point for the exhibition is the arrival of the Portuguese in Asia in the early sixteenth century, when global links began to be more firmly established. The end of the nineteenth century ends the exhibition's timespan, at the point when advances in technology accelerated cultural dynamics towards a more familiar modernity. The themes therefore do not aim to indicate the passage of history nor to chronicle important historical periods; nor do we explore the purely commercial function of ports and shipping - topics well surveyed by economic historians and other exhibitions. Rather, we aim to highlight important cultural features of the Asian ports that resonate with contemporary cosmopolitan life in Asia. Indeed, many aspects of cultural life of the present are not new at all, but are deeply rooted in Asia's forgotten past.

1. Divergences

Networks of people

Asia's long history of mercantile networks and ethnic diasporas form one of the main conceptual foundations of this exhibition. Networks based not only on place of origin, but also on family, type of business, or specialized craft provide completely different ways of understanding Asian history. The exhibition draws attention to this dynamic through the lives of people from different worlds and periods, who lived or worked in places far removed from their place of birth. Here are a few striking examples:

Cornelia van Nijenroode (ca 1629–1691) was born in Hirado, Japan, to a wealthy Dutch merchant and his Japanese wife. She spent her early adulthood in Batavia and died in The Hague. Leonard Blussé's remarkable account of her life allows us to assemble some fascinating objects connected to her.[3]

Georg Franz Müller (1646–1723), who was born in Alsace, spent twelve years in island Southeast Asia as a soldier of the Dutch East

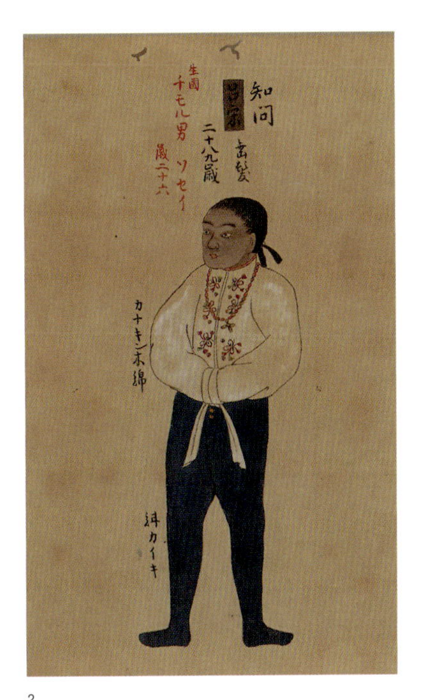

2
Detail of Cat. 94.

India Company before he returned to Europe, where he served in a Swiss Catholic monastery. During his years in Asia, he created a remarkable album of watercolours which record observations of the people he encountered in the Dutch colonial world. He also collected a few artefacts, which he bequeathed to the monastery of St Gallen in Switzerland.[4]

Duzhan Xingying (1628–1706) was a Buddhist monk and painter from Putian, in China's Fujian province, who moved to Nagasaki and became the celebrated fourth abbot of Mampuku-ji, the centre of the Ōbaku sect of Zen Buddhism founded by his master and fellow Fujian native, Yinyuan Longqi. Duzhan's work inspired developments in Japanese seventeenth-century painting and tea ceremony philosophy.

Abdullah bin Abdul Kadir (1797–1854), a Malacca native of mixed Indian, Arab, and Malay ancestry, began his career as an itinerant *munshi* (teacher and translator) for colonial officers, the most well-known of whom was Thomas Stamford Raffles. He travelled widely in the Malay world, was a devout Muslim who assisted the publication work of Christian missionaries, and is celebrated for his autobiography, *Hikayat Abdullah*, regarded as the first example of modern Malay prose.

The **Nattukottai Chettiar community** from India's Tamil Nadu region formed wide networks across Asia in the nineteenth century. The treasury of the Sri Thendayuthapani temple in Ho Chi Minh City, Vietnam, comprises devotional jewellery fashioned of gold and precious stones, donated to the temple in the late nineteenth and early twentieth century. Now stored in Singapore, the collection is a rich visual expressions of intra-Asian networks.

The handscrolls depicting foreigners in Nagasaki also reveal the peripatetic journeys of individuals from Zhangzhou, Ambon, Timor, Goa, Batavia, and Manila, among other cities. For example, natives of Timor are described as coming from Macau and Luzon (Manila) (fig. 2). Child trafficking was also rampant. Many of the foreigners depicted are children, ranging in age from nine to thirteen years.

These disparate examples of the dispersion of individuals provide a clear argument against simplistic notions of East meeting West, or of the influence of one civilization on another. Rather, these individual histories emphasize the heterogeneous, multi-directional nature of human encounter. These diasporas, at the same time, also engendered a certain kind of familiarity across several cities, which often had the same kinds of itinerant merchants and migrant communities and, accordingly, the same kind of physical or built landscape. In various ways, these environments created a shared culture that lay outside the traditionally defined categories of China, India, or the Portuguese (or Dutch or Spanish or British) empires – although it derived from these and other sources.

Networks of goods
The case of Indian cottons for global markets

By the sixteenth century, the globalized marketplace witnessed the wide dispersion of goods from many sources. Asian ports were flooded with these products. Porcelain and textiles were perhaps the most voluminously manufactured and traded, and therefore also the most studied. The dispersion of resist- and mordant-dyed cottons from India, which were in high demand in Europe, Southeast Asia, and Japan, is a perfect expression of a network created by a single type of product. This exhibition highlights the manner in which Indian cotton had become a global commodity by the eighteenth century, mainly distributed through Dutch trading channels (although there was also a significant shadow channel of trade conducted by other European and Asian merchants that was not officially condoned by the Dutch). The Dutch East India Company, or the VOC, had fine-tuned the process of supply and demand, and was able not only to distribute Indian cottons globally, but also to specifically tailor the textiles to regional markets. The extensive lexicon of trade terms for the array of Indian textiles demonstrates the complexity and scale of the market. Many of these terms entered the vocabularies of Asian and European languages. Moreover, many local communities devised their own terms for different types of cloths. In rural and urban communities, the function of these cloths could vary from district to district.

The competitive commercial responses to this lucrative global trade intensified in the eighteenth century and resulted in several significant developments, including the invention of machine-printed cottons in Europe, the rise of the batik industry in Java, and the production of an imitation of Indian chintz in Japan referred to as *wa-sarasa* (和更纱).

In fact, Japanese scholars and collectors first understood the global nature of Indian chintz. *Sarasa*, from the Indian word meaning "colourful", was the VOC trade term for a type of resist-dyed Indian cotton and also a part of Malay vocabulary, but in Japan it connoted any kind of resist-dyed or printed cloth. In Japanese studies, Indian chintz, European and Asian machine-printed cottons, Indonesian batik, and Japanese stencilled or block-printed cottons all fall under the category of *sarasa*. Although Matiebelle Gittinger and John Guy have both furthered knowledge of the global reach of Indian cotton, scholars in general pay little attention to the international responses to this textile.[5] In the same way, scholars of Indonesian batik almost always underestimate the links with Indian chintz, preferring to see batik as a pure, native Indonesian art form. Yet the visually evident interconnections, and the increasing documentation of the dynamic period from the late eighteenth century, when other markets initiated competitive responses to Indian chintz, are creating a wider and more enduring network of printed cotton cloth and its origins.[6]

4

Portrait of a woman.
August Sachtler. Singapore, 1860s.
Albumen print, 30.3 x 22.8 cm.
Mr and Mrs Lee Kip Lee.

5

Three men in a garden.
John Thomson, British.
Singapore, 1860s. Albumen print,
23.5 x 18 cm.
Mr and Mrs Lee Kip Lee.

3
Singapore people.
G. Riemer. Singapore, 1875.
Albumen stereoview print
(published by J. F. Stiehm, Berlin, 1884).
Mr and Mrs Lee Kip Lee.

2. Convergences

Multi-ethnic itinerants and residents

The high visibility of people from various parts of the world converged into one urban centre is a key feature of port cities in Asia. The exhibition augments the voluminous historical accounts on this subject with visual records that attest to the mix of foreigners from the seventeenth century to the late nineteenth century.

Writers passing through Singapore in the nineteenth century often remarked on the vibrant mixture of people wearing fashions from different parts of the world (fig. 3). Singapore therefore exemplifies the intensely multiracial aspect of port cities. This is attested to by the images of pioneer Singapore photographers such as August Sachtler, John Thompson, and Gustave Lambert, who were active in the 1860s (figs. 4, 5). These photographs provide the basis for a display of the diverse costumes that might have been worn in Singapore in the late nineteenth century, just before modern fashion made styles more global and homogenous (Cats. 5–18).

The multiracial character of port cities is already evident in the seventeenth-century album of Georg Franz Müller, with its depictions of the peoples of the Dutch East Indies; two eighteenth-century handscrolls illustrating foreigners in Hirado, Japan; the drawing of a Batavian wedding by the Lutheran minister Jan Brandes; an early nineteenth-century Cantonese watercolour depicting a Parsi and European merchant amid a group of Chinese; and the prints of August van Pers (see Andaya fig. 13).

These manifestations of the multicultural nature of port cities provide important information about origins, objects, society, and dress, but ultimately they can merely hint at the complex convergence and circulation of people. Their art and culture, including luxury goods, cuisine, music, sport, entertainment, religion, fashion, and literature, will be subject to historical investigation for a long time to come.

Batavia
Rethinking the emporium of international goods

Imported goods were distributed widely in Asian port cities, and were remarkable for quantity and variety. For example, inventories of the estates of the multiracial residents of Malacca and Batavia reveal that individuals accumulated belongings that were imported from Europe, Japan, China, Burma, Sulawesi, Java, Sri Lanka, India, among other places.[7] People of different cultures owned many of the same types of objects. For example, European-style paintings circulated among residents of all ethnicities.[8] Moreover, auction records indicate that the city's multiracial residents bought up the used goods on the secondary market. This evidence undermines our present notions of race-based ownership and patronage of goods. Therefore, objects described as representing Portuguese or Dutch or British taste may in reality have had a much wider and less definable market. This exhibition presents a sample selection of the wide variety of imported and locally made objects typically found in any prosperous eighteenth-century Batavian residence, regardless of ethnic origin, including items that were already antiques at that time. The global origin of possessions in a household is therefore hardly a new phenomenon, but in fact one that is already a few centuries old.

The traditional art historical approach typically focuses on the creators and first owners of an object. However, equally important if often ignored in the life of an object is its re-use, repurposing, recycling, and copying. Another phenomenon of imported objects in port cities is the cultural re-assignment in the description of the object by patrons and later collectors, seen, for example, in the term *nyonyaware* (Chinese porcelain made for the Peranakan market), or for the plethora of local names in different languages across Asia for the same export object.

The convergence of goods from around the world in a port, and their circulation within the city, their transformation from new possessions, to second-hand items, and finally to antiques, to their re-purposing (often used in a manner unrelated to the original intent of their maker), their re-imagining, re-naming, or reinterpretation by subsequent owners, and even to their re-export, altogether suggest the necessity of a re-evaluation of so-called "export" objects in Asia.

3. Integrations

The ceaseless, usually random, encounters between people from different communities instigated a multiplicity of marital and cultural outcomes. The innumerable consequences of encounter and the wide spectrum of developments have historically been difficult to track, and apart from very obvious cases, most examples of such integrations have passed unnoticed.

The problem largely stems from the historical prejudices in many communities against mixed marriages. Up to the first half of the twentieth century, mixed race communities we often viewed highly negatively, and were thought to have been tainted with immorality and instability. Academic circles tended to ignore hybrid communities, especially as they fell out of favour with emerging nations, although in recent decades, scholars have begun to study hybrid communities as "legitimate" cultural groups.[9]

Hybrid objects – objects that do not belong to any orthodox cultural category, that are made by craftsmen of diverse ethnicities for customers from equally diverse communities, exhibiting patterns and motifs drawn from a wide spectrum of cultures – have also suffered the same fate. In the first place, hybrid objects are relatively difficult to trace and study. Unless they can be found in sufficient quantities and have visual and written documentary evidence supporting their origin and function, such objects usually have no known provenance and history, and could be the result of adventitious, one-off experimentations, specific orders for a single client, improvisations, and types that might have enjoyed brief popularity in only one area or neighbourhood. They also might have been made or copied by craftsmen of diverse ethnic backgrounds, decorated with popular motifs selected arbitrarily, without any cultural connection to patron or maker.

Compounding these problems is the fact that academic and museum-based art history is often still preoccupied with the original, the authentic, the pure, the national. Museum displays usually centre around royal courts, religious institutions, and intellectual elites. Meaning and symbolism are often of prime importance. Objects are selected to illustrate characteristics that define a style, a community, or a nation. This often means that supposedly impure cultural traits are glossed over, and artefacts without some profound cultural (usually courtly) symbolism and function are relegated to low art or cast as artefacts of popular culture. The larger proportion of artefacts that do not fit the intended idea are excluded, regarded as quaint or debased, while those with some exceptional aesthetic or cultural merit are grouped under the category of exotica.

This traditional academic approach has unfortunately also been adopted in recent studies of hybrid art, which have sought to define and ring-fence the subject, thereby ignoring any other examples that might not fit the proposed definition. For example in Harmen Veldhuisen's study of batik made by European and Eurasian workshops, the criterion of selection was the European surnames of the makers.[10] The cultural influence from the Asian or Arab ancestors of the Eurasian batik-makers was generally ignored. Asian batik-makers who produced the same style and quality of batik were also excluded. Ho Wing Meng's books on Straits Chinese material culture focused primarily on artefacts that were popular with collectors and available in the local antique markets, and many types of objects that were not unique or exceptional to the Peranakan community were excluded.[11]

6
Cat. 165.

In the context of the diverse and inconsistent cultural life of port cities, however, hybrid objects were in fact plentiful and quotidian. Even from the viewpoint of the dominant culture, such objects were not exotic in the slightest. Sünne Juterczenka and Gesa Mackenthun advocate the mathematical concept of "fuzzy logic" as an alternative to the terms "hybridity" and "creolization", because the variable degrees of values inherent in fuzzy logic offer a more cogent conceptual framework to analyse the results of encounters.[12] The so-called inconsistencies of, or spectrum of results in, hybrid artefacts are therefore perfectly consistent within the framework of fuzzy logic. The hybrid and the exotic are the new normal.

This exhibition highlights a few categories that explore the nuances of integration in port cities. Through the medium of painting and photography from the eighteenth to early twentieth century, various hybrid communities in Asia are presented in order to explore diverse Eurasian and intra-Asian encounters. The subjects, visual techniques, and styles, as well as details of clothing, jewellery, and belongings, present a complex and variable mixture of cultural influences. More specifically, this section of the exhibition highlights the evolution of the European-style chair, as well as jewellery, metalwork, and betel nut boxes (fig. 6). Such artefacts represent the wide and erratic range of responses to the meetings of culture. They also reflect the widespread integration or "copying" of the popular by skilled Asian and European craftsmen who could produce work in any cultural style.

The legitimacy of the ambiguous

One common factor underlying many objects in this exhibition is the uncertainty concerning exact origins, despite the fact that they (or similar examples) have been the subject of scholarly research. Quite often, identifying the specific location or region has been based purely on conjecture, although the conclusions have become accepted as fact.

The only realistic approach would be to accept the ambiguity of origins, until solid evidence emerges. In the worst-case scenario, the precise provenance remains unknown, but perhaps this might be less relevant in the paradigm of Asian material culture proposed here.

Networks as culture

7
Plaque: The Christ Child as Navigator.
Unknown, possibly Macau or Manila, early 17th century. Ivory, 12 x 8.2 cm. Asian Civilisations Museum [2015-00220].

It has proved impossible to pinpoint the origins of this and similar plaques. Mostly attributed to Chinese artists, at least one example seems to have been carved in Sri Lanka or India (British Museum).

In many cases, the European trading networks wanted certain types of objects with little regard as to where and by whom they were made. Most famously, the insatiable global market for high-quality porcelain led the Dutch to Japanese kilns when Chinese sources dried up in the middle of the seventeenth century. Similarly, the demand for Christian figures carved of ivory led to the flourishing of artistic workshops across Asia. Facilitated and traded by Portuguese and Spanish trading networks, exquisite ivory figures were produced in such scattered centres as Goa, Sri Lanka, Guangzhou, Fujian, the Philippines, and perhaps Nagasaki and Ayutthaya as well. These small sculptures of the Virgin, the Crucified Christ, and the Christ Child look tantalizing similar. Although modern art historians believe they can distinguish the specific sources of production through documented historical collections in Spain, Mexico, and Portugal; as well as through analysis of specific stylistic features, many confusions remain.

This is especially true of ivories deduced to have East Asian facial features, and therefore attributed to Chinese carvers (fig. 7). But is this necessarily the case, and where were these artists based? Guangzhou and Fujian province in China were active ivory-carving centres, and there appears to have been numerous Chinese carvers working in Manila. Were Chinese ivories shipped to Manila for export to the New World, and did other ethnic groups take up the "Chinese" style of ivory carving? Smaller numbers of ivory carvings have been speculatively attributed to Japan and Thailand, but these closely resemble the Chinese examples. We are forced to concede that the cultural identity and the geographical location of these artists remains for the most part unknown. Indeed, these issues may not matter as much as the idea of styles and consumer systems shared over networks that spanned considerable distances.[13]

The same factors come into play with ebony furniture, which was produced in Sri Lanka and southern India for the Portuguese, Dutch, and British. Not only is there clear evidence that such pieces were made in the Subcontinent for export, but also that ebony workers were brought to Batavia, as reported in 1682.[14] Certain styles have been classified as Dutch and British (and even in some cases as Portuguese), but these categories are not absolute, nor were the consumers of the furniture only European.[15] Ebony or black wood furniture became exceptionally popular throughout Southeast Asia and southern China, and it appears that such examples were made in many ports throughout the region.

One of the most remarkable examples of this instability is an object that seems at first glance a product of seventeenth-century Japan. This Christian shrine is lacquered in rich black, decorated in gold, and inlaid with mother-of-pearl in the style called Namban, as it was meant for the international market (fig. 8). Historians have always assumed that these works were produced exclusively in southern Japan and exported through Nagasaki and other Japanese ports. However, the illustrated example displays technical differences in the layering of mother-of-pearl and the thickness of the gold decoration, which suggests that it was produced in Macau rather than in Japan.[16] This was the result of the dispersal of the Jesuit artistic workshops in Japan after the outlawing of Christianity in the 1620s, which forced many artists to resettle in Macau.

These objects are products of networks which connected port cities across Asia. Although they may have characteristics connected with standard cultural classifications (Chinese, Japanese, Indian, and so forth), they equally lie outside these stylistic spheres. Indeed, the terms sometimes applied to hybrid works of art fall short: sculptures made by Chinese artists resident in Manila, for example, are poorly described as "Hispano-Filipino"; just as "Dutch market" inadequately conveys the new forms developed for the thriving multicultural community of Batavia.

This ambiguity, together with the subtle, manifold, and variable nuances of each object, the range of commercial responses to popular types (competing, copying, counterfeiting, and mass-producing), the wide-ranging distribution of bulk and luxury goods, and the common demand across regional markets, all suggest the urgent need for a new art historical approach to the past, where such characteristics can be rehabilitated and validated.

The art and culture of port cities celebrate human enterprise at its most unbridled and egalitarian. The spirit of unabashed experimentation, creative improvisation, and nonconformist expression resonate with contemporary global art and culture. The freewheeling, impermanent cultural environment of Asia's port cities in the past few hundred years has therefore never been more vitally significant and relevant to understanding the present.

8

Hanging oratory.
Unknown, possibly Japan or Macau,
17th century. Lacquered and painted wood,
30.2 x 23.5 x 4 cm.
Asian Civilisations Museum.

1. Reid 2015.
2. For example, Blussé 1985; Blussé and Fernandez-Armesto 2003; Sutherland 1989, Broeze 1989.
3. Blussé 2002.
4. Schmuki 2001. Mŭller 1669.
5. Osumi 1957 connects the *sarasa* for Europe, Indonesia, Thailand, and Japan, and also includes Indonesian batik. Gittinger 1982, Guy 1998.
6. Lee 2014.
7. Lee 2014, pp. 290–308.
8. North 2010.
9. Reid 2010.
10. Veldhuisen 1993.
11. Ho 1976; Ho 1983; Ho 1987; Ho 1994.
12. Juterczenka and Mackenthun 2009.
13. Chong 2016.
14. Nieuhoff 1682, p. 205.
15. Amsterdam 2003, pp. 207–10; Singapore 2013, p. 106.
16. Communication from Pedro Cancela de Abreu, September 2015. The mother-of-pearl is thinner than normally found on Namban objects, and placed on the final layer of lacquer, rather than set into the lacquer layers.

SECTION DES DAGUERRÉO TYPOMANES
SECTION DES DAGUERRÉOTY POLÂTRES
200
ÉPREUVE R
SANS SOLEIL!
APPAREIL PORTATIF POUR LE VOYAGE
Maurisset
Chez Bauger R. du Croissant 16
LA DAGUERRÉO

POTENCES A LOUER POUR MM. LES GRAVEURS
MAISON SUSSE FRÈRES
ÉTRENNES DAGUERRÉOTYPIENNES POUR 1840
ÉPREUVE DAGUERRIENNE SUR PAPIER
SYSTÊME DU DOCTEUR DONNÉ
SORTIE
FENÊTRES A LOUER
OURNÉE
13 MINUTES
250
APPAREIL POUR LES PORTRAITS DAGUERRÉOTYPÉS.
COMPLET 300F
Imp. d'Aubert & Cie

YPOMANIE

1
Woodbury & Page.
Batavia roadstead, around 1865.
Albumen photograph, 19.4 x 24.5.
National Gallery of Australia.
Ships had to anchor in the "roadstead" offshore. In 1857, photographers Walter Woodbury and James Page and their baggage arrived by a flat-bottomed lighter at this dock in the Harbour Canal.

Getting into the picture
Photographers and their customers in the port cities Goa to Manila, 1840–1900

Gael Newton

From Java I will most likely go to Borneo and from there to Manilla to China and then to India, that is if I live so long. I shall be able to write a voyage around the world when I arrive home.

Walter B. Woodbury, a British photographer, on preparing to depart in May 1857 from work on the Australian goldfields for a world tour in search of fresh opportunities. He got no further than the Dutch East Indies, where he founded the long running firm of Woodbury & Page in 1857. Woodbury ran it with two of his brothers until the 1880s, and they made the largest and most widely distributed nineteenth-century images of colonial-era Indonesia (fig. 1).[1]

Photography ships out

On 8 December 1839, the Paris journal *La caricature* published a cartoon by Théodore Maurisset titled *La daguerreotypomanie*, which satirized the recent craze for the photography process perfected by a well-known painter and the owner-designer of diorama shows in Paris and London, J. L. M. Daguerre (1787–1851). Only months before, the French government had bought Daguerre's patent for producing highly detailed mirror-like images on polished metal plates, and gifted the technical secrets "free to the world" on 19 August. Daguerre's marketing campaign over 1838 and 1839 had whipped up such anticipation that by the time the method was revealed, the world was wild for photography.

Each daguerreotype was unique and needed to be cased to protect its vulnerable surface. A rival British process on paper called "photogenic drawing" was released by W. H. F. Talbot (1800–1877) in January 1839. This was a method for making impressions of objects on sensitized paper by direct contact or photographic images via a camera lens, and

2

3

4

was open to amateurs to attempt. It was easier and cheaper to make but less detailed and not quite as dramatic a performance as seeing a daguerreotype plate develop. The daguerreotype promised to deliver the world in miniature – albeit without colour or action.

In Maurisset's image we see swirling crowds of fashionable Parisians seeking all manner of photographic experiences while out of work artists are shown committing suicide in despair. Trains and steam ships loaded up with cameras can be seen in the background of *La daguerreotypomanie*, and in the foreground a jolly young gent in black suit carries a daguerreotype camera, tripod, and chemicals (fig. 2), captioned as his "appareil portatif pour le voyage" (traveller's portable camera outfit), under his arm, and Daguerre's manual of instructions tucked in his pocket. This is probably the first-ever depiction of a travelling photographer. Maurisset was not fantasizing: photography was on the move. Expensive Daguerre-Giroux brand cameras had been dispatched on a world cruise via South America in September, and the Paris optician N. P. Lerebours was marketing a cheaper apparatus by October.

In the first half-century or so of photography, several hundred such men, mostly from Europe, would take photography to the port cities of Asia in the hopes of making their fortune (fig. 3). Some swapped their black suits for the all-white suiting adopted by foreign residents in Asia. Their arrival inspired both foreign residents and native-born to take up photography as hobby or profession. The first Asian photographers would arise in the port cities.[2] The earliest photographs surviving photographs of Asia and Asians are in the daguerreotypes of 1844 by French trade negotiator Jules Itier (fig. 4).

2
Théodore Maurisset.
La daguerreotypomanie
(detail of plate on pp. 43–44.).
From *La caricature* (1839).

3
G. R. Lambert & Co., Singapore & Delhi.
***William Hancock, Chinese Maritime Customs Service official, posing with unidentified man*, Sumatra, January 1891.**
Albumen photograph, 17 x 11 cm.
Edward Bangs Drew Collection,
Harvard-Yenching Library,
Harvard University.

4
Jules Itier.
***Vue redressée de la Praia à Macao,* Octobre 1844.**
Daguerreotype, 9.3 x 14.3 cm.
Musée français de la Photographie,
Bièvres.

Photography makes port

5
Isidore van Kinsbergen.
Malay family, around 1865.
Albumen silver photograph,
19.5 x 15.2 cm.
National Museum of Singapore
[2011-00701-028].

In January 1840 a daguerreotype camera was advertised by Thacker and Company in Calcutta, the earliest known sale in Asia. The instrument was possibly the one used by Irish physician, pharmacologist, and inventor Dr William O'Shaughnessy (1808–1889) in February to make daguerreotypes. The doctor had already made photogenic drawings in October the previous year. By the end of the year British trade firms in Calcutta included six employees listed as "artist, photographer".[3]

Dr O'Shaughnessy's efforts are the earliest known success in making photographs in Asia. He may have experimented a year or so later with Talbot's superior calotype process of 1841, which allowed for multiple printing of photographic images on paper using a paper negative. The watercolour paper used for the process resulted in a more graphic image than the fine detail and tonal range of the daguerreotype. Portraiture was not possible at this time due to the need for excruciating long exposures. The great advantage of the paper process was that it enabled mass production of prints. The calotype would prove markedly popular in British India in the 1850s, particularly for recording antiquities, but was rarely used elsewhere in Asia. The first photographic images travelled back along the seaways to Europe as soon as they were made, and were often publically exhibited. There was no means to reproduce photographs in the press until the 1880s, but redrawn as engravings and lithographs, the new imagery educated the international public and at the same time fixed stereotyped perceptions of Asia which linger into the present day. [4]

The photographers of mid- to late nineteenth-century Asia followed the patterns set in the previous century by itinerant painters from Europe who went East in search of sales and commissions. Portuguese, Dutch, and British artists in particular sought out the Asian colonial possessions of their nations but the field was open to all. Some would never return, ending their days in far-distant ports like the Prussian-born Jacob Janssen (1779–1856), who left home in 1807, and travelled via Copenhagen, Lisbon, and Boston then studied with an Italian artist in Philadelphia. Janssen worked chiefly as a topographical artist in Rio de Janeiro, Calcutta, Singapore, and Manila in the 1830s before settling in Sydney in 1840. Janssen's generation of travelling artists soon faced competition from itinerant and local photographic artists. For example, the Dutch-Flemish engraver and opera singer Isidore van Kinsbergen (1821–1905), who arrived in Batavia in 1851 with a French Opera Company and stayed on to become a pioneer photographer in the 1850s (fig. 5) and theatre director in his later years.[5]

Asia on a metal plate and paper pages

Despite Maurisset's accurate description of the instant export of cameras and photographers, there was no Parisian-style *daguerreotypomanie* in any of Asia's port cities. Indeed there are scant references even to the very first accounts of the invention of photography around 1840 in foreign-language colonial papers. This lack obscures how effectively travellers on fast ships brought the latest news out of Europe within a few months. By the time cameras or photographers arrived there was no need to explain what photography was. Local newspapers and published reports, while few, show that by the mid-1840s daguerreotype activity of some sort had happened in most of the major Asian ports.

Eyewitness reports of first sightings of photographs being made in Asia are rare.[6] A number of visiting naval expeditions to Asia in the 1840s had cameras aboard, but nothing is known of their use. The 1849 autobiography of Malacca-born Indian *munshi* Abdullah bin Abdul Kadir, a teacher of Malay and translator in Singapore, recorded seeing a doctor from an American warship make a daguerreotype view of Singapore between mid-1840 and 1841.[7] Remarkably, two caches of daguerreotypes from Southeast Asia survive: one from a French trade negotiator and amateur photographer Jules Itier and the other by Adolph Schaefer, one of the first generation of professional photographers at work in Asia (figs. 6, 8).

Jules Itier (1802–1877) was a well-educated administrator with scientific and natural history interests. He travelled in Asia for two years as chief commercial negotiator on the French government legation sent to China to conclude the first Franco-Chinese trade agreement signed at Whampoa in October 1844. While in Singapore, the French Legation stayed at the London Hotel of Gaston Dutronquoy, who hailed from French-speaking Guernsey in the Channel Islands. Dutronquoy was part-showman. He had a theatre on the premises and had advertised a daguerreotype portrait service in December 1843 and January 1844. Self-taught, Dutronquoy seems not to have succeeded with his photographic venture and must have viewed Itier's successful plates with not a little envy.

On his journey, Itier succeeded in making his first Asian views and portraits in Singapore, then in Guangzhou, Macau, Manila, Saigon, as well as Galle in Sri Lanka, on the way home.[8] These are the earliest extant photographic images from these lands. Itier's own account, *Journal d'un voyage en Chine en 1843, 1844, 1845, 1846* (published in Paris between from 1848 to 1853), reveals his curiosity about plants and people, and describes some of his photographic operations. Itier was undaunted by the difficulties of securing portraits. His surviving plates include people in the street, Europeans, and the first known portraits of Chinese sitters.

6
Jules Itier.
***Vue de la villa à commerce a Singapour*, 1844.**
Daguerrotype, 12.6 x 15 cm.
National Museum of Singapore
[2005-00445]

Most revealing is the exchange Itier reported of his photographing Lam Qua (1801–1860), a renowned painter servicing the foreign community

Portrait du peintre chinois Lam-Qua, miniature, peint par lui-même. — Collection particulière de M. Itier.)

7

8

in Guangzhou. Lam Qua was a master of Western academic realist oil painting who had exhibited at the Royal Academy in London. He was keen to see the "admirable apparatus that can draw by itself and that so intrigued the painters of Canton". He sat for a daguerreotype portrait and then a few days later presented Itier with a painted miniature copy of the daguerreotype as a memento, which Itier exhibited with his own plates on his return to France in 1847 (fig. 7).[9]

Itier used some of his own plates as the basis for illustrations in his own publication but did not circulate his works more widely. His archive remained in the family home in southern France unseen until its rediscovery in the 1980s. But even at the tiny scale and murkiness, in his daguerrotypes there is a vivid sense of what the place and people actually looked like, with which no print or painting can compare.[10]

The first instance of government-commissioned photography in the Asia was not in British India, as might be expected from the early interest of the new medium, but in 1841 at the instruction of the Dutch Ministry of Colonies in Amsterdam. They equipped medical officer Jurriaan Munnich (1817–1865) to test out the process in the tropics. Unfortunately his results were not satisfactory. The longer-term object of the Ministry of Colonies had been to see if the antiquities, especially the Borobudur complex in central Java, could be photographed. In 1843 they agreed to advance funds for better equipment for a second attempt by German professional Adolph Schaefer. The *Javasche Courant* of 22 February 1845 reported on his "beautiful and numerous" portraits, which had ended the "despair that Daguerre's art could ever be exercised fully in the tropics". Schaefer succeeded too in photographing Borobudur in 1845. A magnificent set of daguerreotypes of Borobudur and of antiquities in the Batavian Society of Arts and Sciences collection were sent back to the Netherlands but little used. Schaefer stayed on in Asia but was bankrupt by 1849. The Leiden University library holds his plates (fig. 8).[11]

Another significant route for the transmission of the daguerreotype into southern Asia was via the Pacific. During the 1850s, numerous daguerreian artists trained or came to Asia via North America. They moved form port to port as there were not enough customers to stay put. New arrivals tended to advertise that they had the latest American refinements, indicating a general rise in the perception of American technological know-how. This was possibly because daguerreotypes had a poor reputation, and were thought to fade and deteriorate in tropical conditions.

Most daguerreotypists worked for short periods, made only modest incomes, and left little personal trace. One who we do know about through his memoir *Reseminnen från Södra och Norra Amerika, Asien och Afrika* (published in Stockholm in 1886) is Swedish adventure-traveller Cesar von Düben (1819–1888). He set out in 1843 for lands he had read about as a child and returned home in 1858. Already a

7
Portrait of Chinese painter Lam Qua, miniature painted by himself.
Print. Collection de M. Itier in *L'Illustrations: journal universel,* no. 182, vol. 7 (22 August 1846), p. 393. HathiTrust and University of California Libraries.

8
Adolpf Schaefer.
Sculpture of Hindu god Karttikeya, 1845.
Daguerrotype.
Leiden University Library.

competent draughtsman, Düben took up photography in Philadelphia in 1849, practised in Mexico, America, China, the Philippines, Hong Kong, Singapore, Burma, India, and Indonesia. While many failed, Düben successfully used photography for a decade to support himself (fig. 9). Düben's reminiscences, along with Itier's journal, are the only personal glimpses of contact with locals by daguerreotype photographers in Southeast Asia.[12]

We will never know how happy residents in the port cities were with their metal plate images as so few portrait daguerreotypes survive. However, the arrival of the new paper photographs in the late 1850s turned a trickle of images made in Asia into a tide that flowed not only from port to port but also back into Europe and America. The wet-plate process gave superior clarity of detail and when paired with brilliantly detailed albumen (egg white printing papers), provided the means for photographic images and photographers to multiply and circulate worldwide. It was this process that would create the visual heritage of nineteenth-century photography of the people and places of the port cities of Asia.

The wet-plate process allowed for easier and cheaper portraiture, and for the production of views for sale. Advertisements bristle with claims of the superiority of the new paper prints, and offer large arrays of formats from stereographs and bound albums to the modest-priced small portraits on card mounts known as *cartes de visites* that could be collected into albums. Soon studios were offering prints of the people, the city, and surroundings for pasting into albums. For convenience, studios also had their own albums. Within a decade, locals and visitors alike could easily and relatively cheaply insert their own images into the new travel and family albums alongside professional views from home and abroad. The novelties of the era were showpiece panoramas, but these were expensive, had relatively few buyers, and could only be made by the best photographers. Photographers could now compete with the established styles and genres of the graphic arts but with the added value that their images showed what the places and people actually looked like.

The wet-plate era would also preserve the work of the first Asian-born photographers, and in several rare cases, their self-portraits. The latter include a suave self-portrait as a painter, showing the fascinating Hindu professional Hurrichand Chintamon in Bombay (Mumbai) around 1865 (fig. 10).

9

10

9
Cesar Düben.
***Prest van Madras*, plate after original daguerreotype of around 1855.**
Photolithograph in Reseminnen från *Södra och Norra Amerika, Asien och Afrika* (Stockholm 1886). National Gallery of Australia Research Library.

10
Hurrichand Chintamon.
***Self portrait as a painter in Bombay,* around 1860.**
Albumen photograph.
Collection Hugh Ashley Rayner, Bath.

11
Frederik Feibig.
***Dwelling of an English gentleman* [Garden Reach], Calcutta, around 1851.**
Hand-coloured, salted paper photograph. British Library.

Bombay masterworks on paper

In the 1850s and 1860s, British India provided the most fertile ground for early photography on paper, to judge by the number of enthusiasts. The most remarkable figure in early paper photography in South and Southeast Asia was the German artist-lithographer Frederick Fiebig, who was based in Calcutta in the 1840s. He took up calotype photography around 1849 and made hundreds of images in Calcutta, Chennai, and Sri Lanka in the early 1850s (fig. 11). Fiebig made a practice of hand-colouring many of his prints. When Fiebig sold the East India Company in London some five hundred prints in 1856, he told the directors of how "during my travels in India I employed my leisure time in taking photographic views of the principal buildings and other places of interest at Calcutta, Madras, the Coromandel Coast, Ceylon, Mauritius, and the Cape of Good Hope".[13] His output was over a thousand prints, a huge collection to have assembled in such hot climates in three or four years. Photography in the field was not for the faint-hearted in the formative decades of the medium in Asia.

Fiebig's focus was on the comforting vision of the white, neoclassical buildings of the British Corporation and missionaries taking a rightful place alongside the past glory of ancient Hindu and Islamic antiquities. The photographer's challenges compared to the artist are apparent in Fiebig's charming scenes. The photographer has to undertake artistic posing, arranging people and other objects, before making his image. The photographer cannot compile his scene at leisure from various life sketches. Fiebig's images have a characteristic array of carefully positioned Indian figures dispersed throughout the space. No matter how hard they try to avoid it, there is often a person who does not follow direction and is caught clearly looking at the camera. That is the charm of the abundance of extras that come with the photographic view.

Fiebig's photographs show the messier reality of Calcutta and the jostle of new and dilapidated in the port. His visit to Madras in 1852 was reported in *Illustrated Indian Journal of Arts* in February 1852. While local artists were grateful to have had lessons in the useful process, the reviewer warned that photography was a "dangerous aid to the artist, as it teaches him to regard nature too much in her everyday common garb, when she is in general common place, vulgar, and minute." This warning did little to discourage local photographers, particularly among the East India Company's technical ranks. Fiebig's East India Company collection survives in the British Library.[14]

In the 1850s and 60s, East India Company officers undertook extensive journeys beyond the ports, recording the Mughal, Hindu, and Buddhist antiquities of India. They exhibited their works in various locations locally and internationally. Back in port, practitioners of the new medium usually belonged to one of the photographic societies that were the first to be formed anywhere outside Europe. The Calcutta society, founded in 1854, had two hundred members, both Indian and foreign, by 1858. A Bombay society was formed in 1855 and the one in Madras followed in 1856.

The officer class of amateur photographers tended to favour antiquities. One of the earliest and most elegant applications of photography on paper in the Asian ports was initiated in the mid-1850s on the lively metropolis of Bombay, by William Johnson and William Henderson. They were founder members of the Bombay Photographic Society. Both were East India Company civil servants who parlayed their amateur interest into professional practice. Johnson had arrived in Bombay by 1848 and worked as a clerk. He opened a daguerreotype studio in 1852, but shifted to the wet-plate by 1854. Johnson was a founder and secretary, and oversaw the society's monthly issues of the *Indian Amateur's Photographic Album*, with three original prints per issue. Henderson had begun work as a Military Board Office clerk in 1840, and was in partnership with Johnson from 1857 to 1859, then in his own business until 1866.

Johnson and Henderson compiled some of the first published albums in Asia: their *Costumes and Characters of Western India* and *Photographs of Western India* appeared between 1855 and 1862. This album included a three-panel panorama of Calcutta from Saint Thomas Cathedral (made around 1855 and measuring 20 by 78 cm) - one of the earliest in this format in Asia. The large format and rich tones of the wet-plate photographs enabled a real panorama of scenes and images of the many races and cultures (fig. 12). The prints had a veracity of raw detail that drew on all the established genres of topographical views, and occupation and costume studies. Above all else the photographs powerfully captured a sense of the activity on the docks, in particular the Parsee merchants involved in the cotton trade. Not all the images in the published albums were by Johnson or Henderson.

12

12
Willliam Johnson.
The cotton ground, Colaba, Bombay, around 1858.
Albumen photograph.
Photographs of Western India. Volume II. Scenery, Public Buildings, &c. 19 x 24 cm.
National Gallery of Australia.

13
William Johnson.
Goanese Christians, around 1855-62.
Albumen silver photograph, no. 23 in *The Oriental Races and Tribes, Residents and Visitors of Bombay: A Series of Photographs with Letter-Press Descriptions by William Johnson Bombay Civil Service* (uncov.), London, 1863.
DeGolyer Library, Southern Methodist University, Dallas, Texas.

14
William Johnson and William Henderson.
Vallabhácharya Mahárájas, 1856.
The Oriental Races and Tribes, Residents and Visitors of Bombay: A Series of Photographs with Letter-press Descriptions by William Johnson Bombay Civil Service (uncov.), no. 23, v 1, "GuJarât, Kutch, and Kâthiawâr", London,1863.
Albumen silver photograph, letterpress on card, 26 x 18 cm.
National Gallery of Australia.

13

14

In his 1863 book on Bombay, *Mumbaiche Varnan*, Goan writer Govind Narayan (1815–1865) gave an early description of the craze for photography among British and Indian residents. He noted: "There are thirty to forty men who can produce such pictures. Many youths have mastered this process and have established workshops in their houses to capture the images of their near and dear ones". He picked out Dr Narayan Daji Lad and Harischandra (Hurrichand) Chintamon as experts among the Hindus, and William Johnson as the most famous among the British. He noted that Johnson planned to present his series of men and women of the castes of Bombay to "the Empress", that is, Queen Victoria.[15] In that very year, Johnson was in London publishing his two-volume *The Oriental Races and Tribes, Residents and Visitors of Bombay* (1863, 1866). It is the earliest ethnographic publication in Asia to be illustrated with photographs.

From nothing in the mid-1840s, within a decade, hundreds of large-format photographic prints documented the urban character and social structure of India. British residents are not included in these "types", but private albums and official portraits were also creating a record of this class.

The Oriental Races and Tribes, Residents and Visitors of Bombay sought to convey the racial diversity of India's most populous city, and is a tour de force of montages of studio portraits matched with appropriate outdoor shots of urban, domestic, or rural scenes, including proud Rajput princes, Brahmin, Muslim, and Parsee men and women. The Johnson and Henderson images included both the usual types and also a good cross-section of ethno-religious groups, including Goan Catholics, or "Portuguese Christians", seen with males in Western suits and the women in Goan-style saris who had immigrated to Bombay by the 1850s (fig. 13). Their Western-influenced education and experience of higher caste Hindu and Goan Catholics in the Portuguese colonial administration led to significant roles in teaching English and Latin, as well as service in the British navy as seamen or musicians on the expanding commercial shipping lines.

One Goan Christian was Dr Narayan Daji Lad (1828–1875), the first Indian faculty member at Grant Medical College, who had a long career as a photographer in Bombay, including tours with the governor of Bombay. He and his brother Dr Bhau Daji Lad were active in the cultural and political life of Bombay, including organisations such as the Photographic Society. Narayan complained in the press about Johnson's piracy of his images, such as the photograph of Vallabhacharya Maharajas in *Oriental Races and Tribes* (fig. 14). Another photograph in Johnson's album appears to be by Hurrichund Chintamon, who ran the first Indian professional studio in Bombay. Chintamon is an intriguing character yet to be studied in depth. He lived in England for periods from 1874 to the 1880s, as an agent for the Gaekwad of Baroda, and devoted much time to spiritual matters, including for the Theosophy movement.[16] The Johnson and Henderson portfolios remain impressive over a century and a half later. Whether Johnson was responsible for the montage work is not known. It possibly inspired a later catalogue of all Indian races that John Forbes Watson and John William Kaye compiled – an eight-volume study with over four hundred photographs of different castes and races entitled *The People of India*, which appeared between 1868 and 1875. The feel of the two projects is not comparable: the latter being a utilitarian directory of castes, while a sense of a vibrant multiracial, multi-faith community exists in the Johnson tome.

By the early 1860s, a new generation of wet-plate view and portrait photographers were setting up across the network of Southeast Asian ports. As Johnson departed for London, his countryman Samuel Bourne, an ambitious amateur photographer turned professional from Norfolk in England, arrived. He had been a clerk but had determined to come to India and make his name with landscape and topographic photography. He formed a partnership with Charles Shepherd, and their firm Bourne and Shepherd would become one of the major distributors of late nineteenth- and early twentieth-century Indian imagery. Bourne's great achievement was his Himalayan journeys in the 1860s, but the firm also presented panoramic visions of the gleaming English colonial architecture of the port cities (fig. 15).

While providing views for sale, most photographers also made portraits and house calls at fine homes, creating a less visible record of private life, including in some cases for elite Indian customers.

15

17

16

Expanded views

One of the first of the new wet-plate firms to arrive in Java in 1857 was the partnership of Walter B. Woodbury (1834–1885) and James Page. The pair had started their partnership in the Australian goldfields during 1855–57, but found that there was far too much competition. They gave up hopes of a permanent studio in Melbourne and planned to travel from port to port and end up in Valparaiso. On arrival in Batavia, however, they were delighted to find that there were only three other photographers. With the superior technique offered by the wet-plate cartes and the luscious cased photographs on glass called ambrotypes, they were besieged by an extraordinary variety of customers. The clientele included, as Walter wrote home to his mother in 1858, "part Dutch Chinese Malay half casts, Maduresse Bandanese Arabs countesses, chevaliers Javanese and a variety of others."

Woodbury established a charming suburban home and a purpose-built studio in Batavia (fig. 16). He organized extensive excursions to the sites of antiquities such as Borobodur, but also started the first sale of images of Malay and Chinese types, and of the sultans of central Java and their exotic dancers. The Chinese quarter fascinated Woodbury and proved popular as a subject throughout the firm's long operations into the 1880s. He sold stereographs of these subjects to London opticians and publishers Negretti and Zambra in 1861 (fig. 17). Woodbury and Page brought out the first album of views of Batavia in 1864, and sold distinctive albums packaged in portfolios labelled *Vues de Java* for decades. Page died early of a tropical fever and Walter returned home in 1863 to fame as the inventor of the Woodburytype, having set up his firm to continue under brothers Henry and then Albert.

Until its decline in the late 1880s, Woodbury and Page and the army of native assistants and other operators, such as Henry Schuren who went on to play a significant role in Bangkok and Manila in the 1870s, produced thousands of beautifully toned, highly detailed, elegantly composed images that remain well-represented in most overseas archives.

When he arrived in Batavia, Woodbury made no mention of one of the other established photographers. Isidore van Kinsbergen (1821–1905), the Belgian-Flemish artist, theatre performer, and singer, made a name for himself in the 1860s and 1870s photographing the ancient monuments of Java. These photographs were exhibited widely. He also made dramatic tableaux of royalty and natives that were much used in late nineteenth-century publications. Kinsbergen was a true immigrant: he lived a long life in Java and was deeply involved in the foreign community. His tableaux portraits and figure studies have a certain theatrical flair but also demonstrate his connection with his sitters.

The same 1860s patterns can be seen in the Straits Settlements, where the Austrian August Sachtler, who had a substantial career in an Austrian government expedition to Japan in the early 1860s, arrived with his

15
Samuel Bourne.
Calcutta. Panoramic view from Ochlerlony Monument.
Prince of Wales Tour of India 1875–6 (vol.3), 1863–76
Albumen print.
Royal Collection Trust [RCIN 2701702]

16
Woodbury and Page.
Woodbury and Page studio in Batavia.
Albumen photograph.
KITLV (Royal Netherlands Institute of Southeast Asian and Carribbean Studies), Amsterdam.

17
Woodbury & Page.
Chinese Quarter, around 1866.
Albumen photograph.
National Museum of Singapore [2008-05831-010]

The Chinese were required to live just outside the south gate of the city, in an area that became known as Glodok. Today, Glodok is the centre of Jakarta's Chinatown.

brother Carl Herman and opened a studio in Singapore in 1862. He also worked with the Dane Kristen Feilberg (1839–1919), who ran their Penang branch. The Sachtlers were the first studio in Singapore, and made the earliest panoramas of the city in 1863, where both died in 1874 after a decade of successful operation (fig. 18). Like Woodbury and Page and Bourne and Shepherd, the Sachtlers used a base in one port to make photographic expeditions to nearby regions to build inventories. August Sachtler and Feilberg made the first album of views of Singapore, and images of the cultural mix of peoples, referred to as "types", along with the first albums of Sarawak. Feilberg produced impressive, mammoth plates of the wild, unknown regions and of the Batak peoples in Sumatra.

18
Attributed to Sachtler & Co.
***Boat Quay on Singapore River, Court House under Renovation, Harbour Master's Office and Town Hall*, around 1874.**
Four panel panorama on successive pages of a French travel album assembled around 1885; albumen photographs.
Musee Guimet, Paris.

The Scottish photographer John Thomson, who started up in Singapore in 1863 also branched out by working in Penang for ten months to capitalize on a more exotic, older port. He found his true calling as an outdoor photographer on the streets of Penang. Thomson's technical ability allowed him to present his subjects, as in his Durian sellers (fig. 19), with some sense of naturalness. Through this engagement with the locals, Thomson more or less invented the idea of the "travel photographer" as a social reporter rather than merely a merchant of souvenirs. Thomson pioneered the modern model of a photojournalist in the 1870s with the publication of his four-volume *Illustrations of China and its People: A Series of Two Hundred Photographs* (London, 1873–74), using the autotype process.[17]

19

By the mid-1860s, there were established studios in most major Asian ports who sent their operators out on the streets and offshore, and into the hinterlands for views and native types. These photographs were marketed in the ports and also sent abroad for armchair travellers, as well as for publication. While most studios in Southeast Asia were run by European photographers, in India, Hong Kong, and Java, Asian photographers set up studios catering to both European and indigenous clients. For example, Afong and Pun Lun in Hong Kong had studios of their own by the 1860s. Pun Lun operated as well in Singapore and Shanghai.

By the 1880s, in the main port city centres, a few larger enterprises grew to become veritable palaces of photography.[18] These had showrooms hung salon-style floor to ceiling with almost life-sized portraits in heavy wood frames, along with cases of accessories offering multiple choices

20

21

19

John Thomson.
***Malays selling durians*, Singapore or Penang 1862.**
Albumen photograph, carte de visite.
Collection Janet Lehr, New York.

20

Rudolf Jasperacher, photographer,
G.R. Lambert & Co.
***Kling family*, 1886-88.**
Albumen photograph.
Private collection.

21

Premises of Ohnnes Kurkdjian, Surabaya, 1910.
Gelatin silver photograph, KITLV (Royal Netherlands Institute of Southeast Asian and Carribbean Studies), Amsterdam.

for presentation, and a range of prices that could deliver everything from massive albums to miniature lockets. Going to the right photographer for one's social set became a factor. These patterns continued on into the early twentieth century, with packaged tour groups visiting studios to buy souvenirs and postcards.

The most singularly elaborate and far-reaching enterprise was that of the German professional Gustave Lambert (1846–1907), who established his long-running firm in Singapore in 1877.[19] By the 1890s, it had a building decorated with a fancy portico on Orchard Road, as well as a studio in the centre of the city. Lambert served as official photographer to the king of Siam, and his studio cards were emblazoned with medals won at international exhibitions. Branches were set up in Deli, Sumatra, while numerous operators, mostly from Middle Europe, began or passed though the Lambert and Company.

Like Woodbury and Page, who always stamped their work, G. R. Lambert and Company maintained strict quality control, and consequently their prints are usually easily identified through their rich tone, detail, and assured composition. Alexander Koch became owner-manager after Lambert's return to Germany in 1885, and he developed an inventory of some three thousand views of Southeast Asia, including what amounts to a racial atlas of the region. Images from Sumatra and remoter regions were probably made in those locations, but Singapore's many races provided enough Kling, Chinese, Malay, Thai, and mixed-race sitters, who were set up with minimal props (fig. 20). Unlike William Johnson, little effort was expended on appropriate backdrops.

In Surabaya, the firm of Armenian Ohnnes Kurkdjian (1851–1903) dominated the tourist image of Java until well into the 1930s (fig. 21). For a generation of travellers, readers, and collectors worldwide, the output of these major firms defined the tropics and the cultural images of the peoples of South and Southeast Asia.

As they emerged in the late 1880s, many of the firms were run as Chinese family businesses and were patronized by all classes of customers. A spirit of fun is apparent in the many rickshaw comedies played out by Europeans in the Pun Lun studio (fig. 22). What the Chinese operators thought of this is not recorded.

All studios used their card mounts as advertising and all aspired to a European model with crests, mottos, medals, and promises of "negatives kept". The cards of Tan Tjie Lan, who ran a busy studio in Batavia, revealed his many credentials. Increasingly in the period from the 1880s to the early 1900s, locals step into the frame of the studio, often presenting themselves as mirror images to the "types" anonymously shown in the earlier decades.

The views of the docks made in the 1890s by Tan Tjie Lan and G. R. Lambert and Company of Batavia and Singapore respectively (figs. 23, 24) are revealing of other changes, as photography became easier and could capture action. In the Lambert photographs of Tanjong Priok we see a huge crowd of Indian and Chinese dockworkers who were probably there to load coal. They have been marshalled to stop for the photographer. They look up to where the camera is, high above them, awaiting the signal that they must stop, look up – and then get on with their work. With new cameras and faster films, shooting people as they moved about was much more successful. When Tan Tjie Lan photographed the *S.S. Reale* departing Batavia for the outer islands, the photographer is amongst the action. No one has been ordered to stop; the viewer is down on the level of the port and its citizens.

22
Pun Lun, Singapore.
Rickshaw pantomime.
Albumen or pop cabinet card.
Photoweb collection.

23
G.R. Lambert & Co.,
Alexnder Koch director.
***American sailing ship, Tanjong Pagar dry dock*, 1892.**
Albumen silver photograph, 21.7 x 27.2 cm.
National Museum of Singapore
[1993-00285-022].

24
Tan Tjie Lan.
***KPM inter-island boat Reael from Aceh at Tanjung Priok dock, Jakarta*, around 1895.**
Albumen silver photograph, 18.2 x 24.7 cm.
National Gallery of Australia.
Verso of a Tan Tjie Lan cabinet card.

25
Unknown French amateur Kodak photographer.
Singapore Vue prise de L'Hotel de L'Europe and ***Shanghai votre serviteur en brochette i* [self portrait], around 1895.**
Gelatin silver photographs.
Photoweb Collection.

22

23

24

25

What can be glimpsed through the range of standard topographical and souvenir images from Asian ports is the daily life largely hidden from official cultural images, but also the modernization of Asian port cities linked by the maritime trade routes of the Western nations across the globe. The full history of the photographers of Asia, including the few women practitioners from the late nineteenth to the mid-twentieth century, such as Thilly Weissenborn in Java, and indeed the Asian photographers who fanned out across Asia, is yet to be written.

A considerable number of Indian-operated studios catering more or less exclusively to local customers also appeared. These evolved unique forms of coloured and decorated images that reflected Indian art and religious practice as pilgrimage testaments. Personal family photography using the Kodak cameras and processing agents, the growth of popular photographic travel guides, and the mass-market publications in the late nineteenth and early twentieth century eventually brought an end to the local studios providing the bulk of portraits and views as original prints for albums. Then it became more about putting yourself in the picture in the family album. It became more about saying I live here, I visited there, or being personally taken by the author on a journey through images in a book (fig. 25).

The role of professional views and events photography moved to the picture press rather than the original print purchased as a souvenir from a photography studio or hotel. Whatever their personal success in their new profession or hobby, the images left by the pioneer immigrant and local photographers who first set up in the ports of Asia are now the treasured visual heritage of the lands they visited as well as their homelands.

Like novels and other art forms, photographs are contrived and opinionated. They are theatrical set-pieces governed by the many technical limitations of their lack of colour, action, and atmosphere, which painters could easily include or invent to suit. The photographic medium's compensating privilege is that as the viewer bends to peer into the image there is that frisson of a moment and a space shared over time. This is what the photographer chose to see that day, that moment; this is how the subject was cast and performed for the camera. Regardless of original purpose or client, once a photograph was made and circulated and finally archived, it shaped how subjects and viewers were seen and how they understood themselves, and how they are seen by future generations. In the early twenty-first century, the legacy of the pioneer photographers in the port cities of Asia is being rediscovered, exhibited, and revalued, and subject to new scholarly analysis.

26

Albert Honiss.
Fort Santiago the Pasig River wall of the Intramuros [walled city] looking east to the opening into Manila Harbour and roadstead where ships await unloading, around 1880.
Four-panel panorama, albumen photographs over three album pages. Musée Guimet, Paris.

1 Woodbury to Ellen Lloyd, 4 May 1857; in Elliott 1996, letter no. 15.

2 There are many "Asias" and no single directory exists. British Library curator John Falconer has placed his extensive "Biographical dictionary of 19th century photographers in South and South-East Asia" online: www.luminous-lint.com. See also Falconer 1987. For Sri Lanka, see New Delhi 2015. For Indonesia see Rotterdam 1989, including listings of 471 photographers, of which just under half are Chinese and a quarter Japanese. Terry Bennett's histories of photography in Japan, China, and Korea are a major source of directories, and his current work on Indochina will also provide a future directory for that region. Information of Thai photographers is available largely in Thai,, but Joachim Bautze's *Unseen Siam: Early Photography*, 1860–1910 is forthcoming. There is no Philippine directory as yet, but see Gael Newton's Southeast Asian national surveys and individual artist entries in Hannavy 2008 and in Ghesquière 2016.

3 Chris Furedy, "British Tradesmen of Calcutta 1830–1900: A Preliminary Study of Their Economic and Political Roles" in Sealy 1981, pp. 48–62.

4 In 1849–50, French baron Alexis de Lagrange (1820–1880) used French printer Louis-Désiré Blanquart-Évrard's paper negative process to record Indian monuments, five of which were published in Blanquart-Évrard's *Album photographique de l'artiste et de l'amateur* of 1851. Dr John Murray in Calcutta made superb mammoth plate calotype negatives of Indian antiquities, which he published in London in 1858; these were some of the earliest Asian photographs on view in Europe. See Sabeena Gadihoke, "Indian subcontinent, 1839–1900" in *The Oxford Companion to the Photograph*, edited by Robin Lenman (Oxford, 2006). Only a few Indian daguerreotypes returned to Europe and they were mostly used as the basis for engraved illustrations in pictorial papers such as *The Illustrated London News*.

5 Chris Furedy,"British Tradesmen of Calcutta 1830–1900: A Preliminary Study of Their Economic and Political Roles", in Sealy 1981, pp.48–62. *www.yorku.ca/furedy/papers/ko/BTC1830.doc*, accessed online June 2015.

6 The local paper reported that a daguerreotype camera was successfully demonstrated in Sydney on 15 May 1841. See Wood 1996. The first professional, George B. Goodman from London, arrived the following Christmas; a Japanese merchant imported a daguerreotype camera into Yokohama in 1841, although he was unable to get satisfactory results; but the Daimo took an interest later in 1857 and a portrait survives.

7 I am grateful to independent scholar Raimy Che-Ross of Canberra for advice that the ship may have been the East India Company steam frigate *S.S. Sostiros*, which was sent to Amoy in 1841. The US Navy's Africa Squadron under Matthew Perry passed through Asia from 1842, and in 1843 daguerreotypist George R. West was part of the American Treaty of Wanghia, led by Congressman Caleb Cushing on *USS Brandywine*, and signed in Macau on 3 July 1844. West (around 1825–59) became the first photographer to work in China in 1844, and opened the first studio in Hong Kong in 1845. The preparations listed taking a daguerreotype camera. See p. 137, http://www.forgottenbooks.com/readbook_text/Americans_in_Eastern_Asia_1000282628/153, accessed 10 May 2015.

8 Singapore-based French scholar Gilles Massot has proved that the daguerroetype of a temple in Pondicherry attributed to Itier was taken on 22 July 1844, by civilian mission member Natalis Rondot, and is the earliest securely dated photograph of India. See "Jules Itier and the Lagrené Mission" in *History of Photography* 39, no. 4 (November 2015), pp. 319–47.

9 See "Towards a History of the Asian Photographer at Home and Abroad: Case Studies of Southeast Asian Pioneers Francis Chit, Kassian Céphas and Yu Chong", in Newton 2016.

10 Gilles Massot has made a detailed study of Itier's itinerary and life, see Massot 2015.

11 See Gael Newton, "Silver streams Photography arrives in Southeast Asia 1840s–1880s", in Canberra 2014, p. 16.

12 See Rotterdam 1989 and Canberra 2014.

13 There is no monograph on Fiebig; see John Falconer, "Frederic Fiebig" in Hannavy 2008.

14 See "Photography in Madras", *Illustrated Indian Journal of Arts*, pt 4 (February 1832), p. 32; and "Envisioning the Indian city: Researching cross cultural exchanges in colonial and post-colonial India" [eticproject.wordpress.com].

15 *Govind Narayan's Mumbai: An Urban Biography from 1863*, edited by Murali Ranganathan (New Delhi, 2009), p. x.

16 The most extensive discussion is Sandra Hapgood, *Early Bombay photography* (Mumbai 2015).

17 See Canberra 2014 for Woodbury & Page, van Kinsbergen, Sachtler, and Thomson in context.

18 These were run by Europeans, with the exception of Felix Laureano (1866–1952), a Filipino who ran successful studios in the Philippines, India, and Barcelona.

19 Falconer 1987.

Einheimischer Batavischer

ürger und seiner Frau.

Ball at Singapore, in celebration of the anniversary of the settlement.
Unknown artist.
Singapore, 1854.
Wood engraving.
Published in *Illustrated London News*, 22 April 1854.
National Museum of Singapore [XXXX-01426]

Thirty-fifth anniversary of Singapore as a British settlement. Prominent members of the Asian community were invited, including Hoo Ah Kay (Whampoa), at the left, with a fan.

Dressing badly in the ports
Experimental hybrid fashion

Peter Lee

Fashion in the port cities of Asia, especially in Southeast Asia, has always been motivated by improvisations and experimentations. Textiles and garments were produced locally but imported varieties were even more plentiful. In her study of the Dutch trade in Indian chintz, Ruurdje Laarhoven uncovered the diversity and quality of cotton cloth, both plain and patterned, which was in demand throughout Asia in the seventeenth and eighteenth centuries.[1] Chinese and European textiles, as well as those produced all over island Southeast Asia, were also distributed along the same trade routes.

The confluence of international and local woven goods in every port produced a range of cultural reactions. Clothing and textiles of diverse origins were worn together. Anthony Reid describes this phenomena as the "experimentations of the fifteenth and sixteenth centuries", which were widely recorded by European commentators of that period.[2] He also made the important distinction: "the difference in dress between rich and poor, servant and master, king and commoner, was less marked than in pre-industrial Europe, where each man's station, and even vocation would be read in the prescribed style of dress".[3] Patterns (woven, resist-dyed, block-printed, or needlework) were copied and integrated, and those that were common in one type of textile were duplicated in another (for example, ikat patterns were reproduced in block-printed cottons). With this global array of textiles circulating in Asia, it was natural that sumptuary laws were imposed in some Asian port cities.[4]

The Malay writer Munshi Abdullah, who grew up in the more sartorially liberal environment of colonial Malacca in the late eighteenth century and later moved to Singapore, found sumptuary laws distasteful. When he was told that in Pahang it was forbidden to wear shoes, yellow clothing and fine muslin, he remarked:

> "I thought about what was said for awhile, and smiled, because I had heard such stupid and useless regulations, because they were all petty matters that had become wrong and forbidden. Why not forbid birds to fly above the palace, and mosquitoes to bite the Raja and bugs to hide in his pillow?... As to matters that should rightly be forbidden and can be beneficial to mankind, nothing is said..."[5]

Where sumptuary laws existed in port cities, they were not strictly enforced, which exacerbated the inconsistency and variations in fashion, even within communities. Members of the same household might even dress in different kinds of costumes or fashions.

What Anthony Reid describes as "experimentations" are in fact unrestrained, arbitrary, heterogeneous, and individualistic ways of dressing up. This essay aims to propose that these experimentations were not isolated to the fifteenth and sixteenth centuries, but in fact continued into the twentieth century – and with all probability existed before the fifteenth century. This prolonged period of so-called "experimentation" also provides a significant historical parallel with current developments in European and global fashion. The old paradigm of European fashion has generally been pivoted on what Roland Barthes describes as its "tyrannical nature" – the cultural leadership of royal courts, intellectual elites, the film industry, fashion designers and fashion capitals.[6] But the ascendancy of the internet has totally undermined this influence, and has encouraged the rise of the kind of individual experimentation Reid describes in Asia in the fifteenth and sixteenth centuries. Therefore, what is considered unprecedented about global fashion at present, has in fact very deep roots in Asian port cities. The overwhelming visual and written evidence of this not only suggests the need for a revision of global fashion history, but also of the conventional concepts of immutable, traditional Asian "ethnic costume".[7]

Showing off

Ever since the sixteenth century, travellers passing through Asian port cities have remarked on the ostentatiousness of the clothing and jewellery worn by their residents. Anthony Reid was among the first modern scholars to draw attention to this sartorial extravagance in Southeast Asia.[8] In colonial port cities, dressing up was rampant across all levels of society. The travelogue of Dutch writer Jan Huyghen van Linschoten, provides detailed accounts of the showy attire of the women of Goa in the late sixteenth century:

> When they go to church or make other visits, they put on very costly clothing, and gold Bracelets and bands on their arms, and costly gemstones and pearls; from their ears hang straps full of jewels and gems.[9]

François Pyrard, who visited Goa in the early seventeenth century, noted that even the servants of wealthy women in Goa were "richly attired in silk of all colours ..."[10] Christoph Schweitzer observed the same kind of conspicuous luxury in late seventeenth-century Colombo in his somewhat satirical description of a wealthy mestiza woman:

> ...and her hair, which reached down to the ground, was coated everyday with coconut oil and then coiled up, just as the horses' tails are dressed in Germany. She wore a short white camisole of the finest linen, fastened with gold buttons, through which her black skin could be seen. Below this camisole, the stomach to the navel was exposed, just a handspan-length. She wore white linen which reached from the navel to the ground, and over that a beautiful coloured silk. She hung a large rosary made of gold and ivory around her neck.[11]

As Anthony Reid noted, men were equally flamboyant and fastidious when it came to dressing up, revealing how in the *Sejarah Melayu*, Seri Wak Raja, the fifteenth-century Bendahara of Malacca, "would don his sarong and he would undo it twelve or thirteen times until he got it to his liking. Then would come the jacket and head-cloth, and the process with the sarong would be repeated with them until they too were to his liking..."[12]

Dandyism was still prevalent in the early nineteenth century, even among people who could scarcely afford it, as Munshi Abdullah noted on a journey to Pahang in the Malay Peninsula:

> I observed of the people that perhaps only about one in ten did any work; the rest of them loafed about all day in poverty and vice.... About half of them had the habit of solely being concerned with looking smart, wearing fine jackets and trousers, but there was not one who wanted to find his way in life.[13]

Mixing and matching

Men's clothing was a hybrid construct; European jackets and waistcoats were particularly fashionable. François Valentijn's report of eighteenth-century men's fashion in Ambon, perfectly expresses this experimentation:

> I have seen some of the kings, who were wearing very beautiful clothes on Sundays. The old King of Kilang, called Domingos Koëlho, have I often seen wearing a coat [*rok*] of black velvet with big golden buttons, and a fine black hat of the old-fashioned kind called *brantemmers* [fire-buckets], with an expensive golden hatband, worth three or four hundred guilders, and a cane with a beautiful gold knob in his hand.

> Others I have seen in gold brocade, in silver or gold linen, or in silk, or in beautiful fine blue gingham clothing, while at home they also wear beautiful and very fine chintz kebayas. This is not only the costume for Christian men, but also most of the Muslims wear costumes like the common Ambonese, with the difference that the Moors wear a turban or *dastar*, which is a red, blue, or white narrow strip which they wind around their head. One also sees that these same also wind three to four strings very thickly around their head; but they seldom use green-dyed bands, because it is said that only those who are descended from Mohammed are allowed to wear green. The Moors also wear bajus of silk, gingham, or of other silk material in various colours, also chintz kebayas, which is very similar to what other Ambonese wear.[14]

His account also reveals how different ethnic communities adopted similar clothing styles, indicating that transnational and "trans-ethnic" fashion was already present in island Southeast Asia in the eighteenth century.

The eighteenth-century handscrolls depicting foreign figures from the Matsura Historical Museum (see Cats. 93, 94) are extraordinarily detailed documentations of the types of global travellers landing in Nagasaki. The diversity and idiosyncrasy of the clothing worn by ordinary people from various parts of the world are among the handscrolls' most remarkable aspects. One of the scrolls even has detailed descriptions of every item of clothing worn by the foreigners (see Cat. 94).

On another (Cat. 93), a barefoot man described as coming from "Kalapa" (Batavia) is dressed in a chintz jacket derived from a Western type, together with what appears to be pink pantaloons and a Western hat with a narrow rim, with a ceramic decanter in one hand and a tobacco pipe in his mouth (fig. 1). Although the components of his outfit are derived from European models, the way he has assembled them is not conventional. Similarly, another man described as a "Kaffir" (a colonial and derogatory term for an African), pairs a flowery European style chintz jacket with a white loin cloth, sports a gold earring and holds a globular long-necked bottle in his right hand (fig. 2).

On yet another scroll (Cat. 94), a native of Zhangzhou, Fujian, identified as Lin Weizheng, aged 33, is shown dressed in a fine Chinese surcoat unconventionally worn together with a pair of trousers tucked into leggings, more typically seen on foot soldiers. His gold necklace and headband emphasize the nonconformity of the entire outfit (fig. 3). A dark-skinned mestizo dandy perhaps from Goa (described as coming from *Canarim*, which is in fact the language spoken in the region) is dressed in a somewhat rakish early nineteenth-century European manner, with a white summer jacket, flamboyant waistcoat, ruffled shirt, and narrow trousers (fig. 4). His gold earring, headband, and waist sash further accentuate his unorthodox and individual style.

1
Kalapa Man [Batavian man] (咬留吧人).
Detail of Cat. 93.

2
Kaffir Man [African man] (カフリ人).
Detail of Cat. 93.

3
A male native of Zhangzhou, named Lin Weizheng, aged 33 (生國漳州男,名林為政,歳三十三).
Detail of Cat. 94.

4
A male native of Canarim [Goa], named Caetano, aged 20 (生國カナリイン男,名カイタアノ,歳二十).
Detail of Cat. 94.

1

2

3

4

Jan Brandes's illustrations of Batavian life provide rare insights into Dutch colonial society in Asia, and to the improvisations in dressing that was part of daily life. Five illustrations by Brandes depict his son Jantje dressed only in an *oto* (an archaic Hokkien or Javanese term for a Chinese lozenge-shaped bib or apron, known in Mandarin as a *doudou*, 兜兜), which was common for Batavian children of all nationalities up to the age of four of five years, and combined with a Batavian chintz *baju* and sarong, or a Western-style jacket and breeches (fig. 5). Brandes's painting of a Batavian wedding (Cat. 64) is remarkable for the discordant array of costumes and fashions worn by the hosts, guests, slaves, and musicians. Both European and Batavian fashions were acceptable at such events. One guest in particular combines a chintz petticoat (*saia*) with a long kebaya (fig. 6). The puffed sleeves of the kebaya were more common for undress or *deshabillé*, the informal style of European fashion in the eighteenth century, but clearly it was acceptable in formal Batavian events. Even ceremonial robes were improvised in Batavia, as demonstrated in an 1860s photograph of a Batavian bride, who wears a Chinese style skirt with a *baju kurung*, a garment worn in island Southeast Asia, Sri Lanka, and Goa. The bride's jewels are also a riotous mixture of Chinese, indigenous, and hybrid forms (fig. 7). In the late nineteenth century, Chinese fashion in Batavia continued to be self-determined. Wealthy women there and in other port cities of Java developed their own preferences for robes and skirts in European patterned velvet, which they wore with Western-style gold and diamond jewellery sold in ateliers owned by Europeans, Indians, Chinese, and others (fig. 8).

The same kind of arbitrary, hybrid fashion was common in Goa and Manila in the eighteenth and nineteenth centuries (figs. 9, 10). Not only were garments and footwear of different origins combined together in one outfit, new shapes and textile patterns that synthesized diverse influences were also developed. In Mindanao the local *datus* (chiefs) dressed in diverse hybrid styles combining Western, Malay, and Chinese garments (fig. 11). In the early twentieth century, plain or laced cotton blouses, sometimes with puffed sleeves, became the symbol of emancipated women in the West, and was also adopted in various parts of Asia. Queen Saovapha Phongsri of Thailand frequently dressed in this type of lacy blouse with leg-of-mutton sleeves together with a court-style *chong kraben* (a skirt cloth wrapped in a manner resembling trousers), together with European silk stockings and shoes (fig. 12).[15] In the same period, Indian women in Singapore combined this style of blouse with a sari (fig. 13).

Perhaps the ultimate hybrid garment in island Southeast Asia was the kebaya, worn by men and women, made from every textile imaginable from cotton and silk to velvet and polyester, and evolving in cut and design through the centuries.[16] Hybrid styles also influenced other urban centres. In the palaces of Surakarta and Jogjakarta in the nineteenth century, velvet jackets embroidered with gold thread were adapted from European court dress, and worn with special batiks dyed on European machine-made cotton. In the same era in Kandy, Sri Lanka,

5
Jantje and Flora in an upstairs room (detail).
Jan Brandes.
Watercolour over pencil sketch on paper, 1784.
Rijksmuseum, Amsterdam [NG-1985-7-2-4]

6
Detail of Cat. 64.

7
Bride in her finery
Woodbury and Page. Batavia, around 1860s.
Albumen print.
Mr and Mrs Lee Kip Lee, Singapore.

8
Woman in a velvet robe and diamond jewellery.
Batavia, around 1880–1900.
Albumen print.
Mr and Mrs Lee Kip Lee, Singapore.

5

6

7

8

9

10

11

12

13

9

Mestizo of Canarim.
Workshop of Goa, 1785–1800.
Museu Quinta das Cruzes-Funchal, Madeira Tecnopolo.

10

Un mestizo de Manila.
Justinian Asuncion, mid-19th century.
Lithograph.
Asian Civilisations Museum [2014-01425].

11

A group of *datu* (chiefs) from Mindanao, on board a steamship.
Manila, around 1880.
Albumen print.
Mr and Mrs Lee Kip Lee collection, Singapore.

12

Queen Saovapha Pongsri of Siam,
Bangkok, around 1900.
Photograph.
Mr and Mrs Lee Kip Lee collection, Singapore.

The queen is dressed in what was popularly referred to as *seua khaen moo ham* (leg-of-ham blouse) and a *chong kraben* (skirtcloth).

13

An Indian family.
Lee Brothers studio. Singapore, around 1915.
Gelatin silver print.
National Archives of Singapore.

The women wear blouses popularly referred to in America as waistshirts, together with their saris.

aristocrats wore a puff-sleeved jacket and hat inspired by seventeenth-century European fashion, together with skirt cloths imported from the Coromandel Coast.

The improvisations and experimentations in fashion are epitomised in nineteenth-century photographs of Singapore, where members of different communities are portrayed in clothing that defies all notions of strict adherence to conventions. The Malays in Singapore, many of who had mixed origins and came from various parts of the archipelago, dressed in styles that combined different fashions. Women might wear black or white lace veils imported from France, sarongs from southern India, Java, or Sumatra, and tailor their kebayas from imported printed cottons (fig. 14). European jackets were fashionable with Malay gentlemen (fig. 15). Ottoman fashion was worn by people of various ethnicities including Arabs, Turks, Armenians, and Muslims from the Malay Archipelago, and variations of an outfit comprising a long robe and waist sash together with a vest and long overcoat was the equivalent of a Western suit in the Islamic world. In Singapore, these robes were often worn in combination with Malay garments, and it is often impossible to ascertain the ethnic origin of subjects dressed in this manner in early photographs (fig. 16). Japanese women in nineteenth-century Singapore also seem to have taken liberties with how they presented themselves. In one example from the 1880s, the subject wears a silk summer kimono with a simple, narrow *obi* (kimono sash) and has left out the customary silk *obi-jime* (obi cord) and scarf-like *obi-age* (obi bustle), while adding two small jewelled brooches at the front of her kimono, reminiscent of the *kerosang* brooches worn with a kebaya (fig. 17). Rather than *geta* (clogs), she wears *kasut manik* (beaded slippers), fashionable among women of the Malay, Indian, Eurasian, Chinese, and Peranakan communities.

Photographs of Peranakan family groups are exemplary expressions of this kind of sartorial "disorder". In one image from the early 1860s, male members of the family are dressed in *baju lokchuan* comprising trousers, shirt, and jacket, which is derived from southern Chinese fashion. But their headwear (skullcap, boater) and footwear (Chinese shoes, Western-style leather shoes) are diverse. The young standing boy is dressed in Chinese attire, the seated girl in a pan-archipelago style *baju panjang* and sarong, and the infant in quintessential Victorian garb comprising a bonnet and lacy frock (fig. 18). In another group photograph, the men are dressed in various types of Western suits typically worn in the tropics (fig. 19). The seated woman wears an old fashioned *baju panjang* tailored from the latest European printed organdie together with a fine batik cotton sarong from Java. The younger girl to her left is dressed in a modern white cotton lace kebaya made popular by European and Eurasian women in the Dutch East Indies, and a batik cotton sarong. The older girl, however, is dressed in a hybrid Chinese style silk blouse and skirt heavily trimmed with lace inserts. The three women are all in fact wearing completely unrelated styles of fashion, emphasizing how strict codes of dressing were not important for this and many families.

14

15

16

14

Portrait of a woman.
G. R. Lambert. Singapore. around 1900.
Albumen print.
Mr and Mrs Lee Kip Lee, Singapore.

15

Sultan Abu Bakar of Johor, Straits Settlements Governor Sir Frederick Weld and his wife, and Malay dignitaries at the unveiling of Stamford Raffles's statue (detail).
Singapore, 1887.
Albumen print.
Mr and Mrs Lee Kip Lee, Singapore.

The sultan is dressed in a Western suit, but the Malay dignitaries wear diverse combinations of Western and Malay garments.

16

A gentleman in Ottoman-inspired robes.
August Sachtler. around 1860s.
Albumen print.
Mr and Mrs Lee Kip Lee, Singapore.

17

18

19

17

Woman in a kimono.
Singapore, around 1880s.
Albumen print.
Mr and Mrs Lee Kip Lee, Singapore.

18

A Peranakan family.
August Sachtler. Malacca, around 1860s.
Albumen print.
Mr and Mrs Lee Kip Lee, Singapore.

19

A Peranakan family.
Lee Brothers Studio. Singapore,
around 1915.
Gelatin silver print.
National Archives of Singapore.

Negative reactions

Hybrid dressing was disparaged by many foreign observers, especially from the late eighteenth century, when racism and imperialism became increasingly endemic. The opinion of the British captain W. C. Lennon on the dressing of Catharina Johanna Koek, wife of Malacca's last Dutch governor, Abraham Couperus, in 1795, is typical of the thinking of colonial elites:

> Madam Couperus was dressed in the most unbecoming manner possible. A mixture between the Malay and the Portuguese. Her outward garment being made exactly like a shift, she looked as if she reversed the order of her dress altogether. Her hair was drawn so tight to the crown of her head, and the skin of her forehead so stretched, that she could scarcely wink her eyelids.[17]

The remarks of Lord Minto, Governor General of India, about women belonging to the Batavian elite are equally scornful:

> An elderly Batavian lady's upper garment is a loose coarse white cotton jacket fastened nowhere but worn with the graceful negligence of pins and all other fastenings and constraints of a Scotch lass, an equally coarse petticoat, and the coarsest stockings, terminating in wide, thick-soled shoes; but by standing behind her you find out her nobility, for at the back of the head a little circle of hairs is gathered into small crown, and on this are deposited diamonds, rubies, and precious stones often of great value.[18]

Munshi Abdullah's initial reaction to encountering Lord Minto himself, as he set foot in Malacca amid great pomp and ceremony, provides a fascinating counterpoint:

> Moreover, when I saw Lord Minto's appearance and disposition I was astonished because I had expectations about his appearance, his elegance, his height, and his dressing. This reminds me of a Malay saying: hearsay is more beautiful than actuality. I had to bite my forefinger. I saw a thin man past middle-age, with soft manners and a sweet expression. I felt he did not have the ability to lift 20 katis [about 12 kg]. That was how delicate he was. As for his clothing, he wore a black wool jacket and black trousers, and there was nothing else worthy of reporting.[19]

In the late nineteenth century, the sight of Dutch women in the Indies wearing sarong and kebaya still generated condescending comments among visitors. The American art collector Isabella Stewart Gardner, who visited Java in 1883, found this style intolerable: "And everywhere people (Dutch) in the strangest clothes since Eden … the women (ladies?) in Sarongs, no heeled slippers, loose white jackets (absolutely nothing else) and hair down their back." She noted that, "at 8 o'clock, however, dinner comes and people then appear *clothed*." On a separate occasion she also remarked on "people in frightful deshabillé" and how she was "thoroughly disgusted with Dutch people and their clothes."[20] Enslaved to what Roland Barthes describes as fashion's tyranny, privileged women like Gardner would voluntarily restrain themselves in uncomfortable corsets and bustles, and subject themselves to the dictates of Paris and other fashion capitals.

Some Western visitors did not hold such views about Dutch women in sarong and kebaya. Arbot Reid, a British observer, remarked in an article in the *Straits Times*:

> Who are we that we should impose our insular peculiarities upon the dress of the world, or endeavour to judge the custom of our Netherlands neighbour by our own narrow island prejudices.[21]

In the early twentieth century, such Western criticisms of "impure" fashions infiltrated Asian attitudes to their own way of dressing. In 1904 the *Straits Times* reported on the proceedings of a debate organised by the Chinese Literary and Debating Society in Kuala Lumpur about the dress of Straits Chinese women. One speaker proposed that: "The *kabaya*... was not graceful, nor neat, nor even suitable. As regards the *sarong*, its vulgarity was beyond question. On the whole, the costume of the Straits Chinese ladies was a disgrace to modern civilisation." Another speaker commented that "No Chinese lady who had any loyalty or reverence for the mother-country should discard its customs; and as fashion was a custom, it should not be set aside in favour of an alien fashion – the fashion of a non-descript costume." Yet another passed judgement on sarong and kebaya as "bastard garments of an inferior civilisation".[22]

Rehabilitating Asian fashion

None of these negative opinions mattered to the multicultural residents of port cities. From the seventeenth to nineteenth centuries, fashion could not have been more egalitarian, unfettered, and exuberant. What is also remarkable is the way that fashion trends were disseminated, democratically through merchants, peddlers, textile, and clothing shops.[23] Social life in a port city usually centred on family and community. Fashion was therefore not led by the usual arbiters, such as royal courts and intellectual elites, and later fashion designers and journalists. Unlike nineteenth-century European society, there was no bourgeoisie determinedly emulating the ways of their social superiors. Rather, people and communities seemed to possess much more freedom in the way they chose to appear, and when styles, textiles, and garments became viral, they primarily occurred through word of mouth. Asian fashion is currently still heavily entrenched in referencing a mythological past, where dressing well meant dressing "properly", conforming to dictated notions of correct forms. The whole concept of traditional fashion is in fact a modern one, which developed from the early twentieth century, and became more contagious from the mid-twentieth century in the era of Asian nationalism. But historical documents are increasingly revealing how these notions are far from historical fact. In the port cities of Asia, dressing "badly" was in fact the only way to dress well.

1 Laarhoven 1994.

2 Reid 1988, p. 88–89.

3 Reid 1988, p. 85.

4 See Lee 2014 p. 59–61 on Batavian sumptuary laws, and Reid 1988, vol. 1, p. 89, on dress restrictions in Banjarmassin.

5 "... tiada boleh pakai kasut dan pakaian kuning dan kasa nipis, sekalian itu larangan sekali-sekali. Maka apabila sahaya dengar akan larangan yang tersebut itu, maka berfikirlah sahaya sejurus sambil tersenyum sebab menengarkan `adat bodoh dan sia-sia itu, sebab sekalian itu perkara yang kecil-kecil menjadi salah dan larangan. Mengapa tiada dilarangkan burung terbang dari atas istana itu, dan mengapa tiada dilarangkan nyamuk memakan darah, dan pijat-pijat di bantal raja itu... Maka seperti perkara yang patut dilarangkan itu dan boleh menjadi kebajikan kepada segala manusia itu didiamkan..." Abdullah 1838 in Sweeney 2005, p. 43.

6 Barthes 1983, p. 263.

7 See Lee 2014, p. 29.

8 Reid 1988, pp. 85–90.

9 Linschoten 1596, p. 47: "wannerse ter Kercken gaen ofte eenige visitatie doen, hebben seer costelijcke cleederen aen, ende gouwe *Braselettes* ende *Manilias* aen haer armen, ende costelijcke ghesteenten ende Peerlen; aen haer ooren hanghen riemen vol Juweele ende Cleynodien ..." Also see Linschoten 1885, p. 206.

10 Pyrard 1619, pp. 105–6: "richement vestuës de soye de toutes couleurs, avec un grand crespe fin par dessus, qu'ils appellent *Mantes*; mais elles ne sont habillees à la mode de Portugal, et ont de grades pieces de soye qui leur servent de cotillon; Elles ont aussi juppes de soye fort fines, qu'ils appellent *Bajus*."

11 Schweitzer 1688, pp. 111–12: "und weil sie ihre biß auff den boden hangende Haar mit Klapper-Oel von Kochus Nussen gemacht alle Tag schmierte ihre Haar wicklete sie auff wie in Teuschland die Pferdt auff geschwäntzt werden, truge ein, einer Spannen langes weisses Camisol von feinstem Leinwand mit güldenen Knöpfflein fornen zugethan, durch welches ihre schwartze Haut wol zusche war unter disem Camisol, war der Bauch mit dem Nabel ein Spannen lang bloß, under dem Nabel biß auss den Boden, hatte sie Erstlich ein weis von Leinwad darüber ein schönes gefärbtes Seiden Kleidlen umbgewickelt, sie hatte ein grosses Pater Noster von Gold und Helffenbein umb den Hals hangen." Fayle 1929, p. 256.

12 Reid 1988, p. 85. "...barulah ia berkain; dua tiga belas kali dirombaknya, belum baik, dibaikinya. Sudah itu maka berbaju dan berdestar, itu pun demikian juga; dua tiga belas kali diikat, dirombak, belum baik diperbaikinya; bersebai pun demikian juga." *Sejarah Melayu* 1979, p. 183.

13 "Maka adalah sahaya lihat orang-orangnya dalam seratus barangkali sepuluh sahaja yang ada bekerja; dan yang lain itu lalai sahaja sepanjang hari, dalam hal miskin dan jahat... Dan yang ada setengah tabiat mereka itu hendak berchantek sahaja, memakai kain baju dan seluar yang bagus-bagus, tetapi tiada ia mau mencari jalan kehidupannya." Abdullah 1838 in Sweeney 2005, p. 21.

14 "Ik heb'er onder de Koningen gezien, die Sondags al vry pragtig uytgedoscht waren. Den ouden Koning van Kilang, Domingos Kôelho genaamd, heb ik menigmaal met een rok van swart fluweel, met groote goude Knoopen, bekleed gezien, dragende een fy nen swarten hoed, van die ouderwetze slag, die men brantemmers noemt, met een kostelyken gouden Hoeband, wel 3 of 400 gulden weerdig, en een Rotang, van een schoone goude Knop voorzien, in de hand. Andre heb ik dus in 't goud Broccade, in 't Silver, of goud Laken, of in Zyde, of in schoone fyne blauwe Ginganse kleederen gezien; waar nevens zy in huis ook wel fraeje en zeer fyne Chitse Cabaejen hebben. Dit is de dracht der Mannen onder de Christenen niet alleen; maar ook gaan de meeste Mohhammedaanen gelyk de gemeene Amboineesen gekleed, zynde het eenig onderscheid, dat de Mooren een Tulband of Distar dragen, zynde een roode, blauwe, of witte smalle band, dien zy om 't hoofd winden. Men ziet 'er ook, die de zelve zeer dik en met 3 of 4 strengen om 't hoofd winden; dog zelden die ze groen-geverwd hebben; alzo men zegt, dat die kleur alleen den genen, die van Mohhammed's bloed zyn, toegelaten is te dragen. Ook dragen de Mooren wel Badjoe's van Zyde, Ginggang, of van andere Zyde Stroffen van verscheide kouleuren, ook Chitse Cabaejen, komende in hun verdere dragt meest met de andre Amboineesen over een." Valentijn 1724, vol. 2, p. 169.

15 See Woodhouse 2012 on Thai hybrid fashion and photography.

16 Lee 2014.

17 Captain W. C. Lennon, "Journal of expedition to the Molucca Islands" in India Office Records: IOR H/441, pp. 17–18; Harrison 1985, p. 32.

18 Minto 1880, pp. 305–6.

19 "Syahdan apabila aku melihat rupanya dan sifatnya Lord Minto itu ta'jublah sangat hatiku karena kusangkakan bagaimanakah rupanya dan tampannya dan tinggi besarnya dan pakaianny. Maka teringatlah aku seperti umpamaan Melayu katanya: indah khabar dari rupa. Maka kugigit telunjukku. Maka adalah sifatnya kulihat orangnya telah lalu separuh 'umur dan tubuhnya kurus dan kelakuannya lemah lembut dan air mukanya manis. Maka adalah rasa hatiku, tiada boleh ia mangangkat dua puluh kati; begitu lembut orangnya. Maka pakaiannya kulihat baju sakhlat hitam dan seluarnya hitam, tiadalah apa yang lain yang hendak kusebutkan."Abdullah 1849 in Sweeney 2008, p. 309.

20 Chong 2009, pp. 261, 262, 269.

21 "Under Which Flag?", *Straits Times*, 9 June 1894, p. 3.

22 *The Straits Times*, 12 April 1904, p. 4.

23 Lee 2014 discusses the cultural dynamics of how sarong and kebaya styles became popular in island Southeast Asia. The influence of kebaya, batik, and jewellery retailers are examined in Lee 2014, pp. 72–73, 119–20, 206, 212–15, 270–72, 285–86.

貴

People of Nineteenth-Century Singapore

Singapore in the nineteenth century attracted traders and migrants from many parts of the world. The sight of people belonging to several nationalities in one town, dressed in their own fashions, was common in several Asian port cities throughout history. This mixture attracted the attention of travellers, writers, and photographers – and became part of Singapore's identity. Singapore's multicultural society therefore has long historical precedents in the region.

1

1
View of the Padang in Singapore
John Turnbull Thomson,
British (1821–1884)
Singapore, 1847
Watercolour on paper, 34.5 x 52 cm
National Museum of Singapore [HP-0055]

2

2
Four women
Singapore, 1860s
Albumen prints, each approx. 9.3 x 6.1 cm
Mr and Mrs Lee Kip Lee

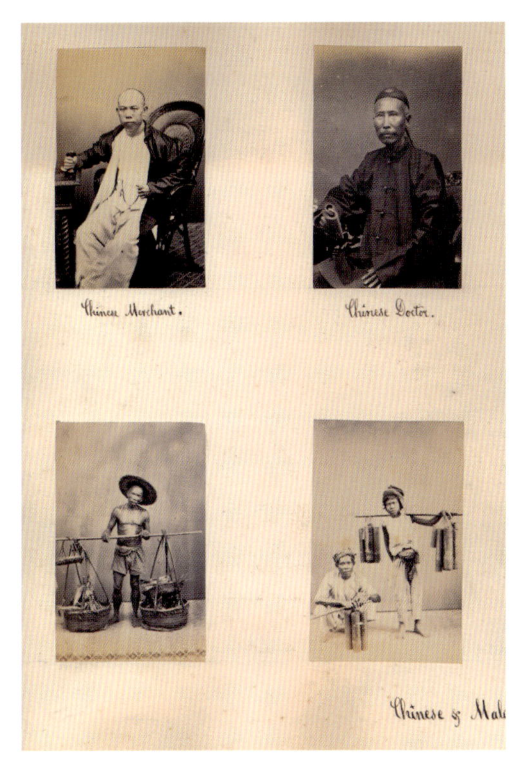

3

4

3
Singapore residents
Singapore, 1860s
Albumen prints, each approx. 9 x 5.8 cm
Mr and Mrs Lee Kip Lee

4
Portrait of a woman
G. R. Lambert & Co.
Singapore, 1880s
Albumen print, 27.4 x 21 cm
Mr and Mrs Lee Kip Lee

5

5
Baju panjang
Penang, late 19th century
Cotton (batik), 154 x 120 cm
Asian Civilisations Museum [1999-01012]

Sarong
Dutch East Indies, late 19th or 20th century
Cotton, 114.4 x 91.5 cm
Peranakan Museum,
Gift of Mr and Mrs Lee Kip Lee [2011-02192]

6

Suit
Europe, 1860s
Linen, vest: 56 x 48 cm;
jacket: 80 x 140 cm; pants: 113.5 x 46 cm
Asian Civilisations Museum,
Gift of Mr and Mrs Lee Kip Lee

7

7
Bisht
Unknown, early 20th century
Wool, 126 x 153 cm
Asian Civilisations Museum,
Gift of Mr and Mrs Lee Kip Lee

8

8
Garo
China, early 20th century;
embroidered in Gujarat, India
Silk, cotton (embroidered), 123.5 x 466.5 cm
Asian Civilisations Museum,
Gift of Mr and Mrs Lee Kip Lee

9

Shirt
China, late 19th century;
tailored in Indonesia
Silk, 77 x 159 cm
Mr and Mrs Lee Kip Lee

Pants
China, late 19th century;
tailored in Indonesia
Silk, 79 x 102 cm
Mr and Mrs Lee Kip Lee

10

10
Angarkha
Southern India, late 19th
or early 20th century
Cotton, 121.5 x 191 cm
Gift of Mr and Mrs Lee Kip Lee

11

11
Kebaya
Europe; tailored in Indonesia,
late 19th or 20th century
Cotton, 85 x 141.5 cm
Peranakan Museum,
Gift of Mr and Mrs Lee Kip Lee [2011-01138]

Sarong
Signed: Lien Metzelaar
Java, late 19th or early 20th century
Cotton (drawn batik), 104.8 x 213.5 cm
Peranakan Museum [2011-02025]

12

12

Dress
India, 19th century; tailored in Europe,
1850s or 1860s
Cotton (ikat), 142.5 x 121 cm
Mr and Mrs Lee Kip Lee

13
Kimono
Japan, late 19th century
Silk, 150 x 106 cm
Asian Civilisations Museum,
Gift of Mr and Mrs Lee Kip Lee

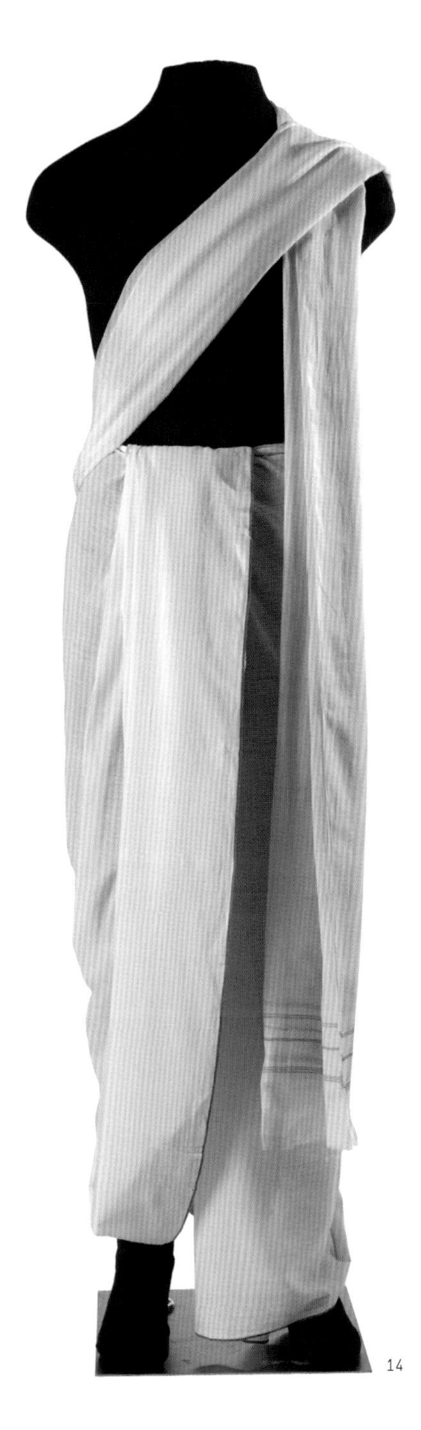

14

14
Dhoti and *angabastra*
India, Bombay (Mumbai), early 20th century
Cotton (machine-made: The Finlay's Group of Mills),
dhoti: 204 x 252 cm; angabastra: 107 x 366 cm
Asian Civilisations Museum,
Gift of Mr and Mrs Lee Kip Lee

15

15

Sari

India, Tamil Nadu, Kanchipuram,
late 19th or early 20th century
Silk, silver, 720 x 111.5 cm
Asian Civilisations Museum,
Gift of Mr and Mrs Lee Kip Lee

16

Baju kurung
Europe; tailored in Sumatra, late 19th century
Cotton, 104 x 51.5 cm
Asian Civilisations Museum, Gift of Mr and Mrs Lee Kip Lee

Sarong
Southern India or Indonesia, late 19th or 20th century
Cotton, 105 x 101 cm
Asian Civilisations Museum, Gift of Mr and Mrs Lee Kip Lee

Veil
Europe, late 19th century
Lace, 116 x 57 cm
Asian Civilisations Museum, Gift of Mr and Mrs Lee Kip Lee

17
Ao
China, late 19th century
Silk, 102 x 145 cm
Asian Civilisations Museum

Qun
China, late 19th century
Silk, 91 x 131 cm
Asian Civilisations Museum

18

Baju
Malay Peninsula, late 19th
or early 20th century
Silk, gold thread, 72.8 x 169.5 cm
Asian Civilisations Museum,
Gift of Mr and Mrs Andy Ng [1997-04226]

Trousers
Sumatra, late 19th orearly 20th century
Silk, 104 x 63 cm
Mr and Mrs Lee Kip Lee

Sarong
Southern India or Indonesia,
late 19th or early 20th century
Cotton, 104 x 98 cm
Peranakan Museum,
Gift of Mr and Mrs Lee Kip Lee [2011-02214]

18

19

View of Singapore from Mount Wallich
Percy Carpenter, British (1820–1895)
Singapore, 1856
Oil on canvas, 97.9 x 203.5 cm
National Museum of Singapore [HP-49]

19

Divergence
Moving, selling, copying

—

The movement of people across the oceans and the circulation of goods, technologies, and ideas from different origins have shaped the development of port cities. One important feature was the widespread presence of certain kinds of people and trade goods. In the seventeenth century, starting up life and trade in a new country took place against enormous odds, but was common in port cities. The impact on material culture was tremendous. Competitive traders immediately produced close copies and cheap imitations, which were also circulated in the region, leading to the acceleration of international styles and fashions.

Even 500 years ago, success was usually the result of cheap or slave labour and unethical trade practices. Devastating social upheaval was created all along the trail of goods, from production sites to cargo routes to markets. Kingdoms and cities rose and fell in the scramble for commercial leadership. The acceleration of these unstable dynamics through the centuries facilitated what we understand as modernity and globalization, and led to the development of popular culture.

New place, new life

The histories of communities and people starting new lives in new environments are as old as port cities. This transnational dynamic underlines the development of port cities. Following some lives and paths reveals that port cities were as intensely globalized and multicultural in the past as they are today.

20

Abdullah bin Abdul Kadir (1797–1854)

Hikayat Abdullah is the autobiography of Abdullah bin Abdul Kadir. Born in Malacca, he had a strict Muslim upbringing and scholarly education. He wrote the book in Jawi, and gives vivid accounts of everyday life and politics in Singapore, Malacca, and the region. He lived in Singapore and died while on a pilgrimage in Jeddah, Saudi Arabia. For his literary contributions, Abdullah has been called the "Father of modern Malay literature".

20
Hikayat Abdullah
Abdullah bin Abdul Kadir (1797–1854)
Mission Press
Singapore, 1849
Printed book, 25.2 x 40 cm (open)
National Library Board, Singapore
[BRN: 4079095]

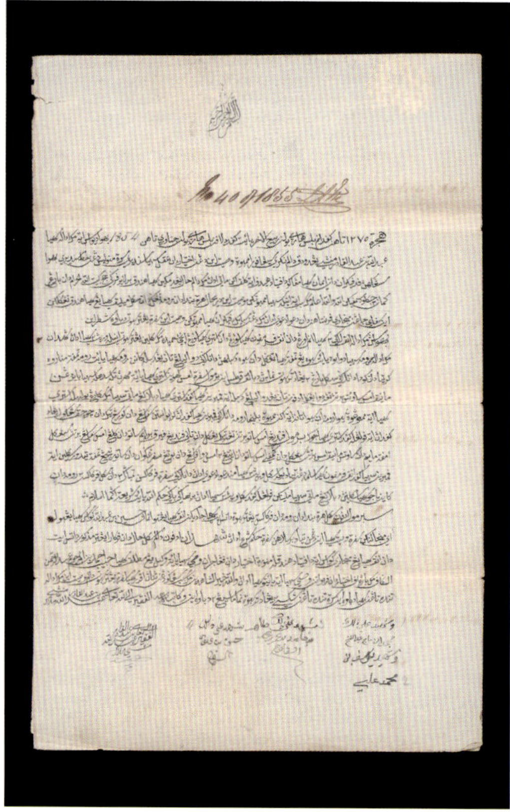

21

22

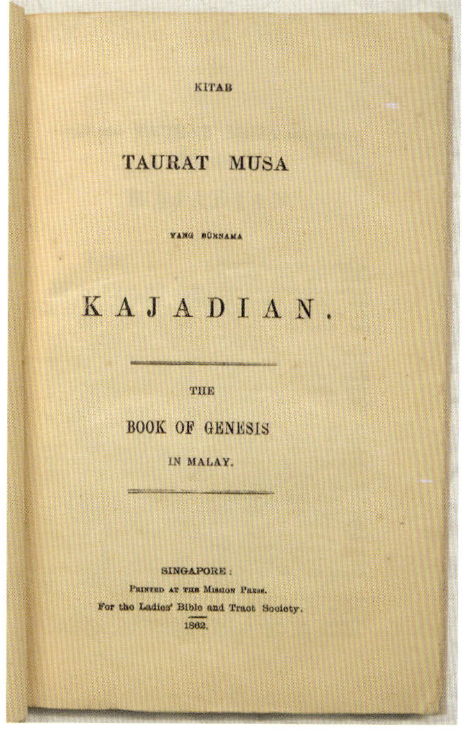

KITAB

TAURAT MUSA

YANG BŬRNAMA

KAJADIAN.

THE

BOOK OF GENESIS

IN MALAY.

SINGAPORE:
PRINTED AT THE MISSION PRESS.
For the Ladies' Bible and Tract Society.
1862.

23

21

Page from the last will and testament of Abdullah Kadir
Singapore, 1855
Ink on paper, 32.8 x 44 cm
National Museum of Singapore,
Gift of Mr John Koh [2000-05663]

22

Translations* from Hakayit Abdullah *with comments by J. T. Thomson
Henry S. King & Co. London, 1874
Printed book, height 21 cm
National Library Board, Singapore [4118867]

23

***Kitab Taurat Musa Yang Bŭrnama Kajadian* (The Book of Genesis in Malay)**
Benjamin Peach Keasberry, John Stronach, London Missionary Society and Ladies' Bible and Tract Society Singapore, 1862
Printed book, 21.2 x 13.6 cm
National Library Board, Singapore [12921117]

The Chettiars of Saigon

The Chettiar community has been trading beyond their homeland in the region of Chettinad, Tamil Nadu, for centuries. More than 200 years ago they built temples in Saigon. Links were strengthened when Pondicherry (near Chettinad) and Saigon became part of the French colonial empire. The treasuries of temples were filled with gold jewellery offered to deities as devotional embellishments.

24

24

Necklace (*kasumalai*)
India, Tamil Nadu, after 1884; exported to Vietnam
Gold (French 10-franc coins),
silver (Austro-Hungarian 1-forint coins), length 50.2 cm
Saigon Chettiar's Temple Trust Ltd [A-006],
on long-term loan to Indian Heritage Centre, Singapore

25

Arch for a deity (*prabhavali*)
India, Tamil Nadu, 20th century; exported to Vietnam
Gold, 108.5 x 63 cm; weight 6,213.6 grams
Saigon Chettiar's Temple Trust Ltd [A-011],
on long-term loan to Indian Heritage Centre, Singapore

26
Ceremonial spear
India, Tamil Nadu, 20th century; exported to Vietnam
Gold, emeralds, rubies, diamonds,
50.8 x 5.7 x 0.7 cm; weight 471.9 grams
Saigon Chettiars' Temple Trust Pte Ltd [A-003]

27
Parrot ornament
India, Tamil Nadu, 20th century; exported to Vietnam
Gold, emeralds, rubies and diamonds,
17.5 x 2.79 x 11.46 cm; weight 262.2 grams
Saigon Chettiars' Temple Trust Pte Ltd [A-040]

28

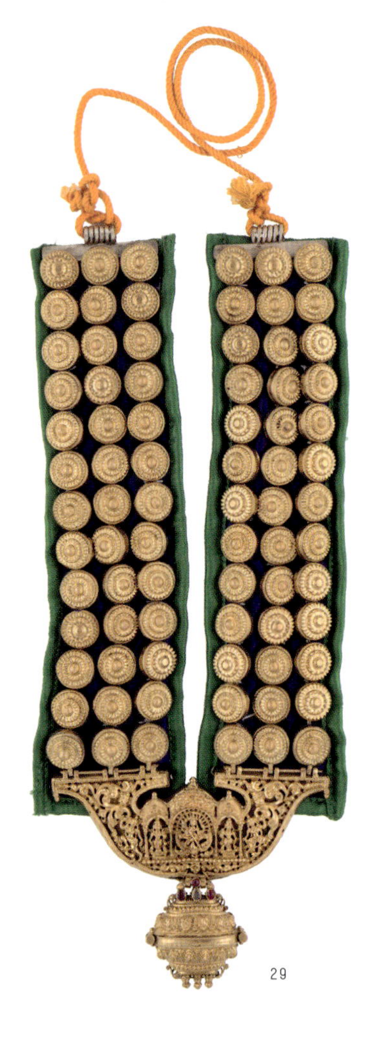

29

28

Necklace (*gowrishankaram*) with rudraksha-seed prayer beads and pendant
India, Tamil Nadu, 20th century; exported to Vietnam
Gold, rudraksha seeds, length 56 cm
Saigon Chettiar's Temple Trust Ltd [A-027],
on long-term loan to Indian Heritage Centre, Singapore

29

Necklace with neckband
India, Tamil Nadu, 20th century; exported to Vietnam
Gold, length 88 cm
Saigon Chettiar's Temple Trust Ltd [A-028],
on long-term loan to Indian Heritage Centre, Singapore

Georg Franz Müller (1646-1723)

Alsace native Georg Franz Müller was a soldier employed by the Dutch East India Company (Vereenigde Oost-Indische Compagnie, or VOC). He left the Netherlands in October 1669 and arrived in Java in August 1670. He served as a soldier in Batavia and other island outposts for twelve years. He studied the Malay language and kept an illustrated diary of his travels, making innumerable illustrations of the people of various nationalities that he saw in Batavia and the islands, as well as exotic flora and fauna, and even mermaids and other imaginary beings. He left Batavia in 1682 and reached Europe the following year. He worked for a Roman Catholic church in Rorschach, Switzerland, and left all his diaries and rare artefacts from his travels to the church.

30
Coin purse
Indonesian Archipelago, 1669-1682
Silk (embroidered),
metal clasp, 15 x 14.5 cm
Stiftsbibliothek, St Gallen, Switzerland

31
Basket
Sulawesi, 1669-1682
Reed, 21.8 x 13 cm
Stiftsbibliothek, St Gallen, Switzerland

32
Shoes
Batavia, 1669-1682
Silk, cotton, leather, length 21.5 cm
Stiftsbibliothek, St Gallen, Switzerland

33
Teapot
China, 17th century
Ceramic, 14 x 8 x 8 cm
Stiftsbibliothek, St Gallen, Switzerland

34
***Reisebuch* (travel diary)**
Georg Franz Müller
Southeast Asia, 1670–1682
Bound volume, 13 x 19.5 cm
Stiftsbibliothek, St Gallen, Switzerland

Manges tanges
Durioen

Ein Persianer
Ein Japanner
Ein Orientalisch Tarter

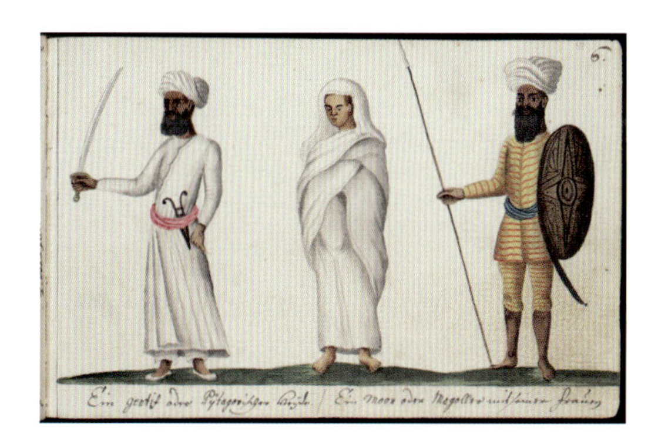

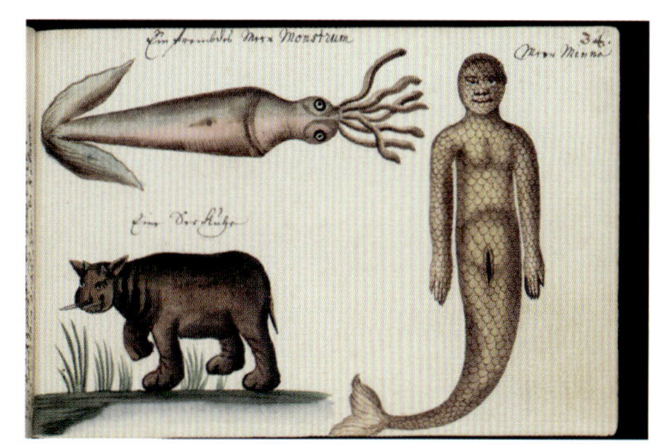
Monstrum

34

Duzhan Xingying (1628-1706)

Duzhan Xingying 独湛性莹 (Dokutan Shoei in Japanese) was a Chinese monk from Putian, Fujian province, who was a disciple of Zen master Yinyuan Longqi 隱元隆琦, also a Fujian native from Fuqing.

In 1654 Duzhan followed Master Yinyuan to the port of Nagasaki together with 30 other disciples. Seven years later, Yinyuan established the Ōbaku 黃檗 sect of Zen Buddhism, centred on Mampuku-ji, a new temple in Uji near Kyoto, and became its first abbot. Duzhan was part of this new sect, and he established a Zen temple, Shosan Horin-ji, in Hamamatsu in 1664. He later became the fourth abbot of Mampuku-ji. His predilection for chanting earned him the nickname Chanting Monk Dokutan. In addition to his Zen teachings, he also made paintings for display in temples and monks' quarters. He lived to a venerable age and passed away in retirement in Uji.

The practices of the Ōbaku sect were closely linked with Pure Land Buddhism and Ming culture. Duzhan had a strong inclination towards Pure Land philosophy and also held strong Confucian attachments to his parents and ancestors.

Information in this section provided by Nishigori Ryosuke, specialist on Ōbaku sect of Zen Buddhism (Nishigori 2016).

35

35

Portrait of Dokutan Shoei
Painting by Fujiwara Tanenobu;
Inscribed by Eppō Doshō
Japan, dated 1706
Hanging scroll, ink and colours on paper,
133.9 x 65.6 cm
Obakusan Mampuku-ji,
Kyoto Prefecture, Japan

The portrait shows the monk in old age, perhaps made for his funerary ceremony. The artist worked for the court of Maeda Tsunanori (1643-1724), who ruled over the Kaga domain (present day Ishikawa and Toyama Prefectures). There are five portraits of Dokutan executed by Fujiwara in the temple collection. The calligraphy is by Eppō Doshō (1655-1734), 25th disciple of Dokutan Shoei. He was born in Hangzhou, Zhejiang province, China, and came to Nagasaki in 1686. He wrote of his respect and admiration for his mentor in the inscription.

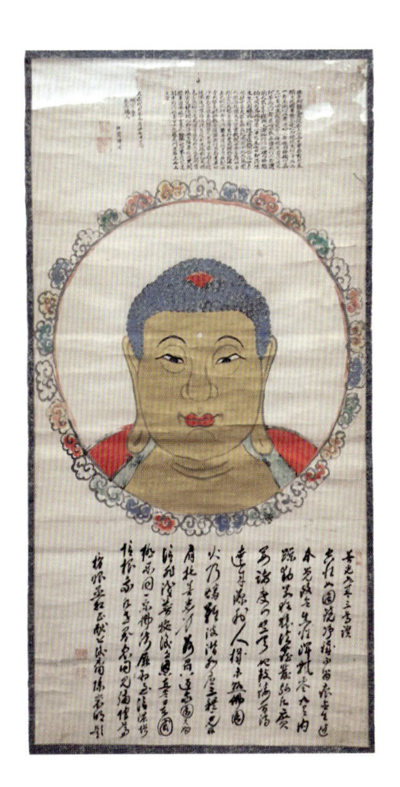

36

36

Seishi Bosatsu, Amida Nyorai, and Kannon Bosatsu
Painted and inscribed by Dokutan Shoei
Japan, 1698
Hanging scrolls, ink and colours on paper,
left: 182 x 94 cm; centre: 183 x 93 cm; right: 185 x 94 cm
Obakusan, Mampuku-ji, Kyoto Prefecture, Japan

At left, Dokutan has drawn his mother as a diminutive worshipper with Seishi Bosatsu, and at right, his father with Kannon Bosatsu.

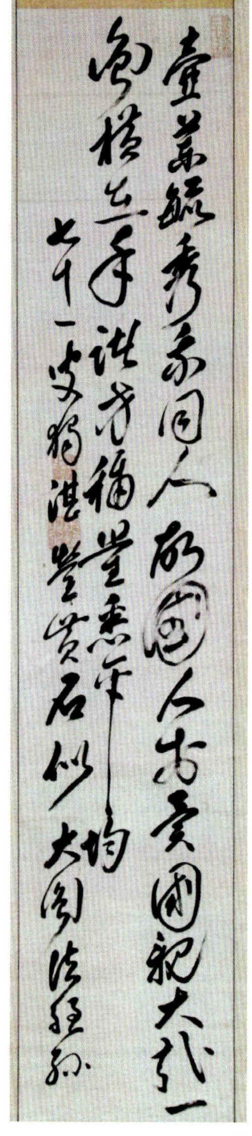

37

新春吉旦恭惟
本師老和尚慈祉萬安舊冬結制蒙
慈蔭平安所賜松隱寒山二集時々捧誦受
教不忘茲者為初山創建已經兩年尚從簡便今妥
再構佛殿恭請
老和尚開山為此方福田願 慈心廣大乞垂許可光崇檀
信福庇將來下情無任感戴之至 上
本師老和尚 方丈
嗣法徒 性瑩 和南百拜

38

37

Calligraphy for Daien
Running script by Dokutan Shoei
Japan, dated 1698
Ink on paper, 133 x 26 cm
Obakusan Mampuku-ji, Kyoto Prefecture, Japan

38

New Year greeting: Dokutan Shoei to his master Ingen Ryuki (Yinyuan Longgi)
Japan, dated 1666
Ink on paper, 108 x 28.5 cm
Obakusan Mampuku-ji, Kyoto Prefecture, Japan

This letter by Dokutan Shoei comes from a scroll of New Year greetings sent to Master Ingen Ryuki. He established Mampuku-ji in 1661, and retired in 1664. The greetings also express well wishes for his retirement.

Cornelia van Nijenroode (around 1629-1691)

Cornelia van Nijenroode was the daughter of Cornelis van Nijenroode, a wealthy Dutch merchant in Hirado, Japan, and Surishia, a Japanese geisha. After her father's death in 1632, the young heiress was sent to Batavia, where in 1652 she married Peter Cnoll, an equally wealthy Dutch merchant. His death in 1672 left her immensely rich. Her marriage four years later to Johan Bitter and his subsequent attempts to take over her inheritance led to one of the most famous and acrimonious international court cases of the 17th century. It was fought all the way to The Hague, where Cornelia passed away.

39

39

Pieter Cnoll, Cornelia van Nijenroode and their Daughters
Jacob Coeman, Dutch (1632-1676)
Batavia, 1665
Oil on canvas, 141.5 x 202.5 x 5.6 cm
Rijksmuseum, Amsterdam [SK-A-4062]

40

40

Table screen sent by Cornelia van Nijenroode to her family in Hirado
China or Batavia, 17th century
Wood, 33 x 16.5 x 16 cm
Private collection, Hirado, Japan

The first Dutch East India (VOC) trading outpost in Japan was on the island of Hirado, off the coast of Kyushu. In 1638, *Sakoku* (national isolation policy) was implemented by the authorities. The Dutch were ordered to transfer all trading activities to the small island of Dejima in Nagasaki harbour. Before the VOC closed its branch in Hirado in 1639, it was recorded that about 40 mixed-race Japanese people were deported to Batavia.

This group included people of all ages who were related to the foreigners by marriage or by blood. In 1660 a series of letters were sent from the deported people to their families to Japan, conveying well wishes and a longing to return home. Some are still today in Hirado. They have come to be known as "Jakarta letters".

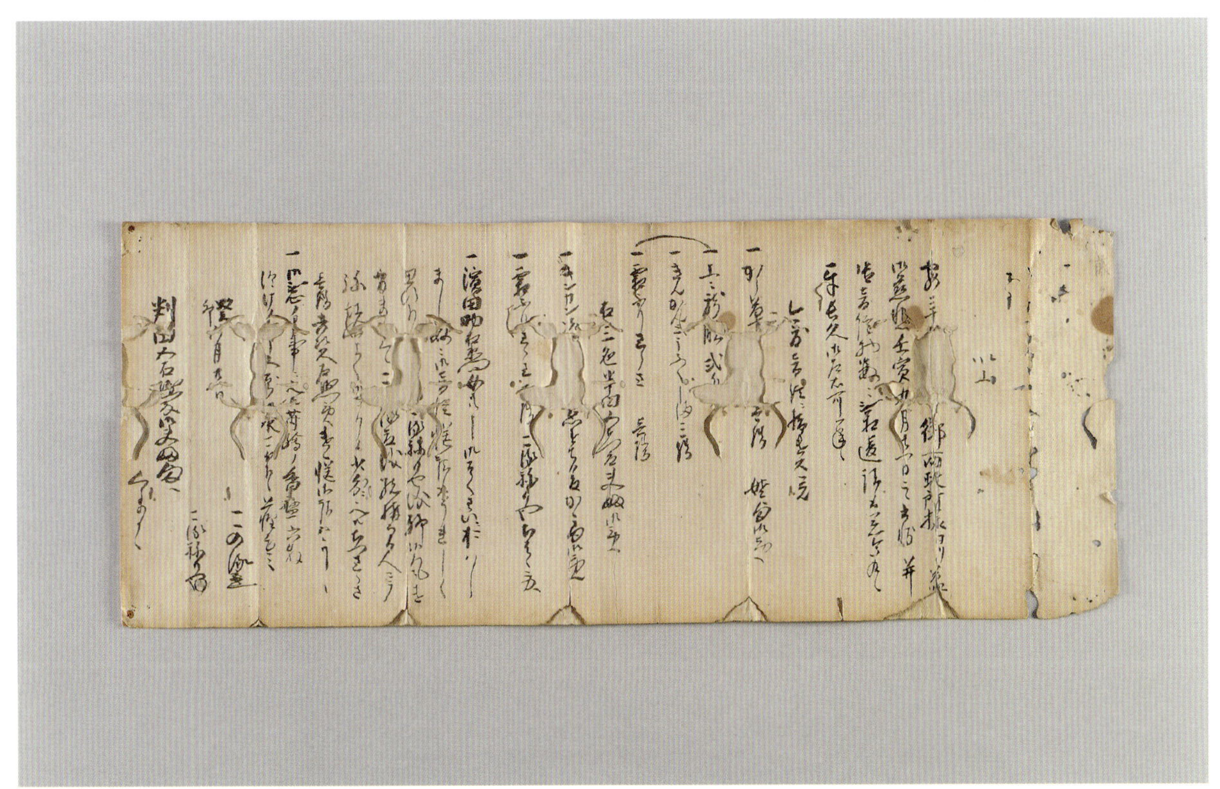

41

41

Jakarta letter: Cornelia van Nijenroode to her family in Hirado
Cornelia van Nijenroode
Batavia, dated 1663
Ink on paper, 20.5 x 52 cm
Matsura Historical Museum, Hirado, Japan

In this letter, Cornelia sends her regards to her relatives, and describes gifts to them from Batavia. The gifts included *sarasa* (Indian cotton), boxes of gingham fabric, and lacquered objects.

42

This letter was written on a patchwork cloth wrapper made up from cotton and silk. The identity of Koshoro is unclear and there is no other record. The author wrote in a poetic prose, confessing longing for home.

Japan, I love thee I love thee, as you stand transiently
Moreover, the heart, too, clings to the thought of a lost hometown
A rugged heart is in tears, sobbing
Further to the side of an indifferent trance
Even if service is wickedly pilfered, but only slightly
With a belief that service can be brought to its knees, at the sight of a single tea caddy
Oh Japan, I love thee, I love thee

Koshoro

A visiting elderly woman

42
Jakarta letter: Koshoro to relatives in Hirado
Batavia, 17th century
Ink on cotton and silk, 21 x 21 cm
Matsura Historical Museum, Hirado, Japan

Everybody wants one
Circulating rarities and replicas

One characteristic feature of consumption in Asian port cities was the widespread circulation not only of raw materials and consumables, but also manufactured goods. The same kinds of porcelain, textiles, and items made of wood, metal, and exotic substances found their way to every port city. Of these, textiles was one of the most heavily traded goods.

The massive demand for such cargo created the most natural commercial responses: copies, in varying degrees of quality, were made by competing manufacturers. Designs were often tweaked to satisfy different regional consumers.

Indian trade textiles present a perfect case study of this dynamic.

Indian Textiles

Because of the massive global demand for Indian cotton cloth – especially the variety known by the terms chintz and *sarasa* – copies were made by competing manufacturers, in varying degrees of quality. Designs were also tweaked to satisfy different regional consumers.

Itinerant artisans of different nationalities, collaborating with migrant and local middlemen, also satisfied consumer demands and began producing goods in any style for any market. Southeast Asian batik and European printed cottons are among the dynamic responses to the widespread popularity of Indian cottons.

43

44

45

43
Inner robe with standing tree design
India, Coromandel Coast, 17th or 18th century; tailored in Japan, Edo period
Cotton, 137.5 x 131 cm
Matsuzakaya Collection, Japan

44
Under robe with flowering plants design
India, Coromandel Coast, 17th or 18th century; tailored in Japan, Edo period
Cotton, 138 x 131 cm
Matsuzakaya Collection, Japan

45
Outer robe (*Tozan*)
India, Coromandel Coast, 17th or 18th century; tailored in Japan, Edo period
Cotton, 137.5 x 130 cm
Matsuzakaya Collection, Japan

46

46

Jacket (*Jinbaori*)
India, Coromandel Coast, 18th century; tailored in Japan
Cotton, 112.5 x 128 cm
Matsura Historical Museum, Hirado, Japan

47

47
Ceremonial cloth (detail)
India, Coromandel Coast, 18th century;
traded to eastern Indonesia
Cotton, 90 x 110 cm
Mr and Mrs Lee Kip Lee

48

48
Dress
India, Coromandel Coast,
18th century; tailored in Europe
Cotton (mordant and resist dyed), 90 x 110 cm
Mr and Mrs Lee Kip Lee

49

49

Kebaya
India, Coromandel Coast, 18th century;
tailored in Lampung, South Sumatra
Cotton, (mordant and resist dyed), 142 x 161.4 cm
Asian Civilisations Museum,
Gift of Mr Lee Kip Lee [2011-00091]

50

Baju
India, Coromandel Coast, 18th century;
tailored in Lampung, South Sumatra
Cotton, (mordant and resist dyed), 143 x 114.5 cm
Mr and Mrs Lee Kip Lee

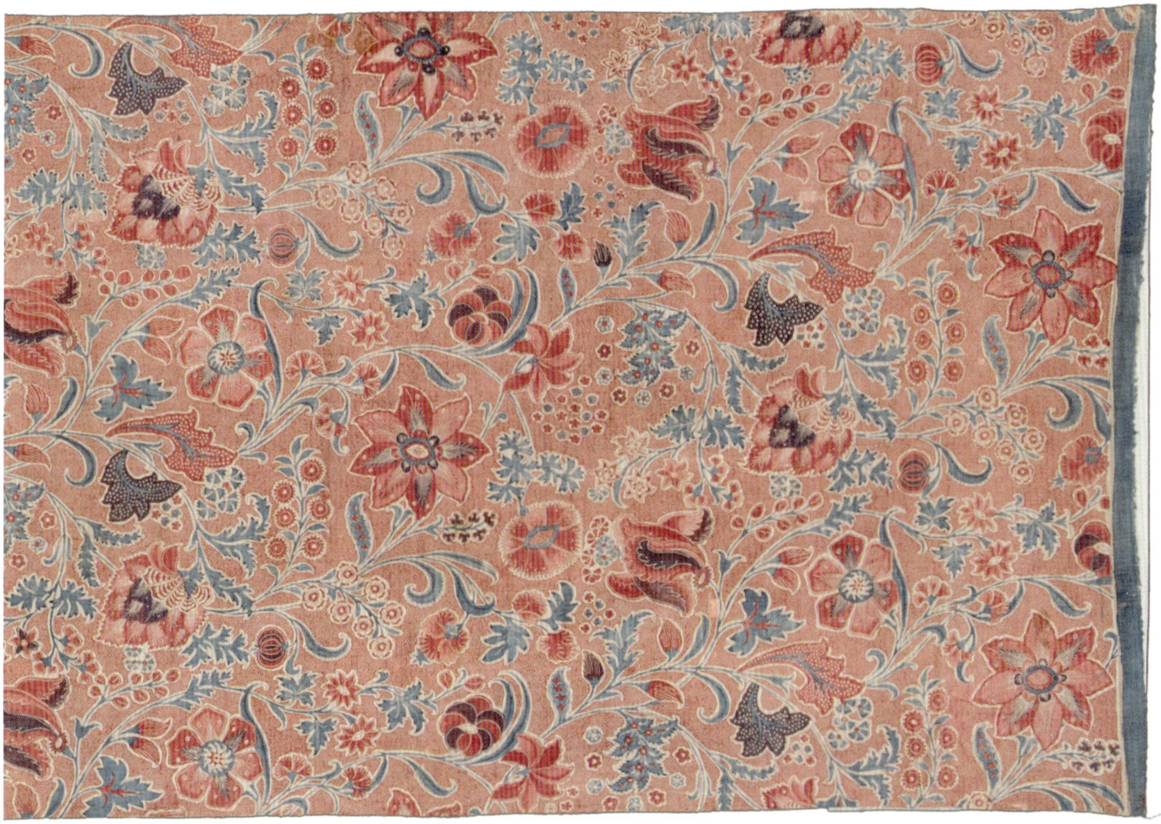

51

51

Ritual hanging (detail)
India, Coromandel Coast, 18th century
Cotton (mordant and resist dyed), 448 x 92.1 cm
Asian Civilisations Museum [2009-01881]

52

53

52

Tea ceremony utensils and pouches
India, Coromandel Coast, 18th century;
tailored in Japan, late 18th century
Cotton (mordant and resist dyed),
tea cover (red), 10 x 9 cm; teacups: 3.5 x 6.5 (each);
tea containers (white): 7 x 6 cm; (blue): 10 x 6 x 5 cm;
tea pewter (oval): 3 x 13 x 8.5 cm;
cups: China, 18th century; Porcelain, each diameter 6.3 cm
Mr and Mrs Lee Kip Lee

53

Tea ceremony mat (*fukusa*)
India, Coromandel Coast, 18th century; traded to Japan
Cotton (mordant and resist dyed), 40.5 x 40.5 cm
Mr and Mrs Lee Kip Lee

54

54

Yardage fabric
India, Coromandel Coast, 18th century; traded to Japan
Cotton (mordant and resist dyed), applied gold, 166 x 141.6 cm
Mr and Mrs Lee Kip Lee

55

56

55

Pouch and bag
Europe, 19th century; traded to Japan
Cotton (machine printed)
Pouch: 16 x 9.5 cm (closed)
Bag: 19 x 27 x 25 cm
Mr and Mrs Lee Kip Lee

56

Baju
Europe, 19th century; tailored in Sumatra
Cotton (machine printed), 72.5 x 126.4 cm
Asian Civilisations Museum,
Gift of Mr and Mrs Lee Kip Lee [2011-00099]

57

57
Yardage fabric (detail)
Europe, 19th century; traded to Indonesia
Cotton (machine printed), 145 x 268 cm
Mr and Mrs Lee Kip Lee

58

58

Sarong
Europe, 19th century; traded to Indonesia
Cotton (machine printed), 110 x 90 cm
Mr and Mrs Lee Kip Lee

59

59
Sarong (detail)
Java, North Coast, late 19th century
Cotton (drawn batik), 110 x 90 cm
Asian Civilisations Museum [T-0834]

60

60

Kain panjang (detail)
Java, North Coast, late 19th century
Cotton (drawn batik), 90 x 110 cm
Mr and Mrs Lee Kip Lee

61

61

Sutra cover
Europe, 19th century; tailored in Sri Lanka
Cotton (machine printed), 44 x 15–16 cm
Mr and Mrs Lee Kip Lee

62

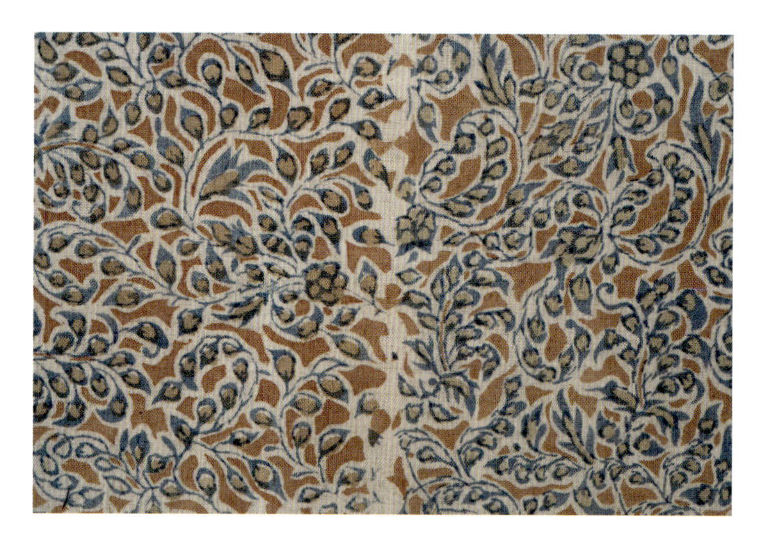

62

Futon cover

India, late 19th century; stenciled and tailored, probably in Nagasaki
Cotton (stencil dyed, a type known as *wa-sarasa*, 和更纱), 183 x 148.5 cm
Mr and Mrs Lee Kip Lee

Convergence
Owning, Collecting, Commissioning

—

The convergence of people and goods created cities characterized by diverse architecture and a demand for imported goods. Conspicuous consumption, social competition, and a globalized marketplace gave rise to shared tastes. Luxury goods and fashions were unifying forces both within and among Asian ports. In other words, members of various communities residing in a city often wanted the same prestige objects.

In port cities one could see an array of imported goods from the whole of Asia. Considered exotic in the West, these objects were an integral and conventional aspect of life in the port cities of Asia.

Batavia, cultural capital of Asia

Batavia (present-day Jakarta), established in 1618 as the capital of the Dutch colonial empire in Asia, was perhaps the most important Asian port city in the 18th century. The flood of Asian goods into one metropolis was probably unprecedented in history. International goods circulated widely among the city's multiracial residents, and were resold and repurposed for centuries, and many continue their existence as the antiques of today.

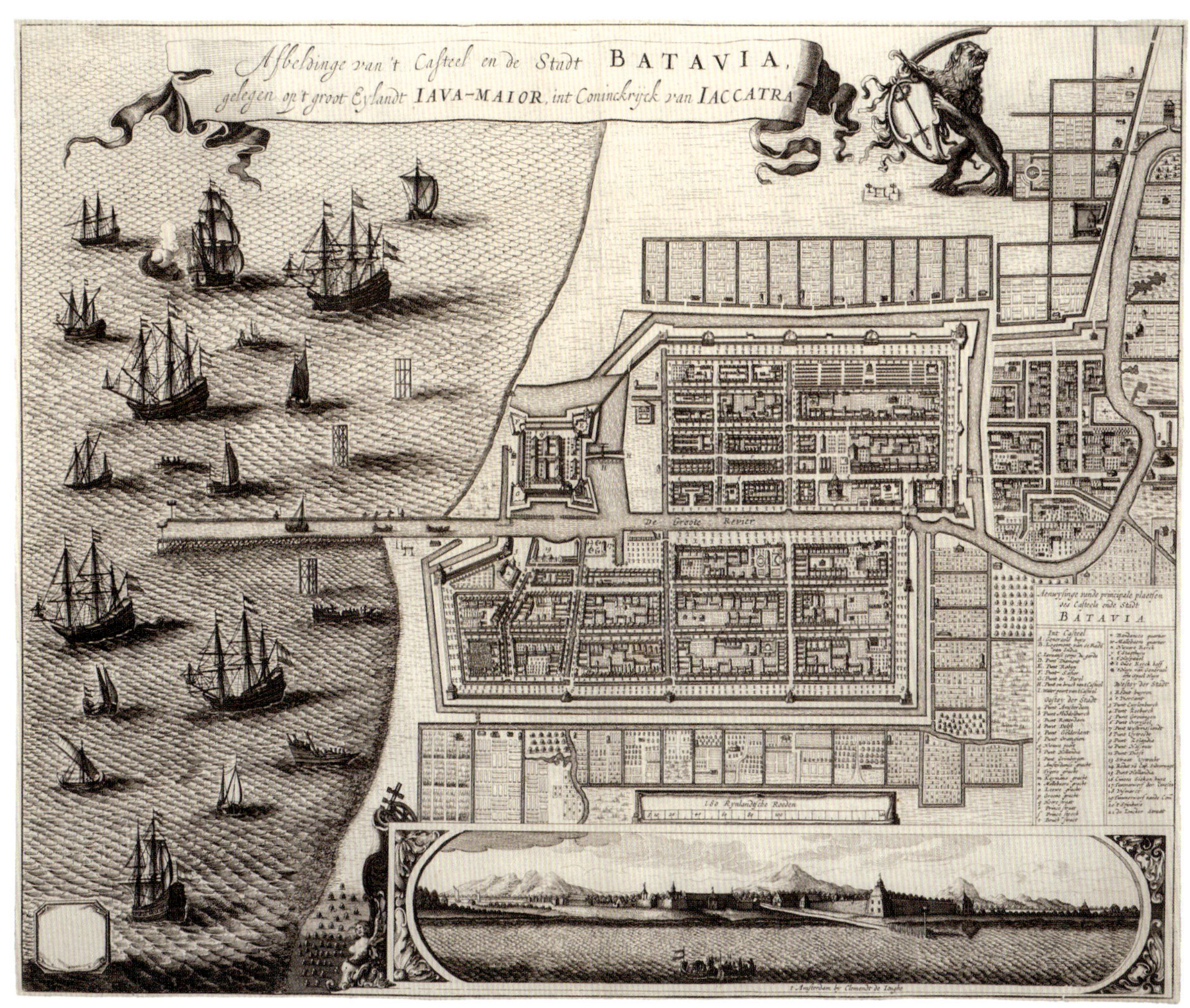

63

63

City plan of old Batavia
Clemendt de Jonghe, Dutch (1624–1677)
Batavia, 1660s
Engraving, 40.5 x 49.5 cm
Asian Civilisations Museum [2016-00358]

64

64

Hollands bruidsfeest te Batavia
(Dutch wedding feast at Batavia)
Jan Brandes, Dutch (1779–1785)
Batavia, 1779–85
Watercolour on paper, 20.1 x 33 cm
Rijksmuseum, Amsterdam [NG-369]

65

65

Cabinet with Adam and Eve
Sri Lanka, late 17th century
Ivory, teak, silver mounts, 45.2 x 55 x 34.3 cm
Asian Civilisations Museum [2015-00188]

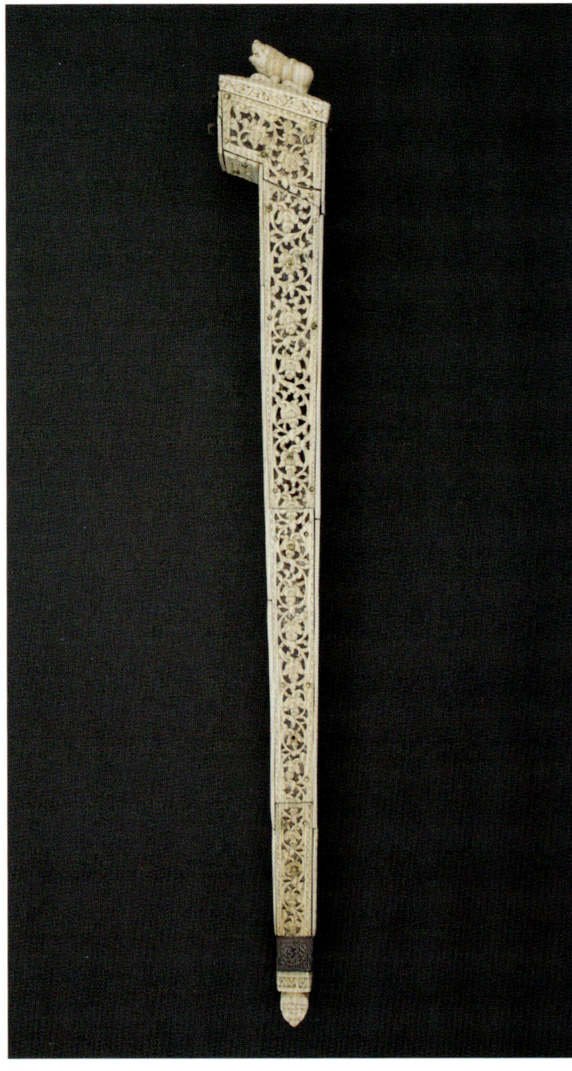

66

66

Double pipe case
Sri Lanka, mid-17th century
Ivory, wood, mica, 6.5 x 5.7 x 52 cm
Asian Civilisations Museum [2011-01494]

67

67

Bible box
Sri Lanka, early 18th century
Ivory, gold mounts, 4.7 x 12.5 x 9.9 cm
Asian Civilisations Museum [2014-00917]

68

68

Commemorative salver
Batavia, 1740 or later
Silver, diameter 30 cm
Asian Civilisations Museum [2014-00930]

69

69

Decanter case with bottles
Batavia and India, Gujarat, 18th century
Wood, silver mounts, velvet lining,
Box: 17.8 x 25 x 16.1 cm; bottles: each 13.5 x 6.2 x 6.4 cm
Asian Civilisations Museum [2013-00749]

70

70

Ewer and salver
Batavia, around 1700
Silver
Ewer: 34 x 28cm, weight 2074.83 gm;
salver: 7.5 x 58 cm, weight 3290 gm
Gemeentemuseum, The Hague, Netherlands [0154599]

71

72

73

71

Coffee pot
China, 18th century
Silver, gilded silver knob, height 15 cm
Asian Civilisations Museum,
Gift from Mr and Mrs Lee Kip Lee [2012-00185]

72

Pair of wine cups
China, 18th or 19th century
Silver, each 3 x 7 x 5.5 cm
Asian Civilisations Museum,
Gift of Mr and Mrs Lee Kip Lee [2012-00196]

73

Keris
Java, Yogyakarta, mid- or late 19th century
Wood, antler, gold, steel, diamonds, length 50 cm
Asian Civilisations Museum [2016-00028]

74

74

Spirit keg: Dutchman sitting on a barrel
Japan, Arita, early 18th century
Porcelain, 36 x 22 x 12.5 cm
Asian Civilisations Museum [2015-00203]

75

76

75

Coffee pot
Porcelain: Japan, Arita, late 17th or early 18th century
Mount: Probably Germany, 18th century
Porcelain and gilded metal, height 27 cm
Asian Civilisations Museum

76

Shaving basin
Japan, early 18th century
Porcelain, 27 x 26.5 x 7 cm
Asian Civilisations Museum [2015-00035]

77

78

77
Bowl
China, 16th century
Porcelain, 12 x 7.1 cm
Asian Civilisations Museum,
Gift of Mr Loh Teh Soon [1996-00092]

78
Covered jar
Batavia, 1736–1795
Porcelain, 21 x 21 cm
Asian Civilisations Museum [2012-00388]

79

80

81

79
Bowl
China, 1752
Porcelain, 7.4 x 15 cm
Asian Civilisations Museum [C-1257]

80
Pair of cups
China, 1752
Porcelain, 5 x 8.3 cm (each)
Asian Civilisations Museum [C-1258]

81
Dish with landscape scene
China, around 1750
Porcelain, 2.4 x 22.9 cm
Asian Civilisations Museum [C-1252]

82

83

82
Kendi
China, Qing dynasty (Kangxi period, 1662–1722)
Porcelain, 14.5 x 4.3 x 22.3 cm
Asian Civilisations Museum [C-0110]

83
Jar
China, Qing dynasty
Glazed stoneware, height 65.5 cm
Asian Civilisations Museum [C-1409]

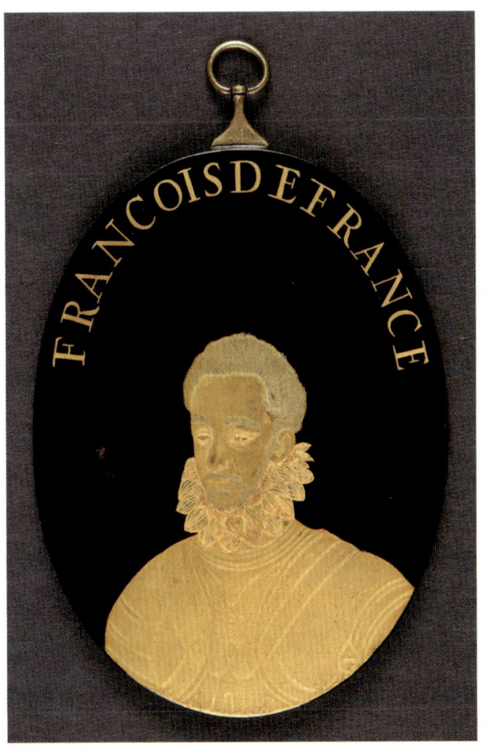

84

84
Medallions
Japan, late 18th century
Lacquer, gold, copper, each 12.2 x 9.2 cm
Asian Civilisations Museum [2015-00478, 2015-00479]

85

85
Screen
Batavia, 18th century
Wood, 190 x 280 x 82 cm
National Museum of Indonesia, Jakarta

86

87

86
Pair of gueridons
Batavia, 18th century
Wood, each 112.5 x 64 cm
Jakarta History Museum

87
Chair
Batavia, 18th century
Wood, 121 x 49 x 67.5 cm
Jakarta History Museum

88

88
Cabinet
Probably central Java or Batavia,
late 18th to early 20th century
Hardwood, 131.5 x 72 x 30.5 cm
Asian Civilisations Museum [2015-00206]

89

Tric-trac table
Batavia, mid-18th century
Amboyna wood, 60 x 129 x 74 cm
Asian Civilisations Museum [2015-00034]

90

90

Miniature cabinet
Batavia, late 17th or early 18th century
Ebony, 44 x 38.5 x 24 cm
Mr and Mrs Lee Kip Lee

91

91

Wall clock (*stoelklok*)
Netherlands, Friesland, 1790
Painted and gilded wood, metal,
66 x 32 cm (without chimer)
Asian Civilisations Museum

92

92

Figure of a Dutch woman
China or Batavia, late 17th century
Wood, height 17.5 cm
Asian Civilisations Museum [2015-00050]

Mayhem
The dark side of dynamic encounters

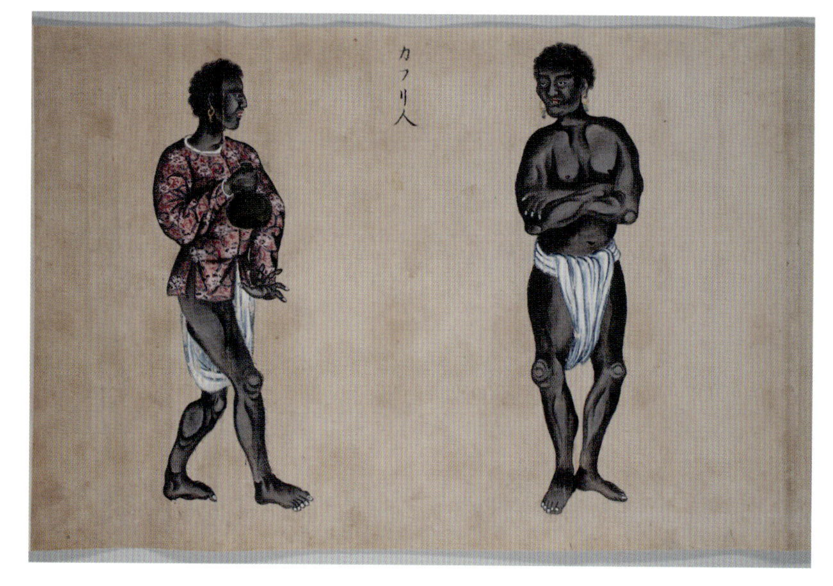

Underlying the extraordinary mingling of people was the brutality of travel and urban life. Piracy was a constant threat, and death on the seas was common. Among warring nationalities, the plunder of cargo was justified as legal booty.

Commerce was often conducted through violence, and involved harming competitors, forcing agreements, and avoiding government control. Trade and armed force went hand in hand.

Slavery and human trafficking, universally practised for most of recorded history, had great impact on urban society, intensifying the racial diversity of the population. Political and religious leaders also put into place ruthless measures to control the population. This did not stop violent upheavals and the whole range of human vices, including murder, gambling, prostitution, and illicit drugs.

93

93

Scroll depicting foreigners (detail)
Japan, Nagasaki, 17th or 18th century
Ink and colours on paper, 27.8 x 823.2 cm
Matsura Historical Museum, Hirado, Japan

This scroll depicts 42 foreigners, including Dutch, British, Portuguese (Macanese), Goan (Kaffir), Spanish (Luzon), Moors, Thai (Siam), Batavian, and several different Chinese ethnicities, genders, and age groups. Because of trade relationships with the Europeans and the Chinese, many such paintings were produced in Japan during this period.

According to the records from the family archive set up by Matsura Kiyoshi, aka Matsura Seizan, (1760–1841), this scroll was painted by artists in Nagasaki during the time of Matsura Atsunobu, 31st-generation Matsura clan lord (1684–1756).

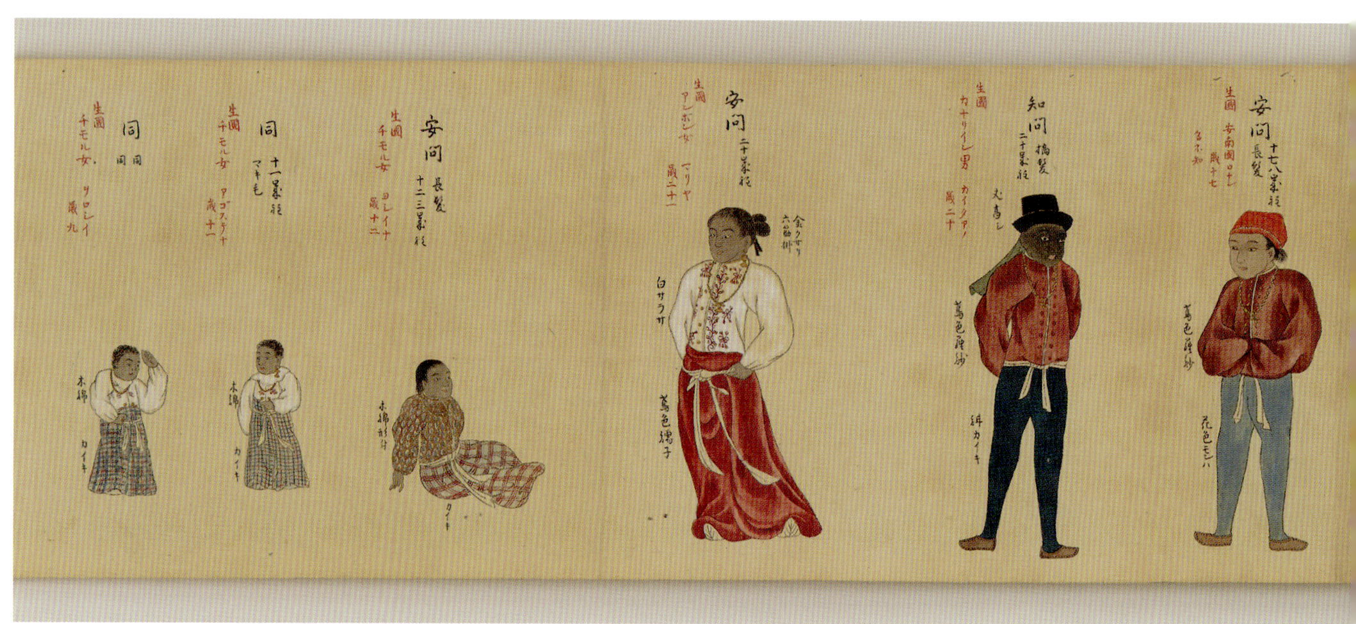

94

Two scrolls depicting foreigners detained in Hirado (detail)
Japan, later 18th or early 19th century
Ink on paper, each 29.5 x 107.5 cm
Matsura Historical Museum, Hirado, Japan

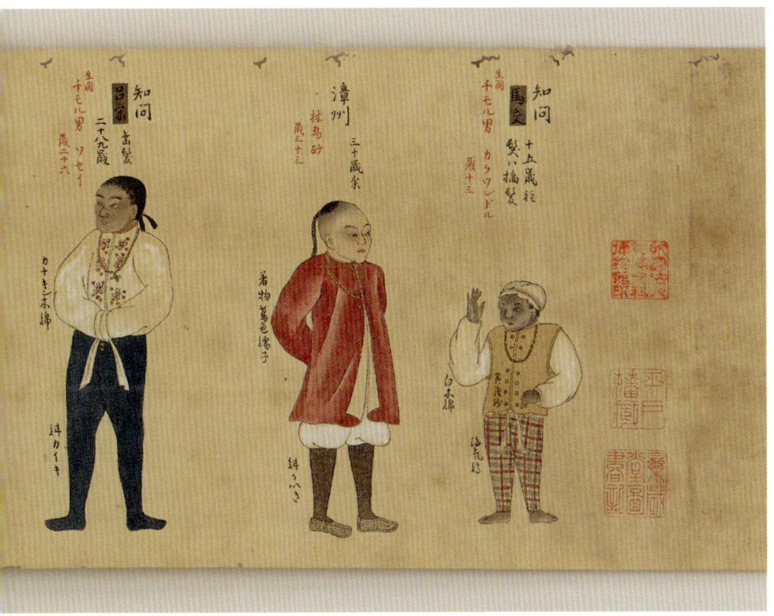

94

The Japanese artist labelled each person, noting origin, attire, and even stated age. The people appear cosmopolitan, mixing the fashions of Asia and Europe.

The annotations list the people as born in Zhangzhou, Timor, Goa, Ambon Island, and Annam, et al. A few are recorded as natives of Goa or Timor, but arrived from Manila and Macau, showing the movements of people in Asia in the period. One scrolls notes weapons confiscated. The scrolls are recorded in the Matsura clan collection as early as 1808.

Between 1633 to 1866, Japan blocked travel as part of a strict national isolation policy. This even applied to Japanese who were travelling or living abroad. Only a few Dutch and Chinese merchants were authorized to enter Japan during the trading season.

95

95

Chinese and foreign merchants in Canton.
Possibly by Puqua (active around 1800)
Canton, early 19th century
Gouache on paper, 23 x 325 cm
Asian Civilisations Museum

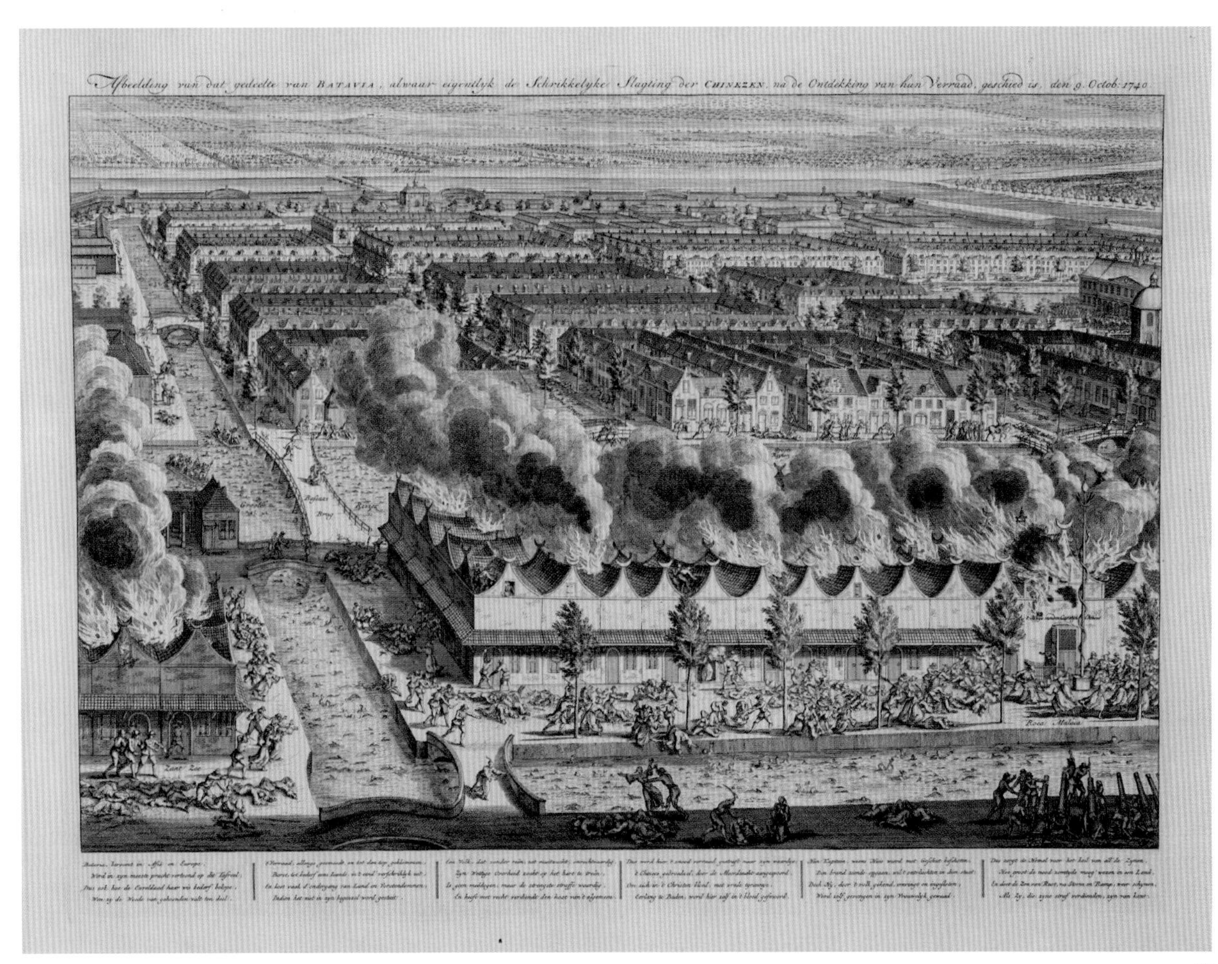

96

96

Depiction of the part of Batavia where the terrible slaughter of the Chinese took place on 9 October 1740.
Amsterdam, Bernardus Mourik, (1740).
Engraving, 43.5 x 56 cm,
after B. Mourik by A. van der Laan.
Mr and Mrs Lee Kip Lee.

97

97

Chinese coolies aboard the *S.S. Andrea Rickmers*, which arrived in Singapore from Swatow on 11 October 1906
Singapore, 1906
Photograph, 12 x 17.3 cm
Mr and Mrs Lee Kip Lee

98

98

Interior of the Ghee Hin Kongsi lodge
Singapore, late 19th century
Watercolour on paper, 41.8 x 30.9 cm
National Museum of Singapore [1996-02889]

99

99

Woman at her bath
G. R. Lambert & Co.
Singapore, around 1900
Albumin print, 23 x 28 cm
Mr and Mrs Lee Kip Lee

100

100

Chinese courtesans and their client
August Sachtler & Co.
Singapore, around 1860s
Photographs, left: 20.6 x 17.3 cm; right: 20.4 x 15.8 cm
Mr and Mrs Lee Kip Lee

The child might be William Henry Ker (1837–1864), son of William Wemyss Ker (1802–1874), a partner of Paterson, Simons and Co., one of the earliest and largest trading firms in the Straits Settlements. In 1859, the firm opened a dry dock on newly acquired land, which in 1900 became known as Keppel Harbour.

European children raised in the colonies were often looked after by local or Chinese attendants and nannies. The presence of the attendant in this painting is noteworthy, perhaps intended to add exotic appeal to the depiction of an English boy who grew up in a Southeast Asian colonial setting.

The artist Aaron Edwin Penley was better known for watercolours, and exhibited regularly at the Royal Academy exhibitions in London.

101

101

Portrait of a boy with a Chinese servant
Attributed to Aaron Edwin Penley, English (1806-1870)
Probably London, mid-19th century
Oil on canvas, 60.9 x 45.9 cm
Asian Civilisations Museum [2016-00214]

Integration

—

The diversity of Asian ports made for, not only racially hybrid communities, but also hybrid forms. In environments where migrants lived far from their motherlands, traditions were often forgotten, or adapted to the new environment.

Increasingly, itinerant craftsmen of different nationalities offered their services in every port of call, and they were adept at producing work in a range of styles. Very similar objects, for example furniture and ivory carvings, were produced in India, Southeast Asia, and China. It is often impossible to determine precisely where objects come from. They were produced for a common clientele in many different ports.

Mixed race communities

Perhaps the most visible aspect of integration in port cities was the mixed race community. European and other travellers in Asia from the seventeenth through nineteenth century never failed to report on the existence of hybrid communities. What they observed were the most apparent kind of hybridized groups, combining European and Asian ancestry. But since ancient times, Asians have been mixing with other Asian nationalities, and these groups often escaped the notice of foreign travellers.

103

102

102

Bramina woman
India, Goa, around 1785-1800
Oil on canvas, 65 x 50 cm
Private Collection, Portugal

103

Gentio de Angraça. Gentia de Pano
India, Goa, around 1785-1800
Oil on canvas, 65 x 50 cm
Private Collection, Portugal

104

105

104

Portrait of a man in *barong tagalog*
Signed and inscribed, lower left:
Nacio en 2_ de Agosto _ 1736 y fallecio
en 2[4?] de Abril de 1836
Severino Flavier Pablo
Philippines, Manila, 1836
Oil on canvas, 84 x 62 cm
Mr Jamie C. Laya

105

Portrait of a mestiza woman
Possibly Severino Flavier Pablo, Filipino (1800–1870)
Philippines, Manila, 1800s
Oil on canvas, 84 x 62 cm
Mr Jamie C. Laya

106

107

106
Woman with a dog
Jan Daniël Beijnon, Eurasian,
born in Batavia (1830–1877)
Batavia, 1873
Signed and dated, lower left: J. D. Beijnon. 1873
Oil on canvas, 58 x 44 cm
Mr Jan Veenendaal

107
Woman in a rocking chair
Jan Daniël Beijnon
Batavia, 1869
Signed and dated, lower right: J. D. Beijnon 1869
Oil on canvas, 56 x 46 cm
Mr Jan Veenendaal

108

111

108

A Straits Chinese family
Singapore, early or mid-20th century
Photograph, 18.8 x 24.4 cm
National Museum of Singapore
[1995-00548]

109

A Jawi Peranakan family
Singapore, 23 November 1932
Photograph, 24 x 29.5 cm
Family of Ahmad Mohamed Ibrahim

110

A Eurasian family in Singapore
Lee Brothers Studio
Singapore, 1910–1925
Photograph, 16.5 x 21.5 cm
National Archives of Singapore,
Lee Brothers Studio Collection [138839]

111

Members of a Surabaya shooting club
Java, 1871
Albumen print, 28.5 x 35.5 cm
Mr and Mrs Lee Kip Lee

The chair in Asia

The evolution of the European chair in Asia began with the arrival of the Portuguese in India in the early sixteenth century. The flows of cultural influences from Europe and all over Asia, the movement of itinerant carpenters of various nationalities, and the exuberant technical and cultural exchanges between multi-ethnic craftsmen in Asia's cosmopolitan ports, created many different hybrid chair forms over the centuries.

113

112

114

112
Chair
Goa, 16th or 17th century
Teak, 108 x 60 x 58 cm
Mr and Mrs Lee Kip Lee

113
Chair
India, Coromandel Coast, late 17th century
Ebony, 98.5 x 56.3 x 48.4 cm
Asian Civilisations Museum [2011-00717]

114
Chair
Batavia, 18th century
Wood, 154 x 76 x 63.5 cm
Gereja Sion (Portuguese Zion Church), Jakarta

115

116

117

115
Chair
Batavia, 18th century
Wood, 136 x 49 x 55 cm
Gereja Sion (Portuguese Zion Church), Jakarta

116
Armchair
India or the Dutch East Indies, 19th century
Rattan, rosewood, brass, 91 x 56 cm
Asian Civilisations Museum [2014-00472]

117
Chair
Singapore, early 20th century
Teak, 111 x 47.5 x 46 cm
Asian Civilisations Museum,
Purchased with funds from
Mr and Mrs Lee Seng Tee [2004-01026]

119

118

120

118

Armchair
Japan, Yokohama, around 1900
Katsura wood, 86 x 64 x 53 cm
Asian Civilisations Museum,
Gift of Mr and Mrs Lee Kip Lee

119

Burgomeister chair
Sri Lanka, Colombo, 1750–1775
Mara wood, 76.5 x 68.5 x 55 cm
Mr and Mrs Lee Kip Lee

120

Armchair
China, Canton, late 19th century
Wood, mother-of-pearl, 102.8 x 67.3 x 53.4 cm
Asian Civilisations Museum [2005-01476]

Jewellery

India had long held the position as the leading global consumer of precious stones. The Deccan region produced the finest diamonds, and in the seventeenth century, Goa rivaled Golconda as an important gem-cutting centre. The quest for gems saw the increasing presence in Asia of international merchants and jewellers, including those from Europe. As portable wealth, jewellery was horded by the wealthy residents of port cities. The demand attracted workshops and retailers from all over the world, creating hybrid forms for a varied clientele.

122

121

123

121
Tiara-comb
India, Goa, late 18th century
Gold, silver, copper, rock crystal, 14.5 x 19 cm
National Museum of Ancient Art, Lisbon [1233 Joa]

122
Hair ornament
India, Goa, 17th century
Gold, glass, 7 x 3.8 cm
National Museum of Ancient Art, Lisbon [1357 Joa]

123
Hair ornament
India, Goa, 19th century
Gold, copper, resinous mass, 15.3 x 8.2 cm
National Museum of Ancient Art, Lisbon [1351 Joa]

Jewellery of Goa
Luísa Penalva

All kinds of traditions, customs, and faiths converged in Goa. This is the background for understanding the extraordinary set of 349 jewels, which, for 50 years, had been hidden away in bank vaults in Goa. In 2012 they were entrusted to the Museo Nacional de Art Antica in Lisbon, where they were studied for the first time.

The jewellery was inspired by Indian symbols and aesthetics, European sensibilities, and the beliefs of the Christian community in the territory. They date from the seventeenth to the nineteenth century, revealing typologies, customs, and traditions previously ignored, and reflecting the combination of influences and tastes over the years in Goa. A set of thirteen pendants of the Child Christ as Salvator Mundi, nine of which are hung on necklaces (Cat. 127), are special. Worn by the Christian women of Goa, these pendants - some modest, others displaying sophisticated goldsmith's work and European taste - are a perfect example of the cultural and religious mixing in which the motif of the Child Christ is harmoniously combined with that of the Child Krishna. And two gold scapulars decorated with Our Lady of Sorrows (Cat. 126) testify unequivocally to the existence in Goa of Christian

124

125

126

124
Hair ornaments
India, Goa, 18th or 19th century
Gold, silver, glass, diameters 4.3 cm
National Museum of Ancient Art, Lisbon [1354 Joa]

125
Earrings
India, Goa, 19th century
Gold, glass, diameters 1.3 cm
National Museum of Ancient Art, Lisbon [1369 Joa]

126
Scapular: Seven Sorrows of the Virgin Mary
India, Goa, late 18th century
Gold, silver, 4 x 4 cm
National Museum of Ancient Art, Lisbon [1282 Joa]

commissions to local craftsmen for symbolic ornaments.

Paintings depicting *braminas* and *bayés* (see Cats. 102, 103) wearing necklaces with pendants, as well as other ornaments represented in the jewellery collection, show that such jewels frequently formed part of a large and magnificent whole, reflecting the multicultural nature of Goan society with their fascinating bright colours and rich appearance.

The set comprises bracelets (*bandio*) (Cat. 130), hair ornaments (*quegoda*) (Cats. 122–124), the typical bracelets of affronting animal heads (*kada*) (Cat. 129), and necklaces composed of a succession of coins (*moedanchen-gantlem*) (Cat. 128).

A comb tiara (*dantoni*), dating from the late eighteenth century, is one of the two most important pieces in the whole group (Cat. 121). Decorated with profuse foliage, shaped and pierced over a silver background, this piece stands out because of the richness of its magnificent gold work, with the prominent large flowers creating a three-dimensional sense of movement that enhances the pierced work of the gold over silver. But it is the large two-headed eagle (*gandabherunda*), with its spread wings, that is the most imposing feature of the comb, which would have held in place the hair of a Goan Christian woman, serving as a symbol of the protection of her virtue as a married woman.

127

128

127
Choker with pendant of Child Christ as Salvator Mundi
India, Goa, 17th or 18th century
Gold, silver, rock crystal, glass, silk, cotton, length 36.2 cm; pendant: 7.5 x 5.6 cm
National Museum of Ancient Art, Lisbon [1252 Joa]

128
Necklace
India, Goa, 18th century
Gold, resinous mass, cotton, length 42 cm
National Museum of Ancient Art, Lisbon [1335 Joa]

The study of these pieces, inspired by the presence of the extremely rare motifs and subject matter, has shown the world the extraordinary beauty and richness of Goan jewellery. The portraits of Goans, Hindus, and Christians, dating from the nineteenth century, which are now to be found scattered among public and private collections, both in Portugal and abroad, together with the bound edition of the drawings of Manuel da Cunha Maldonado and Joaquim António Roncon, have helped in the identification of the jewels, as well as in their distribution throughout the intricate Goan social topography, making them yet another fundamental instrument for the recognition of Goa's cultural and religious multiplicity.

More information can be found in the Museu Nacional de Arte Antiga exhibition catalogue: *Splendours of the Orient. Gold Jewels from the Old Goa* (Lisbon 2014).

129

130

129
Pair of bracelets
India, Goa, 19th century
Gold, resinous mass, 6.8 x 7 cm
National Museum of Ancient Art, Lisbon [1303 Joa]

130
Pair of bracelets
India, Goa, 19th century
Gold, silk, resinous mass, 30.5 x 34 cm
National Museum of Ancient Art, Lisbon [1316 Joa]

Jewellery from the Philippines

—

131

Jewellery from the Straits Settlements

—

132

133

131

Tiara-comb (*pienetas*)
Philippines, mid-19th century
Gold and tortoiseshell, 5 x 7.5 cm
Asian Civilisations Museum

132

Buckle
Indonesia, early 20th century
Gold, diamonds (brilliant cut), 10 x 7.5 cm
Mr and Mrs Lee Kip Lee

133

Buckle
Singapore, Malacca, or Penang,
late 19th or early 20th century
Gold (with Chinese mark), 5.5 x 9 cm
Mr and Mrs Lee Kip Lee

134

135

134
Brooch
Van Arcken & Co.
Batavia, late 19th century
Gold, diamonds (brilliant cut), 13 x 8.5 cm
Mr and Mrs Lee Kip Lee

135
Brooch
Indonesia, early 20th century
Gold, diamonds (rose cut), 9.5 x 6.5 cm
Mr and Mrs Lee Kip Lee

136

137

136

Pendant
Java, Solo, late 19th century
Gold, diamonds (mine cut), 7.5 x 3.8 cm
Asian Civilisations Museum [2000-05601]

137

Brooch
Probably China, Guangdong, 19th century
Gold, tortoiseshell or horn
Box: 2.8 x 12 x 8.4 cm
Asian Civilisations Museum [2011-00685]

138

Kebaya pins
Indonesia, early 20th century
Gold (with Chinese mark), each 1 x 4 cm
Mr and Mrs Lee Kip Lee

139

Hairpin
Sri Lanka, late 19th or early 20th century
Gold, silver, zircon, length 12.5 cm
Mr and Mrs Lee Kip Lee

140

Hairpin
Penang or Rangoon, early 20th century
Gold, diamonds (rose cut), length 15 cm
Mr and Mrs Lee Kip Lee

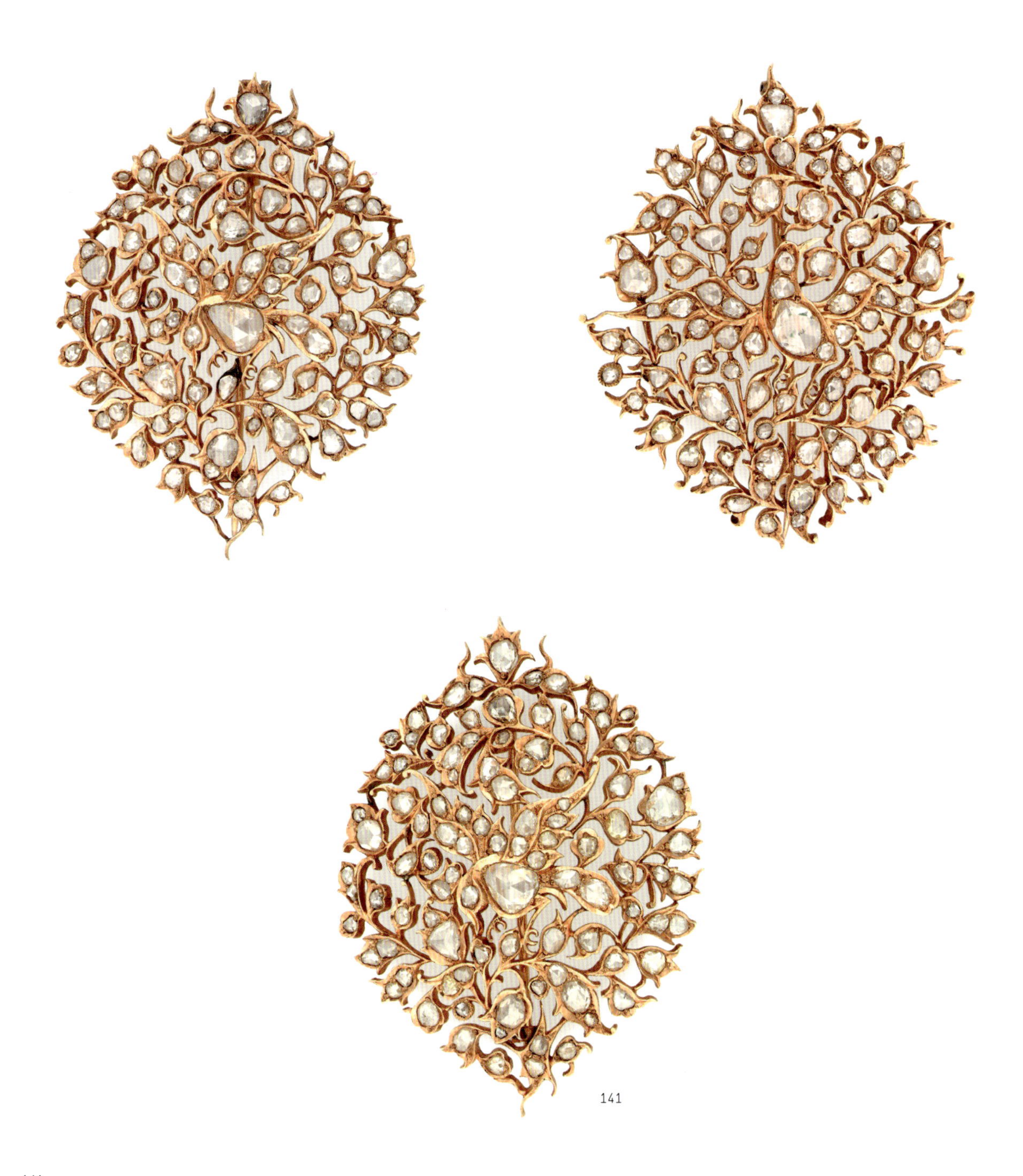

141

141

Kerosang set

Singapore, early 20th century

Gold, diamonds (rose cut), each 8 x 6 cm

Mr and Mrs Lee Kip Lee

142

143

142

***Kerosang* set**
Straits Settlements, 1920s or 1930s
Gold, 35 x 3.6 cm
Asian Civilisations Museum [2002-00421]

143

***Kerosang* set**
Singapore or Malacca,
late 19th or early 20th century
Gold and diamonds
left: 7.8 x 5.9 cm
middle: 3.1 x 2.2 cm
right: diameter 3.7 cm
Asian Civilisations Museum,
Gift of Mr Edmond Chin [2002-00508]

Gold and silver

Itinerant gold- and silversmiths travelled widely and passed on their skills to the local communities wherever they went. Craftsmen were forcibly moved from other countries to develop the industry in colonial ports, often working for employers of other nationalities. All these factors contributed to the development not only of similar styles of metalwork throughout Asia (often making it impossible to identify where they were made), but also of extremely hybrid designs and forms. In the 18th century, Dutch colonial settlements instituted a system of hallmarks, which documented the place of manufacture. Even so, the designs were often extremely mixed.

144

144

Bowl with cover
Inscribed on base:
顺娘 (Hokkien: Soon Neo; Mandarin: Shun-niang)
友娘 (Yew Neo; You-niang)
Batavia, 1700–25
Silver, 10.7 x 16 x 12 x 6 cm
Rijksmuseum [BK-1994-77]

145

145

Charger: Dutch East India Company gift to Sultan Adipati Cakraningrat of Bangkalan, Java, in 1783
Java, around 1783
Gold, diameter 25.5 cm
National Museum of Indonesia, Jakarta

146

147

148

146

Rosewater sprinklers
China or Batavia, late 18th century
Silver (filigree), each 22.7 x 8 cm
Asian Civilisations Museum [2012-00523, 2012-00524]

147

Rosewater sprinklers
China, late 19th or early 20th century
Gilded silver, each 29.5 x 8 cm
Asian Civilisations Museum,
Gift of Mr and Mrs Lee Kip Lee [2012-00175]

148

Rosewater sprinklers
Batavia, late 17th or early 18th century
Silver, each 36 x 10 cm
Mr and Mrs Lee Kip Lee

149

149
Toilette mirror
China, late 19th
Silver, enamels 34 x 19 x 13 cm
Asian Civilisations Museum,
Gift of Mr and Mrs Lee Kip Lee [2012-00201]
Marks: 大興 (Da Xing) ; 紋銀 (wen yin, or "fine silver")

150

151

152

150

Sireh box
Java, 18th century
Enamel, silver (filigree),
precious stones, 9.5 x 21 x 15 cm
Museum Nasional Indonesia, Jakarta

151

Tobacco box
Europe and Sulawesi, 18th century
Gold (filigree) from Sulawesi;
glass (faux lapis lazuli) from Europe, 2.5 x 8 cm
Mr and Mrs Lee Kip Lee

152

Snuff bottle and container
Unknown, possibly Indonesia, 18th or 19th century
Gold, glass, possibly horn, bottle: 7 x 3.7 cm;
container: diameter 4 cm, height 1.5 cm
Mr and Mrs Lee Kip Lee

153

153
Royal Acehnese dagger
Sumatra, Aceh, early or mid-19th century
Gold, wood, steel, black coral, bone, diamonds, sapphires, enamel; length of dagger 32 cm; with sheath 38 cm
Asian Civilisations Museum [2011-01616]

154

155

154
Casket
Unknown, possibly Goa or Indonesia, 17th or 18th century
Silver (filigree), 10 x 11.5 x 8 cm
Mr and Mrs Lee Kip Lee

155
Casket
Unknown, possibly Goa or Indonesia, 17th or 18th century
Silver (filigree), 16 x 22.5 x 15 cm
Mr and Mrs Lee Kip Lee

156

157

158

156

Casket
Indonesia, 18th century
Tortoiseshell, silver, 8.5 x 17.5 x 11.5 cm
Mr and Mrs Lee Kip Lee

157

Casket
Manila, 18th century
Tortoiseshell, silver mounts, 12.5 x 18.8 x 9.5 cm
Mr and Mrs Richard and Sandra Lopez

158

Casket
Manila, 18th century
Silver, 10.9 x 16.5 x 9.5 cm
Mr and Mrs Richard and Sandra Lopez

The cult of betel

The origins of betel chewing are still uncertain, but it was evidently widespread in maritime Asia as a pastime by 1500 among both urban and tribal communities. In the port cities, it was fashionable among all ethnic groups - Asian, European, and Eurasian - and an important component of social interaction and ritual.

The chew is prepared from betel nut (seed of the areca palm), betel leaf (from a vine, *Piper betle*), lime powder (calcium hydroxide), and sometimes tobacco. Elaborate boxes with many small containers were created for storage and presentation of the ingredients.

The betel box is perhaps the quintessential hybrid object, appearing in innumerable forms and styles, and drawing from a wide variety of cultural influences. Until the early 20th century, it could also be the ultimate luxury object, made of the costliest materials.

159

159

Betel set
Stoneware: Vietnam, 15th or 16th century;
mounts: Makassar, Gowa, Sulawesi,
late 19th or early 20th century
Gold, silver, niello, and porcelain, box: 11.5 x 21 x 21 cm
Asian Civilisations Museum [2014-00945]

160

160
Betel containers and tray
Batavia, early 18th century
Silver, height 12 cm (spittoon)
Mr and Mrs Lee Kip Lee

Three "Padang" betel boxes

Covered betel boxes with corners that rise to a point circulated widely in the Malay Archipelago in the eighteenth century, and have designs and materials so disparate that it is most likely the shape was adopted by craftsmen of various backgrounds for patrons of equally diverse origins. These have somehow become associated with the Padang Highlands, perhaps because one example has been documented among the heirlooms of Datuk Pemuncak Alam nan Sati, who lived there (Jasper and Pirngadie 1927, p. 43 and 45.). But there are similar examples among the heirlooms of the Sultanate of Bima, in eastern Indonesia, and the possessions of Pieternella van Hoorn, who lived in Batavia in the early eighteenth century (See KITLV image 41097, Brommer 2015, p. 95.).

Cat. 161 has motifs on the silver fittings inspired by baroque designs, such as those by Daniel Marot. Cat. 162 is in silver made in Batavia with incised decoration that might be inspired by Chinese or Javanese silverwork. Cat. 163 in a gold alloy also has a baroque inspired design, related to carvings on ebony and ivory colonial objects from Sri Lanka.

161

162

163

161
Betel box
City mark and date letter of Batavia
Batavia, 1705
Ambon wood, silver mounts,
8.5 x 19.5 x 13.5 cm
Mr Jan Veenendaal

162
Tobacco Box
Maker's mark: K
Batavia, around 1700–28
Silver, 4 x 13.8 x 7.8 cm
Gemeentemuseum, The Hague,
The Netherlands

163
Betel box
Indonesia, 18th century
Gold, 8.5 x 22 x 12 cm
Mr and Mrs Lee Kip Lee

166

166

Betel box
Indonesia, 18th century
Mother-of-pearl, mastic,
silver fittings, 10 x 23.5 x 15 cm
Mr and Mrs Lee Kip Lee

167

Betel box with lime container
Possibly Canton, early 19th century
Gilded silver, 8 x 20.5 x 14 cm
Mr and Mrs Lee Kip Lee

168

169

168

Spittoon
China, Changsha kilns, 9th century
Porcelain, height 14.1 cm
Asian Civilisations Museum [2015-1-00521]

169

Spittoon
Indonesia, early 19th century
Silver, height 14.5 cm, diameter 14 cm
Mr and Mrs Lee Kip Lee

170

171

172

170
Spittoon
China, 17th or 18th century
Porcelain, height 7.8 cm
Asian Civilisations Museum [2015-00065]

171
Spittoon
Guangzhou, China, 18th century
Copper, enamel, 12 x 11 x 8 cm
Asian Civilisations Museum,
Gift of Amir and Soha Mohtashemi [2014-01069]

172
Spittoon
Sumatra, 19th century
Silver, 16.7 x 17 cm
Asian Civilisations Museum [2013-00739]

Manges tanges

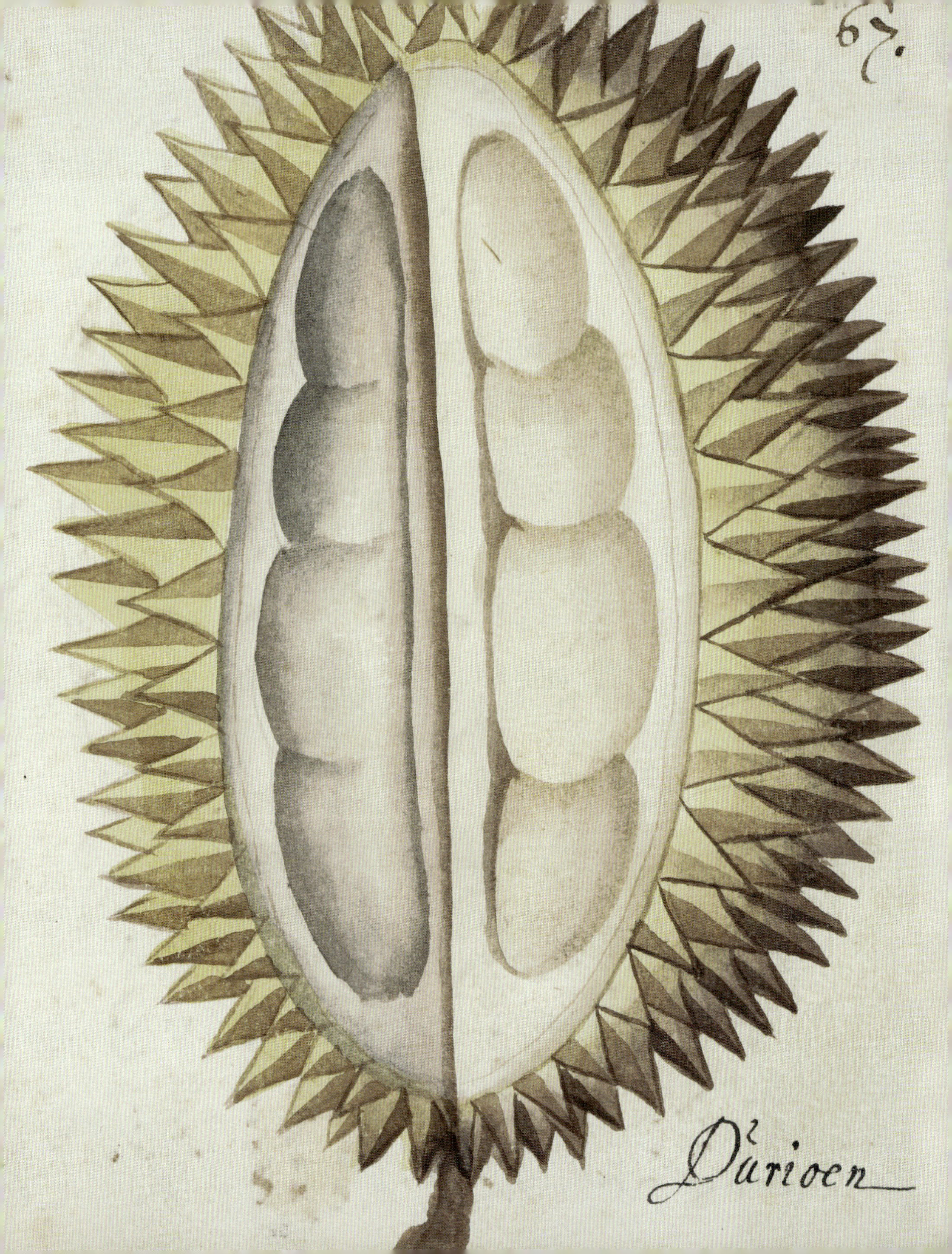
67.
Durioen

Bibliography

Abu-Lughod 1989
Janet L. Abu-Lughod. *Before European Hegemony: The World System, A.D. 1250–1350*. Oxford, 1989.

Amsterdam 2002
The Dutch Encounter with Asia, 1600–1950. Exh. Rijksmuseum, Amsterdam, 2002.

Andaya and Andaya 2015
Barbara Watson Andaya and Leonard Y. Andaya. *A History of Early Modern Southeast Asia*. Cambridge, 2015.

Andaya and Andaya 2016
———. *A History of Malaysia*. 3rd edition. London, 2016.

Andaya 2012
Barbara Watson Andaya. "'A people that range into all the countries of Asia': The Chulia trading network in the seventeenth and eighteenth centuries" in *The Trading World of the Indian Ocean, 1500–1800*, edited by Om Prakash (New Delhi, 2012), pp. 305–36.

Andaya 2016
———. "Rivers, oceans and spirits: Gender and water cosmologies in Southeast Asia." *TRaNS: Trans-Regional and -National Studies of Southeast Asia 4*, no. 2 (July 2016), pp. 239–63.

Ang See 1990
Teresita Ang See. *The Chinese in the Philippines. Vol. 2: Problems and Perspectives*. Manila, 1990.

Aslanian 2011
Sebouh David Aslanian. *From the Indian Ocean to the Mediterranean: The Global Trade Networks of Armenian Merchants from New Julfa*. Berkeley, 2011.

Barros 2012
Amândio Jorge Morais Barros. "The Portuguese in the Indian Ocean in the first global age: Transoceanic exchanges, naval power, port organization and trade" in *Oceans Connect: Reflections on Water Worlds across Time and Space*, edited by Rila Mukherjee (Delhi, 2013).

Barthes 1983
Roland Barthes. *The Fashion System*. New York, 1983. [Translation of Système de la mode (Paris, 1967).]

Blussé 1985
Leonard Blussé. *Strange Company: Chinese Settlers, Mestizo Women and the Dutch in VOC Batavia*. Dordrecht, 1985.

Blussé 1999
———. "Chinese century: The eighteenth century in the China Sea region." *Archipel* 58 (1999).

Blussé 2002
———. *Bitter Bonds: A Colonial Divorce Drama of the Seventeenth Century*. Princeton, 2002.

Blussé and Fernandez-Armesto 2003
Leonard Blussé and Felipe Fernandez-Armesto, eds. *Shifting Communities and Identity Formation in Early Modern Asia*. Leiden, 2003.

Boxer 1966
C. R. Boxer. *The Dutch Seaborne Empire, 1600–1800*. London, 1966.

Boyajian 1993
James C. Boyajian. *Portuguese Trade in Asia under the Habsburgs, 1580–1640*. Baltimore, 1993.

Broeze 1985
Frank Broeze, ed. *Brides of the Sea: Port Cities of Asia from the 16th to the 20th Centuries*. Sydney, 1989.

de Bruijn and Raben 2004
Max de Bruijn and Remco Raben, eds. *The World of Jan Brandes, 1743–1808*. Zwolle, 2004.

Bulley 1992
Anne Bulley. *Free Mariner: John Adolphus Pope in the East Indies, 1786–1821*. London, 1992.

Canberra 2014
Garden of the East: Photography in Indonesia, 1840s–1940s. Exh. National Gallery of Australia, Canberra, 2014. By Gael Newton et al.

Carpio 2005
Rustica C. Carpio. *The Shanghai of My Past*. Manila, 2005.

Chattopdhyay 2005
Swati Chattopdhyay. *Representing Calcutta: Modernity, Nationalism and the Colonial Uncanny*. London, 2005.

Chin 1998
James Kong Chin. "Merchants and other sojourners: The Hokkiens overseas, 1570–1760." PhD dissertation: University of Hong Kong, 1998.

Chong 2016
Alan Chong. "Christian ivories by Chinese artists: Macau, the Philippines, and elsewhere, late 16th and 17th centuries" in Singapore 2016, pp. 204–7.

Chong and Murai 2009
Alan Chong and Noriko Murai, eds. *Journeys East: Isabella Stewart Gardner and Asia*. Exh. Isabella Stewart Gardner Museum, Boston, 2009.

Cortesão 1944
Armando Cortesão, ed. *The Suma Oriental of Tome Pires*. 2 vols. New Delhi, 1990. Reprint of 1944 edition.

Deb 1905
Binaya Krishna Deb. *The Early History and Growth of Calcutta*. Calcutta, 1905.

Dhiravat 2016
Dhiravat na Pombejra. "Catching and selling elephants: Trade and tradition in seventeenth-century Siam" in *Modern Southeast Asia, 1350–1800*, edited by Ooi Keat Gin and Hoang Anh Tuan (London, 2016), pp. 192–209.

Disney 2009
A. R. Disney. *A History of Portugal and the Portuguese Empire*. Vol. 2: The Portuguese Empire. Cambridge, 2009.

Dy 2014
Aristotle C. Dy. "The Virgin Mary as Mazu or Guanyin: The syncretic nature of Chinese religion in the Philippines." *Philippines Sociological Review* 62 (Jan. 2014), pp. 41–63.

Earns 2012
Lane Earns. "Nagasaki: Fusion point for commerce and culture (1571–1945)" in *Places of Encounter*, edited by Aran Mackinnon and Elaine McClarnand Mackinnon (Boulder, 2012), vol. 2.

Elliott 1996
Alan F. Elliott, ed. *The Woodbury Papers: Letters and Documents held by the Royal Photographic Society*. Melbourne, 1996. [Papers of Walter B. Woodbury]

Falconer 1987
John Falconer. *A Vision of the Past: A History of Early Photography in Singapore and Malaya: The Photographs of G. R. Lambert & Co., 1880–1910*. Singapore, 1987.

Faure 1996
David Faure. "History and culture" in *Guangdong: China's Promised Land*, edited by Brian Hook (Hong Kong, 1996), pp. 1–30.

Fayle 1929
C. Ernest Fayle, ed. *Voyages to the East Indies: Chrstopher Fryke and Christopher Schweitzer*. London, 1929.

Felix 1966
Alfonso Felix. *The Chinese in the Philippines*. Vol. 1. Manila, 1966.

Ghesquière 2016
Jérôme Ghesquière, ed. *La photographie ancienne en Asie*. Paris, 2016.

Gipouloux 2011
François Gipouloux. *The Asian Mediterranean: Port Cities and Trading Networks in China, Japan and Southeast Asia, 13th–21th Century*. Cheltenham, 2011.

Gittinger 1982
Matiebelle Gittinger. *Master dyers to the world, technique and trade in early Indian dyed cotton textiles*. Washington, Textile Museum, 1982

Guy 1998
John Guy. *Woven Cargoes: Indian Textiles in the East*. London, 1998.

Hannavy 2008
John Hannavy, ed. *Encyclopedia of Nineteenth-Century Photography*. 2 vols. New York, 2008.

Hapgood 2015
Susan Hapgood. *Early Bombay Photography*. Mumbai, 2015.

Harrison 1985
Brian Harrison. *Holding the Fort: Malacca Under Two Flags, 1795–1845*. Kuala Lumpur, 1985.

Hirado 2010
Shito Hirado: nenpyou to shidan. Matsura Historical Museum, Hirado, Japan, 2010.

Ho 1976
Ho Wing Meng. *Straits Chinese Silver*. Singapore, 1976.

Ho 1983
———. *Straits Chinese Porcelain: A Collector's Guide*. Singapore, 1983.

Ho 1987
———. *Straits Chinese Beadwork & Embroidery: A Collector's Guide*. Singapore, 1987.

Ho 1994
———. *Straits Chinese Furniture: A Collector's Guide*. Singapore, 1994.

Juterczenka and Mackenthun 2009
Sünne Juterczenka and Gesa Mackenthun, eds. *The Fuzzy Logic of Encounter: New Perspectives on Cultural Contact*. Münster, 2009.

Kaufmann and North 2014
Thomas DaCosta Kaufmann and Michael North. "Mediating cultures" in *Mediating Netherlandish Art and Material Culture in Asia* (Amsterdam, 2014).

Kobe 2000
Hirado – Matsura kei meihou ten. Asahi Shimbunsha, Kobe, 2000.

Laarhoven 1994
Ruurdje Laarhoven. "The power of cloth: The textile trade of the Dutch East India Company (VOC), 1600–1780." Dissertation: Australia National University, 1994.

Lee 2014
Peter Lee. *Sarong Kebaya: Peranakan Fashion in an Interconnected World, 1500–1950*. Singapore, 2014.

Linschoten 1596
Jan Huyghen van Linschoten. *Itinerario, voyage ofte schipvaert, van Jan Huygen van Linschoten naer Oost ofte Portugaels Indien inhoudende een corte beschryvinghe der selver landen ende zee-custe*. Amsterdam, 1596.

Linschoten 1885
———. *The Voyage of John Huygen van Linschoten to the East Indies*. Edited by Arthur Coke Burnell and P.A. Tiele. London, 1885.

Lisbon 2014
Museu Nacional de Arte Antiga. *Splendours of the Orient: Gold Jewels from Old Goa*. Lisbon, 2014.

Massot 2015
Gilles Massot. "Jules Itier and the Lagrené Mission." *History of Photography* 39, no. 4 (2015) pp. 319–347.

Matheson 1972
Virginia Matheson. "Mahmud, sultan of Riau and Lingga (1823–1864)." *Indonesia* 13 (April 1972), pp. 119–46.

Meilink-Roelofsz 1962
M. A. P. Meilink-Roelofsz. *Asian Trade and European Influence in the Indonesian Archipelago*. The Hague, 1962.

Minto 1880
Countess of Minto. *Lord Minto in India: Life and Letters of Gilbert Elliot, First Earl of Minto from 1807 to 1814*. London, 1880.

Müller 1669
Georg Franz Müller. "Reisebuch des Elsässer Weltreisenden Georg Franz Müller (1646–1723)." Manuscript, 1669–82. Stiftsbibliothek St Gallen [Cod. Sang. 1311]: http://www.e-codices.unifr.ch/de/list/one/csg/1311

Nagasaki 2016
Kenkyu Kiyou 10 (2016). Special issue. Nagasaki Museum of History and Culture.

Negrón 2015
Ninel Valderrama Negrón. "Nostalgia of the empire: The arrival of the portrait of Ferdinand VII in Manila in 1825." *Rio de Janeiro* 10, no. 2 (2015). http://www.dezenovevinte.net/uah2/nvn_en.htm.

Newbold 1839
T. J. Newbold. *Political and Statistical Account of the British Settlements in the Straits of Malacca*. 2 vols. Kuala Lumpur, 1971. Reprint of 1839 edition.

New Delhi 2009
Govind Narayan's Mumbai: An Urban Biography from 1863. Edited by Murali Ranganathan (New Delhi, 2009), p. x.

New Delhi 2015
Imaging the Isle Across: Vintage Photography from Ceylon. Exh. National Museum, New Delhi, 2015.

Nieuhof 1682
Joan Nieuhof. *Gedenkwaerdige zee en lantreize door de voornaemste landschappen van West en Oostindien*. Edited by Hendrik Nieuhof. Amsterdam, 1682.

Nishigori 2016
Ryosuke Nishigori. *Torai Ōbakusō Dokutan Shoei to sono kaiga* in Nagasaki 2016.

North 2010
Michael North. "Production and reception of art through European company channels in Asia" in *Artistic and Cultural Exchanges Between Europe and Asia, 1400–1900*, edited by Michael North (Farnham, UK, 2010), pp. 89–107.

Osumi 1957
Tamezo Osumi. Kowatari Sarasa, vols. 1-6. Tokyo, 1957] [大隅為三, 古渡更紗, 第1-6, 東京: 美術出版社, 1957]

Pearson 2006
Michael N. Pearson. "Littoral society: The concept and the problems." *Journal of World History* 17, no. 4 (2006), pp. 353–74.

Pieris 2009
Anoma Pieris. *Hidden Hands and Divided Landscapes: A Penal History of Singapore's Plural Society*. Honolulu, 2009.

Pyrard 1619
François Pyrard. *Voyage de François Pyrard, de Laval. Vol. 2: Seconde partie du voyage de François Pyrard depuis l'arrivée à Goa iusques à son return en France*. Paris, 1619. [babel.hathitrust.org]

Raben 1996
Remco Raben. "Batavia and Colombo: The ethnic and spatial order of two colonial cities, 1600–1800. PhD dissertation: University of Leiden, 1996.

Reid 1988
Anthony Reid. *Southeast Asia in the Age of Commerce, 1450–1680*. Vol. 1: The Lands Below the Winds. New Haven, 1988.

Reid 2010
———. "Hybrid identities in the fifteenth-century Straits" in *Southeast Asia in the Fifteenth Century: The China Factor*, edited by Geoff Wade and Sun Laichen (Singapore, 2010), pp. 307–32.

Reid 2015
———. "Early modernity as cosmopolis: Some suggestions from Southeast Asia" in Delimiting Modernities: Conceptual Challenges and Regional Responses, edited by Sven Trakulhun and Ralph Weber (London, 2015), pp. 123–42.

Rotterdam 1989
Toekang potret: 100 jaar fotografie in Nederlands Indië, 1839–1939. 100 Years of Photography in the Ditch Indies, 1839–1939. Exh. Museum voor Volkenkunde, Rotterdam, 1989.

Schmuki 2001
Karl Schmuki. *Der "Indianer" im Kloster St. Gallen: Georg Franz Müller (1646–1723), ein Weltreisender des 17. Jahrhunderts*. St. Gallen, 2001.

Schweitzer 1688
Christoph Schweitzer. *Journal- und Tagebuch seiner sechsjährigen Ost-Indianischen Reise: angefangen den 1. Dec. 1675 und vollendet den 2. September 1682*. Tübingen, 1688.

Sejarah Melayu 1979
Sulalatus Salatin, Sejarah Melayu. Edited by A. Samad Ahmad. Kuala Lumpur, 1979.

Seshan 2012
Radhika Seshan. *Trade and Politics on the Coromandel Coast: Seventeenth and Early Eighteenth Centuries*. Delhi, 2012.

Singapore 2013
Devotion and Desire: Cross-Cultural Art in Asia: New Acquisitions. By Alan Chong, Pedro Moura Carvalho, et al. Singapore, 2013.

Singapore 2016
Christianity in Asia: Sacred Art and Visual Splendour. Edited by Alan Chong. Exh. Asian Civilisations Museum, Singapore, 2016.

Skinner 1982
C. Skinner, ed. *Ahmad Rijaluddin's Hikayat Perintah Negeri Benggala*. The Hague, 1982.

Sutherland 1989
Heather Sutherland. "Eastern emporium and company town: Trade and society in eighteenth-century Makassar" in Broeze 1989, pp. 97–128.

Sweeney 2005
Amin Sweeney. *Karya Lengkap Abdullah bin Abdul Kadir Munsyi*, Jakarta, 2005.

Trocki 2006
Carl A. Trocki. *Singapore: Wealth, Power and the Culture of Control*. London, 2006.

Trocki 2011
———. "Singapore as a nineteenth century migration node" in *Connecting Seas and Connected Ocean Rims*, edited by Donna R. Gabaccia and Dirk Hoerder (Leiden, 2011).

Veenendaal 1985
Jan Veenendaal. *Furniture from Indonesia, Sri Lanka and India during the Dutch Period*. Delft, 1985.

Veenendaal 2014
———. *Asian Art and Dutch Taste*. The Hague, 2014.

Veldhuisen 1993
Harmen C. Veldhuisen. *Batik Belanda, 1840–1940: Dutch Influence in Batik from Java: History and Stories*. Jakarta, 1993.

Wolters 1999
O. W. Wolters. *History, Culture and Region in Southeast Asian Perspectives*. Revised edition. Ithaca, 1999.

Wood 1996
R. Derek Wood. "The voyage of Captain Lucas and the daguerreotype to Sydney." *Journal de la Société des océanistes* 102 (1996), pp. 113–18.

Woodhouse 2012
Leslie Woodhouse. "Concubines with cameras: Royal Siamese consorts picturing femininity and ethnic difference in eary 20th-century Siam." *Trans Asia Photography Review 2*, no. 1 (Spring 2012). [http://hdl.handle.net/2027/spo.7977573.0002.202]

Zhang and Sen 2014
Zhang Xing and Tansen Sen. "The Chinese in South Asia" in *Routledge Handbook of the Chinese Diaspora*, edited by Tan Chee-Beng (New York, 2014), pp. 205–26.

LENDERS

- Family of Ahmad Mohamed Ibrahim, Singapore
- Gemeentemuseum Den Haag, Netherlands
- GPIB Jemaat "SION" DKI Jakarta
- Indian Heritage Centre, Singapore
- Jakarta History Museum
- Matsura Historical Museum, Hirado
- J. Front Retailing Archives Foundation, Inc., Nagoya
- Mr Jan Veenendaal, Belgium
- Mr and Mrs Lee Kip Lee, Singapore
- Mr and Mrs Richard and Sandra Lopez, Manila
- Mr Jaime C. Laya, Manila
- Museu Nacional de Arte Antiga, Lisbon
- Museum Nasional Indonesia, Jakarta
- National Archives of Singapore
- National Library of Singapore
- National Museum of Singapore
- Obakusan Mampuku-ji, Kyoto Prefecture
- Private Collection, Japan
- Private Collection, Portugal
- Rijksmuseum, Amsterdam
- Saigon Chettiar's Temple Trust Pte Ltd, Singapore
- Stiftsbibliothek St Gallen, Switzerland

Acknowledgements

We received support from all the authors; special thanks to Leonard Y. Andaya and Barbara Watson Andaya, Gael Newton, Luísa Penalva, Nishigori Ryosuke, Alan Chong, and Peter Lee.

We are grateful to all the lenders to the exhibition. Many curators and museum personnel generously facilitated research and loans: GPIB Jemaat "SION" DKI Jakarta: Elisabeth Makaminan; Jakarta History Museum: Sri Kusumawati; Museum Nasional Indonesia, Jakarta: Intan Mardiana, Wahyu Ernawati; Fukuoka Art Museum: Iwanaga Etsuko; J. Front Retailing Archives Foundation Inc.: Kato Junichi, Sakakibara Mami; Matsura Historical Museum, Hirado: Okayama Yoshiharu, Sakamoto Kumiko; Obakusan Mampuku-ji, Kyoto: Tanaka Chisei, Hisatsune Shinryu; Gemeentemuseum Den Haag, The Hague: Ap Gewald, Jolanda Zonderop; Rijksmuseum, Amsterdam: Cindy van Weele, Wendela Brouwer; Ayala Museum: Kenneth C. Esguerra, Aprille P. Tijam; Museu Nacional de Arte Antiga, Lisbon: Anal Kol, Conceição Borges de Sousa; Indian Heritage Centre, Singapore: Nalina Gopal; National Archives of Singapore: Eric Chin; National Library of Singapore: Tan Huism; National Museum of Singapore: Szan Tan, Daniel Tham, Ong Shihui; Saigon Chettiar's Temple Trust Pte Ltd, Singapore: Ashwin Muthiah, Chocku Chockalingam; Stiftsbibliothek St Gallen: Silvio Frigg, Franziska Schnoor.

Generous sponsorship towards the exhibition was provided by Port of Singapore Authority, Singapore Cruise Centre, and the Japan Foundation.

We are also grateful for the support and advice of the family of Djohan and Diah Pangestuti, Guus Röell, Shiraishi Akiko, Aurora Martinez, Ramon N. Villegas, Yuhashi Michiko, Matsura Akira, and Matsura Osamu. Valuable assistance was provided by Japan Creative Centre: Ito Misako, Matsunaga Kazunori.

At the Asian Civilisations Museum, we thank:

Richard Lingner, Clement Onn, Naomi Wang, Denisonde Simbol, Ling Li Li, Noorashikin Zulkifli, Loh Pei Ying, Farrah Ismail, Nguyen Thi Tu Linh, Lau Sue Ann, and Ilaria Obata.

pp. 8–9, Print in Philippus Baldaeus, *Naauwkeurige Beschryvinge van Malabar en Choromandel, Derzelver aangrenzende Ryken, en het machtige Eyland Ceylon* [...]. Amsterdam, Janssonius van Waesberge en Van Someren, 1672. Courtesy of Mr and Mrs Lee Kip Lee.

pp. 28–29, Detail of Cat. 19.

pp. 42–43, Theodore Maurisset. *La daguerreotypomanie*, lithograph from *La caricature* (1839). George Eastman House, Rochester, New York.

pp. 62–63, Detail of Cat. 93.

pp. 80–81, Detail of Cat. 77.

pp. 210–211, Detail of Cat. 34.

p. 219, Details of Cats. 39, 102, 104, 106.